Analytical Ultracentrifugation in Biochemistry and Polymer Science

Analytical Ultracentrifugation in Biochemistry and Polymer Science

Edited by

S.E. Harding
University of Nottingham

A.J. Rowe
University of Leicester

J.C. Horton
University of Nottingham

ROYAL SOCIETY OF CHEMISTRY

ISBN 0-85186-345-0

A catalogue record for this book is available from the British Library

Published by The Royal Society of Chemistry,
Thomas Graham House, Science Park, Cambridge CB4 4WF

Printed in England by Redwood Press Ltd., Melksham, Wiltshire

Preface

The history of science is full of examples of the development of new techniques and methodology which served their purpose and were then discarded as the amount of useful information to be obtained became saturated, or more powerful or more convenient procedures became available. In the field of macromolecular science this has *almost* been true of analytical ultracentrifugation. Since its inception by T. Svedberg and co-workers in the 1920's it has *almost* undergone such an evolution. However this book is no funeral for this technique. On the contrary it is the consequence of an exciting re-birth and re-growth that is now well underway.

Without doubt the great boom period for the analytical ultracentrifuge was in the period 1950-1970 with virtually every biochemistry laboratory and many chemical laboratories having access to this facility in the form of the famous Spinco/Beckman Model E Ultracentrifuge or its MSE or MOM equivalent. The focus of attention was on molecular weight and related phenomena and gross conformation determinations (*via* the sedimentation coefficient). However by 1970 in, for example, the field of protein biochemistry simpler to use alternative procedures were available for routine molecular weight determinations - namely gel electrophoresis and gel chromatography; for conformation studies biochemists were no longer interested in crude ellipsoid of revolution representations: analytical ultracentrifuges became discarded everywhere leaving only a handful being actively used.

By the middle of the 1980's biochemists started to realise - in large numbers - the limitations of gel electrophoresis and gel chromatography for molecular weight determinations - particularly for the characterisation of macromolecules that contain *more* than one subunit, or for *polydisperse* systems of macromolecules (such as glycoconjugates and polysaccharides), or for *interacting* systems of macromolecules or for precise measurements on *newly engineered* macromolecules. This was coupled with a growing appreciation that more can be done with sedimentation and molecular conformation data than the use of simple ellipsoid shapes: together with an appreciation of the limitations of X-ray crystallography and high-resolution nuclear magnetic resonance for representing solution conformations of intact, larger macromolecules such as immunologically active antibodies. In parallel to this growing demand for basic solution information on these systems in the field of protein biochemistry has been a similar demand in polymer science: in this latter field, light scattering techniques have been the method of choice for solution characterisations - almost to the point of obsession - and yet for many cases such techniques are intrinsically unsuitable, largely through problems of almost insurmountable requirements of solution clarity from dust and aggregates. Thus the potential of modern analytical ultracentrifuge procedures as an alternative is now clearly recognised.

Resulting from this re-awakening of interest, the growing demand for basic solution information on macromolecular systems in biochemistry and polymer science has had two clear effects. One has been the appearance of National Centres or "Facilities" for this type of measurement (*e.g.* at Storrs in the USA and at Nottingham/Leicester in the UK). The second has been the development and launch by Beckman Instruments of a new analytical ultracentrifuge (the "Optima XL-A") with full computer control and data capture/analysis facilities: this has made the technique and the range of important physical measurements (from simple solution homogeneity checks, sedimentation coefficient and molecular weight determinations right through to the characterisation of molecular weight distributions, subunit structures, interaction constants and sophisticated hydrodynamic bead model representations of molecular conformation in solution) fully accessible to the biochemist and polymer scientist. The availability of a new hardware platform in the form of the Beckman XL-A has thus neatly dovetailed with concurrent advances in information technology to ensure that the old inhibitory factors so far as analytical ultracentrifuge work is concerned - namely tedium of data capture and mathematical complexity of data analysis - have quite simply ceased to apply.

What has been lacking however is any up-to-date guide for potential users of the technology. To what extent are 'classical' methods still applicable? In what areas are advances currently being made? These issues need to be presented to an audience many of whom will have grown up in science with a knowledge (if any) of analytical ultracentrifugation based on 'historic textbook discoveries'. Approaches by the Editors of this volume to a worldwide roster of experts met with an enthusiastic response, and led to the 33 Chapters contained in this volume, in which the present 'state of the art' is presented to a new audience and to older researchers alike.

The book divides naturally into four parts. Part I deals with a description of the state of the art concerning instrumentation including a detailed consideration of optical systems and data capture and analysis procedures. Part II focuses on sedimentation equilibrium methods for obtaining molecular weight, polydispersity and interaction information. Part III considers transport methods (both sedimentation and diffusive) and includes a detailed consideration of rigid and flexible particle conformation analysis, polydisperse and interacting systems and concentration dependence phenomena. In Part IV, we consider a wide range of applications to macromolecular systems in biochemistry and polymer science, including seed and membrane proteins, polysaccharides and glycoconjugates and synthetic macromolecular systems.

We would like to acknowledge Dr. John Mitchell and Dr. Sandra Hill at the University of Nottingham School of Agriculture for allowing us access to their Apple Macintosh and, in particular, its printing facilities. A special thanks also to our old friends and colleagues Dr. Michael Creeth and Dr. Paley Johnson who have been an inspiration to us in the past and to whom this book is dedicated.

This is the first book of its type for nearly two decades (and over 50 years since Svedberg and Pedersen's classical book[§] appeared). It is our view that with the growing resurgence of interest in the technique it will be something less than two decades before the next one appears.

Steve Harding[*]

Arthur Rowe[†]

John Horton[*]

The National Centre for Macromolecular Hydrodynamics
Universities of Nottingham[*] and Leicester[†]
May, 1992

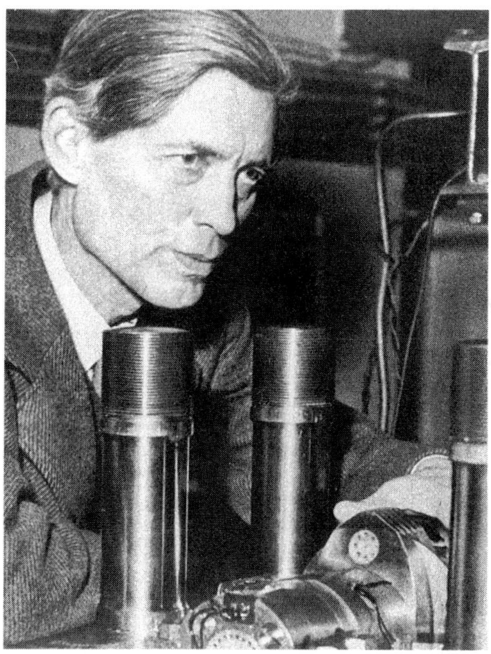

Thé Svedberg

§ "The Ultracentrifuge", Oxford University Press and "Die Ultrazentrifuge", Steinkopff, Dresden, 1940.

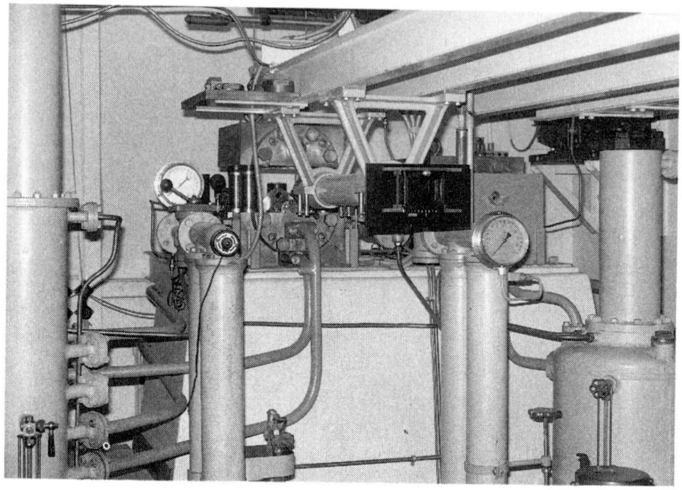

(Part of) Svedberg analytical ultracentrifuge

Beckman / Spinco Model E

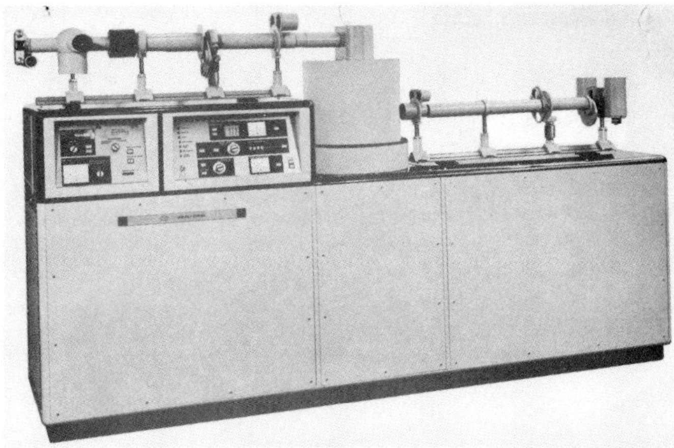

MSE Mk II analytical ultracentrifuge

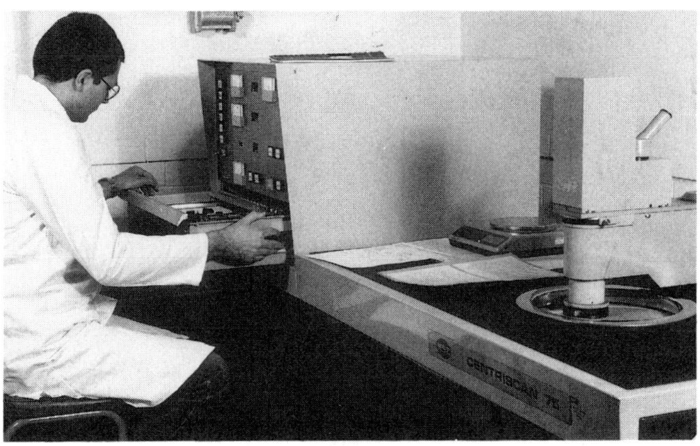

MSE Centriscan analytical ultracentrifuge

Contents

Part I: Instrumentation

Part II: Equilibrium Methods

Part III: Transport Methods

Part I: Instrumentation

1

Is There a Future for the Ultracentrifuge?

By H.K. Schachman

DEPARTMENT OF MOLECULAR AND CELL BIOLOGY, W.M. STANLEY HALL, UNIVERSITY OF CALIFORNIA, BERKELEY, CA 94720, U.S.A.

1. INTRODUCTION

After six decades of exciting research activity with the ultracentrifuge, the pace has slowed considerably. Rarely now do we find the insightful theoretical treatments, innovative experimental techniques and imaginative applications which appeared in the scientific literature repeatedly during the period from the mid 1920s to the 1980s. To many who had used the ultracentrifuge to investigate the size and shape of macromolecules of biological interest it seemed as if the instrument was a tool of the past, and it seemed unlikely that the ultracentrifuge would have a future. What happened? Were the techniques of sedimentation velocity and sedimentation equilibrium replaced by others which were more convenient, more accurate, more discriminating and more sensitive? Were research workers no longer interested in the purity and molecular weights of the newly isolated proteins? Had the biologists, who previously were the principle users of ultracentrifuges, become content with merely "yes or no" answers in their studies of interactions between proteins and nucleic acids? Would filter binding assays, Sephadex columns and polyacrylamide gels continue to satisfy them? Or would shifting goals in the 1990s cause them to be interested in equilibrium constants as well as enthalpy and entropy changes in these interactions?

Scientific history is replete with examples of physico-chemical techniques which served a valuable purpose in one period of frenetic research activity only to be replaced by another newly developed or dramatically improved method. But those research workers interested in the association-dissociation equilibrium of oligomeric proteins or the measurement of the strength of interactions between dissimilar macromolecules have found that no better technique has yet appeared. The ultracentrifuge is still the tool of choice. Hence, as shown in this volume, there are exciting new experimental techniques and theoretical treatments which are just beginning to surface and doubtless will be used by molecular and cell biologists with friendly computer programs, and a smaller, convenient, accurate and sensitive ultracentrifuge.

This introductory chapter summarizes in brief form the glorious past and then anticipates the future by indicating what one "old-timer" would like to have in the ultracentrifuge of the future.

2. THE GLORIOUS PAST

The Period Before 1930

Svedberg's initial interests in the development of the ultracentrifuge were focused on its use in studying the polydispersity of colloidal gold suspensions. In 1923, while a visiting professor at the University of Wisconsin, Svedberg, with Nichols, constructed the first instrument containing an optical system for analyzing the redistribution of particles in the centrifugal field. One year later, after his return to Uppsala, he and Rinde recognized the hazards of convective flow in sedimentation velocity experiments and designed the sector-shaped cell. Meanwhile they were using an absorption optical system to measure the movement or redistribution of particles throughout the cell. After a few disappointing experiments with egg albumin, Fåhraeus proposed studying hemoglobin even though Svedberg was apparently apprehensive because its molecular weight based upon its iron content should have been only about 16,700. Molecules of that size would have been difficult to study at the centrifugal fields available at that time. According to Pedersen, Svedberg was awakened in the middle of the night by a telephone call from Fåhraeus who was watching the ultracentrifuge experiment and shouted, "Thé, I see a dawn." Indeed, as Svedberg found upon rushing to the laboratory there was a marked decrease in the absorbance at the top of the cell.

Thus began a new era in protein chemistry and the ultracentrifuge became the tool of choice for determining the molecular weight of proteins, their size and shape and the homogeneity of preparations. Dissatisfaction with the absorption optical system and the need to measure the darkening of the photographic plate as a function of position in the ultracentrifuge cell led to Lamm's great contribution in the development of the scale method for determining the gradient of refractive index as a function of distance from the axis of rotation. The Lamm scale method which exploited the bending of light in regions of the cell where there was a change in refractive index with distance was very accurate, but it was so tedious to use that very rapidly considerable effort was expended to develop a self-plotting optical system.

Throughout this period of the late 1920s, theory was also being formulated for the determination of molecular weight from measurements of the sedimentation coefficient and diffusion coefficient. The famous Svedberg equation based on the sedimentation velocity method and hydrodynamic considerations had its counterpart in the sedimentation equilibrium expression derived initially by equating the transport toward the bottom of the cell by sedimentation with that in the opposite direction by diffusion. Later this same equation was obtained from thermodynamic principles. Along with this basic and indispensable theory was Lamm's derivation of the famous continuity equation in 1929 and approximate solutions of it which accounted for the shape of boundaries in sedimentation velocity experiments.

Excitement about the potential of the ultracentrifuge for the study of biological systems was world-wide, and within a few years Uppsala became a leading centre for research on the physical chemistry of biological macromolecules. For young postdoctoral fellows and senior investigators interested in proteins Uppsala had become the Mecca.

The Decade from 1930 to 1940

From the point of view of workers using the ultracentrifuge, the design and construction of the schlieren optical system containing a cylindrical lens had the greatest impact. It is of interest that two "self-plotting" optical systems were developed independently by Philpot and Svensson, on the one hand, and by Longsworth on the other. Both yielded superb patterns of moving boundaries represented by the concentration gradient (more properly the refractive index gradient) as a function of radial distance. As soon as properly ground cylindrical lenses became available, the Philpot-Svensson system became the one of choice over the ingenious, mechanical scanning method of Longsworth. Schlieren patterns from sedimentation velocity experiments on numerous proteins dominated the biochemical journals in the 1930s.

To the biochemist this period was a crucial one for the acceptance of the view that proteins were molecules. Svedberg, himself, visualized them as colloidal aggregates. But he recognized the significance of their homogeneity as demonstrated in sedimentation velocity experiments. Indeed, on the basis of his studies on hemocyanins with molecular weights of the order of 10^6 and other proteins with molecular weights from 3.5×10^4 to 2.1×10^5, he proposed "that the numerous proteins are built up according to some general plan which secures for them only a very limited number of different molecular masses and sizes when present in aqueous solution." Although his numbers proved later to be untenable and the range of proteins studied too limited, his provocative ideas and data constituted the first evidence for the concept that proteins are composed of subunits.

Physical chemists at that time were beginning to use hydrodynamic theories for prolate and oblate ellipsoids of revolution in order to interpret the frictional coefficients deduced from sedimentation velocity experiments in terms of the shapes and hydration of the macromolecules. It was a substantial triumph for both the experimental studies and the theoretical treatments when the size and shape of tobacco mosaic virus was predicted from the sedimentation coefficient and the intrinsic viscosity before the rod-like virus particles were observed by electron microscopy.

By the mid 1930s, the oil-turbine ultracentrifuge in Svedberg's laboratory and a few other laboratories in the world were in routine operation and studies of albumin-globulin ratios in serum were the focus of much research. Of even greater significance for the future was the construction of air-driven ultracentrifuges by Beams and Pickels. These important developments in the decade between 1930 and 1940 are summarized in Table 1.

Table 1. Ultracentrifugation from 1930-1940.

- Molecular weights of proteins
- Air turbine ultracentrifuge with air-supported rotor
- Sedimentation of flexible chain polymers
- Frictional ratios and shapes of molecules
- Hydration of proteins
- Prediction of size and shape of tobacco mosaic virus
- Schlieren optical system with cylindrical lens
- Solution of continuity equation
- Analysis of serum
- Proteins as molecules
- Subunits in proteins

The Decade from 1940 to 1950

By 1940 extensive studies of a variety of macromolecules had been made by ultracentrifugal methods. Tables of sedimentation coefficients, diffusion coefficients and frictional coefficients of proteins and viruses were already being compiled. In that year the great monograph, 'The Ultracentrifuge' by Thé Svedberg and Kai O. Pedersen was published. Included in that extremely influential monograph which I termed "The Bible" was a significant section on the air-driven ultracentrifuge written by Bauer and Pickels. Both an elementary theoretical treatment as well as a more detailed approach were included along with extensive sections on experimental measurements. Some shorter sections in this seminal monograph dealt with polydispersity and the sedimentation behaviour of linear high polymers.

Table 2. Ultracentrifugation from 1940-1950.

- "The Bible"
- Molecular constants of proteins
- Shapes of proteins
- Demise of absorption optics
- Disappearance of sedimentation equilibrium
- Transient states - approach to equilibrium
- Electrically-driven ultracentrifuge
- Analysis of mixtures
- Inverted refractive index gradients - lipoproteins

As seen in Table 2 which summarizes the principal developments in the period from 1940 to 1950, this decade saw the demise of absorption optics because of the widespread acceptance of the schlieren optical system. Similarly, the technique of sedimentation equilibrium was abandoned because attaining equilibrium in the ultracentrifuge cells required such long experiments (days and even weeks) and few instruments could operate continuously for such long periods without some malfunction. Ironically, both absorption optics and sedimentation equilibrium were revived at a later time to play important roles in the application of the ultracentrifuge for studies of nucleic acids, viruses, proteins and small molecules.

In the late 1940s anomalies were discovered in the analysis of mixtures of sedimenting compounds and the interpretation was provided by Johnston and Ogston. An extreme form of the Johnston-Ogston effect was observed in the sedimentation analysis of serum where the density of the solution caused the flotation of lipoproteins thereby complicating immensely the schlieren patterns because of inverted refractive index gradients.

Without much doubt, the most significant contribution to ultracentrifugation during this period was the design and construction of the electrically driven ultracentrifuge by Pickels and his colleagues. The availability of an extremely reliable commercial instrument coupled with the establishment of grant programmes in the United States led in the following decade to an almost explosive proliferation in the number of research workers using the ultracentrifuge as their principal tool. It is difficult for workers in the 1950s and later to appreciate that, prior to the availability of a commercial instrument in 1947, only about one-half dozen oil turbine ultracentrifuges and a slightly larger number of air-driven instruments were in operation throughout the world. By 1980 there were about 2,000 Spinco ultracentrifuges used routinely throughout the world.

The Decade from 1950 to 1960

In 1953 I had an opportunity to visit Uppsala and meet some of the pioneers in the early developments and was astonished to hear some of them express concern that all of the great research with the ultracentrifuge had been accomplished. From then on, according to this pessimistic view, there would be just more of the same - a molecular weight and sedimentation coefficient for another protein. How wrong they were. As seen by the large number of entries in Table 3 for the decade from 1950 to 1960, the full impact of the commercial availability of the Model E ultracentrifuge was just beginning to be felt.

Analysis of the transport of material by sedimentation showed that it was the movement not of the peak of the schlieren curves but rather of the second moment of the gradient curves that defined the sedimentation coefficient. A return to the continuity equation and solutions of it for describing the shape and broadening of boundaries due to diffusion led to the incorporation of the concentration dependence of the sedimentation coefficient. Disappearing from the literature was the erroneous term "sedimentation constant" for it was shown repeatedly with many different macromolecules that there was a significant concentration dependence. When this was incorporated into the continuity equation, the self-sharpening of boundaries was shown theoretically and it became clear that analysis of homogeneity from the spreading of the boundary was more complex than had been appreciated.

Table 3. Ultracentrifugation from 1950-1960.

- Second moment of gradient curve
- Rigorous thermodynamic theory
- Synthetic boundary cell
- Differential sedimentation
- Concentration dependence of sedimentation coefficient
- Analysis of boundary spreading
- Diffusion coefficients and heterogeneity
- Self-sharpening of boundaries
- Zone or band centrifugation
- Ultracentrifuge with magnetically supported rotor

- Moving partition separation cell
- Interference optical systems
- Re-birth of sedimentation equilibrium - short columns
- Double-sector ultracentrifuge cells
- Renaissance of absorption optics - DNA
- Density gradient sedimentation equilibrium
- Mechanism of replication of DNA
- Analysis of reversibly interacting systems
  - Self associations
  - Mixed associations
- Discovery of ribosomes
- Analysis of charge effects

Because of the increasing interest in applications of the ultracentrifuge, theoreticians re-examined the thermodynamic treatment for sedimentation equilibrium with the result that much more rigorous formulations were developed. Meanwhile it was recognized that the time to attain sedimentation equilibrium, as had been pointed out much earlier for experiments in gravity, varied inversely with the square of the height of the liquid column. Thus, decreasing the volume of solution in the ultracentrifuge cell led to a considerable shortening of the time required to attain equilibrium. Moreover, highly accurate molecular weights could be attained with these short columns. Thus we witnessed the rebirth of sedimentation equilibrium.

Biochemical research during this period turned somewhat toward nucleic acids, and, especially with DNA, it was apparent that concentrations much lower than those that could be detected by schlieren optics were required to minimize the complications resulting from non-ideality of the solutions. This led to the renaissance of absorption optics virtually in the form described by Svedberg and his colleagues. This crude, tedious optical system was used in density gradient sedimentation equilibrium experiments and in the elegant study demonstrating the semi-conservative mechanism of replication of DNA.

There was a proliferation of different types of ultracentrifuge cells in this period including synthetic boundary cells, moving partition separation cells, layering cells for band centrifugation and double sector cells. The latter were essential for the implementation of the Rayleigh interference optical system which after years of study and improvement became the optical system of choice for the reincarnation of the sedimentation equilibrium method. This combination of short columns and the Rayleigh interferometer restored the sedimentation equilibrium method to its position as the premier method for the determination of molecular weights ranging from 10^2 to 10^6.

During this decade Beams described his progress in developing an ultracentrifuge with a magnetically supported rotor, but the field continued to be dominated by the electrically driven instrument which proved extremely reliable, versatile and adaptable.

The Decade from 1960 to 1970

In the period from 1960 to 1970 there was explosive activity in the use of the ultracentrifuge for the study of DNA and proteins. The density gradient sedimentation equilibrium method was employed to measure the buoyant densities of innumerable DNAs and correlations quickly emerged relating the density to C+G contents relative to A+T. In the meantime, the advances in experimental techniques stimulated theoretical treatments for multicomponent systems. In this way preferential interactions in concentrated cesium chloride solutions were shown to be important. Similarly, protein chemists used the more rigorous theoretical approaches to evaluate the molecular weights of subunits of proteins in concentrated solutions of urea or guanidine hydrochloride. Some of these developments, along with others, are summarized in Table 4.

Table 4. Ultracentrifugation from 1960-1970.

- Multicomponent systems and preferential interactions
- Buoyant densities of DNA
- High-speed sedimentation equilibrium - meniscus depletion
- M_n, M_w and M_z
- Split-beam photoelectric optics scanning absorption optics
- $\bar{v}$ determination by sedimentation equilibrium; D_2O
- Non-ideality in sedimentation equilibrium
- Constituent sedimentation coefficients and molecular weights
- Band sedimentation
- Analysis of self-associating systems
- Pressure effects and volume changes in interactions

The establishment of sedimentation equilibrium as the method of choice for determining molecular weights was demonstrated conclusively during this period with the development of the high-speed or meniscus depletion technique. By using much higher centrifugal fields than had been employed previously, it was possible to reduce the solute concentration at the meniscus to essentially zero. In this way the displacement of the interference fringes yielded absolute values for the concentration. Data from such experiments permitted the calculation of number average, M_n, weight average, M_w, and z-average, M_z, molecular weights thereby providing important information about the homogeneity or heterogeneity of the proteins. This sedimentation equilibrium technique employing the Rayleigh interference optical system has become the dominant method for analyzing association-dissociation equilibria of oligomeric proteins.

By the early 1960s, workers who had been using the ultraviolet absorption optical system to study DNA were expressing considerable discontent because of the absence of direct viewing of the progress of the experiment and because of the lack of flexibility. Some experiments were total failures because the selected speed for the rotor was too high or too low for the material being examined and this knowledge became available only after the experiment was completed. In other experiments, inadvertent errors in selecting exposure times or in developing procedures led to useless photographic plates. On occasion the concentration of macromolecules was too high or too low, and accuracy was lost because of the limited range in the response of the photographic plate to varying light intensity. These frustrations led to major efforts to redesign the primitive absorption system by introducing modern technological techniques. The resulting split-beam scanning absorption optical system proved convenient to use, sensitive and accurate. Moreover, the incorporation of a monochromator which permitted studies over a broad range of wavelengths in the visible and ultraviolet regions of the spectrum provided discrimination permitting differentiating or distinguishing between various macromolecules based on their absorption spectra. With this system, by merely changing the wavelength of the incident light, it was possible to exploit chemical differences among the various constituents.

This photoelectric scanning absorption optical system which simulated a double beam spectrophotometer proved ideal for analyzing chemically interacting systems. In this way, for example, the interaction of methyl orange with bovine plasma albumin was studied quantitatively by both sedimentation velocity and sedimentation equilibrium techniques. The theoretical treatment for such systems dealt with constituent sedimentation coefficients and molecular weights.

During this period it was shown that simultaneous sedimentation equilibrium experiments on a protein in both H_2O and D_2O (or preferably $D_2{}^{18}O$) could yield both partial specific volumes and molecular weights. Moreover, only microgram quantities of protein were required because of the sensitivity of the split beam photoelectric scanning absorption optical system. Hence, the essential parameter, partial specific volume, was now amenable from ultracentrifuge experiments without recourse to difficult measurements requiring much larger amounts of material.

The Decade from 1970 to 1980

By 1970 there were about 1,500 ultracentrifuges in operation throughout the world. Most of them were used very productively in biochemistry and molecular biological laboratories for the characterization of polypeptide hormones, proteins, nucleic acids and viruses in terms of their shapes, molecular weights and interactions. The photoelectric scanning absorption optical system, the schlieren optical system and Rayleigh interference optics were being used in diverse ways. Sensitivity was achieved with absorption optics, and that system was employed for extremely dilute solutions. Convenience came from the direct viewing provided by the schlieren optical system. Rayleigh interference optics yielded the greatest accuracy and was used primarily for sedimentation equilibrium experiments. In the meantime, as seen in Table 5, further improvements were made in the optical systems. Also, there were additional refinements in the theory of sedimentation. These led to the removal of uncertainties and ambiguities in the interpretation of data. Finally, there were new techniques such as difference and differential sedimentation and active enzyme centrifugation.

Table 5. Ultracentrifugation from 1970-1980.

• Preferential interactions	• Lasers and Rayleigh interference optics
• Rebirth of hydrodynamic theory - frictional ratios	• Pulsed laser interferometry
• Rediscovery of spinning top - airfuge	• Self-associating systems - refined treatments
• Difference sedimentation equilibrium	• Simulation of transport processes for interacting systems
• Difference and differential sedimentation	• Active enzyme centrifugation
• Density and viscosity measurements in ultracentrifuge	• Vidicon and multi-channel analyzer - absorption optics
• Microtubule assembly	• Computer interactions with ultracentrifugation

Developments in hydrodynamic theory resulted in treatments of frictional coefficients for realistic models of oligomeric proteins. Thus different arrangements of subunits could be considered along with linear, circular and supercoiled DNA. During this period, the ultracentrifuge was linked to the computer. Various types of transport problems involving chemically interacting species were simulated. Tedious calculations previously performed by research workers were transferred to the computer. As a result, many programs are now available for evaluating sedimentation equilibrium experiments on self-associating systems and mixed associations.

Just as there had been frustration earlier over the original ultraviolet absorption optical system, by this time workers were complaining about the inadequacy of the split beam photoelectric scanning absorption system. Hence research workers turned to the vidicon and a multi-channel analyzer as a replacement for the moving photomultiplier tube. The potential of this approach has still not been demonstrated. The light source used in the Rayleigh interference optical system for some years was replaced by a pulsed laser source in some laboratories. Its superiority both in terms of intensity and monochromaticity led to a considerable enhancement in the quality of the interference patterns. Since rotor precession was a problem with many of the ultracentrifuge drives, some workers aimed for a pulsed laser of such intensity that a fringe pattern could be obtained during a single revolution of the rotor. This goal was not achieved, but shorter exposure times were attained.

By 1980, the pace of research on the ultracentrifuge, both theoretical and experimental, had slowed demonstrably. Even its application for the study of biological systems had diminished. Hence in the next decade only a small number of laboratories were still using the ultracentrifuge, and only a handful of investigators considered the instrument as its primary or indispensable tool. Replacement parts became difficult to find and workers yearned for a new instrument.

3. RENAISSANCE OF THE ULTRACENTRIFUGE

To those who have been struggling for the past ten years with their "ancient" ultracentrifuge, news of the development of a new instrument was an answer to their prayers. No longer would they spend seemingly endless hours trying to locate light sources, cell parts, new rotors, photographic plates or rebuilt electric drives. Others who had not been exposed to the ultracentrifuge in the past were not aware that it could provide answers to questions research workers were reluctant to ask. Is their transcription factor a dimer or trimer and how is the association-dissociation equilibrium affected by various ligands? If an amino acid at a location far from the interface between chains is replaced by site-directed mutagenesis, will the energy of the interchain interaction be affected and by how much?

It was commonplace to hear expressed, "If only we had a small, quiet, convenient, reliable and versatile ultracentrifuge we could do so many experiments that cannot be done now." We need a direct drive of virtually unlimited lifetime. If this is obtained, there would not be a need any longer to change drives periodically and to check the alignment of the optical system. Of course, there must be no rotor precession which plagued workers in the past. It is ironic that the photoelectric absorption optical system was used in some laboratories to test the electric drive just being installed in the ultracentrifuge. By positioning the photomultiplier tube at a position on the image of the cell corresponding to the air-liquid meniscus, one could detect the movement of the image back and forth and measure both the amplitude and period of the movement of the rotor. With some drives this movement caused such blurring of the interference fringes that precision and accuracy were seriously compromised. In a new ultracentrifuge containing a direct drive mounted below the rotor, this precession should be much less. Speed control was not a problem with the electrically driven ultracentrifuge after the mechanical gear box which controlled the speed was replaced by an electronic mechanism. Clearly, rapid changes in speed are to be avoided so as to minimize convective disturbances. Modern technical advances permit temperature measurement and control without the "medieval" troublesome needle dipping into a mercury cup which had been used for so many years. Not only can we look forward to a sensible radiometer temperature sensing system but also the heating or cooling of the rotor must be accomplished so that a uniform temperature is maintained throughout the rotor.

Temperature control is important especially for titanium rotors which are now in vogue because they can be used at higher centrifugal fields than the rotors made of aluminum alloys. It should be recognized that the thermal conductivity of titanium is much smaller than that of aluminum, and gradients of temperature can lead easily to convective disturbances in the cell. This does not constitute a problem in sedimentation equilibrium experiments but it can be serious in sedimentation velocity studies.

The new ultracentrifuge just now appearing in laboratories has only an absorption optical system which produces superb plots of absorbance *vs.* radial distance in both sedimentation velocity and sedimentation equilibrium experiments. By using the ultracentrifuge with light of short wavelength (220 nm or even lower) workers will be able to study proteins at concentrations of only a few μg/ml. This sensitivity should be coupled with accuracy far in excess of that obtained with the earlier split beam photoelectric scanner. Clearly the data for absorbance (concentration) *vs.* distance must be subjected to smoothing procedures with a sliding window so that derivative curves representing the concentration gradient as a function of distance can be available for the research worker to view as the experiment progresses. It is indisputable that mere viewing of a derivative pattern is more informative than a superficial examination of an integral curve. We see asymmetry more easily in a peak than in an s-shaped curve. Hence there is great reluctance to abandon the schlieren system. But if we can obtain good derivative patterns from the absorption system and use the integral curves for analysis of the data by a computer, we can consider the schlieren optical system a technique of the past.

Although the schlieren optical system is dispensable, the ultracentrifuge must include a Rayleigh interference optical system. Such a system provides remarkable precision and accuracy. Moreover, it is needed for those macromolecules and small molecules which exhibit little absorbance of either visible or ultraviolet light. In addition, the interference optical system has the advantage that most substances (on a weight/volume basis) cause about the same increment in refractive index when dissolved in aqueous solutions. Thus this system is especially valuable for mixtures since the number of fringes produced by each species is strictly related to their relative amounts.

With both the interference and the absorption system in operation more or less simultaneously, workers should have the best of both worlds: accuracy, sensitivity, convenience, and discrimination.

Since the ultracentrifuge of the future is to be linked directly to the computer, much of the tedium and accompanying fatigue of workers should be eliminated. Photography of patterns will no longer be necessary and measuring fringe deviations on plot readers will be eliminated. Laborious calculations will be automatic and there will be no excuses for short cuts by using non-rigorous procedures. Hence, in sedimentation velocity experiments, the transport equation will be used to give accurate sedimentation coefficients for pure substances and weight average sedimentation coefficients for mixtures of both non-interacting and interacting systems. Boundary spreading measurements should yield diffusion coefficients and indications of homogeneity or heterogeneity. Because of the sensitivity of the absorption optical system, experiments will be conducted on macromolecules at very low concentrations; as a consequence, artificial sharpening of boundaries can be minimized. This advantage of accuracy at low concentration is also applicable to sedimentation equilibrium experiments which will be performed on very dilute solutions thereby reducing complications from thermodynamic non-ideality.

Almost certainly the ultracentrifuge will be the dominant tool for studying self-associating systems. Much of this work will involve sedimentation equilibrium with the use of both optical systems in conjunction with sophisticated computer programs to test models involving monomers, dimers and higher oligomers. In addition, it is likely that sedimentation velocity experiments will help to differentiate among various types of equilibria. By integrating across the boundary where disturbances of the chemical equilibria occur because of separation of the various species, workers can evaluate unambiguous weight-average sedimentation coefficients which can also be used to assess various models. For interacting systems involving different components, such as a protein and a nucleic acid, the combination of the two optical systems should constitute a powerful tool. Here one would evaluate constituent sedimentation coefficients and constituent molecular weights.

Both the absorption and interference optical systems permit the direct measurement of the small difference in the sedimentation coefficients between two samples. If the two compartments of a double sector cell are filled with two solutions rather than a solution and a solvent, the resulting pattern will represent the difference in absorbance (or refractive index) as a function of distance from the axis of rotation. Hence the addition of a ligand to one sample which promotes a conformational change affecting the sedimentation coefficient will in turn lead to a difference pattern that is readily analyzed to give directly the percent change in sedimentation coefficient. This technique should be capable of measuring a change of 1% in sedimentation coefficient to an accuracy of about 1% or 2%. An analogous difference sedimentation equilibrium experiment might prove useful in measuring very accurately the effect of ligands in shifting the association-dissociation equilibria in self-associating systems.

Biochemists will find that the extinction coefficient of their newly isolated favourite protein can be measured most readily by using the ultracentrifuge with the two optical systems. Merely layering solvent over a dilute solution of protein (1-4 mg/ml) in a synthetic boundary cell will permit the measurement of the absorbance and the number of fringes. The latter quantity yields with high accuracy the concentration in mg/ml and the absorbance can then be converted into the extinction coefficient (on a weight/volume basis). Shapes of macromolecules, especially in terms of ligand-promoted changes, are likely to be the subject of renewed interest based on measured sedimentation coefficients and molecular weights yielding frictional coefficients which can be interpreted in terms of various hydrodynamic models. Undoubtedly some workers will use the ultracentrifuge to measure the compressibility of macromolecules and others will measure volume changes resulting from both self-associations and mixed associations.

It will be fun to see new cells, modified theories, the use of time-dependent concentration changes, and unexpected applications to biological systems. We can hardly wait.

SELECTED REFERENCES

1. R.L. Baldwin and K.E. van Holde, 'Advances in Polymer Science', Springer-Verlag, Berlin, 1960, pp. 451-511.
2. T.J. Bowen, 'An Introduction to Ultracentrifugation', Wiley-Interscience, London, 1970.
3. J.M. Creeth and R.H. Pain, *Progr. in Biophysics and Molec. Biol.*, 1967, *17,* 217.
4. B. Elzen, Ph.D. Thesis, University of Twente, 1988.
5. H. Fujita, 'Foundations of Ultracentrifugal Analysis', Wiley, New York, 1975.
6. H.K. Schachman, 'Ultracentrifugation in Biochemistry', Academic, New York, 1959.
7. H.K. Schachman, 'The Beckman Symposium on Biomedical Instrumentation', The Rockefeller University, New York, 1986, pp. 37-58.
8. T. Svedberg and K.O. Pedersen, 'The Ultracentrifuge', Clarendon, Oxford, 1940.
9. J. Vinograd and J.E. Hearst, 'Progress in the Chemistry of Organic Natural Products', Springer-Verlag, Wien, 1962, pp. 372-422.
10. J.W. Williams (ed.), 'Ultracentrifugal Analysis in Theory and Experiment', Academic, New York, 1963.
11. J.W. Williams, 'Ultracentrifugation of Macromolecules: Modern Topics', Academic, New York, 1972.
12. J.W. Williams, K.E. van Holde, R.L. Baldwin and H. Fujita, *Chem. Revs.*, 1958, *58,* 715.
13. D.A. Yphantis (ed.), 'Advances in Ultracentrifugal Analysis', *Ann. N.Y. Acad. Sci.*, 1969, *164*.

2

The Optima XL-A: A New Analytical Ultracentrifuge with a Novel Precision Absorption Optical System

By Robert Giebeler

BECKMAN INSTRUMENTS, SPINCO DIVISION, 1050 PAGE MILL ROAD, PALO ALTO, CALIFORNIA 94304 U.S.A.

1. INTRODUCTION

A new analytical ultracentrifuge - the Optima XL-A - has been designed that includes a novel precision absorption optical system offering greatly extended performance and ease of use. Radial scans can be accomplished at wavelengths between 190 and 800 nm with a range of up to three absorption units, and in addition, wavelength scans can be done at specified radii in the same manner as a spectrophotometer. This new optical system is designed into a new ultracentrifuge that provides a very smooth-spinning, precession-free induction drive, and precision and highly stable temperature control. Here we review some of the important features of this instrument, whose development was announced some time ago[1], beginning with a close look at the optical system.

2. OPTICS

The improved performance of this absorption optical system is provided by a new design encompassing a highly stable, UV-enhanced xenon flashlamp as the source, a holographically-generated, toroidally-curved diffraction grating especially designed for the application, a new approach for imaging the sample, and lastly, a supersensitive photomultiplier tube for detection (Fig. 1).

The xenon flashlamp is fired as the sector of interest passes directly over the detector, and the duration of the flash is about one microsecond with an exponentially decaying tail. Data is taken one sector at a time in this manner, where the maximum firing rate of the tube is 100 flashes per second, corresponding to once every ten revolutions at 60,000 rev/min. The timing for firing of the tube is generated by detecting the passage of a magnet mounted in the bottom of the rotor with a Hall effect sensing device. The time period is measured between successive magnet index pulses, and a delay time is determined based on that period and the position of the sector and cell of interest with respect to the magnet. Errors in delay time for tube firing are corrected during initial instrument calibration by varying delay time about the nominal value to ensure that the flash is being fired when the sector is directly above the detector.

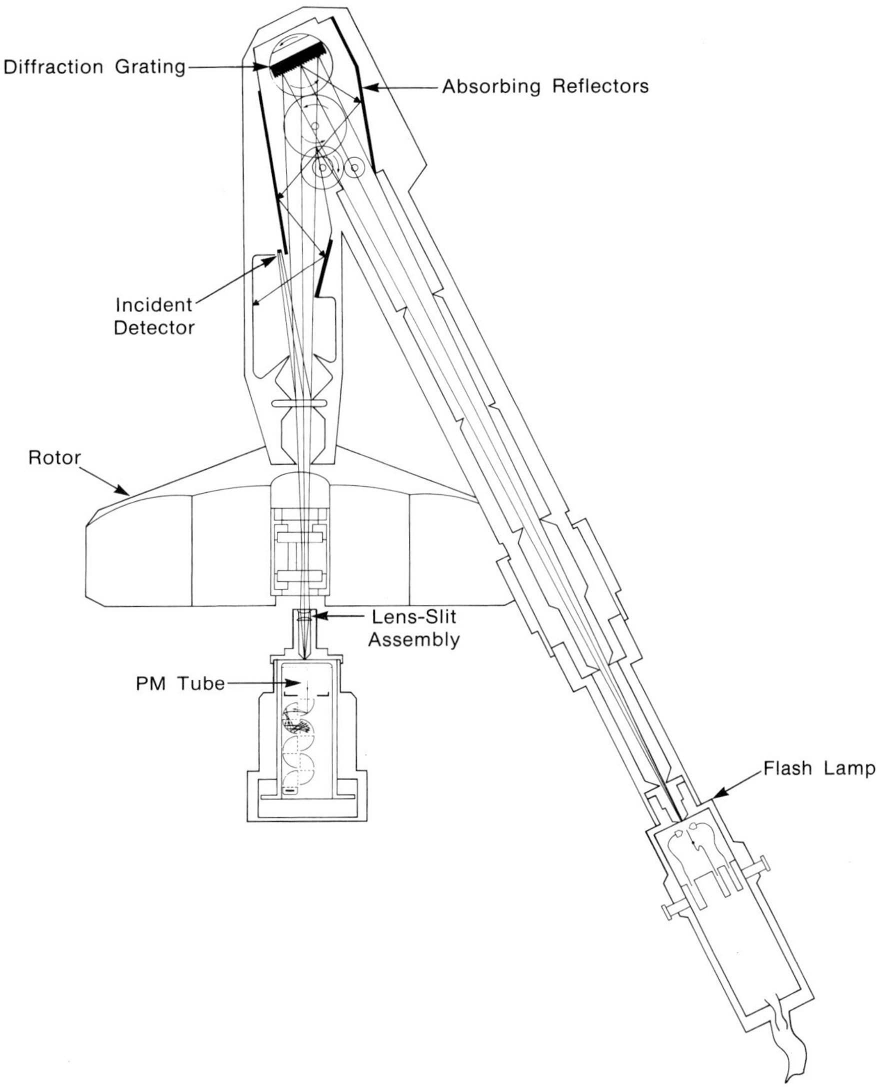

Figure 1. Tangential cross-section of the XL-A. The optical system of the XL-A is depicted in the diagram. Both the flash lamp and the photomultiplier tube, with its associated lens-slit assembly, are permanently mounted in the bottom of the centrifuge housing. The piece holding the diffraction grating may be installed or removed by means of a bayonet mount located by the flash lamp. See the text for details.

Light from the xenon flashlamp is projected from a 1 mm circular aperture which then impinges on the toroidally-curved diffraction grating. This grating is aberration-corrected so that the system has a nominal 2 nm band pass between 200 and 800 nm with very low stray light. The grating is rotated by a precision gear train to select or scan wavelengths in this range. This diffraction grating is additionally blazed by ion bombardment to intensify the first-order light efficiency in the UV range, which also decreases chromatic stray light.

All light not passing out of the exit aperture of the monochromator is absorbed by a series of semi-reflective and non-reflective absorbing filters to prevent light at wavelengths other than that of interest from being reflected back to the diffraction grating and then out of the monochromator. Stray light levels at 210 nm are typically 0.1%.

Light leaving the diffraction grating then impinges on an 8% reflector which reflects this light back to a solid-state detector which is mounted in the virtual focal point of the monochromator system. This detector monitors the intensity of light incident on the sample in the cell assembly. This intensity value, in conjunction with the transmitted intensity value as measured by the PM tube, is used to normalize the pulse-to-pulse variations in the light intensity emitted from the flashlamp. This is necessary because the reference and sample sectors in the cell are measured separately. The exit aperture of the monochromator prevents illumination of the sector not being measured. This eliminates spectral stray light and the resulting non-linearity in absorption measurement.

Light passing through the mid-plane of the sector being measured is then refocussed by the camera lens assembly in the plane of the slit above the PM tube detector (Figs. 2 and 3). This slit serves two purposes in the system. The length of the slit shown in the T-plane (Fig. 3) is the band pass determining aperture for the monochromator, which is about 2 nm, as determined by benzene vapour scans. Effectively, a band of light only about half the width of the sector is used, so that time jitter in the firing of the flashlamp does not cause noisy absorption data. The dispersion of wavelength in this plane of constant radius in the sample does not degrade data quality, because the sample is homogeneous at a constant radius. The width of this slit, shown in the R-plane, is the spatial resolution aperture for radial scans and is about 25 μm wide (Fig. 2).

The function of the other dimension of curvature on the diffraction grating is to collimate the light in the T-plane that illuminates the sample in the cell assembly (Fig. 2). Very parallel light with low spatial dispersion is necessary for illuminating the sample in order to get good quality data in high absorbing samples near the meniscus because of the limited depth of field of the camera lens. In order to ensure that this parallel light is also parallel to the meniscus in the sample, the source position can be adjusted. The image of the sample in the cell is scanned with the lens/slit assembly which moves as a unit. There is an additional aperture above the camera lens in this assembly to limit the angular diversion of light coming from the sample to ±4°, regardless of radial position. This minimizes spatial stray light due to scattering from

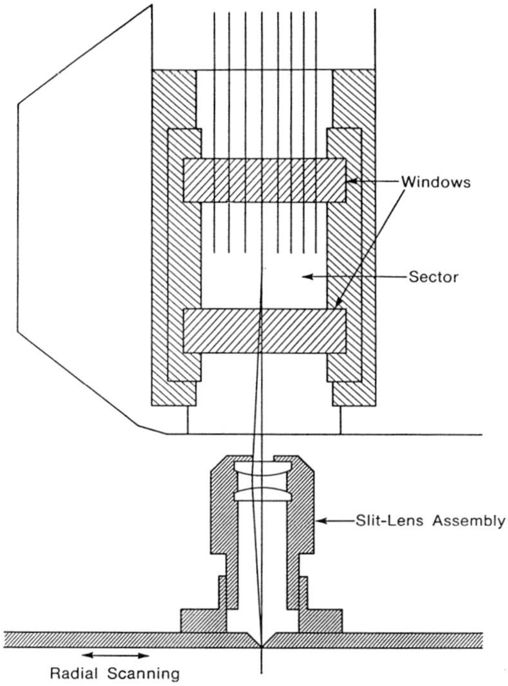

Figure 2. Radial scanning of the sample. A portion of the rotor is shown in cross-section, with the sample cell and its associated quartz windows. Light transmitted through the sample impinges upon a slit-lens assembly, which is displaced radially to scan the sample. The width of the slit is 25 μm.

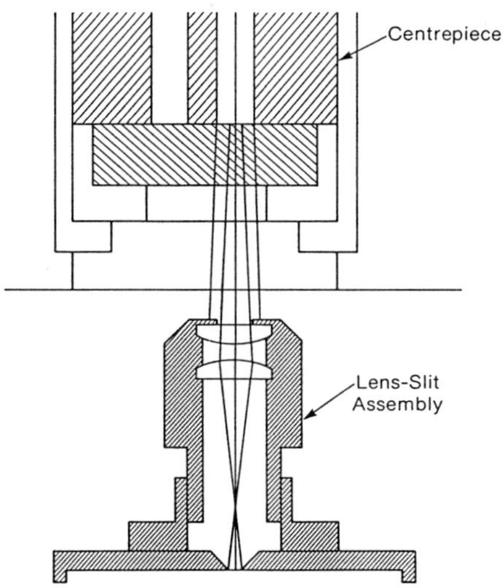

Figure 3. Camera lens-slit imaging. A plane within the sample sector is imaged by lenses in the moving slit assembly. Angular diversion of light from the sample is limited to ±4° to reduce stray light detection.

being detected and, at the same time, allows a generous refraction by the sample due to gradient salts that may be used (Fig. 3). The focal length of this camera lens assembly is adjustable to compensate for a variation in the index of refraction of the glass used at different wavelengths. However, it is rarely necessary to refocus because of the quality of collimation of light illuminating the sample and the relatively small amount of light ray steering that the camera lens accomplishes as a result of moving with the slit assembly.

As a result of the lens moving with the slit, this slit/lens assembly undergoes motion which is directly proportional to the image size of the cell. The active area of the PM tube is larger than this distance and, therefore, remains stationary. Inhomogeneity in the illumination field of the sector in the cell and in the sensitivity of the PM tube is all normalized in the ratioing of sample and reference sectors, provided the slope of these variations is not excessive. An excessive slope in the illumination field will degrade data quality, because the light detected at a given radial position enters the sample and reference sectors at slightly different radial positions when a high differential gradient exists between the two. The image size in the plane of the slit is actually a half of the cell object size. This improves the depth of field of the camera lens assembly.

The last optical element is the blocking filter (Fig. 3) that is placed in the beam path above the slit when operating above 400 nm. This filter blocks the second and third orders of light coming from the diffraction grating that would otherwise appear as stray light and cause excessive non-linearity of absorption measurements while scanning above 400 nm.

3. DATA ACQUISITION SOFTWARE

Implementation of the above-mentioned optical system is made possible by a very software-intensive data acquisition system. This software controls the accurate firing of the flashlamp, the positioning of the grating and the radial detector to the exact position of interest, the calculation of sample absorption based on signals from the incident and transmitted detectors, and lastly, the calibration of the firing delay time, radial position, and wavelength.

As previously discussed, firing of the flashlamp is controlled in software by measuring the period of one rotor revolution and delaying the firing of the tube for a time that equals that needed to position the cell and sector of interest in the optical path. This timing is very critical and needs to be precise to at least one part in 2,000.

Variability of components in the system (time jitter in the index pulse detection system and the firing physics of the xenon flashlamp) necessitates recalibrating delay time with an offset correction before each run. The delay time calibration software offsets the strobe firing delay to a point equivalent to the centre of the peak of the sector transmitted intensity.

Both the wavelength position of the diffraction grating and the radial position of the detector are adjusted by DC motors and detected by precision potentiometers.

Complex Proportional-Integral-Differential Servo algorithms in software control the motors to position radius and wavelength. Calibration of wavelength and radial position again is accomplished in software. After initial factory calibration at several different wavelengths, a repetitive single point wavelength calibration is done before each scan, based on locating the 230 nm emission peak from the xenon flashlamp. Similarly, before each scan, radial accuracy is calibrated based on scanning the reference edges in the counterbalance similar to the method used in the Model E. The software locates these edges by monitoring the transmitted intensity detected by the PM tube, and then corrects all subsequent radial position data. The user can enter, *via* the operator interface, the exact numerical values of these reference edges of the counterbalance in use. These radial values are not compensated for rotor speed.

The absorption of the sample as compared to the reference at each specific wavelength and radial position is calculated from data from the incident and transmitted detectors for both sample and reference sectors. Sample and reference sector transmitted intensity is normalized by the incident intensity detected simultaneously, eliminating the effect of flash-to-flash intensity variations from the xenon flashlamp, except while doing wavelength scans. These normalized intensities are then used to calculate sample absorption.

The intensity values used in this calculation are also compensated for PM tube and Programmable Gain Amplifier (PGA) gains that existed during the reading. These intensity values represent the real light intensity that the PM tube is detecting which, as will be discussed later, serve as a valuable diagnostic tool. The PM tube voltage and PGA settings are adjusted based on sample and reference intensities to maintain signal levels near the maximum at the input to the analogue-to-digital conversion block to maintain maximum resolution (Fig. 4). The PM tube voltage is adjusted with PGA = 1 when signal is near maximum, and then the PGA is set based on sample sector intensity to maintain this analogue signal near the maximum (PM tube voltage remaining constant). This provides signal detection over an overall six decade range with near constant resolution and good linearity.

The detected light intensities are also corrected for base-line offsets. With no light entering the PM tube, the PM tube voltage and Programmable Gain Amplifier settings are adjusted throughout their operational ranges, and the resulting intensities are measured and used in a base-line correction algorithm to establish zero light intensity, irrespective of gain settings, PM tube noise levels, or component offsets.

Data quality is enhanced by data averaging. Up to 50 readings can be averaged together per radial or wavelength data point. This can be done in either of the two different scan modes of continuous or step scan. In the continuous scan mode, the radial slit position moves continuously, and readings between successive data point positions are averaged together. In the step scan mode, radial position is set to each successive radial position and remains in that position as the readings are taken that are used in that average. The advantage of the continuous scan mode is that it is more rapid and also averages the optical aberrations in the cell windows, thereby decreasing the apparent noise level. The advantage of the step scan mode is that more data points can be averaged together, which is necessary when operating down at very low light

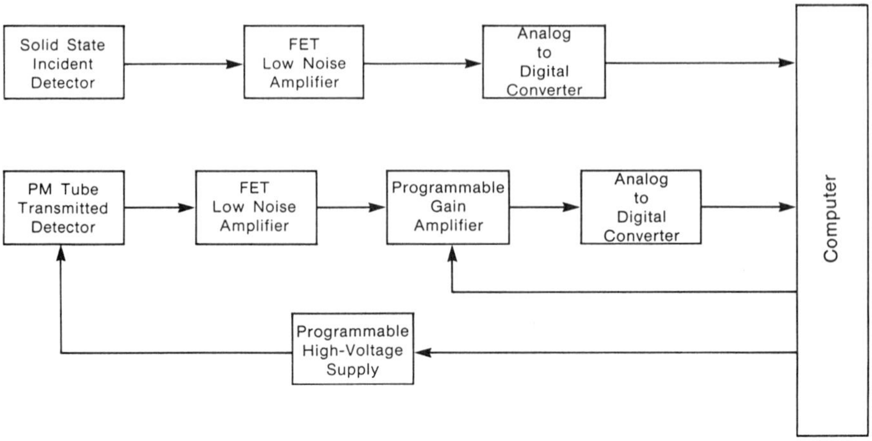

Figure 4. Block diagram: light signals. The sequential processing of the light signal by elements of the system is shown.

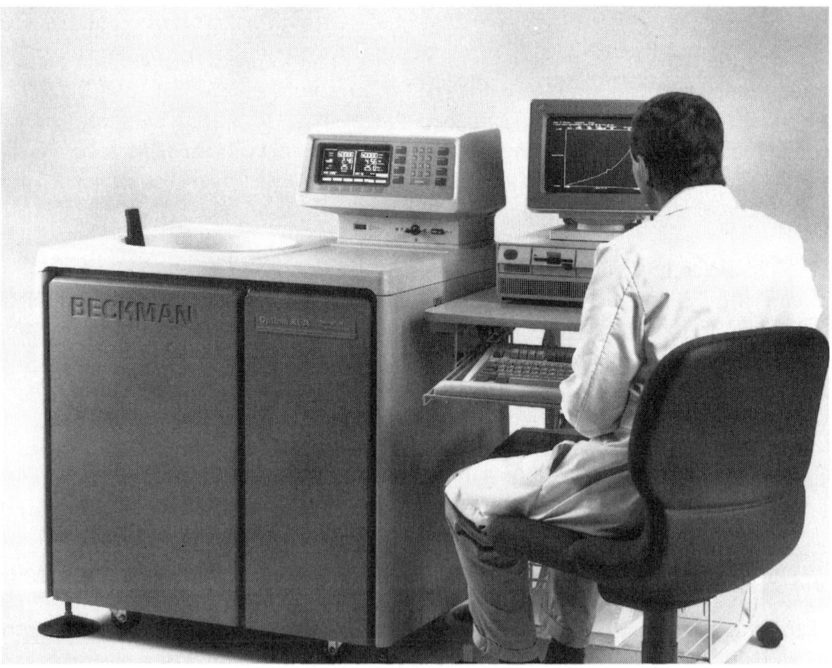

Figure 5. The Optima XL-A analytical ultracentrifuge.

levels to enhance the PM tube signal-to-noise ratio. It is also necessary to use the step scan mode on a wavelength scan so that the wavelength is exactly the same for sample and reference sectors, because the emission spectrum from the xenon flashlamp has very high peaks at specific wavelengths.

4. THE ULTRACENTRIFUGE

The new ultracentrifuge (Fig. 5) that encompasses this new optical system provides many improvements over the Model E ultracentrifuge.

The drive comprises an induction motor that operates under computer control to provide rapid, controllable acceleration rates, vibration-free operation, precision control and that exhibits virtually no precession while at speed. In addition, lifetime expected while using analytical rotors is above 100 billion revolutions, and overall reliability is vastly improved. The unique dynamic damping in this drive system allows all asynchronous precession to dampen out to virtually zero within a matter of minutes, even if induced.

Temperature control is considerably more stable, provides more rapid cool-down and heat-up rates, is thermally more uniform, and has equivalent accuracy as compared to the Model E. This control system uses an isothermal radiometer temperature-sensing system to sense the temperature of the rotor that is emissivity-independent and view factor-corrected in software. Heating and cooling of the rotor are accomplished by the refrigeration can that surrounds the rotor, which is in turn heated and cooled by thermoelectric modules. This environment is very isothermal, and at equilibrium, irrespective of speed or temperature, rotor temperature is within about one degree of the refrigeration can temperature.

The control system that regulates rotor temperature, as monitored by the radiometer, is highly software-intensive. This software encompasses triple proportional-integral-differential control algorithms and proportional integral smoothing algorithms. In addition, radiometer view factors are measured during rotor cool-down to allow more rapid rotor cool-down and more accurate temperature monitoring during cool-down. While at equilibrium, refrigeration can temperature fluctuation does not typically exceed one degree, and the corresponding rotor temperature fluctuation is less than 0.02°C.

The new ultracentrifuge is overall much easier to use and more reliable as a result of this microprocessor-intensive design. This includes improved serviceability as a result of the microprocessor-based system diagnostics.

5. ROTORS AND CELLS

The new analytical ultracentrifuge is designed around the existing cell design used in the Model E. It can run all double-sector 12 mm centrepieces and six-channel equilibrium centrepieces. For ideal results, newly manufactured windows should be used that are made of more pure quartz than windows available previously and have correspondingly less absorption down below 220 nm. In addition, window holders have been redesigned to provide a higher sector view factor all the way to the bottom of the sector. New cell housings should be used with anodised screw rings to reduce stray light.

The new titanium analytical rotor used in this ultracentrifuge is four-place. Additional six- and eight-place rotors are contemplated. This new analytical rotor is lighter and has more uniform stress levels throughout the rotor, providing for a more uniform adiabatic temperature change throughout the rotor as it accelerates. In addition, rotor growth is larger than earlier analytical rotors, requiring different corrections of radial calibration, depending on speed.

6. ANALYTICAL OPERATOR INTERFACE

The user of the new analytical ultracentrifuge controls the instrument through an IBM-PC based operator interface. This operator interface provides the ability to set up run conditions for the ultracentrifuge that alternatively can be done *via* the instrument-based operator interface, to set up the scan conditions for either manual or automated scanning, to review the data generated by scanning, to store and recall this data from the data base, and lastly, to diagnose the analytical system functionality and manually control functions.

The format for this operator interface is a series of pulldown menus. The operator has the choice of either manual scanning, or automatic scanning at a regular time interval. In the manual scanning mode, the operator has the additional choice of performing radial scans at specified wavelengths or wavelength scans at specified radii. Radial scans can be accomplished at up to three different wavelengths in succession in a semi-automated fashion. Wavelength scans can be done in succession between two designated radii at a given delta radius for each successive scan.

The automatic mode allows the operator to scan all cells specified at a specified time interval automatically, with a specified start and stop time. In the manual mode, the data last taken can be immediately displayed. In the automatic mode, any data from any particular scan in the series can be retrieved from the database. Each scan is automatically named in a manner that incorporates key identifiers of the run.

In reviewing data from scans just completed, several options are provided. The data of primary interest to the experimenter is absorption data. The other data, intensity data, which shows the sector intensity from which absorption is calculated, and incident data, which shows the comparison of the incident and transmitted detectors, can serve as useful diagnostics. For example, by monitoring reference sector intensity, one can easily determine if unexpected excessive reference buffer absorption exists, which if present can increase stray light and seriously degrade linearity of absorption measurement.

Presentation of data allows the option of intensity scales being either logarithmic or linear, and in addition a cursor is provided to read off the specific numerical data from the plot. Absorption scales have several display options, including A *vs.* r, ln A *vs.* r^2, and dA/dr *vs.* r.

Sample absorption, when selected from the menu, is calculated as an average of the successive absorption calculations when multiple data points are being averaged to aid in the calculation of standard deviation. This is in contrast to the absorption calculation that can be done on the IBM directly from averaged intensity data. If the operator is averaging large numbers of data points to reduce system noise (*i.e.*, with a highly absorbing reference) and is also operating near a zero absorption baseline, then the absorption data calculated directly from averaged intensities is more accurate, due to the inability to take the log of a negative ratio. Absorption data can be reduced to the subset of interest and stored in this configuration.

The utilities menu allows directories to be created and changed for a specific user's database. The identification of scans in the designated directory is provided by the first four letters of the scan name. Additional diagnostic tools are provided in the service menu. From this menu, various parameters such as wavelength and radius can be manually adjusted. Other types of diagnostic scans such as time scans for checking drive precession, and delay scans for checking flashlamp firing calibration can be performed. In addition, calibration parameters can be modified manually from this menu.

7. CONCLUDING REMARKS

Starting with the work of Svedberg early in this century, the analytical ultracentrifuge has provided a powerful approach to the characterisation of proteins and other macromolecules. In the past decade, as more and more proteins have been produced in pure form through recombinant DNA techniques, the demand for a tool with the capabilities of the analytical ultracentrifuge has increased dramatically. Yet the only available instruments in general use have been ancient commercial Model Es, or modifications of these instruments upgraded by individual workers. At best these instruments have been unreliable, at worst they have been difficult to use. The new Optima XL-A solves this problem by giving a reliable, accurate, easy-to-use instrument. It will serve as a new tool in the repertoire of every biochemist.

REFERENCE

1. H.K. Schachman, *Nature*, 1989, *941*, 259.

3

Experience with the MOM 3180 Analytical Ultracentrifuge

By E. Görnitz and K.-J. Linow

INSTITUTE OF POLYMER CHEMISTRY, KANTSTR. 55, TELTOW-SEEHOF, GERMANY O-1530

1. INTRODUCTION

In the polymer research laboratories of today the method of analytical ultracentrifugation - which is referred to in standard texts of macromolecular chemistry as one of the standard methods for estimating molecular weights of natural and synthetic polymers - can only rarely be found alongside modern computer coupled gel permeation chromatography (GPC)-devices and light scattering photometers. Certainly one of the reasons for this situation is based on the complexity of operation and data handling and in the tendency to malfunction of these big old machines. There are now just two commercially available analytical ultracentrifuges: the Optima XLA very recently launched from Beckman Instruments (see Chapter 2) and also the 3180 analytical ultracentrifuge from the Hungarian Optical Works MOM. Our experience with using MOM analytical ultracentrifuges goes back more than 20 years particularly with the air turbine driven MOM G120. The MOM factory started producing analytical ultracentrifuges in 1959 and since that time has continually worked on the further development of these machines and until very recently has been the only commercial manufacturer of an ultracentrifuge assigned specifically for analytical use and which can also be used for preparative work.

The newest model from Budapest - which we have had in our laboratory now for three years - is the MOM 3180 Analytical Ultracentrifuge (Fig. 1). It has a completely new drive mechanism and is smaller (the floor area requirement being 1.40 m $\times$ 0.75 m) and much more convenient for operation than its predecessors.

2. THE DRIVE SYSTEM

MOM 3180 rotors are, in common with preparative ultracentrifuges (but unlike the Beckman Model E), bottom loading in the armoured vacuum chamber. The chamber is opened from the top, thus the rotor is easy to load or unload from the drive shaft. The *drive* is taken from a normal 1.1 kW asynchronous motor, which very quickly acquires full speed. The heart of the drive mechanism is a special hydraulic clutch, which facilitates a soft acceleration of the rotor. Speed control is achieved by variation of the quantity of oil in the clutch. A digital electronic regulation system, which is

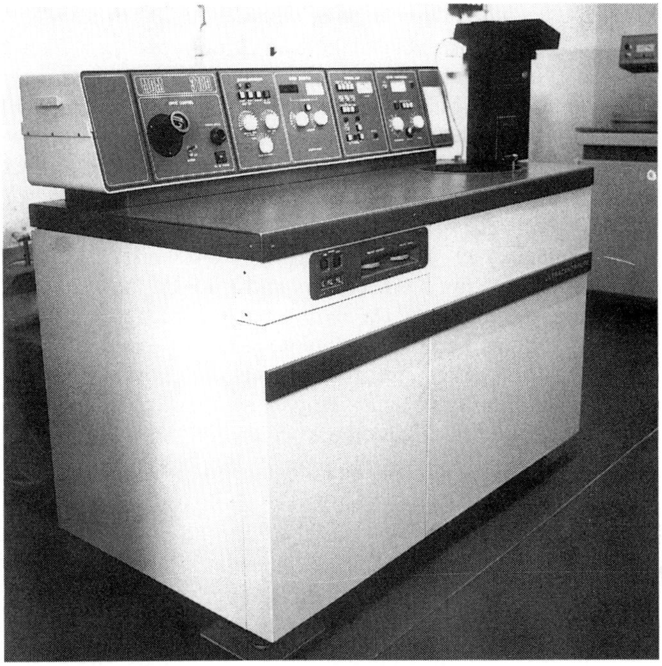

Figure 1. The MOM 3180 analytical ultracentrifuge.

controlled by an odometer system emanating from the drive axle of the rotor, ensures, with the help of a small gear pump, that the quantity of oil in the clutch is at such a level that the transmitted torque maintains the desired rotor speed. The transmission of force from the clutch to the drive shaft of the rotor takes place by the combination of a toothed plastic belt and an acceleration gear-box with an overall transmission ratio of 1:25. The high speed bearing for the rotor shaft in the chamber bottom is constructed in such a way that no oil from the outer case of the bearing can enter the vacuum chamber. Indeed, during our work with the apparatus we had no problems from oil mist which can lead to dirty optical surfaces in the chamber (this is particularly serious for ultraviolet absorption measurements).

The digital electronic speed regulation system permits selection between 39 discrete rotor speeds from 2,000 rev/min to 65,000 rev/min and with an accuracy of ±30 rev/min. Acceleration of the standard two-place rotor from zero to 60,000 rev/min takes place in ~6 min in the case of our machine. An electronic safety system prevents overspeeding and operation errors. Thus, by means of photoelectric registration of light impulses reflected from a safety coding ring at the bottom of every rotor, the drive is taken off automatically, if the speed becomes higher than the maximum speed value of this rotor.

3. VACUUM AND TEMPERATURE SYSTEMS

The *vacuum* is produced by a two stage rotary valve pump and for higher rotor speeds by an additional oil diffusion pump.

An exact *temperature control* of the rotor is highly important for reliable analytical working. In our apparatus the inner wall of the vacuum chamber is cooled to an adjustable value by means of a refrigeration system. The temperature is measured by a thermistor, dipping into a blackened tube projecting from the rotor top and in this way permitting the accurate detection of the radiation temperature of the rotor. The measured signal controls an electronic device which regulates the heat output of a small heater arranged in the vicinity of the rotor surface. This system regulates the temperature of the rotor in the range from 0°C to +50°C with an absolute accuracy of ±0.1°C and with a constancy of regulation of ±0.05°C. At working temperatures quite different from room temperature it is necessary to thermostat the rotor (with the filled cell) before placing it in the MOM 3180 in order to avoid long thermal equilibration times. An additional heating insert in the vacuum chamber makes the apparatus suitable for measurements at elevated temperatures up to 150°C, this being of particular importance for investigations of polymer melts or synthetic polymers soluble only at high temperatures: for these measurements a special high temperature rotor is available. Thus it is fair to say that the overall temperature range for analytical measurements is from 0°C to 150°C.

4. OPTICAL SYSTEMS

The *optical system* of the MOM 3180 contains, like its predecessor models, Philpot-Svensson optics and Rayleigh interference optics employing a 200 W high-pressure mercury vapour lamp with 546 nm interference filter as the light source. An additional feature is the "combined" optical system, permitting the simultaneous recording of schlieren and interference pictures on one photograph by placing a particular combination of an optical grating with a slit in the optical path. In our investigations this "combined optics" has proved its worth, not only for the evaluation of synthetic boundary experiments (for the determination of solute concentrations or diffusion coefficients), but also in sedimentation equilibrium experiments. The optical system of the centrifuge can be very easily changed even during the course of the experiment by rotating a disc containing the interference grating, the combined slit-grating and three slits of different width for schlieren measurements.

Photographic recording of the schlieren and interference pictures onto roll-film of size 6 cm × 4.5 cm is performed by means of an automatic camera operation which allows large variation in controlling exposure times and intervals. The magnification factor from sample-cell to film is about two, both in the horizontal and vertical directions. The optical conversion factor of the Philpot optics (*i.e.* the distance between the collimating lens and the plane of the schlieren diaphragm) is about 75 cm.

As an optional extra accessory an ultraviolet absorption optical system is available. For this facility an independent light path is used and this requires two additional quartz windows in the bottom and the lid of the chamber. A low-pressure mercury vapour lamp with 254 nm interference filter serves as a light source and a plate camera for uv sensitive photographic plates as a uv detector. Here a major disadvantage of this apparatus becomes apparent for all scientists working in the field of biochemistry and needing absorption in the uv range as a main tool for the detection of sedimentation behaviour, because the absence of a photoelectric scanner makes the measurement highly complicated. Besides the mandatory densitometric evaluation of the photographs, the relation between concentration of the solution and blackening of the photographic plate has to be estimated for every system investigated by a series of calibration measurements. On the other hand it is an advantage of the independent light path of the uv absorption optics that schlieren and/or interference pictures and uv absorption can be recorded simultaneously. [The ease of conversion between optical systems was also a feature of the old MSE Mark II and Centriscan analytical ultracentrifuges - now no longer commercially available - see *e.g.* refs 2 and 3 - the latter having a photoelectric scanning system for both schlieren and absorption optics.]

For our work in the field of polymer research the problem of the absence of a scanning absorption system was not of such importance since most of the synthetic polymers investigated had no appreciable absorption in the usable ultraviolet region of the spectrum. However we can use the absorption optical system to address a completely different kind of problem that one is often faced with in the field of polymer technology, namely for the analysis of the particle size of polymer latices. For this purpose we replaced the uv light source with a He-Ne laser, coupled, for the simplification of the adjustment, *via* a fibre optic. The camera was eliminated and replaced by a photoelement connected to a recorder for registration of light intensity as a function of time. If an analytical cell is used, of which one quartz window is covered by an apertured stop with the slit perpendicular to the direction of sedimentation, one will find, when the sedimentation boundary passes by, an increase of the light intensity because of the decreasing turbidity of the particle suspension. From the shape of the intensity-time-curve the particle size distribution can be calculated, taking into account the dilution effect in the sector-shaped cell and the particle size dependence of turbidity by the help of Mie scattering coefficients. Thus this method, which was described in detail in ref. 1 and has been perfected by other polymer scientists[4,5], could be adapted to our apparatus without too much difficulty. This facility allows the determination of particle size distributions in the range between 50 nm to 5 μm. For smaller particles - which contribute practically no turbidity - the particle size distribution can be estimated by the conventional determination of the distribution of sedimentation coefficients using the schlieren method, as has been shown earlier[6].

5. MOM 3180 ACCESSORIES

The *accessories* of the MOM 3180 consist of a relatively large selection of rotors (all made from aluminium alloys) and analytical cells. Only the standard two-place rotor can be used up to 65,000 rev/min. In addition, there is a two-place rotor (maximum angular speed, ω_{max}≈50,000 rev/min) useful for work with lower solute concentration in cells with larger path lengths (up to 30 mm), a four-place rotor (ω_{max}≈50,000 rev/min) for measurements with wedge window cells in multi-cell schlieren or interference runs, a six-place rotor (ω_{max}≈40,000 rev/min) for a set of double sector interference cells with wedge-shaped window holders and centrepieces (-1°, -0.5°, 0°, +0.5°, +1° and one counterbalance) for simultaneous generation of the interference patterns of five samples. There is, in addition, an especially heavy four-place rotor (ω_{max}≈20,000 rev/min) to facilitate extremely smooth running at low rotor speeds and the above mentioned two-place high temperature rotor (ω_{max}≈45,000 rev/min, T_{max}=150°C). In addition there are a number of swinging bucket and fixed angle rotors for preparative applications.

The great number of analytical cells with centrepieces made from materials of different mechanical stability and chemical resistance (aluminium, polyamide and filled Epon) permits the application of all established techniques in analytical ultracentrifugation and known from work on the famous Beckman Model E apparatus. There are centrepieces for single and double sector cells, synthetic boundary cells, band forming cells for the zonal and density gradient technique, mechanical separation cells and Yphantis-type multi-channel equilibrium cells. Cell windows from quartz and sapphire are available.

It must, however, be stressed that the use of multi-cell runs for the generation of several schlieren or interference pictures at one photograph by means of wedge-shaped windows or centrepieces can only be an unsatisfactory substitute of a real scanning or multiplex operation, which is not available with the MOM 3180: judging from our experience it is a very involved procedure to achieve a fairly satisfactory quality of interference records from such a five-cell interference run.

6. SEDIMENTATION EQUILIBRIUM OF POLYMERIC SYSTEMS

In our investigations in the field of synthetic polyelectrolytes besides conventional measurements of sedimentation velocity[7] the various techniques of sedimentation equilibrium[8] become more and more interesting. Such aqueous polymeric systems - needing the addition of low-molecular electrolyte for the screening of electrical charges at the molecular chain - are unfortunately mostly prone to aggregation, complicating the characterization of the molecular parameters. It has recently been shown[9] that by low-speed equilibrium measurements in combination with synthetic boundary experiments at very low rotor speeds it is possible to estimate the molecular weight of the molecularly dispersed polymer as well as the amount and averaged weight of the aggregates: in these experiments, which employ both the interference and combined schlieren and interference systems, the apparatus has proved its value with regard to smooth running (with the heavy rotor a run is free from vibrations even with speeds as low as 2,000 rev/min) and the reliability of the machine at long running times (up to several days).

7. THE FUTURE OF THE MOM 3180

To conclude then, we think it is fair to say that because of the reliability and the wide range of applications of the MOM 3180, it is a helpful tool for the characterization of polymer solutions and dispersions especially by routine techniques, including sedimentation velocity and short-column sedimentation equilibrium measurements. However, to push it well beyond the limits of rival techniques - such as light scattering - a number of demands arise with regard to higher efficiency of data output from analytical runs from this instrument. There is first of all the requirement of supplying the uv absorption optical system on the MOM 3180 with a photoelectric scanner. Further and particularly for such applications as sedimentation equilibrium or density gradient experiments (often needing a long series of time consuming measurements), a multiplex device for the operation with six- or eight-place rotors will be necessary.

With regard to the drive system, a continuous and programmable adjustment of the rotor speed is desirable, which would be helpful for investigations of broad distributions of molecular weights or particle sizes by sedimentation velocity measurements and for the technique of overspeeding to shorten the time to achieve sedimentation equilibrium. But the most important question for the future of the instrument - and for the method of analytical ultracentrifugation generally - seems to be, how to inexpensively solve the problem of an on-line data registration and transfer from the optical systems to a computer.

REFERENCES

1. W. Scholtan and H. Lange, *Kolloid-Z. Z. Polym.*, 1972, *250*, 782.
2. T.J. Bowen, 'An Introduction to Ultracentrifugation', 1970, John Wiley, London.
3. C.H. Emes in 'Centrifugation: a Practical Approach' (ed. D. Rickwood), 1978, pp. 173-194, IRL Press, London.
4. W. Mächtle, *Angew. Macromol. Chem.*, 1988, *162*, 35.
5. H.G. Müller, *Colloid Polym. Sci.*, 1989, *267*, 1113.
6. K.-J. Linow, K.-H. Goebel and H.-J. Schneider, *Acta Polymerica*, 1981, *32*, 660.
7. G.I. Timofejeva, S.A. Pavlova, Ch. Wandrey, W. Jaeger, M. Hahn, K.-J. Linow and E. Görnitz, *Acta Polymerica*, 1990, *41*, 479.
8. P.M. Budd, in 'Comprehensive Polymer Science', G. Allen and J.C. Bevington (eds.), Vol.1: Polymer Characterization, Pergamon Press, 1989, pp. 199 and 215.
9. E. Görnitz, K.-J. Linow, Ch. Wandrey and H. Dautzenberg, in preparation.

4

The Preparative Ultracentrifuge as an Analytical Tool

By Geoffrey J. Howlett

RUSSELL GRIMWADE SCHOOL OF BIOCHEMISTRY, UNIVERSITY OF MELBOURNE, PARKVILLE 3052, VICTORIA, AUSTRALIA

1. INTRODUCTION

Ultracentrifuges have been used extensively to study the molecular weight and interactions of proteins in solution. The majority of such studies have used the analytical ultracentrifuge equipped with interference, schlieren or absorption optics to determine the concentration distributions formed during centrifugation. Protein size, self-association and the binding of light absorbing ligands to proteins can be readily analysed using this approach. These techniques are limited however by the sensitivity of the detection system (usually requiring protein concentrations greater than 10 μg/ml) and, in the case of complex mixtures of proteins, by the difficulty of distinguishing the distributions of the individual components.

Methods developed for the preparative ultracentrifuge allow the analysis of the tube contents at the conclusion of centrifugation. Sensitive and specific assays may therefore be used to analyse the sedimentation behaviour of particular components. In sucrose gradient sedimentation velocity studies[1] a linear sucrose gradient is used to stabilize the sedimenting zones and to prevent stirring of the tube contents during deceleration of the rotor. The use of this method for the characterization of detergent solubilised membrane proteins is described in Chapter 25 of this volume. More recently, methods have been devised for the analysis of the sedimentation behaviour of macromolecules and the binding of ligands using the bench-top air-driven ultracentrifuge, the TL-100 ultracentrifuge and other preparative ultracentrifuges[2-5]. The small tubes used in these instruments offer significant advantages over the longer column lengths used for sucrose gradient experiments, allowing sedimentation equilibrium to be achieved within a reasonable time (24-48 hr).

Ligand-binding studies and molecular weight studies using the air-driven ultracentrifuge are based on the observation that under suitable conditions of density stabilization (*e.g.* a self-forming gradient using 5 mg/ml bovine serum albumin) it is possible to stop the rotor and to determine the concentration distributions formed during centrifugation directly[6-8]. Methods were devised[7] for the analysis of ligand binding to proteins which were suitable even for relatively small proteins such as ribonuclease (M~14,000). Initially, sedimentation equilibrium methods were devised which measured the fraction of a specific macromolecule remaining in an upper fraction of the tube and compared this value to the values obtained using a range of proteins of known molecular weight[6]. A significant advance was the use of techniques to sequentially fractionate the tube contents at the conclusion of centrifugation. These methods generated complete sedimentation profiles suitable for analysis using conventional sedimentation equilibrium theory[8]. This Chapter describes the methods which have been developed for the preparative ultracentrifuge and their applications for both ligand-binding and molecular weight studies.

2. EXPERIMENTAL

The Air-Driven Ultracentrifuge

The air-driven ultracentrifuge or "Airfuge" (Beckman Instruments, Inc., Palo Alto, U.S.A.) is a desk-top air-turbine centrifuge with a rotor approximately 4 cm in diameter which can be lifted and driven by ordinary laboratory compressed air. The six sample tubes, of capacity 175 µl, are oriented at an angle of 18° to the axis of rotation. For a tube containing 150 µl of solution the radial distances from meniscus and base of the solution to the centre of rotation are approximately 1.1 cm and 1.45 cm respectively. Rotor speed is measured using a light-sensitive transistor and frequency counter[9], stroboscopically[7] or using a commercially available attachment (Beckman Instruments). The rotor can attain maximum operating speed in about 30 s. Deceleration from top speed involves an initial coasting period of approximately 2 min, followed by mechanical braking for a similar period. The speeds used for sedimentation equilibrium experiments are generally in the range 30,000 rev/min - 50,000 rev/min and are achieved using air pressures of 5-10 p.s.i. The minimum rotor speed for stable operation is approximately 25,000 rev/min

The temperature of the Airfuge is not controlled directly but stable temperatures may be achieved by placing the instrument in a temperature controlled room. Measurements of the temperature of the rotor chamber during centrifugation indicate that, after an equilibration period of about 2-3 h, the temperature is constant and slightly above ambient. In a typical experiment, using a rotor speed of 35,000 rev/min and an Airfuge located at 4°C, the temperature of the rotor chamber after the initial equilibration period and the temperature of the rotor contents after 24 h of centrifugation is approximately 9°C [10].

A plastic cap is used to seal the centrifuge tubes in the rotor. Light greasing of the cap prevents significant evaporation of the tube contents during centrifugation. The individual tubes in the rotor are connected by an air space. In the case of samples containing H_2O and D_2O, mixing of the solvent occurs *via* this airspace by "distillation" causing convection. Small plastic caps fitted to each centrifuge tube have been used to prevent this mixing and to allow sedimentation equilibrium experiments using D_2O and H_2O samples in the same rotor[11].

The TL-100 Ultracentrifuge

The recent development of the bench-top TL-100 ultracentrifuge (Beckman Instruments) has provided an alternative to the use of the air-driven ultracentrifuge for sedimentation equilibrium studies[5]. The small TLA-100 rotor for this centrifuge holds

20 tubes at an angle of 30° to the vertical. The capacity of the tubes (200 μl) is similar to the capacity of the tubes used in the air-driven ultracentrifuge. The increased angle of the tubes to the vertical leads to slightly longer column lengths for the same sample volume. A major advantage of the TL-100 ultracentrifuge is the availability of temperature control. A current limitation of the air-driven ultracentrifuge is the inability to centrifuge at the low speeds required for sedimentation equilibrium studies of macromolecular complexes of molecular weight in excess of 500,000. The low speed capability of the TL-100 ultracentrifuge has the potential to overcome this limitation. A swinging bucket rotor (TLS-55) is available for the TL-100 ultracentrifuge. This rotor, with suitable tube adaptors, has also been used for sedimentation equilibrium studies of macromolecules[5].

Density Stabilization

An essential part of sedimentation equilibrium experiments using preparative ultracentrifuges is the use of a solute for density stabilization. Dextran (molecular weight fractions 10,000, 40,000 or 70,000) or bovine serum albumin at initial concentrations of 5 mg/ml may be used for this purpose. These solutes form concentration gradients which stabilize the tube contents during centrifugation and braking of the rotor. Higher concentrations of Dextran or bovine serum albumin (in excess of 10 mg/ml) are undesirable since significant non-ideality effects are observed under these conditions[10]. Bovine serum albumin, as well as being a density stabilizer, offers the additional advantage, that it prevents non-specific absorption of samples to surfaces.

Time to Reach Equilibrium

Theoretical considerations indicate the time taken to reach sedimentation equilibrium varies in direct proportion to the square of the column length and inversely to the diffusion coefficient of the sedimenting solute[12]. The short column lengths used in air-driven ultracentrifuge studies (approximately 3 mm) favour the attainment of equilibrium. Convective flow in the centrifuge tube as the sedimenting species collide with the inclined wall may also reduce the time to reach equilibrium[6]. In practical terms the time required to reach equilibrium should be determined experimentally. Experiments with a number of ^{125}I-labelled proteins and the air-driven ultracentrifuge, using solution volumes of 150 μl and the A100/18 rotor at 35,000 rev/min, indicated no significant difference between data obtained after centrifugation for 22 or 30 hr [10], indicating that equilibrium was reached after 22 hrs.

Fractionation of Tube Contents

Several methods have been employed for the fractionation of the tube contents at the conclusion of centrifugation. Early experiments[8] involved the use of a micromanipulator and microcapillary tubes to take sequential sample volumes of 10 μl. The radial position of each aliquot was then determined from a consideration of tube geometry. The subsequent development of a microfractionator (Beckman Instruments) allowed disposable plastic pipette tips to be used to take successive aliquots. Using this method the volume of each aliquot is calculated from the weight of the aliquot and the solution density. The time taken to fractionate each tube using the microfractionator is approximately six minutes. Studies with ^{125}I human α_1-acid glycoprotein[10] indicate the distributions are stable during the time taken to fractionate several tubes.

The most significant improvement in fractionation procedures involved the development of an automatic fractionation device[13]. This device can collect laminar fractions 0.1 mm high (approximate volume 1 μl) with a precision of better than ±2 %. Attri and Minton[13] demonstrated that the automatic fractionator provides concentration gradients of radiolabelled macromolecules that are comparable in accuracy and resolution than those obtained for optically absorbing solutes using the optical scanner of the analytical ultracentrifuge. The commercial development of this equipment[5] will greatly facilitate studies of macromolecular interactions using preparative ultracentrifuges.

Optical Scanning

A non-destructive method for determining concentration gradients within the centrifuge tube has also been described by Attri and Minton[14]. This technique uses quartz centrifuge tubes which may be used, with suitable adaptors, in either the A110/30 fixed angle rotor of the air-driven ultracentrifuge, the SW-41 swinging bucket rotor for conventional ultracentrifuges or the TLS-55 swinging bucket rotor of the Beckman TL-100 benchtop ultracentrifuge. Experiments with γ-globulin indicate good agreement between results obtained by optical scanning compared to results obtained by fractionation of the tube contents[14]. While the use of optical scanning has the potential for greater accuracy, the approach is limited to cases where the solute of interest may be detected optically and independently of the presence of other solutes.

Geometry of the Tube

Analysis of sedimentation equilibrium data obtained using the preparative ultracentrifuge requires the conversion of concentration data *versus* aliquot number to be converted to concentration as a function of radial position. The relationship between the volume of each aliquot and the radial position in the centrifuge tube has been considered for the air-driven ultracentrifuge[8] and more generally for the air-driven and TL-100 ultracentrifuges[3,15].

The tube is considered as a half sphere (radius a), joined to a cylinder (Fig. 1). The radial position with respect to the centre of rotation (r_c) represents the inner position of the junction between the sphere and the cylinder and θ is the angle of inclination of the tube to the vertical. The volume (V) at radial position r is given by

$$V(r) = \int_{r_m}^{r} A(r)\ dr \qquad (1)$$

where A(r) is the cross-sectional area of the solution perpendicular to the radial path. For $r \leq r_c$

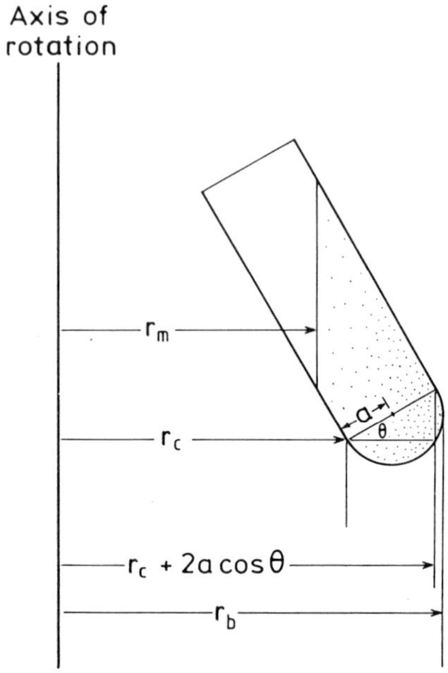

Figure 1. Geometry of the preparative ultracentrifuge tube. The tube is considered as a half-sphere (radius a) joined to a cylinder inclined at an angle (θ) to the vertical. The radial positions of the inner (r_c) and outer junctions are indicated. The distance of the meniscus (r_m) and base of the tube (r_b) to the axis of rotation are also shown.

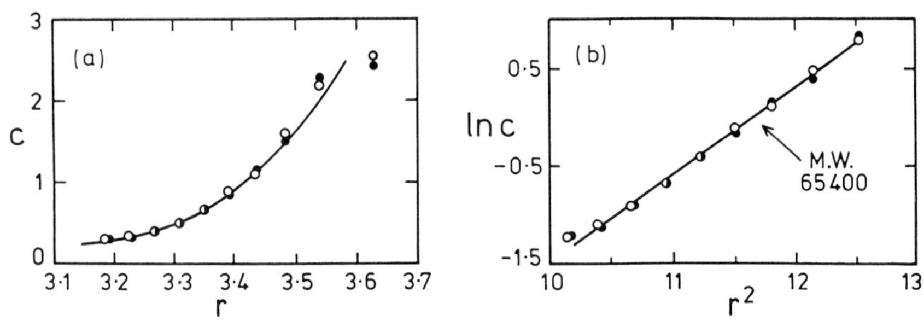

Figure 2. Sedimentation equilibrium distributions of ^{125}I-labelled bovine serum albumin obtained using the TL-100 ultracentrifuge (S. Longano and G.J. Howlett, unpublished). The distributions were formed by centrifugation of samples containing ^{125}I-labelled bovine serum albumin (650,000 cpm/ml) for 48 h at 15,000 rev/min and 10°C. The buffer used was 5 mM phosphate buffer, pH 7.4, 150 mM NaCl containing 5 mg/ml of unlabelled bovine serum albumin. The data from duplicate tubes (indicated by the open and closed symbols) are given in the form of (a) c *vs.* r and (b) ln c *vs.* r^2, where c is the concentration of radioactivity relative to the initial concentration and r is the radial position in the tube (cm).

$$A(r) = \pi a^2 \csc\theta \qquad (2a)$$

For $r_c < r < r_c + 2a\cos\theta$

$$A(r) = a^2 \left\{ \csc\theta \left[\frac{\pi}{2} + \alpha\sqrt{1-\alpha^2} + \sin^{-1}\alpha \right] + \right.$$

$$\left. (1 - \alpha^2\cos^2\theta) \left[\frac{\pi}{2} - \beta\sqrt{1-\beta^2} - \sin^{-1}\beta \right] \right\} \qquad (2b)$$

For $r > (r_c + 2a\cos\theta)$

$$A(r) = \pi a^2 (1 - \alpha^2\cos^2\theta) \qquad (2c)$$

where

$$\alpha = 1 - \frac{r - r_c}{a}\sec\theta \qquad (2d)$$

and

$$\beta = \frac{\alpha\sin\theta}{\sqrt{1 - \alpha^2\cos^2\theta}} \qquad (2e)$$

Values of V(r) as a function of radial distance for the TLA-100 rotor are given in Table 1.

Table 1. Relationship between tube volume and radial position in the TLA-100 rotor[a]

Radius (cm)	Volume (μl)	Radius (cm)	Volume (μl)
3.70	1.88	3.35	84.20
3.65	6.82	3.30	101.16
3.60	14.57	3.25	118.15
3.55	24.87	3.20	135.13
3.50	37.40	3.15	152.11
3.45	51.78	3.10	169.09
3.40	67.55	3.05	186.07

[a] The volumes were calculated by numerical integration using eq. (1), increments of Δr of 0.01 and values of $\theta = 0.5236$ rad, $r_c = 3.314$ cm and $a = 0.2325$ cm.

3. DETERMINATION OF MOLECULAR WEIGHT

At sedimentation equilibrium the concentration of an ideal solute, c_i, as a function of radial distance in the centrifuge cell, r, is given by the relationship

$$c_i(r_2) = c_i(r_1) \exp(\sigma_i (r_2^2 - r_1^2)) \tag{3}$$

where

$$\sigma_i = \frac{M_i(1-\bar{v}_i\rho)\omega^2}{2RT} \tag{4}$$

Thus

$$\frac{d \ln c_i}{d(r^2)} = \sigma_i \tag{5}$$

where M_i and $\bar{v}_i$ are the molecular weight and partial specific volume of solute i, respectively, R is the gas constant, T the absolute temperature, ω the angular velocity and ρ the solution density. Determination of the concentration distribution at sedimentation equilibrium provides a value for $(d \ln c_i)/(dr^2)$ and hence a value for M_i using appropriate values for the other parameters in eq. (4).

Analysis of the experimental data obtained from a sedimentation equilibrium experiment using ^{125}I-labelled bovine serum albumin is presented in Table 2. For this experiment the sample was sedimented for 48 hr at 15,000 rev/min and 10°C. At the conclusion of centrifugation aliquots were removed sequentially. The weight of each aliquot together with the solution density was used to determine the volume of each aliquot. The radial position of the midpoint of each aliquot (r) was calculated from eq. (2). The data in the form of the concentration of radioactivity (relative to the initial concentration) in each aliquot (c) *versus* r are presented in Fig. 2a. Data from duplicate tubes yield values of 0.927 and 0.917 for the slope of the ln c *vs.* r^2 plot of the data (Fig. 2b). Using a value of 0.735 mg/ml for the partial specific volume of bovine serum albumin[16], the measured values for the solution density, angular velocity and temperature a value of 65,400 is calculated for the molecular weight using eq. (4). This value is in good agreement with the value (66,200) calculated from the amino acid sequence.

4. PROTEIN-PROTEIN INTERACTIONS

Self-Association

Pollet *et. al.*[8] used the air-driven ultracentrifuge to analyse the self association of insulin. Molecular weight values of 5,600, 9,800 and 22,000 were obtained using initial insulin concentrations of 1×10^{-10} mol l^{-1}, 2×10^{-5} mol l^{-1} and 2×10^{-4} mol l^{-1} respectively. The ability to examine sedimentation equilibrium behaviour at the sub-micromolar level significantly extends the range of concentrations amenable to study using the analytical ultracentrifuge[3].

Other examples of the application of the air-driven ultracentrifuge for the study of protein self-association are presented in Fig. 3. The results in Fig. 3a refer to the behaviour of concanavalin A [2,17]. Analysis of this data in terms of a dimer-tetramer

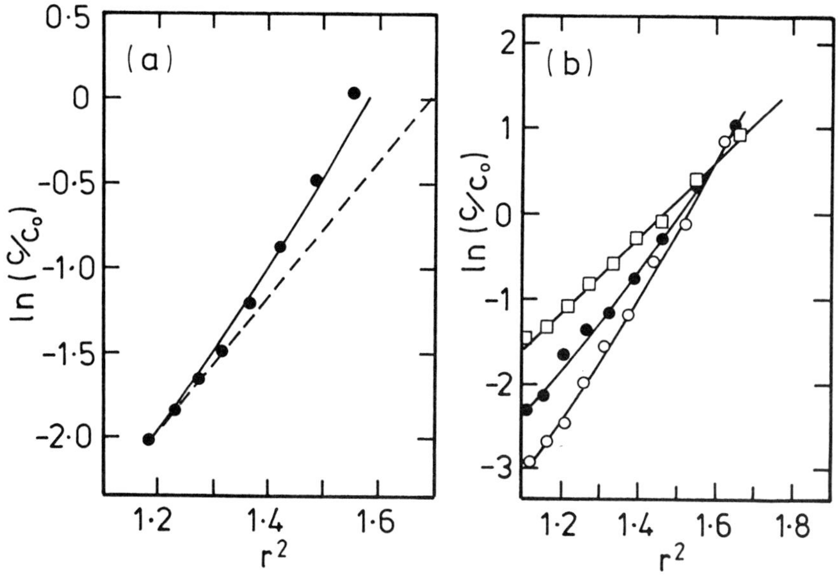

Figure 3. (a) The sedimentation equilibrium behaviour of ^{125}I-labelled concanavalin A [17]. The buffer used was 0.25 M potassium phosphate, pH 7.0 containing 5 mg/ml bovine serum albumin. The solution of concanavalin A (2 mg/ml) was centrifuged 22 h at 36,670 rev/min and 10°C using the air-driven ultracentrifuge. The broken line was calculated using eq. (3) and values for the molecular weight and partial specific volume of dimeric concanavalin A. The solid line was calculated using eqs. (6-9) assuming a dimer-tetramer equilibrium governed by an equilibrium constant of 1.6×10^4 $l.mol^{-1}$.
(b) The effect of NAD^+ and tyrosine on the sedimentation equilibrium behaviour of chorismate mutase dehydrogenase from *E. coli*[19]. The buffer used was 50 mM Tris-HCl, pH 7.5 containing 10 mM 2-mercaptoethanol, 5 mg/ml dextran T40, and 5 % v/v glycerol. The enzyme solution (initial concentration 0.86 mg/ml) was centrifuged for 22 h at 31,440 rev/min and 10°C. The data obtained in the absence of added ligands (open squares) yielded a value of 76,700 for the molecular weight of the enzyme. The data obtained in the presence of 0.5 mM NAD^+ (•) and in the presence of 0.05 mM NAD^+, 0.05 mM tyrosine (o) have been analysed in terms of a dimer-tetramer equilibrium. The solid lines were calculated using eqs. (6-9) and values of 0.1×10^6 $l.mol^{-1}$ and 2.3×10^6 $l.mol^{-1}$ for the association constants, respectively[19].

Table 2. Sedimentation equilibrium distribution of ^{125}I-bovine serum albumin using the TL-100 ultracentrifuge[a]

Aliquot volume (µl)	Cumulative volume (µl)	Mid volume (µl)	Radius (cm)	Concentration (c/c_0)
13.4	13.4	138.50	3.1901	0.2988
13.5	26.9	125.05	3.2297	0.322
13.5	40.4	111.55	3.2694	0.403
13.7	54.1	97.95	3.3095	0.5077
14.0	68.1	84.10	3.3503	0.6616
14.0	82.1	70.10	3.3922	0.8519
14.2	96.3	56.00	3.4363	1.1639
14.4	110.7	41.70	3.4845	1.4905
14.6	125.3	27.20	3.5400	2.2841
19.9	145.2	9.95	3.6277	2.4389

[a] The conditions used for the experiment are described in the caption to Fig. 2. The weight of each aliquot taken at the conclusion of centrifugation was determined and the volume calculated from the density of solution. The mid-volume of each aliquot was calculated starting from the base of the tube. The radius of the mid-volume of each aliquot was calculated using eq. (2). The concentration refers to the concentration of radioactivity in each aliquot (c) relative to the initial concentration (c_0).

equilibrium was performed as follows. The total concentration (in g per unit volume) at the meniscus (r_m) is equal to the sum of the concentrations of dimer $c_D(r_m)$ and tetramer $c_T(r_m)$

$$\bar{c}(r_m) = c_D(r_m) + c_T(r_m) \tag{6}$$

The equilibrium constant is defined as

$$K = \frac{c_T(r_m)}{c_D^2(r_m)} \tag{7}$$

Combination of eqs. (6) and (7) yields

$$\bar{c}(r_m) - c_D(r_m) - Kc_D^2(r_m) = 0 \tag{8}$$

Taking the experimental value for (r_m) and an assumed value of K it is therefore possible to solve eq. (8) to determine $c_D(r_m)$ and then to obtain $c_T(r_m)$ using eq. (6). The entire concentration distribution can then be determined using eq. (3) written for dimer and tetramer.

$$\bar{c}(r) = c_D(r_m)^{\sigma_D(r^2-r_m^2)} + c_T(r_m)^{\sigma_T(r^2-r_m^2)} \tag{9}$$

This procedure can be repeated using different values of K until the theoretical curve provides the best fit to the experimental data. The solid line drawn through the data in Fig. 3a was calculated assuming a dimer-tetramer equilibrium governed by an association constant of 2×10^4 l mol^{-1}. This estimate of the association constant is in good agreement with the value (5×10^4 l mol^{-1}) obtained using the analytical ultracentrifuge[18].

The results in Fig. 3b describe the sedimentation equilibrium behaviour of the enzyme chorismate mutase prephenate dehydrogenase from *E. coli*[19]. The concentration distribution for this enzyme was determined from enzyme activity measurements. The molecular weight obtained for the enzyme alone (78,000) indicates the presence of a dimer. The addition of one of the substrates (NAD^+) increases the average molecular weight (106,000) while in the presence of both NAD^+ and tyrosine the molecular weight increases still further (136,000). The solid lines drawn through the data were calculated, using the procedures described above, for a dimer-tetramer equilibrium favoured by the ligands NAD^+ and tyrosine. The model provides a quantitative mechanism for the inhibition of the enzyme by the end-product, tyrosine in which both tyrosine and NAD^+ bind preferentially to the less active, tetrameric form of the enzyme[19].

Heterogeneous Interactions

The use of the air-driven ultracentrifuge for studies of the interactions of dissimilar proteins was initially described for the salt-dependent interaction between ovalbumin and lysozyme[20]. Other examples include the interaction of spectrin with ^{125}I-calmodulin[21], bovine serum albumin with ^{125}I-calmodulin[22], concanavalin A with glycoproteins[17] and the interaction of a monoclonal antibody with the influenza virus neuraminidase[23].

The analysis of such interactions may be considered in two categories depending on the relative molecular weights of the interacting species. Where the difference is large the smaller species may be considered as a low molecular weight ligand and conditions chosen so that only the unbound ligand remains at the meniscus at sedimentation equilibrium. Theoretical curves for such a system are presented in Fig. 4. The system described is for an antibody (M~150,000) interacting with a ligand (M~15,000). The distributions were calculated assuming a sector shaped geometry and using equations originally derived to describe the interaction of a low molecular weight ligand (L) with a macromolecule[7]. The profiles for the individual species indicate that the antibody (A) and antibody complexes (AL and AL_2) are depleted from the meniscus leaving unbound L. Also shown in Fig. 4 (broken line) is the calculated distribution for a sample containing only L. The ratio of the concentration of L at the meniscus in the presence and absence of added antibody (indicated by the arrows in Fig. 4) provides a measure of the fraction of free L relative to the total amount of L.

Further analysis of the interaction requires consideration of the effects of centrifugation on the total amount of free L in the centrifuge tube. These effects have been shown to be small for low molecular weight ligands[7]. The effects of centrifugation on the system described in Fig. 4 may be considered by determining the concentration of free L as a function of the concentration of bound L and constructing a theoretical "Scatchard" plot. The results in Fig. 5 indicate that accurate values are obtained for the stoichiometry of the interaction and an overestimate (1.5-fold) of the association constant. This theoretical treatment allows the experimentally determined equilibrium constant to be corrected for the effect of centrifugation on the concentration of free ligand[23]. As indicated by results in Fig. 5 the correction factor becomes smaller as the molecular weight of L decreases.

The results in Fig. 6 describe the sedimentation equilibrium behaviour of radiolabelled lysozyme in the presence and absence of monoclonal D1.3 [24]. Analysis of the profile obtained in the absence of antibody yielded the expected value for the molecular weight of lysozyme. The decrease in the concentration of lysozyme at the meniscus in the presence of antibody is an indicator of the extent of the interaction of the antibody with lysozyme. The results of several experiments in which the fraction of lysozyme remaining at the meniscus was determined using different initial concentrations of lysozyme and antibody yielded the equilibrium constant for the interaction. The value of 2.0×10^9 $l.mol^{-1}$, obtained after correction for the effects of centrifugation described above, may be compared with a value of 0.5×10^9 $l.mol^{-1}$ obtained from fluorescence measurements[25].

The second category of protein-protein interaction which may be considered is where the molecular weights of the interacting species are similar. In this case it is not possible to choose conditions where the meniscus is selectively depleted of one of the species. An indirect method to characterize the interaction is required[5]. An example of such a system is the interaction of bovine serum albumin (M~66,200) with calmodulin (M~17,000). Sedimentation equilibrium profiles for ^{125}I-calmodulin in the presence and absence of bovine serum albumin are presented in Fig. 7. The increase in the slope of the ln c *versus* r^2 data in the presence of albumin is indicative of an interaction

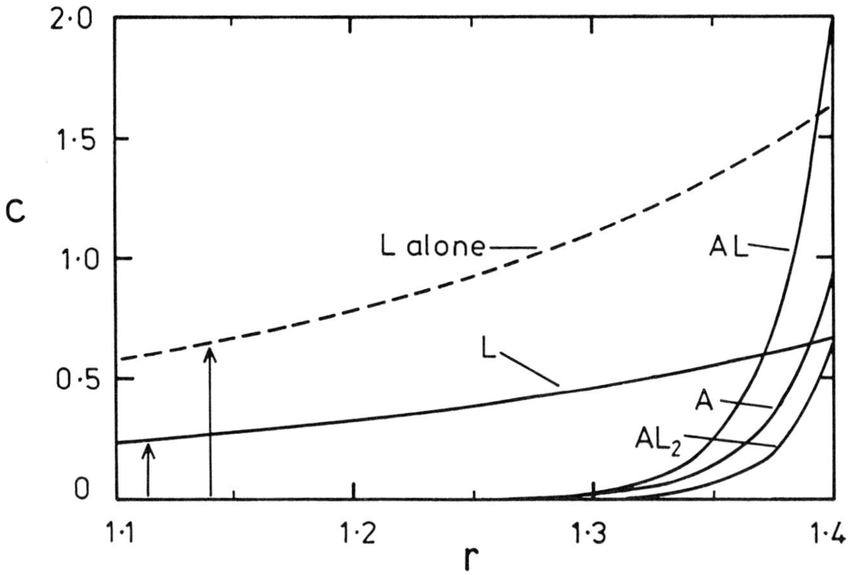

Figure 4. Computer-simulated sedimentation equilibrium distributions for an antibody-antigen interaction. The system described is for an antibody, A (M~150,000) interacting with a lower molecular weight protein antigen (M~15,000). The partial specific volumes for A and L were taken as 0.73 ml/g. It was assumed there was no volume change on interaction. The distributions were calculated assuming a sector shaped geometry using equations described previously[7]. Other parameters assigned were the angular velocity (4000 rad/sec) and temperature (10°C). The initial concentrations of total A and total L were taken as 0.6×10^{-9} M and the intrinsic association constant taken as 1×10^{9} $l.mol^{-1}$. The concentration distributions for L, AL and AL_2 have been normalized to the total initial concentration of L and the concentration of free A normalized to the total initial concentration of A. The broken line has been computed for a system containing L alone.

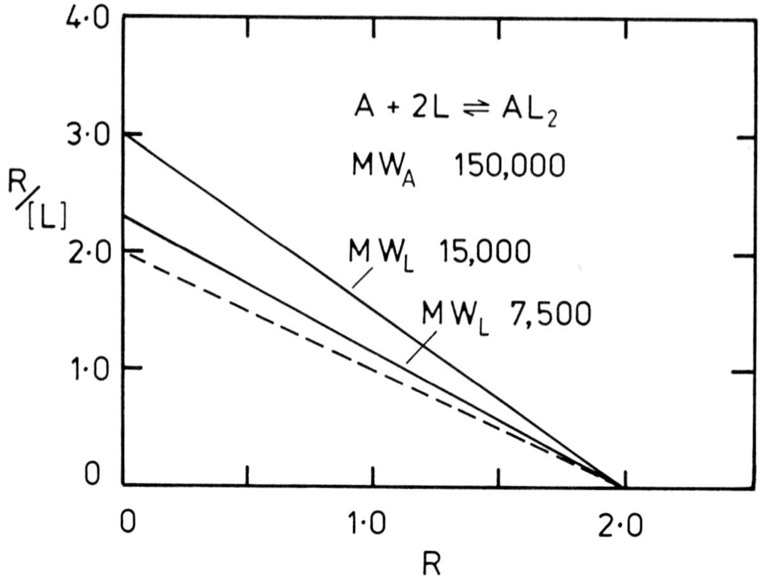

Figure 5. Theoretical "Scatchard" plots for an antibody-antigen interaction. The upper line was calculated[7] for the system described in Fig. 4. The concentration of L at the meniscus in the presence and absence of added antibody (indicated by the arrows in Fig. 4) was used to calculate the average concentration of free L in the centrifuge cell. The number of moles of bound ligand per mole of antibody (R) was calculated as a function of the average nanomolar concentration of free ligand, [L]. Also shown in the figure is a plot for the case where the molecular weight of L is 7,500. The broken line describes the plot for the case where the molecular weight of L is negligible. The slope and intercepts of the broken line yield the correct values for the stoichiometry and association constant for the interaction.

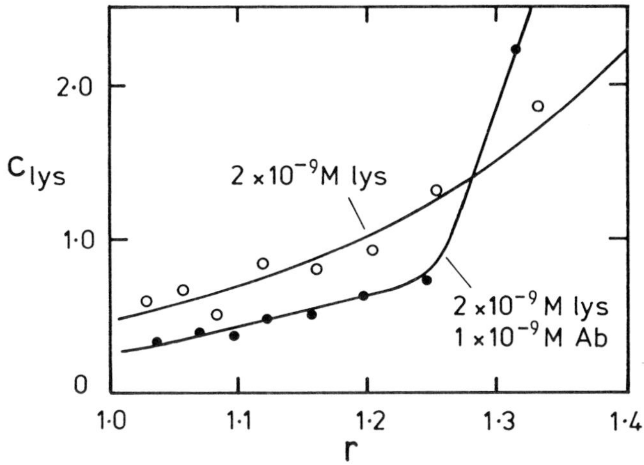

Figure 6. Sedimentation equilibrium distributions for ^{125}I-labelled lysozyme in the presence (•) and absence (o) of monoclonal antibody D1.3 [24]. The initial concentrations of lysozyme and antibody were 2×10^{-9} M and 1×10^{-9} M respectively. The samples, prepared in phosphate buffered saline containing 5 mg/ml bovine serum albumin, were centrifuged at 40,700 rev/min and 10°C for 24 h using the air-driven ultracentrifuge. The concentration (c_{lys}) of radiolabelled lyzozyme (determined at the conclusion of centrifugation) relative to the initial concentration of radiolabelled lysozyme is plotted as a function of radial distance (r) in the air-driven ultracentrifuge tube (unpublished data, T. M[c]Inerney, C. Gruen, G. Howlett and D. Jackson).

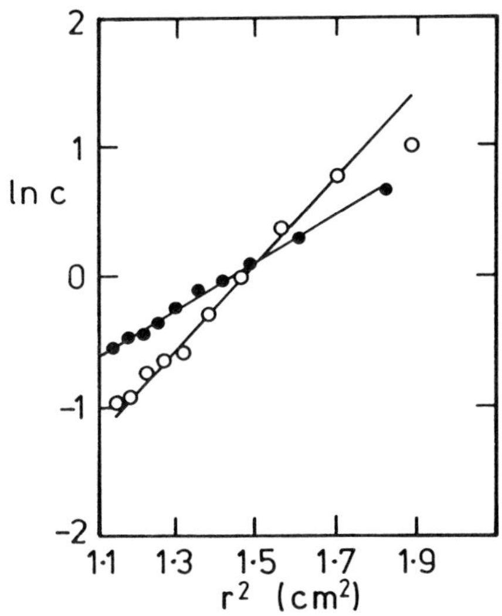

Figure 7. Sedimentation equilibrium profile of ^{125}I-calmodulin in the presence and absence of unlabelled bovine serum albumin[22]. The samples were prepared in 25 mM Tris-HCl, pH 7.5, pCa 6.68. Final concentrations in the tubes were 209 nM ^{125}I-calmodulin, 5 mg/ml dextran T40 (•) and 209 nM ^{125}I-calmodulin, 5 mg/ml bovine serum albumin (o). Centrifugation using the air-driven ultracentrifuge was at 38,520 rev/min for 24 h at 10°C.

between the two proteins. The equilibrium constant for the interaction may be calculated as follows. Assuming a single 1:1 complex, the total concentration of calmodulin at the meniscus, [T], is given by

$$[T] = [C] + [BC] \tag{10}$$

where [C] is the concentration of free calmodulin and [BC] is the concentration of the complex. The weight average reduced molecular weight at the meniscus is given by

$$\sigma_{av} = \frac{d \ln [T]}{d(r^2)} = \frac{\sigma_C[C] + \sigma_{BC}[BC]}{[T]} \tag{11}$$

The value of d ln $[T]/d(r^2)$ is available from the experimental data, σ_C and σ_{BC} may be calculated using eq. (2) and the known values of the molecular weight and partial specific volumes of the species, assuming no volume change on interaction. Eqs. (10) and (11) may be solved for [C] and [BC]. Under conditions where there is a large molar excess of B compared to C, the concentration of B at the meniscus, [B], may be determined at the conclusion of centrifugation allowing the equilibrium constant to be calculated as $K = [BC]/([C]\times[B])$. A value of 2.1×10^4 $l.mol^{-1}$ was obtained for the interaction between albumin and calmodulin[22]. The interaction of calmodulin with bovine serum albumin indicates the need for caution, in some instances, in the use of albumin to provide density stabilization during centrifugation.

Non-ideality Effects

A less specific form of protein-protein interaction occurs at high concentrations of protein due to the effects of excluded volume or non-ideality. These effects are a significant factor in the determination of colloidal osmotic pressure, solubility and chemical equilibria. Non-ideal effects need to be considered when analysing the sedimentation equilibrium behaviour of concentrated solutions of macromolecules. Experiments with ^{125}I-labelled bovine serum albumin in concentrations of unlabelled albumin ranging from 20-80 mg/ml, indicated significant reductions in the slope of the ln c *vs.* r^2 plots as the concentration of protein increased. Using procedures similar to those used to analyse dimer-tetramer equilibria (eqs. (6-9)) it is possible to analyse such non-ideality effects to obtain the appropriate non-ideality constants[10].

A more detailed analysis of the effects of non-ideality to reveal hidden associations of protein is provided by the work of Muramatsu and Minton[26]. These studies showed that although the sedimentation equilibrium behaviour of bovine serum albumin could be adequately accounted for on the basis of excluded volume effects the results with ovalbumin and aldolase were indicative of weak self-association occurring at the higher protein concentrations.

5. CONCLUDING REMARKS

Preparative ultracentrifuges such as the air-driven ultracentrifuge and TL-100 ultracentrifuge provide a simple and relatively inexpensive method for the analysis of macromolecular interactions in solution. The ability to stop the rotor and to analyse the concentration distribution directly using a range of sensitive assays, including radioactivity, substantially extends the concentration range for sedimentation equilibrium experiments. Ligand binding experiments may be performed by selectively removing ligand acceptor complexes from the meniscus leaving unbound ligand. The ability to use sedimentation equilibrium theory to estimate the magnitude of effects due to centrifugation provides a sound basis for such experiments. Changes in the weight average molecular weight of a radiolabelled species in the presence of unlabelled components provides a powerful approach to the investigation of the interactions between proteins and other macromolecules in solution. The development of a commercially-available automatic fractionating device will facilitate the application of the preparative ultracentrifuge for the analysis of such interactions.

REFERENCES

1. R.G. Martin and B.N. Ames, *J. Biol. Chem.*, 1961, *236*, 1372.
2. G.J. Howlett, 'Molecular and Chemical Characterization of Membrane Receptors' (J. Craig and L. C. Harrison, eds), Alan R. Liss, New York, 1984, Vol 3, p. 179.
3. R.J. Pollet, 'Methods in Enzymology' (C.H.W. Hirs and S.N. Timasheff, eds), Academic Press, 1985 ,vol 117, p. 3.
4. G.J. Howlett, 'Methods in Enzymology' (G. Di Sabatino ed.), Academic Press, 1987, vol 150, p. 447.
5. A.P. Minton, *Anal. Biochem.*, 1989, *176*, 209.
6. M.A. Bothwell, G.J. Howlett and H.K. Schachman, *J. Biol. Chem.*, 1978, *253*, 2073.
7. G.J. Howlett, E. Yeh and H.K. Schachman, *Arch. Biochem. Biophys.*, 1978, *190*, 809.
8. R.J. Pollet, B.A. Haase and M.L. Standaert, *J. Biol. Chem.*, 1979, *254*, 30.
9. G.J. Howlett and G. Markov, *Arch. Biochem. Biophys.*, 1980, *202*, 507.
10. G.J. Howlett, P.W. Dickson, H. Birch and G. Schreiber, *Arch. Biochem. Biophys.*, 1982, *215*, 309.
11. G.J. Howlett, H. Birch, P.W. Dickson and G. Schreiber, *Biochem. Biochem., Res. Commun.*, 1982, *105*, 895.
12. K.E. Van Holde and R.L. Baldwin, *J. Phys. Chem.*,1958, *62*, 734.
13. A.K. Attri and A.P. Minton, *Anal. Biochem.*,1986, *152*, 319.
14. A.K. Attri and A.P. Minton, *Anal. Biochem.*,1983, *133*, 142.
15. S.C. Longano, J. Knesel, G.J. Howlett and G.S. Baldwin, *Arch. Biochem. Biophys.*, 1988, *263*, 410.
16. J.C. Lee and S.N. Timasheff, *Biochemistry*, 1974, *13*, 257.
17. G.J. Howlett, P.J. Roche and G. Schreiber, *Arch. Biochem. Biophys.*, 1983, *224*, 178.
18. D.F. Senear and D.C. Teller, *Biochemistry*, 1981, *20*, 3076.
19. G.S. Hudson, G.J. Howlett and B.E. Davidson, *J. Biol. Chem.*, 1983, *258*, 3114.
20. R.G. Clarke and G.J. Howlett, *Arch. Biochem. Biophys.*, 1979, *195*, 235.
21. A. Hussain, G.J. Howlett and W.H. Sawyer, *Biochem. Biophys., Res. Commun.*, 1984, *122*, 1194.
22. A. Hussain, G.J. Howlett and W.H. Sawyer, *Anal. Biochem.*, 1985, *145*, 217.
23. D.C. Jackson, G.J. Howlett, A. Nestorowicz and R.G. Webster, *J. Immunol.*,

1983, *130*, 1313.

24. A.G. Amit, R.A. Mariuzza, S.E.V. Phillips and R.J. Poljak, *Science*, 1986, *233*, 747.
25. E.S. Ward, D. Gussow, A.D. Griffiths, P.T. Jones and G. Winter, *Nature*, 1989, *341*, 544.
26. N. Muramatsu and A.P. Minton, *J. Molecular Recognition*, 1989, *1*, 166.

5

Methods for Off-line Analysis of Sedimentation Velocity and Sedimentation Equilibrium Patterns

By A.J. Rowe*, S. Wynne Jones#, D.G. Thomas* and S.E. Harding#

NATIONAL CENTRE FOR MACROMOLECULAR HYDRODYNAMICS, *UNIVERSITY OF LEICESTER, LEICESTER LE1 7RH, U.K. AND #UNIVERSITY OF NOTTINGHAM, SUTTON BONINGTON, LE12 5RD, U.K.

1. INTRODUCTION

The equilibrium or transient sedimentation velocity distribution of particles dispersed in an aqueous solute situated in a centrifugal accelerative field is routinely studied by means of either an optical trace recorded photographically or by chart recorder traces. Rayleigh interferometric fringe patterns have been widely used to record *equilibrium* distributions, in which the displacement of the parallel fringes is directly related to particle concentration differences. In this Chapter we will outline how we have developed a simple but highly efficient frameshift algorithm for automatic interpretation of these patterns[1]: results obtained from extensive use and further definition of this algorithm confirm its validity and utility.

We will also describe the use of algorithms we have developed for the interpretation of Fresnel fringe patterns yielded by an alternative optical system: Schlieren optics: these more complex patterns involving non parallel fringes can be analysed successfully, subject to certain conditions, with a precision similar to that obtained using Rayleigh interference optics. For the analysis of sedimentation equilibrium *or* sedimentation velocity records from chart paper rather than photographic output (as applies for example, to the use of the scanning absorption optics system on the Beckman (Palo Alto, U.S.A.) Model E or MSE (Crawley, U.K.) Mk II analytical ultracentrifuges or the scanning absorption and knife edge Schlieren optics on the MSE Centriscan ultracentrifuge), the use of a simple digitising table is sufficient to give acceptable precision combined with convenience of use. Some reference to these features will also be made in this Chapter.

Although the theory of sedimentation transport and equilibrium has been dealt with elsewhere in this volume, it is still instructive at this stage to consider some basic formulae which indicate whether it is concentration or concentration gradient distribution a particular data capture regime is designed to record.

The basic equation governing the distribution of particles at *sedimentation equilibrium* in a centrifugal field is given at infinite dilution by[3]

$$\frac{dc}{dr} = \sigma c r \tag{1}$$

where c is the concentration of particles at radial position r, and σ is a "reduced" molecular weight (*cf* Chapter 7), given by

$$\sigma = \frac{M_r\,(1-\bar{v}\rho)\omega^2}{RT} \tag{2}$$

where M_r is the molecular weight, or, as biochemists like to use, the "relative molecular mass", $\bar{v}$ the partial specific volume ($\approx$ reciprocal density) of the particles, ρ the density of the fluid, R the gas constant, ω the angular velocity and T the temperature (in degrees Kelvin).

Eq. (1) is frequently used in its "integrated" form:

$$\frac{\Delta \log c}{\Delta(r^2)} = \frac{\sigma}{2} \tag{3}$$

Normally the parameter M_r will be the object of study. It may itself be a function of r, either directly as a result of polydispersity or indirectly as a result of depending upon c (which itself will vary with r), through non-ideality or associative phenomena.

The basic purpose of any analysing optical system is thus to record a pattern capable of being interpreted to yield *either* c *or* dc/dr as a function of r, at the stable, equilibrium distribution.

By contrast, *sedimentation velocity* is a transport method, and a time (t)-dependent series of distributions must be logged in this case. If the interface which is being monitored between solution and solvent is at a radial position r_b, then the sedimentation coefficient s (velocity per unit field) is given by

$$s = \frac{1}{\omega^2 r_b}\left(\frac{dr_b}{dt}\right) \tag{4}$$

Again it is often more convenient to use the integrated form:

$$s = \frac{1}{\omega^2}\left(\frac{\Delta \log_e(r_b)}{\Delta t}\right) \tag{5}$$

from which s is readily determined from the regression of $\log_e(r_b)$ on t.

2. DATA CAPTURE IN OFF-LINE MODE

Historically only a small minority of workers have used wholly on-line systems for data capture in the analytical ultracentrifuge. This has been related to the cost and complexity of the systems required. It may now be taken for granted, however, that on-line capture will be normal for scanning absorbance work, particularly with the advent of the new Beckman Optima XLA Analytical Ultracentrifuge, which incorporates as standard the logging of data to disk (formatted as ASCII files). Almost certainly the same will apply to Rayleigh Interferometric and to Schlieren patterns, as and when the modern instrument incorporates the appropriate optical systems. However, in the meantime - which is likely to extend for some years to come - it will be necessary (other than in laboratories possessing specialised devices) to evaluate distributions recorded using these optics in off-line mode. We consider now optimal procedures for doing this, and note that the interpretative algorithms involved may well be equally useful for analysis of on-line data.

Rayleigh Interferometric and Schlieren patterns are invariably recorded on photographic film. Modern film backings give adequate dimensional stability, although care must be exercised to avoid serious temperature changes during processing. 'Fine grain' emulsions offer sufficient (line) resolution in almost every case, normally around 100 cycles per mm. Absorption and refractometric scanning patterns are recorded onto chart paper. Because of the large spatial magnification used, resolution is not a problem in this case, although for several reasons ('shot noise' from sampling, window absorption and scattering, electronic noise) the attainable precision is normally lower than with the two other optical systems.

Analysis to Yield c *vs.* r Data for Equilibrium Distributions

The classical approach has been to use Rayleigh interference optics to give a pattern in which the displacements of the fringes in a direction (z) normal to radial is a linear function of the concentration increment at the radial position in question. The fringes are of course equi-spaced and parallel, and hence a scan across them in the z direction yields a sinusoidal intensity function whose phase is a measure of (the non integer part of) the fringe shift.

We have developed a simple but fast and stable algorithm for deriving the phase shift from the intensity function[1]. The latter is logged from the photographic record of the fringe pattern, using a commercial scanning densitometer, the LKB (Bromma, Sweden) 2202 laser densitometer. Then if Q fringes are contained within the window analysed, an iterative frameshift is performed within the data set, to maximise the Fourier coefficient of order Q. The method is thus a null method, which searches for the frameshift which will set the phase term to zero[1]. This algorithm, unlike earlier approaches in this area, yields estimates for the fringe increment whose precision is not a function of the latter parameter itself. The precision of the recorded fringes may be gauged subjectively from Fig. 1 and using this system a precision of f/500 (f is a single fringe increment) is in our experience attainable[1].

Analysis to Yield dc/dr *vs.* r Data

The earliest optical method used to analyse distributions within the ultracentrifuge cell was the 'Schlieren' optical system, in which an analysing diaphragm is inserted into the back focal plane of the camera lens employed to image the cell. Shadows or other traces are produced whose displacement, again in the z direction, is proportional to the first derivative of solute concentration with respect to radial distance. Other than in the earliest work, a 180° phase plate has been used as the analysing diaphragm. The resulting single trace is rather broad as compared to an interference fringe (Fig. 2).

It has been universally considered that the precision with which this trace can be interpreted falls well short of what can be achieved using Rayleigh interference optics. Subjectively this is understandable. The Schlieren trace appears relatively broad, and only a single trace is yielded, thus making unavailable the reduction in noise/signal normally achieved from multiple records. Yet the principal optical components of the two optical systems are identical, and are used at the same working aperture. Insofar as distinctive components are introduced in either method, there is no reason to suppose that these limit the information transfer function, which one would expect to be very similar in both cases, given adequate interpretative algorithms.

We have therefore researched the possibility of developing the interpretation of Schlieren records to a much higher level than heretofore. There are a number of practical reasons for doing this. As detailed below, we find that with suitable developments of the methodology, and subject to certain relatively minor reservations, results from the Schlieren optical method can indeed be interpreted with a precision approaching those obtained by the Rayleigh interference method. The basis of this is the recording and interpretation of the more complex *Fresnel* fringe patterns generated by the Schlieren optical diaphragms. Several approaches to the interpretation of Fresnel fringe patterns can be defined. It seems likely that an optimal approach has yet to be delineated, but results to date are more than adequate to demonstrate the potential of work in this area.

3. FRINGE SHIFTS IN RAYLEIGH INTERFEROMETRIC PATTERNS

Our off-line data capture algorithm (ANALYSER written in Turbo Pascal) from c *vs.* r records has already been described in some detail[1,2]. ANALYSER[1], which is based on a frameshift Fourier algorithm, has now been upgraded to version ANALYSE2 which permits the analysis of data from a two-dimensional data acquisition system as opposed to a series of individual one-dimensional scans. Sophisticated search procedures have been incorporated to ensure that the system reproducibly and stably

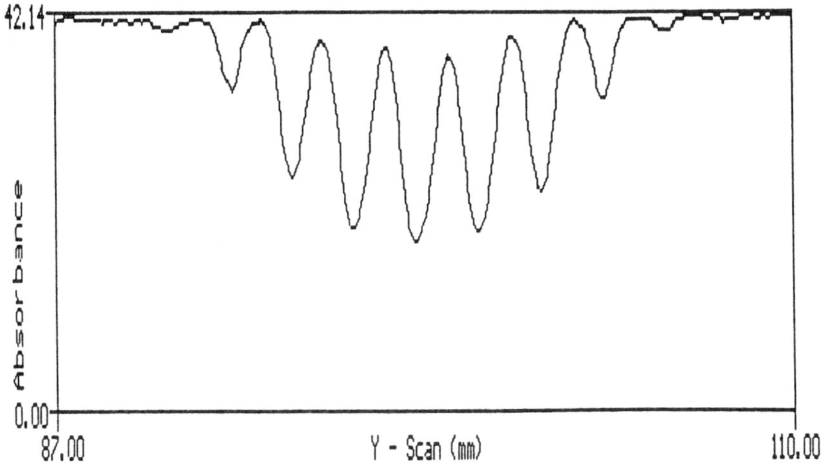

Figure 1. Digitised optical density values output from an LKB Ultroscan XL 2222 (two-dimensional scanner) scanning at a single radial position. 575 values were logged in this case at each of 175 radial positions in the cell, and these form the data set for subsequent analysis. Data for sedimentation equilibrium experiment on lipase, using Rayleigh interference optics.

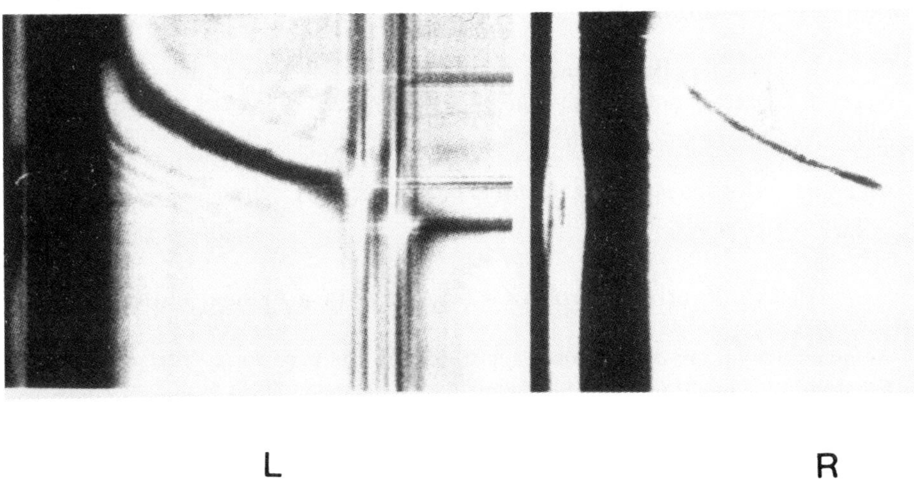

Figure 2. Phase plate Schlieren records of a solution at sedimentation equilibrium in an ultracentrifuge cell (MSE Mk II Analytical Ultracentrifuge). The solution column is 2 mm long in real space. From original negatives conventionally (L) and correctly (R) exposed.

finds the correct fringe intensity maximum in what is now a full two-dimensional record. As it is now possible to analyse data at up to 200 radial positions from a single experiment, rigorous tests can be performed to assess such factors as sample homogeneity and interactions (Fig. 3). The completed system is now in intensive use, and results on many systems have fully justified our initial estimates[1] of the precision attainable. What we would now like to focus on for the purpose of this Chapter is a new off-line data capture system for optical *Schlieren* records of *dc/dr vs.* r.

4. REFRACTOMETRIC OPTICS AND FRESNEL FRINGE PATTERNS

Refractometric Optics

The presence of an analysing diaphragm in the back focal plane of the camera lens imaging a cell, in this case located in an ultracentrifuge rotor, leads to the formation of a 'Schlieren' pattern in the image plane. The presence of a cylindrical lens results in a pattern in the image plane in which the z deflection of the trace at any radial position is linearly related to the refractive index change (dc/dr) in the conjugate locus in the cell plane (Lloyd[3]). Although any physical form of diaphragm can in principle be employed, it has long been customary to use a 180° phase plate (Wolter[4]), which causes no loss of transmitted light and has been considered on general principles (rigorous analysis appears not to have been performed) to maximise the information transfer function[3]. As a pure phase plate records no signal as dc/dr > 0, a thin line of metal evaporated onto the half wave step is normally added[3]. The latter *only* produces a trace for zero or very low dc/dr values. The optics of the transition region of dc/dr have yet to be been defined.

The resulting trace shows a well defined but rather broad line (Fig. 1). This represent the zero'th order fringe of an often poorly resolved Fresnel pattern. We have addressed ourselves to the definition at high resolution of the co-ordinates of this pattern, both by location of the zero'th order fringe, and by an alternative approach in which we derive and apply a relationship between Fresnel fringe *spacing* at defined r and the corresponding dc/dr and Δc values.

Definition of the Zero'th Order Fringe in a Fresnel Pattern

As noted above, the conventional approach here has been to record the Schlieren pattern using a modified phase plate diaphragm. The pattern is in principle symmetric with respect to the zero'th order fringe, which is located in the centre of the line trace. However, as was noted by Rowe and Khan[5], a simple knife edge diaphragm has a better established optical theory, and can yield results of a precision equal to that given by a phase plate. We have therefore explored the use of a simple knife edge diaphragm to generate the Fresnel pattern.

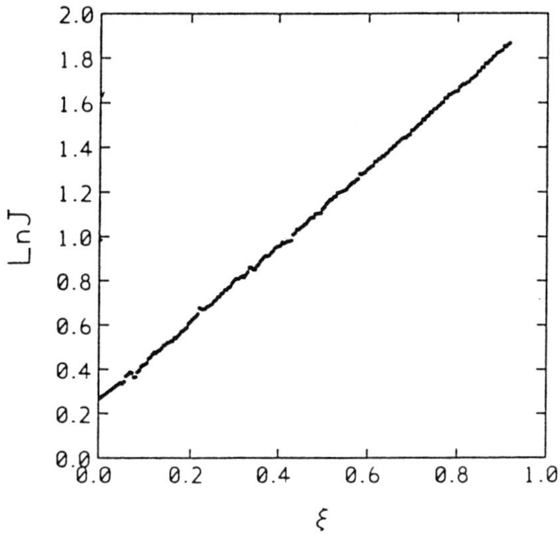

Figure 3. Plot of the logarithm of the solute concentration (expressed in absolute fringe numbers, J) *versus* the radial displacement (squared) parameter ξ. Data from a low speed sedimentation equilibrium experiment on recombinant hirudin, loading concentration 0.8 mg/ml. From the slope a weight averaged molecular weight of (7080±100) is computed (from sequence = 6964).

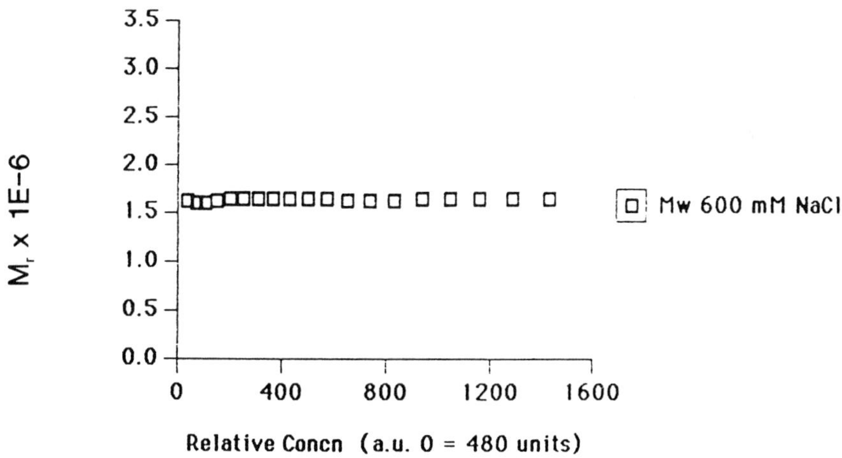

Figure 4. $M_{r,w}$ values computed (from an equilibrium solute distribution recorded using Schlieren optics) for the protein dynein at a cell loading concentration of 0.5 mg/ml. These point weight average values are computed by numerical integration of the dc/dr values to give c values, the constant of integration being found by a numerical manipulation based upon the equivalence of harmonically related averages.

We have in all cases adopted certain modified procedures for setting up of the ultracentrifuge cells and optics designed to maximise the information transfer function. These will be described elsewhere. They are not relevant to the interpretative algorithms described below, but are indispensable if results of the precision described are to be obtained in practice.

The Zero'th Order in a Phase Plate Generated Fresnel Pattern

The zero'th order in this case is simply the maximum density (minimal intensity) in the z scan across the line pattern (Trautman and Burns[6]). Unfortunately as normally recorded the line is rather broad and the precision in the estimation of the position of the maximum not high. This is the conventional reason behind the general opinion that Schlieren optics are inherently of low precision.

Several experimental procedures can be adopted which largely circumvent the problem of the width and lack of definition of the optical trace. The experimental procedure most relevant to the present discussion is the use of longer photographic exposures to bring the optical record *near to the centre of the line trace* into the linear part of the gamma curve of the recording material. It seems not to have been generally appreciated (though Lloyd[3] commented briefly on the matter) that giving a 'normal' exposure as gauged for the whole photograph seriously degrades the transfer function in the critical spatial region.

Locating the zero'th order in a suitably exposed photographic image (*i.e.* highly *over* exposed with respect to 'non-information') by analysis of successive radial z scans results in a very smooth data set. Fig. 4 below illustrates typical final results computed from data measured in this way.

These results are comparable to those which would be obtained using Rayleigh interferometric optics together with the Fourier algorithms described above, and would correspond to a precision of at least f/300 in the latter terms. There are however certain qualifications which must be made concerning the absolute *accuracy* as distinct from the *precision* of the results. As noted above, for low values of dc/dr, the optical behaviour of a compound phase plate is far from well defined. This can be circumvented by avoidance of such conditions. More seriously, it seems not to have been appreciated that the physical properties of the phase plate and in particular its phase angle are critical.

We have computed intensity distributions in the image plane for a phase plate of various angles, using the Cornu Spiral construction[6]. The treatment given by Trautman and Burns[6] assumes a phase angle of 180°. We have extended this to the more general case, and as the tabulated values for the Cornu Spiral are in some cases of insufficient resolution we have computed the co-ordinates from the relation[7]:

$$F(x) = \left(\frac{2}{\pi}\right)^{0.5} \left\{ x + \frac{jx^3}{3} + \left(\frac{j^2}{2!}\right)\left(\frac{x^5}{5}\right) + \ldots + \left(\frac{j^n}{n!}\right)\left(\frac{x^{2n+1}}{2n+1}\right) \right\}^{0.5} \tag{6}$$

for n = 0, 1, 2, ... and the spiral is plotted in the complex plane ($j=\sqrt{-1}$).

The results are shown in Fig. 5 below for phase angles of 180°, 150° and 120°.

It is clear that the location of the true geometrical edge coincides with the minimum of the intensity distribution *only* for the case of a 180° phase angle. Moreover the pattern is not symmetrical for other phase angles. Thus the wavelength used is critical, and must be tuned to the particular phase plate used, a precaution which has not formed part of normal practice. Longer photographic exposures are now needed, to compensate for the reduced light intensity in monochromated light.

The Zero'th Order from a Knife Edge Pattern

A simple knife edge used as a diaphragm results in a 'shadow' pattern, with a set of Fresnel fringes (Fig. 5). Although actually of slightly smaller amplitude than those generated by a phase plate[6], the lower dynamic range of the image means they are frequently better registered.

The location of the zero'th order fringe can be computed by measurement of (say) the first and second order minima, and from the knowledge that their location (z) with respect to the zero'th order is given for order i by

$$z = \{4(i+1) - 0.5\}^{0.5} \tag{7}$$

In practice only the first two or three orders can be measured in a z scan (Fig. 6).

Zero'th Order Determined from the Fresnel Fringe Spacing

It is possible to use the values of the fringe spacings in the z direction to evaluate Δc *directly* instead of by computation of the location of the zero'th order. This is because of the defined relationship[8] between fringe spacing and the second derivative of the refraction (and hence concentration) gradient, from which it follows at once that

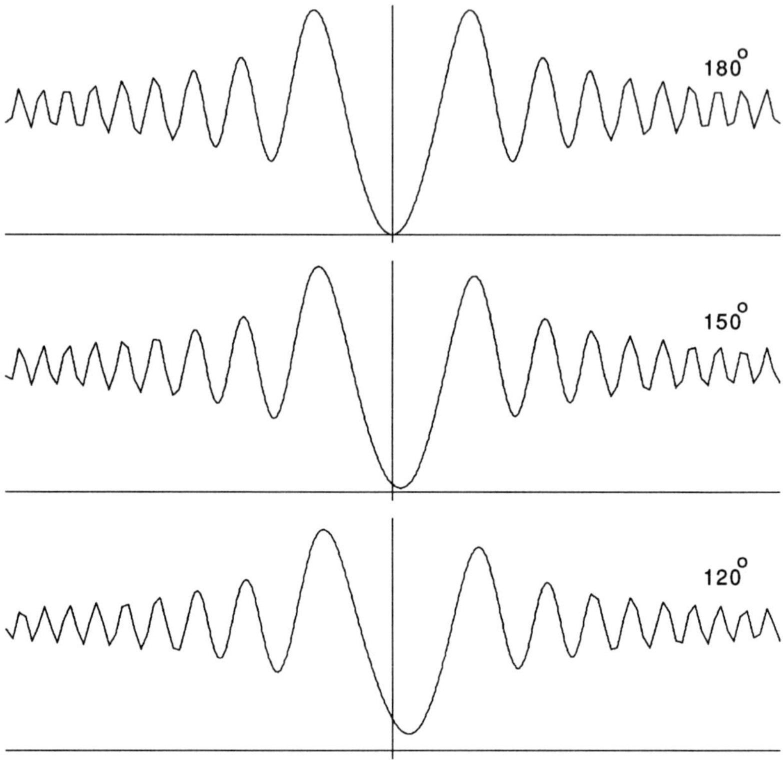

Figure 5. Computed intensity distribution of the Fresnel pattern yielded by a phase plate of phase angle 180°, 150° and 120°. The location of the true geometrical edge is shown by the vertical line in each case.

$$\Delta c = \int_{r_a}^{r_b} \int_{r_a}^{r_b} \left\{ \sum_{j=1}^{j=n-1} \frac{\left[\dfrac{z_{i+1} - z_i}{\{4(i+1) - 0.5\}^{0.5}}\right]}{\left[\dfrac{(4i-0.5)^{0.5}}{n-1}\right]} \right\} dr\, dr \tag{8}$$

This equation defines the relationship between Fresnel fringe spacings and Rayleigh fringe shifts. The double integration is highly favourable with respect to error reduction in the data set, achieving at least an order of magnitude of diminution, greater if the summation can be effected by measurement of multiple fringes. A constant of integration is required for the first integration. This is in fact the zero'th order spacing, but as an independent estimate of this constant can be obtained from each radial scan, errors in its estimation are not serious.

We have evaluated several procedures for determining the z spacings in a multiple fringe pattern. Direct fitting of the Fresnel function by a least squares algorithm has been implemented, but is not totally successful. This is primarily because the intensities in a Fresnel pattern, ranging from true zero upwards, cannot possibly be recorded photographically within the linear part of the gamma curve of the emulsion. The true pattern is thus convoluted with an envelope function, which as the Fresnel amplitude/intensity function is anharmonic, cannot readily be deconvoluted as with Rayleigh patterns[1]. Furthermore, as noted above, although up to ten or more Fresnel fringes can be discerned by eye, the intensity scans give only two or three clearly defined maxima/minima (Fig. 6).

Thus this approach, whilst in principle superior to the simple evaluation of the zero'th order, has yet to be developed to its full potential.

Data Capture from Chart Records

This is particularly simple. No method can equal the convenience and ease of use of a digitising table linked to a microcomputer. A spatial resolution of at least ten cycles per mm over the active area is required, which is readily attainable with most commercial tables. The use of a stylus *or* cursor with cross-hairs is largely a matter for individual taste. We have for many years used a BIT-PAD linked to an Apple II microcomputer - its current upgrade is an Apple Macintosh linked to a Wacom tablet. A problem for some users will be that no single format is standard: for those not feeling confident about writing their own system the simplest approach is to contact an existing analytical ultracentrifuge user owning a system mounted on a hardware platform matching the enquirer's.

(a)

(b)

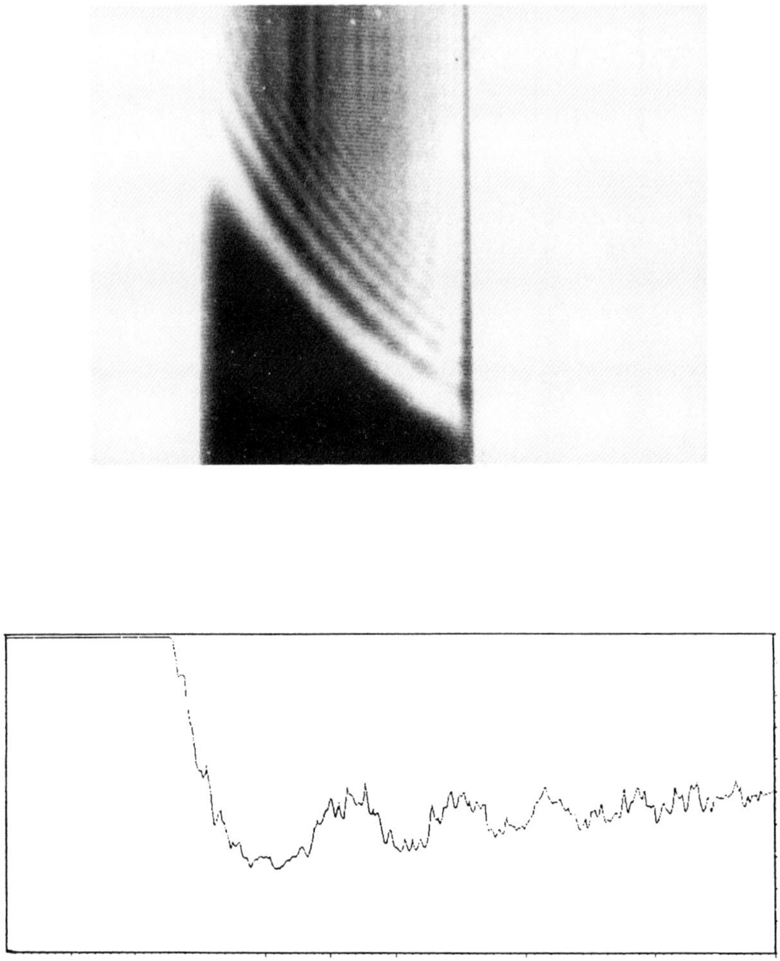

Figure 6. (a) A part of a Schlieren pattern recorded with a knife edge diaphragm from a solution of an enzyme (chloramphenicol acetyl transferase) at sedimentation equilibrium. Centrifugal direction is from left to right; (b) Vertical (z) scan across the above Fresnel fringes. The location of the centre of the peaks in the scans can be determined by established procedures[1]. The resulting precision is found to be of the order of 1% to 2% of the fringe spacing.

Particular points which need to be borne in mind by those writing their own software are:

For *velocity* runs, the programme must incorporate a check as to whether the original data is charted or photographic. This is because the time interval is *non-constant* in the former case, due to the differential time taken for the scanner to reach the boundary position.

For *equilibrium runs*, consideration must be given to the fact that intervals in the radial direction cannot be taken as wholly constant, due to the limitations of the hand-eye combination and possibly of the resolution of the table. It may be desirable to incorporate an algorithm transforming the data set to equal spacing in r (or usually better r^2).

In both cases, care should be taken that both raw and derived data are presented to the user *and* made available for further analysis and graphical presentation. The simplest way of achieving the latter aims is to store ASCII files which can be read into a commercial spreadsheet, unless the user wishes to undertake the (perfectly feasible but not particularly simple) task of writing dedicated presentation software.

4. CONCLUSIONS

Off-line evaluation offers a simple approach which is not instrument-dependent. For some time to come it will be used as the only rapid way to evaluate certain types of data, and the simple digitising table/microcomputer combination offers a simple and universal approach of an accuracy and precision acceptable for many purposes. Very high precision can however be obtained, where the original record is photographic. The evaluation of relative solute concentrations within an ultracentrifuge cell by Rayleigh interferometric fringe shifts can, for example, using optimal procedures and interpretative algorithms, yield a data set with a precision approaching f/500, where f is a single fringe shift[1]. The alternative Schlieren optical system can under identical experimental conditions (*i.e.* same solute concentration and optical path length) approach the Rayleigh fringe level of precision. In terms of Rayleigh fringe shifts, the direct evaluation of the zero'th order by a phase plate diaphragm attains a precision equivalent to f/300 in Rayleigh interference terms, though with some danger of systematic error, and by knife edge diaphragm attains f/150 to f/200. By the use of the fringe spacings and the transformation noted above (eq. (6)) a precision of close on f/300 is attainable in principle. This latter equation offers a formal proof of the transformation required to relate Rayleigh fringe increments to Fresnel fringe spacings.

These refractometric (Schlieren) optics have a number of advantages over Rayleigh interferometric optics. Their alignment is much simpler, and window distortions are a much less serious problem. The widely held supposition that their precision is much inferior to Rayleigh optics lacks a theoretical basis and is now shown to be untrue in practice. Given more advanced two-dimensional analysis of the recorded fringe patterns to enable up to ten Fresnel fringe spacings to be analysed, the precision of refraction and hence concentration increment determinations should be essentially the same by either method, and it will be possible to choose the simpler refractometric system when experimental conditions so dictate.

ACKNOWLEDGEMENTS

We are indebted to the Science and Engineering Research Council (U.K.) for support of this work.

REFERENCES

1. S.E. Harding and A.J. Rowe, *Optics and Lasers in Engineering*, 1988, *8*, 83.
2. S.E. Harding and A.J. Rowe, *Biochem. Soc. Trans.*, 1987, *17*, 1046
3. T. Bowen and A.J. Rowe, 'An Introduction to Ultracentrifugation', Wiley, London, 1970.
4. P.H. Lloyd, 'Optical Methods in Ultracentrifugation, Electrophoresis and Diffusion', Oxford, 1974.
5. Von H. Wolter, *Ann. der Physik*, 1950, *VI.7*, 182.
6. A.J. Rowe and G.M. Khan, *Rev. Sci. Instr.*, 1971, *42*, 1472.
7. R. Trautman and V.W. Burns, *Biochem. Biophys. Acta*, 1954, *14*, 26.
8. A. Papoulis, 'Systems and Transform with Application to Optics', pp. 70-73, McGraw Hill, Maidenhead, England, 1968.

6

On-Line Data Acquisition and Analysis from the Rayleigh Interferometer

By Thomas M. Laue

DEPARTMENT OF BIOCHEMISTRY, UNIVERSITY OF NEW HAMPSHIRE, DURHAM, NEW HAMPSHIRE 03824, U.S.A.

1. INTRODUCTION

In order for the intrinsic merits of analytical ultracentrifugation to be fully realised, the acquisition of data must be automated and simplified. Nowhere is this more plain than for the Rayleigh interference optical system. The interferometer provides a cell image in which the refractive index difference between sample and reference at each radial (horizontal) position is given by the vertical displacement of a set of evenly spaced horizontal fringes (Fig. 1)[1,2]. Manual acquisition of fringe displacements is extremely tedious. However, the Rayleigh image is well-suited for automated data acquisition[3,4].

Several rapid and precise instruments for acquiring fringe displacements from photographs have been described[3-6]. Nevertheless, only a few automated plate readers were built, partly because of their construction costs and complexity, but also because of the general decline in the use of the analytical ultracentrifuge. Moreover, these systems still required that an interference photograph be taken, aligned and read. As centrifugation was used only occasionally, most experimenters chose to use the more easily operated absorbance scanner. In many cases there was no obvious loss in information. Thus, before rushing headlong into the construction of an automated Rayleigh interferometer, it is worthwhile to assess whether or not refractive optics serve any useful purpose in modern centrifugation.

Why Bother with Interferometry?

The primary measurement in analytical ultracentrifugation is the concentration as a function of radial position. Absorbance and interference (refractive) optical systems provide complementary, not redundant, capabilities for determining the concentration distribution. Neither method alone can satisfy the wide range of applications of analytical ultracentrifugation.

The selectivity of the absorbance system cannot be matched by the interferometer. Moreover, the absorbance system often is able to detect lower concentrations of solute than can be measured refractively. But it is the concentration, not the absorbance, that is needed for analysis. For determining the sedimentation

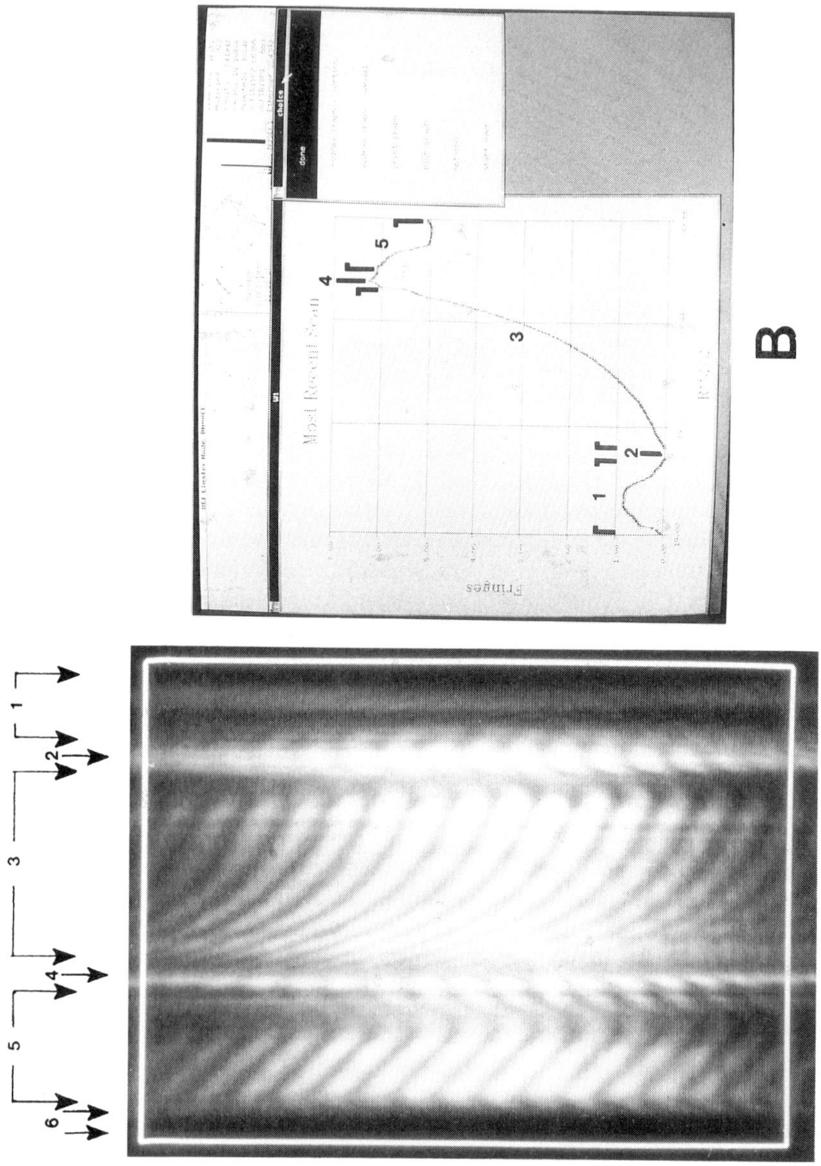
Most Recent Scan
Fringes
done
choice
1
2
3
4
5
6
A
B

Figure 1. The televised presentation of a magnified portion of a Rayleigh image from the Model E analytical ultracentrifuge and the resultant graph of fringe displacement as a function of radial position. Photograph of Rayleigh interference image (**A**) as displayed on the television monitor of the on-line interferometer. Because there is no viewing mirror to reverse the image, sedimentation is from right to left when presented on the monitor. The corresponding graph (**B**) of the fringe displacement is reversed so that lower radii are to the left. The box in the image defines the area analyzed to produce the graph. The vertical white stripes delineate the radial region to be examined and the horizontal white stripes define the portion of the diffraction envelope to be used in the Fourier analysis. Going from right to left in the image, is a region devoid of fringes (**1**), the air-liquid interface (**2**), the solution-solvent fringes (**3**), the interface between the solution and the FC-43 fluorocarbon (**4**), the fluorocarbon-solvent fringes (**5**) and a second region devoid of fringes (**6**). The corresponding regions are marked in the graph. The image was made using a 15 mW He-Ne source and a 12 mm thick, short column centrepiece so that the solution column (region **3**) is 500 μm from meniscus to base[9]. The sample is prothrombin (0.5%) in 100 mM NaCl, 50 mM Tris (pH 7.6), 1 mM benzamidine and 2 mM Ca^{2+} examined at 16000 rev/min. Under these conditions, the solution reached equilibrium in less than 60 minutes. The image is obtained in $^{1}/_{30}$th second and the Fourier data reduction and production of the graph requires about 10 seconds. Sixteen fringes were used in the Fourier analysis at a centre-to-centre spacing of about 27 pixels. Six fringes are crossed in going from the meniscus to the base and there are approximately 75 data points over this region. The graph in panel **B** may be edited using a mouse and pointer, corrected by blank subtraction and the resultant data analyzed immediately.

coefficient or the molecular weight of a single solute that obeys Beer's law, this ordinarily poses no problem. Even for such simple systems, however, turbidity, fluorescence, phosphorescence and adhesion to the cell windows can add complications. Likewise, refraction of the light beam can contribute to the apparent absorbance, particularly in regions containing steep concentration gradients. The need for sufficient light to make accurate measurements and for narrow slits to define the radial position can lead to undesirable compromises.

More detailed analysis of absorbance data must rely on the accuracy and constancy of extinction coefficients. However, extinction coefficients often differ profoundly for otherwise similar solutes (*e.g.* protein absorbance at 280 nm) and can change significantly with variation in solvent conditions or solute aggregation. For systems containing more than one absorbing species, ill-posed and often intractable deconvolution of absorbance data is needed to obtain the concentrations required for analysis. Moreover, many solutes do not exhibit significant absorbance except in the far ultraviolet, where absorbance by solvent components may interfere. Finally, the useful absorbance range of any spectrophotometer is limited fundamentally by source intensity, source stability, stray light and detector noise. Thus, relying solely on absorbance detection can restrict the utility of the ultracentrifuge.

Despite its non-selectivity and somewhat lower sensitivity, for many cases refractive detection offers several advantages over absorbance measurements. For the Rayleigh system, it is the refractive increment (dn/dc) which relates the fringe displacement to solute concentration. The refractive increment is nearly identical for materials of similar composition, regardless of their aggregation state, and nearly constant over a wide range of solvent conditions[7]. Fringe displacements are not affected by sample turbidity, fluorescence or phosphorescence. Thus, the precision and accuracy of the Rayleigh system cannot be achieved by absorbance systems. The insensitivity of the refractive increment to solute aggregation makes interference optics the method of choice in studying associating chemical systems. Furthermore, so long as there is sufficient light to produce detectable fringes, any solute concentration can be measured with equally good accuracy. Thus, interferometry is useful for examining high concentration solutions or solutions absorbing light strongly in the ultraviolet. Radial resolution is typically an order of magnitude better than with the absorbance system and very steep gradients can be traced accurately[8].

It is clear that both optical systems are needed for the analytical ultracentrifuge to satisfy all of its uses.

2. DESCRIPTION OF THE ON-LINE INTERFEROMETER HARDWARE

A prototype on-line interferometer has been constructed using a Beckman Model E analytical ultracentrifuge as a foundation (Fig. 2-I, A). The basic optics of the Model E Rayleigh system have been retained, with major changes being made to the light source and the detector. For the on-line system, the AH-6 mercury arc light source is replaced with a laser (Fig. 2-I, D) and the photographic plate is replaced by a solid-state television camera (Fig. 2-I, F). Both the laser and the camera are operated by computer (Fig. 2-I, B) so that their functions can be synchronized to one another and to the spinning rotor. The computer is also used to acquire images, calculate the concentration distribution, store information, and edit and analyse the data. A detailed description of one on-line system is provided below, but it should be emphasized that any number of laser sources, cameras, computers and interfaces could be used satisfactorily. It is more important that the critical design criteria be understood so that advantage can be taken of the rapid advances in electronics and optics. Indeed, since the system described here was constructed, two other laboratories (David Yphantis[10] and Walter Stafford[11]) have constructed on-line interferometers. Their systems employ the essential features described here, but differ in detail. Included here are descriptions of these variants whenever it seems useful.

Light Source

The AH-6 mercury arc light suffers many problems that make it unsuitable for use with on-line data acquisition. Lasers are a superior source for interferometry, and there are several descriptions of laser sources for the Model E[12,13]. Using a pulsed laser results in clearer images and facilitates the use of multi-hole rotors[4,12,13]. Gas lasers (either helium-neon or pulsed argon) have been effective in these applications, but there have been great strides in opto-electronics, so that now laser diodes are an appealing option as a source. They provide ample light, are small, inexpensive, have a long life and can be pulsed. Typically, the coherence length is a few centimetres, which is less than is available from He-Ne lasers, but is far better than that of the AH-6 lamp. Currently, the shortest wavelength available for commercial laser diodes is 670 nm, which means that there is some loss in sensitivity compared to the AH-6 light source, since

$$\frac{dn}{dc} = \left(\frac{dn}{dc}\right)_0 (0.94 + 2\times10^6.\lambda^{-2})$$

where $(dn/dc)_0$ is the refractive increment at 578 nm and λ is the wavelength of interest in Å [7]. However, laser diode technology is advancing rapidly and shorter wavelength laser diodes will be available in the near future. Laser diodes differ from gas lasers in that the beam emerges as a diverging ellipse instead of being circular and collimated. However, it is a simple task to expand and collimate the beam using commercially available prisms and lenses.

For the system described here, either continuous He-Ne gas lasers (15 mW model 124b, 25 mW model 127, Spectra-Physics, Mountain View, CA, U.S.A.) or 5 mW to 30 mW, 780 nm pulsed laser diodes modules (LDM-4412 diode mount, ILX Lightwave Corp., Bozeman, Montana, using a Mitsubishi ML6702A diode) were used. The latter system included a modulatable power source (model LDX-3620) and thermo-electric cooling (model LDT-5910). Cooling is essential for gain-guided laser diodes to maintain a unimodal (*i.e.* single wavelength) output, but it probably will not be required for the more recently developed index-guided laser diodes. Tests indicate that lasers of 5 mW are sufficient, though higher power lasers can be useful for generating high intensity pulses of short duration (100 ns - 1000 ns), thus eliminating the need for slits on the cell (discussed below).

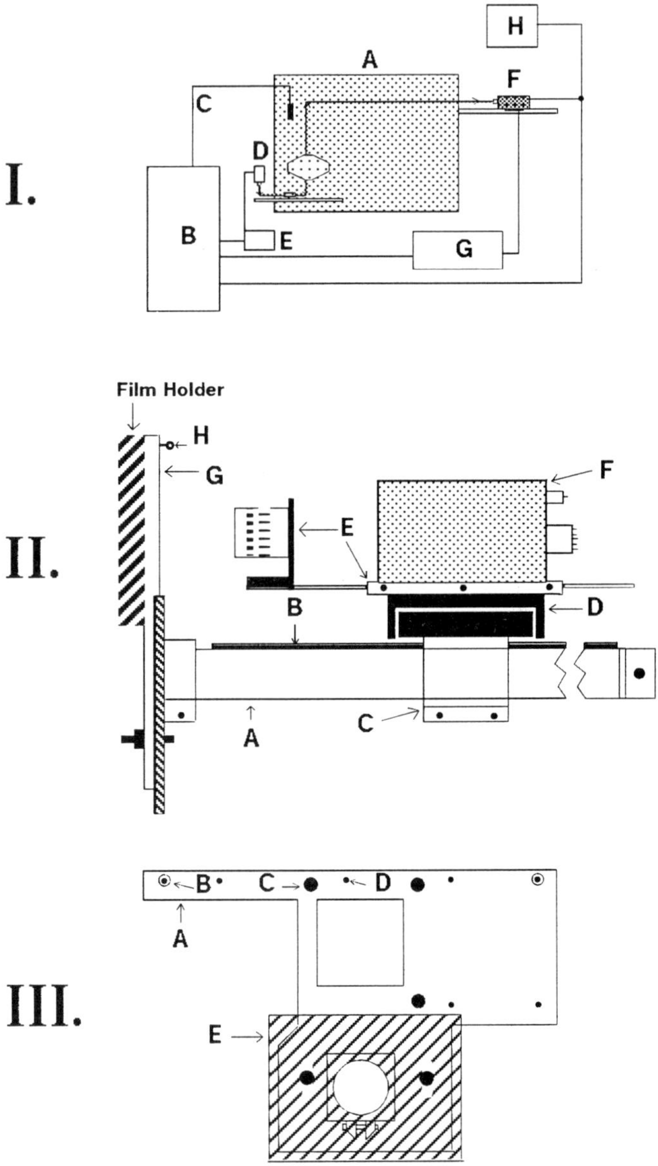
I.
H
A
F
C
D
B
E
G
II.
Film Holder
H
G
F
E
B
D
A
C
III.
B
C
D
A
E

Figure 2. Design of the on-line interferometer hardware. Panel **I** is a block diagram identifying and placing the major components: **A.** Model E analytical ultracentrifuge with the optical path noted, **B.** VAXlab computer, **C.** sensor providing a pulse for each rotor revolution[15], **D.** laser light source mounted vertically with its beam deflected by prisms[12], **E.** laser controller to synchronize light bursts to the spinning rotor[16], **F.** CCD television camera and relay lens mounted on a boom (detailed in panel **II**), **G.** interface between the computer and the camera, **H.** television monitor. Panel **II** shows the television camera mount in greater detail: **A.** 1.5" diameter boom, 26" long which serves as the optical track, **B.** $^1/_8$" high by $^7/_{16}$" wide by 24" long rail for alignment of the camera mount carrier, **C.** camera mount carrier which holds the motorized stage, **D.** the motorized linear stage, **E.** extendable lens holder and lens mounted in a groove so that the distance between the lens and the television sensor can be varied, **F.** the television camera, **G.** the adapter plate which attaches the camera mount to the centrifuge, **H.** attachment so that guy wires running from the adapter plate to the end of the camera boom may be used to stabilize the camera mount. Subsequently it has been found that the guy wires are not absolutely necessary. Panel **III** is a detailed view of the adapter plate which holds the television camera mount to the film holder: **A.** $^1/_4$"-thick aluminium plate which attaches to the back of the film holder so that the window in the centre of the plate is aligned with the door at the end of the optical track on the Model E, **B.** the guy wire attachments, **C.** and **D.** are mounting screws which hold the plate to the back of the film holder and which use the existing screw holes in the film holder, **E.** plate with collar to hold the camera boom.

Synchronizing the Light Source to the Spinning Rotor

Pulsed lasers permit the use of multi-cell rotors without using wedged-windows. However, their use requires that timing circuits be used to ensure that light pulses consistently illuminate the same arc of the rotor. Electronic designs that use phase-lock-loop[4,14,15] and rate-multiplying[16] circuits have been described. Both systems are able to produce a laser pulse that hits the same rotor position to within one milliradian. However, the latter design is simpler, more reliable and has been adapted to use counter/timer microcomputer interfaces. In this case, the computer program replaces the function of the rate-multiplier circuit. The added flexibility offered by the computer makes this the method of choice.

In this scheme, a single pulse is required each revolution and the period between pulses (*i.e.* one rotation) is measured by counting with a high-frequency clock. The value in the counter is multiplied by a fraction corresponding to the portion of a revolution to wait before pulsing the laser. A counter is loaded with the resulting value and decremented by the same clock used to determine the rotor period. When the count equals zero, the laser is pulsed. Two pairs of counters are operated so that a light pulse is produced each revolution, and the timing of the pulse is based on the period of the previous revolution[16]. Thus, even rapid rotor acceleration can be tracked. Propagation delays in the electronics (usually only a few μsec) are compensated for by delaying the start of decrementing for a period equal to the total delay. A method for setting the compensation time has been described[4,14]. The duration of the laser pulse is determined in the same fashion so that a constant position and constant arc are illuminated. If the pulse is of very short duration, then the Rayleigh slits on the cell may be eliminated[4], thus facilitating the simultaneous use of the scanner and interference optical systems.

Solid State Cameras as Detectors

A television camera replaces photography with the on-line interferometer, which raises concerns about comparative resolution and sensitivity. Rapid advances have been made in television cameras in the past decade, largely in response to consumer demand for home video equipment. These cameras typically use a charge-coupled device (CCD) sensor in place of the vidicon tube commonly employed in the past. There are several advantages in using CCD sensors rather than vidicon tubes for this application. The major advantage is geometric stability. This results in light sensitive elements (pixels) that are of a well-defined size, shape and spacing on CCDs. By contrast, the pixel on a vidicon is a generated quantity, dependent on timing and electron beam diameter, rather than being a geometric entity. The currently available high-resolution CCD arrays have photosensitive areas that correspond in size to the conventional $^1/_2$" or $^2/_3$" vidicon. The resulting solid-state cameras are small, inexpensive and rugged. CCD sensors are well suited to use with bright laser light in that they do not "burn" (*i.e.* suffer permanent damage), do not "bloom" (*i.e.* spill electrons to surrounding pixels) easily and exhibit almost no "memory" (*i.e.* image retention from scan to scan). They are less light sensitive than vidicons, but this is easily overcome in the current application by using brighter sources.

It is possible with many cameras to synchronize the camera to an appropriate computer interface by using a "slow scan" option. However, our experience has been that there is no real advantage to using slow scan signals for this application. Using standard video signals reduces the cost of both the camera and the computer interface considerably. There are two principle standards currently in wide-spread use, RS-170 (based on a 60 Hz frame rate and used in the U.S. and Japan) and PAL (based on a 50 Hz frame rate and used in Europe). The arrays available using either standard are satisfactory, but not ideal.

The major drawback to existing high resolution CCD-based cameras is their sensor size. A moment's regard will provide insight to the problem. Conformance to the $^{1}/_{2}$" or $^{2}/_{3}$" camera tube format standard means that the sensors are nominally 6-8 mm wide by 4-6 mm high with about 600-750 pixels (depending on the manufacturer) in each row. On the other hand, the nominal distance between counterweight reference holes is 16 mm and the Model E optics provide an approximate 2× magnification. Thus, the cell image at the photographic plane is approximately 32 mm wide. This means that either the image must be demagnified at least four-fold from its current size to fit on the sensor or the camera must be moved radially, acquiring only a portion of the total cell image at any one time. Demagnification results in lost resolution, about 25 μm/pixel in the cell frame of reference, compared to the 10 μm or less per radial increment typically used in reading interferograms. Thus, demagnification limits the maximum gradient that can be followed and reduces the number of data points available from some of the most useful cell types (*e.g.* short column equilibrium cells[9]). For these reasons, it was decided in the present application to retain the magnified image and move the camera, even though this results in more complexity. A system that uses a demagnified image has been developed by Walter Stafford[11], and will be described in a forthcoming publication. Likewise, the commercialization of high-definition television will result in larger sensors with more pixels per row and corresponding improvements in computer interfaces. These systems promise to nearly double the current resolution, eliminating the need to move the camera.

For the system described here, two cameras were used with equivalent results. The first is a Phillips 56475 (Phillips Industrial Devices, Inc. Eindhoven, The Netherlands) imaging module which uses an Amperex NXA1031/03 frame transfer CCD. The image area is 6 mm horizontal by 4.5 mm vertical. There are 610 pixels per row and 492 pixels per column. Each pixel is 9.9 μm wide by 18.6 μm tall. The specified minimum detectable sensor illumination is approximately 8 nW/cm^2 and saturation occurs at 4 $\mu W/cm^2$, with a signal to noise ratio of 40-60dB over the full range of illumination. The second camera is a Cohu CCD camera model 4810 (Cohu, Inc., San Diego, CA, U.S.A.). The image area is 8.8 mm horizontal by 6.6 mm vertical (corresponding to the $^{2}/_{3}$" camera tube pickup). There are 754 pixels per row and 488 pixels per column. Each pixel is 11.5 μm wide by 27.0 μm tall. The minimum sensitivity is about 12 nW/cm^2 and saturation occurs at 5 $\mu W/cm^2$, with a signal to noise ratio near 50 dB over the full range of illumination. This camera had the glass coverslip over the sensing area removed so as to eliminate the Newton's rings resulting from the highly coherent laser light. Both cameras provide RS-170 standard output.

It should be noted that sensor manufacturers tend to use several different specifications for illumination sensitivity. Nearly all CCD sensors have the same basic sensitivity per μm^2 of pixel, since this is determined by the properties of silicon (roughly 40% quantum efficiency at 633 nm). The only exception to this are arrays that use charge-injection device (CID) sensors, where metallization of the faceplate decreases sensitivity approximately two-fold. To help in comparing camera specifications, it is useful to know that 1 lux = 1 lumen/m^2, 1 phot = 1 lumen/cm^2, 1 candle = 4π lumen, 1 candle-metre = 1 lux and 1 foot-candle = 1 lumen/ft^2. Finally, 1 lumen (at 556 nm) equals 1.61 mW.

The Camera to Computer Interface

The operation of the television camera, video signal digitizing and temporary data storage are provided by a "frame grabber" interface. Suitable interfaces are available from many companies and for several common laboratory computers. The system described here uses a Data Translation DT2651 interface (Data Translation, Inc., Marlborough, MA, U.S.A.) attached to the Q-bus backplane of a VAXlab (Digital Equipment Corp., Maynard, MA, U.S.A.). This interface converts the RS-170 analogue video signal into an array of 8-bit numbers. It also allows for the attachment of up to 4 cameras and contains two on-board buffer-memories, each capable of holding a 512-pixel by 512-pixel by 8-bit image. Standard RS-170 synchronizing signals, either supplied separately or superimposed on the video signal (*i.e.* composite video), can be provided by the interface or by the camera. Alternatively, separate lines are provided for the horizontal sync, vertical sync and for the pixel clock to allow slow-scan operation. Composite video is provided for connecting a television monitor.

In our current arrangement, the camera is used to provide an analogue, composite video signal. This choice significantly reduces the cost and complexity of the camera and television monitor since it eliminates the need for scan converters. However, using composite video leads to limitations in the radial resolution beyond that calculated from the sensor specifications. The problem is that the interface digitizes the incoming analogue signal at 512 equally spaced intervals during the horizontal sweep period. Meanwhile, the same horizontal scan of the camera contains intensities from more than 512 pixels (*e.g.* 610 pixels for the Phillips and 754 for the Cohu). Thus, the true radial resolution is only $(512/610)\times100$ = 84% or $(512/754)\times100$ = 68% of that expected based solely on the sensor specifications. Clearly, there is no one-to-one correspondence between the pixels on the sensor and the video data array.

The use of the analogue signal also leads to the possibility for geometric distortion. This will occur if there are problems with synchronizing the timing of the interface to the composite video (*e.g.* poor sync separation, bad phase-lock-loop operation, *etc.*) or if there is significant distortion from the video amplifiers (*e.g.* overshoot, poor settling time). The former problem is usually greatest at the start of a frame when the phase-lock-loop is readjusting to the vertical sync timing. The latter problem would be most apparent when there are abrupt intensity changes along a horizontal sweep. Each of these will lead to a further reduction in the radial resolution.

Undoubtedly, an optimal system would achieve the highest radial resolution and least potential for distortion by using slow-scan operation of the camera and interface to insure a one-to-one correspondence between the pixels and the resulting data array, and a scan converter to provide the video signal for a monitor. To date, the added complexity has not been necessary since the radial resolution is determined largely by the image magnification and image distortion seems negligible. It will become essential, however, to use the slow-scan scheme if non-magnified images and larger sensors are used to image an entire cell at once.

Computer Systems

Two computers are used in the operation of the on-line interferometer, one to process the images and one to perform the real-time chores associated with synchronizing the laser to the spinning rotor. The first computer is interfaced to the television camera and performs the Fourier analysis that yields the fringe displacement. This computer must provide mass storage capabilities since it is easily possible to acquire 10 Mbytes or more of information in a day. With this high acquisition rate, it is useful to have some sort of archival storage (removable discs or magnetic tape) available. Variants of the on-line interferometry system have been built by us using a VAXlab running VMS (Digital Equipment Corp., Maynard, MA, U.S.A.), using a Macintosh IIplus[11] (Apple Computer, Inc., Cupertino, CA, U.S.A.) and using a 80386-based PC running DOS[10] (Zenith Computer, Inc., Benton Harbor, MI, U.S.A.). A maths co-processor is essential for rapid image analysis.

A second, less powerful IBM PC compatible is used to operate the counter-based laser controller (above). A maths co-processor is required. The programs on both computers must allow for interrupt-driven communications over serial lines, thus increasing the complexity of the software. As hardware and software become more powerful, the two computers eventually could be merged into one.

For the system described here, a MicroVAX II-based VAXlab running the VMS operating system serves as the image analysis and storage computer. It also operates the motorized camera stage and determines the rotor speed. This computer has 13 Mbytes of memory, 220 Mbytes of disc storage and archival storage provided by either $^1/_2$", 9-track tape or tape cartridges. It is connected by ethernet to three other VAX-based workstations, thus allowing the CPU-intensive task of data analysis to be conducted simultaneously and separately from data acquisition. Originally the system operated with a continuous He-Ne laser source, thus limiting acquisition to a single cell. For multi-hole operation, an 8 MHz, 8086-based Zenith Z-151 IBM PC compatible with 640 kbytes of memory and 8087 maths co-processor serves as the laser controller. The interface design and operating software (written in Basic) were modifications of that supplied by The University of Connecticut Centrifuge Facility[10].

3. ALIGNMENT AND CALIBRATION

Arrangement of the Camera

The camera (Fig. 2-II, F) is positioned about 25 cm behind the photographic plane of the centrifuge and arranged so that each column of pixels corresponds to a radial position. The lens (75 mm, f/1.8) is removed from the camera and is mounted on an extendable rod (Fig. 2-II, E) or an extension tube. Positioned in this fashion, it acts as a relay lens so that the fringes at the photographic plane are magnified and projected onto the CCD sensor. The extent of magnification can be calculated from simple lens equations. Typically the lens and camera are positioned so that the overall magnification is about 2.5×, including the 2× magnification afforded by the Model E optics. This results in a little more than 3 mm of the cell (in cell co-ordinates) being viewed at any time. The camera is mounted on a motorized precision linear stage and position encoder (model 16328, Oriel Corp. Stratford, CT, U.S.A.) whose position is under computer control *via* an RS-232-coupled interface (model 18011, Oriel Corp., Stratford, CT, U.S.A.). This stage has a positioning reproducibility of ±0.3 μm and a maximum cumulative error of 2 μm over the full 50 mm of travel.

The camera is mounted on a boom (Fig. 2-II, A) that serves as a tilt table and optical track. The boom is attached by a collar (Fig. 2-II, C) to an adapter plate (Fig. 2-III) that is, in turn, bolted to the photographic plate holder. Slots in the adapter plate allow the boom assembly to be shifted up or down, thus permitting the image to be centred vertically on the camera sensor. The collar holding the boom can be loosened so that the boom may be rotated, permitting the horizontal motion of the stage to be aligned with the radial axis of the image. Guy wires (Figs. 2-II, H and 2-III, B) are connected from the boom to the centrifuge frame to provide additional support.

The alignment and focusing of the Rayleigh system is performed as has been described previously[2], with adjustments to take account of the laser light source[12,17]. Two modifications to these procedures were used. The first is that the relay lens and television camera were positioned on the boom and focused on a ruled scale mounted at the plane of the photographic plate. The relative position of the relay lens and camera were adjusted so that a 7 mm portion of the scale spanned the array. Then, using vacuum fringes, the boom was adjusted vertically so that light from the central diffraction envelope of the interference system was roughly centred on the sensor and adjusted rotationally until the fringes were approximately parallel to the rows of pixels. At this point, the television camera can be used in place of film for the remaining steps of the focusing procedure. The second change, suggested by Walter Stafford, was the use of a standard incandescent lamp and a 633 nm bandpass filter (model P10-633-S, Corion Corp., Holliston, MA, U.S.A.) in place of the "kinetic diffuser" needed to overcome the laser light's temporal coherence[12,17]. The light bulb and filter combination provide a superior image of the ruled disc during the focusing of the camera lens. These two changes significantly shorten the time needed to focus the centrifuge optics. For the work reported here, the $^2/_3$ plane of the cell, used to minimize Wiener skewing[2,8], was brought into focus at the photographic plate holder. In this way, direct comparison could be made of the data from the on-line system and from photographs. Once focused, none of the centrifuge optics need to be moved, except for shifting the position of the camera and cylinder lenses on the Model E when longer pathlength cells are used[2].

The final alignment consists of tipping the television camera so that the motion of the camera stage is parallel to the radial direction. The approach used was first suggested by David Yphantis. The procedure can be done either with or without a rotor in the chamber. In this method, the cross-covariance array is generated by mathematically "sliding" an intensity profile (Fig. 3A) taken at the inner counterweight reference holes past an intensity profile taken at the outer reference holes. The resulting graph of the cross-covariance (Fig. 3B) has a single maximum at the number of pixels needed to bring the two profiles into alignment. Other peaks occur at a spacing equal to the centre-to-centre distance between the fringes. The camera is rotated on the boom and the cross-correlation re-determined until the image is within one fringe of perfect alignment. Any residual rotational error is corrected for during data acquisition (below). Evidence from over 200 experiments indicate that the rotational alignment is constant under all experimental conditions so long as the mask or cylinder lens are not rotated.

A second, and separate test is made to insure that the rows of pixels are aligned with the radial axis. In essence, this tests for the alignment of the array with respect to the axis of the motorized stage. What is desired is that the fringes observed with no rotor in the chamber be exactly parallel to the rows of pixels. To test this, the fringe displacement is measured across the image, as described below, and the slope determined. To date, no adjustment to the camera has been necessary, indicating that the array is well aligned with the camera case and that the case is flat with the motorized stage. Should there be a significant slope to the fringes, shims can be placed between the camera case and the motorized stage until a zero slope is obtained. At this point it would be worthwhile to check the entire rotational alignment. As above, this alignment is stable from run-to-run.

Calibration of the Radial Position

Once the camera is aligned, each column of pixels on the sensor, hence, each column of intensities stored in the computer, corresponds to a unique radial position. The apparent radial position in the intensity array is given by the column index, which ranges from 1 to 512. Both the motorized stage position and the column index must be considered in the calculation of the radial position. Increments of the stage position and of the column index are related by independent magnification factors to increments in radial position in the cell co-ordinate system. Both magnification factors are determined using the ruled disc supplied for focusing the Rayleigh optics.

For both the determination of both magnification factors, the ruled disc is inserted into the holder, ruled side down, along with the appropriate spacers to bring it to the $^2/_3$ height and arranged in the optics as though the camera lens were being focused[2,12,17]. The incandescent light and bandpass filter are used for illumination (above) and the rotor aligned so that a clear image of the 1 mm-spaced lines appears on the television monitor. At the magnification used here, three lines should be in the

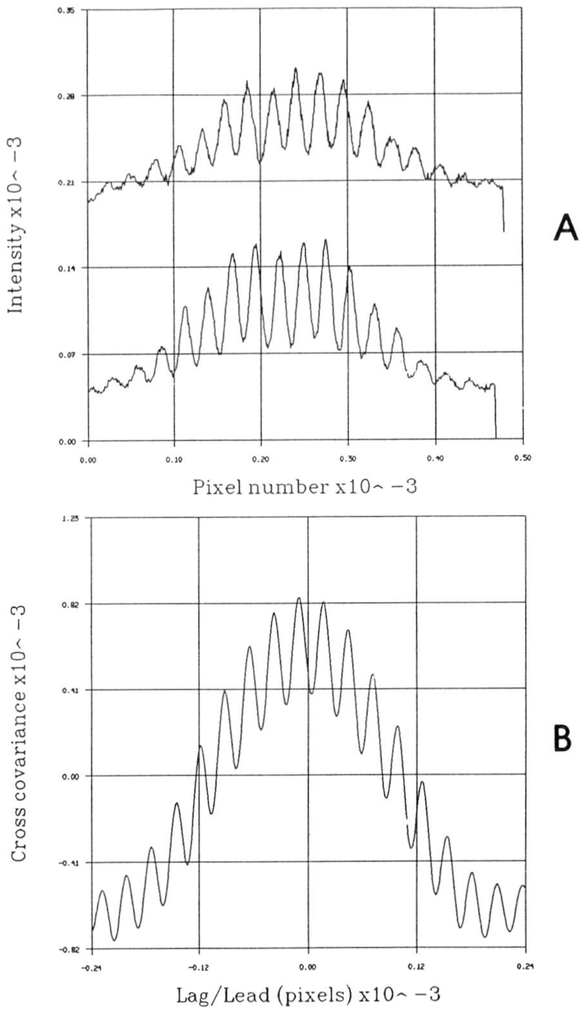

Figure 3. Rotational alignment of the image is facilitated by using intensity profiles (**A**) made at each of the reference holes. The top intensity profile in panel **A** was made at the inner reference pattern and the lower profile at the outer reference pattern. The abscissa in this graph is the row index for the 479 pixels that constitute the intensity profiles. If the image were perfectly aligned, the maxima in the two profiles would exactly coincide. Panel **B** is the cross covariance function that is generated from the images in **A**. A lag indicates that the intensity profile for the inner reference edge is shifted to the left with respect to the outer reference edge, *i.e.* the image must be rotated clockwise. The magnitude of the cross covariance reaches the maximum when the images are aligned. For the case shown here, the image is misaligned by 10 pixels, or -0.37 fringe. This rotation can be corrected for in software (see text).

image at any given time. Programs were written that permit the location of a vertical white stripe to be monitored as it was positioned in the image on the monitor. The vertical line should be exactly parallel to the lines on the ruled disc if everything is aligned properly. To start, the white stripe is placed at column 256 (the centre of the sensor) and the motorized stage moved until the left edge of one of the ruled-disc lines is adjacent to the stripe. The distance needed to move the stage to align each of the lines on the ruled disc with the stripe is noted. An average is taken of several such measurements. The ratio of the average increment in the stage position (in millimetres) per ruled line is the magnification factor for the motorized stage (M_m). This magnification factor is a unitless ratio, since the ruled lines are precisely 1 mm apart.

To obtain the magnification factor at the camera sensor, the increment in the pixel column index needed to move the vertical white stripe to the left edge of each of the ruled lines in the image is determined. This measurement is repeated at each of several positions of the motorized stage (*i.e.* not just when one of the ruled lines is positioned at column index 256). The ratio of the average increment in the column index per ruled line is the magnification factor for the camera sensor (M_c), and has units of pixels/mm. The value of M_c should be determined as a function of the column index as a test for spherical aberration or electronic distortion affecting the horizontal scan.

It is the inverse of M_m and M_c that appear in the calculation of the radial position. Since the motorized stage encoder is calibrated in microns, it is convenient to use the inverse of the camera sensor magnification factor (M_c^{-1}) in units of microns per column index increment (or microns per pixel). The position of any column in the array, relative to the approximate centre of the array, can be calculated from its index as $(256\text{-}i)\cdot M_c^{-1}$. Note that the image is "reversed" on the sensor (Fig. 1), so that the right hand side (*i.e.* larger values of i) correspond to lower radii.

The final information needed to determine the radial position is the position of the counterweight reference edges. These are determined by moving the motorized stage until the image of the reference edge is in the field of view, then moving a vertical white stripe until it is aligned with the inner edge (*i.e.* that nearest the cell image) of the image. The two reference edge positions are then calculated to be:

$$ref_a = E_a \cdot M_m^{-1} + (256 - i_a) \cdot M_c^{-1} \tag{1}$$

$$ref_b = E_b \cdot M_m^{-1} + (256 - i_b) \cdot M_c^{-1} \tag{2}$$

where ref_a is the position in microns corresponding to the centripetal reference edge, E_a is the motorized stage position returned by the encoder, M_m^{-1} is the inverse of the magnification factor for the motorized stage, i_a is the column index and M_c^{-1} is the inverse of the camera magnification factor. These same definitions apply to the positions at the outer reference edge, designated as ref_b, E_b and i_b. The midpoint between ref_a and ref_b is 6.5 cm from the centre of rotation, so that the radial position can be calculated from any encoder position and column index as:

$$r = 6.5 + \{E \cdot M_m^{-1} + (256 - i) \cdot M_c^{-1} - (ref_a + ref_b)/2\} \cdot 10^{-4} \tag{3}$$

The factor of 10^{-4} converts the position from μm to cm.

4. FRINGE DISPLACEMENT MEASUREMENT

In the Rayleigh interferometer, monochromatic illumination of a pair of parallel slits yields fringes that are equally spaced at the image plane (Fig. 1), regardless of the refractive index difference between the solvent and solution or the gradient in refractive index[1,3,8]. The constant spatial interval between fringes is evident in an intensity profile (Fig. 4) and a discrete Fourier analysis of the profile reveals a single, rather narrow maximum at a spatial frequency corresponding to this interval (Fig. 5). The dominance of this maximum means that the hundreds of intensities constituting the profile can be replaced by a single descriptor, the spatial frequency of a sine wave. The relative vertical displacement of the fringes requires a second descriptor, which, is obtained as the phase of Fourier analysis. However, before going further it is worthwhile to consider how the Fourier analysis is used in the on-line interferometer.

The on-line system uses a single-frequency, discrete Fourier analysis conducted at the spatial frequency of the fringes. Under these circumstances, the displacement of the maxima in the intensity profile relative to the maxima in the sine function equals the phase, **P**, of the Fourier analysis. Moreover, **P** varies linearly with the fringe shift. Thus, an integral fringe displacement results in a 2π change in **P**, and the fractional fringe displacement equals $\mathbf{P}/2\pi$. Moreover, the Fourier analysis is an effective filter against noise at all frequencies other than that being examined. Thus, the effects of defects in the optics and the sensor are reduced dramatically.

In order to conduct the single-frequency, discrete Fourier analysis rapidly, 512 values of the sine and cosine functions are stored in look-up tables. The values in these tables are calculated as $\sin(i\cdot 2\pi/ppf)$ and $\cos(i\cdot 2\pi/ppf)$ where i ranges between 1 and 512 and *ppf* is the number of rows of pixels corresponding to the centre-to-centre spacing of the fringes. Two sums are calculated in this analysis, one corresponding to the product of the pixel intensities and the sine function, the other to the product of the intensities and the cosine function:

$$\mathrm{SSUM} = -\sum_{j=\mathrm{bottom}}^{j=\mathrm{top}} \sin(j)\cdot I(j) \tag{4}$$

$$\mathrm{CSUM} = -\sum_{j=\mathrm{bottom}}^{j=\mathrm{top}} \cos(j)\cdot I(j) \tag{5}$$

where I(j) is the intensity and the row index of the pixel, j, serves as the pointer to the sine and cosine tables. This index, j, ranges in value from the top of the region used for the analysis to the bottom of this region (Fig. 1) and is incremented by 2 since the image is a 2:1 interlace. The sums are taken over an integral number of cycles of the sine and cosine functions. Normalization of the sums is not necessary, but can be achieved by dividing the sums by (bottom-top)/4. These sums can be calculated rapidly (< 10 ms using 479 intensities).

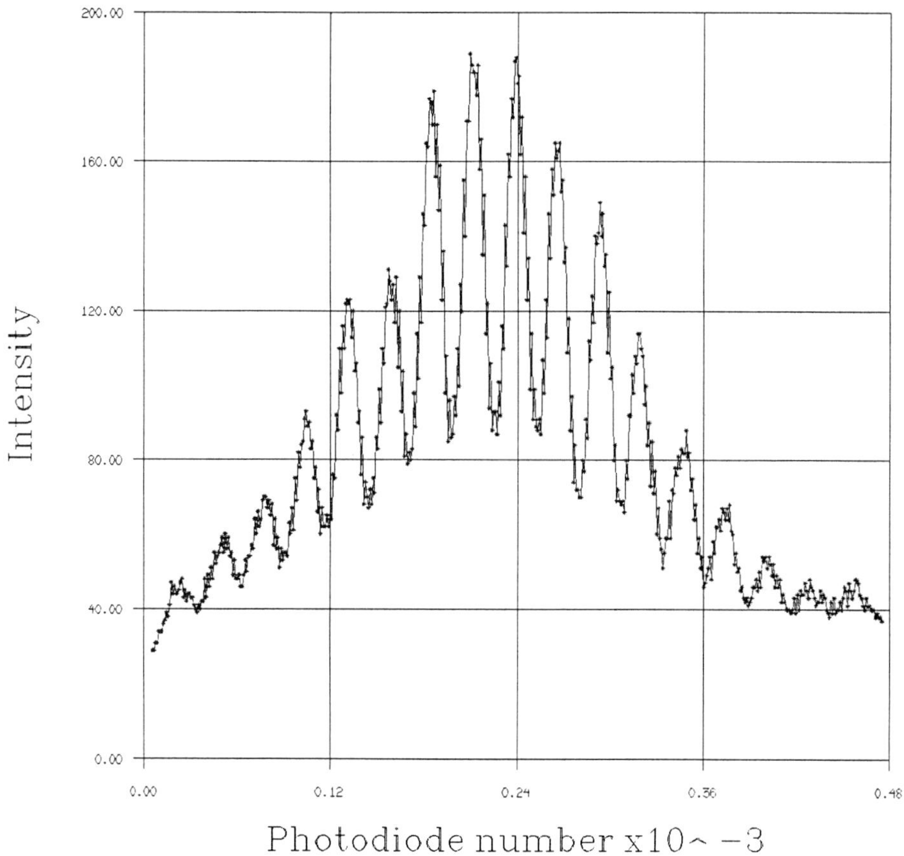

Figure 4. Intensity profile as used for the Fourier analysis. In this example, there are 470 pixels across 17 fringes, or 27.6 pixels (labelled photodiodes) from centre-to-centre of each fringe. This pattern was made using a continuous laser (15 mW He-Ne) and, consequently, the fringe envelope shows the single-slit diffraction background, resulting in a maximum modulation depth of 50%[1,3]. Use of a pulsed laser eliminates this background, resulting in intensity profiles which are more pleasing aesthetically, but which do not result in any detectable improvement in the precision of the fringe displacement measurement. The use of a pulsed laser has other advantages, which are described in the text.

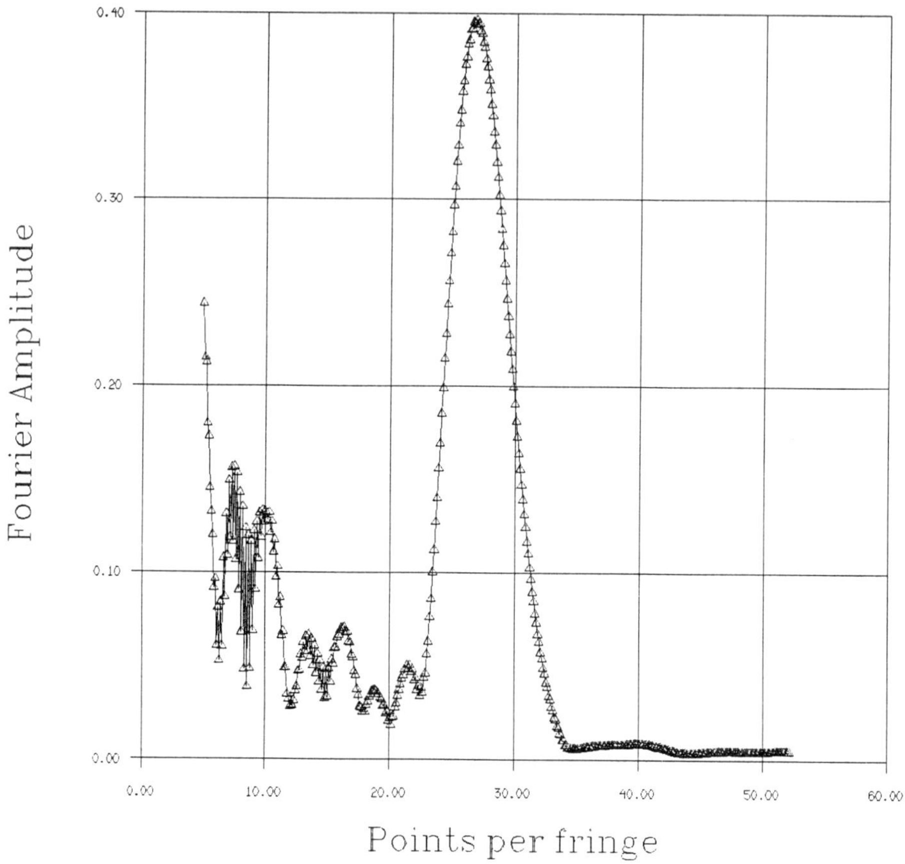

Figure 5. The amplitude of the Fourier analysis (ordinate) as a function of the spatial frequency, *ppf* (abscissa), for the fringe profile shown in Fig. 4 and assuming 17 fringes are in the diffraction envelope. The amplitude, **A**, is a measure of the goodness of fit between the spatial frequency of the fringes and the frequency used in the analysis (see text). Tests indicate that the fringe displacement measurements are most precise when the ppf having the largest **A** is used for the analysis[4].

Two values are calculated from these sums. The first is the Fourier amplitude **A**, calculated by adding the squares of the sine and cosine sums. The second is the Fourier phase **P**, calculated as the arctangent of the ratio of the sine sum and the cosine sum:

$$\mathbf{A} = \sqrt{\mathrm{SSUM}^2 + \mathrm{CSUM}^2} \tag{6}$$

$$\mathbf{P} = \arctan\left[\frac{\mathrm{SSUM}}{\mathrm{CSUM}}\right]$$

It is important to note that a full four-quadrant arctangent function is required and that most programming languages only supply a two-quadrant arctangent. The four-quadrant phase can be calculated from the two-quadrant form by testing SSUM. If SSUM is less than or equal to 0, then **P** should be adjusted by adding 2π. The fractional fringe displacement, Y(j) at position j is $\mathbf{P}/2\pi$.

Determining the Optimum Value for *ppf*

The most accurate fringe displacement measurements are obtained when the wavelength of the Fourier analysis exactly matches number of pixels, ppf, corresponding to the centre-to-centre spacing of the fringes[4]. The units of *ppf* are pixels per fringe. The Fourier amplitude is a measure of the image quality and of the goodness of fit between the spatial frequency of the fringes and that used for the analysis. Since the fringe spacing depends on the cell mask, it is necessary to determine *ppf* for each cell mask. This is done by calculating the Fourier amplitude as a function of *ppf*. In order for the Fourier analysis to be correct, an integral number of cycles (fringes) must be used. This is accomplished by counting the number of fringes, *Nf*, visible in the diffraction envelope (Fig. 1 and Fig. 4). The product of ppf and *Nf* is the number of pixels used for the analysis. In the on-line system presented here, the user may select a portion of the envelope containing an integral number of fringes. Alternatively, the entire envelope may be used. There is evidence that the latter choice increases the accuracy of the analysis.

A range of values of *ppf* is entered by the user, and **A** calculated for values of *ppf* at intervals of $1/Nf$ and a graph presented to the user (Fig. 5). The value of *ppf* giving the greatest value of **A** is used for the Fourier analysis.

Determining the Total Fringe Displacement

During operation of the on-line system, the fractional fringe displacement is calculated at each radial position. In order to reconstruct the image, then, it is necessary to convert the fractional fringe displacements to total cumulative fringe displacements. This is done by determining the fringe displacement increment between adjacent data points, *i.e.* $\delta\mathbf{Y}=\mathbf{Y}(j)-\mathbf{Y}(j-1)$, and summing $\delta\mathbf{Y}$ going from the top to the bottom of the cell image. The first data point has $\delta\mathbf{Y} = 0$, regardless of the true fringe displacement, reflecting the fact that only relative concentrations are available from the Rayleigh system[1]. Calculating the sign of the $\delta\mathbf{Y}$ is a little tricky when the fractional fringe displacement is near the crossover between one and zero and there is little gradient in the image. This could lead to situations where $\delta\mathbf{Y}$ is off by an integral fringe. An algorithm must be included to determine whether $\delta\mathbf{Y}$ needs to be adjusted by one. This is done by first testing whether the absolute value of $\delta\mathbf{Y}$ is greater than 0.5 fringe. If not, then the jth fringe displacement, $\mathbf{Y}(j)$, is set equal to $\mathbf{Y}(j-1)-\delta\mathbf{Y}$. If the absolute value of $\delta\mathbf{Y}$ is greater than 0.5, and $\delta\mathbf{Y}$ is positive, then $\mathbf{Y}(j) = \mathbf{Y}(j-1)+1-\delta\mathbf{Y}$, else $\mathbf{Y}(j) = \mathbf{Y}(j-1)-1-\delta\mathbf{Y}$.

Notice that the test on $\delta\mathbf{Y}$ limits the maximum gradient that can be tracked by the on-line interferometer to 0.5 fringe per column of pixels. At the magnification currently used on our system, the radial increment (in cell co-ordinates) is about 6.7 µm per column, leading to a maximum traceable gradient of 75 fringes/mm. This is ample for all but the most steep gradients, where limitations in accuracy due to Wiener skewing become a concern[1,8]. Because of limitations in the optics and electronics, the actual limiting gradient may be less than this.

Rotation of the Data

The procedure for adjusting the rotation of the motorized mount is described above. The effects of any residual rotation can be minimized by subtracting a straight line drawn between the central fringe of each of the counterweight reference patterns. The amount of rotation is determined as described above and each fringe displacement is then adjusted by: $(\Delta Y/1.656)(r-r_0)$, where ΔY is the difference in fringes needed to align the reference edges and $r-r_0$ is the distance in cm from the first data point. The value of 1.656 is the distance between the reference edges plus 200 µm, to account for the fact that the images used for determining the rotation are away from the inner edge of the reference holes. In principle, the radial increment should be adjusted as well. However, for rotations of less than four fringes, the correction is negligible. The fringe displacement adjustment can be added to the data during collection or it may be applied subsequently using the data editor.

5. DESCRIPTION OF THE ON-LINE INTERFEROMETER SOFTWARE

Overview

Software has been written by the various laboratories for the three computer systems described above. For the system described here software was written in DEC FORTRAN (Digital Equipment Corp., Maynard, MA, U.S.A.) and runs under the VMS operating system. There are four major tasks that must be performed by the on-line system: (1) alignment and calibration of the acquisition system, (2) reduction of the image to an array of fringe displacements as a function of radial position, (3) editing of the data and (4) analysis of the data. The software is modular so that these functions actually run as separate programs, albeit as subtasks from the parent program which serves as the user interface. This allows upgrades to the various component programs to be incorporated into the real time system without the need to recompile and re-link the entire program. Data are passed between component programs as disc files. The analysis programs were developed by David Yphantis[10] and they will be described only briefly with references to more complete descriptions.

Development of software to operate the television camera interface was facilitated by using the DT-IRIS subroutine library (Data Translation, Inc., Marlborough, MA, U.S.A.). Graphics and serial line communications make use of subroutines supplied as part of the VAXlab software package (Digital Equipment Corp., Maynard, MA, U.S.A.). Emphasis is placed on using graphs to display data to the experimenter. A graphics subroutine for the VAX (MVAG2) was developed that permits the display, blank subtraction and editing of x-y data[18]. This subroutine can handle up to 20 data sets simultaneously and provides for a number of useful manipulations of the data. Provision is made for hard-copy output to a variety of devices.

For sedimentation velocity analyses that require that derivatives be taken with the data, a cubic parametric spline is used to smooth the data and reduce it to a functional form[18]. The algorithms in the FITPACK subroutine package (P. Dierckx, Katholieke Universiteit Leuven, Leuven, Belgium) were used in the development of these programs. A graphics interface is incorporated so that the investigator can monitor the progression of the smoothing process and intervene by adding knots to the spline, if desired. In addition to smoothing the data, the size of the data file is reduced considerably since only the three parameters describing each knot plus those describing the end points are saved in the output file. Typically only 20 knots or so are needed to describe the shape of a distribution containing 512 sets data points.

File Format

All programs use a common file format containing a header and data arranged as three columns of numbers. The first column is $r^2/2$ in cm^2, the second is the fringe displacement and the third column is the Fourier amplitude, **A**. A test is made to see if a line contains three columns of numbers and, if not, it is treated as header information. Thus, the header can be of arbitrary length and comments can be placed throughout a file. The data are read in free format so that the columns do not have to be aligned in particular fields. The only exception to this are the files for the splines. These files contain a header followed by the parameters describing the knots, and must be read by a different subroutine.

Data Editing

Data editing is performed using REEDIT, written by David Yphantis. Versions are available for VMS and for DOS-based computers[10]. REEDIT is a graphics-based editor which allows for blank subtraction, data rotation, data truncation and removal of individual data points. A circle is drawn around each data point, with the diameter of the circle dependent on the inverse of the Fourier magnitude. A large circle indicates a data point that has a low magnitude, and thus should be eliminated from analysis. Data points are not actually deleted from the edited data set. Instead, the Fourier amplitude, ordinarily a positive value, is set negative, acting as a flag to all other programs so that the point is excluded from analysis. Provision is made to reset the amplitudes of such data points so that it is possible to examine different portions of the same concentration distribution.

When parametric spline smoothing is used, editing is performed within the program generating the spline since the function must be recalculated after each round of editing. The program used for this purpose will be described in a forthcoming communication[18].

A useful program has been developed by David Yphantis (MATCH) which automates the optimum alignment of two concentration distributions. Linear translation of one distribution is performed to minimize the residual error in the difference between the two distributions. Graphs of the residual errors are provided. This program is extremely useful in determining whether equilibrium has been reached. Moreover, the magnitude of the abscissa translation required to bring data sets into alignment provides an estimate of the extent of rotor precession.

Analysis of Sedimentation Equilibrium Experiments

The edited, blank-subtracted data sets are analyzed using NONLIN[19]. The current version analyses the combined information from up to 15 data sets taken at different rotor speeds, different radii and cell pathlengths. Scanner data and Rayleigh data can be co-analysed. The data are fit to various models that can include either homogeneous or heterogeneous interacting species and non-ideality. The homogeneity of a sample can be tested, as can the homogeneity of an interaction[19]. Models can be changed during the course of an analysis.

Point-average molecular weights can be calculated using BIOSPIN[20]. This program is used primarily for diagnostic purposes and to provide an experimenter with a graphical view of how a chemical system is behaving.

Analysis of Sedimentation Velocity Experiments

The edited, blank-subtracted, smoothed data sets currently are analysed by a variety of programs being developed. Recent efforts have been focused on four areas: (1) the production of sedimentation coefficient distribution functions[18,21,22], (2) extrapolation to infinite time to minimize the effects of diffusion, thus sharpening distribution functions[21], (3) the development of diagnostic methods for determining the suitability of performing more detailed analysis on a sedimenting boundary[18] and (4) providing more detailed analysis of individual boundaries. All of these are being developed for on-line use and will be described in forthcoming publications. The ability of the on-line interferometer to produce accurate dc/dt curves and their transformation to sedimentation coefficient distribution functions are described by W. Stafford in Chapter 20 of this volume.

6. PERFORMANCE OF THE ON-LINE SYSTEM

Fig. 1 shows the televised image from the on-line system described here and the resultant graph of concentration as a function of radial position ($r^2/2$) as it is presented to the experimenter. By using a short column centrepiece[9], equilibrium is achieved in about 45 minutes. The graph is available within 10 seconds from the time the image is captured. These data can be edited and analysed from within this same program. Thus, on-line acquisition and analysis, coupled with short column methods, results in an immediacy that ordinarily is not associated with sedimentation equilibrium. From the standpoint of an experimenter's or a technician's morale, the performance of the on-line system is superb. However, the more important issue is the accuracy of the concentration distribution. To this end, only two issues must be addressed: (1) the precision of the fringe displacement measurement and (2) the accuracy of the radial position measurement.

The precision of the fringe displacement measurement can be addressed several ways. One is to tip a set of otherwise flat fringes purposely, then determine the linearity of the fringe displacement as measured by the on-line system[4]. This method is useful since there is systematic error in the determination of the fringe displacement using the Fourier method (discussed with regard to the Walsh analysis by Laue, 1981)[4]. The error is manifested as oscillations about the true displacement and is periodic with the phase. The extent of the error is affected in a complex way by image quality, the number of fringes used in the analysis, the shape of the central diffraction envelope and the mismatch of the Fourier frequency and the fringe spacing. By using all of the fringes in the diffraction envelope and using the optimum value of ppf for the Fourier analysis, the peak-to-peak error is kept below 0.01 fringe, with an rms of a few thousandths of a fringe. Often, however, this is the dominant residual noise in the analysis of high-quality data.

A somewhat more realistic test of the precision of the fringe displacement is to examine the reproducibility of fringe displacement measurements using a real sample (Fig. 6). This error will include the effects of precession on the fringe displacement measurement. In the case shown here, there are 9-10 fringes in a solution column 500 μm long (Fig. 6B). Two scans were made 10 minutes apart and subtracted from one another. The residual error (Fig. 6A) shows several features of interest. Overall, the error is quite small, except at the ends of the cell where it is >0.02 fringe. The residuals are sloped when examined over the entire cell. When fitted to a straight line, and neglecting the few points at the top and bottom of the cell, the rms error about the line is about 0.001 of a fringe, indicating that the reproducibility of the on-line system is excellent. The increased error at the ends of the cell result from the mismatch of the meniscus and fluorocarbon base caused by precession. Likewise, the slight tip to the residual error results from precession: subsequent scans will show less tip or a tip of approximately the same magnitude only of opposite sign to that shown here. When all points are included in the calculation and the slope of the error is neglected, the rms error is 0.009 fringe. Analysis of several images using the MATCH program (above) indicates that our drive precesses with a magnitude of about 10 μm, consistent with the error seen here. Thus, the precision of the fringe displacement measurements by the on-line interferometer currently is limited by rotor precession to about 0.01 fringe.

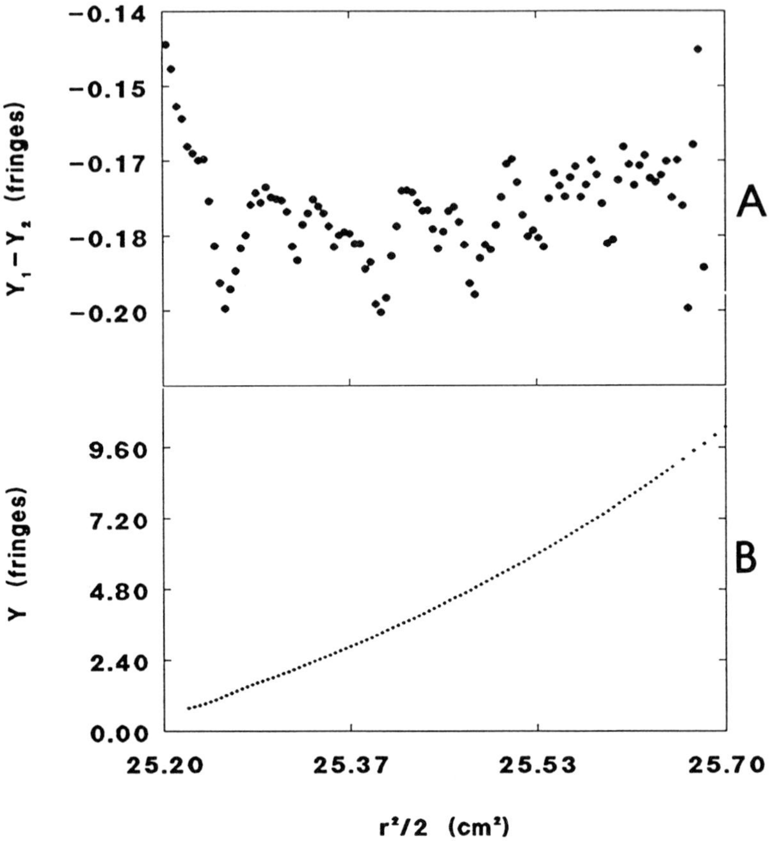

Figure 6. Reproducibility of the real time system on subsequent scans. Shown in panel **A** is the difference between fringe displacement measurements made from television images made about 10 minutes apart. The data used in the measurement are shown in panel **B**, and come from a 500 μm long, short-column experiment[9] in which a high concentration sample of bovine prothrombin (10 mg/ml) was spun at 12000 rev/min, resulting in a fringe displacement of over 10 fringes and a maximum gradient of over 25 fringes per mm. The difference plot (panel **A**) exhibits an offset of -0.18 fringe, due to the arbitrary assignment of $\delta \mathbf{Y}=\mathbf{0}$ for the first data point in each scan. The rms error over all of the data points in panel **A** is 0.009 fringe. However, there is increased error in the points at the top and bottom of the cell, as well as systematic variation in the residuals throughout the column which arise from rotor precession (see text). If the points at the top and bottom of the cell are excluded and the slope to the remaining residuals is taken into account, the reproducibility is better than 0.002 fringe.

This level of precision is consistent with our experience over the last 18 months of use.

Newton's rings often are observed when using television cameras to view laser-derived images. This problem is similar to that described by Williams[12]. The source of the rings can be traced to the glass coverslip over the television sensor. Even though the rings had no visible effect on the determination of the fringe displacements, a camera was purchased which had the coverslip removed. This camera provides images which are more aesthetically pleasing, but does not reduce the residual noise in fringe displacement determinations. Apparently, the filtering provided by the Fourier analysis is sufficient to remove this source of error.

The accuracy of the radial position determination is affected by the accuracy of the motorized linear stage, the accuracy of the pixel positioning on the sensor, the accuracy of the two magnification factors and the accuracy of the reference edge positions. Of these factors, the error in the magnification factors seem to be the most serious. At present M_m can be determined with a precision of ± 0.001, resulting in ± 2.5 µm error in the radial position measurement. The error in M_c is ± 1 column of pixels per mm. Thus, the error in the determination of the radial position in cell co-ordinates depends on the overall magnification of the image on the sensor. For the system described here, there are about 150 columns of pixels per mm in the cell frame of reference, leading to an M_c^{-1} of 6.67 µm/pixel. The total error in r (in µm) is estimated to be 256 times M_c^{-1}, when M_c^{-1} is calculated using either 149 or 151 columns of pixels per mm. This calculation leads to an maximum estimated error of ± 10 µm at either edge of the image and less error towards the centre of the image. This error is approximately the same as the uncertainty caused by rotor precession. Use of reduced magnification at the sensor results in greater uncertainty in r.

Overall, the precision of the data from the on-line system is better than that from manual acquisition and is approximately the same as that from automated plate readers[3-6]. Rotor precession limits the accuracy of both the fringe displacement and radial position measurements.

7. PROSPECTS AND CONCLUSION

In order for analytical ultracentrifugation to regain its image as a useful method that can be conducted rapidly and with certainty, the automation of the data acquisition is essential. The on-line Rayleigh interferometer presented here fulfils this requirement. However, the extreme rapidity with which precise data can be acquired extends analytical ultracentrifugation two other ways. First, it makes possible the accomplishment of far more complicated experimental protocols than could be considered if photographs were used. For example, it is reasonable to determine the molecular weight of a dozen or more samples in less than an hour using short column methods[9], NONLIN[19] and the on-line interferometer. This permits a rapid means of surveying buffer conditions, performing titrations and testing sample homogeneity. The on-line system is useful for examining unstable compounds, where often it is possible to monitor sample stability during the course of an experiment. Likewise, it is now a relatively simple prospect to gather the large quantity of high precision data necessary to characterize interacting systems fully. Finally, it has proven to be indispensable for the investigation of heterogeneous associations, where a large number of conditions need to be examined to assure the accuracy of the model[23].

The second way the on-line interferometer extends analytical ultracentrifugation is by permitting the development of entirely new methods. For example, the ability to acquire fringe displacement measurements in a few seconds allows, for the first time, the rapid determination of dc(r)/dt. By transforming the radial axis to the apparent sedimentation coefficient, it is possible to calculate the differential sedimentation coefficient distribution function on-line (see Chapter 20). Likewise, it becomes possible to monitor, and possibly correct for, the effects of rotor precession by acquiring a large number of data sets in a relatively short period of time. It is quite likely that other methods will be developed that rely on the capabilities of the on-line interferometer.

ACKNOWLEDGEMENTS

This work was supported by grants BBS 86-15815 and DIR 90-02027 from the National Science Foundation. The author thanks Philip Olsen and Theresa Ridgeway for help with the figures, as well as Philip Olsen and Daryl Lyons for conducting the many experiments used to test the on-line system.

REFERENCES

1. E.G. Richards and H.K. Schachman, *J. Phys. Chem.*, 1959, *63*, 1578.
2. E.G. Richards, D.C. Teller and H.K. Schachman, *Anal. Biochem.*, 1971, *41*, 180.
3. D.J. DeRosier, P. Munk and D.J. Cox, *Anal. Biochem.*, 1972, *50*, 139.
4. T.M. Laue, Ph.D. Dissertation, 1981, University of Connecticut, Storrs, CT.
5. D.C. Teller, *Anal. Biochem.*, 1967, *19*, 256.
6. J.H. Richards and E.G. Richards, *Anal. Biochem.*, 1974, *62*, 523.
7. G.E. Perlman and L.G. Longsworth, *J. Am. Chem. Soc.*, 1948, *70*, 2719.
8. D.A. Yphantis, *Biochemistry*, 1964, *3*, 297.
9. D.A. Yphantis, *Ann. N.Y. Acad. Sci.*, 1960, *88*, 586.
10. D. Yphantis, 'Cell and Molecular Biology', University of Connecticut, Storrs, CT, 06268, U.S.A.
11. W.F. Stafford, 'Muscle Research', Boston Biomed. Res. Inst., Boston, MA, 02114, U.S.A.
12. R.C. Williams, Jr., *Anal. Biochem.*, 1972, *48*, 164.
13. C.H. Paul and D.A. Yphantis, *Anal. Biochem.*, 1973, *56*, 52.
14. D.L. Rockholt, C.R. Royce and E.G. Richards, *Biophys. Chem.*, 1976, *5*, 55.
15. T.M. Laue, R.A. Domanic and D.A. Yphantis, *Anal. Biochem.*, 1983, *131*, 220.
16. D.A. Yphantis, T.M. Laue and I.A. Anderson, *Anal. Biochem.*, 1984, *143*, 95.
17. D.G. Rhodes, Ph.D. Dissertation, 1982, University of Connecticut, Storrs, CT, U.S.A.
18. D.B. Hayes and T.M. Laue, *Biophys. J.*, 1990, *57*, 573a.

19. M.L. Johnson, J.C. Correia, D. Yphantis and H.E. Halvorson, *Biophys. J.*, 1981, *36*, 575.
20. D.E. Roark and D.A. Yphantis, *Ann. N.Y. Acad. Sci.*, 1969, *164*, 245.
21. K.E. Van Holde and W.O. Weischet, *Biopolymers*, 1978, *17*, 1387.
22. M.A. Donovan and T.M. Laue, *Biophys. J.*, 1990, *57*, 256a.
23. E.A. Luckow, D.A. Lyons, T.M. Ridgeway, C.T. Esmon and T.M. Laue, *Biochemistry*, 1989, *28*, 2348.

7

Computer-Aided Interpretation of Analytical Sedimentation Data For Proteins

By Thomas M. Laue, Bhairavi D. Shah, Theresa M. Ridgeway and Sandra L. Pelletier

DEPARTMENT OF BIOCHEMISTRY, SPAULDING LIFE SCIENCES BUILDING, UNIVERSITY OF NEW HAMPSHIRE, DURHAM, NEW HAMPSHIRE 03824, U.S.A.

1. INTRODUCTION

This Chapter will focus on the *interpretation* of sedimentation data for purified proteins, glycoproteins or lipoproteins exhibiting relatively simple behaviour in aqueous buffers. Proper interpretation of analytical sedimentation data can provide estimates of molecular weight, subunit stoichiometries, association constants, thermodynamic nonideality, molecular sizes and relative structural asymmetry for a wide variety of macromolecules[1-7]. Estimates of the reduced apparent molecular weight (σ, Table 1) or the apparent sedimentation coefficient (s^*) only require knowledge of the concentration distribution, rotor speed and the time from the start of the experiment[3,8-12]. However, full interpretation of sedimentation data requires knowledge of ancillary information such as the solution density and viscosity, as well as the protein's partial specific volume. Presented here are the methods used to make estimates of this essential ancillary information when explicit data are not available. Following this are the equations used to extract structural information from sedimentation data. The applications and limitations of the methods are discussed.

Naturally, proper data interpretation requires good data acquisition. Many practical suggestions are available in books[8,10], operating manuals[11,12] or reviews[9] that describe the acquisition of sedimentation velocity[9-11], sedimentation equilibrium[9-12,14] and diffusion data[9,10]. Due to the limited scope of this Chapter, the reader should consult the appropriate portions of this book concerning the sedimentation of non-proteins or the analysis of chemical systems exhibiting more complex behaviour[15-17]. Even in these cases, many of the approaches described here are appropriate starting points for interpreting the data.

2. BACKGROUND

Information Needed

Regardless of whether velocity or equilibrium experiments are conducted, the values obtained by direct analysis of the data must be adjusted for the particular conditions of the experiment. To demonstrate how the parameters of interest are determined from experimental data, the "working" equations are presented. For diffusion, Fick's first law for a single ideal component may be written using the Einstein-Sutherland approximation[18]:

$$J_D = -D\frac{dc(r)}{dr} = -\frac{RT}{N_0 f}\frac{dc(r)}{dr} \tag{1}$$

where J_D is the flux (g/cm^2-sec) due to diffusion, D is the diffusion coefficient (cm^2/sec), R is the gas constant, T is the absolute temperature (K), N_0 is Avogadro's number, f is the translational friction coefficient (g/sec) and (dc(r)/dr) is the weight concentration gradient at a given radial position (Table 1). Measurement of D is sufficient to calculate f, and methods for doing this from sedimentation data are described in this volume and elsewhere[9-11].

The flux due to sedimentation (J_s) in a two component system is the product of the mass concentration of solute and its velocity:

$$J_s = c\frac{dr}{dt} = cs\omega^2 r \tag{2}$$

where c is the mass concentration of solute particles, dr/dt is their velocity (cm/sec), s is the sedimentation coefficient (sec), ω is the rotor's angular velocity and r is the distance from the centre of rotation. The substitution of $s\omega^2 r$ for dr/dt is appropriate* since s is defined as the velocity of a particle divided by the acceleration ($\omega^2 r$). As will be shown in the next section, it is the measurable parameter s that is of immediate interest[20-24].

The Apparent Sedimentation Coefficient is Determined Directly from the Data Rearrangement of eq. (2) reveals that the sedimentation coefficient can be determined from the slope of a graph of ln r *versus* $\omega^2 t$, where r is the distance from the centre of rotation to the midpoint of the concentration boundary. (More appropriately, the second moment of the boundary should be used[2], but for a single component the difference in s is negligible.) The interpretation of s begins by observing that:

$$s = \frac{M(1-\bar{v}\rho)}{N_0 f} = \frac{M_b}{N_0 f} \tag{3}$$

---

* There are limitations to this substitution for solutions that exhibit highly non-Newtonian viscosity.

Table 1. Symbols used in this Chapter.

Symbol	Definition	Units
s^* or s	Measured or apparent sedimentation coefficient. Uncorrected for density, viscosity or solute concentration (eq. (3)).	sec
$s_{20,w}$	Sedimentation coefficient corrected for density and viscosity, but not solute concentration (eq. (18)).	sec
s_0	Maximum sedimentation coefficient predicted assuming protein is an anhydrous, close-packed sphere (eq. (3) using f_0 from eq. (20)).	sec
k_s	Empirical constant taking account of effect of c_p on s.	$l\ g^{-1}$
s^0	Sedimentation coefficient corrected to zero solute concentration, but not for solvent viscosity and density.	sec
$s^0_{20,w}$	Sedimentation coefficient corrected for density, viscosity and solute concentration.	sec
σ	Reduced apparent molecular weight $= M(1-\bar{v}\rho)\omega^2/RT$ (eq. (4)).	cm^{-2}
J_D	Flux due to diffusion (eq. (1)).	$g\ cm^{-2}\ sec^{-1}$
D	Diffusion coefficient (eq. (1)).	$cm^2\ sec^{-1}$
D_0	Maximum diffusion coefficient predicted assuming protein is an anhydrous, close-packed sphere (eq. (1) using f_0 from eq. (20)).	$cm^2\ sec^{-1}$
J_s	Flux due to sedimentation (eq. (2)).	$g\ cm^{-2}\ sec^{-1}$
R	Gas constant, conventionally in cgs units (8.3144×10^7).	$g\ mol^{-1}\ K^{-1}$
N_0	Avogadro's number (6.023×10^{23}).	mol^{-1}
N_i	Number of residues of ith type.	unitless
N_t	Total number of amino acids in a protein (eq. (10)).	unitless
$N_{aromatic}$	Total number of aromatic amino acids + histidine (eq. (10)).	unitless
$N_{carbohydrate}$	Total number of saccharide units (eq. (10)).	unitless
c	Weight or mass concentration.	$g\ l^{-1}$
c(r)	Mass concentration at radial position r.	$g\ l^{-1}$
c_p	Mass concentration of protein.	$g\ l^{-1}$
C	Molar concentration.	$mol\ l^{-1}$
C_i	Molar concentration of the ith component.	$mol\ l^{-1}$
ω^2	Square of the rotor angular velocity (rpm $\pi/30)^2$.	sec^{-2}
M_r	Molecular weight (used in a generic sense).	$g\ mol^{-1}$
M_p	Molecular weight of protein in a complex (eq. (6)).	$g\ mol^{-1}$
M_d	Molecular weight of a denaturant (eq. (10)).	$g\ mol^{-1}$
M_b	Buoyant mass $= M(1-\bar{v}\rho)$ (eq. (3)).	$g\ mol^{-1}$
$G(s^*)$	Integral sedimentation coefficient distribution function (fractional value of $c(s^*)/c_1$).	unitless

$g(s^*)$	Differential sedimentation coefficient distribution function = s^* $dG(s^*)/ds$.	unitless
a_i to f_i	Coefficients of polynomial fit (Tables 3 and 4).	Depends on context
T	Temperature (units depend on context).	K or °C
T_{max}	Temperature of maximum density for water or heavy water.	°C
$\Delta\rho_{C_i}$	Density increment of solution containing component i at molar concentration C and at 293.15 K (eq. (13)).	g ml^{-1}
ρ_T	Density of pure water at experimental temperature (eq. (11) or (12)).	g ml^{-1}
$\rho_{T,b}$	Solution density at the experimental temperature (eq. (14)).	g ml^{-1}
ρ_{max}	Maximum density, observed at T_{max}.	g ml^{-1}
τ	Temperature difference between experiment and T_{max}.	°C
F	Empirical constant dependent on water isotope types.	unitless
$\bar{v}$	Partial specific volume of the protein at 25°C (eq. (7)).	ml g^{-1}
$\bar{v}_T$	Partial specific volume of the protein at temperature T (eq. (8)).	ml g^{-1}
v_p	Empirical specific "packing volume" based on Teller, *et al.*[30,33] and equal to 0.7654 ml g^{-1}.	ml g^{-1}
$\bar{v}_3$	Partial specific volume of a denaturant (eqs. (9) and (10)).	ml g^{-1}
ϕ_2'	Apparent isopotential partial specific volume (eq. (9)).	ml g^{-1}
δ_3	Solvation of protein by denaturant (eq. (9)) expressed as g-denaturant g-protein^{-1}.	unitless
δ_1	Hydration of protein by water (eq. (9)) expressed as g-H_2O g-protein^{-1}, estimated using eq. (22).	unitless
g_3	Denaturant contribution to solvent composition (eq. (9)) expressed as g-denaturant g-H_2O^{-1}.	unitless
η_T	Viscosity of pure water at temperature T (eq. (15)).	centipoise
$\Delta\eta_{C_i}$	Viscosity increment of solution containing component i at molar concentration C (eq. (16)).	centipoise
$\eta_{T,b}$	Viscosity of solution at the experimental temperature (eq. (17)).	centipoise
R_0	Radius of anhydrous sphere of equivalent volume to protein of given M_r and $\bar{v}$ (eq. (19)).	cm
R_p	Empirical packing radius of sphere equivalent volume to protein of given M_r (eq. (19)).	cm
f_0	Minimum frictional coefficient assuming protein is an anhydrous, close-packed sphere (eq. (20)).	g sec^{-1}

f^*	Observed frictional coefficient obtained from rearrangement of eq. (3), supplying s^* and M_b.	g sec^{-1}
$f^0_{20,w}$	Frictional coefficient obtained as described for f^*, except using $s^0_{20,w}$ in place of s^*.	g sec^{-1}
F_o	Frictional coefficient ratio for anhydrous oblate ellipsoid having an axial ratio of a/b and calculated using Perrin's equations (eq. (26)).	unitless
F_p	Frictional coefficient ratio for anhydrous prolate ellipsoid having an axial ratio of a/b and calculated using Perrin's equations (eq. (25)).	unitless
$a/b_{o,m}$	Maximum axial ratio for anhydrous oblate ellipsoid (eq. (24)).	unitless
$a/b_{p,m}$	Maximum axial ratio for anhydrous prolate ellipsoid (eq. (23)).	unitless
X	Fractional difference between s_0 and $s^0_{20,w}$ (eqs. (23) and (24)).	unitless
H_i	Hydration of amino acid type i expressed in mole-H_2O mole-amino-acid^{-1} (eq. (22)).	unitless
$\delta_{1,m}$	Maximum degree of hydration (g-H_2O g-protein^{-1}) assuming spherical protein (eq. (21)).	unitless
$\delta_{1,p}$	Estimated degree of hydration (g-H_2O g-protein^{-1}) assuming prolate axial ratio of a/b (eq. (27)).	unitless
$\delta_{1,o}$	Estimated degree of hydration (g-H_2O g-protein^{-1}) assuming oblate axial ratio of a/b (eq. (28)).	unitless

where M is the solute's anhydrous molecular weight (g/mole), $\bar{v}$ is its partial specific volume (ml/g), ρ is the solvent density (g/ml), N_0 is Avogadro's number, M_b is the buoyant molecular weight and f is the translational frictional coefficient.

For a hydrodynamically ideal molecule, all information concerning the shape and size of the molecule is expressed through f. Changes in f are propagated inversely to s. Eq. (3) shows that knowledge of M and s are sufficient to determine f, even in the absence of determining D. This latter approach to the determination of f is preferred because good estimates of M often are available and since the determination of D from sedimentation data can be difficult.

For solutions containing two or more non-interacting proteins, and in cases where the boundaries are well resolved, the analyses described below may be applied to each component's boundary. It must be kept in mind, however, that the effects of solute-solute interaction are not always innocuous and they can produce measurable alterations in the shape and position of the boundaries[20]. For cases where there is considerable heterogeneity in a sample, it is useful to examine the integral sedimentation coefficient distribution function, $G(s^*)$ or its derivative $g(s^*) = s^* dG(s^*)/ds^*$ (see W. Stafford III, Chapter 20 of this volume).

The Reduced Apparent Molecular Weight, σ, is Directly Available from Sedimentation Equilibrium

Equating eqs. (1) and (2), one obtains the fundamental equation describing sedimentation equilibrium of a single, thermodynamically ideal protein:

$$\sigma = \frac{d \ln\{c(r)\}}{d\left\{\frac{r^2}{2}\right\}} = \frac{s\omega^2}{D} = \frac{M(1-\bar{v}\rho)\omega^2}{RT} \qquad (4)$$

where σ is the reduced apparent molecular weight[14,25], c(r) is the weight concentration at radial position r, and the other terms are as described in Table 1. While σ may be determined from the slope of a graph of ln c as a function of $r^2/2$, sedimentation equilibrium data is analysed best by nonlinear least squares procedures, which can accommodate more complicated cases involving association and nonideality[2,10,14,26].

$\bar{v}$, η and ρ are Needed for Further Interpretation of s^* and σ

s^* and σ are directly accessible from sedimentation data. However, more often than not, it is the anhydrous molecular weight or "relative molecular mass", M_r and the frictional coefficient (f) that are of interest to the experimenter since these are required to learn about the size and shape of a molecule. Rearrangement of eq. (4) yields:

$$M_r = \frac{\sigma RT}{(1-\bar{v}\rho)\omega^2} \qquad (5)$$

which shows that calculation of M_r from σ will require $\bar{v}$, ρ, ω^2 and T. Similarly, rearrangement of eq. (3) shows that $\bar{v}$ and ρ are required to determine f from M_r and s. Further interpretation of f requires that the solvent viscosity (η) be taken into account and that estimates of f be made for anhydrous spheres of identical molecular weight and density. These calculations permit, through still more calculations, the estimation of the limits on a protein's hydration and asymmetry. While none of the calculations is difficult, it is useful to assemble all of them into a computer program to facilitate data interpretation. Such a program not only eliminates the tedium associated with routine analysis, it also provides a consistent and reliable means of quantifying hydrodynamic observations.

The remainder of this Chapter will focus on: (1) defining the parameters needed for the interpretation of sedimentation data, (2) presenting the equations and algorithms needed for calculating the parameters and (3) discussing the rules and limitations for their use.

3. MOLECULAR WEIGHT

It should be noted that a "generic" relative molecular weight, M_r, is used throughout this Chapter, even though various treatments of sedimentation equilibrium data will yield the well-defined n-, w-, z- and higher molecular weight averages[14]. It is important to realise that such averages come from primary measurements and are not "relative" (except to ^{12}C) in the same sense as molecular weights estimated from gel electrophoresis, gel-permeation chromatography or other secondary methods. Still, it is a common practice of journals to insist that M_r be used as the abbreviation for molecular weights determined from sedimentation equilibrium, and this practice is maintained here.

As shown in Fig. 3, three methods should be available for supplying the molecular weight: (1) M_r entered explicitly by the user, (2) M_r calculated from σ and (3) M_r calculated from a user-entered composition. Each is described below.

Explicit Entry of M_r

Molecular weight estimates obtained from sequence analysis, gel electrophoresis, gel filtration or other methods may be entered explicitly. This user-supplied molecular weight should have priority over all others in subsequent calculations.

Molecular Weight from σ

From eq. (5) it can be seen that entry of σ along with the rotor speed, temperature, ρ and $\bar{v}$ is sufficient for the calculation of M_r. Estimates of $\bar{v}$ can be calculated from the protein's composition and ρ from the buffer composition (below).

In addition to the total particle molecular weight (eq. (5)), the buoyant molecular weight ($M_r\,[1\text{-}\bar{v}\rho]$) and an estimate of the unconjugated protein molecular weight, M_p, should be provided. The latter value is calculated as:

$$M_p = \frac{\sigma RT}{\left\{ (1-\bar{v}_p\rho) + \sum \delta_i \, (1-\bar{v}_i\rho) \right\} \omega^2} \tag{6}$$

where $\bar{v}_p$ is partial specific volume of the protein, δ_i is stoichiometry of ligand binding for the ith component (in g-ligand per g-protein), and $\bar{v}_i$ is its partial specific volume[6,34]. Estimates of M_p are particularly useful in the analysis of detergent-solubilized proteins or proteins examined in the presence of denaturants[35-38]. This is discussed in more detail below.

Molecular Weight from Composition

The *minimum* molecular weight is estimated by summing up the g/mol of each of the protein's components. The amino acid composition is entered as mol-residue per mol-protein. Entries should be provided for the twenty common amino acids, as well as for glx and asx (average of the values for gln and glu and asn and asp respectively) and unk, the average value of the twenty amino acids, for use when the residue identity is unknown. The use of these average values will decrease the accuracy of subsequent calculations.

Table 2 presents M_r, $\bar{v}$ and degree of hydration (H_i) for the amino acids, as well as for several common protein-associated components. It is important to note that the values of M_r for the amino acids and carbohydrates have been adjusted to reflect the loss of H_2O associated with the formation of amide, ester or glycosidic linkages. Other components have *not* been corrected in this fashion. It must be realized that M_r calculated in this fashion can be in serious error if the composition is based on DNA sequence data and significant post-translational modification of the protein occurs or if the protein exists as an oligomer.

4. PARTIAL SPECIFIC VOLUME

The partial specific volume ($\bar{v}$) is a measure of the change in the volume (in ml) of the solution per gram of solute. For typical proteins, any errors in $\bar{v}$ are magnified three-fold in the determination of M_r [9]. Therefore, good estimates of $\bar{v}$ are necessary. An excellent discussion of partial specific volumes, along with an extensive tabulation of values, may be found in the review by Durschlag[6]. Several methods are available for the measurement of $\bar{v}$, but these typically require relatively high concentrations of solute, may require specialized equipment and definitely require careful analytical techniques for their success[39,40].

An alternative to measuring $\bar{v}$ is to estimate its value based on the protein's composition. Comparison of measured and estimated $\bar{v}$ indicate that the error typically is less than 1% [40]. Somewhat larger errors are observed for proteins that have an unusual amino acid composition, proteins that are conjugated with carbohydrate, proteins that exhibit preferential hydration or proteins that have a non-globular structure[6,40,41]. Even under these circumstances, the error is typically less than 3% if the correct composition is used[6,40].

Table 2a. Amino acids[a] 25°C

Amino acid	M_r	$\bar{v}$ (ml g^{-1})	H_i mol-H_2O mol-aa (pH 6-8)	H_i mol-H_2O mol-aa (pH 4)
Ala	71.1	0.74	1.5	1.5
Arg	156.2	0.70	3	3
Asn	114.1	0.62	2	2
Asp	115.1	0.60	6	2
Asx[c]	114.6	0.61	4	2
Cys	103.2	0.63[b]	1	1
2Cys	204.3	0.63	-	-
Gln	128.1	0.67	2	2
Glu	129.1	0.66	7.5	2
Glx[c]	128.6	0.665	4.8	2
Gly	57.1	0.64	1	1
His	137.2	0.67	4	4
Ile	113.2	0.90	1	1
Leu	113.2	0.90	1	1
Lys	128.2	0.82	4.5	4.5
Met	131.2	0.75	1	1
Phe	147.2	0.77	0	0
Pro	97.1	0.76	3	3
Ser	87.1	0.63	2	2
Thr	101.1	0.70	2	2
Trp	186.2	0.74	2	2
Tyr	163.2	0.71	3	3
Unk[c]	119.0	0.72	2.4	2
Val	99.1	0.86	1	1

[a] Based on values from Cohn and Edsall as presented in Durschlag[6]

[b] Based on value from Zmyatnin as presented in Durschlag[6]

[c] Values for Asx and Glx are averages of acid and amide forms. Values for Unk are average of all amino acids.

Table 2b. Carbohydrates at 20°C[a]

Carbohydrate	M_r	$\bar{v}$ (ml/g)
Fructose	180	0.614
Fucose	164	0.671
Galactose	180	0.622
Glucose (calculated)	180	0.622
(0.5M)	180	0.623
(3M)	180	0.638
Hexose	180	0.613
Hexosamine	179	0.666
Sucrose (0.05 M)	342	0.613
(0.1-0.2 M)	342	0.616
(1 M)	342	0.620
Lactose (0.1 M)	342	0.606
(0.4M)	342	0.610
Mannose	180	0.607
Methyl-pentose	165	0.678
N-Acetyl-galactose	221	0.684
N-Acetyl-glucose	221	0.684
N-Acetyl-hexosamine	221	0.666
N-Acetyl-neuraminic acid	308	0.584
Raffinose[b]	486	0.608
Sialic acid	308	0.584

[a] Values of $\bar{v}$ excerpted from Durschlag[6]
[b] Determined at 25°C

Table 2c. Denaturants at 20°C[a]

Denaturant	M_r	$\bar{v}$ (ml/g)
Guanidine HCL (lim $c \rightarrow 0$)	96	0.70
(1 M)	96	0.73
(2 M)	96	0.743
(4 M)	96	0.754
(6 M)	96	0.763
Urea (lim $c \rightarrow 0$)	60	0.735
(1 M)	60	0.745
(8 M)	60	0.763

[a] Values of $\bar{v}$ excerpted from Durschlag[6]

Table 2d. Detergents 20°C[a] (cmc = critical micelle concentration).

Detergent	$\bar{v}$ ml/g
Brij-56 (above cmc)[b]	0.955
DDAO Dodecyl-dimethyl-amine oxide	1.122
Lubrol PX (above cmc)	0.958
Lubrol WX (above cmc)[b]	0.929
DOC sodium deoxycholate (below and above cmc)	0.779
SDS sodium dodecyl sulphate (below cmc)	0.814
(above cmc)[b]	~0.86
(above cmc in H_2O)[b]	0.854
(above cmc 0.1 M NaCl)[b]	0.863
Triton N-101 (above cmc)[b]	0.922
Triton X-100 (above cmc)	0.913
Tween-80 (above cmc)[b]	0.896

[a] Values of $\bar{v}$ excerpted from Durschlag[6]
[b] Determined at 25°C

Table 2e. Ions at 20°C[a]

Ion	$\bar{v}$ ml/g
$^1/_2\ Ca^{2+}$	-0.220
Cs^+	0.160
K^+	0.022
Li^+	-0.130
Na^+	-0.061
	-0.288[b]
Cl^-	0.507
SO_4^{2-}	0.310

[a] Values of $\bar{v}$ excerpted from Durschlag[6]
[b] Different buffer conditions. See Perkins *et al.*, *J. Mol. Biol.*, 1981, *150*, 69.

Table 2f. Organic Acids at 20°C[a]

Acid	$\bar{v}$ ml/g
Acetic acid (0.1 M)	0.856
(1M)	0.861
(10M)	0.918
Butyric acid (0.1 M)	0.944
(1M)	0.967
(10M)	1.036
Propionic acid (0.1 M)	0.885
(1M)	0.904
(10M)	0.991

[a] Values of $\bar{v}$ excerpted from Durschlag[6]

Table 2g. Miscellaneous at 20°C[a]

Substance	$\bar{v}$ ml/g
Acetyl-CoA	0.638
Actinomycin[b]	0.786
Adenosine	0.638
(in 0.2 M NaCl)	0.62
ATP	0.44
Carbamyl phosphate	0.48
CTP	0.44
Ethanol	1.18
FMN[c]	0.574
Glycerol (10%)	0.767
(20%)	0.768
(30%)	0.770
(40%)	0.772

[a] Values of $\bar{v}$ excerpted from Durschlag[6]
[b] Determined at 5°C
[c] Determined at 25°C

The partial specific volume usually is estimated using the method of Cohn and Edsall[4-6,40]:

$$\bar{v} = \frac{\sum n_i M_i \bar{v}_i}{\sum n_i M_i} \tag{7}$$

where n_i is the number of moles of the ith component and M_i is its molecular weight. Values of $\bar{v}$ for the amino acids and several protein-associated components are presented in Table 2. The values in this table are for solutions at 25°C (unless noted otherwise), but $\bar{v}$ can be adjusted over a temperature range of 4-45°C using[6]:

$$\bar{v}_T = \bar{v}_{20} + 4.25\times10^{-4}\ (T-298.15) \tag{8}$$

where $\bar{v}_T$ is the partial specific volume at temperature T and $\bar{v}_{20}$ is the value calculated using eq. (7).

There is no computational method to correct $\bar{v}$ for the effects of pH. However, it has been observed that $\bar{v}$ varies only slightly with pH unless denaturation occurs, at which point $\bar{v}$ decreases by about 3% [6]. This decrease in $\bar{v}$ with pH parallels other changes in the protein structure (*e.g.* unfolding as monitored by viscosity or by circular dichroism).

Buffer composition and preferential solvation also can effect $\bar{v}$ [6], but, except as noted below, these are usually neglected.

All of the partial specific volumes presented in Table 2 have been determined for unconjugated molecules. For a few non-polar ligands, this approximation has been shown to be reasonable, however, caution should be exercised when calculating partial specific volumes for lipoproteins[35]. Likewise, greater uncertainty results from the use of $\bar{v}$ determined for unconjugated carbohydrates[41].

The Use of Denaturants and the Effects of Preferential Hydration

It is often useful or necessary to measure the molecular weight of proteins in denaturants. In this case, preferential binding of the denaturant to the protein will affect the partial specific volume and must be taken into account. Under these circumstances, $\bar{v}$ must be replaced by the apparent isopotential partial specific volume ϕ_2' (the nomenclature follows that of Durschlag, where component 2 is the protein)[3,6]. In general, increasing ionic strength results in a linear increase in ϕ_2', with the slopes following the Hofmeister series[6]. However, preferential hydration can lead to nonlinear changes in ϕ_2' [6].

Special provisions may be made in the program for estimating ϕ_2' of simple proteins in solvents containing 8 M urea or 6 M guanidinium chloride using described methods[6,37,38,42]. In these cases ϕ_2' replaces $\bar{v}$ in eq. (5). Values of ϕ_2' can be determined as described[42] or they can be estimated from[37]:

$$\phi_2' = \bar{v} + \left\{\frac{1}{\rho} - \bar{v}_3\right\}(\delta_3 - g_3\delta_1) \qquad (9)$$

where ϕ_2' is the calculated value of the isopotential partial specific volume, $\bar{v}$ is from eq. (7), ρ is the solvent density, $\bar{v}_3$ is the partial specific volume of the denaturant, δ_3 is the number of grams of denaturant bound per gram of protein, δ_1 is the hydration in grams of water per gram of protein (described below) and g_3 is the number of grams of denaturant per gram of water. For 6 M guanidine HCl at 20°C, the values are ρ = 1.1418 g/ml, $\bar{v}_3$ = 0.763 ml/g and g_3 = 1.007 g guanidine HCl per g H_2O [42]. For 8 M urea at 20°C the corresponding values are ρ = 1.1152 g/ml, $\bar{v}_3$ = 0.763 ml/g and g_3 = 0.752 g urea per gram of water[37]. For both urea and guanidine HCl, δ_3 is calculated by assuming one molecule of denaturant is bound per pair of peptide bonds and one to each aromatic side chain, including histidine[37,42]:

$$\delta^3 = \frac{M_d}{M_p}\left\{\frac{N_t - 1}{2} + N_{aromatic}\right\} \qquad (10)$$

where N_t is the total number of amino acids, $N_{aromatic}$ is the number of aromatic residues, M_d is the molecular weight of the denaturant (M_d = 95.54 for guanidine and 60 for urea) and M_p is the subunit molecular weight for the protein.

Note that erroneous values of ϕ_2' may be calculated if the overall protein composition has significant levels (>10% wt/wt) of non-amino acid constituents[43]. However, empirical evidence suggests that reasonable estimates for ϕ_2' are obtained for glycoproteins in 6 M guanidine HCl by assuming each carbohydrate moiety binds one guanidine molecule and by assuming δ_1 is not affected by the presence of the carbohydrate[43]. In this case, eq. (10) is modified by adding the term $[M_d/M_p]\times N_{carbohydrate}$ to account for the guanidine binding. There has not been sufficient systematic experimental evidence to justify the use of a similar analysis for 8 M urea denaturation of glycoproteins[37].

Non-denaturants may benefit from similar analysis. Timasheff and co-workers have described calculations of ϕ_2' for simple proteins in solvents containing varying amounts of NaCl, Na_2SO_4, $MgSO_4$, glycine, α-alanine, betaine and CH_3COONa[38]. It should be noted that such calculations require knowledge of the interaction (*i.e.* solvation) parameters (δ_1 and δ_3) for the particular salt, and that it is incorrect to attempt these calculations using interaction data for a different salt[38].

5. SOLVENT DENSITY

The density is simply the mass (in g) of one ml of solvent and is dependent on temperature and composition. The solvent density may be corrected to the experimental temperature using the equation of Kell for pure water (ρ_T), as presented in the CRC Handbook of Chemistry and Physics[44] and modified to yield cgs units:

$$\rho_T = 1.000028\times10^{-3}\left\{\frac{999.83952 + 16.945176T}{1 + 16.879850\times10^{-3}T}\right\}$$

$$- 1.000028\times10^{-3}\left\{\frac{7.9870401\times10^{-3}T^2 + 46.170461\times10^{-6}T^3}{1 + 16.879850\times10^{-3}T}\right\}$$

$$+ 1.000028\times10^{-3}\left\{\frac{105.56302\times10^{-9}T^4 - 280.54253\times10^{-12}T^5}{1 + 16.879850\times10^{-3}T}\right\} \tag{11}$$

where T is the temperature in °C and the multiplicative factor 1.000028×10^{-3} converts the units from $kg\text{-}m^{-3}$ to $g\text{-}ml^{-1}$. Densities calculated using this equation are good to at least five decimal places over the 0°C-100°C temperature range.

The temperature correction for solvents containing heavy water (either D or ^{18}O) may be made using the methods described by Steckel and Szapiro[45]:

$$\rho_T = \rho_{max}\left\{1 - F\tau^2\left[1.74224 + \frac{482.502}{\tau + 77.861}\right]\times10^{-6}\right\} \tag{12}$$

where ρ_T is the density at temperature T in °C, ρ_{max} is the maximum density (observed at temperature T_{max}), τ is the temperature difference between that of the experiment and the temperature of maximum density (*i.e.*, $T\text{-}T_{max}$) and F is an empirical constant. Eq. (12) may be used for all water species by substituting in the appropriate values from the following table. Note that in cases where ^{18}O is present, isotopic purity is about 98%, and it is 100% when D is present. Values for H_2O reflect the natural isotopic abundance of ^{18}O and D. Likewise, O and H reflect natural isotopic abundance in D_2O and $H_2{}^{18}O$. Eq. (12) is valid from the melting temperature (T_{mp}) to approximately 80°C. Simple mixing rules can be used when different isotopes of water are combined[45].

Isotope	T_{max} °C	ρ_{max} g/ml	F	T_{mp} °C ±0.02
H_2O	3.98_6	1.0000000	1.0	0.00
$H_2{}^{18}O$	4.30_5 ±0.02	1.11255±0.00003	1.0	0.28
D_2O	11.24 ±0.04	1.10602±0.00003	1.0555	3.81
$D_2{}^{18}O$	11.46 ±0.03	1.21691±0.00008	1.0555	4.02

Note that use of either eq. (11) or (12) assumes that the solvent's temperature dependence is dominated by that of pure water or its isotopes. While this is true for solutions containing moderate concentrations of other components, significant errors can be introduced by using either equation with solvents containing high solute concentrations.

The Effect of Buffer Composition on ρ

The density of a solution can be estimated by summing the density increments calculated for each component[8]. The density increments may be calculated using the tabulated values of ρ_{C_i} as a function of C_i (molar concentration of the ith component) fit to the polynomial:

$$\Delta\rho_{C_i} = (a_i + b_iC_i^{1/2} + c_iC_i + d_iC_i^2 + e_iC_i^3 + f_iC_i^4) - 0.998234 \tag{13}$$

where $\Delta\rho_{C_i}$ is the dependent variable, C_i is the independent variable and the coefficients a_i to f_i are the fitted parameters. Values of a_i to f_i for a variety of common buffer components are presented in Table 3. These coefficients were determined using least squares procedures[46]. The inclusion of the $C^{1/2}$ term improves the fit at low concentrations for some ionic solutes. In principle, a_i is the density of pure water at 20°C (0.998234 g/ml) [44], but substantially improved fits are obtained when this term is allowed to vary. Thus, eq. (13) is used as an empirical fitting function, and, as such, no physical interpretation of the parameters should be made. For each component, the parameters from the most parsimonious polynomial providing an adequate fit are presented. Adequacy of the fits were judged by the following criteria: (1) the rms error of the residuals should correspond to the last decimal place of the tabulated values; and (2) graphs of the residuals as a function of either the dependent or independent variable should exhibit no systematic deviations with magnitudes greater than the fourth decimal place (Fig. 1a). For alcohols, deviations in the third decimal place were allowed since these data typically are of lower precision (Fig. 1b).

The concentration range over which each set of coefficients is valid and the maximum error also are presented in Table 3. In general, polynomials fit the density data for simple salts far better than for organic salts and alcohols. Systematic deviations are observed in fitting ρ for alcohols (Fig. 1b), with that for ethanol being the worst case. Even though the magnitude of the errors is within the last decimal place of the tabulated data, it is clear that the polynomial function provides only a marginally adequate model for these substances. For most components, the precision of the fit using the data in Table 3 is better than 6×10^{-5} g/ml. An error of this magnitude in ρ will propagate as a 0.3% error in M_r for a protein having a $\bar{v}$ of 0.75 ml/g and a true M_r of 130,000. This level of error would be acceptable for most research, but could interfere in more demanding cases (*e.g.* determining association constants for cases where large stoichiometries are observed).

Using density increments, the solution density is predicted to be:

$$\rho_{T,b} = \frac{(0.998234 + \Sigma\Delta\rho_{C_i})\,\rho_T}{0.998234} \tag{14}$$

Table 3. Coefficients for the power series approximation of the density

$$\rho = a + b\times10^{-3}\, C^{1/2} + c\times10^{-2}\, C + d\times10^{-3}\, C^2 + e\times10^{-4}\, C^3 + f\times10^{-6}\, C^4$$

Solute	Form	a	b	c	d	e	f	C_{max} M	$10^4\times$Max. error g/ml
Acetate	acid	0.9990	-4.04	1.023	-0.9156	0.67	-2.403	16.0	18
	trichloro	0.98774	-	7.6173	2.8618	-19.437	239.421	4.0	6
Ammonium	chloride	0.99787	1.3717	1.5743	-3.8795	-	-	4.7	0.7
	sulphate	0.99828	-	7.835	-59.751	4.0581	-	3.8	1
Calcium	chloride	0.99741	4.5424	8.5102	-1.5852	-	-	5.0	2
Cesium	chloride	0.99850	-1.7323	12.9622	-0.8867	-0.5153	10.0485	7.0	1.3
Magnesium	chloride	0.99801	-	7.7423	-2.8939	2.1536	-	4.0	4
	sulphate	0.99763	3.9236	11.547	-4.8526	3.0613	-	2.9	0.6
Phosphoric	acid	0.99834	-	5.3345	-3.0937	-	-	0.6	0.6
Potassium	bromide	0.99793	1.6329	8.303	-0.5438	0.1187	-	4.6	0.6
	chloride	0.99791	1.5586	4.5573	-0.7041	-0.2539	10.4034	3.9	0.6
	hydroxide	0.99842	-	4.7917	-1.1602	0.2968	-	12.0	6
	phosphate mono-basic	0.99815	-	9.7583	-7.4352	-	-	0.8	0.8
	phosphate di-basic	0.99837	-	14.5429	-7.3137	-	-	0.5	0.8
Sodium	acetate	0.99623	12.4776	2.2804	9.2092	-27.6174	267.596	4.2	1.6
	chloride	0.99828	-	4.1261	-1.0580	0.6097	-	6.2	1
	citrate tri-basic	0.99497	1.5782	-	-27.741	-	-	1.8	7
	hydroxide	0.99624	7.0838	3.8735	-0.8969	0.1253	-	14.0	3.5
	phosphate mono-basic	0.99826	-	8.838	-3.8893	5.4158	-41.9615	5.0	0.9
	phosphate di-basic	0.99794	3.4466	13.197	-7.0728	-	-	0.4	0.8

	phosphate tri-basic	0.99843	-	18.6772	-7.5401	-	-	0.5	1
Alcohols	methanol	0.99678	-3.5931	-0.3603	-0.0548	-	-0.1766	24.0	9
	ethanol	0.99662	-	-0.6422	-	-0.1895	-	17.0	17
	1-propanol	1.00593	-27.587	-1.3306	-3.8094	2.8667	-8.8572	13.8	10
	2-propanol	1.00369	-20.974	-0.6123	-1.7529	0.3473	-	12.7	14
Other	EDTA	0.99824	-	17.8212	-16.271	-	-	0.18	0.6
	Ethylene glycol	0.99823	-	0.7561	0.1273	-0.1259	-	10.0	0.6
	Sucrose	0.99822	-0.0346	13.2207	-1.8117	-	-24.855	4.0	1.2
	Tris buffer	0.99846	-1.3079	3.0305	0.2876	-0.6279	-	3.5	0.6
	Urea	0.99835	-0.4476	1.528	0.5425	-1.449	9.824	9.0	0.6
	Guanidine HCl	0.99831	-	2.557	-2.7133	-	-	6.2	3.8

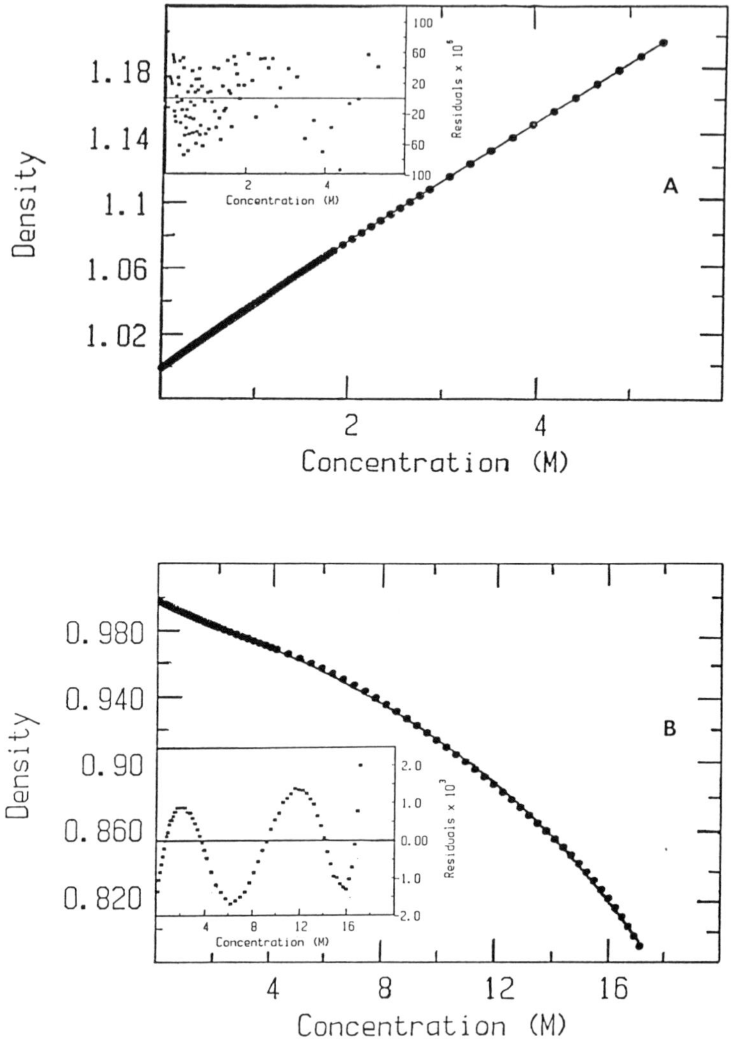

Figure 1. Shown here are representative graphs of the concentration dependence of the density for NaCl (**A**) and ethanol (**B**). The lines through the data points were drawn using the coefficients presented in Table 3 for the polynomial fitting function. The inset in each panel show the residuals of the fit as a function of concentration. Both fits exhibit systematic variation in the residuals, with that for ethanol being an order of magnitude greater than that for NaCl. Still, the error in the determination of ρ is in the least significant digit of the tabulated data. The error for ethanol is the worst of all cases examined (Table 3). Even this level of error would have only minor affects on a molecular weight determination (see text).

where $\rho_{T,b}$ is the calculated density, $\rho_T/0.998234$ is the temperature correction factor, and $\Sigma\Delta\rho_{C_i}$ is the sum of the density increments. In automating the use of eq. (14), density increments should be calculated only for buffer components with non-zero concentrations. This avoids the inclusion of residual errors (Σ [a_i-0.998234]) at zero solute concentration.

Use of eq. (14) assumes that there is no volume change upon mixing. Moreover, this calculation neglects density changes resulting from component redistribution during sedimentation, as well as neglecting the presence of protein and ignoring solution compressibility. For dilute aqueous solutions these assumptions will not result in serious error.

6. VISCOSITY

Viscosity is a measure of the energy dissipated in a fluid in order to maintain flow. In sedimentation, the flow of interest is the movement of the protein through the solvent. For a fixed amount of energy, the greater the solvent's viscosity, the slower the protein will move. Therefore, solvent viscosity will affect both sedimentation velocity and the time needed to reach equilibrium[2,14,17]. Excellent reviews are available describing methods for determining viscosity and for interpreting viscosity data[47,48]. The viscosity is dependent on both the temperature and the solution composition. A correction for the temperature dependence of η can be made using[49]:

$$\eta_T = 1.002 \log^{-1}\left\{\frac{1301}{998.333 + 8.1855(T-20) + 5.85\times10^{-3}(T-20)^2} - 3.302\right\}$$

Range 0°C ≤ T ≤ 20°C

$$\eta_T = 1.002 \log^{-1}\left\{\frac{1.3272(20-T) - 1.053\times10^{-3}(T-20)^2}{T+105}\right\}$$

Range 20°C ≤ T ≤ 100°C (15)

Note that these equations are formulated for use with the centigrade temperature scale.

The Effect of Buffer Components on η

As was done for the density estimate, the viscosity of a solution may be estimated using viscosity increments[8]. Table 4 presents the coefficients for the polynomial series describing the concentration dependence of the viscosity. Note that the *maximum* error is listed in this table. The root mean square error is typically one-half of these values. As with the density increment, the viscosity increment should be calculated only for components with non-zero concentration.

$$\Delta\eta_{C_i} = (a_i + b_iC_i^{1/2} + c_iC_i + d_iC_i^2 + e_iC_i^3) - 1.002 \qquad (16)$$

The solvent viscosity is estimated from the viscosity increments as:

Table 4. Coefficients for the power series approximation of the viscosity.

$$\eta = a + b\times10^{-3}\, C^{1/2} + c\times10^{-2}\, C + d\times10^{-3}\, C^2 + e\times10^{-4}\, C^3 + f\times10^{-6}\, C^4$$

Solute	Form	a	b	c	d	e	Max. Conc. M	Max. error
Acetate	acid	1.0002	0.678	11.803	2.3071	-0.9677	8.79	7.7
	trichloro	1.0108	-88.69	0.5012	1.4585	66.04	3.92	3.4
Ammonium	sulphate	0.9954	11.99	16.637	28.73	96.50	3.72	3.6
Calcium	chloride	1.0310	-202.71	60.415	-210.19	1027.2	1.083	1.9
Cesium	chloride	1.0076	-52.68	3.2944	-10.892	32.260	1.916	0.88
Magnesium	chloride	0.9970	18.500	38.404	37.382	444.48	1.911	0.50
	sulphate	1.0000	-16.565	66.225	198.99	817.18	1.557	7.0
Phosphoric	acid	0.9979	-13.132	26.463	22.713	69.597	5.117	5.9
Potassium	chloride	0.9902	39.626	-5.3310	13.294	-3.1543	3.742	1.8
	hydroxide	1.0384	-134.59	21.302	-13.012	25.109	10.558	22.0
	phosphate mono-basic	1.0082	-70.419	36.881	0.0146	144.65	0.786	0.83
	phosphate di-basic	1.0000	-3.1676	37.897	25.712	1435.8	0.491	0.46
Sodium	acetate	1.0171	-89.188	4.1142	39.567	122.05	2.112	4.9
	chloride	1.0045	-11.248	9.9661	0.1800	30.711	5.326	8.6
	hydroxide	1.1202	-445.64	61.627	-71.304	152.77	7.576	33.0
	phosphate mono-basic	1.0083	-45.986	43.878	40.559	463.85	2.387	4.0
	phosphate di-basic	1.0000	-9.6435	57.917	317.22	0.0000	0.408	0.45
	phosphate tri-basic	0.9867	162.29	46.442	1481.0	-3355.2	0.535	0.78

Alcohols	methanol	1.0150	-26.331	11.567	0.9280	-3.5699	12.209	19.0
	ethanol	1.0062	20.588	13.721	48.909	-47.870	8.488	17.0
	1-propanol	1.1412	-478.79	69.171	-26.356	-8.9613	6.741	12.0
	2-propanol	1.1401	-387.52	54.348	58.036	-89.135	6.742	30.2
Other	Ethylene glycol	1.0265	-115.81	24.791	-4.4436	20.079	10.406	5.7
	Tris buffer	1.0555	-340.23	80.249	-187.66	895.27	3.657	20.0
	Urea	1.0135	-46.746	7.2998	-3.1993	7.7571	8.665	3.3

$$\eta_{T,b} = (1.002 + \Sigma\Delta\eta_{C_i})\,\frac{\eta_T}{1.002} \tag{17}$$

where $\Delta\eta_{C_i}$ is the viscosity increment for component i at molar concentration C_i and η_T is calculated using eq. (15).

It must be realized that the viscosity of a sample may be affected by low concentrations of macromolecular solutes. Thus, the estimate from eq. (17) is valid only at zero protein concentration. Viscosity must be measured for solutions of highly asymmetric molecules or solutions containing moderate to high concentrations of any macromolecular solute.

7. CONVERTING s^* TO $s^0_{20,w}$

The sedimentation coefficient provides a rigorous hydrodynamic descriptor of a molecule. However, to be most useful in this regard, and prior to further interpretation, s^* must be standardized and corrected for concentration effects on s^*. By convention, s^* which has been standardized to conditions corresponding to pure water at 20°C and extrapolated to zero protein concentration is designated $s^0_{20,w}$. Because of the solvent-independent nature of $s^0_{20,w}$, it describes quantitatively the fundamental hydrodynamic properties of the protein and it is this value which is most useful in comparing the sedimentation behaviour of different proteins. Moreover, $s^0_{20,w}$ must be used in calculating the limits of hydration and estimating the asymmetry of a macromolecule. Finally, comparisons of $s^0_{20,w}$ determined for a molecule in solvents of differing composition can yield unique information concerning changes in macromolecular interactions, shapes, sizes and hydration. The corrections account for the solvent density (ρ) and viscosity (η), and the protein's partial specific volume ($\bar{v}$). ρ and $\bar{v}$ affect s^* by their presence in the term $(1-\bar{v}\rho)$ (eq. (3)) and η affects s^* inversely through f.

Standardizing s^*

s^* is corrected to standard conditions of water at 20°C ($s_{20,w}$) using:

$$s_{20,w} = s^* \frac{(1-\bar{v}\rho)_{20,w}\;\eta_{T,b}}{(1-\bar{v}\rho)_{T,b}\;\eta_{20,w}} \tag{18}$$

where the subscripts refer to the experimental temperature (T, 20 = 20°C) and buffer conditions (b, w = water), respectively. Note that $\bar{v}$ in the numerator refers to the value at 20°C, whereas $\bar{v}$ in the denominator should be corrected for temperature (above) and should be corrected for specific ion effects whenever possible[6]. Values for water at 20°C are $\rho_{20,w}$ = 0.9982 g/ml and $\eta_{20,w}$ = 1.002 cp.

The Concentration Dependence of s^*

Prior to further analysis, $s_{20,w}$ should be determined for a number of samples at different initial protein concentrations. A graph then should be constructed to extrapolate $s_{20,w}$ to zero protein concentration. The nomenclature for a standardized, concentration-corrected (*i.e.* lim c→0) $s_{20,w}$ is $s^0_{20,w}$. In cases where well resolved boundaries are being examined, such corrections should be made for each component individually, if at all possible. The determination of $s_{20,w}$ at various solute concentrations may reveal either increases or decreases in s with increasing solute concentration. A decrease in $s_{20,w}$ with increasing c is expected on hydrodynamic grounds[21,22]. A rather weak dependence (about a 1% change in s per mg/ml) is observed for spherical molecules, whereas asymmetric molecules will exhibit a much stronger concentration dependence[23].

Increases in $s_{20,w}$ with increasing solute concentration or the appearance and growth of faster-sedimenting boundaries at increased solute concentrations is diagnostic for mass-action association. The full analysis of sedimentation velocity data from interacting systems is reviewed in this volume and elsewhere[17,24].

In certain cases, the effect of protein concentration on s_{obs} can be taken into account using $s^0_{20,w} = s_{20,w}/(1-k_s c_p)$, where the protein concentration (c_p) is in mg/ml and k_s is an empirical coefficient equal to 0.009 ml/mg for spherical proteins[21,22,27,28]. Much larger values of k_s are expected for asymmetric proteins and the value of 0.009 should be used only if it is known that the protein is spherical or nearly so. For a consideration of the variation in k_s with axial ratio, the reader is referred to Chapter 21 of this volume.

8. THE IMPORTANCE AND DETERMINATION OF R_0

The starting point for hydrodynamic analysis from sedimentation data is the calculation of the expected radius for a rigid, impermeable and incompressible sphere having the identical volume as the protein of interest. This quantity is called the "equivalent radius" and is designated R_0. As will be explained below, there is some ambiguity as to exactly how R_0 should be determined.

There have been several methods proposed for the calculation of shape or hydration data based on the hydrodynamic data available wholly[27,29] or in part[28-32] from the analytical ultracentrifuge. For this Chapter, it will be assumed that only sedimentation data are available, and methods that rely on knowledge of the viscosity increment[31,32] or the radius of gyration[32] will not be discussed here. While this places severe limitations on the interpretation, it is reasonable for situations where either material or instrumentation is limited. Should sufficient material and equipment be available for supplementary measurements (radius of gyration or viscosity increment), then it would be wiser to conduct the additional experiments needed to complete the hydrodynamic interpretation[28,31,32].

In order to account for the fact that the protein is displacing solvent rather than occupying a vacuum, the value of R_0 often is calculated as in eq. (19a):

$$R_0 = \sqrt[3]{\frac{3M_r\bar{v}}{4\pi N_0}} \tag{19a}$$

$$R_p = 6.72\times10^{-9}\, M_r^{1/3} \tag{19b}$$

However, there is no theoretical justification for the use of $\bar{v}$ in this calculation[9] except that $\bar{v}$ reflects, in large part, the volume of solvent displaced by the protein. It is often found that R_0 calculated using eq. (5) is too small, thus leading to over-estimates of the axial ratios (below)[30,33]. This is not surprising since $\bar{v}$ includes solute-solvent interactions other than solvent displacement. Significant errors can be introduced if, for example, there is considerable electrostriction of the solvent by a highly charged protein[6,34]. Moreover, this calculation underestimates the contribution surface roughness (rugosity) makes to the protein hydrodynamics[30].

Since there is no theoretical justification for eq. (19a), it has been suggested that this calculation can be made using an empirical relationship based on X-ray crystallographic structures[28,33]. Teller has described two such apparent radii[33]. The first R_p, evaluated as shown in eq. (19b), is determined from the packing volume in protein-protein complexes. A second estimate of an effective radius, $R_s = 0.9405\pm0.001\ M_r^{1/3}\times10^{-8}$ cm, accounts for the surface roughness of the protein. For monomeric proteins, $R_s=1.4R_p$, and use of R_s in hydrodynamic calculations leads to absurdly low estimates of the hydration and axial ratio (unpublished data). Other authors have used R_p as an estimate of the anhydrous radius, but have neglected R_s, apparently for similar reasons[28]. Use of these relationships typically yield larger values than R_0 and, consequently, smaller estimates of the axial ratios.

Regardless of whether eq. (19a) or (19b) is used, R_0 and R_p are considered to be the minimum radii possible for the anhydrous protein[28-33]. All evidence to date suggests that R_p provides better estimates of the anhydrous radius.

The predicted minimum frictional coefficient is calculated from the Stokes-Einstein relationship using either R_0 or R_p [18]:

$$f_0 = 6\pi\eta R_0 \tag{20a}$$

$$f_0 = 6\pi\eta R_p \tag{20b}$$

D_0 and s_0 are the expected *maximum* values when f_0 is used in their calculation by eqs. (1) and (3), respectively.

9. ESTIMATION OF PROTEIN HYDRATION AND ASYMMETRY

One cannot determine accurate, unambiguous values for both the degree of hydration, δ_1 and the ellipsoid of revolution axial ratio (a/b) from sedimentation experiments alone[27-30]. (N.B. The symbol w is sometimes used instead of δ_1 [32].) However, maximum values for these parameters may be calculated if s_{obs}, M_r and $\bar{v}$ (only for R_0, eq. (19a)) are available. In addition, if one of the parameters (δ_1 or a/b) is known independently, estimates of the second parameter can be made. However, at the outset it must be realized that these parameters, especially that of (a/b), are merely descriptors for models and may bear little relevance to the actual molecular structure.

There are two sources of error that must be addressed. First, the calculation of (a/b) and δ_1 will rely on an estimate of R_0 (or R_p) and, as discussed above, the equations for these are empirical. Secondly, this radius is used to calculate f_0 using eq. (20). However, eq. (20) is derived assuming complete "stick" boundary conditions (*i.e.* a monolayer of solvent moves with the macromolecule). If complete "slip" boundary conditions are assumed (*i.e.* there is no interaction between the solvent and the macromolecule), eq. (20) should be rewritten as $f_0 = 4\pi R_0$ [18]. Thus, even though all evidence suggests that a purely "stick" boundary condition prevails, this assumption increases the uncertainty in the f_0 calculation.

Hydration-δ_1

Water of hydration is considered to be water that differs from the bulk phase due to its interaction with the protein. Our knowledge of the degree of hydration for proteins is poor, largely because the values δ_1 determined are dependent on the techniques used[28-33,39-43,51-54]. The problem is that different methods will report different amounts of water "under the influence" of the protein because the methods are intrinsically sensitive to dissimilar properties of the water. Most methods distinguish two different realms of hydration water, that directly interacting with the protein (primary hydration shell) and that merely "influenced" by being in contact with the primary shell (secondary hydration)[52]. It is not clear whether one, both or some combination of these two shells corresponds to the hydration that influences sedimentation behaviour. This ambiguity complicates the estimation of the asymmetry due to the common effects hydration and shape have on hydrodynamics[28,29]. If one assumes that the protein is spherical, then the maximum degree of hydration ($\delta_{1,m}$) can be calculated. Alternatively, one can estimate δ_1.

Calculation of $\delta_{1,m}$ The predicted *maximum* degree of hydration (g-H_2O per g-protein) can be calculated in different ways depending on whether eq. (19a) or (19b) is used to estimate the radius needed for f_0 (eq. (20a) or (20b)) and depending on how the protein volume is calculated. If one follows the more "traditional" pathway of calculating R_0 using eq. (19a) and f_0 from eq. (20a), then:

$$\delta_{1,m} = \frac{\left\{\left[\frac{f}{f_0}\right]^3 - 1\right\}\bar{v}}{0.998234} \tag{21}$$

where f is the frictional coefficient calculated from $s^0_{20,w}$ and M_r, f_0 is the minimum frictional coefficient from eq. (20a) and $\bar{v}$ is the partial specific volume of the solute. On the other hand, $\delta_{1,m}$ can be calculated from the empirical equations of Teller[33] by substituting f_0 calculated using eqs. (19b) and (20b) and substituting v_p=0.7654 ml/g for $\bar{v}$.

Estimating δ_1 It has become common practice to specify a value of 0.25 or 0.3 for δ_1, with little justification. For unconjugated proteins, typical values of "hydrodynamic hydration" range between 0.2-0.6 g-H_2O/g-protein[55]. It would seem that glycosylation should tend to increase this value, although there is little evidence in the literature on it. For collagen, a fibrous protein, δ_1 has been estimated to be 0.48 g-H_2O/g-protein[53].

One estimate of the degree of hydration can be made based on the amino acid composition and using the method of Kuntz[51]. This method relies on the amount of "unfreezable" water being an accurate reflector of hydration as it relates to hydrodynamics. The calculation is simply:

$$\delta_1 = \frac{18}{M_r} \sum (H_i N_i) \tag{22}$$

where H_i is the hydration from Table 2 (in mol H_2O per mol of amino acid i) and N_i is the number of residues of type i. The quantity $18/M_r$ converts the value from the relative molar scale to the relative mass scale. Estimates of δ_1 calculated using eq. (22) are not adjusted for the effects of pH, but are reported to be valid to within 10% above pH 6. Values for H_1 at pH 4 reflect the lower hydration expected of the carboxyl groups when they are protonated.

Use of eq. (22) assumes that all of the amino acids are exposed to solvent, which for globular proteins clearly cannot be the case. The expectation is that δ_1 would tend to be overestimated. Although there have been studies correlating predicted hydration using eq. (22) with "non-freezable" water[51,54], there has been minimal systematic exploration of whether these values correlate with hydrodynamic properties[54]. However, empirical evidence suggests that δ_1 calculated with this method is "better" than anticipated, and the values of δ_1 are useful in determining ϕ_2' from eq. (9) [42].

Should an estimate of the degree of hydration be available, the corresponding values for $(a/b)_p$ and $(a/b)_o$ are calculated from eqs. (23) and (24) (below) using $X=[(f/f_0)/([\bar{v}+\delta_1]/\bar{v})^{1/3} - 1]$, where δ_1 is the degree of hydration expressed as g-H_2O/g-protein, $\bar{v}$ is the partial specific volume of the protein and f/f_0 is the ratio of the experimentally determined frictional coefficient to that for the equivalent anhydrous sphere (both at 20°C). Again, Teller's[33] empirical estimates of R_p for evaluating f_0 and his estimate of v_p=0.7654 ml/g can be substituted for $\bar{v}$ in this calculation.

Ellipsoid Asymmetry Prediction

Calculation of $(a/b)_{o,m}$ and $(a/b)_{p,m}$ The maximum axial ratio of the prolate ellipsoid of revolution, $(a/b)_{p,m}$ and the oblate ellipsoid of revolution, $(a/b)_{o,m}$ can be determined using a power series approximation of the tabulated data for (a/b) as a function of (f_0/f) [18]. In this case, the coefficients of the corresponding power series in $(s_0/s^0_{20,w} - 1)$ describing (a/b) were determined using nonlinear least squares analysis[46]. For the prolate case:

$$\left[\frac{a}{b}\right]_{p,m} = 1 + 2.346X^{1/2} + 8.297X + 8.4X^2 - 0.4589X^3 + 0.0314X^4 \quad (23)$$

and for the oblate case:

$$\left[\frac{a}{b}\right]_{o,m} = 1 + 2.975X^{1/2} + 8.48X + 13.295X^2 - 3.219X^3 + 0.0996X^4 \quad (24)$$

where $X = (s_0/s^0_{20,w} - 1) = (f^0_{20,w}/f_0 - 1)$, and s_0 or f_0 are calculated as described above. Over the range $1 < (a/b) < 200$, these functions are within ±0.01 of the true axial ratio and certainly introduce no error beyond that in estimating s_0 or f_0 (Fig. 2).

User-Entered Axial Ratio Should the user supply an axial ratio (as major-axis/minor-axis), the corresponding values of the degree of hydration for both the prolate and oblate ellipsoid models can be calculated. To do this, the frictional coefficient ratios (prolate = F_p and oblate = F_o) corresponding to the user-supplied (a/b) may be calculated using the formulas of Perrin[18]:

$$F_p = \frac{\left\{1 - \left[\frac{b}{a}\right]^2\right\}^{1/2}}{\left[\frac{b}{a}\right]^{2/3}} \ln\left(\frac{1 + \left\{1 - \left[\frac{b}{a}\right]^2\right\}^{1/2}}{\left[\frac{b}{a}\right]}\right) \quad (25)$$

$$F_o = \frac{\left\{\left[\frac{a}{b}\right]^2 - 1\right\}^{1/2}}{\left[\frac{a}{b}\right]^{2/3}} \tan^{-1}\left(\left\{\left[\frac{a}{b}\right]^2 - 1\right\}^{1/2}\right) \quad (26)$$

These values then are used to calculate the corresponding degree of hydration:

$$\delta_{1,p} = \left[\left\{\frac{1}{F_p}\frac{f}{f_0}\right\}^3 - 1\right]\frac{\bar{v}}{0.998234} \quad (27)$$

$$\delta_{1,o} = \left[\left\{\frac{1}{F_o}\frac{f}{f_0}\right\}^3 - 1\right]\frac{\bar{v}}{0.998234} \quad (28)$$

where $\delta_{1,p}$ is the predicted degree of hydration for the prolate model, $\delta_{1,o}$ is that for the oblate model and $\bar{v}$ is the protein's partial specific volume. As before, f_0 can be calculated using either eq. (20a) or (20b) and v_p can replace $\bar{v}$.

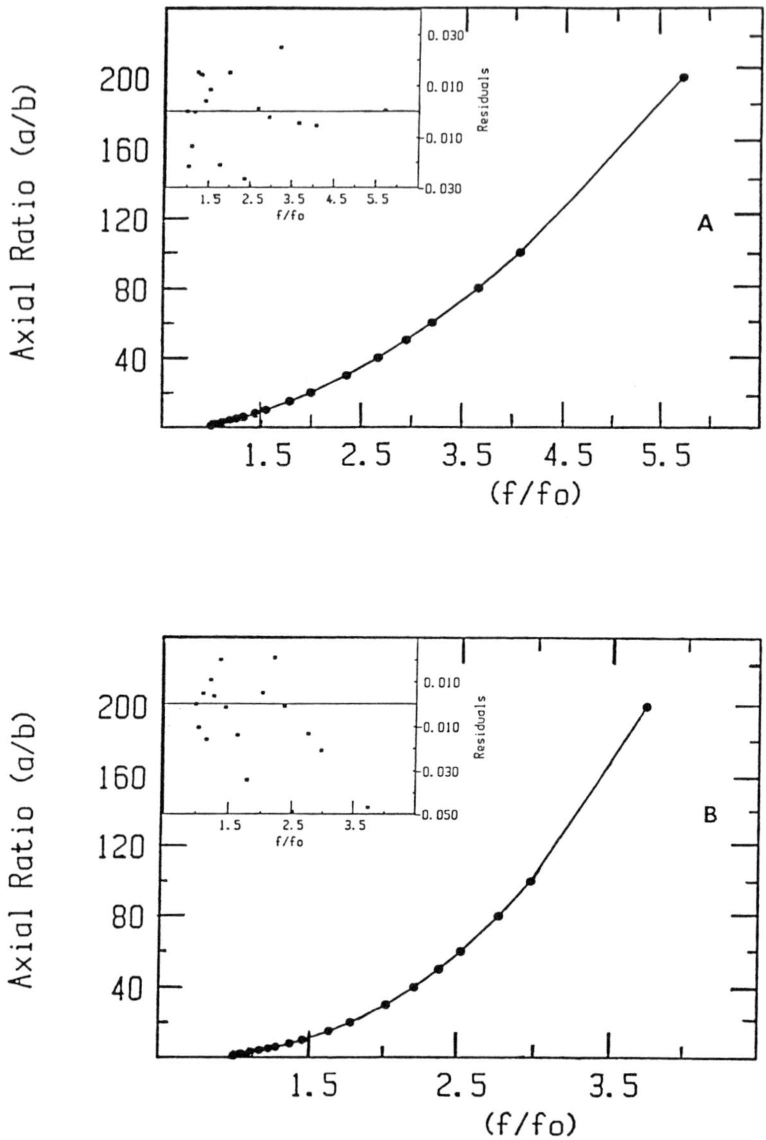

Figure 2. Presented here are the graphs of the axial ratio as a function of f/f_0 for a prolate (**A**) and oblate (**B**) ellipsoid of revolution. Data points are from Cantor and Schimmel[18]. The lines were drawn using the polynomials presented in the text. The insets show the residuals as a function of f/f_0 and demonstrate that the polynomials fit the data to within ±0.05 over the range of axial ratios from 1 to 200.

10. DISCUSSION

Studies of the structure of macromolecules and their complexes are often aided by thermodynamic and hydrodynamic data available from analytical ultracentrifugation. These techniques are amongst the most versatile, rigorous and accurate for estimating useful physical parameters for a wide range of chemical systems. The equations describing sedimentation are derived from first principles[1-3]. All terms in the mathematical equations that describe sedimentation behaviour are either readily and accurately estimable[4,5] or experimentally determinable[6,7]. The quantity and quality of the information furnished is great given the relatively small effort needed to perform the experiments. The sheer number of calculations and their interrelations, however, can seem overwhelming to the non-expert. For the full power of analytical sedimentation to be realized it is necessary for many of the routine calculations to be automated.

The use of appropriate polynomial functions to represent tabulated data permits the construction of a program that uses readily available experimental parameters (temperature, buffer composition, *etc.*) in the computations needed for the interpretation of analytical sedimentation data. Proper design of the program permits the maximum interpretation of data while requiring only a minimum amount of experimental observations. At the same time, those calculations which cannot be made with certainty should not be performed, thus preventing overinterpretation of data. The program should be designed so that user-supplied information always takes precedence in calculations. Thus, as more experimental data is available, more rigorous interpretations may be made.

All computations can be carried out using eight-digit precision. The algorithms presented here can be implemented using any of the common programming languages or spreadsheets (Lotus, *etc.*). These programs offer an attractive alternative since spreadsheets provide a consistent environment for both the development and the use of the program. Moreover, the templates are easily extended to include a variety of cases.

Data Entry and Default Values

The algorithms should be constructed to be flexible with regard to using the available information. Default values must be supplied (either *per se* or calculated) for all required parameters. For maximum flexibility, the user should be allowed to change any default value, which means that error checking must be supplied to insure that the values entered are reasonable. The ability to change default values permits the user to conduct a rapid, computerized appraisal of the effects that can be expected from various experimental protocols. For example, by having the program provide estimates of σ using eq. (4), an experimenter quickly can determine which rotor speeds will provide reasonable values of σ for subsequent data analysis[14].

For dilute aqueous solutions, the following default values are suggested: temperature (T) 293.15 K, density (ρ) 0.9982 g/ml and viscosity (η) 1.002 centipoise. Both ρ and η can be adjusted for temperature and buffer components as described below. When the program is started, all concentrations, including the protein, are set to zero. The degree of hydration (δ_1) is set to 0.0 g-H_2O/g-protein and the ellipsoid axial ratio (a/b) is set to 1.0. The correction factor for adjusting the sedimentation coefficient to zero protein concentration is set to 0.009 l/g, a reasonable value for non-interacting spherical proteins[21,27,28]. A value of 0.735 ml/g may be used as a default for $\bar{v}$. No default is specified for M_r, and a software flag should be set if there is no M_r available (Fig. 3). Should the protein's composition be known, the program should calculate default values for M_r, $\bar{v}$, R_0, f_0, D_0 and s_0 for the hypothetical anhydrous spherical particle[27-33].

Establishing a Priority for Values

It is essential that the program be clear about which values it is using in its calculations. While this may seem obvious, there are cases where ambiguous parameters might become a problem unless a well-defined priority is established. For example, M_r is required by the program for the estimation of the shape and size of a protein. However, there are three distinct ways M_r can be provided (Fig. 3): (i) a value for M_r can be entered directly by the user, (ii) σ can be supplied, which will require the program to calculate M_r from eq. (5), or (iii) M_r can be calculated from a user-supplied composition (amino acid, carbohydrate, *etc.*). A problem arises when sufficient information is available to provide M_r in more than one way and the estimates of M_r differ significantly from one another. This problem is alleviated by setting the priority for which M_r is to be used. Similar choices are experienced with the other terms ($\bar{v}$, ρ, η, f_0, *etc.*) needed for the interpretation of sedimentation data.

For maximum flexibility it is recommended that precedence should be given to any user supplied values (*e.g.* M_r, $\bar{v}$, ρ, *etc.*), since this suggests that an experimentally determined value is available. Likewise, priority should be given to any directly determined experimental value which requires only corrections to be made (*e.g.* σ should be given precedence in determining M_r over a value calculated solely from composition data). Values based solely on calculations should be given lowest priority.

Limitations

There are several limitations to the predictive procedures described here. The limitations can be divided into two categories. The first category includes those limitations that can be overcome by extending the principles used here to include more sophisticated algorithms. For example, modelling proteins as ellipsoids of revolution is most applicable to the examination of monomeric, globular proteins, and is less useful in the detailed examination of oligomeric structures. To overcome this limitation, either more exact or more general models should be used (see *e.g.* Chapter 18 of this volume). However, the use of such models requires that the experimenter already have some structural information available. In these cases, provided that sufficient material of high enough purity is available, there are other methods which can provide more detailed structural information. That is to say, the real power of analytical ultracentrifugation is to provide an initial, rough idea of the asymmetry or

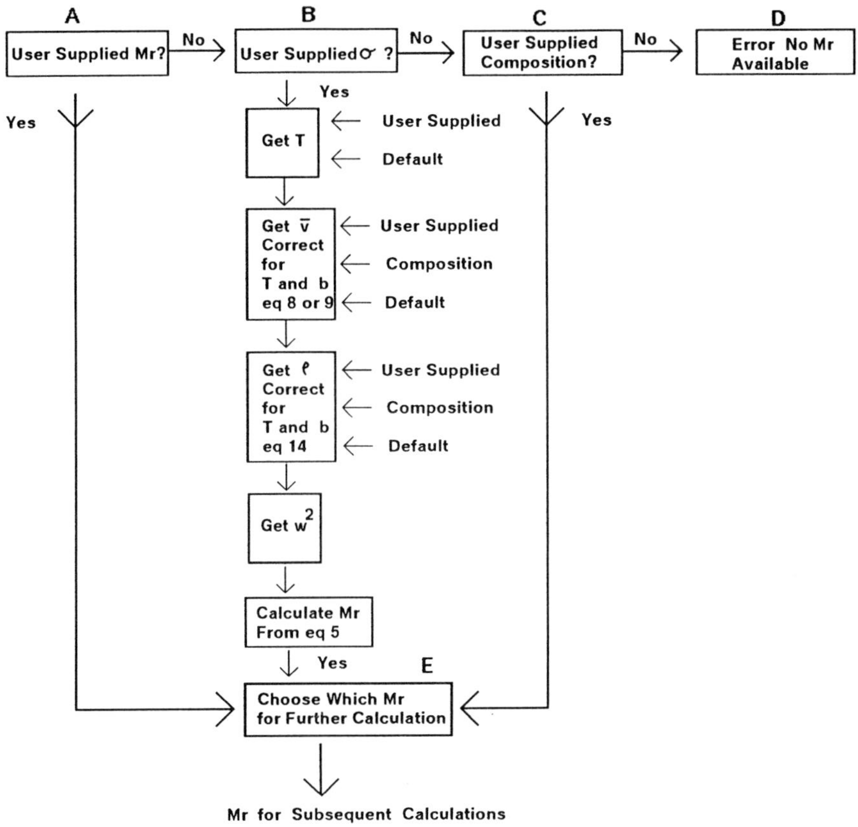

Figure 3. Three ways are available to supply the program with M_r, and a choice must be made (step **E**) as to which value will be used in subsequent calculations. It is suggested that M_r entered directly (path **A**) be given highest priority, that M_r calculated from σ (path **B**) be given second highest priority and that M_r from composition (path **C**) be given lowest priority. An error condition results (state **D**) if no M_r is available, and the program must guide the user into supplying sufficient information for M_r to be evaluated. Note that similar logic must be supplied for $\bar{v}$, ρ and η. Moreover, these terms will be affected by T which, in turn, will affect the calculation of M_r from σ. Thus, a system of software flags must be tested and set appropriately every time new information is provided so that M_r will be updated as the user amends ancillary information (*e.g.* $\bar{v}$,T or ρ).

hydration of a protein. If other structural data exist, then difference sedimentation velocity[56,57] methods can provide useful information concerning the *changes* in structure. The interpretation of data from these methods is outside the scope of this Chapter, but can be found elsewhere[9-12,32,56,57].

The second category of limitations includes those that are fundamental to the use of predictive schemes. Without compelling reason, none of the calculated or predicted values should be used as a substitute for experimental determinations. Many of the equations are derived assuming a two-component solution, even though experiments routinely are performed in buffers of three or more components. On theoretical grounds it is clear that this assumption is not correct[3]. However, over a reasonable protein concentration range and when thorough dialysis has been performed against solvents of moderate pH (*i.e.* where the net-charge to molecular weight ratio of the protein is less than about 1:5000 mol/g) and moderate ionic strength (0.05 M - 0.5 M), the sedimentation behaviour of many proteins is described adequately[2,3]. It is still important that the experimenter be aware that neglect of complexities resulting from multicomponent solutions can lead to misinterpretation of data[3,13]. For example, due to non-additive volume changes on mixing, densities calculated using eq. (14) may be in serious error when two or more solutes are present in moderate concentrations. This same problem can arise for mixtures containing both salts and a high mole fraction of non-ionic components. Indeed, a program will calculate predicted densities for solutions that can never be created! Thus, it must be emphasized that such a computer program cannot replace experimental determinations.

In the Future

The Beckman Optima XL-A analytical ultracentrifuge can be equipped to provide interpretable experimental data (s^* and σ) during the course of an experiment. It is reasonable to include real-time data interpretation as part of this machine. The equations and algorithms presented here form the starting point for such a program. A preliminary version of a program that implements many of the formulas described above has been written for the IBM-PC and is available from the authors.

ACKNOWLEDGEMENTS

This work has been supported in part by a Grant-in-Aid #850-781 and #870-1092 from the American Heart Association and Grants BBS 86-15815 and DIR 90-02027 from the National Science Foundation. This is publication #1542 from the New Hampshire Agricultural Experiment Station. The authors thank Daryl A. Lyons and Hope M. Snazelle for their help in fitting the viscosity data and in preparation of this manuscript.

REFERENCES

1. R.J. Goldberg, *J. Phys. Chem.*, 1953, *57*, 194.
2. J.W. Williams, K.E. Van Holde, R.L. Baldwin, and H. Fujita, *Chem. Rev.*, 1958, *58*, 715.

3. E.F. Cassasa and H. Eisenberg, *Adv. Prot. Chem.*, 1964, *19*, 287.
4. E.J. Cohn and J.T. Edsall, 'Proteins, Amino Acids and Peptides as Ions and Dipolar Ions', Rheinhold, New York, 1943, Chapter 4, p. 157.
5. T.L. McMeekin and K. Marshall, *Science*, 1952, *116*, 142.
6. H. Durchschlag, 'Thermodynamic Data for Biochemistry and Biotechnology', H.-J. Hinz, ed., Springer-Verlag, New York, 1986, Chapter 3, p. 45.
7. D.W. Kupke, 'Physical Principles and Techniques of Protein Chemistry', S.J. Leach, ed., Academic Press, New York, 1973, Part C, Chapter 1, p. 1.
8. T. Svedberg, 'The Ultracentrifuge', T. Svedberg and K.O. Pederson, eds., Johnson Reprint Corporation, New York, 1940.
9. H.K. Schachman, 'Methods in Enzymology', Volume IV, S.P. Colowick and N.O. Kaplan, eds., Academic Press, New York, 1957, Chapter 2, p. 32.
10. H.K. Schachman, 'Ultracentrifugation in Biochemistry', Academic Press, New York, 1959.
11. C.H. Chervenka, 'A Manual of Methods for the Analytical Ultracentrifuge', Spinco Division of Beckman Instruments, 1970, Palo Alto, California.
12. R. Eason, 'Centrifugation - a Practical Approach', D. Rickwood, ed., second edition, IRL Press, 1984, Chapter 8, p. 251.
13. G. Cohen and H. Eisenberg, *Biopolymers*, 1968, *6*, 26.
14. D. Yphantis, *Biochemistry*, 1964, *3*, 297.
15. W.F. Harrington and G. Kegeles, 'Methods in Enzymology Volume XXVII', C.H.W. Hirs and S.N. Timasheff, eds., Academic Press, New York, 1973, Chapter 13, p. 306.
16. W.F. Harrington, *Fractions*, 1975, *1*, 10.
17. H. Fujita, 'Foundations of Ultracentrifugal Analysis', J. Wiley and Sons, New York, 1975.
18. C.R. Cantor and P.R. Schimmel, 'Biophysical Chemistry', Part II, W.H. Freeman and Co., San Francisco, 1980, Chapter 10, p. 539.
19. C.R. Cantor and P.R. Schimmel, 'Biophysical Chemistry', Part II, W.H. Freeman and Co., San Francisco, 1980, Chapter 12, p. 643.
20. J.P. Johnston and A.G. Ogston, *Trans. Faraday Soc.*, 1946, *42*, 789.
21. J.M. Creeth and C.G. Knight, *Biochem. Biophys. Acta*, 1965, *102*, 549.
22. C.R. Cantor and P.R. Schimmel, 'Biophysical Chemistry', Part II, W.H. Freeman and Co., San Francisco, 1980, Chapter 11, p. 591.
23. C.R. Cantor and P.R. Schimmel, 'Biophysical Chemistry', Part II, W.H. Freeman and Co., San Francisco, 1980, p. 613.
24. J.R. Cann, 'Interacting Macromolecules: The Theory and Practice of Their Electrophoresis, Ultracentrifugation and Chromatography', Academic Press, New York, 1970, Chapter III, p. 93.
25. D.A Yphantis and D.F. Waugh, *J. Phys. Chem.*, 1956, *60*, 623.
26. M.L. Johnson, J.C. Correia, D.A. Yphantis and H.R. Halvorson, *Biophysical J.*, 1981, *36*, 575.
27. D.C. Teller, 'Methods in Enzymology', Volume XXVII, C.H.W. Hirs and S.N. Timasheff, eds., Academic Press, New York, 1973, Chapter 14, p. 346.
28. T.F. Kumosinski and H. Pessen, 'Methods in Enzymology', Volume 117, C.H.W. Hirs and S.N. Timasheff, eds., Academic Press, New York, 1985, Chapter 11, p. 154.
29. J.L. Oncley, *Ann. N.Y. Acad. Sci.*, 1941, *41*, 121.

30. D.C. Teller, E. Swanson and C. De Häen, 'Methods in Enzymology', Volume 61, C.H.W. Hirs and S.N. Timasheff, eds., Academic Press, New York, 1979, Chapter 8, p. 103.
31. H.A. Scheraga and L. Mandelkern, *J. Am. Chem. Soc.*, 1953, *75*, 179.
32. S.E. Harding, 'Dynamic Properties of Biomolecular Assemblies', S.E. Harding and A.J. Rowe, eds., Royal Society of Chemistry, Cambridge, 1989, Chapter 2.
33. D.C. Teller, *Nature (London)*, 1976, *260*, 729.
34. H. Höiland, 'Thermodynamic Data for Biochemistry and Biotechnology', H.-J. Hinz, ed., Springer-Verlag, Berlin, 1986, Chapter 2, p. 17.
35. J.H.C. Steele, C. Tanford and J.A. Reynolds, 'Methods in Enzymology', Volume 48, C.H.W. Hirs and S.N. Timasheff, eds., Academic Press, Orlando, Florida, 1978, Chapter 2, p. 18.
36. J.A. Reynolds and D.R. McCaslin, 'Methods in Enzymology', Volume 117, C.H.W. Hirs and S.N. Timasheff, eds., Academic Press, Orlando, Florida, 1985, Chapter 3, p. 41.
37. V. Prakash and S.N. Timasheff, 'Methods in Enzymology', Volume 117, C.H.W. Hirs and S.N. Timasheff, eds., Academic Press, Orlando, Florida, 1985, Chapter 4, p. 53.
38. T. Arakawa and S.N. Timasheff, 'Methods in Enzymology', Volume 117, C.H.W. Hirs and S.N. Timasheff, eds., Academic Press, Orlando, Florida, 1985, Chapter 5, p. 60.
39. D.W. Kupke, 'Physical Principles and Techniques of Protein Chemistry', Part C, S.J. Leach, ed., Academic Press, New York, 1973, Chapter 1, p. 1.
40. S.J. Perkins, *Eur. J. Biochem.*, 1986, *157*, 169.
41. R.A. Gibbons, 'Glycoproteins', Part A, A. Gottschalk, ed., Elsevier, Amsterdam, 1972, p. 31.
42. J.C. Lee and S.N. Timasheff, *Biochemistry*, 1974, *13*, 257.
43. J.C Lee and S.N. Timasheff, 'Methods in Enzymology', Volume 61, C.H.W. Hirs and S.N. Timasheff, eds., Academic Press, Orlando, Florida, 1979, Chapter 4, p. 49.
44. R.C. Weast, ed., 'CRC Handbook of Chemistry and Physics', 67th edition, CRC Press, Inc., Boca Raton, Florida, 1986, p. F-5.
45. F. Steckel and S. Szapiro, *Trans. Faraday Soc.*, 1963, *59*, 331.
46. M.L. Johnson and S.G. Frasier, 'Methods in Enzymology', Volume 117, C.H.W. Hirs and S.N. Timasheff, eds., Academic Press, Orlando, Florida, 1985, Chapter 16, p. 301.
47. J.T. Yang, 'Recent Advances in Protein Chemistry', Volume 16, C.B. Anfinsen, M.L Anson, K. Bailey, and J.T. Edsall, eds., Academic Press, New York, 1961, p. 323.
48. J.F. Johnson, J.R. Martin and R.S. Porter, 'Physical Methods of Chemistry', A. Weissberger and B.W. Rossiter, eds., Volume 1, Part VI, Wiley, New York, 1977, p. 63.
49. R.C. Weast (ed.), 'CRC Handbook of Chemistry and Physics', 67th edition, CRC Press, Inc., Boca Raton, Florida, 1986, p. F-37.
50. K.E. van Holde, 'Physical Biochemistry', second edition, Prentice-Hall, New Jersey, 1985, p. 117.
51. I.D. Kuntz, *J. Am. Chem. Soc.*, 1971, *93*, 514.

52. M. Lüscher-Mattli, 'Thermodynamic Data for Biochemistry and Biotechnology', H.-J. Hinz, ed., Springer-Verlag, Berlin, 1986, Chapter 11, p. 276.
53. G.M. Mrevlishvili, 'Thermodynamic Data for Biochemistry and Biotechnology', H.-J. Hinz, ed., Springer-Verlag, Berlin, 1986, Chapter 5, p. 148.
54. I.D. Kuntz and W. Kauzman, *Adv. Protein Chem.*, 1974, *28*, 239.
55. H. Pessen and T.F. Kumosinski, 'Methods in Enzymology', Volume 117, C.H.W. Hirs and S.N. Timasheff, eds., Academic Press, New York, 1985, Chapter 14, p. 219.
56. E.G. Richards and H.K. Schachman, *J. Am. Chem. Soc.*, 1957, *79*, 5324.
57. M.W. Kirschner and H.K. Schachman, *Biochemistry*, 1971, *10*, 1900.

8

Data Acquisition and Analysis Systems for the Absorption Optical System of the Analytical Ultracentrifuge

By Marc S. Lewis

BIOMEDICAL ENGINEERING AND INSTRUMENTATION PROGRAM, NATIONAL CENTER FOR RESEARCH RESOURCES, NATIONAL INSTITUTES OF HEALTH, BETHESDA, MD 20892, U.S.A.†

1. INTRODUCTION

Analytical ultracentrifugation first became a viable scientific discipline in the 1920s with the pioneering work of Svedberg and his co-workers, and it became a practical research tool in the late 1940s with the introduction of the Model E by the Spinco Division of Beckman Instruments. The contemporary era of analytical ultracentrifugation using computer data acquisition systems with the absorption optical system began with the work of Spragg and Goodman[1], who appear to have been the first to publish on the application of the digital computer as a data acquisition and analysis system. This was followed by studies by Crepeau, Edelstein, and Rehman[2] and by Crepeau, Hensley, and Edelstein[3]. The international symposium "Fifty Years of the Ultracentrifuge", held in 1975[4], presented more sophisticated systems developed by Williams[5]; by Spragg, *et al.*[6]; by Crepeau, Conrad, and Edelstein[7]; and by Cohen, *et al.*[8] All of these systems used computers of the class of the PDP–8E, the Nova 1220, or the Honeywell H516. The limited memory space in these machines represented a very severe constraint and created significant problems for data analysis as well as for data acquisition. These limitations have been overcome by the use of a variety of personal computers that have significantly more memory and speed, are easier to use, and are less expensive than earlier computers. The personal computers now available permit greatly simplified data acquisition systems, and they permit the same computer used to acquire the data to be used for sophisticated analytical procedures and, in addition, the production of publication quality graphs.

At this time, with a renewed interest in the application of analytical ultracentrifugation to the problems of biochemistry and molecular biology, discussion

† The opinions expressed herein are entirely those of the author. Mention of specific software or hardware does not constitute endorsement of that product by the Department of Health and Human Services, the National Institutes of Health, the National Center for Research Resources or the Biomedical Engineering and Instrumentation Program.

of data analysis systems appears quite appropriate. Since the new Beckman Instruments Optima™ XL–A analytical ultracentrifuge is now available, utility of the discussion of data acquisition systems for existing ultracentrifuges is open to question. However, data acquisition boards and computers suitable for their use and for data analysis are now so inexpensive and their use is so simple that it is worthwhile to acquire such systems for existing Beckman Model E ultracentrifuges unless these instruments are to be retired immediately following the acquisition of new XL–A analytical ultracentrifuges. Much useful work can still be done with upgraded Model E analytical ultracentrifuges, and Beckman Instruments appears to intend to support them to the extent that they will have parts available for at least the next several years. Assuming that no other application can be found for it in the laboratory, only the data acquisition board would lose utility when a Model E was finally retired; an appropriately chosen computer would be equally useful with the XL-A.

It has not been possible to evaluate all of the data acquisition boards and all of the data analysis systems available today. The choices made are based on the more limited selection available a few years ago that appeared most suitable at the time; products available since then have not offered sufficient advantage to warrant change. Thus, omission of a specific product here does not imply that the product is considered deficient in any way but rather indicates that a suitably comprehensive evaluation has not been possible; obtaining experimentally useful results has been of greater concern than evaluating every possible means of obtaining them. The reader should consider what is presented here as a starting point for developing optimal systems for his/her own applications.

2. DATA ACQUISITION SYSTEMS

Two approaches to the problem of data acquisition can be used. The more elegant, difficult, and expensive approach is to have direct acquisition from the photomultiplier tube using analogue-to-digital converters operating either directly from the output of the photomultiplier tube or following some level of amplification of the photomultiplier tube signal. The papers of Williams[5] and of Cohen, *et al.*[8] describe such systems. In this approach, signal sorting, multiplexing, and conversion to absorbancies as a function of radial position and cell number are performed primarily by software. It is difficult to advocate such a system at this time when quite good results, obtained by direct digitization of the output from the Model E absorption scanner electronics, can be accomplished using any of a number of modestly priced analogue-to-digital boards designed to work with a variety of different personal computers. Such systems also require appropriate software; some operational systems and their associated software are discussed below.

Two types of analogue-to-digital converters have been used in data acquisition systems for the Model E analytical ultracentrifuge. The most commonly used has been the 12-bit analogue to digital (A-to-D) converter where the signal sampling takes place at discrete intervals under software control. Two of the systems described here use this method. The other approach is to use a system such as the MetraByte (Kiethley MetraByte/Asyst/DAC, Taunton, MA 02780) CHROM-1 board which converts voltages to frequency and then counts pulses for a period of time under software control. Such a system has the advantage that the sampling system acts as an integrator and tends to minimize the effects of random noise. With the system described by Sackett *et al.*[9], the voltage from the scanner output is first passed through an interface box that inverts the voltage and multiplies the voltage by 2.5 to match the input to the CHROM-1 input range. The board is protected by a diode that provides a voltage clamp to prevent the voltage from going below -0.13 V. The interface box also contains a triggering circuit that initiates acquisition at the outer reference position; acquisition termination is also software controlled.

Excellent results were obtained using a sample interval of 20 ms (50 Hz) and the normal scan rate to obtain a data density of approximately 2000 points per centimetre of cell radius. Preliminary data processing was performed using Lotus 1-2-3 on the acquisition computer; this consisted of averaging successive scans to minimize the effects of rotor precession and conversion of voltages to absorbancies using experimentally obtained conversion factors. The MLAB* data analysis system was then used to analyse the data. In the past this was done on the NIH§ DEC–10 mainframe computer, but with the termination of use of the DEC-10 at the NIH, data are now being analysed using the new version of MLAB operating on 80X86 computers.

Experience with the other data acquisition boards clearly indicates that a data density of 2000 points per centimetre of cell radius is not at all necessary and that equally good results can be obtained using the rapid scan rate with a data density of approximately 425 points per centimetre, which corresponds to a radial distance of 0.024 millimetres per data point. With the commonly used photomultiplier slit width of 0.10 millimetres and an approximate magnification factor of 2, the photomultiplier is averaging the absorbancy over a radial distance of 0.05 millimetres in the cell, so a data density corresponding to half that distance appears quite sufficient. The investigator choosing to use this data acquisition board should also keep in mind that more rapid data acquisition minimizes the effects of rotor precession, saves time, particularly when multiple scans are made, and minimizes acquisition time distortion of a moving boundary in sedimentation velocity experiments.

One of the 12-bit A-to-D data acquisition systems that has been used in this laboratory is the G W Industries MacADIOS II board in a Macintosh II computer[10]. A precision voltage divider is used so that the A-to-D converter can sample ±2.0 volts on the ±1.0 volt setting in the acquisition software. The 12-bit A-to-D converter thus gives a reading uncertainty of just under ±0.1 mV corresponding to approximately ±0.0007 absorbance units. The rapid scan rate is used and 45,000 voltage samples are taken in the 18 seconds required for the scan; these are averaged in groups of 100, and a position number, mean voltage value, and standard deviation are recorded. This gives 212 data points per centimetre of radius in the cell or approximately 0.05 mm of radius per data point. With the commonly used photomultiplier slit width of 0.10 mm and a magnification factor of approximately 2, the photomultiplier is viewing this same radial distance at any given time. A very minor change in the acquisition program

---

* MLAB: Modelling Laboratory - see later in the text.

§ NIH: National Institutes of Health

permits doubling the data acquisition rate to give a data density of 425 points per centimetre. The recorded data are then edited using a program written specifically for that purpose in which position numbers are converted to radial positions, voltages are converted to absorbancies using appropriate conversion factors, unnecessary data are discarded and an edited data file is written on the hard disk. As stated above, the data were transmitted to a DEC–10 mainframe computer for more detailed analysis using the MLAB program, but are now transferred to an AST Premium/286 computer for analysis with the new version of MLAB operating on 80X86 computers. Acquisition control software is furnished by GWI in several languages, and the acquisition software and the editing software used in this laboratory are written in Think Pascal.

Studies on standards of known absorbance demonstrate that with this system, the standard error of data is approximately 0.003 absorbance units at an absorbance of 0.000 and is 0.008 absorbance units at an absorbance of 1.000 at a wavelength of 280 nm. The error is essentially a linear function of absorbance. The use of the rapid scan rate minimizes the effects of rotor precession; repetitive scans have usually been reproducible within their RMS error. The MacADIOS II board has the additional advantage that a number of different daughter boards can be added to the mother board, including a 16-bit A-to-D acquisition board should one wish to have more precision and digital I/O boards, should one wish to use an external A-to-D converter for direct acquisition from the photomultiplier tube. The investigator considering the use of this system should keep in mind that the computer interface for the new Beckman XL-A is designed for an 80X86 type computer for both data acquisition and control and that rewriting the interface so a Macintosh II can be used will not be a trivial undertaking even for an experienced Macintosh programmer. The advantages of the Macintosh II, and in particular, the Macintosh IIci and the Macintosh IIfx, are such that this is a possibility worth considering as long as one is aware of the magnitude of the problems involved and has the resources to deal with them.

The other 12-bit A-to-D data acquisition that has been used in this laboratory is the MetraByte DAS-8 board in an AST Premium/286 computer[11]. The system is quite similar to the system used on the Macintosh II, except that 90,000 voltage samples are taken during the rapid scan with averaging of every 100 samples, giving a data density of 425 points per centimetre of radius in the cell, and the standard deviations are not calculated. The data are then edited and written on the hard disk as before. MetraByte also furnishes acquisition control software in several languages, and the acquisition software and editing software were written in Turbo Pascal.

The performance of this system has been the same as the performance of the system on the Macintosh II. It has the additional advantage in this laboratory that the same computer is used for data acquisition, data editing, and data analysis using MLAB. Most of the recent work in this laboratory has been done using this MetraByte DAS-8 A-to-D system because of its intrinsic compatibility and convenience. The use of MLAB for data analysis will be described later.

An alternative approach when using MetraByte data acquisition boards is to use Asyst for control of data acquisition, editing, and data analysis. Asyst has been successfully applied to ultracentrifugal research in other laboratories.

3. DATA ANALYSIS SYSTEMS

The investigator with ultracentrifugal data in need of analysis has various options: data analysis programs can be written using published algorithms, programs written by other investigators can be used, or suitable commercial data analysis software systems can be used. MLAB, the major software system described in this section, is very effective for the task at hand and is quite user friendly. Only the use of MLAB is described in some detail here; while other software is available and certainly warrants most serious consideration, lack of experience with these other systems precludes effective comparison here.

There are two absolutely mandatory features required for any software used to analyse ultracentrifugal data: it must be capable of performing non-linear least-squares curve-fitting and it must have good graphics capabilities. Non-linear least-squares curve-fitting is required for fitting data in the form of concentration as a function of radial position. For the very simple problem of determining the molecular weight of a monodisperse ideal solute, the common practice of fitting the natural logarithm of the concentration as a function of the square of the radius results in an unacceptable transformation of the distribution of the error of the concentration data values and will usually result in erroneous estimates of the values of the molecular weight and its standard error. Additionally, there does not appear to be any particularly good way to weight the data to avoid these problems.

If one is dealing with problems of multiple components, non-ideality, or interacting systems, then non-linear least-squares curve-fitting to appropriate mathematical models is even more essential if the results are to be at all meaningful. The only other way to deal with these problems is to do a running fit of a linear model to a limited portion of the data to obtain apparent weight-average and z-average molecular weights. If the meniscus is depleted, then the number-average molecular weight can in addition be readily calculated. These average molecular weights as functions of concentration can then be used to determine the proper model and the parameter values describing it. While such methods have been successfully used in the past, they have the drawback that they impose an error introducing procedure between the basic data and the analysis of the model. In addition to being more time consuming and more difficult, such procedures are inherently less accurate, and it is very difficult to obtain meaningful error estimates for the various parameters.

Good graphical capabilities are essential if publication quality plots are to be prepared. Even more important, during the modelling procedure, plots of the fit of a model to the data and also plots of the distribution of the residuals for that model are invaluable aids in determining the appropriateness of a specific model for fitting the data. Other features, while highly desirable, may not always be quite as essential. They are discussed in the description of the use of the system.

MLAB is a computer program whose name is an acronym for Modelling LABoratory. It was initially intended to run on DEC-10 and DEC-20 mainframe computers and was designed and programmed at the NIH and has been continually and significantly enlarged and enhanced since 1973[12]. A new, completely revised and enhanced version of MLAB has been written in the C language to run on 80X86 type computers and is available from Civilized Software, Inc. (Bethesda, MD 20815). It is this version that is discussed here.

MLAB is an interactive system for mathematical modelling. The major components of the MLAB system are a curve-fitting program using a modified Marquardt-Levenberg method, a robust differential equation system solving program that permits using differential equation defined models, more than 300 mathematical and statistical functions and operators, a collection of routines for the construction and display of graphs, and mechanisms for saving information between computing sessions. Because MLAB is an interactive system, it has the inherent disadvantage of being slower than a compiled system. This disadvantage is important primarily if one is analyzing a large quantity of data with only one model. On the other hand, with an interactive interpretive shell, the data manipulation and mathematical and statistical calculations for developing and testing models are much more convenient and more rapid than when using a compiled system. MLAB is a command-line system, and the user communicates with the system by typing appropriate commands. These are usually executed immediately; those that are not will prompt the user for additional information. A text database of information may be accessed by typing HELP.

The application of MLAB to the analysis of ultracentrifugal data is well illustrated by describing the analysis of data from a homogeneous ideal macromolecular species. The analysis is performed by typing the following commands:

```
DATA=READ("C:\DIRNAME\FILENAME.SCA")
TYPE DATA ROW 1:11
DELETE DATA ROW 1:11
MAT=SHAPE(NROWS(DATA)/2,2,DATA)
FCT OD(MV)=6.203E-8*MV^2+6.734E-4*MV-.243
MAT COL 2=OD ON MAT COL 2
FCT C(R)=CB*EXP(A*M*(R^2-RB^2))+E
A=(1-VBAR*RHO)*(2*PI*RPM/60)^2/(2*8.314E7*T))
CB=1;  M=50000;  E=0;  RB=7.19
FIT(M,CB,E), C TO DATA
```

The first command above reads a file named `FILENAME` with the extension `SCA` denoting scan data in the directory `DIRNAME` on drive `C:` and places it in an MLAB one-column matrix named `DATA`. The second command causes the appearance of the run information on the screen, and the third command removes that information from `DATA`. `DATA` now has the form of a vector with alternating radial position values and millivolt values from the scanner. The fourth command reshapes `DATA` into the two-column matrix `MAT` with one column for radial values and the other for millivolt values. `NROWS` is a function that returns the number of rows in `DATA`, and it is used in this command to instruct MLAB to reshape `DATA` into a two-column matrix with half as many rows as before. The fifth command defines a function `OD` that transforms millivolt values to absorbancies; the parameters in the function were derived from calibration experiments. The sixth command transforms the millivolt values of column 2 of MAT to absorbancies. The seventh command defines a function C that describes the concentration distribution of a homogeneous ideal solute. The terms have their usual meaning; E is used here to denote a small baseline error. The eighth command defines the value of the term A; numerical values are used for VBAR, RHO, RPM and T; PI is defined in MLAB. The ninth command defines the initial parameter estimates and the value of RB. The tenth command causes MLAB to perform the actual fitting and present the obtained parameter values on the screen. The MLAB output also includes the estimated standard deviations and dependency values of the parameters, the sum of squares, the root-mean-square (RMS) error, the determination coefficient (R^2), and the percent data error.

If one anticipates doing a number of fits of this type, then significant time can be saved by creating a so-called DO-file that would consist of commands 2 through 10 and that could be named DOFILE1. It could be saved in the same directory as MLAB; then all that would be necessary would be to type the first command, followed by the command DO DOFILE1.

If one is dealing with a problem of non-ideality, then the root finder in MLAB is useful. In this case, the fitting function is an implicit function. It is written:

```
FCT C(R)=ROOT(X,0,CB,CB*EXP(A*(M/(1+B*M*X))*(R^2-RB^2))-E-X)
```

The root finder operator returns a value of X, lying between zero and CB such that the value of the last expression in the parentheses is zero. M, CB, E, and B, the virial coefficient, are the fitting parameters in this model. The root-finder can be equally well applied to more complex models.

A graph of the results with default options is easily created by typing the following two commands:

```
DRAW MAT LINETYPE 0 POINTTYPE CIRCLE PTSIZE .01
DRAW POINTS(C,6.70:7.15!100)
```

LINETYPE 0 instructs MLAB to draw no line; POINTTYPE CIRCLE instructs MLAB to draw circles (there are seven line types and twenty-one point types available); PTSIZE .01 means that the point will be .01 times the width of the frame the graph is drawn in. The arguments of the operator POINTS tells MLAB to draw the function C, using the fitting parameters, for 100 equally spaced values of R between 6.70 and 7.15, and since no line type is given, draw it with a solid line. If desired, the graph then may be sized with the commands:

```
WINDOW 6.7 TO 7.2, 0 TO .6
IMAGE 1.75 TO 6.25, 1.5 TO 5.5 INCHES
```

The first command sizes the graph in terms of Cartesian co-ordinates; the second command sizes the graph in screen dimensions with the lower left-hand corner denoted as (0, 0). There are additional commands used for removing or adding labels, titles, and tick marks and for otherwise formatting the graph, all of which are useful in preparing a publication quality graph. The graph may be seen at any time using the command VIEW; the RETURN key then restores MLAB to the command mode. A typical example of such a graph as well as a graph of the residuals is shown in Fig. 1. While we wish to have both graphs together ultimately, we must create them separately. This is done with the commands:

```
FRAME 0 TO 9, 0 TO 5.75 INCHES
W1=W
BLANK W1
```

The first command delineates a region of the screen in which the graph will appear; the second command assigns the name W1 to the default window W; the third command blanks the graph until it is restored with an UNBLANK command. The graph of the residuals is created with the commands:

```
DMAT COL 1=MAT COL 1
DMAT COL 2=MAT COL 2-C ON (MAT COL 1)
L=6.7:7.2!2&'0
DRAW L
WINDOW 6.7 TO 7.2, -.02 TO .02
IMAGE 1.75 TO 6.25, 6.0 TO 7.75 INCHES
FRAME 0 TO 9, 5.75 TO 8 INCHES
W2=W
```

With the exception of the third command, the intent of these commands should be obvious; the third command creates a 2×2 matrix with the x co-ordinates of 6.7 and 7.2 in column 1 and the y co-ordinates of 0 and 0 in column 2 for the purpose of drawing the horizontal line L. When the graph of the residuals has been completed, the first graph is restored with the command UNBLANK W1, and the combined graphs are saved in a plot file with the command PLOT, which creates a file named MLABPSL.N, where N is an identifying integer. After exiting from MLAB, the graph may be drawn by a postscript laser printer with the DOS command PRINT MLABPSL.N. Before exiting, all of the analysis can be saved with the command SAVE IN FILENAME, where FILENAME is a chosen name. This command creates a file named FILENAME.SAV, which can be used again in MLAB with the command REUSE FILENAME. When the EXIT command is given, a file named MLAB.LOG is created that contains all of the commands that have been given by the analyst and the responses (with the exception of graphs) made by MLAB. This may be printed with the DOS command PRINT MLAB.LOG. This log will be automatically deleted and a new log begun at the initiation of the next MLAB session.

For purposes of comparison, Fig. 2 shows data obtained from a Beckman XL-A analytical ultracentrifuge. The data from the XL-A were analysed by essentially the same procedure as those from the Model E. The data from the Model E have a somewhat better RMS error, 0.004 *vs.* 0.006 absorbancy units, but the data from the XL-A have somewhat better distributed residuals. These figures suggest that these two instruments can produce data of essentially comparable quality under the experimental

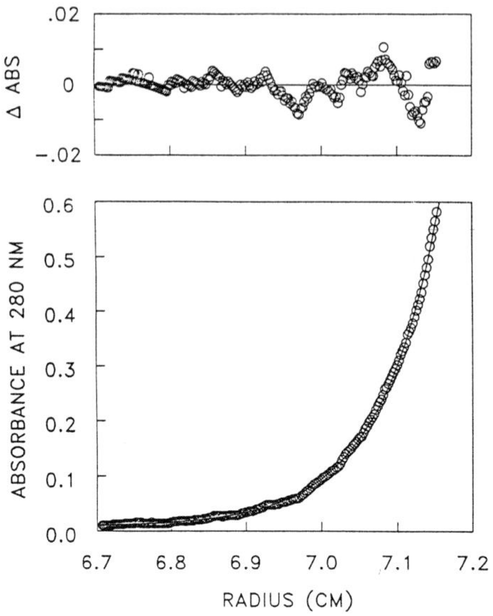

Figure 1. Lower panel: concentration distribution of the Interleukin II receptor at equilibrium at 20,000 rev/min and 20° in a Beckman Model E analytical ultracentrifuge. The weighted fit is for an ideal monomer as a model. Upper panel: distribution of the residuals for this fit. The RMS error is 0.004 absorbancy units.

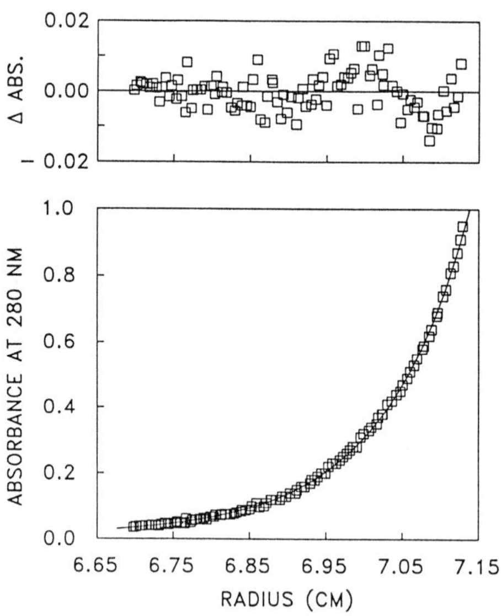

Figure 2. Lower panel: concentration distribution of myoglobin at equilibrium at 20,000 rev/min and 20° in a Beckman XL–A analytical ultracentrifuge. The weighted fit is for an ideal monomer as a model. Upper panel: distribution of the residuals for this fit. The RMS error is 0.006 absorbancy units.

conditions used here, which are scanning at 280 nm and maximum absorbancies of less than 1.0. The XL-A can, however, produce good data at scanning wavelengths as low as 200 nm and with absorbancies up to at least 2.0, experimental conditions not attainable with the Model E that offer new and important opportunities to the investigator.

In these examples, examination of the distributions of the residuals indicates that it would be appropriate to weight the data during the fitting procedure. The proper weight for any given datum value is the reciprocal of its variance. If variances as a function of concentration are available, then it is simple to construct a weighting vector that is a function of concentration. An alternative to this is to use the `EWT` function provided in MLAB. This function constructs a non-parametric model of the data by repetitively smoothing it, then estimates the variances of the points from their deviations from this model and constructs the appropriate weighting vector. MLAB has the capability to simultaneously fit several models with both global (shared) and local parameters. It is particularly important to use weights when performing simultaneous fits of two or more data sets with different magnitudes and, thus, different error structures so that the data set or sets of greater magnitude do not unduly affect the obtained parameter values.

As an example of simultaneously fitting multiple data sets with weights and also having the convenience of using user-defined functions within functions, let us consider fitting two data sets for a reversible monomer-dimer equilibrium characterised by the dimerisation constant, k. We define the function:

```
FCT U(R)=EXP(A*M1*(R^2-RB^2))
```

Then the two fitting functions are:

```
FCT CI1(R)=CB1*U(R)+EXP(LNK12)*(CB1*U(R))^2+E1
FCT CI2(R)=CB2*U(R)+EXP(LNK12)*(CB2*U(R))^2+E2
```

and the fitting command is:

```
FIT(LNK12,CB1,CB2,E1,E2), CI1 TO DATA1 WITH WT EWT(DATA1),\
                          CI2 TO DATA2 WITH WT EWT(DATA2)
```

The parameter `LNK12` is global to both data sets while `CB1` and `E1` are local to `DATA1` and `CB2` and `E2` are local to `DATA2`. The backslash specifies that the command continues on the next line. The returned sum of squares, RMS error, R^2 and percent data error will be for the simultaneous fit and will reflect the magnitude of the weights generated by `EWT`.

In this situation, it is important to fit for the value of ln k for three reasons. First, this is the desired parameter for thermodynamic calculations. The value ln k is on an absorbancy concentration scale and must be converted to a molar concentration scale by adding $\ln(E_1/2)$, where E_1 is the molar extinction coefficient of the monomer and it is assumed that $E_2 = 2E_1$. Second, fitting for ln k guarantees that the value of k will always be greater than zero and thus be physically meaningful. While this could be achieved by the use of constraints, this formulation of the model gives a better shaped sum-of-squares surface. Finally, if one fits for k and plots the sum of squares as a function of that parameter value, keeping all other parameters fixed, a curve that is asymmetrical about the optimal parameter value will result. When fitting for ln k, this curve will be parabolic and almost perfectly symmetrical about the optimal parameter value. Since a parabolic curve, symmetrical about the optimal parameter value, would be obtained for a model that is linear in the parameters, has normally distributed error and where the weight for each data point is the reciprocal of the variance of that point, this becomes a heuristic validation that the obtained parameter value is probably close to a maximum-likelihood estimate and that the standard deviation estimate is also probably more nearly correct.

There are aspects of MLAB that detract from its performance. Plots can be printed only by using postscript printers. The convenience of using a print screen command to produce a notebook quality graph is not available. Of greater importance is the 640k memory limitation of the DOS operating system. MLAB is such a large program that there are only approximately 200k of working memory available for computation. Since curve-fitting requires additional memory space, this limits the total number of points in the data matrices that can be analysed. The upper limit of 600 to 1000 points can be troublesome if one is simultaneously analyzing more than two or three data sets.

Finally, in common with all other non-linear least-squares curve-fitting programs known to this author, MLAB gives parameter error estimates that are only linear approximations. Thus, for some models, they may be grossly in error and it is necessary to evaluate the sum-of-squares as a function of each parameter to make a meaningful evaluation of the standard errors that MLAB returns. This evaluation of the sum-of-squares fit can be done in MLAB, and a contour map of the sum-of-squares surface for any pair of parameters can be drawn and evaluated. MLAB is equally well suited for analysis of error by Monte Carlo, Jackknife and Bootstrap methods.

The description of MLAB given here has been, by necessity, very limited. MLAB has been in use in this laboratory for more than fifteen years, and it has been used for almost every type of ultracentrifugal experiment possible. It has never failed to provide effective analysis for any ultracentrifugal experiment and has been equally effective for a wide range of other computational problems involving mathematical modelling.

REFERENCES

1. S.P. Spragg and R.F. Goodman, *Ann. N.Y. Acad. Sci.*, 1969, *164*, 294.
2. R.H. Crepeau, S.J. Edelstein and M.J. Rehman, *Anal. Biochem.*, 1972, *50*, 213.
3. R.H. Crepeau, C.P. Hensley, Jr. and S.J. Edelstein, *Biochem.*, 1974, *13*, 4860.
4. 'Proceedings of the International Conference "50 Years of the Ultracentrifuge"', M.S. Lewis and G.H. Weiss, editors, *Biophys. Chem.*, 1976, *5*, 1.

5. R.C. Williams, Jr., *Biophys. Chem.,* 1976, *5,* 19.
6. S.P. Spragg, W.A. Barnett, J.K. Wilcox and J. Roche, *Biophys. Chem.*, 1976, *5,* 43.
7. R.H. Crepeau, R. Conrad and S.J. Edelstein, *Biophys. Chem.*, 1976, *5,* 27.
8. R. Cohen, J. Cluzet, H. Cohen, P. Male, M. Moignier and C. Soulte, *Biophys. Chem.*, 1976, *5,* 77
9. D.L. Sackett, R.E. Lippoldt, C. Gibson and M.S. Lewis, *Anal. Biochem.*, 1989, *180,* 319.
10. M.S. Lewis, *G W Industries Application Notes*, No. 13, 1990.
11. A.P. Minton and M.S. Lewis, unpublished.
12. G.D. Knott, *Comput. Programs Biomed.*, 1979, *10,* 271.

9

A Fitting Function for the Analysis of Sedimentation Velocity Concentration Distributions

By Arun K. Attri and Marc S. Lewis

LABORATORY OF CELL BIOLOGY, NATIONAL HEART, LUNG AND BLOOD INSTITUTE AND BIOMEDICAL ENGINEERING AND INSTRUMENTATION PROGRAM, NATIONAL CENTER FOR RESEARCH RESOURCES, NATIONAL INSTITUTES OF HEALTH, BETHESDA, MARYLAND 20892, U.S.A.

1. INTRODUCTION

The use of simple data acquisition systems for recording data from the absorption optical system scanner of the Beckman Model E analytical ultracentrifuge using analogue-to-digital acquisition boards in 80286 or 80386 based computers[1,2] or in Macintosh II computers[3] has provided data of precision superior to that obtainable by manual digitization of the strip chart recordings from the scanner and has provided many more data points for analysis than was practical with that procedure. The new Beckman Optima™ XL-A analytical ultracentrifuge may be expected to provide data that are at least as good. Software for the analysis of the concentration distributions obtained from sedimentation velocity experiments has been developed for both types of computer[4,5]. These programs locate the radial position given by the square root of the second moment of the concentration gradient curve and also locate the radial position corresponding to the concentration that is half the plateau region concentration. The program for 80286 or 80386 computers gives the apparent diffusion coefficient as well as the sedimentation coefficient. These programs give weight average sedimentation coefficients for disperse systems but give a meaningful diffusion coefficient only when the solution is monodisperse.

An alternative approach, which can be used for either monodisperse or paucidisperse solutions, is to fit the concentration gradient curve with an appropriate fitting function. One means of doing this is to use a fitting function based upon a solution of the Lamm equation such as that of Holladay[6]. While such a function has the virtue of being a proper model, it suffers from the defects of being relatively slow from a computational point of view and has difficulty resolving two solute components unless a reasonably well-defined plateau region is present between the gradients of each component[7]. Another means of doing this is to use an arbitrary fitting function. If an arbitrary fitting function is used as a model, however, several important criteria need to be fulfilled. It should accurately locate the radial position corresponding to the square root of the second moment of the gradient curve. It must fit the data well, using the criteria of minimal deviations of the residuals with minimal systematic deviations and small root-mean-square (RMS) errors. It should permit the calculation of an apparent diffusion coefficient. Finally, if possible, it should be able to resolve the gradients of two components relatively early in a sedimentation velocity experiment before there is a well-established plateau region between the gradients of the components.

The fitting function that we propose here is based upon the equation

$$f(x) = \frac{\exp(x)}{1 + \exp(x)} \tag{1}$$

which describes a sigmoidal curve where $0 < f(x) < 1$ for $-\infty < x < \infty$ [8]. In the following sections we will demonstrate how a fitting function based on this equation fulfils the criteria described above.

2. ANALYSIS

The radial position $\bar{r}$ obtained by second moment analysis is defined by

$$\bar{r}^2 = r_p^2 - \frac{1}{c_p} \int_{r_m}^{r_p} c_r \, dr^2 \tag{2}$$

where r_m is the radial position of the meniscus, r_p is a radial position in the plateau region, c_r is the concentration at the radial position r, and c_p is the concentration in the plateau region. It is readily apparent from eq. (2) that the value of $\bar{r}$ is not dependent on the precise location of r_p in the plateau region. Eq. (2) suggests that the appropriate form of the fitting function is given by

$$c(r,t) = \sum_{1}^{m} c_{p,i,t} \frac{\exp(n_{i,t}(r^2 - \bar{r}_{i,t}^2))}{1 + \exp(n_{i,t}(r^2 - \bar{r}_{i,t}^2))} + c_{b,t} \tag{3}$$

where the subscript i is the index for a component and may have values of 1 to m, t as a subscript denotes time dependence, $n_{i,t}$ determines the width of the distribution and is related to the apparent diffusion coefficient, and $c_{b,t}$ is a solute-depleted region baseline correction term. When using this function, $\bar{r}_i$, $n_{i,t}$, $c_{p,i}$, and $c_{b,t}$ will be the fitting parameters whose values are to be adjusted by the non-linear least-squares curve-fitting program.

Considering only one component, substitution of eq. (3) in the integral of eq. (2), evaluation of the integral between limits and rearrangement then gives

$$c_{p,t}(r_{p,t}^2 - \bar{r}_t^2) =$$

$$c_{p,t} \left\{ \frac{\ln\{1+\exp[n_t(r_{p,t}^2 - \bar{r}_t^2)]\} - \ln\{1+\exp[n_t(r_{m,t}^2 - \bar{r}_t^2)]\}}{n_t} \right\} \tag{4}$$

Appropriate algebra then gives, sequentially

$$\exp[n_t(r_{p,t}^2-\bar{r}_t^2)] = \frac{1+\exp[n_t(r_{p,t}^2-\bar{r}_t^2)]}{1+\exp[n_t(r_{m,t}^2-\bar{r}_t^2)]} \tag{5}$$

$$1 + \exp[n_t(r_{p,t}^2-\bar{r}_t^2)]$$

$$= \exp[n_t(r_{p,t}^2-\bar{r}_t^2)] + \exp[n_t(r_{p,t}^2-\bar{r}_t^2)]\exp[n_t(r_{m,t}^2-\bar{r}_t^2)] \tag{6}$$

$$1 = \exp[n_t(r_{p,t}^2-\bar{r}_t^2)]\exp[n_t(r_{m,t}^2-\bar{r}_t^2)]$$

$$= \exp[n_t(r_{p,t}^2-\bar{r}_t^2)] + n_t\,(r_{m,t}^2 - \bar{r}_t^2) \tag{7}$$

$$0 = n_t\,(r_{p,t}^2 + r_{m,t}^2 - 2\bar{r}_t^2) \tag{8}$$

and since the distribution must have finite width, n cannot be equal to zero, and we obtain the requisite condition that

$$r_{p,t}^2 = 2\bar{r}^2 - r_{m,t}^2 \tag{9}$$

Thus, eq. (3) will be an appropriate fitting function and will give a value of $\bar{r}$ that, within the limits of experimental error, is the same as that given by second moment analysis if the function fits the data well and if the condition for the position of r_p required by eq. (9) is obtained. Both of these may be verified by direct inspection of the fit of the function to the data or, more critically, by inspection of a plot of the distribution of the residuals.

The value of the sedimentation coefficient, s, may be obtained from the values of $\bar{r}$ as a function of time by the use of the definition of the sedimentation coefficient

$$s \equiv \frac{1}{\omega^2 r}\frac{dr}{dt} = \frac{1}{\omega^2}\frac{d(\ln r)}{dt} \tag{10}$$

which leads to the alternative fitting functions

$$\bar{r}_t = r_m \exp\{\omega^2 s(t+t_0)\} \tag{11a}$$

$$\ln \bar{r}_t = \ln r_m + \omega^2 s(t+t_0) \tag{11b}$$

where ω is the angular velocity of the rotor, t is the time after the rotor reaches maximum velocity, and t_0 is a fitting parameter that accounts for the time of rotor acceleration. The value of s so obtained must then be corrected to standard conditions of water at 20°C by the usual viscosity and density corrections.

Generally, only a single macromolecular component will be studied. However, if more than one component is known to be present or if the observed concentration distributions suggest the possibility of more than one component, then the multi-component form of eq. (3) must be used, fitting for the fewest components that give acceptable RMS error values and distributions of residuals. Visualization of the distribution of the components is significantly enhanced by a graphical representation of the equivalent of the schlieren pattern. This may be done by plotting the derivative of the concentration given in eq. (3) with respect to radius, using the parameter values obtained during fitting. This derivative is given by

$$\frac{dc_{r,t}}{dr} = \sum_{1}^{m} 2c_{p,i,t}\, n_{i,t}\, r \frac{\exp\{n_{i,t}(r^2-\bar{r}_{i,t}^2)\}}{1+\exp\{n_{i,t}(r^2-\bar{r}_{i,t}^2)\}} \tag{12}$$

The contributions of the individual components may be demonstrated by plotting the individual derivatives of their respective concentration distributions.

Our treatment of the method for obtaining the apparent diffusion coefficient is similar to that of Muramatsu and Minton, except that the fitting function used here permits greater simplicity[9]. For a rectangular cell with an infinitely sharp boundary at zero time, $c_{x,t} = 0$ for $x < 0$ and $c_{x,t} = c_{max}$ for $x \geq 0$. At any later time, t,

$$\frac{dc_{x,t}}{dx} = \frac{c_{max}}{2(\pi Dt)^{1/2}} \exp\left\{\frac{-(x-x_0)^2}{4Dt}\right\} \tag{13}$$

where D is the diffusion coefficient. However, at the mid-point of the boundary, at x = 0,

$$\left[\frac{dc_{x,t}}{dx}\right]_{x=0} = \frac{c_{max}}{2(\pi Dt)^{1/2}} \tag{14}$$

If we assume that the sector-shaped centrifuge cell is approximated well by a rectangular cell, then equivalent equations can be written for the centrifuge cell and we obtain

$$\left[\frac{dc_{r,t}}{dr}\right]_{r=\bar{r}} = \frac{c_{p,t}}{2(\pi Dt)^{1/2}} \tag{15}$$

Eq. (12), written for one component and evaluated at $r = \bar{r}$, gives

$$\left[\frac{dc_{r,t}}{dr}\right]_{r=\bar{r}} = \frac{c_{p,t}\, n_t\, \bar{r}_t}{2} \tag{16}$$

Combining eqs. (15) and (16) then gives

$$n_t \bar{r}_t = \frac{1}{(\pi Dt)^{1/2}} \tag{17}$$

In addition to the assumption regarding the shape of the cell, this equation does not take into account the effect of sedimentation on the broadening of the diffusion boundary.

For the Gaussian-shaped curve of a schlieren pattern, Fujita gives the equation

$$\left\{\frac{A}{H}\right\}^2 = D\frac{2\pi}{s\omega^2}\left[\exp(2s\omega^2 t) - 1\right] \tag{18}$$

which is derived from the Faxén solution of the Lamm equation and where A is the area under the curve and H is the height of the maximum ordinate[10]. Because eq. (12) describes the schlieren pattern, eq. (16) gives $H = c_{p,t}\, n_t\, \bar{r}_t / 2$ since the maximum ordinate will be at $r = \bar{r}$. Eq. (3) is the integral of eq. (12) and, evaluated between the limits of r_m and r_p, gives that $A = c_{p,t}$. If we make the appropriate substitutions and expand the exponent in eq. (18) as a power series, we then obtain

$$\left\{\frac{2}{\bar{r}_t n_t}\right\}^2 = D\frac{2\pi}{s\omega^2}\left[1 + 2s\omega^2 t + \frac{(2s\omega^2 t)^2}{2!} + \frac{(2s\omega^2 t)^3}{3!} + \ldots - 1\right] \tag{19}$$

If we truncate the series after the first term involving time, we obtain eq. (17). Truncation after the second such term gives

$$n_t \bar{r}_t = \frac{1}{\{\pi D t(1 + s\omega^2 t)\}^{1/2}} \tag{20}$$

Substitution of appropriate values for s, ω^2 and t demonstrate that the difference between the values of D obtained using eqs. (17) and (20) is not trivial, the difference being of the order of 10%. The same substitution shows that including third and higher power terms has no significant effect. Data in the form of the values of the product of $\bar{r}$ and n as a function of time may be fitted to give the value of D as a fitting parameter. The value of the diffusion coefficient so obtained then needs to be corrected for temperature and solvent viscosity. In practice it was found to be necessary to account for the time of rotor acceleration, thus eq. (20) must be modified to

$$n_t \bar{r}_t = \frac{1}{\{\pi D t\ (1 + s\omega^2(t + t_0))\}^{1/2}} \tag{21}$$

where t_0 is now also a fitting parameter.

3. EXPERIMENTAL

Materials and Methods

Well-characterized globular proteins whose values of s and D have minimal concentration dependence were used to demonstrate the application of the fitting functions described above. Rabbit muscle aldolase was obtained from Pharmacia; bovine serum albumin fraction V and ovalbumin were obtained from Sigma. The proteins were dissolved in pH 7.4 phosphate buffered saline (PBS) and were chromatographed on 100 cm Sephacryl 200 columns (Pharmacia). In every case, the fraction used was selected from the most concentrated region of the elution profile, which had a band width one-half the width of the elution profile at half maximum height. The selected fractions were then dialyzed against PBS and their concentration

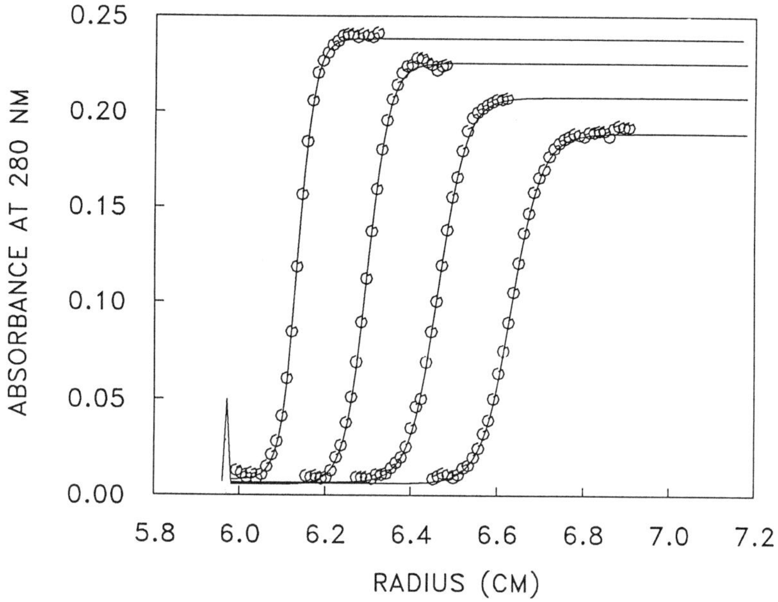

Figure 1. Fits of eq. (3) as a model for concentration distributions obtained from a sedimentation velocity experiment using ovalbumin in PBS at 20°C and 60,000 rev/min. The gradients shown are for sedimentation times of 15, 30, 45 and 60 minutes. More data points than are shown were used for the analysis.

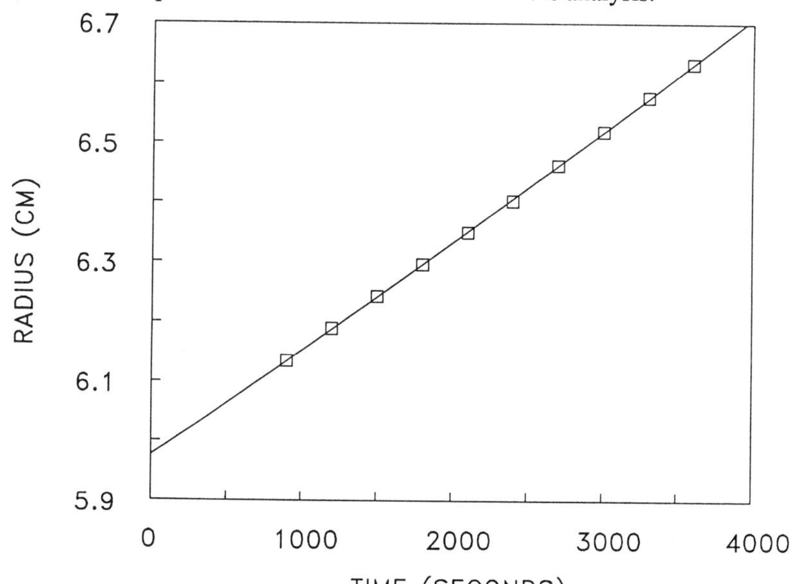

Figure 2. Fit of eq. (11a) to the values of $\bar{r}$, the second moment radial positions of the sedimenting boundaries, as a function of time for the sedimentation velocity experiment described in Fig. 1. The sedimentation times are from 15 to 60 minutes with increments of 5 minutes.

adjusted with dialysate so that their absorbance at 280 nm was between 0.3 and 0.6.

These protein solutions, either singly or in mixture, were examined by sedimentation velocity at 60,000 rev/min and 20°C, using an aluminium-filled epon double-sector cell in a Beckman Model E analytical ultracentrifuge. A MetraByte (Keithley MetraByte/Asyst/DAC, Taunton, MA 02780) DAS-8 analogue to digital board in a 10 MHz AST Premium/286 computer was used as a data acquisition system. Data were acquired in the rapid scan mode, and 900 data points, each of which was the average of 100 data samplings, were acquired in approximately 18 seconds, giving an actual data density of 425 data points per centimetre in the ultracentrifuge cell. Scans for analysis were taken only after a fully depleted region below the meniscus was attained, and data acquisition was terminated while there was still a significant plateau region remaining. Six to ten scans were taken, with the time interval between scans appropriate to the rate of sedimentation.

The data were initially edited in the acquisition program, and the edited data were then analysed by mathematical modelling using non-linear least-squares curve-fitting by means of the PC version of MLAB operating on the acquisition computer (Civilized Software, Bethesda, MD 20815). This is an enhanced version of the original MLAB that was designed to run on DEC-10 and DEC-20 mainframe computers[11].

Results

Fig. 1 shows the fit to the data for the concentration distributions of ovalbumin as a function of time obtained in a sedimentation velocity experiment using eq. (3) for a single component as a model. The quality of the fits is quite satisfactory with a mean value for the RMS error of approximately 0.004 absorbance units. The values of $\bar{r}$ as a function of time obtained from these fits are shown in Fig. 2 where eq (11a) was used as a model to obtain the sedimentation coefficient; a value of $(3.262 \pm 0.003)\times10^{-13}$ seconds was obtained. Fig. 3 shows the values of the products of n and $\bar{r}$ as a function of time obtained when fitting the concentration distributions and fit using eq. (21) to obtain the diffusion coefficient; a value of $(7.03 \pm 0.14)\times10^{-7}$ cm^2 sec^{-1} was obtained.

Fig. 4 shows the distributions obtained with a mixture of rabbit muscle aldolase and bovine serum albumin. These distributions were fit with eq. (3) for two components as a model. Eq. (12) was used as a model for the generation of the individual and the composite curves shown in Fig. 5. The fitting parameters unique to each component were used in a single-component model to generate the individual curves; these same values were then used in a two-component model to generate the composite curves.

The fitting function described here appears to be suitable for determining the second moment radial positions of a sedimenting boundary for a single monodisperse solute and also appears suitable for estimating the diffusion coefficient under these circumstances. While the function appears to resolve two components well, the sedimentation coefficients obtained may have significant error, particularly if their

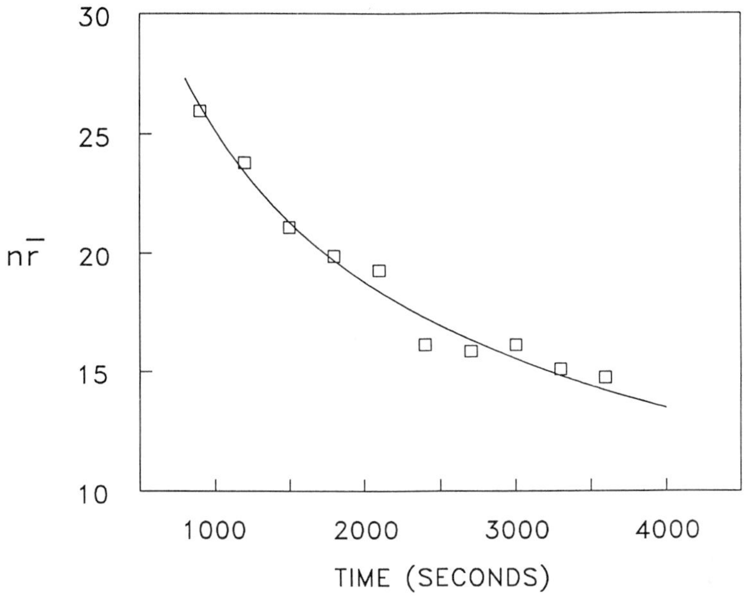

Figure 3. Fit of eq. (21) to the values of the product of n and $\bar{r}$ as a function of time for the sedimentation velocity experiment described in Fig. 1. The value of s obtained from the fit shown in Fig. 2 was used in eq. (21).

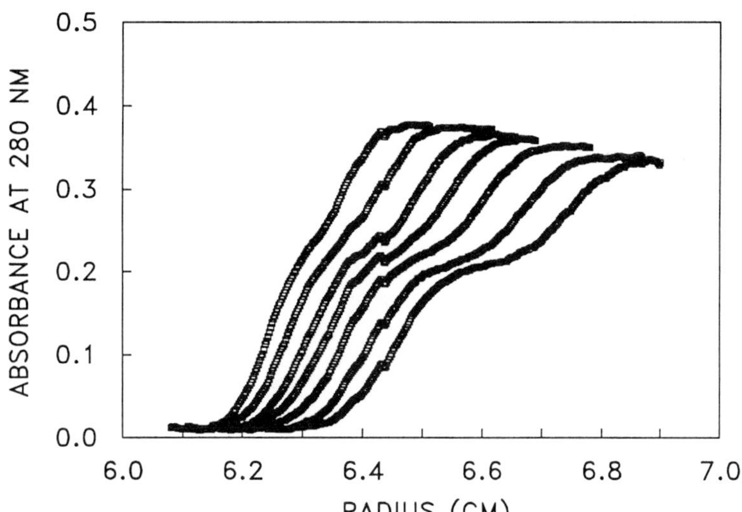

Figure 4. Concentration distributions obtained from a sedimentation velocity experiment using a mixture of rabbit muscle aldolase and bovine serum albumin in PBS at 20°C and 60,000 rev/min. Eq. (3) was used as a two-component model for fitting the data; the fitting lines are not shown. The concentration distributions shown are for sedimentation times of 30, 35, 40, 45, 50, 56 and 62 minutes.

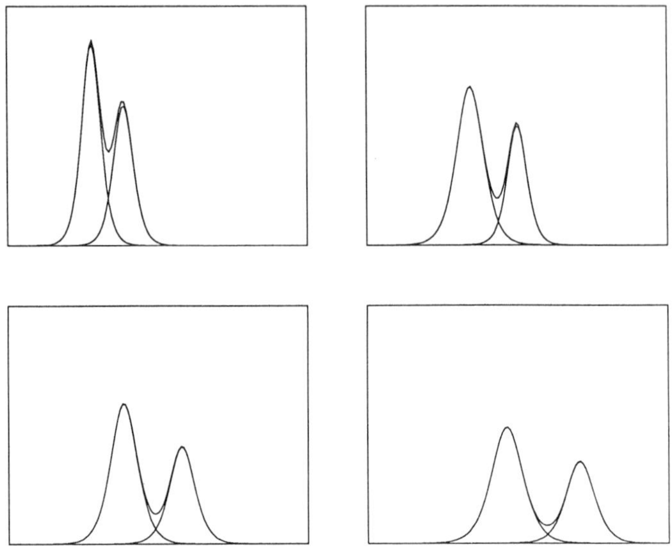

Figure 5. Plots of dc/dr as a function of radius generated by eq. (12) using the fitting parameters obtained when fitting the concentration distributions shown in Fig. 4. The plots shown are for sedimentation times of 30, 40, 50 and 62 minutes.

values do not differ much, and the diffusion coefficients exhibit estimated errors to an extent that these values must be considered unreliable. The function probably will not fit well for the hypersharp boundaries obtained if the sedimenting macromolecule exhibits significant concentration dependence since the fits of steeper gradients exhibit a larger RMS error than the fits of shallower gradients. The radial position of half the plateau concentration should be at least as satisfactory and could be more rapidly obtained. The function would be completely unsuitable for the analysis of the boundaries obtained for heterogeneous solutes or for systems of interacting macromolecules. The proper analysis of such boundaries would require the use of eq. (2) with numerical integration.

REFERENCES

1. D.L. Sackett, R.E. Lippoldt, C. Gibson and M.S. Lewis, *Anal. Biochem.*, 1989, *180*, 319.
2. A.P. Minton and M.S. Lewis, *unpublished.*
3. M. S. Lewis, *G W Instruments Application Notes*, No. 13, 1990.
4. A.P. Minton, *unpublished.*
5. M.S. Lewis, *unpublished.*
6. L.A. Holladay, *Biophys. Chem.*, 1979, *10*, 187.
7. L.A. Holladay, *personal communication.*
8. A.K. Attri, M.S. Lewis and E.D. Korn, *J. Biol. Chem.*, 1991, *266*, 6815.
9. N. Muramatsu and A.P. Minton, *Anal. Biochem.*, 1988, *168*, 345.
10. H. Fujita, 'Mathematical Theory of Sedimentation Analysis', Academic Press, New York and London, 1962, Chapter 2, p 79.
11. G.D. Knott, *Comput. Programs Biomed.*, 1979, *10*, 271.

10

Analysis of Polymer Dispersions with an Eight-Cell-AUC-Mutliplexer: High Resolution Particle Size Distribution and Density Gradient Techniques

By W. Mächtle

POLYMER RESEARCH LABORATORY, BASF AKTIENGESELLSCHAFT, D-6700 LUDWIGSHAFEN, GERMANY

1. INTRODUCTION

A specially adapted analytical ultracentrifuge (AUC) - meaning an AUC which allows the measurement of eight samples simultaneously[1,2] and on which the rotor speed can be increased in a continuous exponential manner from zero to maximum speed[3,4] - is a powerful tool for characterising complex polymer dispersions: Fig. 1 shows a schematic outline of the various types of characterisation that can be performed on such an instrument.

The AUC so adapted provides a multi-purpose apparatus for measuring the *density* ρ and the *size* of macromolecules and microparticles, either in terms of molecular weight M or diameter D. The main characterising feature of such an ultracentrifuge is that the macromolecules and the microparticles can be *fractionated* in *two different* ways:

(i) In the so-called sedimentation- or S-run they can be fractionated according to their *size*, *i.e.* according to their molecular weight M or their diameter D.

(ii) In the so-called density gradient- or DG-run they can be fractionated according to their *density* ρ, in other words, according to the chemical nature of the components.

Fractionation means, that we cannot only just measure *average* values such as average $\overline{M}$, average $\overline{D}$ and average $\overline{\rho}$, but also complete *distributions*, such as molecular weight distributions (MWD), particle size distributions (PSD) and density distributions (DD).

The size range covered by the AUC comprises particle diameters D from 10 nm to 3000 nm and molecular weights M from ~300 g/mol to ~10^{14} g/mol (corresponding to a D~3000 nm particle). The density range is smaller: from ~0.8 g/cm^3 to ~2.0 g/cm^3.

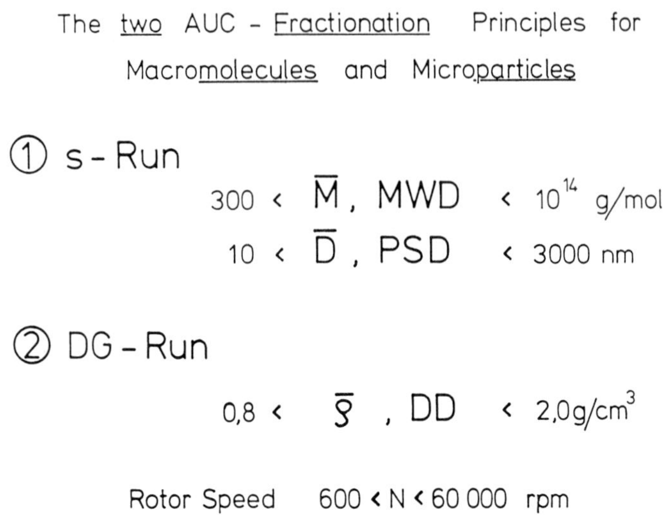

Figure 1. Schematic outline of the possibilities of the AUC to characterize macromolecules and microparticles.

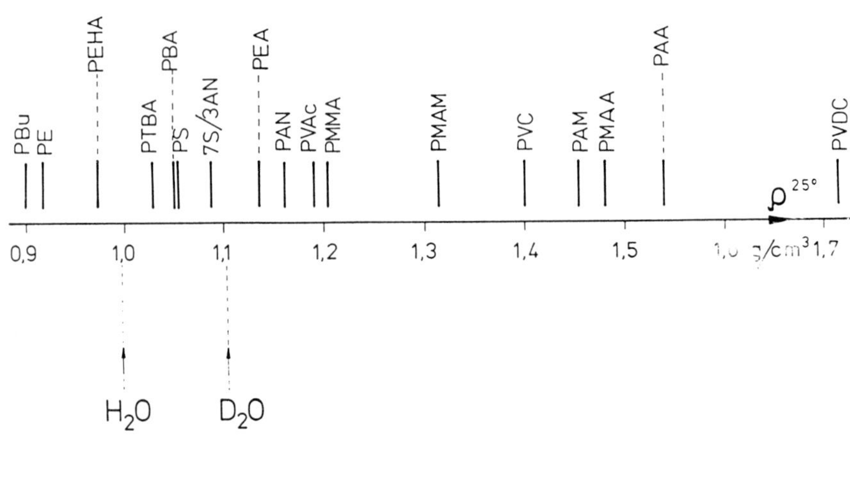

Figure 2. Particle densities of some well known homopolymer dispersions.

The particle size range we can analyse is so large as a result of the fact that we can adjust to the different particle sizes by variation of the rotor speed N from 600 rev/min to 60,000 rev/min. The sedimentation rate of a macromolecule or a microparticle is proportional to the *square* of the rotor speed N, that means, it will sediment 10,000 times faster, if we change N from 600 to 60,000 rev/min.

At first sight the "sedimentation" run seems to be the more important one for microparticles and also for polymer *dispersions* since it yields the particle size distribution. But in our opinion the "density gradient" run[5] is more important since it can reveal a great deal concerning the chemistry of a dispersion. To explain this let us consider Fig. 2. In Fig. 2 some frequently used *homo*polymer dispersions are listed according to their particle densities along a ρ-axis. Modern industrial dispersions are not only homopolymers or simple statistical co-polymers of two homopolymers, but they often are complex mixtures of chemically different dispersions, mixtures of particles and macro*molecules* dissolved in water, like polyacrylic acid and polyvinylpyrrolidone. Often these have a complex structure, for example a core/shell structure. So measurements of the density distribution of an unknown dispersion in a density gradient run yields a lot of information about their chemical structure. For instance: if a chemist is able to incorporate the heavy polyacrylic acid (PAA) into a light polybutylacrylate particle (PBA), the particle density will increase.

This Chapter, which could be retitled, "Characterization of Polymer Dispersions by means of the Analytical Ultracentrifuge" is divided into four remaining sections. In the first, we shall explain the principles of the AUC, the basic experiments and our eight-cell laser multiplexer. The next section will deal with the measurements of particle size distributions (PSD) of chemically uniform dispersions by means of the technique of continuously increasing rotor speed from 0 to 40,000 rev/min. In the next section we shall consider the estimation of chemical heterogeneity, especially of core/shell dispersions. In the last section, systems will be looked at which contain both dispersed *particles* and dissolved macro*molecules*. In this contribution we lay emphasis on examples of applications and not on our instrumentation (details of the latter can be found in refs. 1, 3-8).

2. PRINCIPLES OF THE AUC, EIGHT-CELL LASER MULTIPLEXER AND BASIC EXPERIMENTS

Fig. 3 (left) shows an outline of a standard AUC, a rotor hanging on an axle in a high vacuum chamber and the *analytical optical* light path: a mercury high pressure lamp, vacuum-tight lenses, mirrors and a photographic plate all set up for so-called schlieren optics: we will now consider this in detail.

In Fig. 3 the rotor is seen to bear two cells, a *measuring* cell with quartz windows on top and bottom and a *counterbalance* cell with two radius reference holes. Since the rotor turns very fast, one gets a *superimposed* picture of *both* cells on a photographic plate. There are three such schlieren photographs, taken after a sedimentation time respectively of 50, 140 and 260 minutes. (Later we shall discuss in detail this sedimenting schlieren peak of polystyrene macromolecules dissolved in tetrahydrofuran.)

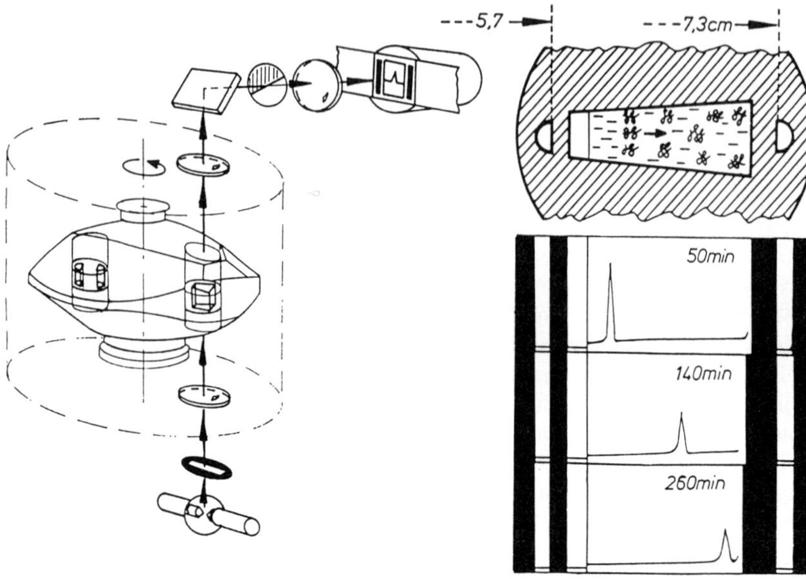

Figure 3. Schlieren optics of an AUC and explanation how schlieren pictures (peaks) come about.

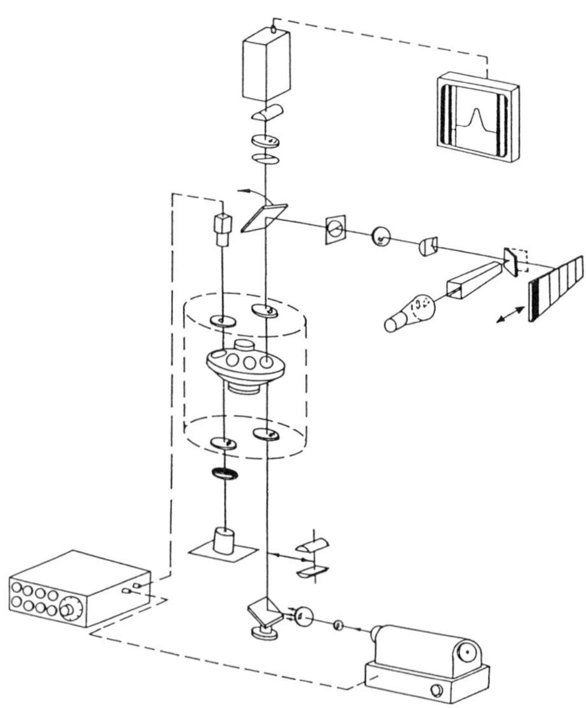

Figure 4. Outline of an eight-cell AUC laser multiplexer for schlieren and interference optics with photographic plate and digital TV camera.

Above these schlieren records in Fig. 3 (right), one sees on the same scale a superimposed top view of the measuring cell and the two radius references holes of the counterbalance cell, which gives the evaluation of the exact radial position of the schlieren peaks. The left line has the radial position of 5.70 cm from the rotor axle and the right line a position of 7.30 cm. Owing to the fact that we can measure the peak radius position as a function of time we are able to estimate the sedimentation coefficient, s, of the peak.

In order to make clear what causes such schlieren peaks let us consider the solution compartment of the measuring cell (Fig. 3, right). On the left of the meniscus one sees the air column, on the right the liquid column. After 50 min the schlieren peak of the macromolecules is far away from the meniscus and completely within the liquid column. On the left of the peak there is the pure solvent of a low refractive index, on the right the solution of a higher refractive index. This difference in the refractive index around the sedimentation front causes the schlieren peak. The illuminated rectilinear slit above the light source is projected through the cell onto the photographic plate. If there is only pure solvent in the cell (*i.e.* there is no difference in the radial refractive index), the image of the slit will be rectilinear again. But if the cell contains a solution showing differences in the radial refractive index, the image of the slit will be distorted to give a "schlieren" peak.

Fig. 3 covers really the outline of the well known Spinco-Beckman Model E AUC. We have actually enhanced the capabilities of this instrument considerably[1,3-8]. Fig. 4 shows our eight-cell laser multiplexer which represents an important improvement, especially for the interesting density gradient technique[5].

We firstly replaced the old *continuously* burning lamp by a *controllable* stroboscopic lamp, *i.e.* by a *modulable* laser. Then we replaced the old *two*-hole-rotor by an *eight*-hole-rotor. Now we can measure eight samples, that means eight macromolecular solutions or eight polymer dispersions *simultaneously*. We can also change from the photographic plate to a digital TV camera or to a photomultiplier. Details of such a construction are explained in refs. 1 and 8.

At this stage we will now stop considering the description of the principles of the ultracentrifuge because it is not necessary for the following to know how a computer controls this complicated system, which has to be gated in a microsecond time range if the rotor is spinning at a speed of 60,000 rev/min. Instead we will describe by means of Fig. 5 the *two* basic AUC experiments and to illuminate this we will consider a polystyrene dispersion as our "example" system.

The two basic experiments are a sedimentation- or S-run, in which, as explained above, we fractionate according to the particle *size*, and a density gradient- or DG-run, in which we fractionate according to the particle *density*. Since we use *uniform* polystyrene particles in this example *no* fractionation can take place, but by means of this example it is easier to explain the principles of the two runs. (Later we shall use non-uniform particles and then fractionation will take place.)

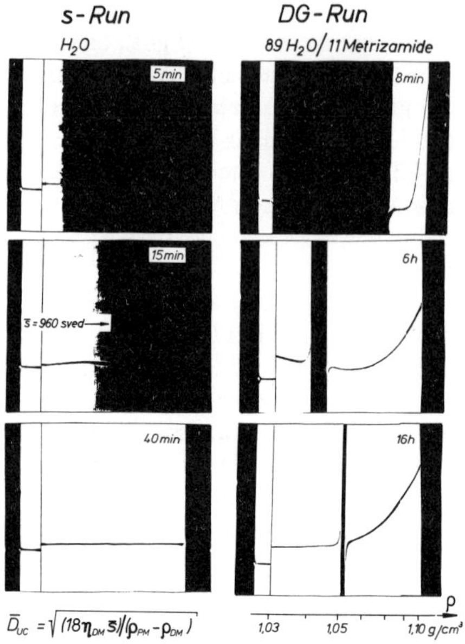

Figure 5. Schlieren photographs of a sedimentation run (left) and a density gradient run (right) of polystyrene dispersion particles with D = 160 nm.

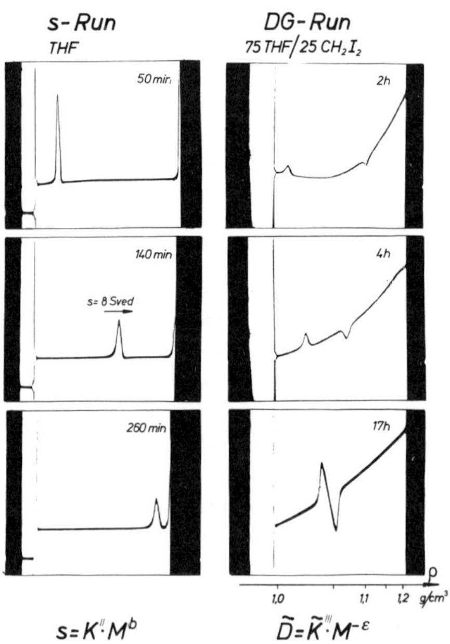

Figure 6. Schlieren photographs of a sedimentation run (left) and a density gradient run (right) of dissolved monodisperse polystyrene macromolecules with M = 1,800,000 g/mol.

For the sedimentation run we use a water-diluted dispersion. The concentration of the polystyrene particles is 1 mg/ml. The rotor speed is low, *i.e.*, 5,000 rev/min. After a sedimentation time of 5, 15 and 40 minutes we take photographs of the sedimenting turbidity boundary of the particles and estimate from these photographs the sedimentation coefficient, s, in Svedberg units (S). The s values of most of these "dispersion" particles are found to be between 500 S and 10,000 S. In contrast to this result the sedimentation velocities of dissolved macro*molecules* are 1,000 times smaller and lie between 0.5 S and 10 S.

For microparticles sedimenting in an AUC, Stokes' Law (see bottom of Fig. 5) is valid and consequently we can calculate the particle *diameter* D_{uc} from the measured s value, the known viscosity η_{DM} and density ρ_{DM} of the dispersion medium and the particle density ρ_{PM}. The result for these polystyrene particles, with s = 960 S, is 160 nm, which is in good agreement with estimates using electron microscopy.

If we want to change from the sedimentation run to the density gradient run[5], shown on the three photographs on the right of Fig. 5, we have to add to the water merely 11 weight percent of metrizamide (this is a water-soluble iodinated sugar of very high density: 2.1 g/cm^3). Then we have to wait 16 hours (*i.e.* overnight) until the equilibrium state is reached. During this time these heavy sugar molecules will be enriched near the cell bottom and so a radial density *gradient* from 1.03 g/cm^3 up to 1.10 g/cm^3 will be built up. At the same time the polystyrene particles sediment and float simultaneously to the position where the gradient density and the particle density are identical. There they gather in a very narrow turbidity band. The position of the radial band yields the particle density within a limit of error of ±0.1% (for example ρ_{PM} = 1.052 g/cm^3 for the polystyrene particles in Fig. 5). It is worth stressing that this position is *independent* of the particle diameter.

Each of the polystyrene particles consists of about 10,000 single polystyrene macromolecules. So, if we put this *aqueous* polystyrene dispersion into an *organic* solvent, such as tetrahydrofuran (THF), which is also water miscible, the particles will be dissolved and the turbidity will disappear (Fig. 6).

The polystyrene *macromolecules* have a molecular weight of 1,800,000 and a coil diameter of 160 nm. Again let us consider both a sedimentation run and a density gradient run. For this purpose we have to choose a rotor speed of 60,000 rev/min, which is typical for macromolecules. The three schlieren photographs of the sedimentation run, on the left, already discussed in Fig. 3, yield a sedimentation coefficient of s = 8 S, a value which can provide the molecular weight M of the dissolved polystyrene molecules *via* the Mark-Houwink-Kuhn-Sakurada (MHKS) relation $s = K'' \times M^b$ (see bottom of Fig. 6). Both constants, K" and b, are known; b is about 0.5.

As before, we can perform a density gradient run in this example by adding 25 weight percent of the heavy solvent di-iodomethane of a density 3.17 g/cm^3 to the light tetrahydrofuran. Again we have to wait for 17 hours until an equilibrium density gradient from 1.0 g/cm^3 to 1.2 g/cm^3 will be produced; again the polystyrene molecules sediment and float simultaneously to the position of equal density, where they stop. The width of this double schlieren peak yields the diffusion coefficient D† which, like the s value, depends on the molecular weight *via* the MHKS relation $D = K''' \times M^{-\varepsilon}$. Again K''' and ε are constants. Like the constant b, the constant ε is about 0.5. This means that the higher the molecular weight, the narrower is the double schlieren peak: this is why particles show an extremely narrow band and the results represented in Fig. 7, from an eight-cell run (measured simultaneously over one night), illustrate this.

One sees the density gradient schlieren photographs of six polystyrene samples with different molecular weights (indicated in Fig. 7). Cell 8 contains a mixture of polystyrenes M~37,000, 1,800,000 and 150,000,000 g/mol. Each single component is clearly seen in cell 8. The polystyrene M~150,000,000 component is simply *intra*particulate *cross-linked* polystyrene latex particles with a diameter of ~100 nm.

It is interesting to note, that in the case of polystyrene 150,000,000 the double schlieren peak is degenerated into an extremely narrow *turbidity* band, because the diffusion coefficient is very small. We call this a *microgel* dispersion.

Of course, we do not use density gradients to measure molecular weights or diffusion coefficients. But we use them to measure chemical heterogeneities as demonstrated in Fig. 8.

There exist many density gradient solvent systems. A common one for polystyrene, polybutadiene and co-polymers of both is a mixture of hexane and chloroform. In that system polystyrene of a molecular weight of 670,000 has an isodensity position near the cell bottom, polybutadiene of a molecular weight of 400,000 has an isodensity position near the meniscus. A co-polymer of 50 styrene and 50 butadiene has a position in the middle of the cell. If these three specimens are mixed in a 1:1:1 mixture, one will get a clear separation, that means a fractionation according to the chemical nature of the three components. With the aid of the particular density (*i.e.* composition) axis at the bottom of Fig. 8, one can identify each component.

The example just considered dealt with dissolved macro*molecules*. If we had considered cross-linked micro*particles* of the same chemical nature, we would have got the same result with the exception that, instead of double schlieren peaks, we would have obtained very narrow bands at the *same* positions.

Having treated the principles of the AUC and the two basic experiments, *i.e.*, the sedimentation run and the density gradient run, we can now consider the measurement of particle size distributions.

---

† Not to be confused with the diameter used elsewhere in this Chapter.

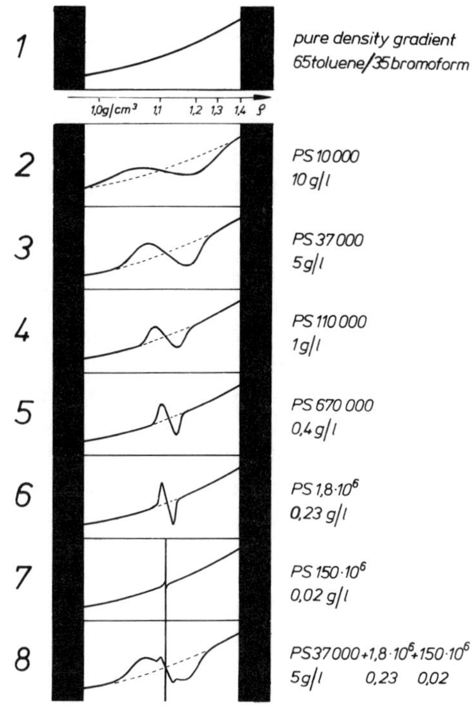

Figure 7. Eight-cells density gradient run of seven different polystyrene samples, measured in one night with an AUC schlieren optics multiplexer.

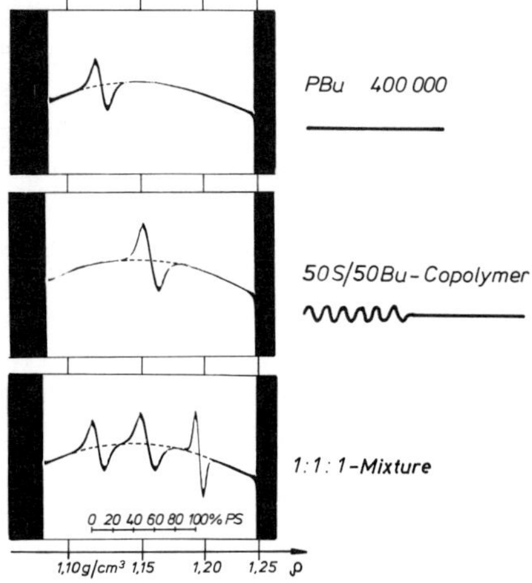

Figure 8. AUC density gradient schlieren photographs of PS, PBu and an S/Bu-copolymer.

3. HIGH RESOLUTION PARTICLE SIZE DISTRIBUTION (PSD) WITHIN AN AUC

At the beginning of this section we would like to stress that only chemically *homogeneous* dispersions will be considered here: that is to say dispersions where *all* the particles have the *same constant* density. Fig. 9 shows a simplified outline of our particular PSD technique[3,4].

In Fig. 9 one sees represented sedimenting particles in an ultracentrifuge cell. In Fig. 5 (left) above, we observed the *same* experiment by means of *schlieren* optics and we saw the sedimenting turbidity front of the dispersion. Now the measurement of the particle size distribution can be observed by means of the so-called *turbidity* optics. In this case the lower cell window is covered by a diaphragm with a small rectilinear slit of 0.2 mm vertical to the rotor radius direction. Parallel monochromatic light passes through the slit and reaches a photomultiplier. When passing through the turbid dispersion the light intensity will be reduced by light scattering from the dispersed particles. The reduced light intensity I is measured on a recorder as a function of the running time, t.

Let us first discuss *mono*disperse particles (dotted lines of Fig. 9). *All* of them will sediment with exactly the *same* velocity. This turbidity front starts from the meniscus at radius position r_m. In the first half of the experiment only the lower intensity (I_0) will be measured. However at the particular moment in time when the sharp turbidity front is passing the slit at radius position r_s, the light intensity will jump to the high value of the pure dispersion medium water I_{DM} and will be constant in the second half of the experiment. From the time when the jump happens we can calculate the diameter of the monodisperse particles by the well-known Stokes' Law equation

$$D = \sqrt{\frac{18\eta DMs}{\rho_{PM} - \rho_{DM}}} = \sqrt{\frac{18\eta DM \ln(r_s/r_m)}{(\rho_{PM} - \rho_{DM})\,\omega^2 t}} \tag{1}$$

Now let us discuss a dispersion with a broad particle size distribution (continuous lines of Fig. 9). In this case a *fractionation* according to the particle size occurs within the sedimenting turbidity front. The front will be smeared owing to the *big* particles moving faster and the small particles moving more slowly. On the recorder a broad *continuous* transition from I_0 to I_{DM} appears. We approximate this continuous curve by a multi-step curve, that means by a superimposition of many monodisperse particle fractions and calculate from this curve the integral mass distribution of the particle diameters (= PSD). From the different transition times t_i we get, by Stokes' Law, the different diameters D_i of the different fractions and from the different measured light intensity jumps ΔI_i we obtain, by the aid of Mie's light scattering theory[3,4,11], the mass percentages m_i of the different fractions. By conservation of mass criteria, m_i summed over i will yield 100 wt %.

Fig. 9 as given is, in fact, a *simplified* picture of our PSD apparatus. In reality we measure seven samples simultaneously with an eight-cell multiplexer and the technique of continuously increasing rotor speed $\omega = 2\pi N$ from 0 to 40,000 rev/min. We measure by a two-beam technique, because cell 8, which also triggers the whole

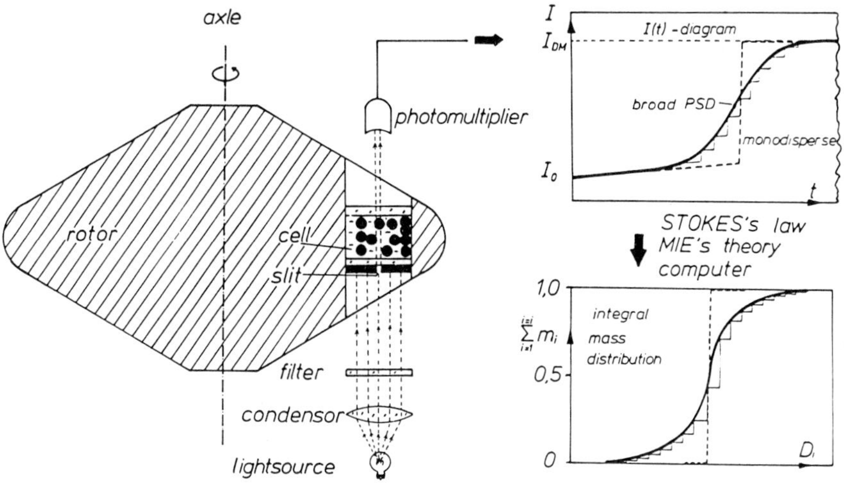

Figure 9. Experimental AUC setup to measure the PSD of a dispersion with a turbidity detector.

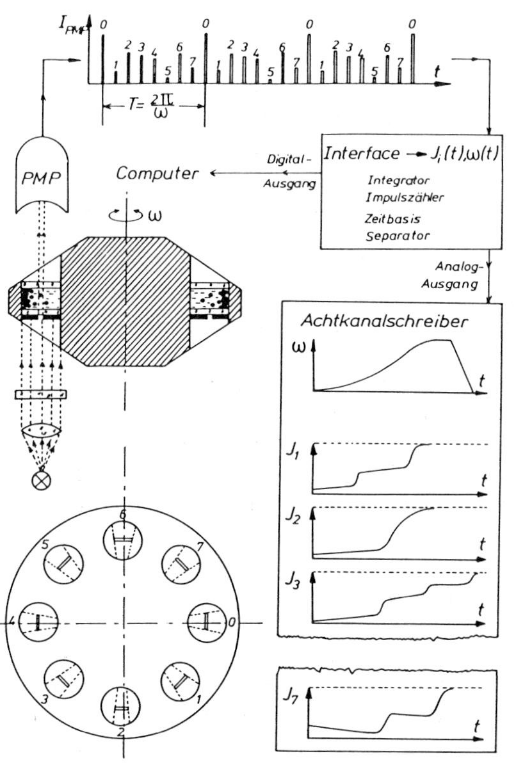

Figure 10. AUC setup to measure the PSD of seven dispersions simultaneously with a turbidity detector.

system, is filled with the pure dispersion medium (DM) water. Fig. 10 shows our eight-cell setup.

Cantow[9] first introduced an ultracentrifuge technique to determine the PSD of polymer latices which again is based on Stokes' Law as well as an earlier paper by Nichols, Bailey, and Kraemer[10] but in addition to this takes into account the Mie light scattering theory for spheres[11]. Thereafter Scholtan and Lange[12] developed an analogous ultracentrifuge technique featuring a rather simple experimental arrangement. This method was further developed by ourselves[3] taking the following steps, outlined in Fig. 10: (a) transition from single beam to two beam technique; (b) increase of the measuring capacity by a factor of seven by introducing an eight-cell rotor and a multiplexer which allows simultaneous measurement of seven samples and a reference cell; (c) transition from the common operation mode using a constant rotor speed N to a time dependent operation mode N(t) where during each run the rotor speed is increased from N = 0 to the maximum N = 40,000 rev/min within one hour always in the same manner. In this case we have to replace the term $\omega^2 t$ in Stokes' Law by the cumulative time integral

$$\int \omega^2 \, dt$$

which the computer calculates. The advantage of this new, nearly exponentially time dependent operation mode is that by the end of each run samples containing very small particles of about 20 nm can be measured together with samples containing very large particles of about 2000 nm. Thus time consuming test runs to find a proper rotor speed N are no longer needed. The PSD of nearly any kind of dispersion is measurable with this simple ultracentrifuge/light scattering device.

The ability to measure several I(t) curves simultaneously is very important for our coupling-PSD technique[4] to measure samples with an extremely broad PSD (Fig. 12) and for our H_2O/D_2O analysis[3] to measure the particle density ρ_{PM} (or the chemical heterogeneity, if present) with the same AUC device. Figs. 11-14 show some application examples of our PSD equipment.

Fig. 11 shows the PSD curves of ten narrow distributed polystyrene dispersions with different diameters, measured in two one-hour runs with our ultracentrifuge/light scattering device. The ten average diameters are shown in Fig. 11: $D_{50\%}$ = 67, 113, 166, 246, 318, 356, 486, 680, 840 and 1220 nm. This is a huge diameter range. These ten dispersions were polymerised in our laboratory.

Now we create a new dispersion with an extremely broad PSD by mixing together all these ten single dispersions in equal parts, 10/10/10/... and so on. Then we measure simultaneously the two I(t) curves at the standard concentration c_s = 0.35 mg/ml and at the increased concentration c_h = 3.5 mg/ml. Fig. 12 (top) shows that in the standard I(t) curve each single component (with the exception of the smallest 67 nm component) is clearly seen as a single step. The 67 nm particles are only perceptible as a tiny, not exactly resolvable step, at t = 4600 s. But in the increased concentration I(t) curve these 67 nm particles show a clear step.

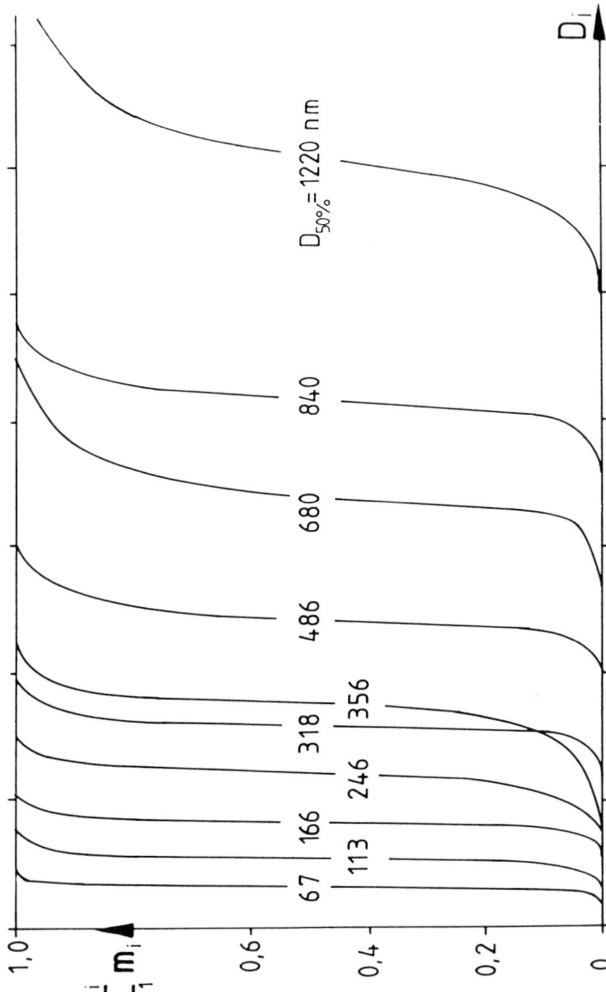

Figure 11. PSD of ten narrow distributed polystyrene dispersions measured in two runs with an eight-cell AUC setup.

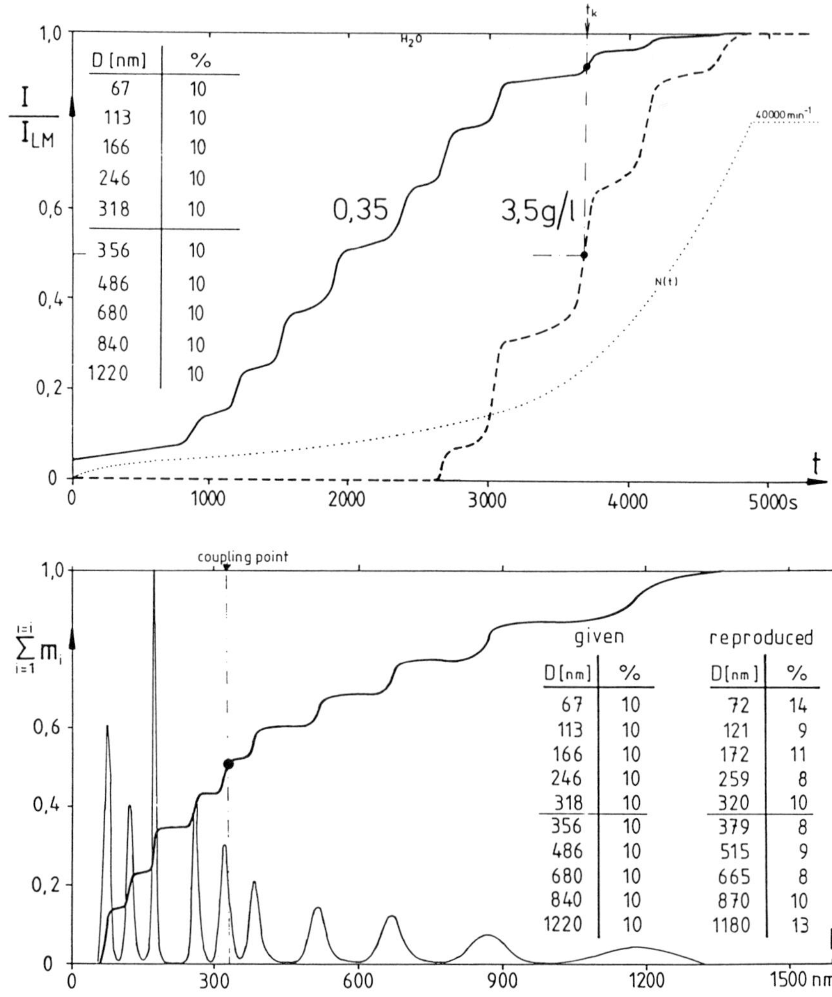

Figure 12. PSD analysis of a ten-component mixture of ten narrow polystyrene dispersions. Above: Simultaneously measured I(t) curves. Below: Resulting coupling-PSD in integral (thick line) and differential (thin line) form.

The resulting coupling-PSD is shown in Fig. 12 (bottom) in two different forms, in the usual integral form and in the differential form. The latter form especially demonstrates the excellent resolving power of our ultracentrifuge/light scattering device. All ten components of this extremely broadly distributed dispersion are resolved down to the base line. This is also valid for the two closely neighbouring components 318 and 356 nm, the diameters of which have a difference of only 12%.

The values in the first table of Fig. 12 (bottom) are given. The values in the second table are the ones reproduced from the ten-component mixture. To within ±5% we reproduce the diameters D and to within ±15% we reproduce the (small) mixing parts.

The next example application of our eight-cell PSD apparatus, Fig. 13, demonstrates, that it is possible to measure particle densities ρ_{PM} using only Stokes' Law. We call this *H_2O/D_2O analysis*[3]. The idea behind it is simply to measure the same particles in the two different dispersion media H_2O and D_2O having different densities ρ_{DM} = 0.99704 and 1.1044 g/cm^3 (at 25°C). The two resulting Stokes equations with the two unknown quantities, D_{PM} and ρ_{PM}, are easily soluble.

In Fig. 13 we choose as an "unknown" dispersion a narrow distributed aqueous polystyrene dispersion with D = 155 nm and the usual solid content of about 50 wt %. We always dilute such an unknown dispersion by roughly 1:1000 with H_2O, with D_2O and with a 1:1 H_2O/D_2O mixture.(The last one we do not really need. It is only for computer control purposes, to decide between sedimentation and flotation.) Then we measure the three corresponding I(t) curves in one AUC run *simultaneously*. Afterwards the computer creates horizontal intersecting lines across these three I(t) curves and for every line it estimates the three *coupled* running time t (or more precisely, the three travelling time integrals $\int\omega^2 dt$). These times only together with Stokes' law yield the particle density ρ *and* the particle diameter D, which are printed on the right hand side of each line in Fig. 13.

In simple terms: the ρ/D table constructed in this manner represents the particle *density* distribution (PDD) and the particle *size* distribution (PSD) of the unknown dispersion. In Fig. 13 we find ρ = *constant* = 1.054 g/cm^3: *i.e.* we are dealing with a chemically homogeneous *polystyrene* dispersion: in addition we can say that this dispersion is narrowly distributed with an average diameter of D = 155 nm. (Further H_2O/D_2O analysis examples, especially of chemically heterogeneous dispersions are considered in ref. 3.)

Fig. 14 shows another application example of our PSD apparatus. Particle size distribution measurements by means of an ultracentrifuge are very helpful to exemplify reaction kinetics of emulsion polymerisation. Fig. 14 in fact shows the variation of the particle size distribution for an acrylic ester co-polymer, obtained by different reaction conditions. By the usual batch (or one step) technique, *viz* with one reactor, we get nearly monodisperse particles with a diameter of 150 nm. As variation number 1 we introduce the pre-reactor between the storage bins and the (main) reactor: all the other reaction conditions remain constant. As a result we get a broadened particle size distribution again with an average diameter of 150 nm.

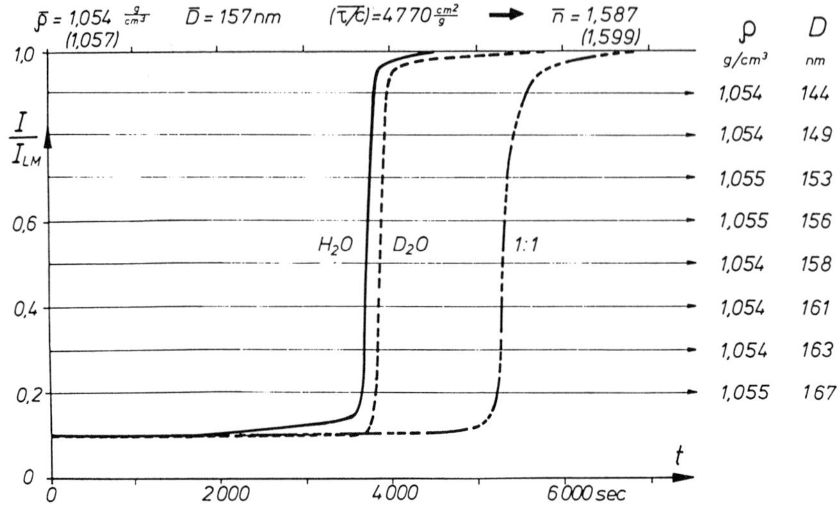

Figure 13. H_2O/D_2O analysis of a narrow distributed polystyrene dispersion with D = 155 nm.

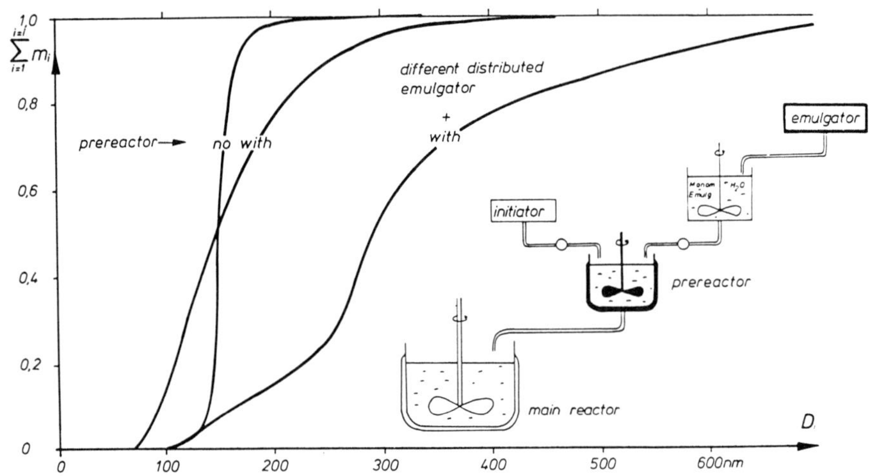

Figure 14. PSD variation by different reaction guidance of an acrylic ester emulsion polymerisation.

In variation number 2 we introduce the upper additional storage bin and put half of the emulsifier into it. With the other half of the emulsifier in the first storage bin, we start the reaction. When 50% of the reaction time has elapsed we open the tap of the additional emulsifier storage bin and the emulsifier will flow into the reacting system. As a final result we get an extremely broad, nearly bimodal, particle size distribution with very big particles up to 600 nm. Indeed, most of the details of the reaction kinetics which happen in this emulsion polymerisation may be explained with the help of the particle size distributions measured in this way.

Having now treated the estimation of high resolution PSD and the H_2O/D_2O analysis of chemically *homogeneous* dispersions we now look at chemically heterogeneous dispersions.

4. CHEMICAL HETEROGENEITY - CORE/SHELL DISPERSIONS

Very often the surfaces of the particles of a simple dispersion, for example of polybutylacrylate, will be modified by another polymer, for instance by grafting a styrene/acrylonitrile-copolymer (abbreviated to "SAN") in the form of a grafting shell. This is done by producing rubber-modified high impact materials. For example, soft polybutylacrylate (PBA) particles will be introduced into a hard continuous matrix of SAN: in this case, the grafting of the SAN to the polybutylacrylate particles causes a very good bonding between the rubber particles and the matrix.

Sometimes it is difficult to carry out such a surface grafting, (*i.e.* the production of core/shell-dispersions) because of the risk that no grafting reaction will take place and that a separate homopolymerisation of the shell material will occur instead.

In Figs. 15-22 we shall explain how to control such a complicated grafting reaction (which yields chemically non-uniform particles) by means of the analytical ultracentrifuge.

Let us consider such a grafting experiment. Fig. 15 gives the scheme for a polybutylacrylate (PBA) dispersion which was made from four different monodisperse dispersions with diameters D_{0i} of about 100, 200, 300 and 400 nm by mixing. The mixing ratios m_{0i} were 40 to 30 to 20 to 10 weight percent.

Onto the surfaces of the PBA particles, the monomers styrene and acrylonitrile (SAN), shown on the right, will be grafted. The subscript "s" means "shell" and "c" means "core". The planned *average* grafting degree "p" is the mass M_s of all the grafting monomers divided by M_s plus M_c, *i.e.* the mass of all the core particles. Not only the diameter D_{0i} of the particles will increase by grafting, but *also* the density of the core particle ρ_c, because the value for the density of the SAN-shell material ρ_s, amounting to 1.08 g/cm^3, is higher than that for the core polybutylacrylate ρ_c (amounting to 1.04 g/cm^3).

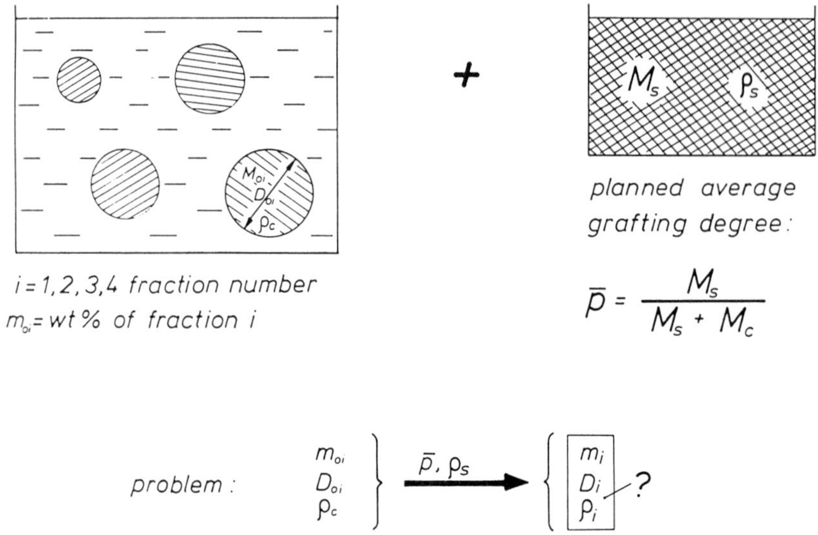

Figure 15. Scheme of a core/shell dispersion grafting reaction.

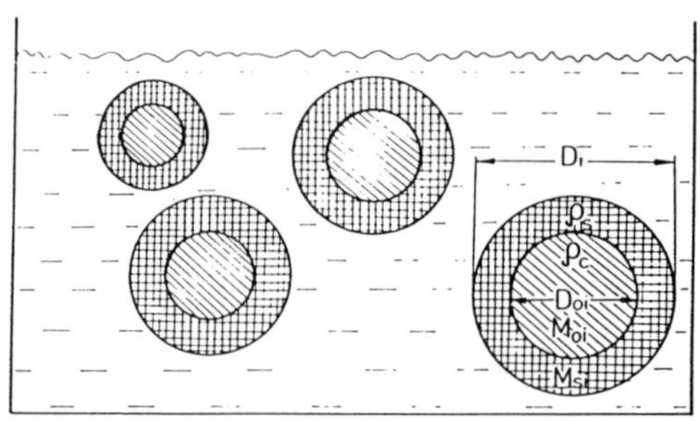

Kind of grafting distribution	mass of the grafting shell
① statistical	M_{si} = statistical
② homogeneous	
- number	M_{si} = const.
- diameter	" $\sim D_{oi}$
- surface	" $\sim D_{oi}^2$
- volume	" $\sim D_{oi}^3$

Figure 16. Theoretical models of a core/shell dispersion grafting reaction.

Consequently our problem is the following: What are, *after* the grafting reaction, the new mass percentages m_i, diameters D_i and particles densities ρ_i of the four starting fractions like? Two grafting models, shown in Fig. 16, are conceivable: a *statistical* and a *homogeneous* grafting degree distribution. In the statistical model the distribution of the grafting shell masses M_{si} over *all* cores is of merely statistical nature. In every homogeneous model all the cores of the same diameter have the same degree of grafting, that means the same grafting shell mass M_{si}. Four homogeneous models are in turn conceivable: a number-, a diameter-, a surface- and a volume-distribution of the grafting degree. In the first model M_{si} is constant for every core, in the second one M_{si} is proportional to the core diameter D_{0i}, in the third one it is proportional to the core surface (*i.e.* to the square of D_{0i}) and in the fourth one it is proportional to the core volume (*i.e.* to the cube of D_{0i}).

In addition to the theory, let us consider ultracentrifuge *experiments* in order to assess which of these models is the correct one. Fig. 17 shows the first experiment. To differentiate between statistical and homogeneous grafting, we pick one of the four ungrafted monodisperse polybutylacrylate dispersions, called PBA 2. Number 2 means that their diameter is about 200 nm (strictly 241 nm) which shows the measured particle size distribution above. In the density gradient below PBA 2 shows, as expected, a narrow turbidity band at the density position 1.038 g/cm^3.

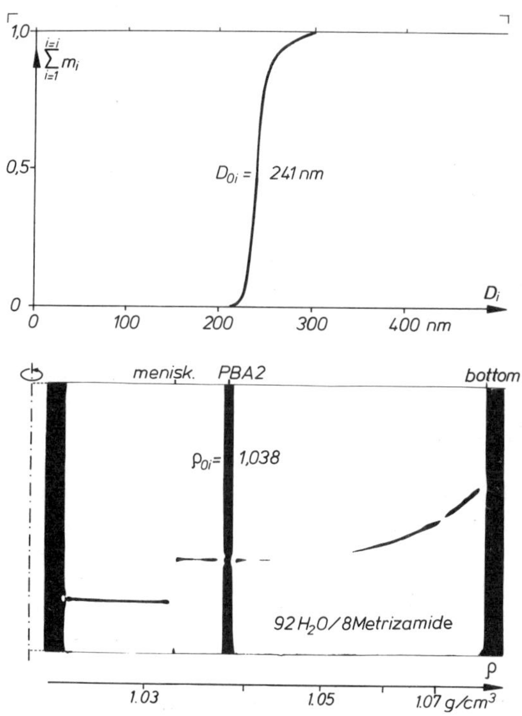

Figure 17. PSD and density gradient of a monodisperse PBA-dispersion.

Now we will graft PBA 2 with SAN. The planned average grafting degree p is 0.5, that means the ratio of the mass of core to the mass of the shell is 1:1. The result is shown in Fig. 18.

In the density gradient below the ungrafted and the grafted dispersions are measured *simultaneously*. The resulting narrow turbidity band of the grafted dispersion proves that the grafting of the SAN is of a *homogeneous* and not of a statistical nature. All grafted particles have exactly the same density (1.059 g/cm^3): this value is the theoretically expected one for a complete homogeneous grafting reaction.

The measured particle size distribution of the grafted dispersion above shows the same result, if it is compared with the particle size distribution of the ungrafted dispersion. Again the new distribution is monodisperse. The diameter increase from 241 to 300 nm is ~25% (= $2^{1/3}$-1), as theoretically expected for a completely homogeneous grafting reaction.

Fig. 19 shows that each of the four single monodisperse polybutylacrylate dispersions was grafted with SAN, in the same manner as PBA 2, and in each case the result was the same: a completely homogeneous grafting reaction. All starting diameters increase at 25% and all starting densities change from 1.039 to 1.059 g/cm^3 by grafting.

Now we do the most important experiment: we mix the four monodisperse starting dispersions, as mentioned before, in the mixing ratio of 40:30:20:10 wt% and afterwards we graft them in *common*. This means that we graft a starting dispersion with a *broad* particle size distribution. The result is shown in Fig. 20.

The dotted curve is the measured broad particle size distribution of the ungrafted starting dispersion, obtained by the 40:30:20:10 wt % mixture of the four monodisperse polybutylacrylate dispersions. The continuous curve is the new particle size distribution after the grafting reaction. Again this curve shows four monodisperse fractions. But the increase of the diameters, which is due to the grafting, is *not* equal and is *not* 25% for all fractions. It is much higher than 25% for the small 100 nm starting particles and much lower than 25% for the large 400 nm starting particles. Moreover the mass percentage of the small 100 nm starting particles has increased from 40% to 50% owing to the grafting. This means that smaller particles are grafted to a greater extent than larger particles.

Another proof is furnished by the density gradient. In this density gradient, the ungrafted and the grafted mixture were measured simultaneously. The four ungrafted fractions of 100, 200, 300 and 400 nm particles are gathered in *one* single narrow band at the expected density position (1.039 g/cm^3). In contrast to this result the four grafted fractions show *four different* bands at four different density positions ρ_1, ρ_2, ρ_3 and ρ_4 which we can attach to the grafted 100, 200, 300 and 400 nm particles.

An exact analysis of all these ultracentrifuge experiments leads to the following conclusion: The grafting of a starting dispersion with broad particle size distribution is *homogeneous* and the distribution of the grafting degree is proportional to the *surface* areas of the different fractions.

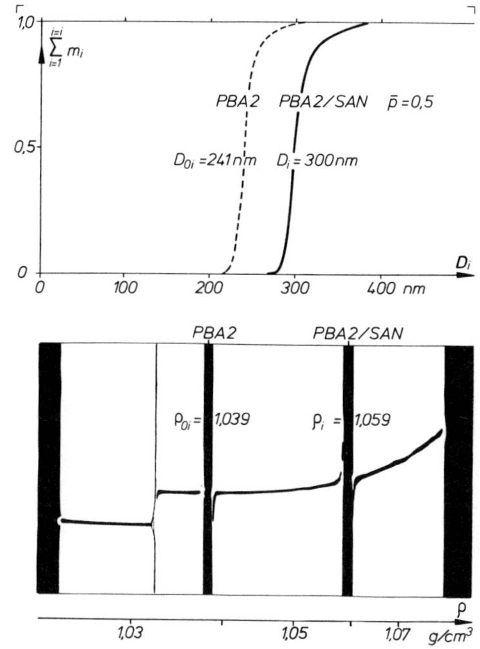

Figure 18. PSD and density gradient of an ungrafted and a SAN-grafted monodisperse PBA-dispersion.

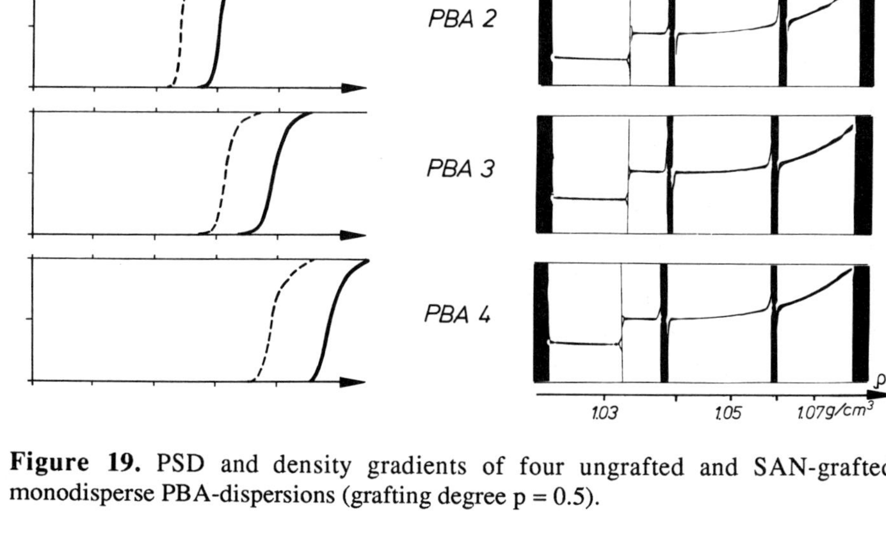

Figure 19. PSD and density gradients of four ungrafted and SAN-grafted monodisperse PBA-dispersions (grafting degree p = 0.5).

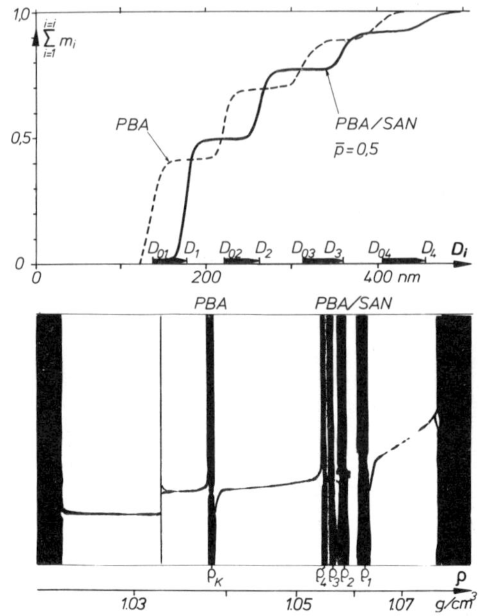

Figure 20. PSD and density gradient of a four-modal ungrafted and a SAN-grafted PBA-dispersion (40:30:20:10 wt % mixture).

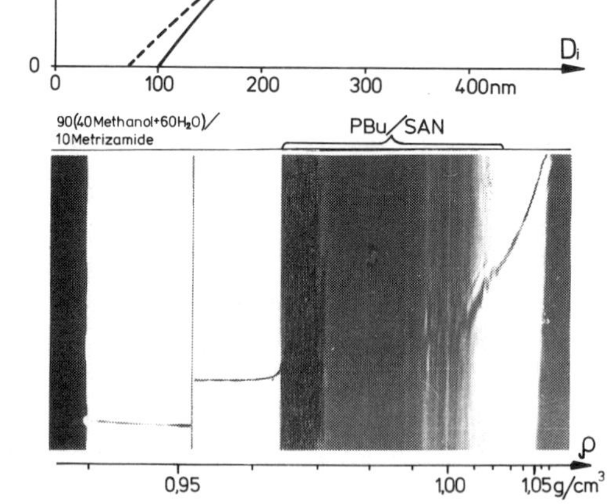

Figure 21. PSD and density gradient of an ungrafted and a SAN-grafted broad distributed PBu-dispersion.

If this conclusion is valid in general, a starting dispersion with a *continuously* broad particle size distribution must have a continuous distribution of the grafting degree, that means a *continuously* broad density gradient band after grafting as well. Let us prove this statement by considering the situation represented in Fig. 21.

A polybutadiene (PBu) dispersion with a continuous and very broad particle size distribution (dotted line) was grafted again with SAN. The planned average degree of grafting was 0.5. The starting diameters covered a continuous range from 70 to 400 nm. Again the resulting particle size distribution is continuously broad (continuous line). The percentage increase of the smaller particles is higher than that of the bigger ones.

The density gradient of the grafted dispersion alone shows a very broad continuous turbidity band. This proves that also in the case of grafting SAN onto polybutadiene particles the grafting is homogeneous and proportional to the surface areas of the particles. Ungrafted polybutadiene particles are not within this density gradient, because their very low uniform density of 0.901 g/cm^3 lies beyond the range of this density gradient.

In Figs. 15-21 we discussed simple core/shell dispersions. In Fig. 22 let us consider now a more complicated situation: this is a core/shell 1/shell 2-dispersion. Onto a basic dispersion, an ethylhexylacrylate(EHA)/styrene(S) co-polymer, with the low uniform particle density of 1.00 g/cm^3 we graft, in a first step, a 30 wt % shell of heavy methylmethacrylate (MMA) with the density of 1.20 g/cm^3. The resulting middle dispersion shows a narrow density band at 1.06 g/cm^3. That means that the grafting degree is uniform as expected, because the starting particle size distribution (Fig. 22, above) was nearly monodisperse.

In a second step we graft a second 30 wt % shell of styrene onto the particles of the middle dispersion. Since the densities of the particles of the middle dispersion and of the grafted polystyrene are the same, the position of the particles of the final dispersion in the density gradient is again 1.06 g/cm^3. But from the measured particle size distributions one can see that, owing to the second grafting step, the particle diameter has increased again.

Having now treated the characterization of chemically heterogeneous core/shell dispersions (*i.e.* systems with dispersed microparticles only), we will now go on to consider systems which consist at the same time of dispersed *particles* and dissolved macro*molecules*.

5. SYSTEMS WITH DISPERSED MICROPARTICLES AND DISSOLVED MACROMOLECULES

As already mentioned in the introduction, modern dispersion systems often consist of *particles* and of water-dissolved macro*molecules* at the same time. On the other hand, each polymer particle of an *aqueous* dispersion can be dissolved in a suitable *organic* solvent, if it is not cross-linked. If it is cross-linked, we get a so-called secondary dispersion of the swollen particles in the organic solvent (we call such cross-linked

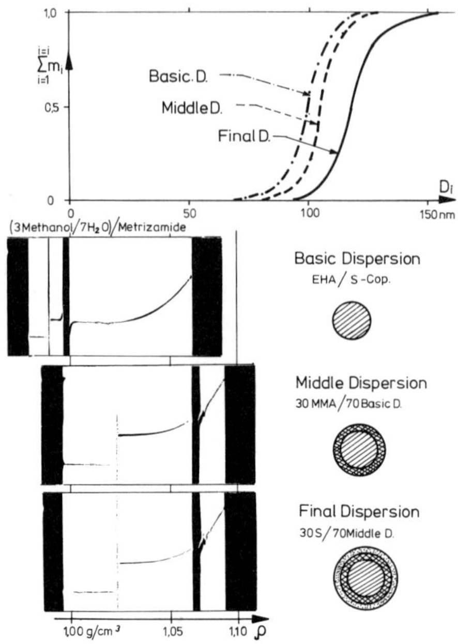

Figure 22. PSD and density gradients after the three polymerisation steps of a core/shell 1/shell 2 grafting dispersion.

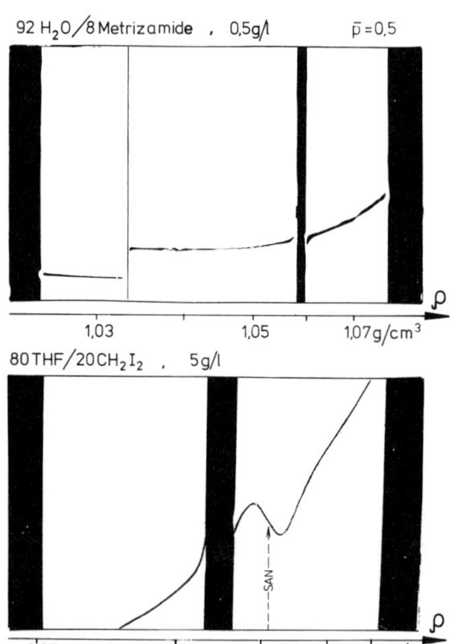

Figure 23. Aqueous and organic density gradients of the SAN-grafted monodisperse PBA 2-dispersion (50 PBA/50 SAN).

dispersion particles, already discussed in Fig. 7, PS 150,000,000, microgels). The analytical ultracentrifuge is the *sole instrument* that allows us to investigate such dispersions in which there exist microparticles and dissolved macromolecules at the same time. In the following we will look at an organic and an aqueous system. First, Fig. 23 shows the organic system.

This monodisperse PBA 2 dispersion, grafted with SAN, and the water/metrizamide density gradient above have already been considered in Fig. 17. We stated there that the SAN grafting is complete and homogeneous. But the question is if all these SAN macromolecules are really bound covalently onto the polybutylacrylate core surfaces or if all of them or parts of them are merely adsorbed or precipitated onto the core surfaces, because SAN macromolecules are not soluble in water.

In order to find an answer to this question we put these grafted particles into the *organic* tetrahydrofuran/di-iodomethane density gradient below, where uncrosslinked polybutylacrylate and SAN are soluble. Since prior to the SAN grafting the starting polybutylacrylate core particles were completely cross-linked we do not expect the double schlieren peak of single *dissolved* polybutylacrylate macro*molecules* at their known density position 1.015 g/cm^3 in this organic density gradient. Instead, as expected, we see the turbidity band of cross-linked *particles* at a density position from which it can be concluded that the particles cannot be of pure cross-linked polybutylacrylate, but of polybutylacrylate covalently *grafted* with 40 wt % - 50 wt % of SAN. On the other hand we see a small separate double schlieren peak of pure dissolved SAN macromolecules: they diffused away from the particle surfaces because they were not bound covalently onto these surfaces.

For the measurement of the amount and of the molecular weight of the ungrafted SAN molecules we put the original aqueous dispersion into pure tetrahydrofuran and carry out a sedimentation run, which is shown in Fig. 24.

During the first minutes of the run the very fast turbidity front of the cross-linked microgel particles sediments completely to the cell bottom. Later on, the slow sedimenting schlieren peak of the ungrafted SAN molecules becomes evident. The sedimentation coefficient of this peak, s_0=5.2 S, yields a molecular weight of 170,000 g/mol of the ungrafted SAN molecules. The schlieren peak area shows the concentration of the dissolved SAN molecules to be ~8 mg/ml. Since we know the overall concentration of the whole dispersion, namely 80 mg/ml, it means that ~10% of the planned 50% SAN are not covalently grafted onto the surfaces of the polybutylacrylate particles. If the real grafting degree were much lower, these rubber particles would yield a bad rubber-modified high impact material.

Let us now leave discussing this organic system and turn over to a very important *aqueous* dispersion system in which microparticles and dissolved macromolecules exist at the same time, *i.e.*, acrylic particles and dissolved polyacrylic acid (PAA) macromolecules.

Fig. 25 shows aqueous water/metrizamide density gradients of three acrylic *homopolymer* dispersions, *i.e.*, polyethylacrylate (PEA) particles with a high density, polybutylacrylate (PBA) particles with a medium density and polyethylhexylacrylate (PEHA) particles with a low density.

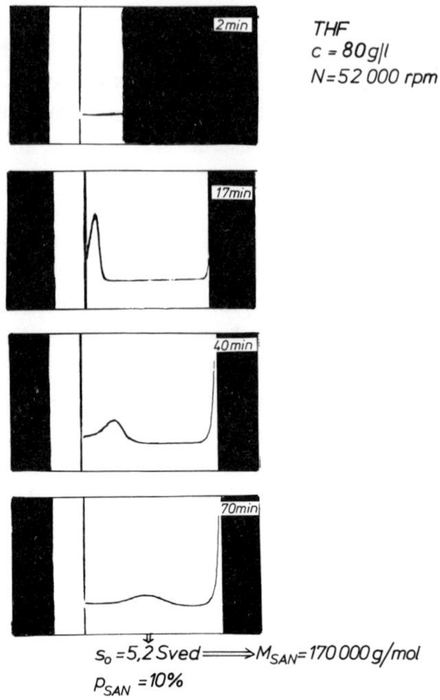

Figure 24. Sedimentation run of the SAN-grafted monodisperse PBA 2-dispersion (50 PBA/50 SAN) in the organic solvent THF.

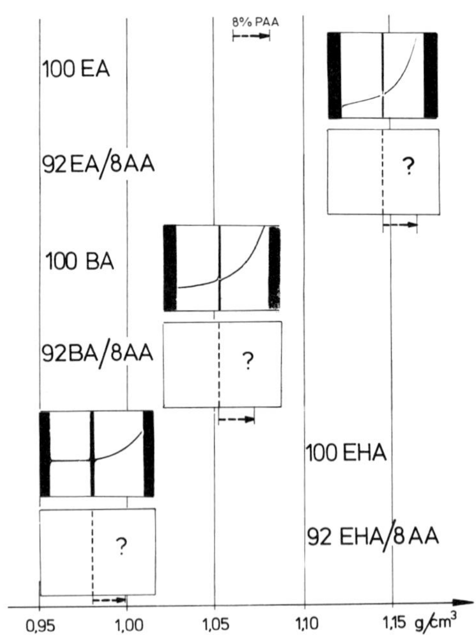

Figure 25. H_2O/metrizamide density gradients of three acrylic homopolyme' dispersions.

Fig. 25 demonstrates the usefulness of density gradient measurements for the differentiation of unknown dispersions according to their chemical structure, as the resolving power of AUC density gradients according to the particle density is very high.

Now we shall introduce 8 wt % of the hydrophilic acrylic acid (PAA) into each of these homopolymers by co-polymerisation.

From Fig. 2 we know that 8 wt % of the heavy acrylic acid, with a density of 1.54 g/cm^3, incorporated into the homopolymers will increase the particle densities considerably. We can calculate this increase on the assumption that the volume is additive and that the acrylic acid is incorporated completely during the co-polymerisation. The result of this calculation is indicated by horizontal arrows in Fig. 25.

But we are not sure that the second assumption - the complete incorporation of the acrylic acid - is really fulfilled, since the monomer acrylic acid and the polymer acrylic acid are water-soluble. So it is possible that during the co-polymer reactions the acrylic acid will not be incorporated into the acrylate chains but will polymerise separately to homo-poly(acrylic) acid. The density gradient measurements of the co-polymers in Fig. 26 show what really happens.

The increase of the particle density of the ethylacrylate/acrylic acid co-polymers is as expected. That means that no separate homopolymerisation of acrylic acid takes place. In the case of butylacrylate/acrylic acid co-polymers the increase of density is smaller than expected. That means that only 5% of the planned 8% acrylic acid are incorporated into the co-polymer. The remaining 3% consist of separate *free* homopolyacrylic acid. It is nearly the same in the case of ethylhexylacrylate/acrylic acid co-polymers. Only 2% acrylic acid are incorporated and 6% are separate free homo-polyacrylic acid.

If there really exists separate water-soluble *free* polyacrylic acid in the two last co-polymer dispersions, a sedimentation run is necessary to furnish the proof and Fig. 27 demonstrates this proof.

We had to choose a very high overall concentration of 80 mg/ml to render visible the small amounts of free dissolved polyacrylic acid eventually present. Because of the high rotor speed of 44,000 rev/min, the very fast particles (the dominant main parts of the six dispersions) sediment together in the first four dispersions and float in the last two dispersions to the bottom or to the meniscus of the cell during the first minutes of the run: there one can see these particles as a thick layer.

After an 80 minutes running time the first sedimenting schlieren peaks of free polyacrylic acid are to be seen, if present. Of course, the free polyacrylic acid is not present in the three homopolymer dispersions. It is also not present in the ethylacrylate/acrylic acid co-polymer, but it is present in the butylacrylate/acrylic acid co-polymer and the ethylhexylacrylate/acrylic acid co-polymer, as predicted by the density gradient experiment. The analysis of the schlieren peak areas reveals for these two cases that 3% and 5%, respectively, of the planned 8% acrylic acid exist as free polyacrylic acid with a molecular weight of about 1000 g/mol - 20 000 g/mol.

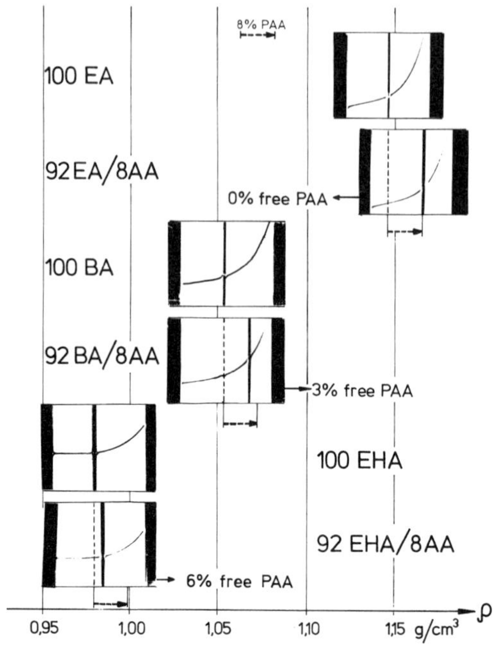

Figure 26. H_2O/metrizamide density gradients of some acrylic homopolymer and co-polymer dispersions.

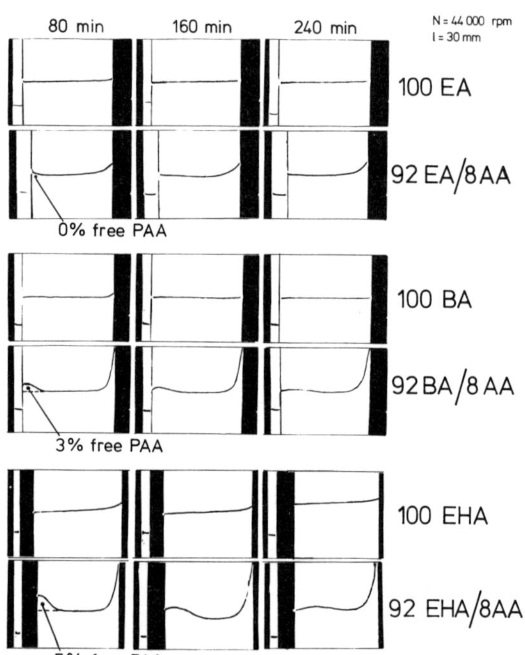

Figure 27. Sedimentation runs of some acrylic homopolymer and co-polymer dispersions at increased concentration c = 80 mg/ml in H_2O.

It is clear that the presence or the absence of free polyacrylic acid in films, made of such dispersions, will influence the performance of these films.

6. CONCLUDING REMARKS

In the preceding sections, we have shown what kinds of polymer dispersion characterisations can be achieved by means of a specially adapted analytical ultracentrifuge. This modernisation (*i.e.* having an eight-cell multiplexer and using the technique of a continuously increasing rotor speed from zero to maximum speed) is essential to perform such a characterisation in an industrial laboratory in a rational manner. Without this adaptation, rapid high resolution particle size distribution measurements and the interesting versatile density gradient technique are too time consuming.

In this Chapter we have not laid any emphasis on instrumentation, (*i.e.* how we adapted our AUC) because we have described that already in detail elsewhere[1,3-8]. Instead, we have concentrated on demonstrating some interesting examples of applications from our daily work in an industrial laboratory. We hope these examples illustrate that the AUC is a very interesting instrument to analyse and characterise polymer dispersions.

REFERENCES

1. W. Mächtle and U. Klodwig, *Makromol. Chem.*, 1979, *180*, 2507.
2. H.G. Müller, *Colloid Polym. Sci.*, 1989, *267*, 113.
3. W. Mächtle, *Makromol. Chem.*, 1984, *185*, 1025.
4. W. Mächtle, *Angew. Makromol. Chem.*, 1988, 162, 35.
5. W. Mächtle, *Colloid Polym. Sci.*, 1984, *262*, 270.
6. W. Mächtle and U. Klodwig, *Makromol. Chem.*, 1976, *177*, 1607.
7. W. Mächtle, Preprints Short Communications IUPAC MACRO, Mainz, 17-21 September 1979, Vol II, pages 731-734.
8. U. Klodwig and W. Mächtle, *Colloid Polym. Sci.*, 1989, *267*, 1117.
9. H.-J. Cantow, *Makromol. Chem.*, 1964, *70*, 130.
10. J.B. Nichols, E.D. Bailey and E.O. Kraemer, 84th Meeting of the Am. Chem. Soc., Division of Colloid Chemistry, Denver, 1932.
11. G. Mie, *Ann. Phys.*, 1908, *25*, 377.
12. W. Scholtan and H. Lange, *Kolloid-Z .Z. Polym.*, 1972, *250*, 782.

11

A Fluorescence Detection System for the Analytical Ultracentrifuge and its Application to Proteins, Nucleic Acids, Viroids and Viruses

By Bettina Schmidt and Detlev Riesner

HEINRICH-HEINE-UNIVERSITÄT DÜSSELDORF, INSTITUT FÜR PHYSIKALISCHE BIOLOGIE, UNIVERSITÄTSSTRASSE 1, 4000 DÜSSELDORF, GERMANY

1. INTRODUCTION

As mentioned earlier in this volume (Chapter 1), some years ago the molecular characterization of proteins, nucleic acids, viruses and other biological compounds used to be determined routinely in the analytical ultracentrifuge[1]. For more specialist cases molecular weight, buoyant density, shape and flexibility are still analysed in the analytical ultracentrifuge, but simpler and less expensive techniques are now applied for routine work: thus, sample purity and molecular weight are, for many systems, analysed by gel electrophoresis and chromatography. Further, determination of the nucleotide composition of nucleic acids by buoyant density centrifugation is no longer needed because the complete nucleotide sequences are known in many cases. For studies of the structure and dynamics of biopolymers other methods like dynamic light scattering, fluorescence decay or neutron scattering have become readily available[2]. Analytical ultracentrifugation is *still* the method of choice for a series of quantitative studies whenever *macromolecular interactions* under a wide variability of solution conditions have to be analysed. Examples are molecular weights of oligo- or multi-subunit proteins under native conditions and protein nucleic acid interaction under equilibrium conditions[3-5].

If the detection system could be improved in respect to sensitivity and selectivity, the analytical ultracentrifuge would be superior over other techniques. In many problems of biochemistry and molecular biology the biological material is available only in minute quantities, and often not in a highly purified form. In the present work we try to show that a fluorescence detection system may overcome a number of the existing limitations and that new types of studies may be carried out by applying this system.

At first, *higher sensitivity* has to be expected from an analytical ultracentrifuge with a fluorescence detection system, because lower detectable concentrations are obtainable with fluorescence as compared with UV absorption. Furthermore, molecules with fluorophores may be recorded *selectively* even if they were present as a minor component in a mixture of other molecules. On the other hand considerable experience is available from the literature on how to label biopolymers with fluorophores. Labelling of proteins may be achieved in most cases by covalently linking fluorescein isothiocyanate (FITC) without losing protein activity[6]. Nucleic acids can be labelled non-covalently by strongly binding such dyes as ethidium bromide or others used in fluorescence microscopy (see *e.g.* ref. 7); fluorescent dyes may be linked covalently to nucleic acids, too, either in enzymatic reactions or by using oligonucleotide synthesis chemistry[8]. Fluorescence-labelled antibodies are widely used in immunology and may even be obtained from commercial sources.

The advantages of an analytical ultracentrifuge with a fluorescence detection system (which in fact could be anticipated from the general literature on fluorescence methods), stimulated us to develop such a system. Indeed the potential of the new system can be outlined with several examples. In what follows it is shown for a well known protein that concentrations which are lower by up to four orders of magnitude as compared to conventional UV absorption are sufficient for studies with fluorescence detection. Similar improvements were achieved in studies on nucleic acids. Applications to dsRNA segments of bovine rotavirus and to viroids are described in which the stiffness of dsRNA was determined. The RNA material for this study could be obtained only by elution after gel electrophoresis but was still sufficient for a study in the analytical ultracentrifuge with fluorescence detection. The equilibrium association constants of the viroid-polymerase interaction and an antibody-antigen binding were derived from evaluating the boundary heights in sedimentation velocity runs. Finally, a new procedure for virus detection with fluorescent antibodies in CsCl density gradient centrifugation is presented. It should be noted that a fluorescence detection system based on similar principles as that of this work was reported by Crepeau and colleagues[9], but - to the best of our knowledge - applications to new problems have not yet been described by these authors. For full details on the instrumentation side of this work the reader is referred to Rappold[10] and Schmidt *et al.*[11].

2. INSTRUMENTAL DEVELOPMENT OF THE FLUORESCENCE DETECTION SYSTEM WITH LASER SCANNING EXCITATION

Optical system

A fluorescence detection system was newly developed for the analytical ultracentrifuge Spinco model E. This optical system was mounted in place of the conventional Schlieren optics. The UV absorption detection system remained unchanged and can be used simultaneously with the fluorescence system. The principle of the fluorescence system is that of a scanning microscope, in which only a very small area of the sample is illuminated by an electron beam or a laser beam, and the intensity of the emitted or reflected radiation is recorded under a wide angle of observation. The spatial resolution is limited by the illuminated area inside the cell, in this case a cross section of about 60 μm.

The basic features of the fluorescence detection system are depicted in Fig. 1. Details of the fluorescence detection optics are shown in Fig. 2. An argon ion laser (Spectra Physics, Series 2000, 2025-05, power supply 265, 5 W total power) is used for fluorescence excitation. Its beam is focused into an entrance aperture ("11" of Fig. 1) of 120 µm. The image of the entrance aperture is formed in the centre of the cell ("18" of Fig. 2) by the projector lens ("8" of Fig. 1). The radial position of the focused beam inside the cell is varied continuously by moving the position table ("10" of Fig. 1) which carries the reflector ("9") and the projector lens ("8"). Thus only those molecules of the solution inside the cell which are located at the radial position of the beam are excited for fluorescence. After passing the cell the laser beam is quenched in a light trap ("7" in Fig. 1). The laser beam is tilted at an angle of 2.3° to the axis of rotation in tangential direction in order to avoid interference between the windows.

Whereas the excitation occurs only in a very small aperture typical for a laser beam, the fluorescence light is emitted in a wide angle. The emitted light which passes the light trap is collected on the photomultiplier ("4") by the collector lens ("5"). The difference in aperture size between excitation and emission is essential for the function of the optical system, because unconventionally, excitation and emission are in the same direction. Scattering effects of the excitation light are suppressed on the photomultiplier by a set of filters ("6"). The filter set contains a combination of an absorption and a bandpass interference filter for both maximum suppression of the excitation and maximum transmission of the emitted light. For detection of the dye fluorescein isothiocyanate (FITC) fluorescence is excited with the laser line 488 nm, the filter set contains a liquid filter of orange G (A488 = 6) and a bandpass interference filter between 530 nm and 570 nm (Microsan, Hamburg, No 35-5552). The dye ethidium bromide is excited with the laser line 514.5 nm, the corresponding filter set contains an edge filter KV 550 (Schott, Mainz) and a bandpass interference filter NAL 600 (Schott, Mainz). The suppression of the intensity of the primary laser beam on the photomultiplier is 10^7-10^8-fold. Because of the light trap ("7"), made from carbon with spectroscopic purity, the filters are not damaged by the primary laser beam.

The photomultiplier R 1104 (Hamamatsu) was chosen for detection of fluorescence emission with very high sensitivity. Optimum sensitivity reaches from 200 nm up to 650 nm.

Signal processing

A signal processing unit was developed in co-operation with Technosystem (Edinghausen, FRG). A pre-amplifier amplifies the signals from the photomultiplier up to values of 10 V. To improve the signal-to-noise ratio, the signal is integrated over the whole period the cell is moving through the laser beam. At the end of the integration the value is digitised and transferred into the central storage of a microcomputer (Apple II plus) for further processing. The electronic device for analogue processing is present twice in order to process the signal of two different cells during the same rotation period of the rotor. Signals of the different cells are selected due to the original multiplexer system of the UV scanner; the light bulbs in the optical pickup had to be replaced by infra-red light in order to prevent stray light in the

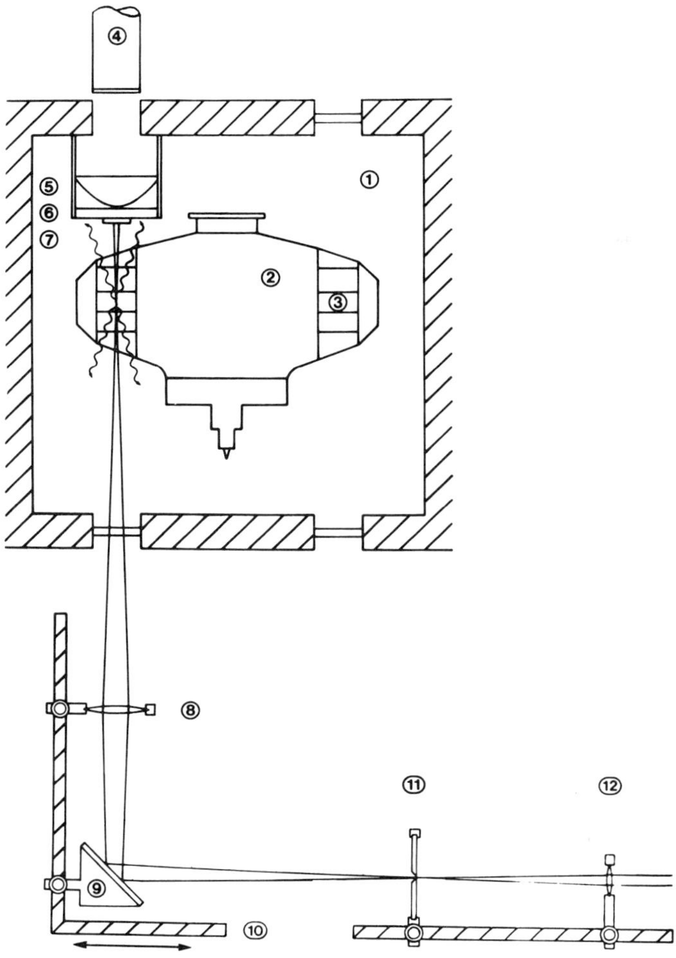

Figure 1. Schematic representation of the fluorescence detection system. 1: vacuum chamber; 2: rotor; 3: cell; 4: photomultiplier; 5: collector lens; 6: set of filters; 7:light trap; 8: projector lens; 9: reflector; 10: positioning table; 11: entrance aperture; 12: collector lens. Portions of the vacuum chamber, positioning table and support of the entrance aperture are drawn true to scale, distances between them are shortened. (From ref. 11.)

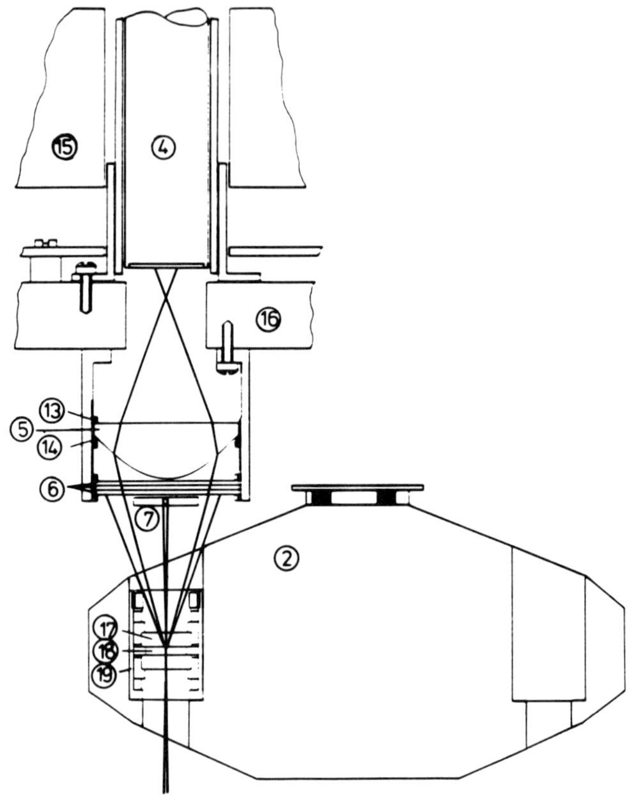

Figure 2. Representation of the fluorescence detection optics (to scale). Components which are shown in Fig. 1 are indicated with the same numbers as in Fig.1. Additional components are: 13: threaded ring; 14: O-ring; 15: drive; 16: steel plate; 17: quartz or sapphire window; 18: sample volume; 19: cell housing. The light trap (7) blocks the primary laser beam. The total fluorescence cone passes the set of filters (6) and the collector lens (5) and is collected on the photomutiplier (3). (From ref. 10.)

fluorescence detection system. Further details of the electronic system can be found elsewhere[10].

3. TEST FOR SENSITIVITY

The improvement in sensitivity due to the fluorescence detection system was demonstrated with two examples: one involving a protein, labelled covalently with a fluorescence dye, and one involving a nucleic acid labelled non-covalently with an intercalating dye.

Sedimentation Velocity of Fluorescence-Labelled Proteins

Fluorescein is widely used as a fluorescence dye binding covalently to proteins. Because of the high absorption coefficient and quantum yield fluorescein labelled proteins can be studied at very low concentration. There is a wealth of experience available from the literature on how to label different proteins with the fluorescein derivative fluorescein isothiocyanate (FITC) [12,13]. The optimal excitation and emission wavelengths of the unbound FITC are 494 nm and 518 nm. The wavelengths are shifted to 505 nm and 521 nm, respectively, if the dye is linked to amino-groups of proteins. For excitation of FITC fluorescence in the ultracentrifuge the strong laser line at 488 nm is most suitable since the line is on the side of the shorter wavelengths compared to the maximum of excitation, and the distance to the maximum of emission was raised to 33 nm. Bovine serum albumin (BSA) was used as a model protein. It was labelled with FITC in a molar ratio of 1:3 (1 mol BSA:3 mol FITC). Details of incubation for the linking reaction may be seen from the legend of Fig. 3.

In Fig. 3 the sedimentation profiles of BSA in five different concentrations are depicted. A three-dimensional presentation of the profiles was chosen with the time-axis in the third dimension running from the back to the front. In the order of the Figs. 3a to e the concentrations were lowered by an order of magnitude from one figure to the next. The increase in noise and background intensity with decreasing concentration is obvious. It does not, however, prevent a quantitative evaluation of the sedimentation coefficient even in the case of the lowest concentration. In Table 1, the concentration independence of the sedimentation coefficients is indicated. It is shown that the sedimentation coefficient does not depend significantly upon the concentration and agrees well with the value of 4.31 S known from the literature[14].

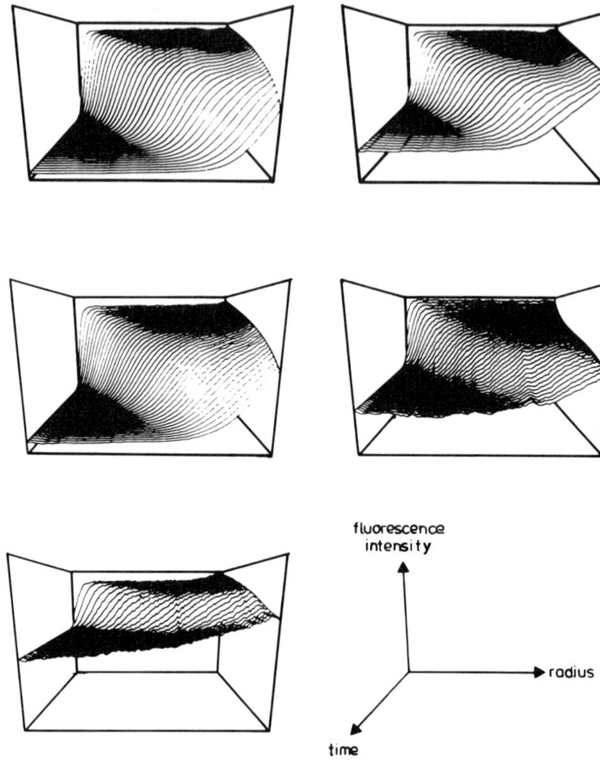

Figure 3. Sedimentation profiles of bovine serum albumin (BSA) recorded with the fluorescence detection system. Labelling of BSA with FITC was carried out with 3×10^{-4} M BSA and 6×10^{-4} M FITC in a volume of 1 ml of 25 mM sodium borate buffer, pH 9.2. Incubation was done for 1 h at 20°C in the dark under gentle stirring. Unreacted FITC was separated on a Sephadex column. Concentrations, sample sizes, cells, rotation and sedimentation coefficients are listed in Table 1. Radial displacement limits are 6.12 cm at the left border and 7.16 cm at the right border. Differences in time between two consecutive scans are 4 min (a-d), 8 min (e). (From ref. 11.)

Table 1. Sedimentation coefficients of bovine serum albumin (BSA) at different concentrations. BSA was labelled with the fluorescence dye FITC and monitored with the fluorescence detection system (*cf* Fig. 3).

BSA (M)	Amount (ng)	Volume (μl)	Optical pathlength (mm)	Rotor speed (rev/min)	$s_{20,w}$ (S)	Fig.
10^{-6}	2600	40	1.5	60,000	4.28	3a
10^{-7}	260	40	1.5	60,000	4.38	3b
10^{-8}	26	40	1.5	52,000	4.31	3c
10^{-9}	5.3	80	3.0	52,000	4.45	3d
10^{-10}	2.1	320	12	52,000	4.32	3e

The lowest concentration was as little as 10^{-10} M, corresponding to 6.5 ng/ml. This is four to five orders of magnitude lower than would be detectable with the UV absorption optics. Since on average three dye molecules were attached to each protein molecule, the concentration of the fluorophore was 3×10^{-10} M. This sensitivity of fluorescence detection is close to that of a good commercial fluorimeter.

Sedimentation Velocity of Fluorescence-Labelled Nucleic Acids

In order to determine the limits of the fluorescence detection system when applied to nucleic acids, double-stranded RNA was labelled with ethidium bromide and analysed in velocity sedimentation runs. The RNA was the double-stranded form of a satellite RNA which is associated as a fifth component with cucumber mosaic virus and therefore called dsCARNA 5 [15]. Its length is 330 base pairs. Ethidium bromide is a fluorescent dye which intercalates between the base pairs of nucleic acids with binding constants between 10^7 M^{-1} and 10^5 M^{-1} depending upon the ionic strength of the solution. Intercalation is saturated with one dye bound for four base pairs. The fluorescence properties of free and bound ethidium bromide are listed in Table 2.

Table 2. Fluorescence properties of ethidium bromide (EB) unbound in solution, bound to DNA and bound to RNA.

Component	I_{max}	λ^{ex}_{max} (nm)	λ^{em}_{max} (nm)	$\Delta\lambda$ (nm)
EB, unbound	2	490	584	94
EB bound to DNA	56	538	580	42
EB bound to RNA	100	532	574	42

λ^{ex}_{max}: wavelength of maximum excitation under experimental conditions
λ^{em}_{max}: wavelength of maximum emission under experimental conditions
$\Delta\lambda$: $\lambda^{em}_{max} - \lambda^{ex}_{max}$
I_{max}: relative intensity with excitation at λ^{ex}_{max} and emission at λ^{em}_{max}

The fluorescence properties clearly demonstrate that ethidium bromide is a very favourable dye for labelling nucleic acids because the absorption coefficients increase drastically due to intercalation and the distance between excitation and emission wavelengths is sufficiently large. Excitation was best with the laser line 514.5 nm which is the most powerful line of the laser.

In Fig. 4 the sedimentation profile of dsCARNA 5 in a concentration of 6×10^{-11} M is shown. This concentration corresponds to a total amount of 530 pg RNA. The concentration of bound ethidium bromide was 2.5×10^{-9} M. In a series of sedimentation runs with decreasing concentration of dsRNA and ethidium bromide the concentrations as applied in the runs of Fig. 4 were the lowest which still allowed us to evaluate the value of s with appreciable accuracy. These numbers have to be compared with 0.5-1 μg RNA which is about the detection limit with the UV absorption optics. The sedimentation coefficient $s^0_{20,w}$ was determined to (6.0±0.3) S. This value is about 23% lower than the value expected from the literature[15]. The deviation is a consequence of the very low ionic strength used in this experiment (20 mM NaCl) and not the influence of the intercalating dye. This was proven in measurements with the UV absorption optics with and without ethidium bromide but substantially higher concentrations of RNA.

Both examples (*i.e.* sedimentation of FITC-labelled BSA and sedimentation of ethidium bromide stained dsCARNA 5) were presented to demonstrate the marked increase in sensitivity of the fluorescence detection system compared to the conventional UV absorption optics, and to show that the profiles obtained by fluorescence detection allowed us to evaluate quantitatively sedimentation coefficients.

4. APPLICATIONS

The Stiffness of Double-Stranded RNA and Viroids

Double-stranded RNA (dsRNA) and viroids were studied in the analytical ultracentrifuge equipped with fluorescence optics for several reasons. The main reason was because only small amounts of material were available in purified form (by eluting bands in a slab gel after electrophoresis). Furthermore, when viroid-polymerase complexes were studied, the viroid band could be stained selectively by ethidium bromide, and the amount of viroids both sedimenting as free molecules and complexed with polymerase could be determined without interference from protein absorption phenomena. (One should note that detailed investigations of viroid sedimentation were carried out well before the use of fluorescence optics was established. Indeed for the sake of this review type of presentation we would like to combine corresponding data on dsRNA with those on viroids, although *only* the data on a dsRNA were obtained exclusively with the fluorescence optics.)

Double-Stranded RNA Double-stranded RNA is found in nature in several forms. Certain groups of viruses, *e.g.* the *reoviridae*[16], the birnaviruses[17], some mycoviruses[18] and bacteriophage ϕ6 [19] contain dsRNA as genomes. From other

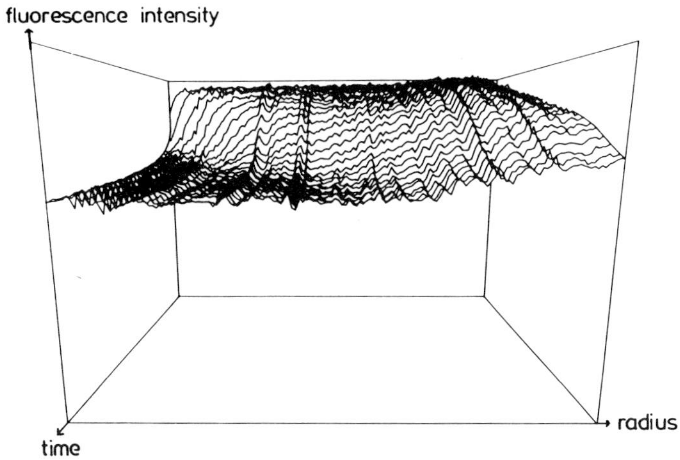

Figure 4. Sedimentation profiles of dsCARNA 5 at a concentration of 6×10^{-11} M (corresponding to 530 pg RNA). Ethidium bromide was used at 1×10^{-7} M. Radial displacement limits are 6.16 cm at the left border and 7.15 cm at the right border. Differences in time between two scans are 5 min.

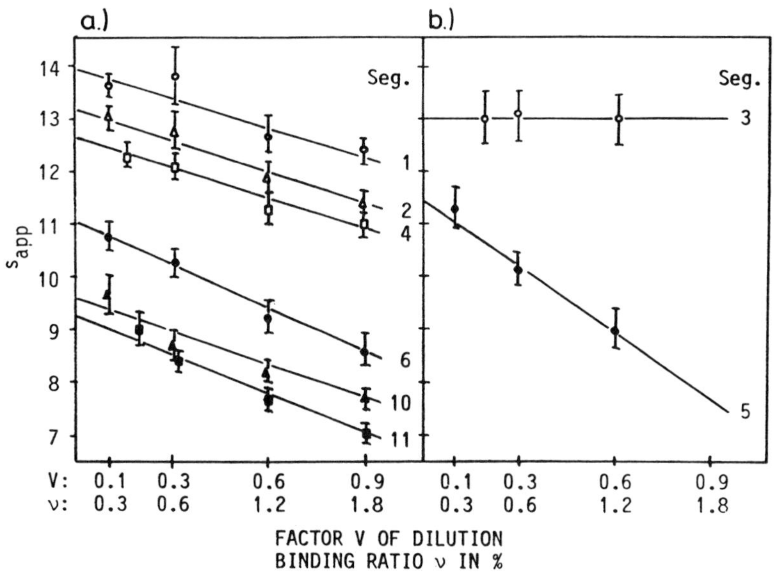

Figure 5. Dependence of the measured s values upon the dilution V of the RNA stock solution upon the binding ratio. The straight lines were obtained from least squares analysis. The s values are corrected for temperature and dialysing buffer. (a) Segments with similar, *i.e.* intermediate dependencies: segments 1, 2, 4, 6, 10, 11; (b) segments with a different dependence: segment 3 and 5. Segments 7-9 were lost during dialysis. The concentration of the stock solutions were the following: 1×10^{-5} M base pairs for segments 1, 2, 5; 7×10^{-6} M base pairs for segments 4, 6, and 3×10^{-6} M base pairs for segment 3, 10, 11. (From ref. 35.)

viruses and bacteriophages with single-stranded genomic RNA, double-stranded forms may be extracted from the infected host. In some plant viruses as cucumber mosaic virus and peanut stunt virus, satellite RNAs are present in both forms, as single-strands and in large excess as double-strands[20]. The formation of RNA double-strands has also some importance as a laboratory technique because it is used in the most sensitive hybridization procedure (see *e.g.* refs. 21 and 22).

Structural and thermodynamic studies on dsRNA have been carried out mostly with synthetic polynucleotides; the results, however, are characteristic for particular periodic sequences used in these studies, and possibly do not hold for natural RNAs. Aspects of the conformation of dsRNA in solution - such as their structural flexibility or "stiffness" - have not been studied so far. Current theories about the stiffness of "worm-like" chain polymers[23-26] allow the calculation of their hydrodynamic behaviour as a function of contour length. The stiffness may be expressed quantitatively as the *persistence length*, which is defined at every given point of the polymer as the tangential distance over which the direction of the chain persists before the direction of the chain is distributed randomly due to bending.

The persistence length of dsRNA was determined from the sedimentation coefficients of the eleven RNA segments of rotavirus, a genus of the *Reoviridae* family[16]. These RNA segments were used because their lengths are known from sequencing[27-33] and electron microscopy measurements[34]. The eleven RNA segments were separated by preparative gel electrophoresis and isolated by elution from gel pieces. The RNA was non-covalently labelled with ethidium bromide at a low dye/polymer ratio. The influences of dye staining, of RNA concentration and of impurities on the measurements could be eliminated. Sedimentation velocity runs were performed in a four-hole rotor TiAN-F at 44,000 rev/min for larger segments and 52,000 rev/min for smaller segments. Single sector centrepieces with an optical pathlength of 1.5 mm and a sample volume of 40 μl were used. The temperature was kept near 20°C. The details of the preparation of dsRNA segments and of the measurements have been described elsewhere[35].

Sedimentation coefficients were determined at varying RNA concentrations and at varying ratios of bound dye per polymer. The concentration of the isolated segments was too low for determination by UV spectroscopy. Therefore, concentrations were evaluated by ultracentrifugation from the height of the sedimenting RNA boundary after staining with a large excess of ethidium bromide. The concentrations of the stock preparations of the isolated segments were determined between 10^{-5} and 10^{-6} M base pairs. From all segments, four dilutions (V), *i.e.* 0.9, 0.6, 0.3, 0.1 of the stock solution, were prepared for the measurements. Total concentrations of dye were chosen to vary the ratio of bound dye per RNA base pair (ν) in the same manner as the RNA concentration. Consequently, in Fig. 5 the s values are given as dependent upon the RNA concentration *and* the ratio of bound dye per base pair. Only 0.3% to 1.8% of all neighbouring base pairs contained an intercalated ethidium. Since both dependencies chosen are proportional, they extrapolate to zero at the same ordinate. Only the lowest concentration of RNA (0.1 times that of the stock solution) was stained with 0.3% instead of 0.2% because of the limit of sensitivity for the fluorescence detection. All s values of one segment were determined in the same run using the four cells of the TiAN-F-rotor.

The results in Fig. 5a show that the dependencies described above of segments 1, 2, 4, 6, 10 and 11 were very similar. Only segment 3 showed a lower and segment 5 a higher dependence (*cf* Fig. 5b). Impurities from the gel matrix after elution of the segments from the polyacrylamide gel could stay with the RNA samples and increase the viscosity of the solutions. This artifact, however, is corrected for by extrapolation to zero concentration of RNA (*cf* Fig. 5). The corrected sedimentation coefficients of all segments studied are listed in Table 3.

Evaluation on the Basis of the Model of Flexible Cylinders For a quantitative interpretation of the experimental s values one cannot apply the model of a rigid cylindrical rod, but one has to take into account the degree of stiffness of dsRNA in the size range under consideration. When dealing with closely related problems concerning double-stranded DNA[36] and the partially base paired viroids[15] the theory of Yamakawa and Fujii[25] has been very successfully applied. According to these authors, polymers were adequately accommodated by the "worm-like" chain model, in which the stiffness is described by Kuhn's statistical length λ^{-1} or the "persistence length" P ($\equiv {}^1/_2\lambda^{-1}$) according to Kratky and Pörod[23]. The following equations describe the dependence of the s value upon the contour length L and diameter d, both counted in units of λ^{-1}.

$$s^0_{20,w} = \frac{M_L\,(1-\bar{v}\rho)}{N_A\;3\pi\eta_0} \times \begin{cases} C_1 \ln(L/d) + C_2 + C_3 L + C_4 L^2 & \text{for } L \leq 2.278 \\ A_1 L^{1/2} + A_2 + A_3 L^{1/2} + A_4 L^{-1} & \text{for } L \leq 2.278 \end{cases}$$

C_1 to C_4 and A_1 to A_4 are *slightly* dependent upon d; the exact relationships are given by Yamakawa and Fujii[25]. M_L was taken as 241.0 Å^{-1} which is the value for a RNA 11-helix with 660 daltons and 2.74 Å helical rise per base pair[37,38]. Values for λ^{-1} and d were obtained from a least square fit of the formulae to the experimental s values. This fit is shown in Fig. 6. The best fit was obtained assuming a persistence length P, or ${}^1/_2\lambda^{-1}$, of 1125 Å and 30 Å for the hydrated diameter, d. In order to estimate the accuracy of the value for the persistence length, the curves with a deviation of ±20% for ${}^1/_2\lambda^{-1}$ were added to the graph. It was estimated that the value of a persistence length of 1125 Å is accurate to within ±100 Å. As outlined above, this value is not influenced by concentration effects of the polymer, nor by the intercalating dye or by impurities. It is, however, characteristic for a particular ionic strength, which was 0.5 NaCl in these experiments. We have chosen the high value of ionic strength because comparable data of DNA and viroids (*cf* below) have been obtained under those test conditions. Furthermore, at high ionic strength the intramolecular repulsion of the negatively charged phosphates of the backbone is shielded and, therefore, the persistence length reflects mainly the structural stiffness from internal forces such as base stacking and hydrogen bonding.

Viroids Viroids are plant pathogens distinguished from viruses by the absence of a protein coat and by their small size. Although different "viroid species" may cause characteristic diseases in different host plants, all viroids have common structural features. They are circular, single-stranded RNA molecules consisting of a

Table 3. Sedimentation coefficients, numbers of base pairs and molecular weights (M_r) for rotavirus RNA segments and ds CARNA 5.

ds RNA	Base pairs	$10^{-6}\times M_r$	$s^0_{20,w}$ (S)
seg. 1	3409	2.25	13.9±0.2
seg. 2	2737	1.81	13.2±0.3
seg. 3	2653	1.75	13.0±0.6
seg. 4	2359	1.56	12.6±0.3
seg. 5	1519	1.00	11.3±0.6
seg. 6	1356	0.89	11.0±0.3
segs. 7-9	1059-1016	0.70	-
seg. 10	751	0.50	9.3±0.2
seg. 11	663	0.44	9.6±0.2
ds CARNA 5	335	0.22	7.8±0.2

The numbers of base pairs of rotavirus RNA segments were determined by sequencing (segments 6-11 [27-33]) and by electron microscopy (segments 1-5 [34]). The $s^0_{20,w}$ values are extrapolated from Fig. 5 for dilution V = 0, intercalated dye per RNA ν = 0. The value for dsCARNA 5 is taken from ref. 43 (*cf* also above).

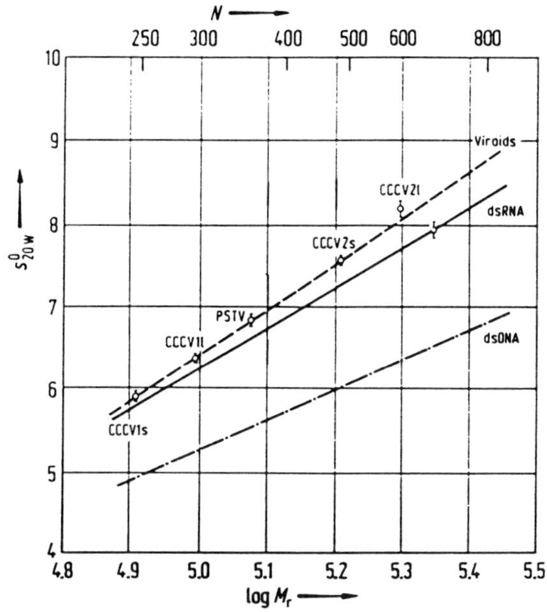

Figure 6. $s^0_{20,w}$ **values for segments from rotaviruses and dsCARNA 5. The fitted lines are calculated with the theory of Yamakawa and Fujii[25] d = 30 Å; (——): $^1/_2\lambda^{-1}$ =** 1125 Å; (---): $^1/_2\lambda^{-1}$ = 850 Å; (·—·—): $^1/_2\lambda^{-1}$ = 1350 Å. (From ref. 35.)

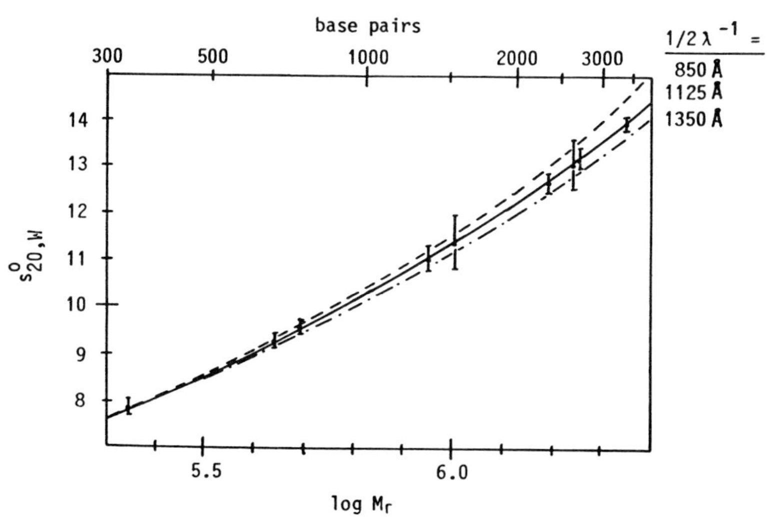

Figure 7. Dependence of $s^0_{20,w}$ upon logarithm of molecular weight M_r for dsRNA, dsDNA, and the viroids potato spindle tuber viroid (PSTVd) and different forms of the coconut cadang-cadang viroid (CCCVd 1s, CCCVd 1l, CCCVd 2s, CCCVd 2l). (Adapted from ref. 41.)

few hundred nucleotides, the smallest having about 240 and the largest about 600 nucleotides. They exist under native conditions as highly base-paired rod-like structures as suggested originally by electron microscopic, hydrodynamic and thermodynamic investigations[39] and this was finally confirmed when the nucleotide sequences were determined[40,41]. Since viroids can have only a very limited coding capacity and as there is no experimental evidence for a viroid-coded translation product, one has to assume that viroid replication and pathogenesis depend completely on the enzyme system of the host. Complex formation between viroids and the replicating polymerase from the host plant will be described in the next paragraph.

The hydrodynamic properties of the potato spindle tuber viroid (PSTVd) and four different viroid-like RNA species causing the cadang-cadang disease of coconut (CCCVd) were studied systematically and evaluated in terms of the "worm-like" cylindrical model described by Yamakawa and Fujii[25]. The derivation of the persistence length from $s^0_{20,w}$ values was carried out in a very similar way to that described above for dsRNA. The experiments were carried out in the Spinco model E analytical ultracentrifuge *without* the fluorescence detection system, but equipped with a high-intensity illumination system[42], a photoelectric scanner and a multiplex system. Four cells in an AN F-Ti rotor were analysed within the same run. The rotor speed was 52,000 rev/min. The temperature was 20°C. (Further details of the experiments may be seen from the original publication[43].) In Table 4 all sedimentation coefficients measured are listed, together with the number of nucleotides and molecular weights M_r. The measurements were carried out in 10 mM sodium cacodylate, 1 mM EDTA, 500 mM NaCl, pH 6.8. The high ionic strength was chosen for the same reasons as discussed above for dsRNA. An experimental error of ±0.1 S was estimated. CCCVd-small is a mixture of two species[44], the uncertainty in the size of CCCVd-large is because the stem used in this work was different from that which was sequenced[44].

Table 4: Sedimentation coefficients of viroids.

RNA	Number of nucleotides	M_r	Average value of s (S)
CCCVd1 - small	246/247	83,500	5.91
CCCVd1 - large	295 (±8)	101,300	6.34
PSTVd	359	121,200	6.82
CCCVd2 - small	492/494	167,000	7.57
CCCVd2 - large	590 (±16)	202,600	8.16

In Fig. 7, the measured sedimentation coefficients are plotted against log M_r. The solid line for dsRNA is taken from the measurements described in the paragraph above; the experimental point is that of dsCARNA 5. One should mention that in the molecular size range below 300 base pairs which is depicted in the figure, the curve for dsRNA (persistence length of 1125 Å, d of 30 Å, M_L of 241 $Å^{-1}$ and $\bar{v}$ of 0.53 cm^3/g) does not deviate significantly from a straight line which would be deduced for the model of a completely rigid rod.

The deviation of the sedimentation coefficients of PSTVd and the cadang-cadang RNAs from the curve of dsRNA could be interpreted as being due to higher flexibility of viroids. Consequently M_L, $\bar{v}$ and d were allotted the values for dsRNA and the persistence length P was taken as the adjustable parameter. The best fit was obtained when the value of P was 300 Å (*cf* dotted line in Fig. 7).

Summary of Hydrodynamic Features of Nucleic Acids in Solution as Derived from Sedimentation Coefficients The results on dsRNA and viroids complete the structural data of nucleic acids in solution. As known from X-ray analysis (*cf* ref. 37), dsRNA is a double-helix in the A-form. It is commonly described as being more compact than the B-form; this means that the helical rise is lower (2.74 Å instead of 3.4 Å) and the molecular mass per unit length is higher (241 $Å^{-1}$ instead of 195 $Å^{-1}$). The persistence length of 1125 Å is nearly twice as great as that of dsDNA in the B-form (500-600 Å) [19,36,46,47]. Thus, the higher compactness of the A-form in comparison to the B-form goes in parallel with the higher rigidity in solution.

The comparison of chain flexibility of nucleic acids is not restricted to the A- and B-form, but could be extended to even more flexible or to more stiff nucleic acids. For viroid RNA, which forms a rod-like structure of an unbranched series of short double helices and small internal loops, a persistence length of 300 Å was found. This suggests that small internal loops induce a high degree of flexibility but do not invalidate the model of an elastic cylinder. The sedimentation data on viroids, dsDNA in the B-form and dsRNA are summarized in Fig. 8, which shows the different behaviour for low and high molecular weights. As the other extreme, for DNA in the Z-form a very high persistence length of 2080 Å was obtained from light scattering data[47].

Viroid-Polymerase Binding as an Example for Protein-Nucleic Acid Interaction

Analytical ultracentrifugation is generally applicable as a method for studying protein-nucleic acid interactions. Although other methods such as filter binding, gel retention, equilibrium dialysis, titration with optical recording, *etc.* are less expensive and/or less laborious, ultracentrifugation is highly favourable if the binding constants of the interaction are low, *i.e.* 10^6 M^{-1} or even lower. This is because it does not shift the binding equilibrium even if the equilibrium is fast compared to the separation process. Furthermore, it does require as a parameter for detection of binding merely a change in the sedimentation constants. Examples of successful applications with detailed discussions of the methodological aspects have been studies on tRNA-synthethase binding[5], oligonucleotide-viroid interactions[48], protein-protein interactions[49] and histone-DNA interactions[50]. So far, this kind of study has been limited to nucleic acid concentrations higher than 10^{-5} M nucleotides due to the sensitivity of the UV absorption optics. Thus, the same increase in sensitivity with the fluorescence detection system as discussed with isolated nucleic acids should be favourable for investigations of nucleic acid-protein interaction. We will describe in the following an example from our viroid research.

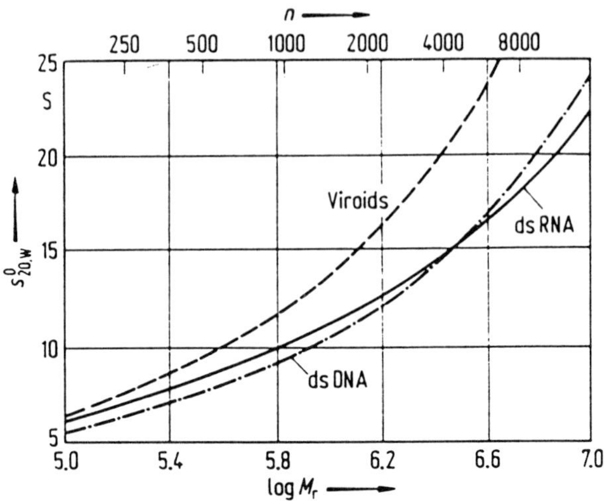

Figure 8. Dependence of $s^0_{20,w}$ on log M_r for dsRNA, dsDNA and viroids. Curves according to the theory of Yamakawa and Fujii.(——) dsRNA: d = 30 Å, $^1/_2\lambda^{-1}$ = 1125 Å [35]; (·—·—) dsDNA: d = 27 Å, $^1/_2\lambda^{-1}$ = 575 Å [19,36,46,47]; (- - -) viroids: d = 29 Å, $^1/_2\lambda^{-1}$ = 300 Å [43]. From ref. 41.

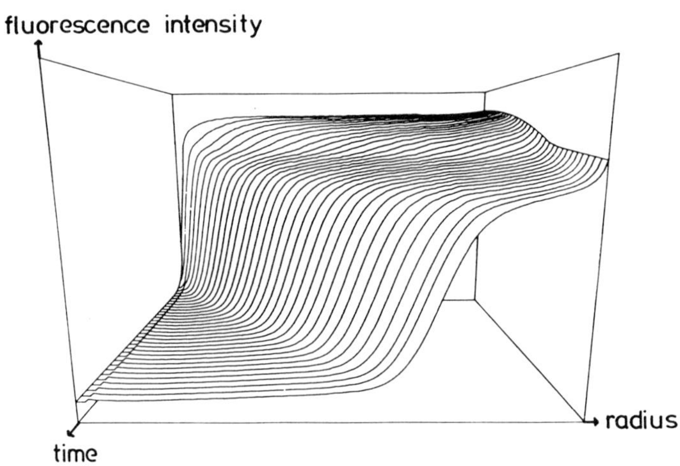

Figure 9. Sedimentation of viroid (slow moving boundary) and viroid-polymerase complexes (faster moving boundary). Viroid and polymerase were used at a concentration of 1×10^{-7} M and 3.3×10^{-8} M. Viroid was stained with 5×10^{-6} M ethidium bromide. Radial displacement limits are 6.16 cm at the left and 7.13 cm at the right border. Sedimentation profiles were recorded every 4.5 min. (From ref. 10.)

While a great deal is known about the structure of viroids (*cf* previous paragraph), relatively less is known of the detailed process and enzymology of viroid replication. As derived from several *in vivo* and *in vitro* studies[51] viroids are replicated by the DNA-dependent RNA polymerase II from the host cell. DNA-dependent RNA polymerase II preparations from both wheat germ and tomato plant tissues are also able to produce full length viroid RNA *in vitro*[52]. We employed a model system composed of purified potato spindle tuber viroid (PSTVd) and DNA-dependent RNA polymerase II isolated from wheat germ to determine the equilibrium association constant and stoichiometry of the viroid polymerase interaction. It would be instructive to now compare the relative affinity of the polymerase for viroid RNA and other RNA and DNA templates.

Determination of Binding Constants The viroid-polymerase interaction was studied using sedimentation velocity experiments monitoring both UV absorption and fluorescence of an intercalated dye. The buffer solution contained 25 mM Tris-HCl, pH 7.9, 25% glycerol, 50 mM NaCl, 6 mM $MgCl_2$, 1 mM EDTA. For monitoring the nucleic acid moiety by fluorescence, ethidium bromide or ethidium homodimer (Fluka, Ulm, Germany) were bound to the viroids non-covalently. (The original data were published earlier[4,10].)

Fig. 9 shows the results of a typical analytical ultracentrifuge experiment to determine an equilibrium binding constant between viroid and polymerase II. Viroid concentration was in three fold excess over the enzyme concentration. In addition to the viroid-polymerase equilibrium the binding of ethidium bromide to viroid has to be considered. In reference titration experiments it was found that neither the binding constant nor the number of binding sites, nor the quantum yield of ethidium bromide changes significantly due to the complex formation with the polymerase. This absence of a counterwise influence was expected, because it was known from electron microscopy that the polymerase molecules bind to the very ends of the rod-like viroid structure[4]. Ethidium bromide was added at 50-fold molar excess over the viroid concentration, *i.e.* at 2.5-fold molar excess over the concentration of binding sites on the viroid. The absolute molar concentration (5×10^{-6} M) was, however, below the reciprocal binding constant ($\approx10^{-5}$ M under the experimental conditions). Thus, the viroid is not saturated with the dye, and the free dye is in excess over the viroid-bound dye. The equilibrium is fast and constant throughout the ultracentrifuge cell, and the viroid-polymerase interaction is not affected by fluorescence staining with ethidium bromide. The free dye does not contribute to the total fluorescence intensity markedly because of its low quantum yield. In summary, the fluorescence intensity of ethidium bromide indicates the relative concentrations of free viroid and viroid sedimenting in a complex with the polymerase.

In addition to the major sedimentation boundary (free viroid) a faster moving boundary of the viroid-polymerase complex is evident from the sedimentation profiles in Fig. 9. After correction for viscosity the sedimentation coefficients $s_{20,w}$ of the free viroid was (6.1±0.3) S and that of the complex (13±1.5) S.

By estimating the relative heights of the sedimenting boundaries and knowing the initial concentrations of viroid and polymerase, an association constant (K_a) can be calculated[5]. The K_a values determined in different experiments are plotted in Fig. 10 as a function of total enzyme concentration. The graph also contains data from earlier studies with the UV absorption optical system[4]. When a 1:1 viroid:polymerase stoichiometry was assumed, this figure showed that the observed equilibrium constant, K_a, was dependent on the total enzyme concentration. Extrapolation by a linear regression to an infinitely dilute enzyme concentration yielded a K_a for the viroid-polymerase interaction of $(2.1 \pm 0.5) \times 10^7$ M^{-1}. The phenomenon of the equilibrium constant decreasing with increasing enzyme concentration is difficult to interpret quantitatively and may come about as the result of the Johnston-Ogston effect in a two component system[53].

Separation of Different Types of Complexes Viroid-polymerase complex formation was studied also under conditions where more than one enzyme molecule could possibly bind to viroid. These conditions were (i) excess of enzyme over viroid and (ii) concentrations higher than K_a. Fig. 11 shows corresponding sedimentation profiles with two-fold (left scan) and six-fold (right scan) molar excess of polymerase over viroid. Free viroid moves with an s value of 6.6 S. The 15 S sedimentation boundary is known from the experiments described above as the 1:1 complex. The 20 S boundary results from partial formation of a complex of two polymerase molecules and one viroid. When a six-fold excess of polymerase was applied a nearly homogeneous boundary of the 2:1 complex with 26S was measured.

The results from ultracentrifugation are in agreement with electron micrographs of the polymerase-viroid interactions[4]. The micrographs showed 2:1 complexes, with the enzyme at both ends of the viroid in addition to 1:1 complexes. Since the 2:1, 26 S complex is observed only at high enzyme:viroid ratios and high enzyme concentration, the equilibrium constant for its formation must be much smaller than the K_a determined here for the 1:1 viroid-polymerase complex. This suggests that the 2:1 complex is not physiologically significant. This also indicates that binding at one of the ends must be non-specific.

The results on the viroid-polymerase interaction which were obtained with fluorescence detection system, were mainly a confirmation of earlier data obtained with UV absorption optics. They showed, however, the advantage of specifically staining one component of the interacting system, in this case the nucleic acid. The other advantage, *i.e.* extension of the concentration range to very low values could not be utilized with this system because of the fairly low affinities, which did not allow us to follow complex formation at very low concentrations.

The biological implications were discussed in detail by Goodman *et al.*[4] and Riesner *et al.*[54] and are not the subject of this Chapter here. One aspect should be emphasized, however, in this context. In electron micrographs, 1:1 complexes, 2:1 complexes and also complexes with other viral RNAs, which did not function as templates for the polymerase, were visible without differentiation of relative affinities. Quantitative sedimentation studies, however, did yield such a differentiation and demonstrated that 1:1 complex formation was sufficiently strong for biological activity

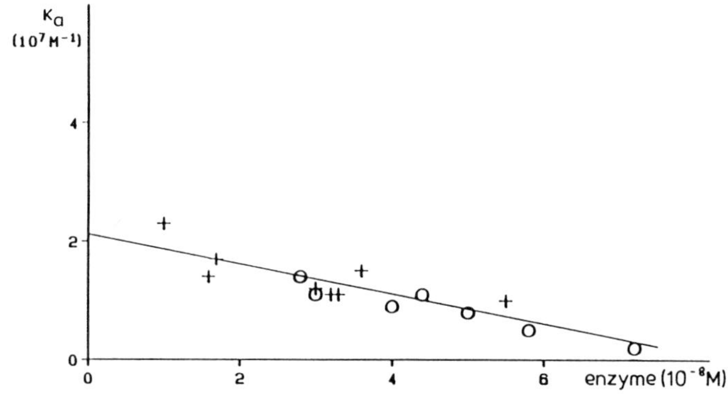

Figure 10. Dependence of the equilibrium constant K_a for viroid-polymerase interaction on enzyme concentration. O: K_a determined from experiments observed with UV optics; +: K_a measured with fluorescence optics. K_a results from extrapolation to zero concentration by linear regression to $(2.1\pm0.5)\times10^7$ M^{-1}. (From ref. 10.)

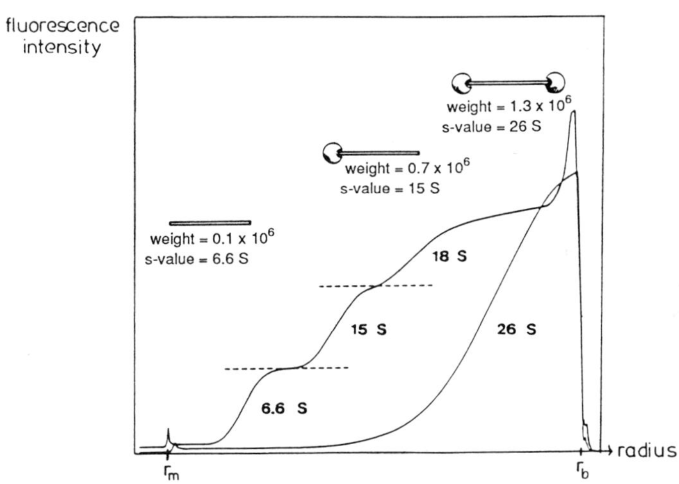

Figure 11. Sedimentation profiles of viroid and viroid-polymerase complexes under excess of enzyme in two different experiments. Both experiments were performed at 20°C and 48,000 rev/min. Radial displacement limits are 5.89 cm at the left and 7.27 cm at the right border. Viroid concentration was 2.2×10^{-7} M and ethidium homodimer was present at a concentration 1.1×10^{-6} M. The polymerase was in twofold molar excess over the viroid concentration in the leftmost scan. Three boundaries are separated by dashed lines (6.6 S, 15 S and 20 S, respectively). When a six-fold excess over enzyme is added (rightmost scan) only one single broader boundary (26 S) is observed. (Adapted from ref. 4.)

whereas the 2:1 complex and the complexes with biologically non-active RNAs were only of marginal affinity.

Antigen-Antibody Complexes as an Example of Protein-Protein Interaction

The binding of antibodies to antigens was investigated as an example of protein-protein interactions. For fluorescence labelling either the specific antibody carries the fluorescence label, *e.g.* FITC, or labelled second antibodies which bind to the Fc-part of the first specific antibody must be utilized for fluorescence labelling. The Fc-part is constant among all antibodies of a given immunoglobulin class from a particular animal species. These second antibodies are commercially available including those with covalently attached fluorescence dyes, *e.g.* FITC. Thus, the labelling reaction is not an extra task for us in investigating each antigen-antibody reaction.

The binding reaction of polyclonal, FITC-labelled antibodies which were produced in goat and are directed against immunoglobulins IgG from mice (goat anti-mouse, GAM_F) was analysed using sedimentation velocity. In these therefore the immunoglobulins from mice (IgG2a) represented the *antigen* and the GAM_F the *antibody* of this antibody-antigen interaction. The GAM_F-fraction consisted of $F(ab')_2$ fragments with both antigen binding sites present, but lacking the Fc-part of the antibodies. The fragments were FITC-labelled in a molar ratio of 3:1 (mol FITC/mol $F(ab')_2$).

In the experiment represented by Fig. 12, the complex formation of 10^{-8} M fluorescence-labelled GAM_F-$F(ab')_2$-fragments and 2×10^{-8} M IgG2a was analysed. The buffer solution contained 140 mM NaCl, 10 mM sodium-phosphate, pH 7.8; rotor speed was 52,000 rev/min. A faster moving component of GAM_F-IgG2a-complex was clearly resolved from the boundary of free $GAM_F \cdot F(ab')_2$. The broadening of this boundary indicated the existence of higher aggregates than the 1:1 complex. The portion of complexed GAM_F resulted directly from the ratio of fluorescence intensities in the fast and the slow boundaries if it was assumed that the quantum yield of FITC is not altered during the binding of GAM_F to the antigen.

In order to test the quality of commercially available GAM_F as antibodies against different IgG-subclasses, *e.g.* IgG2a or IgG1, titration experiments were carried out. The GAM_F concentration was kept constant and the IgG concentration was titrated up to three-fold excess. It was expected that the portion of free GAM_F disappeared with increasing IgG concentration. The results of such titration experiments are shown in Fig. 13. Only 60%-70% of GAM_F could form complexes with the antigen IgG1 (a), with IgG2a (b) or a mixture of both (c). The point of saturation was obtained at a molar ratio (1.2±0.2) IgG/GAM_F. If this ratio is corrected for the fact that 30%-40% of GAM_F do not take part in complex formation at all, the ratio is close to two antigens bound per one active antibody. Thus, our analysis showed that 30%-40% of the commercial antibody fragments were inactive but the rest were active with both binding sites[55].

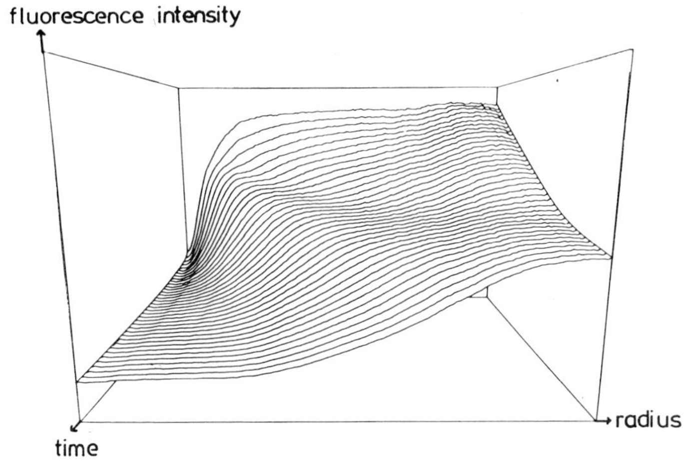

Figure 12. Sedimentation of antigen-antibody complexes (IgG2a-GAM_F). IgG2a and GAM_F were present at a concentration of 2×10^{-8} M. The faster moving boundary results from IgG2a-GAM_F complexes, while the slow moving boundary represents the free GAM_F portion. Radial displacement limits are 6.17 cm and 7.16 cm. The time between two scans amounts 3 min. (From ref. 10.)

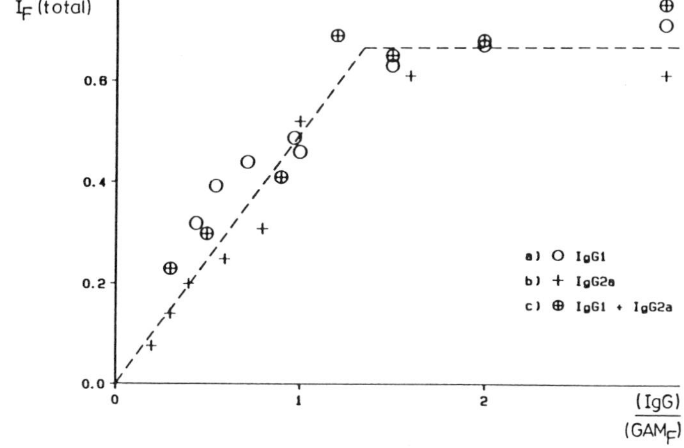

Figure 13. Titration of IgG1 and IgG2a to a constant concentration of GAM_F. Complex formation was evaluated from the different heights of the sedimentation boundaries. The GAM_F concentration was kept to 1×10^{-7} M. The point of saturation was obtained at a molar ratio of (1.2±0.2) IgG/GAM_F. (From ref. 10.)

It was thus shown that antigen-antibody reactions can be studied qualitatively and quantitatively in very low concentrations: in fact, antibody concentrations as low as 10^{-8} M were recorded free of background and with good signal to noise ratio.

Virus Detection in the CsCl Density Gradient

The fluorescence detection system was also applied in order to detect viruses and to determine quantitatively their concentration in crude extracts of biological material, *e.g.* homogenised tissue. Thus, the purpose here is to characterise viruses without extended purification procedures. The method has been established with purified tobacco mosaic virus as a model system which was readily available to us. Several advantages of the combination of analytical ultracentrifugation and fluorescence detection were expected for this purpose: high sensitivity of detection, selective detection of the labelled virus and further increase of sensitivity by concentrating the virus from a large volume into the narrow buoyant band. The procedure presented may compete with the most advanced methods of virus detection applied otherwise, and in addition yields quantitative estimates of concentration, density, homogeneity, *etc.*

Labelling of Virus The method of labelling viruses with antibody-bound fluorescence dyes is a modification of standard techniques in immunology[56]. The principle of this technique is depicted in Fig. 14. Two binding processes are involved in the labelling procedure. A specific monoclonal or polyclonal antibody binds to a coat protein of the virus. In our case a monoclonal antibody (MCA 75) was available[57]. The fluorescence dye is attached covalently to a second antibody, which binds to the Fc-part of the first, the virus-specific antibody. The FITC-labelled polyclonal antibody (GAM_F-$F(ab')_2$ fragment) was the same as used in the experiments described above. In order to obtain a fluorescence label on the virus which is not lost during the time of centrifugation or lost due to the non-physiological solution conditions of several molar CsCl, the binding equilibria were fixed irreversibly by incubation with glutaraldehyde or UV irradiation at 254 nm[58].

Resolution of Virus and Free Antibody in Density Gradient Centrifugation For standard experiments 6×10^8 TMV particles/ml ($\equiv 10^{-12}$ M) were incubated with 10^{-9} M specific and 10^{-9} M second antibody and covalently linked by UV irradiation. The solution was analysed in a CsCl density gradient at 48,000 rev/min with starting density $\rho_0 = 1.325$ g/cm^3.

In Fig. 15a the fluorescence intensity profile after 3.5 h centrifugation time is shown. A narrow virus peak was clearly resolved from the broad background of excess, unbound antibody which exhibited a broad maximum at higher density than that of the virus. The buoyant densities, if determined from the profile of Fig. 15a, should be regarded with care for two reasons. First, the equilibrium distribution was not reached after 3.5 h. Secondly, only the virus labelled by antibodies is observed and the buoyant density of the virus-antibody complex is lower than that of the uncomplexed virus. From the UV absorption data[11] it was estimated, that the difference in buoyant density was 0.008 g/cm^3.

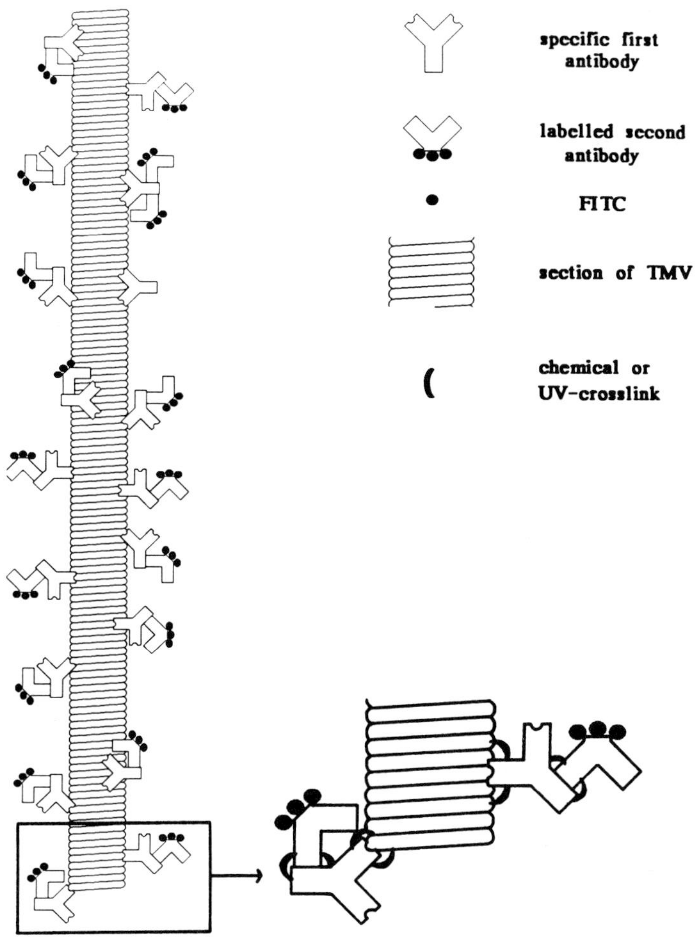

Figure 14. Schematical representation of labelled TMV-antibody complexes used for analysis by analytical ultracentrifugation equipped with fluorescence detection. The complete scheme is drawn true to scale. For standard labelling 3.2 ng TMV was incubated with 12 ng monoclonal antibody in a volume of 80 µl of 1 x PBS (140 mM NaCl, 2.3 mM KCl, 1.5 mM KH_2PO_4, 8.1 mM Na_2HPO_4, 1% BSA, 0.1 % NaN_3, pH 7.2) at 4°C over-night or 2 h at 37°C. After incubation 12 ng of the secondary antibody GAM_F (GAM $F(ab')_2$ fragments carrying FITC) were added, made up to a final volume of 100 µl with 1 x PBS, and incubated again at 37°C for 2 h. Covalent crosslinking of virus with monoclonal antibody and the secondary antibodies GAM_F was achieved by UV irradiation of the solution on ice for 15 min at 254 nm at a distance of 6 cm.

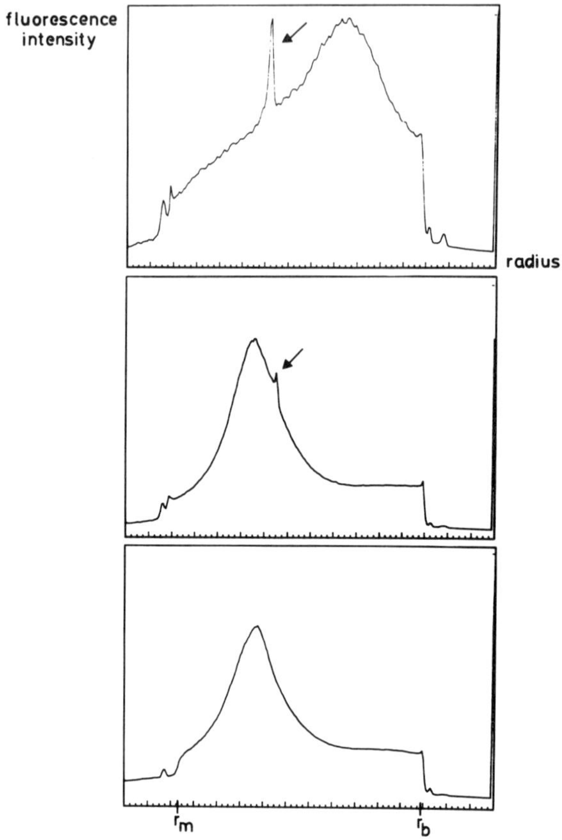

Figure 15. CsCl density gradient centrifugation of fluorescence labelled virus-antibody complexes. Virus-antibody complexes were UV cross-linked and monitored (a) after 3.5 h and (b) after 18.5 h. Concentrations were: 10^{-12} M virus ($\equiv 6\times10^8$ virus particles/ml $\equiv$3.2 ng virus/80 µl, 10^{-9} M specific antibody, and 10^{-9} M second antibody. Rotor speed was 48,000 rev/min, ρ_0=1.325 g/cm^3. In (c) the same amount of virus-antibody complexes was centrifuged but without crosslinking, and monitored after 18.5 h. Complexes are indicated by an arrow. Radial displacement limits are 5.91 cm at the left border and 7.51 cm at the right border. (From ref. 11.)

The profile of the same solution as in Fig. 15a but after reaching the equilibrium gradient is seen in Fig. 15b. A smaller virus peak as compared with that of Fig. 15a was detected on the high density flank of the broad peak of free antibody. One has to conclude that the virus by itself or the virus-antibody complex was not stable in CsCl over the time of 18.5 h. Covalent crosslinking by UV irradiation clearly stabilised the complex but did not prevent disintegration completely.

As a control experiment equal amounts of virus and antibody were centrifuged under the same conditions, but now without UV irradiation or chemical crosslinking. The fluorescence intensity profile after 18.5 h (Fig. 15c) demonstrated that the virus-antibody complex was disintegrated completely and only the concentration profile of the free antibody was detected. From the peak areas of the virus-antibody complexes and of the excess antibodies the amount of fluorescence-labelled antibodies bound to the virus was estimated. After 3.5 h, 4% of the fluorescence-labelled antibody was bound and after 18.5 h, 3%. From these values it could be derived that at the beginning of the run about 40 antibody molecules were bound per virus particle. It follows that only 40 out of 780 binding sites present on TMV[57] were occupied by the combination of a specific first antibody, a fluorescent second antibody, and chemical crosslinking between the three components. In later experiments a similar value for the yield was obtained.

The sensitivity of virus detection as shown in Figs. 15a and 15b was restricted due to the large excess of unbound antibody. Since virus-antibody complexes and free antibody differ considerably in molecular weight, both components were separated by preparative centrifugation. The cross-linked complexes were separated from the unbound antibody by centrifugation at 15,000 g, 4°C for 3 h or at 105,000 g, 20°C for 30 min in an airfuge (Beckman Instruments, Munich) The resulting sediment was resuspended in 50 µl of 1 x PBS (140 mM NaCl, 2.3 mM KCl, 1.5 mM KH_2PO_4, 8.1 mM Na_2HPO_4, 1% BSA, 0.1% NaN_3, pH 7.2). The advantage of the preparative separation step prior to the density gradient is clearly seen from Fig. 16. The same solution as that of Fig. 15 was analysed, but the excess antibody was removed as mentioned. After 3.5 h centrifugation time, a narrow peak at a density ρ of 1.319 g/cm^3 on low background intensity was recorded (Fig. 16). The two peaks at lower densities were artifacts, which were also present in a reference run without virus. According to our experience these peaks although in most cases much smaller, are typical for laser excited fluorescence in CsCl gradients at the meniscus of the solution. The narrow and symmetric form of the virus peak indicates a homogeneous degree of antibody binding and crosslinking to the virus.

Detection of Virus in a Very Crude Tissue Homogenate For testing the applicability of the method for virus detection in crude tissue homogenate, the same low concentration of TMV as above (6×10^8 particles/ml $\equiv 1\times10^{-12}$ M) was added to a tissue homogenate from the conifer "silver fir". Tissue homogenates from conifers are biochemically very difficult because of the many low and high molecular weight compounds which exert by themselves high absorption and fluorescence. Also in an immunoassay, these compounds lead to extreme background problems. It should be noted that these experiments were carried out as part of a research project on virus infections in forest trees.

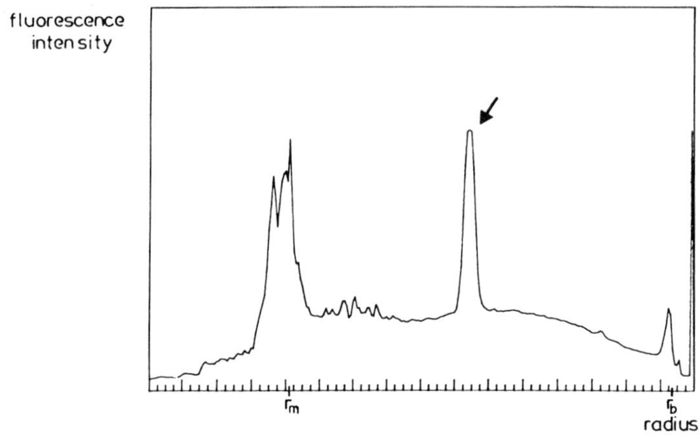

Figure 16. Density gradient centrifugation of virus antibody complexes after UV cross-linking and removal of excess antibody by high speed preparative centrifugation. Labelled virus-antibody complexes at a concentration of 10^{-12} M virus, 10^{-9} M specific antibody and 10^{-9} M second antibody were UV cross-linked and centrifuged before the analytical run for 15 min at 105,000 g. The analytical run was done with the complexes from the sediment and monitored after 3.5 h. The rotor speed was 48,000 rev/min, $\rho_0 = 1.325$ g/cm^3. Radial displacement limits are 5.68 cm at the left border and 7.28 cm at the right border. (From ref. 11.)

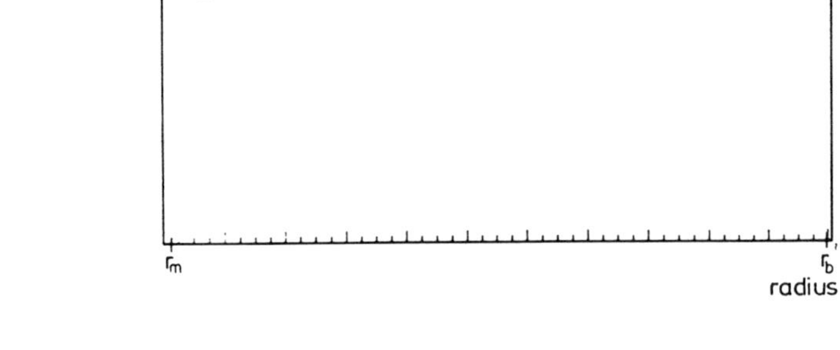

Figure 17. CsCl density gradient centrifugation of 3.2 ng TMV in a crude tissue homogenate from silver fir. TMV was used at the same low concentration as reported in Fig. 15 and Fig. 16. The run was carried out at 44,000 rev/min, $\rho_0 = 1.325$ g/cm^3. Radial displacement limits are 6.09 cm at the left and 7.21 cm at the right border. Virus-antibody complexes at the characteristic buoyant density are indicated by an arrow.

The homogenate was prepared from the needles in a simple phenol/chloroform extraction: as a result, the extract included all primary and secondary plant metabolites like polyphenols, carbohydrates, lipids and nucleic acids but no protein. To compensate for the lack of protein, bovine serum albumin was added to the tissue homogenate at a concentration of 4 mg/ml ($\equiv 10^{17}$ particles/ml). Labelling of viruses with specific first antibody and fluorescent second antibody was carried out within this extract. The fluorescence intensity profile after 8.5 h during CsCl density centrifugation is shown in Fig. 17.

The small peak of virus antibody complexes could be detected at the characteristic buoyant density in the flank of residual antibodies. The high background intensity is caused by fluorescent plant particles in the homogenate. Although the sensitivity of the method was reduced in crude extracts compared to the purified components, 3.2 ng TMV/80 µl could be detected unambiguously and characterized with respect to the buoyant density of CsCl. It was shown that the new method of virus detection can indeed be carried out in crude extracts from the biochemically very difficult tissues of conifers.

Comparison with Other Techniques of Virus Detection One may compare the achievements of the method presented here with those of commonly used techniques in virology. Similarly to the method described in this work, the most advanced techniques are based on specific binding of antibodies to viruses. Sensitive detection of the viruses is achieved either by an enzymatic reaction (Enzyme Linked Immuno Sorbent Assay, ELISA)[56], or by gel electrophoretic separation, consecutive transfer to a membrane and staining with specific antibody (Electro Blot Immuno Assay, EBIA)[60], or electron microscopic inspection of viruses captured on antibody coated grids (Immuno Sorbent Electron Microscopy, ISEM)[61,62]. Whereas in ELISA only the presence or absence of a virus is tested, in EBIA the molecular weights of the antigenic capsid proteins is determined and in ISEM the morphology of the virus-antibody complex may be analysed. The ultracentrifuge method as given here allows the determination of a whole set of physical properties, such as buoyant density, molecular weight, s value and associated parameters derived from all these (although we have not yet systematically assayed the effect of bound antibodies on physical properties).

It is important that the lowest concentrations and total amounts of virus that may still be detected with the different methods be compared. Such comparison of the commonly used procedures was given by Koenig and colleagues[61]: in ELISA 200 µl with 1.5 ng virus/ml, in EBIA 75 µl with 32 ng/ml, and in ISEM 20 µl with 16 ng/ml were traced. The ISEM technique may best be compared to the centrifugation technique as described in this Chapter, because in both cases whole virus particles are analysed. Because the sensitivity of the centrifugation technique has not yet been optimised - and it seems that still lower concentrations would still be detectable - it is fair to state that at present ISEM and centrifugation with fluorescence detection reach the same sensitivity.

5. CONCLUSIONS

In this Chapter we have described the setting up, function and applications of a recently developed fluorescence detection system for the analytical ultracentrifuge. The principle of the optical system is basically that of a scanning microscope, in which only a very small area of the sample is illuminated by an electron beam or a laser beam, and the intensity of emitted or reflected radiation is recorded under a wide angle of observation. The spatial resolution is determined by the illuminated area inside the cell, in this case about 60 μm diameter. One may argue that the system is too expensive to be useful. In fact, the scanning with the laser beam is fairly simple and less expensive than the UV scanner. The only expensive component is the laser light source.

It is obvious from the examples of application that such a system is favourable in respect to *sensitivity* and *selectivity*. The higher sensitivity of fluorescence detection allows us to record sedimentation profiles of proteins present in about 10^4 times lower concentration as compared to that which is required for the conventional UV absorption scanner. It was shown that the sedimentation coefficient of a protein (the example considered here was bovine serum albumin) may be determined if the protein is present in a concentration as low as 2.1 ng/320 μl corresponding to 10^{-10} M. Since three molecules of the dye FITC were attached per protein molecule, the concentration of the fluorophore was 3×10^{-10} M. Thus, the sensitivity of the fluorescence detection system is close that of a good commercial fluorimeter. In the study on nucleic acids, sedimentation profiles could still be evaluated quantitatively, if the concentration was 530 pg/40 μl (*i.e.* about three orders of magnitude less than required for measurement with UV absorption). The nucleic acid was labelled non-covalently by the intercalating dye ethidium bromide. From these investigations the persistence length of natural double-stranded RNA could be derived, and the measurements had to be carried out with isolated gene segments of rotavirus, which indeed were available only in minute amounts. In this example, new results which are of common interest for biochemists and biophysicists, were obtained *only* after the fluorescence detection system was available. It could be shown, that the model of the "worm-like" cylinder was applicable to viroids, dsRNA, dsDNA and Z-DNA in the same manner, but that the adjustable parameter of the persistence length increased by nearly an order of magnitude from the most flexible molecules (*i.e.* viroids) to the stiffest nucleic acids (*i.e.* Z-DNA).

The selective observation of fluorescent labels attached to biopolymers - or of naturally fluorescent molecules - renders possible quantitative studies on these labelled biopolymers even if they are present only as one component in an oligo-component system or as a minor component of a complex mixture. The method needs, however, some additional preparative effort, because the fluorophores have to be attached to the biopolymer. For this purpose a whole variety of procedures is available from the literature, and a series of new labels and attachment reactions have been worked out recently in order to replace radioactive labels by fluorescence labels in molecular biological applications[8]. The advantage of selective labelling of one component is obvious from the applications described above. If protein-nucleic acid or protein-protein interactions are to be analysed, the distribution of the labelled component between the free and the complexed state can be easily evaluated and binding constants and stoichiometries determined. For promotor-polymerase interactions fairly low binding constants, *i.e.*, around 10^7 M^{-1}, can be determined. It is noteworthy that low binding constants of biopolymers are difficult to determine, because the binding equilibrium is fast, and the complexes may dissociate during their measurement. This is particularly serious if the concentrations were analysed by separation techniques such as preparative ultracentrifugation, gel electrophoresis, chromatography, *etc*. In sedimentation runs, however, the binding equilibrium is sustained during the measurements, as pointed out earlier[5].

The achievements of the virus detection procedure were discussed above. We would like to summarise by saying that the favourable features of fluorimetry and centrifugation can clearly be combined. By using antibody-bound fluorophores virus can be detected with high sensitivity and selectivity. Concentrating the virus at the radius of its characteristic buoyant density increases the sensitivity even more and permits us to carry out the analyses with unpurified solutions. It may be regarded as a unique feature of the method that *quantitative* physical characterization of *unpurified* substances may be performed with high sensitivity.

We expect in the future that studies on protein-nucleic acid and protein-protein interactions under equilibrium conditions will be particularly attractive, if one component is labelled by a fluorophore. The whole dynamic range of fluorescence measurements could be used and this would allow the determination of binding constants from very low to very high values (*e.g.*, 10^{11} M^{-1}). Quite generally, fluorescence detection may be superior to absorption, not only because of the higher sensitivity but also because of the outstanding dynamic range.

ACKNOWLEDGEMENTS

We would like to thank Ms. H. Gruber for help in preparing the manuscript. The work was supported by grants from the Deutsche Forschungsgemeinschaft, the Bundesminister für Forschung und Technologie and the Fonds der Chemischen Industrie.

REFERENCES

1. H.K. Schachman, 'Ultracentrifugation in Biochemistry', Academic Press, New York, 1959.
2. A. Neuberger and L.L.M. van Deenen, 'Modern Physical Methods in Biochemistry', Elsevier, Amsterdam, 1988.
3. M.S. Lewis and G.H. Weiss, 'Fifty Years of the Ultracentrifuge', North-Holland Publishing Company Amsterdam, New York and Oxford, 1976.
4. T.C. Goodman, L. Nagel, W. Rappold, G. Klotz and D. Riesner, *Nucleic Acids Res.*, 1984, *12*, 6231.
5. G. Krauss, A. Pingoud, D. Boehme, D. Riesner, F. Peters and G. Maass, *Eur. J. Biochem.*, 1975, *55*, 517.
6. Goldman, 'Fluorescent Antibody Methods', Academic Press, New York, 1968.

7. J.G.J. Bauman, J. Wiegant, and P. van Duijn, *J. Histochem. Cytochem.*, 1981, *29*, 138.
8. L.M. Smith, S. Fung, M.W. Hunkapiller, T.J. Hunkapiller and L.E. Hood, *Nucleic Acids Res.*, 1985, *13*, 2389.
9. R.H. Crepeau, R.H. Conrad and S.J. Edelstein, *Biophys. Chem.*, 1976, *5*, 27.
10. W. Rappold, Ph.D. Thesis, Universität Düsseldorf, 1986.
11. B. Schmidt, W. Rappold, V. Rosenbaum, R. Fischer and D. Riesner, *Colloid Polym. Sci.*, 1990, *268*, 45.
12. T.H. The and T.E.W. Feltkamp, *Immunology*, 1970, *18*, 865.
13. G. Steinbach and H. v. Mayersbach, *Acta Histochem.*, 1976, *55*, 110.
14. C. Tanford, 'Physical Chemistry of Macromolecules', J. Wiley, New York, 1961.
15. J.M. Kaper and M.E. Tousignant, *Endeavour New Series*, 1984, *8*, 194.
16. W.K.J. Joklik, 'The Reoviridae', Plenum Press, New York, 1983.
17. P. Dobos, B.J. Hill, R. Hallet, D.T.C. Kells, H. Becht and D. Teninges, *J. Virol.*, 1979, *32*, 593.
18. K.W. Buck, H.-W. Ackermann, R.F. Bozarth, J.A. Bruenn, X. Koltin, C.J. Rawlinson, R. Ushiyama and H.A. Wood, *Intervirology*, 1984, *22*, 17.
19. Z. Kam, N. Borochov and H. Eisenberg, *Biopolymers*, 1981, *20*, 2671.
20. J.R. Diaz-Ruiz and J.M. Kaper, *Biochim. Biophys. Acta*, 1979, *564*, 275.
21. E. Spiessmacher, H.-P. Mühlbach, M. Tabler and H.L. Sänger, *Biosci. Rep.*, 1985, *5*, 251.
22. D.K. Bodkin and D.L. Knudson, *Virology*, 1985, *143*, 55.
23. O. Kratky and G. Pörod, *Recl. Trav. Chim.*, 1949, *68*, 1106.
24. J.E. Hearst and W.H. Stockmayer, *J. Chem. Phys.*, 1962, *37*, 1425.
25. H. Yamakawa and M. Fujii, *Macromolecules*, 1973, *6*, 407.
26. J. García de la Torre and V.A. Bloomfield, *Q. Rev. Biophys.*, 1981, *14*, 81.
27. G.W. Both, L.J. Siegmann, A.R. Bellamy, N. Ikegami, A. Shatkin and J. Furuichi, *J. Virol.*, 1984, *51*, 97.
28. J. Cohnen, F. Lefevre, M.K. Estes and M. Bremont, *Virology*, 1984, *138*, 178.
29. M.L. Dyall-Smith, T.C. Elleman, P.A. Hoyne, J.H. Holmes and A.A. Azad, *Nucleic Acids Res.*, 1983, *11*, 3351.
30. T.C. Elleman, P.A. Hoyne, M.L. Dyall-Smith, J.H. Holmes and A.A. Azad, *Nucleic Acids Res.*, 1983, *11*, 4689.
31. C.W. Ward, T.C. Elleman, A.A. Azad and M.L. Dyall-Smith, *Virology*, 1984, *134*, 249.
32. H.N. Baybutt and M.A. McCrae, *Virus Research*, 1984, *1*, 533.
33. M. Imai, M.A. Richardson, N. Ikegami, A.J. Shatkin and J. Furuichi, *Proc. Natl. Acad. Sci. U.S.A.*, 1983, *80*, 373.
34. F. Rixon, P. Taylor and U. Desselberger, *J. Gen.Virol.*, 1984, *65*, 133.
35. R. Kapahnke, W. Rappold, U. Desselberger and D. Riesner, *Nucleic Acids Res.*, 1986, *14*, 3215.
36. R.T. Kovacic and K.E. van Holde, *Biochem.*, 1977, *16*, 1490.
37. W. Saenger, 'Principle of Nucleic Acid Structure', Springer-Verlag, Heidelberg, 1984.
38. P. Taylor, F. Rixon and U. Desselberger, *Virus Res.*, 1985, *2*, 175.
39. H.L. Sänger, G. Klotz, D. Riesner, H.J. Gross and A.K. Kleinschmidt, *Proc. Natl. Acad. Sci. U.S.A.*, 1976, *73*, 3852.

40. D. Riesner and J.J. Gross, *Ann. Rev. Biochem.*, 1985, *54*, 531.
41. D. Riesner and G. Steger, 'Landoldt-Börnstein, Numerical Data and Functional Relationships in Science and Technology', New Series, Group VIII, Vol 1, Nuleic Acids, Sub volume d, ed. W. Saenger, p. 194, Springer-Verlag, Berlin, 1990.
42. J. Floßdorf, *Makromol. Chem.*, 1980, *181*, 715.
43. D. Riesner, J.M. Kaper and J.W. Randles, *Nucleic Acids Res.*, 1982, *10*, 5587.
44. J. Haseloff, N.A. Mohamed and R.H. Symons, *Nature*, 1982, *299*, 316.
45. G. Steger, M. Tabler, W. Brüggemann, M. Colpan, G. Klotz, H.L. Sänger and D. Riesner, *Nucleic Acids Res.*, 1986, *14*, 9613.
46. N. Borochov, H. Eisenberg and Z. Kam, *Biopolymers*, 1981, *20*, 231.
47. T.J. Thomas and V.A. Bloomfield, *Nucleic Acids Res.*, 1983, *11*, 1919.
48. U. Wild, K. Ramm, H.L. Sänger and D. Riesner, *Eur. J. Biochem.*, 1980, *103*, 227.
49. R. Girg, R. Jaenicke and R. Rudolph, *Biochem. Int.*, 1983, *7*, 433.
50. H.Eisenberg and G. Felsenfeld, *J. Mol. Biol.*, 1981, *150*, 537.
51. H.L. Sänger, 'The Viroids' ed. Diener, Plenum Publishing Corporation, New York, 1987.
52. H.-R. Rackwitz, W. Rohde and H.L. Sänger, *Nature*, 1981, *291*, 297.
53. J.P. Johnston and A.G. Ogston, *Trans. Faraday Soc.*, 1946, *42*, 789.
54. D. Riesner, J. Harders, R. Hecker, P. Klaff, P. Loss, N. Lukacs and G. Steger, 1991, NATO ASI Series, 'Plant Molecular Biology', in press.
55. D.R. Davies and H. Metzger, *Ann. Rev. Biochem.*, 1983, *1*, 87.
56. M.F. Clarke and A.N. Adams, *J. Gen. Virol.*, 1977, *34*, 475.
57. P. Fischer and F. Kreuzaler, manuscript in preparation.
58. H.J.P. Schoemaker and P.R. Schimmel, *J. Mol. Biol.*, 1974, *84*, 503.
59. M.H.V. van Regenmortel and G. Hardie, *Immunochem.*, 1976, *13*, 503.
60. E.P. Rybicki and M.B. von Wechmar, *J. Virol. Methods*, 1982, *5*, 267.
61. K.S. Derrick, *Virology*, 1973, *56*, 652.
62. R.G. Milne, *Acta Hortic.*, 1980, *110*, 129.
63. R. Koenig, D.E. Lesemann and D.E. Burgermeister, *Phytopath. Z.*, 1984, *111*, 244.

12

Sedimentation Field Flow Fractionation

By E.C. Arner and J.J. Kirkland

INFLAMMATORY DISEASES RESEARCH, THE DU PONT MERCK PHARMACEUTICAL COMPANY, WILMINGTON, DE 19880-0400, U.S.A.

1. INTRODUCTION

Sedimentation field flow fractionation (SFFF) is a relatively new, high resolution technique for the separation of a wide range of particulates, colloids and soluble macromolecules. SFFF is a member of a family of field flow fractionation (FFF) methods which were introduced by Giddings in the mid-1960s[1,2].

All of the FFF methods achieve separation through a common principle that can be thought of as one-phase chromatographic methods. They are similar to chromatography in that separation is achieved by differential elution of the sample components as they pass through a column or channel. However, in FFF there is no stationary phase or solid support. The sample is injected into a very thin, open channel with a continuously flowing carrier liquid (Fig. 1), and components initially are uniformly distributed across the channel. Generation of laminar flow within this narrow channel results in a characteristic parabolic flow profile with maximal flow rate at the centre of the channel and flow decreasing to near zero at the channel walls. An external force field is applied perpendicular to the direction of the flow. This field, which can be a temperature gradient as in thermal FFF, electrical potential as in electrical FFF, hydraulic pressure as in cross-flow FFF or centrifugal force as in sedimentation FFF[3-5] causes redistribution of the sample components with larger or more dense particles forced into slower flow streams near the channel wall. Smaller or less dense particles are forced into faster flow streams further from the wall (Fig. 1). Thus components elute in order of increasing effective mass or molecular weight.

SFFF is the most extensively developed of the various FFF methods. It can be used for materials with a wide range of molecular weights (10^5 to 10^{11}) or a size range of approximately 0.01 μ to 1.0 μ. It also has the highest resolving power, and allows determination of molecular weights directly from physical parameters of the sample components and experimental conditions. SFFF is a comparatively low-sheer and non-interactive technique that provides rapid, mild separations compared with conventional methods based on density equilibrium, sizing gels or chemical interactions. Therefore, it appears to be particularly well suited for the fractionation and isolation of fragile macromolecules and cellular components of interest in biological research. A unique feature of SFFF is that the molecular weight of components can be determined without reference to standards, since this mass-separating method is based on first principles[6].

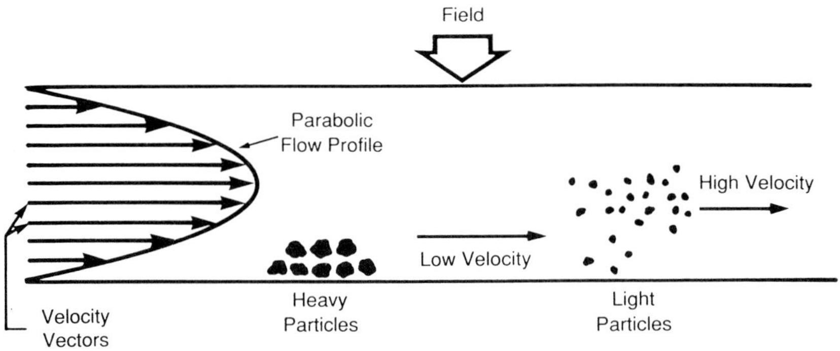

Figure 1. Principle of field flow fractionation.

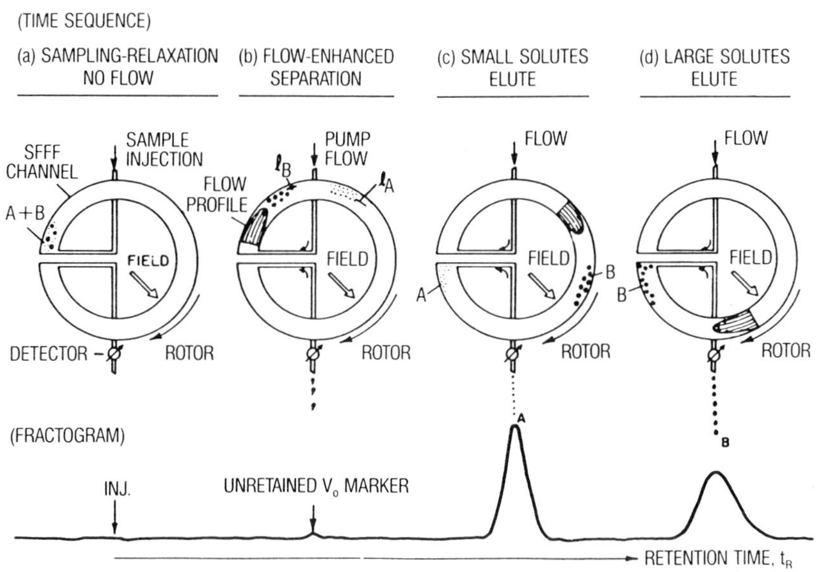

Figure 2. Development of an SFFF separation. Following sample loading, flow is stopped and sample components reach sedimentation equilibrium during the relaxation period (a). Flow is then re-initiated and sample components migrate at different rates depending on the flow rate of the fluid stream carrying them (b). Components elute in order of increasing mass (c and d).

2. THEORY

The theoretical basis for SFFF is briefly reviewed. For more detailed discussion, see refs 6-10.

Separation takes place in a very thin (*e.g.*, 0.25 mm) open, ribbon-like channel, located in the rotor of a centrifuge, through which a carrier liquid flows at a uniform velocity. The sample is injected into the rotating channel as a uniform suspension or solution in the carrier. After the sample has entered the separation channel, carrier flow is stopped. The external centrifugal force, G, perpendicular to the channel is held constant, allowing sedimentation equilibrium to be achieved (Fig. 2a). In this equilibration or relaxation step, the centrifugal force causes sample components more dense than the carrier liquid to be pushed toward the outer wall of the channel. Accumulation of sample concentration near the wall is opposed by normal diffusion of the particles in the opposite direction. The distribution of the solute or dispersed particle is exponential:

$$c = c_0 \, e^{-x/l} \qquad (1)$$

where c is the concentration of the solute at a distance x from the wall and l is the mean thickness. The value of l depends on the velocity, v, due to the applied force and to the normal diffusion coefficient, D, such that $l = D/v$. Distinct species assume different characteristic layer thicknesses within the channel. Smaller particles have higher diffusion rates, and, as illustrated in Fig. 2b, accumulate in a layer (l_a) that extends further from the outer wall than that of larger particles (l_b).

After the prerequisite creation of stratified layers across the channel is accomplished through the development of sedimentation equilibrium, the actual separation is accomplished by the carrier flow. The carrier flow through the channel is resumed, and a parabolic velocity flow profile is formed (Fig. 2b). Therefore, particles closer to the walls are in regions of slower flow and particles further from the walls are in regions of more rapid flow. Carrier molecules and very small particles that are not influenced by the field remain uniformly distributed across the channel. These materials travel in all flow streams at an average velocity equal to the carrier and elute first (Fig. 2b), followed by particles in order of increasing effective mass (Figs. 2c and 2d). Elution is continuously monitored by a detector which measures concentration. In this way, a fractogram is obtained which displays eluted peaks *versus* retention time, similar to a chromatogram (Fig. 2).

At a constant centrifugal force, the molecular weight of a retained component relates to its retention as follows:

$$M = \left\{ \frac{6RTV}{V_0 G W \; (\Delta\delta/\delta_s)} \right\} \qquad (2)$$

where M is the molecular weight of the component (g/mol); R is the gas constant (8.31×10^7 g.cm.s^{-1}.K^{-1}.mol^{-1}); T is temperature (Kelvin); V is the retention volume of the eluted component (ml); V_0 is the dead volume of the channel (ml); W is the thickness of the channel (cm); and $\Delta\delta$ is the density difference (g/cm^3) between the component density δ_s (g/cm^3) and the carrier liquid. G is the force field that is related to rotor speed by:

$$G = \left\{\frac{\omega\pi}{30}\right\}^2 r \tag{3}$$

where ω is the rotor speed in rev/min, and r is the distance (cm) from the centre of the rotor to the channel wall.

Using eq. (2) the molecular weight of a component eluting from an SFFF channel can be calculated from physical parameters that are known (rotor speed, channel dimensions, carrier and particle density, *etc.*) as a function of the observed retention volume. Since shape-dependent frictional coefficients associated with particle movement cancel out, conformational features of the components normally do not effect retention[10]. Therefore, SFFF can, as with conventional sedimentation in the analytical ultracentrifuge (Chapters 13 to 17) directly measure the mass of a component without the need for standards.

An SFFF instrument can be operated at a constant centrifugal force field (rotor speed), or it can be programmed to vary the field. Although the highest resolution is obtained with a constant force field, separations have several limitations in this mode. Constant-field SFFF gives uneven particle size resolution across the fractogram and the range of possible size separation is very narrow. This presents a problem for separating samples with wide mass or size distribution. In addition, the widely broadened peaks of late-eluting components may be difficult to detect. Fig. 3A illustrates these disadvantages by showing fractograms of a mixture of five narrow particle-size distribution polystyrene latex standards fractionated at two different constant force fields. With the lower force field (1710 rev/min), only the largest particle component (peak E) was completely resolved from the other particles. At the higher force field (5360 rev/min), although the two smallest components still overlap, the larger particles are resolved but only at excessively long retention times. This demonstrates the narrow range of constant-field SFFF; the five components cannot be resolved in a reasonable analysis time with a single force field. The non-uniform resolution is demonstrated by the much better resolution of the larger particles than the smaller ones. The excessive broadening of late-eluting peaks is exemplified by peak E from the high force field separation.

A controlled reduction of gravitational field in SFFF is effective in overcoming these problems[11-13]. In this method, called time-delay, exponential-decay SFFF (TDE-SFFF), after sample injection and relaxation, the initial force, G_0, is maintained after the start of the carrier for a time period of τ. Then, the force field is allowed to decay exponentially at a rate corresponding to the same constant, τ. This method has the advantage of allowing the separation of samples containing components with a very wide mass range to be carried out with good resolution in a reasonable time period[7]. The advantages of TDE-SFFF are illustrated in Fig. 3B, showing the fractogram for five polystyrene latex standards that were separated in approximately 20 minutes with easily-detected bands of about equal width.

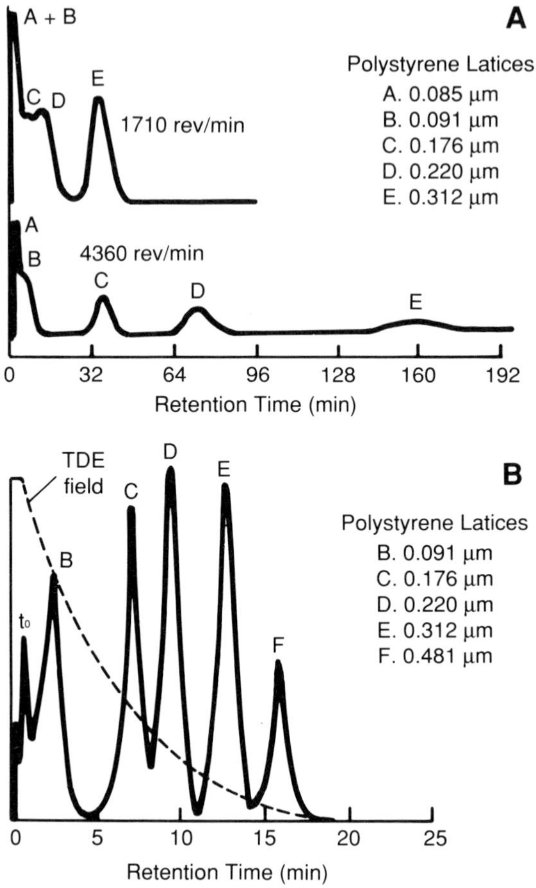

Figure 3. Separation by constant force and time-delay exponential SFFF. (A) Constant force field separations at two rotor speeds. Channel, 57 by 2.54 by 0.0254 cm; mobile phase, 0.1% FL-70 surfactant; flow rate, 2.0 ml/min.; relaxation, 10 min. at ω_0; sample, 25 ml, 0.1% of 0.085 μm, 0.9% of 0.091 mm, and 0.04% each of 0.176 μm, 0.220 μm and 0.312 mm polystyrene latex standards; ultraviolet detector, 300 nm; temperature 22°C. (B) Time delay exponential separation. Conditions as in A with the following exceptions: flow rate, 3.0 ml/min.; initial rotor speed, 10,000 rev/min; decay constant, $\tau = 4.76$ min; sample, 10 μl, 0.99% of 0.09 μm, 0.04% each of 0.176 μm, 0.220 μm, and 0.312 μm and 0.05% of 0.418 μm polystyrene latex standards[21].

TDE-SFFF also enables optimisation of resolution for any portion of the molecular weight range separable by SFFF. The resolution of smaller sample components is optimised by initially using a relatively high force field, held constant for a long time period and then rapidly exponentially decayed (Fig. 4A). Fractionation of the same sample starting with a lower force field, a shorter time at constant field and a longer decay period optimises resolution of the higher molecular-weight components (Fig. 4B). Thus, in a given sample, isolation of a higher molecular-weight component can be accomplished in the same amount of time as a low molecular-weight component, depending on the selection of conditions.

For well-retained samples in TDE-SFFF, the molecular weight (g/mol) and retention time t_R (min) of the components exhibit a simple log-linear relationship:

$$\ln M = \ln A + \frac{t_R}{\tau} \tag{4}$$

where

$$A = 6R_0T\tau \exp\left\{-t_0G_0W\frac{\Delta\delta}{\delta_s}\right\} \tag{5}$$

t_0 is the elution time of an unretained component (channel dead time) (min); τ is the time (min) of the delay and also the time constant of the exponential decay; and G_0 is the force field initially applied.

When the density of the sample is unknown, several SFFF runs with different carrier liquid densities provides data so that both particle size and particle density can be determined[14]. This type of experiment can be carried out in either constant-field or TDE-SFFF mode.

3. COMPARISON OF SFFF WITH OTHER METHODS

Chromatography

Some differences between SFFF and gel permeation chromatography are due to the different type of forces used to induce retention. In gel permeation chromatography, these forces are highly localised at surfaces and phase boundaries. They tend to be very powerful and capable of causing structural disruption of macromolecules. Retention forces in SFFF are more diffuse and locally weak, and thus, normally do not cause alteration of molecular conformation. As the molecular size of sample components increases, the flow process in packed chromatography beds is capable of exerting disruptive shear forces on macromolecules[15]. By contrast, the shear forces induced by the uniform laminar flow of SFFF channels are relatively gentle.

The resolving power of SFFF can be up to five times that of size-exclusion chromatography (SEC)[16]. This difference in resolution is due to the different mechanisms of separation with these two methods. In SEC the separation is confined to the pore volume of the gel, with all components eluting between the total exclusion

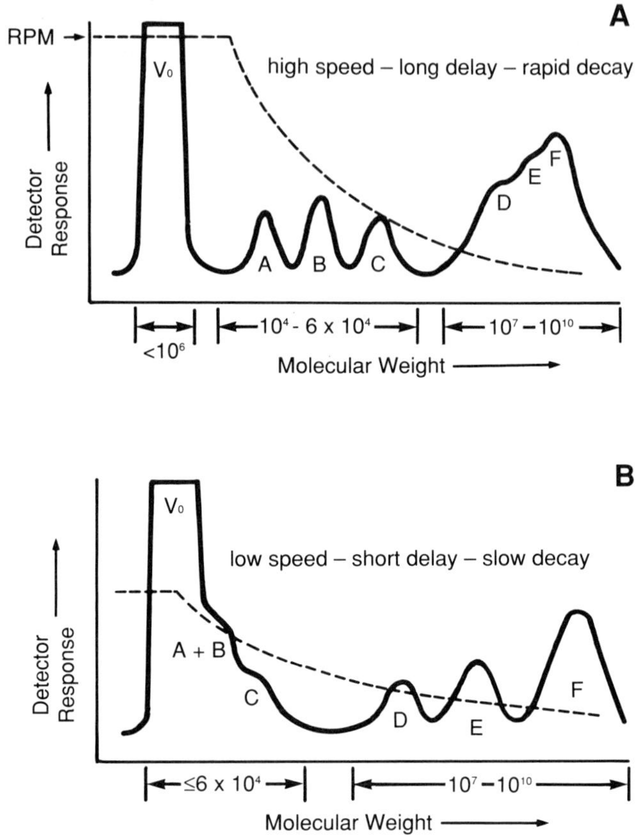

Figure 4. Optimisation of resolution using exponential-decay SFFF. A hypothetical sample containing components with a broad range of molecular weights can be separated to optimise resolution of the lower molecular weight components by using a high initial speed, large delay and rapid decay (A) or to optimise resolution of higher molecular weight components by using low initial speed, short delay and slow decay (B).

and total permeation volumes. However, in SFFF peaks elute well after the unretained solvent peak. Thus, SFFF has the potential for large peak capacity and high resolution in comparison to SEC. Although both SFFF and chromatography have the flexibility of altering the speed of separation by altering flow rate, retention in SFFF is induced by externally-controlled forces. Therefore, with no change in the SFFF channel or equipment, the centrifugal force (rotor speed) can be adjusted to maximize the separation of a particular component. Alternatively, it can be gradually changed in the course of a run (*e.g.*, by TDE-SFFF) to accommodate differing sample components, or to control resolution and speed.

While SEC is not an absolute method without appropriate standards, SFFF separations are based directly on physical first principles. Therefore, it is possible to deduce molecular weights or particle size directly without the need for standards.

Conventional Centrifugation

Conventional centrifugational methods based on sedimentation velocity are affected by the viscosity of the medium and the shape of the solutes. SFFF, on the other hand, is a true sedimentation-equilibrium-based process and, with the use of an appropriate detector, can therefore provide accurate and absolute measurements of solute molecular weight[6]. As an *analytical* tool, it is, in a sense, complementary to conventional sedimentation equilibrium in the analytical ultracentrifuge and provides separation as well. Indeed, as a separation tool, SFFF can provide shorter analysis times and higher resolution than conventional centrifugation techniques, because of different sedimentation distances and separation mechanisms associated with these methods. The sedimentation distances are extremely small in SFFF (~0.25 mm), about two orders of magnitude smaller than those in conventional centrifugation separations. Thus, SFFF provides faster separations. SFFF also has a larger peak capacity than conventional centrifugation. In conventional centrifugation separation methods, separation is also limited by the length of the centrifuge tube, and therefore, this method has a limited peak capacity. Increasing the resolution of the lower density components at the top of the tube will result in a simultaneous loss of resolution of the higher density components at the bottom of the tube.In SFFF, resolution and peak capacity of all components are enhanced by increasing centrifugal force, with the limitations of practical analysis time and maximal field available.

4. INSTRUMENTATION

In practice, the SFFF channel is located within the rotor of a centrifuge. A schematic diagram of the general configuration of SFFF equipment is shown in Fig. 5. Standard liquid chromatography equipment is used to control carrier flow, sample injection and sample detection. A liquid chromatography solvent-metering pump provides carrier flow, and sample injection is accomplished using a microsampling loop-injector valve.

Because this is a continuous-flow system within a centrifuge, an essential part of the assembly is a specially designed, water-cooled, rotating force-seal which allows sample and carrier liquids to be introduced and removed from the channel while

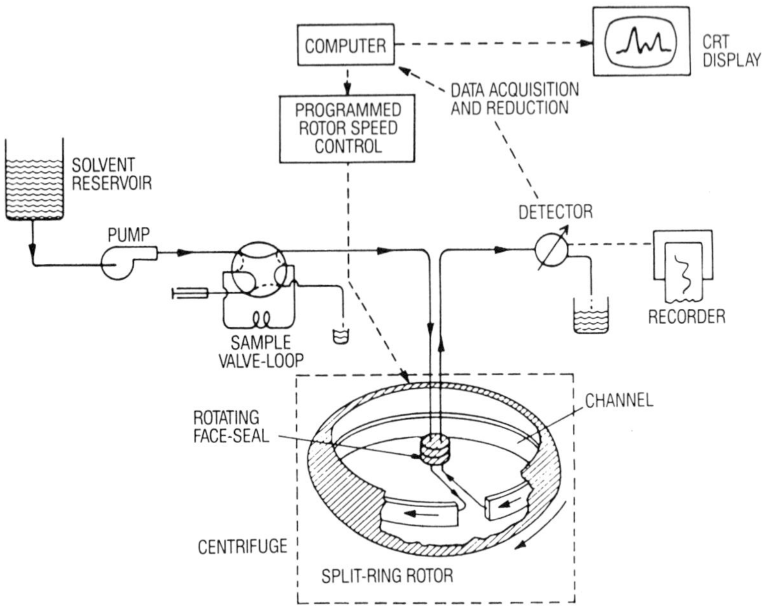

Figure 5. A schematic of SFFF equipment.

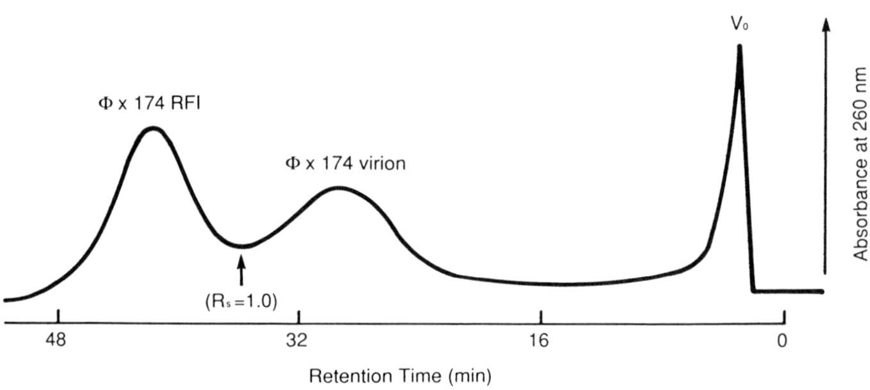

Figure 6. Separation of a DNA mixture. Conditions were: sample, 6 mg of ϕX174 RFI and 6 mg of ϕX174 virion in 100 μl; mobile phase, 0.1 M Tris and 0.1M NaCl (pH 7.6); flow rate, 0.5 ml/min; initial rotor speed, 30,000 rev/min; channel thickness 0.024 cm; and time-delay constant, $\tau = 12$ min.

spinning in the centrifuge[8]. Seals typically are used daily for months without replacement.

Several types of channel configurations have been developed for use in high-speed and ultracentrifuges[8]. One design uses continuous rings of an appropriate engineering plastic, precision fitted to the polished inside surface of a metal rotor bowl. Because of its inherent elasticity the plastic ring can expand with the centrifuge bowl as the force field is increased. An important aspect of the design is that the area inside the rotor and around the plastic ring is filled with liquid of approximately the same density as the carrier fluid. At high gravitational field this minimizes the pressure differences developed, between carrier liquid in the channel and the outside region around the plastic. In this way, stresses on the plastic channel are greatly reduced.

Components eluting from the channel can be detected using a spectrophotometric detector with a flow cell. This is used either for direct measurement of sample absorbance or as a turbidometric detector, depending on sample requirements. A microcomputer or PC is used to control rotor speed during exponential programming, monitor rotor speed, store detector output and compute particle size or molecular weight data.

5. APPLICATIONS

SFFF was initially used for separating and determining the size of a wide range of non-biological particles and colloids of industrial interest[6,8,17-20]. However, the most important applications for this method may be for the isolation and characterization of macromolecules and particulates of biological origin. Because SFFF is a very mild, non-interactive technique which imposes low shear forces on components, it is particularly attractive for work with fragile biomolecules such as nucleic acids and many glycopolymers. In addition, typical separations are complete in 1-2 hours and recoveries of isolated materials are high (80%-100%). Since SFFF separations of biological materials are normally done using conventional buffered-salt solutions as the carrier fluid, isolated materials can often be used without additional clean-up procedures. Further, and as we have stressed above, since sample concentration and molecular conformation do not influence retention, molecular weights can be obtained without reference to standards, thus allowing the determination of molecular weights during the separation process.

Nucleic Acids

Many nucleic acids of interest in biology or of use in molecular biology have molecular weights in the range accessible by SFFF. Separation and characterization of a number of viral and plasmid DNAs in the molecular weight range of $1.5\times10^6 - 33\times10^6$ by SFFF[8,21-23] demonstrate the advantages of this technique for biological applications. Fractionation of a mixture of commercially available ϕX174 RFI DNA ($M \sim 3.5\times10^6$) and ϕX174 virion DNA ($M \sim 1.75\times10^6$) in Fig. 6 illustrates the high degree of resolution which can be obtained with SFFF[22]. If these components were collected separately at the arrow, each fraction would exhibit a purity of 98%. This rapid (<1 hr) separation also demonstrates close agreement between the molecular weights of these DNAs determined by SFFF as compared with reported values. Agreement between literature values for molecular weights and those determined by SFFF have been demonstrated for a wide range of DNAs (Table 1)[8,22-23].

Table 1. Molecular weights of nucleic acids and assemblies thereof as determined by SFFF compared with other techniques.

Sample	Molecular Weight	
	Literature	SFFF
pBR322	2.8×10^6	2.7×10^6
pBR325	3.6×10^6	3.5×10^6
λ DNA	33.0×10^6	33.4×10^6
pDM24D	1.5×10^6	1.3×10^6
ΦX174 RFI	3.5×10^6	3.5×10^6
ΦX174 Virion	1.75×10^6	3.5×10^6

Studies with the cloning vector pBR322 and λ-viral DNA demonstrated that both small, super-coiled DNA and large, linear, fragile λ-DNA can be separated without shear damage. Electrophoresis of λ-DNA samples after being subjected to SFFF showed no alteration in molecular weight nor indication of degradation as compared to untreated control λ-DNA (Fig. 7)[22]. The biological activity of the λ-DNA isolated by SFFF, as assessed by monitoring its ability to lyse bacteria, was equivalent to that of a control sample, illustrating that, SFFF does not alter biological activity of this material. In addition, studies with both pBR322 and λ-DNA showed essentially quantitative isolation yield, indicating that no significant amount of DNA was irreversibly retained during SFFF separation.

SFFF has also shown applicability in a practical recombinant DNA procedure involving the preparative isolation of pBR322 plasmid from crude cellular lysates containing large amounts of contaminating RNA and small amounts of protein. Separation of a single retained peak was obtained, completely resolved from V_0 which contained the RNA and protein contaminants (Fig. 8)[23]. Total purification time including the final SFFF isolation was about three quarters of a day, significantly less than required using conventional isolation techniques.

Liposomes

The use of liposomes in drug delivery systems requires an exact knowledge of their size. Quantitative particle-size distribution analyses of unilamellar and larger multilamellar liposomes have been achieved using SFFF (Fig. 9)[24]. The weight-average diameter obtained by SFFF compared well to the value determined by negative-stain electron microscopy. Again, this low-shear method did not cause a decrease of multilamellar liposomes to smaller particles by shear, nor an increase in the apparent size through aggregation. SFFF has also been used to monitor changes in

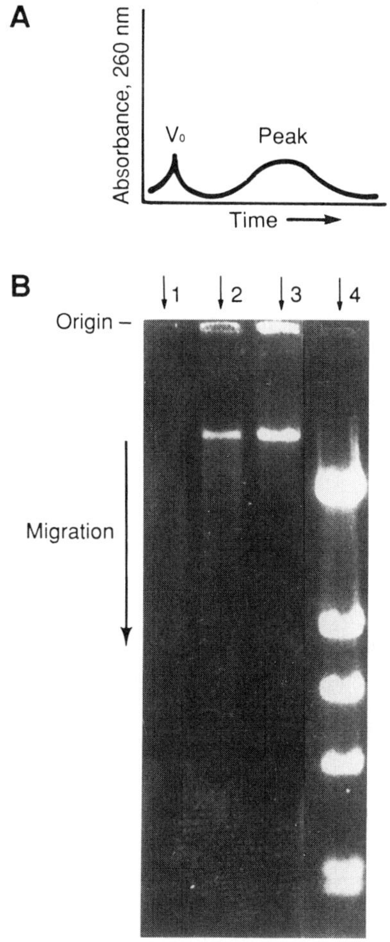

Figure 7. Fractionation and analysis of λ-DNA. (A) Fractogram of λ-DNA. Conditions were: sample, 6 μg λ-DNA in 100 μl; rotor speed, 15,000 rev/min; relaxation time, 10 min; time-delay constant, τ = 10 min; flow rate, 0.5 ml/min; mobile phase, 10 mM Tris, 10 mM NaCl (pH 7.6). (B) Agarose gel electrophoresis of fractionated λDNA. The gel was a 0.4% agarose and running buffer, 4 mM Tris (pH 7.4), 36 mM PO_4, 1 mM EDTA.

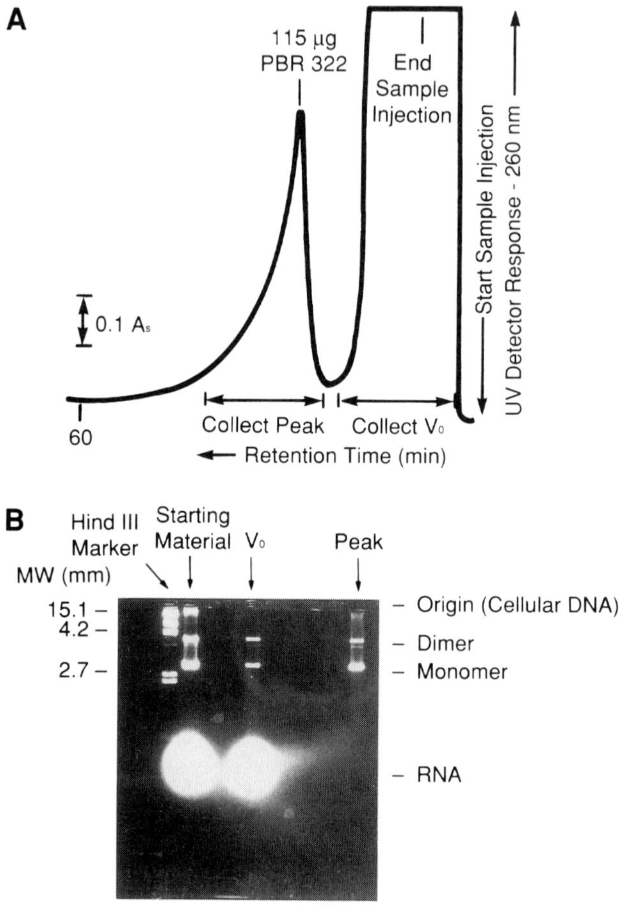

Figure 8. Preparative SFFF isolation and analysis of pBR322. (A) Fractogram of pBR322. Conditions were: sample, 3.3 ml of crude lysate; channel thickness, 0.024 cm; rotor speed, 32,000 rev/min (80,000 g); mobile phase, 0.01 M Tris - 0.1 M NaCl (pH 7.6); flow rate 0.5 ml/min. (B) Agarose gel electrophoresis of pBR322 preparative fractions. Conditions were: 0.7% agarose gel; sample, 0.5 mg each lane.

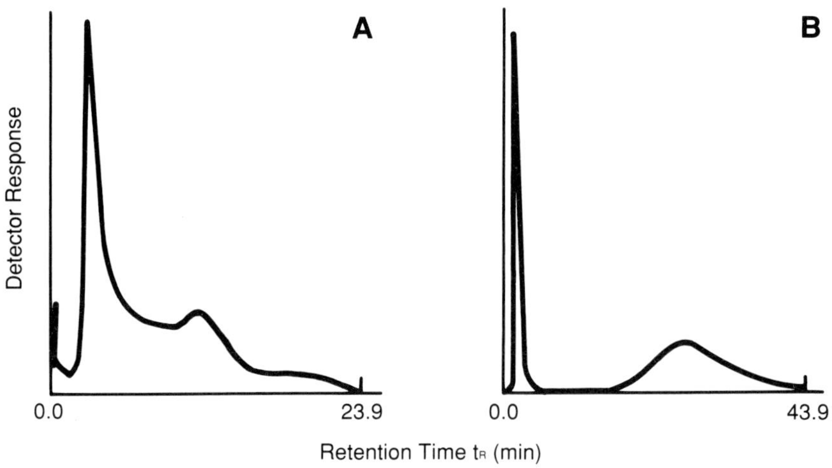

Figure 9. SFFF fractograms for liposome preparations. Conditions were: channel, 57 cm × 2.54 cm × 0.0254 cm; mobile phase, 0.01 M Hepes, 0.142 M NaCl, 0.0055 M NaOH (pH 7.45); flow rate, 2.0 ml/min. (A) Sonicated unilamellar vesicles. Conditions: mobile phase, buffer described above plus sugar (ρ = 1.18 g/cm^3); initial rotor speed, 18,000 rev/min; exponential decay constant, 4.0 min; initial time delay, 6 min; ultraviolet detector, 240 nm. (B) Multilamellar vesicles. Conditions same as for (A) except buffer (ρ = 1.004 g/cm^3); initial time delay, 4.0 min; ultraviolet detector, 350 nm.

vesicle size during the formation of fused unilamellar vesicles of dipalmitoylphosphatidyl-chlorine[25]. SFFF compares favourably with other methods of liposome analysis. Where electron microscopy suffers from possible staining artifacts and gel filtration is limited by poor lipid recoveries and large elution times, SFFF is a rapid and non-destructive method of analysis with quantitative recovery of the vesicles.

Bioparticulates

Studies have demonstrated the ability of SFFF to fractionate viruses including T-2[11,26], QB and P22[27], fD[11] and T-4D[28]. Observed molecular weights were in close agreement with literature values. Starting with lysates of inclusion bodies from gypsy moth larvae, monomeric, dimeric and trimeric forms of the enveloped gypsy moth nuclear polyhedrosis virus have been purified to homogeneity (Fig. 10)[29] in a single SFFF run. In comparison a sucrose density-gradient sedimentation of the same virus sample showed less resolution and required much longer total analysis time. The observed molecular weights for the enveloped monomers and envelope-free monomers corresponded closely with literature values[29].

SFFF has been used to subfractionate streptococcal cell wall fragments with a continuous weight distribution of 10^6-10^8 into samples of more defined, narrower mass range[30]. These fragments are used to induce arthritis in animal models; severity of the arthritis is dependent on fragment size. SFFF improved the degree of fractionation and resulted in time savings of several days over conventional methods. Particle size information on experimental particulates of aggregated albumin for liver-imaging studies has also been obtained[21], as well as the characterisation of other albumin microspheres[31].

High-resolution droplet-diameter profiles have been obtained for perfluorocarbon blood substitutes by SFFF analysis[32]. Since finely-dispersed perfluorocarbon emulsions are more stable than course emulsions and larger particles are cleared from the circulation more rapidly than smaller ones, particle size is the most important characteristic of perfluorocarbon blood substitutes. SFFF separations of perfluorocarbon emulsions, Fluosol-DA 20% and Fluosol-43 (Fig. 11)[32] show a shift in the droplet diameter profiles to larger diameter for both of these blood substitutes in less than two months although the change did not appear to be as rapid for Fluosol-43. These studies suggest that SFFF may have utility for monitoring emulsion stability of blood substitutes.

SFFF has been shown to be a convenient, efficient method for analysis of a wide variety of subcellular particle preparations, including eucaryotic ribosomes[33] as well as mitochondria, microsomes, Golgi membranes or plasma membranes from corn roots[34]. Rat liver microsomal pellet organelle components have also been fractionated[35] and various components identified. Although the microsomal pellet is a very heterogeneous mixture of materials (10^6-10^9 molecular-weight), resolution with SFFF was significantly higher than with conventional methods. These data suggest that SFFF may also prove useful for micro-preparative fractionation of some subcellular particle preparations.

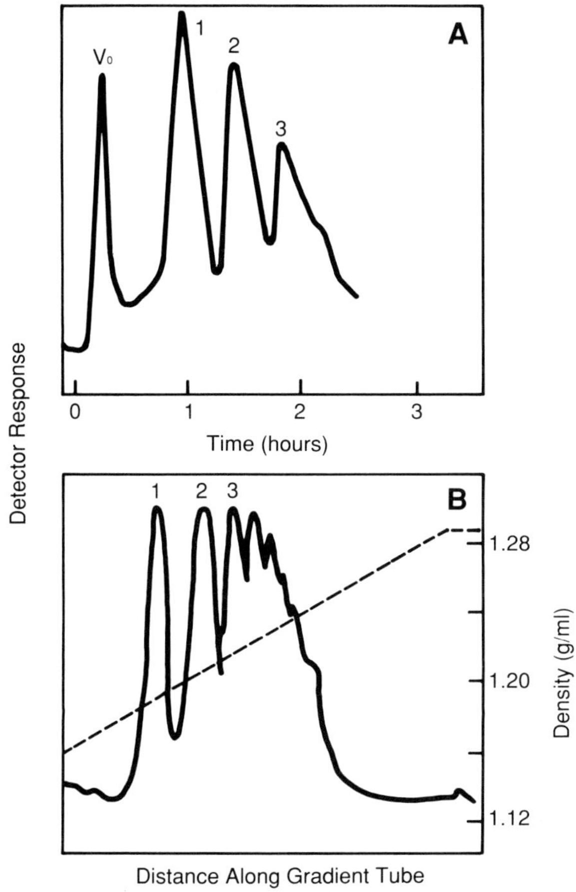

Figure 10. Separation of viral particles from gypsy moth larvae [redrawn from data in ref. 29]. (A) SFFF fractogram. Conditions were: channel, 83.3 cm × 2.0 cm × 0.0254 cm; flow rate, 0.4 ml/min; rotor speed, 1417 rev/min; sample, 500 mg in 100 μl; ultraviolet detector, 254 nm. (B) Sucrose density gradient centrifugation. Conditions were: density, 1.283 g/ml to 1.127 g/ml (sucrose in 0.01 M tris - HAc, 0.001 M EDTA, pH 7.4); centrifugation, 24,000 rev/min for 4 hours; ultra-violet monitor, 254 nm; sample, 3 ml.

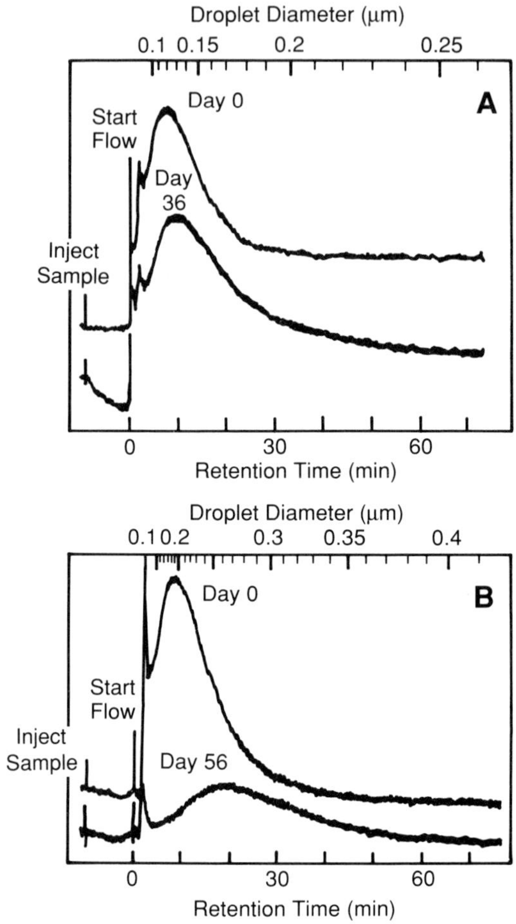

Figure 11. Time variation of fractograms for Flusol-43 and Fluosol-DA 20% by SFFF. (A) Flusol-43 fractograms. Conditions were: flow, 1.0 ml/min; channel; 79.4 by 2.0 by 0.0127 cm; sample, 10 μl; ultraviolet detector, 253.7 nm; field, 86.2 g. (B) Fluosol-DA 20% fractograms. Conditions were as in (A) except field was 21.5 g.

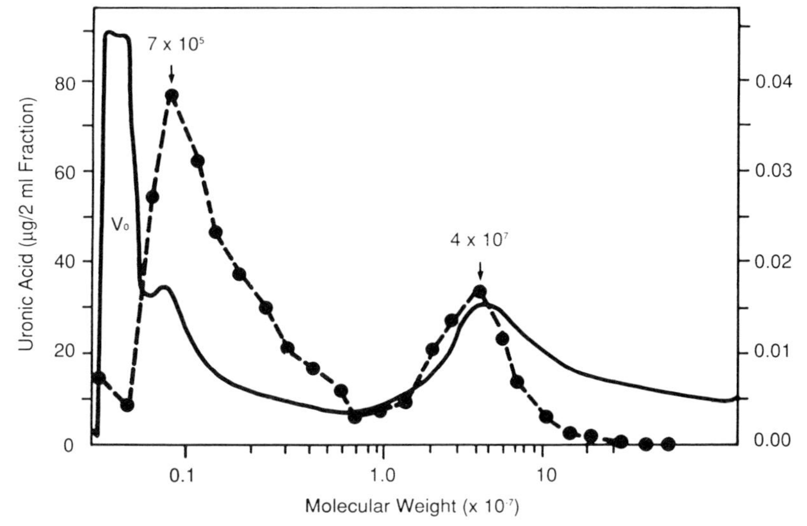

Figure 12. Fractogram of calf nasal cartilage proteoglycans. Solid line is absorbance at 280 nm; dash line represents the uronic acid content. Conditions were sample, 0.4 mg uronic acid in 0.5 ml; mobile phase, 0.05 M sodium acetate, pH 6, 0.15 M NaCl; flow rate, 0.5 ml/min; initial rotor speed, 32,000 rev/min (80,000 g); time delay, 20 min, exponential decay constant, 12 min.

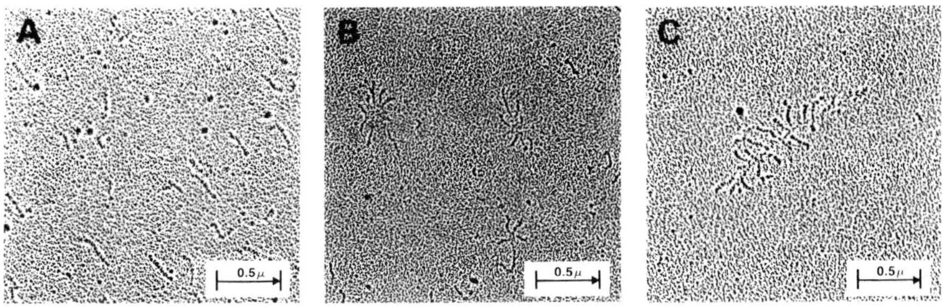

Figure 13. Transmission electron microscopy of calf articular proteoglycans from SFFF fractions. Separation conditions were as in Fig. 12. Fractions were collected at various retention intervals (A) M~1×10^6 - 2×10^6 (B) M~3×10^6 - 7×10^6 and (C) M > 1×10^6.

Proteoglycans

Proteoglycans, the component of cartilage responsible for its shock-absorbing properties, have been rapidly fractionated (1-2 hours) by SFFF[36-37]. In cartilage these complex glycoproteins exist as monomers ($M \sim 0.5\times10^6 - 2.5\times10^6$), and as aggregates with hyaluronic acid ($M \sim 10^7$-10^8). Mixtures of monomers and aggregates from bovine articular or nasal cartilage and rabbit articular cartilage have been fractionated. Observed molecular weight ranges calculated from SFFF correspond closely to those reported in the literature. Fig. 12[36] shows a typical SFFF separation of proteoglycans from calf nasal cartilage using as monitors both UV detection at 280 nm and uronic acid content, a more specific measure of proteoglycans.

Transmission electron microscopy of isolated fractions (Fig. 13) confirm the separation of the monomer and aggregate forms and the preservation of the fragile fern-like structure of the aggregate. SFFF has shown utility in detecting small variations in proteoglycan size distribution due to animal species and age, cartilage type, and exposure to cytokines[36-37]. In comparison to conventional separation methods used for proteoglycans, SFFF is more rapid, and allows recovery of separated components for further analysis without clean-up steps.

6. CONCLUDING REMARKS: ADVANTAGES OF SFFF FOR SEPARATION OF BIOMOLECULES

The applications discussed above have illustrated a number of advantages of SFFF for the fractionation and molecular weight analysis of biological materials. (1) The gentleness and low-shear properties of SFFF allow separation and isolation with little risk for loss of biological activity. (2) Separation time is short and, by adjusting instrument conditions, is generally not dependent on sample molecular weight. A single component can normally be separated from contaminants within about an hour. (3) Resolution and sensitivity have been shown to be adequate for many biological separations. Components differing in mass by 20% can be separated with baseline resolution. (4) The yield and purity of biological materials appear to be at least comparable to those obtained by conventional methods. (5) Molecular weight information can be determined accurately by on-line analysis without the need for reference standards. (6) Isolated materials are recovered ready for use without further preparation.

REFERENCES

1. J.C. Giddings, *Sep. Sci.*, 1966, *1,* 123.
2. G.H. Thompson, M.N. Myers and J.C. Giddings, *Sep. Sci.*, 1967, *2,* 797.
3. J.C. Giddings, M.N. Myers and J.F. Moellmer, *J. Chromatogr.*, 1978, *149*, 501.
4. J.C. Giddings, S.R. Fisher and M.N. Myers, *Am. Lab.*, 1978, *10*, 15.
5. J.C. Giddings, M.N. Myers, K.D. Caldwell and S.R. Fisher, *Anal. Chem.*, 1980, *26,* 79.

6. J.C. Giddings, F.J.F. Yang and M.N. Myers, *Anal. Chem.*, 1974, *46,* 1917.
7. J.J. Kirkland, S.W. Rementer and W.W. Yau, *Anal. Chem.*, 1981, *53,* 1730.
8. J.J. Kirkland, C.H. Dilks, Jr. and W.W. Yau, *J. Chromatogr.*, 1983, *255*, 255.
9. J.C. Giddings, G. Karaiskakis, K.D. Caldwell and M.N. Myers, *J. Colloid Interface Sci.*, 1983, *92,* 66.
10. J.J. Kirkland, L.E. Schallinger and W.W. Yau, *Anal. Chem.*, 1985, *57*, 2271.
11. J.J. Kirkland, W.W. Yau, W.A. Doerner and J.W. Grant, *Anal. Chem.*, 1980, *52,* 1944.
12. W.W. Yau and J.J. Kirkland, *Sep. Sci. Technol.*, 1981, *16*, 577.
13. J.J. Kirkland and W.W. Yau, U.S. Patent 4,285,810, 1981.
14. J.J. Kirkland and W.W. Yau, *Anal. Chem.*, 1983, *55*, 2165.
15. P. Leopairet and F.W. Merrill, *J. Liq. Chromatogr.*, 1978, *1,* 21.
16. A.D. Bangham, D. de Gier and G.D. Freville, *Chem. Phys. Lipids*, 1976, *1*, 225.
17. J.C. Giddings, B.W. Barman and H. Li, *J. Colloid Interface Sci.*, 1989, *132*, 554.
18. J. Janca, D. Pribylova, K. Bouchel, V. Tyrackova and E. Zurkova, *J. Liq. Chromatogr.*, 1986, *9*, 2059.
19. J.C. Giddings, K.D. Caldwell and H.K. Jones, *A.C.S. Symp. Ser.*, 1987, *332*, 215.
20. L.E. Oppenheimer and G.A. Smith, *J. Chromatogr.*, 1989, *461*, 103.
21. J.J. Kirkland and W.W. Yau, *Science*, 1982, *218*, 121.
22. L.E. Schallinger, W.W. Yau and J.J. Kirkland, *Science*, 1984, *225*, 434.
23. L.E. Schallinger, J.E. Gray, L.W. Wagner, S. Knowlton and J.J. Kirkland, *J. Chromatagr.*, 1985, *342*, 67.
24. J.J. Kirkland, W.W. Yau and F.C. Szoka, *Science*, 1982, *215,* 296.
25. R. Dreyer, E. Hawrot, A.C. Sartorelli and P.P. Constantinides, *Anal. Biochem.*, 1988, *175*, 433.
26. J.C. Giddings, F.J.F. Yang and M.N. Myers, *Sep. Sci.*, 1975, *10*, 133.
27. J.C. Giddings, M.N. Myers, K.D. Caldwell and S.R. Fisher, *Methods Biochem. Anal.*, 1980, *26*, 79.
28. K.D. Caldwell, G. Karaiskakis and J.C. Giddings, *J. Chromatogr.*, 1981, *215*, 323.
29. K.D. Caldwell, T.T. Nguyen, J.C. Giddings and M.N. Mazzone, *J. Virol. Methods*, 1980, *1*, 241.
30. A. Fox, L.E. Schallinger and J.J. Kirkland, *J. Microbiological Methods*, 1985, *3*, 273.
31. K.D. Caldwell, G. Karaiskakis, M.N. Myers and J.C. Giddings, *J. Pharm. Sci.*, 1981, *70*, 1315.
32. F.-S. Yang, D. Caldwell, J.C. Giddings and L. Astle, *Anal. Biochem.*, 1984, *138*, 488.
33. J.J. Kirkland and R.M. McCormick, *Chromatographia*, 1987, *24*, 58.
34. S.M. Mozersky, K.D. Caldwell, S.B. Jones and B.E. Maleeff, *Anal. Biochem.*, 1988, *172,* 113.
35. E.H. Harrison and W.E. Bowers, *J. Biol. Chem.*, 1983, *258,* 7134.
36. L.E. Schallinger, E.C. Arner and J.J. Kirkland, *Biochim. Biophys. Acta*, 1988, *966*, 231.
37. E.C. Arner and J.J. Kirkland, *Biochim. Biophys. Acta*, 1989, *993*, 100.

Part II: Equilibrium Methods

13

Equilibrium Sedimentation in Short Solution Columns

By John J. Correia* and David A. Yphantis

DEPARTMENT OF MOLECULAR AND CELL BIOLOGY, UNIVERSITY OF CONNECTICUT, BOX U-125, STORRS, CT, 06269-3125, U.S.A.

1. INTRODUCTION

We present here the details of a rapid and moderately precise method to estimate molecular weights by sedimentation equilibrium. Initially, van Holde and Baldwin[1] demonstrated the feasibility of equilibrium ultracentrifugation in short (1 mm high) columns of solution. Subsequently, a short column technique was proposed[2] utilizing column heights of 0.75 mm in specially designed circular channels. Necessarily, the technique was less precise than conventional sedimentation equilibrium experiments in 3 mm high solution columns; it provided estimates of M_w[†] to ±3% and of M_z to ±10%. The short column technique has had relatively limited use[3-4]; however, recent improvements in the ease and precision of data acquisition along with advances in data analyses have made the short column technique a more useful experimental tool[5-13]. For example, the use of laser light sources and of sapphire windows has overcome some limitations associated with poor fringe resolution and with the use of the schlieren optical system to estimate the refractive index gradient. Early developments suggested this method to be promising for the determination of thermodynamic parameters of association-dissociation reactions and indicated that the method should be useful with large macromolecules. Recently, an application of short column equilibrium sedimentation to the intermediate stages of tobacco mosaic virus protein assembly has appeared[14]. The technique is promising for rapid survey of many samples or for characterization of unstable systems since equilibrium is achieved rapidly in short column experiments, *e.g.* sixteen times faster in a 0.75 mm high solution column than in the 3 mm high solution columns typical of most present-day sedimentation equilibrium experiments[1,15]. Such rapid attainment of equilibrium is useful when surveying purified macromolecules under various conditions. For example, equilibrium is obtained in four to five hours with short column experiments on *E. coli* ribosomes, compared with the 70 hours required in conventional

* Current address: Department of Biochemistry, University of Mississippi Medical Center, 2500 N. State St., Jackson, MS, 39216, U.S.A.

† Abbreviations: BSA, bovine serum albumin; M_w, weight average molecular weight; M_z, z-average molecular weight; M_{z+1}, (z+1)-average molecular weight; r_a, radial position of the meniscus (cm); r_b, radial position of the base (cm); TMV, tobacco mosaic virus; TMVP, tobacco mosaic virus protein.

experiments. Here we describe theoretical, experimental and computational tools that we have used to carry out rapid short column sedimentation equilibrium experiments.

2. MATERIALS AND METHODS

Equipment

The Beckman/Spinco Model E ultracentrifuge used was equipped with Rayleigh interference optics, electronic speed control, a temperature control unit calibrated against a National Bureau of Standards certified thermometer and a TRW model 83AR (or equivalent) pulsed argon ion laser (λ=5145 Å) as the interferometric light source[5]. The use of pulsed lasers as ultracentrifuge light sources has been reviewed[16]. Other modifications to the Model E have been described elsewhere[8,9,13]. Interferograms were measured automatically using a Gaertner microcomparator stage interfaced to a PDP-8/L minicomputer[17]. Fringe phases and the magnitudes of the associated transforms were obtained using a modified Walsh analysis of the digitised light intensities across the interference envelope at each radial position[18] *.

Materials

Crystalline bovine serum albumin (BSA), Lot 1207, was from Pentex Co. and was examined in the following isoelectric buffer[19]: 0.15 M sodium chloride, 0.02 m sodium acetate, 0.03 M acetic acid, pH 4.40. Crystalline bovine pancreatic ribonuclease-A, Lot 16C-8056, was from Sigma Chemical Co. and was examined in the following isoelectric buffer[20]: 0.1 M sodium chloride, 0.035 M K_2HPO_4, 0.004 M KH_2PO_4, pH 7.75.

Samples of BSA or ribonuclease-A were equilibrated in the appropriate buffer by gel filtration on a pre-equilibrated Sephadex G-75 column (0.9 cm $\times$ 100 cm) and eluted with the appropriate buffer. Trailing fractions of the main peak were taken for the ultracentrifuge experiments to avoid any heavy aggregates that may form upon storage. *E. coli* ribosomes were provided by Dr. A.J. Whaba, University of Mississippi Medical Center and were examined in 25 mM malonic acid, 5.20 mM sodium phosphate, with KOH added to attain pH 7.80. The ribosomes were titrated to a specific Mg^{2+} concentration. Malonic acid is a weak metal ion binder and thus buffers the free Mg^{2+} concentration. Ribosome association and stability is strongly dependent on free Mg^{2+} concentration[21]. A titration curve was established to determine the free Mg^{2+} concentration at a specific total Mg^{2+} concentration[22]. Ribosomes were stored frozen at -80°C in 50% glycerol.

Upon receipt of the ribosomes, a static light scattering magnesium titration (fractional association *vs.* Mg^{2+} concentration) was performed on the sample. A value of 1.83-1.87 was observed for the ratio of the scattered light intensity of the ribosomes solution with 6 mM Mg^{2+} to the same solution at 1 mM Mg^{2+}. This value

---

* T.M. Laue and D.A. Yphantis, in preparation.

corresponds to a sample that is fully associated at 6 mM Mg^{2+}. These ribosomes are designated as type A or "tight couples". Sucrose density gradient runs containing 6 mM free Mg^{2+} confirmed this assessment, revealing less than 10% contamination from type B or "loose couples". "Loose couples" are fully dissociated at 6 mM Mg^{2+} and fully associated at 18 mM Mg^{2+}.

Frozen ribosomes were thawed and equilibrated in the appropriate buffer by passing the sample over a pre-equilibrated Sephadex column. To save critical time, sedimentation velocity runs were conducted on the same solutions but with a different Model E at the same time as the equilibrium runs. These solutions were "jumped" to 6 mM Mg^{2+} to ascertain the concentration of "loose couples" present. The partial specific volumes, $\bar{v}$, used in the calculations for these materials were: BSA, 0.7343 ml g^{-1} [23]; ribonuclease 0.695 ml g^{-1} [1]; TMV protein, 0.730 ml g^{-1} [24]. Values for the partial specific volumes of 30 S, 50 S and 70 S ribosomal particles were linearly interpolated to the appropriate temperature[25]. Solvent densities were either measured or estimated from data in the International Critical Tables. No correction was applied for the density increments of the solutes of interest.

Experimental

Meniscus depletion experiments were performed according to the method of Yphantis[15] using external loading cells[26]. Short column cells were filled according to the method of Yphantis[2] and multiple speed runs on the same samples were performed with increasing ultracentrifuge speeds only. All experiments below 10000 rev/min were performed using an An-J rotor to minimize convection. Synthetic boundary experiments were carried out in double-channel capillary type centrepieces supplied by the manufacturer or constructed from standard double sector centrepieces. Plastic needles were used to avoid metal ion contamination. Equilibrium times were estimated[1] and checked empirically by comparing photographs from different times. Blanks were run before and after each experiment[26,27] and compared with one another. If discrepancies were found, then only the "after blank" was used as a correction for the experiment.

Data Analyses

Automated plate reading was performed as described earlier[28]. The optimum parameters used for the Walsh analyses (the number of points per fringe, the order of the transform), the number of flashes of the laser, the delay times of the pulses, the number of fringes scanned and the region of the fringe envelope scanned were investigated separately to maximize data precision[28]. Data files, stored on magnetic tape, were transferred from the PDP-8/L plate reading system to an IBM/370 and analyzed as described below. Interference fringes in the meniscus region tend to be "spotted", probably because of diffraction from the meniscus[29]. These data points tend to be associated with significant deviations and were thus virtually always excluded from any calculations. In addition, points at the bases of solution columns that corresponded to gradients exceeding the gradient cutoff for Weiner skewing[30] or points clearly beyond the limits of sensible resolution were also excluded.

Examination of the first differences, $c_i - c_{i-1}$, (where c_i is the concentration of the ith radial point) was useful in such editing. Data points of poorer quality from the meniscus and base regions were discarded until the first differences were uniform. For short column data this editing was performed with care so as to maintain the central point of the final fringe displacements at the midpoint of the channel. The effects of this editing process on short column analyses are considered below.

Two versions of a data-fitting computer program, ORTHO, were developed: one program uses a three term recursive formula[31] to generate orthogonal combinations of exponentials, while the second program uses the more general Gram-Schmidt orthonormalisation procedure[32]. The first program generates a polynomial of exponentials from the 0th to the nth term, exp $(n\sigma_1\xi)$, where $\xi = r^2/2 - r_a^2/2$, and that includes all n-1 intermediate terms. [For ideal systems sigma, σ_i, the effective reduced molecular weight, is defined as $M_i(1-\bar{v}\rho)\omega^2/RT$, where R is the gas constant, T is the absolute temperature, ω is the angular velocity, $\bar{v}$ is the partial specific volume of the solute, and ρ is the density of the solution[33].] Thus, if one is fitting up to a hexamer, then the dimer, trimer, tetramer and pentamer terms are of necessity also calculated. The error in the estimated coefficients of any term increased dramatically with each additional term because of the lack of orthogonality of exponentials[28]. In general, the fitting of real or of simulated noise-modulated fringe displacements corresponding to a monomer - n-mer system with this routine is not successful: the analysis does not reasonably reflect the components known to be present, rather it introduces significant amounts of spurious components. Consequently, this approach was abandoned for a more general approach. The Gram-Schmidt orthonormalisation allows specific terms to be chosen, *e.g.* just a monomer and a hexamer term, and thus minimizes errors in the coefficients, in the inferred equilibrium constants, and in the derivatives. Both versions of ORTHO are **linear** least squares fitting routines. The user provides apparent sigma values for the exponential functions. If the sigma values used for the fitting procedure correspond to the apparent reduced molecular weights of the actual physical components then the coefficients of the fit should be related to the meniscus concentrations of the n-mers and to the equilibrium constants for the n-mer associations. Under these conditions the procedure would become essentially a nongraphical application of Steiner's[34] method, quite similar to well known linear least squares procedures[35-38]. However, if the sigma values for the fitting procedure are not properly chosen (as the sigmas of the true, physical components), then the coefficients of the exponentials will not have physical significance. Rather than attempting to assign meaning to specific coefficients, we fit the general exponential shape of the curve as a sum of products of coefficients times exponentials of $\sigma_1{}^*\xi$, where $\sigma_1{}^*$ can be considered as an **arbitrary** pseudo "apparent reduced molecular weight". Calculation of the gradient, of the curvature, of the kurtosis, and of the values of σ_z and σ_{z+1} at any point ξ is then performed using differentiation with respect to ξ:

$$c(\xi) = \sum A_i \exp(\sigma_1{}^*\xi) \tag{1}$$

$$\frac{dc(\xi)}{d\xi} = \sum A_i \, \sigma_i{}^* \exp(\sigma_1{}^*\xi) \tag{2}$$

$$\frac{d^2c(\xi)}{d\xi^2} = \sum A_i \, (\sigma_i^*)^2 \exp(\sigma_1^*\xi) \tag{3}$$

$$\frac{d^3c(\xi)}{d\xi^3} = \sum A_i \, (\sigma_i^*)^3 \exp(\sigma_1^*\xi) \tag{4}$$

$$\sigma_z(\xi) = \frac{d^2c}{d\xi^2} \Big/ \frac{dc}{d\xi} \tag{5}$$

$$\sigma_{z+1}(\xi) = \frac{d^3c}{d\xi^3} \Big/ \frac{d^2c}{d\xi^2} \tag{6}$$

The point averages, $\sigma_w(\xi)$, $\sigma_z(z)$ and $\sigma_{z+1}(\xi)$, are defined in the usual manner[39]. In practice most such estimates of the σ_{z+1} are not meaningful and estimates of the $\sigma_z(z)$ are only meaningful near the middle of the solution column. One does have an accurate estimate of the midpoint concentration for short column cells[1,2] provided that the equilibrium speed is sufficiently low: If $\sigma < 1.04$ cm^{-2} for all components, then the midpoint concentration is, approximately, the loading concentration, c_0. Thus, one can also calculate the value of $\sigma_w(\xi)$ at that point:

$$\sigma_w(\xi_{mid}) = \frac{1}{c_0}\frac{dc}{d\xi} \tag{7}$$

This mid-cell value corresponds to the weight average value of σ for the initial solution on the assumption of mass conservation[1,2]. Under the same assumption, the value of $\sigma_z(\xi_{mid})$ corresponds to the z-average value of σ for the initial solution.

Meniscus depletion experiments were analyzed by nonlinear least-squares fitting to appropriate models using a previously described program, NONLIN[40]. Care was taken to examine the observed distributions for evidence of heterogeneity[12] as described in the Results section. The same (or, occasionally, equivalent) solutions were used for parallel short column experiments.

3. RESULTS

Simulations and Testing of the Method

Concentration distributions from short column experiments inherently contain less information than do concentration distributions in longer, conventional columns. Accordingly, the intrinsically poor orthogonality of exponentials can be put to use by fitting short column data to a sum of arbitrary exponentials: The general features of the distributions can be fit accurately enough to estimate the first and second derivatives of the distributions with respect to ξ near the middle of the solution column. This approach was initially tested with synthetic data. Three general types of short column

Table 1. Estimation of σ from synthetic monomer-dimer short column data with program ORTHO.

Terms in the fit	Value of σ_1^* used in the fit (cm^{-2})					
	0.4	0.5	0.6	1.0	1.5	2.0
3	2.35±0.02	2.35±0.03	2.35±0.02	2.34±0.02	2.31±0.03	2.31±0.02
4	2.34±0.03	2.32±0.03	2.32±0.02	2.31±0.02	2.31±0.03	2.30±0.04
5	2.33±0.06	2.33±0.06	2.30±0.07	2.33±0.07	2.32±0.05	2.30±0.04
6	2.33±0.08	2.32±0.09	2.28±0.09	2.33±0.11	2.30±0.09	2.30±0.08

The estimates are the mean ± the standard deviation of replicate fits of the data modulated with different sets of random noise (1.5×10^{-3} mm of fringe displacement). Generating parameters for the data were: $\sigma_1 = 1.5$ cm^{-2}, $K_2 = 1.0$ (mm of fringe displacement)$^{-1}$, $c_0 = 1.0$ mm fringe displacement, $\sigma_{z,mid} = 2.31$ cm^{-2}, $r_a = 6.575$ cm, $r_b = 6.5$ cm.

experiments were simulated: ideal systems (Fig. 1a), heterogeneous systems (Fig. 1b), and self-associating systems (Table 1). These "perfect" data sets were then modulated by addition of random numbers having a Gaussian distribution, a mean of zero and a standard deviation representative of typical experimental error to generate simulated "noisy" data for fitting. At least five such "noisy" data sets were fit for each case so as to provide an estimate of the uncertainty of the estimations of the first and second derivatives of the concentration[12,41]. We present here only results for the estimation of σ_z.

The analyses reported in Fig. 1 used a three term fit:

$$c = A_0 + A_1 \exp(\sigma_1^* \xi) + A_2 \exp(n\sigma_1^* \xi) \tag{8}$$

where n was taken as 2. Comparable results, over slightly narrower ranges of σ_1^*, were found with n = 3 or 4. Initial attempts to optimise the value of σ_1^* and σ_n^* (= $n\sigma_1^*$) by the use of splines were abandoned when it became apparent how insensitive the σ_z estimates were to the choice of the values of the σ_1^*. Any pair of values that spanned the true value of σ_z provided nearly identical results, as shown in Fig. 1. The procedure provides poor estimates of σ_z at low σ_z values when high values are chosen for σ_1^*, the pseudo σ of fitting equation[8]. Similarly, poor estimates of high values of σ_z were found if σ_1^* were chosen too low. Table 1 shows that a three term fit [according to eq. (8)] is adequate to extract reasonable estimates of σ_z values from short column observations.

Estimations of σ_{z+1} require determinations of $d^3c/d\xi^3$. The third derivatives estimated from these simulations were ill behaved, at times becoming even negative. Exclusion of the constant term A_0 gave only slight improvement. However, just as larger σ values improve the estimations of $d^2c/d\xi^2$ and thus of σ_z (*vide infra*), the use of very high speeds may enable estimations of the σ_{z+1} in favourable cases. The largest source of error in the values of the σ_z arises from the estimation of the $d^2c/d\xi^2$ values. The largest error in the σ_w values generally appears to be from determinations of the c_0.

The automated plate reading system can measure fringe displacements at 2 µm intervals, providing several hundred data points across a 0.75 mm solution column. Such sampling density is not practical with manual data acquisition. To test the limitations of the method, real data was edited so as to use every point, every other point, every third, fourth or fifth data point for molecular weight estimations (Table 2). As can be seen, consistent results were obtained with all subsets of the original data sets, suggesting that this approach need not be limited to the use of automated data acquisition. High quality interferograms, however, do appear to be critical for precise estimation of apparent molecular weights. Errors in the fitting parameters appear to increase as, N, the number of data points used, decreases. The uncertainty in gradient estimates appears to be proportional to 1/N and the uncertainty in σ_z to $1/\sqrt{N}$.

As described in the methods, edge effects require the removal of data points evenly from the meniscus and base regions. This approximates, in the limit, the original interferometric method[2] of determining the gradient at the midpoint of the channel from either a two point slope or a quadratic or cubic fit of five to nine points

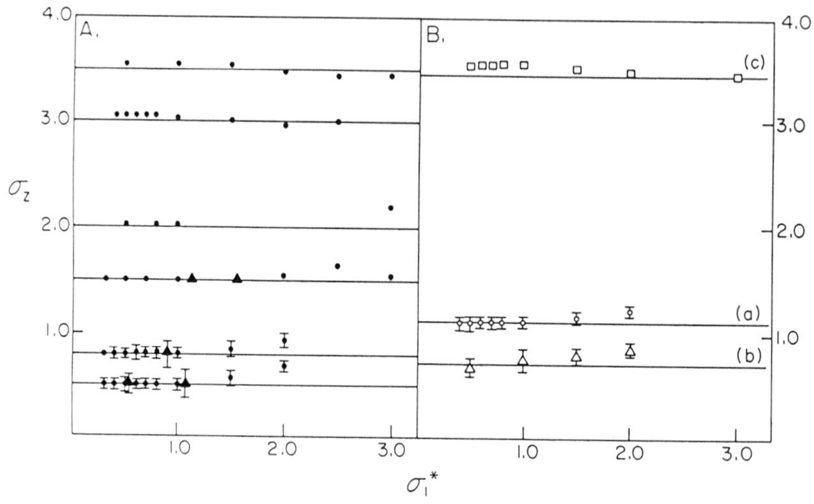

Figure 1. Estimation of σ_z at the midpoint of artificially generated short column data using a three term exponential fitting procedure (see Data Analysis). All data sets (100 equally spaced data points in the region 6.5 cm < r < 6.575 cm, c_0 = 1 mm of fringe displacement) were modulated with randomly distributed noise of standard deviation 1.5 μm fringe displacement, except for the points marked as Δ in panel A where the noise added corresponded to 5 μm. Values of $\sigma_{z,app}$ are plotted *vs.* σ_1^*, the value of the exponential basis vector [see eq. (8)]. The lines represent the true $\sigma_{z,mid}$ values that are being estimated.

(A) Fitting from ideal, single component short column data.

(B) Fitting of short column data for heterogeneous mixtures:

- (a) Two components, σ_i = 0.3 and 1.3 cm^{-2}, $\sigma_{z,mid}$ = 1.11 cm^{-2}, $c_{0,i}$ = 0.5 mm of fringe displacement.
- (b) Nine species, σ_i = 0.0, 0.125, 0.25, 0.375, 0.5, 0.625, 0.75, 0.875, and 1.0 cm^{-2}, $\sigma_{z,mid}$ = 0.71 cm^{-2}, $c_{0,i}$ = 0.111 mm of fringe displacement.
- (c) Nine species, σ_i = 0.0, 0.625, 1.25, 1.875, 2.5, 3.125, 3.75, 4.375, and 5.0 cm^{-2}, $\sigma_{z,mid}$ = 3.44 cm^{-2}, $c_{0,i}$ = 0.111 mm of fringe displacement.

Table 2. The effect of varying the number of points in the fitting of short column data.

c (mg/ml)	Number of points	dc/dξ ($mm_{fringe\ displ}\ cm^{-2}$)	σ (cm^{-2})	RMS (μm)
5.33	143	9.1191±0.0025	1.4890±0.0106	4.04
	72	9.1194±0.0056	1.4905±0.0151	4.13
	48	9.0087±0.0085	1.4842±0.0186	4.05
	36	9.0602±0.0124	1.4821±0.0228	4.37
	29	8.9751±0.0114	1.4860±0.0208	3.60
4.00	143	5.8439±0.0010	1.4728±0.0074	2.38
	72	5.8236±0.0021	1.4737±0.0104	2.38
	48	5.7863±0.0033	1.4724±0.0133	2.43
	36	5.7836±0.0043	1.4733±0.0150	2.50
	29	5.7472±0.0053	1.4725±0.0165	2.41
3.00	158	4.2374±0.0005	1.5426±0.0054	1.50
	79	4.2063±0.0009	1.5395±0.0078	1.53
	53	4.1918±0.0015	1.5390±0.0097	1.59
	40	4.2069±0.0023	1.5359±0.0115	1.65
	32	4.1926±0.0030	1.5359±0.0129	1.66
2.00	173	2.4646±0.0005	1.4794±0.0099	2.10
	87	2.4755±0.0011	1.4818±0.0141	2.12
	58	2.4755±0.0018	1.4797±0.0174	2.11
	44	2.4762±0.0024	1.4811±0.0199	2.14
	35	2.4200±0.0028	1.4795±0.0234	2.17

The data points were selected from across the entire channel by choosing each point, each other point, every third, fourth, or fifth point. The data are from an experiment on TMVP at pH 7.0, 20°C, 4400 rev/min[14].

Table 3. The effect of discarding points from the base and meniscus regions of short column data.

c (mg/ml)	Number of points	dc/dξ ($mm_{fringe\ displ}\ cm^{-2}$)	σ_z (cm^{-2})	RMS (μm)
5.33	183	9.1243±0.0015	1.5002±0.0082	5.71
	143	9.1191±0.0025	1.4890±0.0106	4.04
	133	9.1174±0.0033	1.5035±0.0128	4.08
	123	9.1107±0.0040	1.5311±0.0147	3.88
	113	9.1090±0.0053	1.4580±0.0178	3.80
	103	9.1067±0.0073	1.5635±0.0222	3.74
	83	9.0935±0.0137	1.5638±0.0350	3.43
	63	9.0825±0.0301	1.6047±0.0666	3.05
4.00	185	5.8408±0.0003	1.5010±0.0058	3.01
	153	5.8439±0.0010	1.4728±0.0074	2.38
	143	5.8469±0.0013	1.4800±0.0087	2.36
	133	5.8438±0.0017	1.4961±0.0103	2.32
	123	5.8328±0.0020	1.5177±0.0116	2.17
	113	5.8346±0.0026	1.5478±0.0139	2.09
	93	5.8373±0.0047	1.4682±0.0205	1.88
	73	5.8304±0.0111	1.4653±0.0405	2.02
3.00	203	4.2422±0.0001	1.5399±0.0066	3.51
	158	4.2374±0.0005	1.5426±0.0054	1.50
	148	4.2369±0.0006	1.5469±0.0064	1.50
	138	4.2402±0.0010	1.5403±0.0079	1.52
	128	4.2403±0.0013	1.5262±0.0096	1.53
	118	4.2404±0.0015	1.5181±0.0222	1.50
	98	4.2367±0.0026	1.5661±0.0166	1.40
	78	4.2304±0.0050	1.6396±0.0262	1.16
2.00	195	2.4610±0.0001	1.5024±0.0086	2.62
	173	2.4646±0.0005	1.4792±0.0099	2.10
	163	2.4634±0.0007	1.4900±0.0115	2.11
	153	2.4626±0.0009	1.4965±0.0139	2.17
	143	2.4635±0.0011	1.4774±0.0165	2.18
	133	2.4655±0.0013	1.4235±0.0183	2.02
	113	2.4667±0.0020	1.3033±0.0236	1.75
	93	2.4748±0.0033	1.2916±0.0344	1.65

The actual data used in the analysis are underlined. The minimum number of points corresponds to column heights of 0.25-0.35 mm. These data are taken from the same experiment as Table 2, with TMVP at pH 7.0, 20°C, 4400 rev/min.

symmetrically disposed about the midpoint of the channel. Table 3 illustrates the effect of symmetric editing on the estimation of parameters. The data presented here are from real, rather than simulated, experiments. (Comparable results are seen from noise perturbed ideal synthetic data.) The values of the gradient, $dc/d\xi$, are unaffected. The values of the second derivative, $d^2c/d\xi^2$, and of σ_z are very sensitive to the use of narrower data spans. Errors in these parameters increase much more rapidly by shortening the data span than by thinning out the data while keeping a constant data span (as in Table 2). Modelling studies with synthetic data[28] showed that it is essential to include as wide a data span as possible, despite the lower deviations associated with the higher quality fringes near the central regions of the observation channels and with the possibly better fits over more limited portions of the concentration distributions. The increased error observed in parameter estimates from narrower data subsets explains why the earlier attempts to determine M_z by short column methods gave, at best, estimates with a precision of only 10%. In summary, these short column approaches of determining z-averages work best when as much as possible of the concentration range of the short column data is spanned. The improved precision associated with a wider concentration range is consistent with observations from sliding fit methods[41,42] used to determine apparent molecular weight moments from sedimentation equilibrium distributions. (An algorithm similar to ORTHO could be readily adapted to carry out sliding fits on equilibrium concentration distributions and would be preferable to polynomial smoothing since the exponential terms provide better approximations.)

Testing with Macromolecules

Ribonuclease-A In order to properly evaluate the effectiveness of short column experiments the same protein solutions were analysed by both the ORTHO algorithm and by conventional sedimentation equilibrium experiments[15]. Table 4 presents a summary of three meniscus depletion experiments performed with purified ribonuclease-A. Experiments 1 and 2 were performed on the same samples three days apart, before and after a short column experiment. Experiment 3 was carried out on a separate sample, after a short column experiment. Each channel was analysed individually as a single ideal species. In each case the results suggested a trend with concentration. Joint fits of the data as a single ideal species and as a single non-ideal species were performed. The results are in reasonable agreement with the formula weight of ribonuclease, 13.683 kg mol^{-1}, and with the expected (colligative) second virial coefficient for an uncharged spherical protein, $B = 4\bar{v}/M = 2.03\times10^{-4}$ mol ml g^{-2} [43]. Table 5 presents an analysis of the heterogeneity of these three experiments[12]. The data were truncated so as to span the identical concentration range in each channel of each experiment. A joint fit of the three channels to σ and B provided a "communal" value of σ. This communal value was then held constant and used to fit each channel individually for B_{app}. The fitted values of B_{app} are tabulated for comparison in Table 5. The protein in experiment 1 appears to have been homogeneous by this test: The values of B are all well within the error limits. The proteins in the second and third experiments appeared to be slightly heterogeneous. A channel by channel comparison of the values of B_{app} for experiments 2 and 3 showed these samples to be slightly different. (These samples were approximately the same age, with the solutions in experiment 2 being one day older.) The increase in B_{app} with loading concentration for

Table 4. Summary of meniscus depletion experiments with ribonuclease A at pH 7.75, 20°C, 48000 rev/min.

Exp.	c_0 (mm)	Channel	M (g mol^{-1})	10^4×B (ml mol g^{-2})	RMS (μm)
1	1.60	A	13379 <13355, 13402>	-	2.68
	0.91	B	13362 <13335, 13389>	-	3.95
	0.42	C	13447 <13410, 13485>	-	3.69
		joint	13385 <13358, 13412>	-	3.60
		joint	13616 <13515, 13716>	4.55 <2.83, 6.26>	3.12
2	1.53	A	13366 <13329, 13404>	-	5.47
	0.91	B	13376 <13349, 13402>	-	3.50
	0.42	C	13624 <13565, 13683>	-	4.74
		joint	13406 <13369, 13443>	-	5.00
		joint	13641 <13510, 13443>	4.50 <2.35, 6.62>	4.61
3	1.16	A	13320 <13290, 13350>	-	3.70
	0.58	B	13521 <13497, 13547>	-	3.12
	0.33	C	13595 <13556, 13633>	-	3.52
		joint	13469 <13436, 13501>	-	4.17
		joint	13590 <13511, 13669>	2.51 <1.10, 3.91>	4.06

these two experiments is consistent with contamination by larger species, possibly ribonuclease-A oligomers. Estimation of the amount of contaminating oligomer depends on its size; less than 1% of a dimer or of a trimer would account for our observations. It seems reasonable to suppose that ribonuclease A can be freshly prepared as a homogeneous sample that behaves as a single non-ideal component; with time the solutions may become heterogeneous as a result of slow formation of oligomers. Similar oligomerisations have been observed on prolonged storage[2,44] or on lyophilisation[45] of ribonuclease preparations.

Fig. 2 summarises short column experiments performed with ribonuclease-A. The lines correspond to the predicted behaviour of an uncharged spherical protein, with $\bar{v} = 0.695$ ml g^{-1} and M= 13.683 kg mol^{-1}, whose only non-ideality arises from its excluded volume. The data are consistent with models of either a single ideal or a single non-ideal species. Table 6 presents a summary of attempts to fit multiple channels of the short column data with a non-ideal model using NONLIN[40]. Apparently only the 36000 rev/min data gives a result completely consistent with the high speed experiments in 3 mm columns. Only at this higher speed does the estimated value of the apparent virial coefficient agree with the values reported in Table 4, and only at this speed does the RMS of the fit decrease with the inclusion of B as a fitting parameter. Apparently there is insufficient information in the set of the four short column data sets to differentiate between an ideal and a non-ideal model at the lower speed. At the highest speed, 36000 rev/min, where the curvature and kurtosis of the fringe displacements are significant, one can start to distinguish between the ideal and non-ideal models. At the two intermediate speeds (24000 and 34000 rev/min) the nonlinear least squares approach does provide reasonable apparent molecular weight estimates. The data from 16000 rev/min (the lowest speed) appear to be inadequate to even fix the apparent molecular weight, either by nonlinear least squares fitting or by estimation of σ_z through the second derivative, although the same data does provide good estimates of the σ_w using values of the first derivative of the concentration with ξ and of the measured refractive increments of the initial solutions.

Table 5. Analysis of meniscus depletion experiments with ribonuclease-A for heterogeneity of the solution.

Exp.	Channel	σ (held) (cm^{-2})	$10^3 \times B$ (mm^{-1})	RMS (μm)
1	A	4.2187	4.56 <3.36, 5.97>	2.47
	B	4.2187	4.52 <2.88, 6.16>	3.31
	C	4.2187	4.85 <3.19, 6.55>	3.35
2	A	4.2260	8.78 <4.20, 12.7>	5.28
	B	4.2260	4.77 <2.30, 7.29>	3.34
	C	4.2260	1.86 <-1.56, 5.34>	4.69
3	A	4.2219	6.52 <5.39, 7.65>	3.62
	B	4.2219	2.53 <1.57, 3.50>	2.95
	C	4.2219	-0.14 <-1.36, 1.09>	3.52

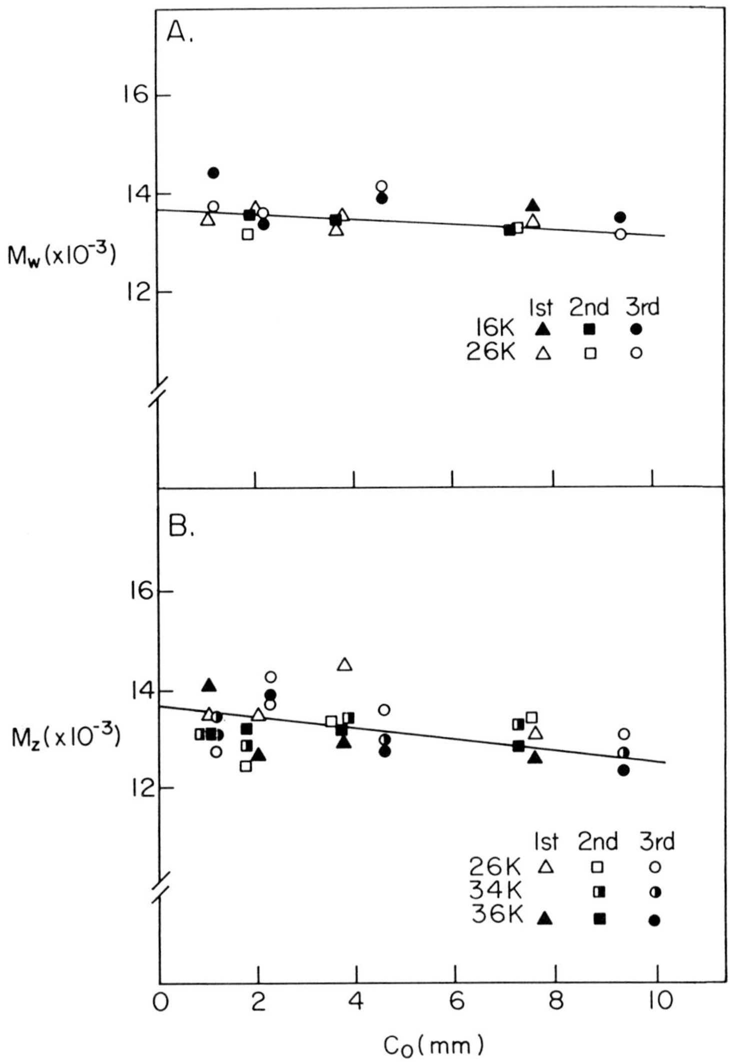

Figure 2. (A) Observed values of the apparent weight average molecular weights, $M_{w,app}$, as a function of the loading concentrations, c_0, from three short column experiments on ribonuclease-A at pH 7.75, 20°C. (B) Values of the observed apparent z-average molecular weight, $M_{z,app}$, *vs.* c_0 for the same three short column experiments.

The lines indicate the behaviour expected for a spherical, uncharged protein of molecular weight 13.683 kg mol^{-1}, with $\bar{v}$, the partial specific volume, taken as 0.695 ml g^{-1}. The slope of these lines reflects non-ideality from excluded volume. The concentration scale is in units of mm of fringe displacement in a 12 mm thick centrepiece using light with a wavelength of 5145 Å. One mm of fringe displacement here corresponds to about 0.84 mg ml^{-1}.

Table 6. Multiple channel fitting of ribonuclease-A short column data to an ideal and a non-ideal model.

Speed (rev/min)	M (g mol^{-1})	B (ml mol g^{-2})	RMS (μm)
16000	17010 <16441, 21135>	-	2.09
	failed		
26000	13378 <13008, 13658>	-	2.49
	13567 <13121, 13892>	4.84 <-14.2, 9.7> $\times 10^{-6}$	2.49
34000	13171 <12998, 13338>	-	3.30
	13389 <13050, 13737>	6.76 <-2.5, 16.2> $\times 10^{-5}$	3.29
36000	13062 <12825, 13294>	-	3.94
	13648 <13247, 14063>	2.65 <1.44, 3.85> $\times 10^{-4}$	3.73

Eight channels from Fig. 2 (exp. 1 and 2) were simultaneously fit with NONLIN[40].

We believe that the failure of the non-ideal fitting of the data at the lowest speed reflects the small curvature present in these data. This tiny curvature is the reason that short column experiments with low values of σ (in this case σ_w = 0.472 cm^{-2}) are useful only for estimations of the M_w. The short column estimates of M_w and M_z appear to be consistent with the known behaviour of ribonuclease-A in solution. Generally, comparisons of values obtained at multiple equilibrium speeds may facilitate interpretation of apparent molecular weights. For example, the absence of any speed dependence of the apparent molecular weights with the intermediate aggregates of TMVP excluded the presence of non-reequilibrating stacks of disks in solution[14,28].

Bovine Serum Albumin Fig. 3 summarises the results of two short column experiments with BSA. Table 7 presents a summary of a meniscus depletion experiment on one of the same solutions of BSA. The three channels were fit individually with NONLIN[40] assuming a single ideal species and the three channels were jointly fit as both a single ideal and as a single non-ideal species. The fitted values are in agreement with the expected molecular weight, 66.296 kg mol^{-1} [46], and the estimated second virial coefficient for BSA, $4\bar{v}/M = 4.4\times10^{-5}$ ml mol g^{-2} [43]. Table 8 shows a test of the heterogeneity of the BSA in the high speed equilibrium experiment. The three channels were edited so as to encompass the same concentration range and then jointly fit as either a single non-ideal species or as a reversibly associating monomer-dimer system. (The model of a monomer-dimer system is not correct; however, it should provide the same quantitative and qualitative information as the non-ideal model[47].) The best estimate of σ from the joint fit was then used to fit each channel individually with the same model. The results were compared and used to infer heterogeneity[12]. As can be seen in Table 8, the values of B_{app} decrease with increasing loading concentration when using the non-ideality model, and, correspondingly, the values of ln K_{app} increase with loading concentration when using the model of ideal association. Channels A and B are almost identical within the estimated confidence limits. However these two channels do differ slightly but significantly from channel C. If a larger species were present its effect would have been in the opposite direction from what is observed: in such a case M_{app} would have been expected to decrease with increasing loading concentration since the data selected for fitting originates from more centrifugal regions of the channels as the loading concentration decreases. The minor discrepancies of the fits of channel C from the fits of the other channels may have their origins in systematic experimental errors. A prime candidate for the source of such errors is the possibility of convective disturbances. Such disturbances are seen from time to time in equilibrium ultracentrifugation[15]. They seem to be correlated with the presence of low concentration regions where there is little stabilization of solute distributions by self-generated density gradients, either from the solute itself or from the redistribution of solvent components. The reasonable agreement of the joint fits presented in Table 7 with the expected values of M and B for BSA is a strong indication of a fairly homogeneous preparation.

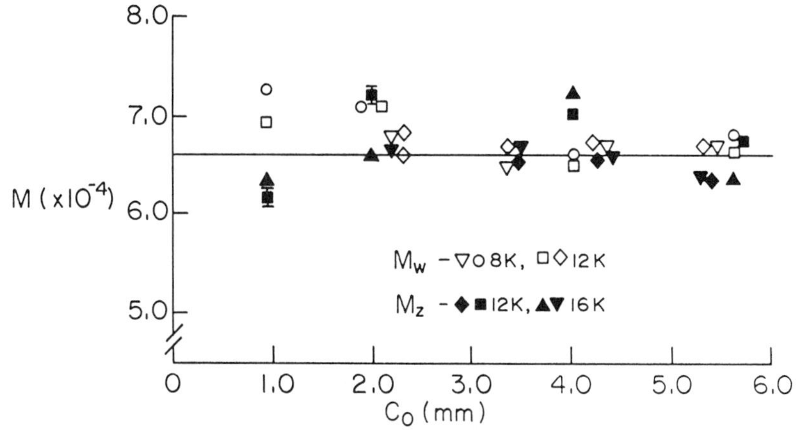

Figure 3. Apparent weight average and z-average molecular weights from short column experiments on bovine serum albumin at pH 4.40, 20°C. The horizontal straight line is the expected behaviour for an ideal monomer with molecular weight of 66.296 kg mol^{-1}. Again one mm of fringe displacement corresponds to a concentration of about 0.8 mg/ml (≡0.08%).

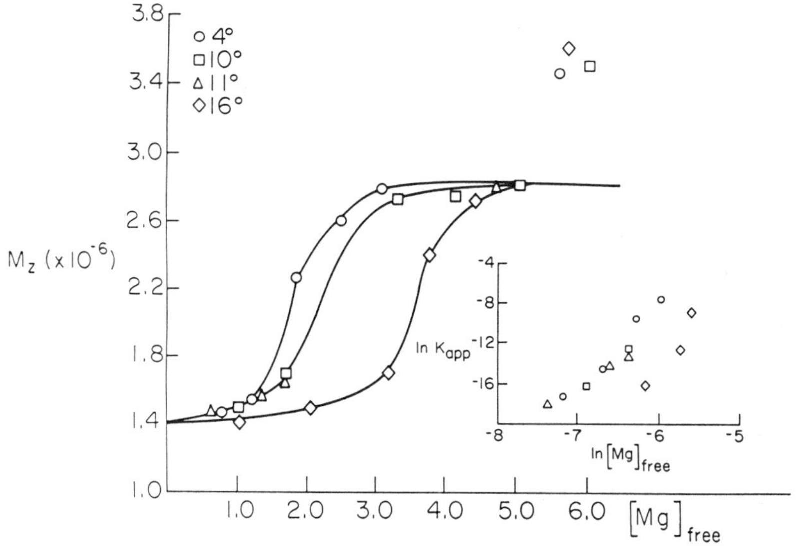

Figure 4. Z-average molecular weight of *E. coli* ribosomes as a function of concentration of Mg^{2+} concentration in millimoles per litre. All short column experiments were carried out at 2400 rev/min at a loading concentration of 2 mg ml^{-1} at the indicated temperatures. The lines indicate the trends of the observations. The insert is a graph of ln K, the logarithm of the apparent association constant, as a function of the Mg^{2+} concentration on a logarithmic scale. These apparent association constants were estimated from the $M_{z,app}$. Only data corresponding to $M_{z,app}$ values under 2700 kg mol^{-1} are presented. At each temperature the slopes of ln K *vs.* ln [Mg^{2+}] provide estimates of the number of Mg^{2+} involved in the association of 30 S and 50 S particles to form 70 S ribosomal particles (see text).

Table 7. Summary of the results from a meniscus depletion experiment with bovine serum albumin at pH 4.40, 20°C, 24000 rev/min.

Channel	c (mm)	M (kg mol^{-1})	$10^5\times$B (ml mol g^{-2})	RMS (μm)
A	1.14	65.4 <65.2, 65.6>	-	4.17
B	0.59	66.3 <66.1, 66.4>	-	3.16
C	0.30	65.2 <64.8, 65.5>	-	2.74
joint		66.0 <65.8, 66.1>	-	3.88
joint		66.4 <66.0, 66.8>	3.40 <0.85, 5.92>	3.82

Table 8. Analysis of a meniscus depletion experiment on bovine serum albumin for heterogeneity of the solutions.

Assuming single-component non-ideal system

Channel	σ_{held} (cm^{-2})	$10^5\times$B (mm^{-1})	RMS (μm)
A	4.5519	5.72 <2.13, 9.38>	3.44
B	4.5519	8.14 <5.11, 11.2>	2.72
C	4.5519	12.9 <9.59, 16.2>	2.92

Assuming ideal reversibly associating system

Channel	σ_{held} (cm^{-2})	$\ln K_{app}$ ($mm_{fringe\ displ}^{-1}$)	RMS (μm)
A	4.382	-3.91 <-4.27, -3.59>	3.80
B	4.382	-4.30 <-4.65, -3.98>	2.57
C	4.382	-5.61 <-6.44, -4.95>	2.67

The results from the short column experiments are presented in Fig. 3. The average value of $M_{w,app}$ is (68.2±2.1) kg mol^{-1} while the average value of $M_{z,app}$ is (66.6±3.2) kg mol^{-1}, in reasonable agreement with the meniscus depletion experiment and with the formula weight of 66.296 kg mol^{-1} for BSA. There is a weak suggestion of non-ideality and no evidence of any aggregates.

***E. coli* Ribosomes** Fig. 4 summarises four experiments with *E. coli* ribosomes at different temperatures. The apparent molecular weight increases with [Mg^{2+}] and the 50% association point shifts to higher [Mg^{2+}] with increasing temperature, as expected. The average $M_{z,app}$* value at [Mg^{2+}] < 1.0 mM is 1460±20

* Calculations were limited to estimations of z-average molecular weights. Speeds under 1500 rev/min would be required to estimate M_w values for *E. coli*

kg mol^{-1}, in agreement with the expected z-average of a 1:1 mixture of particles 1000 (30 S) and 1650 (50 S) kg mol^{-1}, *i.e.* 1410 kg mol^{-1}. The plateau at "full" association corresponds to values of 2700-2800 kg mol^{-1}. Sedimentation velocity experiments (data not presented) show that 100 S particles, presumed to be dimers of 70 S ribosomes, were present in many of the more associated solutions. Estimation of the amount of the 100 S component is difficult because of small peak size in the schlieren patterns and because the 70 S and 100 S boundaries appear to represent a rapidly interacting system. The existence of this further interaction is supported by high values of $M_{z,app}$, *i.e.* 3500 kg mol^{-1}, when $[Mg^{2+}] > 5.5$ mM. These $M_{z,app}$ estimates correspond to 18% of the 100 S dimer and 82% of the 70 S monomer and further suggest that the 70 S dimerisation reaction depends strongly on Mg^{2+} concentration. Previous determinations[25] of the molecular weight of 70 S "washed" ribosomes indicated $M_{z,app}$ values in excess of 3000 kg mol^{-1}. Those experiments were performed in different buffer system at 10 mM total $[Mg^{2+}]$. The magnesium concentrations for 50% association are slightly higher than the values reported by Hui Bon Hoa *et al.*[48] †. This is not surprising for the higher temperature (16°C) where appreciable loose couple formation occurs. Graphs of ln K_{app} *vs*. ln $[Mg^{2+}]$ (inset to Fig. 4) for each temperature should have slopes equal to the number of Mg^{2+} involved in the association[48]. Metal ion binding to 70 S particles is believed to produce charge-neutralization and thus is associated with a (nonspecific) decrease of the electrostatic free energy of assembled subunits[49]. The value of such slopes has been reported previously to vary from 6.0 at 7.5°C to 7.4 at 25°C. In general, observations in the temperature range 4°C-11°C have a similar slope. Variations in slope at higher extents of association or at 16°C may arise from the presence of 100 S particles and/or loose couples. The results summarized in Fig. 4 are thus consistent with previous observations and strongly suggest that the molecular weights observed in short column experiments with *E. coli* ribosomes are reliable.

4. DISCUSSION

The simulations summarized in Fig. 1 and Table 1 demonstrate that the short column approach presented here can provide a powerful and relatively precise means of estimating z-average molecular weight: a wide range of σ values can be accurately estimated with a three term fit of exponential functions of $\sigma\xi$. The experimental results presented (Figs. 2-4) show that it is practical to estimate both weight average and z-average molecular weights for various molecular sizes by this approach. Just as with any other ultracentrifugation technique it is essential that the distributions be stable,

---

ribosomes. This appears to be beyond the lower limit of stable operation of the Model E ultracentrifuge.

† Direct comparison of our results with the light scattering results of Hui Bon Hoa *et al.*[41] is difficult. These authors used a different buffer system, equilibrated their ribosomes with 10 mM Mg^{2+} before making up the solutions and presented results as a function of total $[Mg^{2+}]$. They also used a lower ribosome concentration, *i.e.* 5.0 A_{260} units or approximately 0.3 g/ml. Furthermore, absolute molecular weights were not calculated; rather, relative values of the scattered light intensity were used to infer extents of association.

with no significant convective disturbances. Our best experimental cases, presumably at stable speeds and with convection free equilibria, appear to provide moderately accurate results. The precision of estimation of weight average values of σ can be 1% or better (*e.g.* the standard deviation of the σ_W estimated from the second ribonuclease experiment of Fig. 2 at 16000 rev/min was about 0.2% when compared to the expected non-ideal behaviour; experiment 1 of the same figure at 26000 rev/min gave a standard deviation of 1% from the predicted values). In general[#], 2% is a reasonable estimate of the precision in σ_W if the concentration is high enough. The estimation of σ_Z can be made to better than 4% in reasonably good experiments (the standard deviation of the higher speed values of σ_Z from the expected non-ideal values for ribonuclease is less than 3%). The observed experimental variations most likely arise from occasional convective phenomena, especially at low speed and with loading concentrations under 2 mg/ml, conditions that result in reduced density-gradient stabilization. The major limitation of these methods for molecular weight estimation is the requirement for moderately high protein concentrations (>2 mg/ml) for the equilibrium and (>5 mg/ml) for determinations of refractive increment with synthetic boundary experiments. Estimates of two molecular weight averages M_W and M_Z can facilitate the selection of appropriate models by methods similar to those applied to TMVP assembly[14] or by more traditional two species analysis[41]. Relatively precise estimations of only σ_Z (to obviate the requirement for determination of refractive increment) at high equilibrium speeds can be made with considerably less than 100 µg of solute in a short period of time. Potential applications include the rapid survey of numerous fractions, *e.g.* chromatographic, or of various solvent conditions in a single run, especially since one can run up to five short column cells (when multiplexing of the light source is available[6,8,9]) in a type G six-hole rotor with up to four solutions per cell. This short column procedure has been routinely used in our laboratories for over ten years to study several macromolecules[14,50-53].

ACKNOWLEDGEMENTS

We deeply appreciate the generous provision of computer time by the University of Connecticut Computer Center. We thank Mike Crossin for assistance with the *E. coli* ribosome work and Tom Laue and Brad Chaires for critical reading of this manuscript at various stages. We also thank Raymond Kikas for outstanding technical assistance. This work was supported by NSF Grants PCM 76-21847 and BBS 86-12159 (D.A.Y.).

[#] It should be noted that solute adsorption generally sets limits on the accuracy of the σ_W at low concentrations by violating mass conservation in the solution columns[2]. No mass conservation assumption is needed to estimate the σ_Z. However if there is selective adsorption of solute components then the σ_Z will not reflect the composition of the initial solution loaded into the cell.

REFERENCES

1. K.E. van Holde and R.L. Baldwin, *J. Phys. Chem.*, 1958, *62*, 734.
2. D.A. Yphantis, *Ann. N.Y. Acad. Sci.*, 1960, *88*, 586.
3. I.M. Mackie and J.J. Connell, *Biochem. et Biophys. Acta*, 1964, *93*, 544.
4. H. Utiyama, N. Tayata and M. Kurata, *J. Phys. Chem.*, 1969, *73*, 1448.
5. C.H. Paul and D.A. Yphantis, *Anal. Biochem.*, 1972, *48*, 588.
6. C.H. Paul and D.A. Yphantis, *Anal. Biochem.*, 1972, *48*, 605.
7. T.M. Laue, R.A. Domanik, D.G. Rhodes and D.A. Yphantis, *Biophys. J.*, 1977, *17*, 101a.
8. T.M. Laue, R.A. Domanik and D.A. Yphantis, *Anal. Biochem.*, 1983, *131*, 220.
9. T.M. Laue, D.A. Yphantis and D.G. Rhodes, *Anal. Biochem.*, 1984, *143*, 103.
10. R.A. Domanik, D.A. Yphantis and T.M. Laue, *Biophys. J.*, 1977, *17*, 101a.
11. T.M. Laue, and D.A. Yphantis, *Biophys. J.*, 1979, *25*, 104a.
12. D.A. Yphantis, J.J. Correia, M.L. Johnson and G.-M. Wu, in 'Physical Aspects of Protein Interactions', (N. Catsimpoolas, ed.), Elsevier North Holland, Inc., 1978, p. 275.
13. D.A. Yphantis, T.M. Laue and I. Anderson, *Anal. Biochem.*, 1984, *143*, 95.
14. J.J. Correia, S. Shire, D.A. Yphantis and T.M. Schuster, *Biochemistry*, 1985, *24*, 3292.
15. D.A. Yphantis, *Biochemistry*, 1964, *3*, 297.
16. D.A. Yphantis, Methods in Enzymology vol *61*, (C.W. Hirs and S.N. Timasheff, eds.) Academic Press, New York, 1980, p. 3.
17. T.M. Laue, Ph.D. Dissertation, University of Connecticut, 1981.
18. K.G. Beauchamp, 'Walsh functions and their applications', Academic Press, Inc., London, 1975.
19. G. Kegeles and F.J. Gutter, *J. Am. Chem. Soc.*, 1951, *73*, 3770.
20. A. Rothen, *J. Gen. Physiol.*, 1940, *24*, 203.
21. J.B. Chaires, M. Tai, C. Huang, G. Kegeles, A.A. Infante and A.J. Wahba, *Biophys. Chem.*, 1977, *7*, 179.
22. M. Crossin, Ph.D. Dissertation, University of Connecticut, 1981.
23. M.O. Dayhoff, G.E. Perlmann and D.A. Maclnnes, *J. Am. Chem. Soc.*, 1952, *74*, 2515.
24. R. Jaenicke and M.A. Lauffer, *Biochemistry,* 1969, *8*, 3077.
25. W.E. Hill, G.P. Rossetti and K.E. van Holde, *J. Mol. Biol.*, 1969, *44*, 263.
26. A.T. Ansevin, D.E. Roark and D.A. Yphantis, *Anal. Biochem.*, 1970, *34*, 237.
27. T.A. Horbett and D.C. Teller, *Anal. Biochem.*, 1972, *45*, 86.
28. J.J. Correia, Ph.D. Dissertation, University of Connecticut, 1981.
29. R.C. Williams, Jr., *Anal. Biochem.*, 1972, *48*, 164.
30. H. Svensson, *Optica Acta*, 1954, *1*, 25.
31. G. Forsythe, *J. Soc. Indust. Appl. Math.*, 1957, *5*, 74.
32. P. Davis and P. Rabinowitz, *Association for Computing Machinery,* 1954, *1*, 184.
33. D.A. Yphantis and D.F. Waugh, *J. Phys. Chem.*, 1956, *60*, 23.
34. R.F. Steiner, *Arch. Biochem. Biophys.*, 1952, *39*, 333.
35. L.A. Holladay and A.J. Sophianopoulos, *J. Biol. Chem.*, 1972, *247*, 427.
36. R.H. Haschemeyer and W.F. Bowers, *Biochemistry*, 1970, *9*, 435.

37. M.F. Rohde, S. O'Brien, S. Cooper, and K.C. Aune, *Biochemistry*, 1975, *14*, 1079.
38. K.C. Aune and M.F. Rohde, *Anal. Biochem.*, 1979, *79*, 110.
39. W.D. Lansing and E.O. Kraemer, *J. Am. Chem. Soc.*, 1935, *57*, 1369.
40. M.L. Johnson, J.J. Correia, D.A. Yphantis and H.R. Halvorson, *Biophys. J.*, 1981, *36*, 575.
41. D.E. Roark and D.A. Yphantis, *Ann. N.Y. Acad. Sci.*, 1969, *64*, 245.
42. D.C. Teller, Methods in Enzymology. *27D*. (C.W. Hirs and S.N. Timasheff, eds.), Academic Press, Inc., N.Y., 1973, p. 345.
43. C. Tanford, Physical Chemistry of Macromolecules, John Wiley and Sons, Inc., New York, 1961.
44. D.A. Yphantis, *J. Phys. Chem.*, 1959, *63*, 1742.
45. A.M. Crestfield, W.H. Stein and S. Moore, *Arch. of Biochem. and Biophys., Supplement*, 1962, *1*, 217.
46. M.O. Dayhoff, 'Atlas of Protein Sequence and Structure', National Biomedical Research Foundation, 1976, Vol. *5*, p. 267.
47. W.F. Stafford, III and D.A. Yphantis, *Biophys. J.*, 1972, *10*, 1359.
48. G. Hui Bon Hoa, M. Gaffe and M. Grunber-Manago, *Biochemistry*, 1977, *16*, 2800.
49. A. Wishnia and A.S. Boussert, *J. Mol. Biol.*, 1977, *116*, 577.
50. D.A. Yphantis and T. Arakawa, *Biophys. J.*, 1986, *49*, 499a.
51. T. Arakawa and D.A. Yphantis, *J. Biol. Chem.*, 1987, *262*, 484.
52. D.A. Yphantis and T. Arakawa, *Biochemistry*, 1987, *26*, 5422.
53. T. Arakawa, Y.-R. Hsu and D.A. Yphantis, *Biochemistry*, 1987, *26*, 5428.

14

The Use of the Omega Function for Sedimentation Equilibrium Analysis

By G.B. Ralston and M.B. Morris

DEPARTMENT OF BIOCHEMISTRY, THE UNIVERSITY OF SYDNEY, SYDNEY, NSW 2006, AUSTRALIA

1. INTRODUCTION

Analysis of sedimentation equilibrium measurements in the analytical ultracentrifuge requires the solution to the following problems: (i) can the presence or absence of contaminants be demonstrated? (ii) can the influence of non-ideality be taken into account? (iii) in the case of self-association, can the presence of reversible association reactions be demonstrated? In particular, can the pattern of self-association be assessed and can the thermodynamic parameters governing self-association be estimated? Furthermore, is it possible to analyse self-association reactions even in the presence of thermodynamic non-ideality?

In sedimentation equilibrium experiments, most of the information about the nature of the self-association can be obtained only by the prior choice of a reaction model. However, the advantages of this method are: that precise tests are available for determining whether chemical equilibrium has been achieved; that data can be obtained without perturbing the chemical equilibrium of the sample; and that data are obtained over a wide range of solute concentration from a single experiment. These factors are essential for quantitative, thermodynamic analysis of self-associating systems.

Many methods have been detailed in the literature for analysing the results obtained from sedimentation equilibrium experiments with self-associating solutes (for reviews, see refs. 1 and 2). These methods allow some, or all, of the thermodynamic parameters that define the self-association to be quantified, including the equilibrium constant(s), a measure of the non-ideality of the solute, and the molecular weight of the smallest participating species. For the results of such an analysis to be meaningful, such methods must also include a means for determining whether the solute is heterogeneous, and whether the system has attained chemical equilibrium during the time-course of the experiment.

The most direct approach is through analysis of the concentration as a function of radial distance. The method of Johnson *et al.*[3] employs this strategy to estimate equilibrium constants pertaining to a selected reaction model, the monomer molecular weight and a measure of solution non-ideality from the concentration distribution.

However, this approach does not allow a simple means for demonstrating the absence of contaminating species, or for demonstrating that chemical and sedimentation equilibrium have been attained.

Traditionally, self-association in the presence of non-ideality has often been studied through analysis of the distribution of the apparent weight-average molecular weight, $M_{w,app}$. Three of the more commonly used methods have been: (i) direct fitting of the apparent weight-average molecular weight distribution with the use of non-linear regression[4]; (ii) the Adams approach, which requires model-dependent functions of the apparent weight-average and number-average molecular weights[5-8] and (iii) the use of combined number-, weight- and z-average molecular weights for analysis of discrete association patterns[9].

The use of molecular weight averages has the advantage of being conceptually easy to comprehend. One can often see at a glance that self-association is occurring, and can often make a reasonable judgement whether it may be limited to dimerization or to some other relatively simple monomer-n-mer scheme. Non-ideality can often be detected easily through the presence of a maximum in the $M_{w,app}$ distribution. However, when thermodynamic non-ideality is relatively severe, the maximum value of $M_{w,app}$ may be considerably less than the molecular weight of the largest species in discrete association reactions, leading one to underestimate the stoichiometry in this simple approach. On the other hand, for indefinite self-associations, in which oligomer size increases without limit, there may be no decrease in $M_{w,app}$ with increasing concentration, even when non-ideality is reasonably severe.

The distribution of $M_{w,app}$ allows one to test for attainment of combined chemical and sedimentation equilibrium. Only when equilibrium is achieved, will plots of $M_{w,app}$ superimpose for different operating conditions: for different loading concentrations of solute, different radial positions of the solution column, different angular velocities of the rotor, and different column lengths[2,10,11]. The self-associating solute behaves as a single thermodynamic component[12]. Failure to achieve equilibrium (*i.e.*, if the time to attain chemical equilibrium is long in relation to the time of the experiment), the presence of contaminating species, or association-incompetence of some of the solute material, will all be manifest as a failure of overlap in the $M_{w,app}$ distribution. The presence of contaminants makes the value of $M_{w,app}$ no longer uniquely dependent on the total solute concentration, but instead its value will also be dependent on the radial position of the solution column[12]. In many cases, the type of heterogeneity present may be evaluated by noting the form of the divergence between the $M_{w,app}$ distributions[1,2,13].

However, the extensive data manipulation required in these methods, particularly the numerical differentiation required in the calculation of $M_{w,app}$, tends to magnify errors in the original concentration distribution[1,2,11]. Moreover, the use of the logarithm of concentration leads to non-uniform distribution of errors that can lead to a systematic bias in the $M_{w,app}$ distribution, particularly in the low concentration range of the data. Smoothing of the c(r) *vs.* r data, frequently advised as a means of minimising the magnification of errors[1], improves the appearance of the data, but often superimposes its own error structure, leading to oscillations in the $M_{w,app}$

distributions. Errors in the number average molecular weight, $M_{n,app}$, derived from $M_{w,app}$, will be even greater since $M_{n,app}$ is obtained from $M_{w,app}$ by numerical integration[1,2,11]. The accumulation of errors frequently makes it difficult to decide whether molecular weight distributions from separate experiments lie on a single continuous line, and thus whether equilibrium has been attained and whether heterogeneity is negligible.

An alternative approach utilises the omega function as a sensitive means of demonstrating attainment of equilibrium and the absence of contaminants or of association-incompetent material. Furthermore, the omega function can be used as a means of determining patterns of self-association reactions, of estimating the parameters of self-association, and of quantifying thermodynamic non-ideality.

2. THE OMEGA FUNCTION

The omega function was introduced by Milthorpe *et al.*[14] as a means of analysing sedimentation equilibrium data obtained with self-associating solutes. For a single macromolecular associating component, the omega function, $\Omega(r)$, is defined as[14]:

$$\Omega(r) = \frac{c(r)\exp[\phi_1 M_1(r_F^2-r^2)]}{c(r_F)} \tag{1a}$$

$$= \frac{a_1(r_F)c(r)}{a_1(r)c(r_F)} \tag{1b}$$

where $\phi_1 = (1 - \bar{v}_1\rho)\omega^2/2RT$, with $\bar{v}_1$ the partial specific volume of the smallest polymerising species (or protomer), ρ the solvent density*, ω the angular velocity, R the universal gas constant and T the absolute temperature of the sedimentation equilibrium experiment. The value of ϕ_1 is assumed to be independent of radial distance; a reasonable assumption for an incompressible solution with a single solute component and with no volume change on association. M_1 is the molar mass of the protomer, $c(r)$ and $c(r_F)$ are the total weight concentrations of the associating solute at radial positions r and an arbitrarily chosen reference position, r_F, respectively, and a_1 is the thermodynamic activity of the protomer. From eq. (1a), if M_1 and $\bar{v}_1$ are known and an appropriate reference concentration, $c(r_F)$, is chosen, a plot of $\Omega(r)$ *vs*. $c(r)$ can be drawn from experimental measurements obtained with Rayleigh interference or absorption optics. By definition $\Omega(r) = 1$ at $c = c(r_F)$.

Since the relationship between $a_1(r)$ and $c(r)$ depends on the parameters of self-association and non-ideality, eq. (1b) relates the omega function to the parameters of self association for any selected self-association reaction model.

---

* Wills *et al*. (ref. 15) have shown that the *solvent* density is the correct quantity to use here, rather than the frequently quoted *solution* density; its use allows the determination of the thermodynamic activity of the solute in terms of *molarity*. This correction has come about through a rigorous analysis of the appropriate definition of the chemical potential in sedimentation experiments, and the concentration scale in which it should be expressed.

One of the major features of the omega analysis is that the successive differentiation and integration of the data required by many other methods are avoided, and that, as a result, less error accumulates in the calculations.

By extrapolating the omega function for a self-associating solute to zero concentration, the thermodynamic activity of the self-associating *monomer*, $a_1(r)$, can be determined at the chosen reference point in the centrifuge cell. From eq. (1b) it can be shown that[14]:

$$\lim_{c(r)\to 0} \Omega(r) = \Omega_0 = \frac{a_1(r_F)}{c(r_F)} \tag{2}$$

Since the reference concentration, $c(r_F)$, is known, the thermodynamic activity of the monomer at the reference point, $a_1(r_F)$, can be calculated from the intercept at zero concentration. The value of $a_1(r_F)$ may then be used to determine the monomer activity, $a_1(r)$, *independent of any particular model for the reaction*, at any radial position and hence as a function of the total concentration, $c(r)$[14]:

$$a_1(r) = a_1(r_F)\exp[\phi_1 M_1(r^2-r_F^2)] \tag{3}$$

The $a_1(r)$ *vs*. $c(r)$ data can be fitted with reaction models[14]. In the case of simple reaction schemes where ideal behaviour is assumed, $a_1(r)$ may be approximated closely by $c_1(r)$, and, for an ideal monomer-n-mer reaction:

$$k_{app} = \frac{c(r) - c_1(r)}{[c_1(r)]^n} \tag{4}$$

From plots of $\ln[c(r)-c_1(r)]$ *vs*. $\ln c_1(r)$, one obtains the stoichiometry, n, from the slope, and $\ln k_{app}$ from the intercept[16-18].

In more complex reaction schemes, and when non-ideality must be taken into account, the availability of the thermodynamic activity of the monomer at each point in the cell allows the estimation of the parameters of self-association as well as a measure of solution non-ideality as is described below.

Tests for Heterogeneity

Plots of $\Omega(r)$ *vs*. $c(r)$ provide a sensitive and robust test for determining if the self-association reactions from a series of sedimentation equilibrium experiments have reached chemical equilibrium and if the samples behave as a single thermodynamic component. Under these conditions, the plots will be coincident over their common concentration range when a common $c(r_F)$ is used (provided no volume changes accompany the association reactions)[14].

In practice, one would choose a reference point such that the errors in determining the concentration at that point are minimised. A value near 1.0 g/L is usually convenient; this value is sufficiently high that the relative errors are small, and yet, even with meniscus depletion experiments, the concentration gradient is not sufficiently high to produce large absolute errors.

Data from several different experiments can be combined into a single curve, even if a value of $c(r_F)$ cannot be found that is common to all experiments. Milthorpe *et al.*[14] showed that this could be done through the intermediate calculations of the $a_1(r)$ distribution. However, it is possible to combine omega distributions without reference to the intermediate activities.

For example, if data were available from three experiments in which part of the $\Omega(r)$ distribution of experiment 1 overlaps part of the distribution of experiment 2, a reference concentration, $c(r_F)$, common to experiments 1 and 2, and a concentration c^*, common to experiments 2 and 3, can be selected. The value of Ω^*, referred to the original reference concentration $c(r_F)$, will be common to experiments 2 and 3 at the common concentration c^*. This value of Ω^*, together with the radial position r_3 corresponding to c^* in experiment 3, can then be used to compute the reference radial position, r_F, for experiment 3 through the use of a re-arrangement of eq. (1a):

$$r_F^2 = r_3^2 + \frac{\ln [\Omega^* c(r_F)/c^*]}{\phi_1 M_1} \tag{5}$$

Note that the value of ϕ_1 in this calculation will be that pertaining to experiment 3. With the value of r_F now available for experiment 3, all other $\Omega(r)$ values may be calculated through the use of eq. (1a). This approach allows the combination of data from high-speed meniscus depletion experiments, which give one access to very low concentrations and thus aid extrapolation to infinite dilution, and from low speed experiments which allow more precise measurements at higher concentrations.

Fig. 1 shows omega function curves for a simulated monomer-dimer reaction contaminated by various non-participating contaminants, at three different loading concentrations of 0.1 g/L, 0.32 g/L and 1.0 g/L, calculated by the method outlined by Teller[1] for meniscus-depletion conditions. In Fig. 1a, 20% of the monomer is unable to take part in the reaction. This has the effect of lowering the value of the equilibrium constant estimated from a single loading concentration[1], and so it is important to be able to recognise this problem when it occurs. Failure of the three curves to overlap is most evident at the low concentration end of the distribution, but detectable divergence of the curves also occurs at concentrations above the reference concentration.

When the reaction is contaminated with 20% of irreversible tetramer (perhaps as a result of sulphydryl oxidation) divergence of the three omega curves is clearly seen on either side of $c(r_F)$ (Fig. 1b). The presence of 20% irreversible octamer (Fig. 1c) is clearly evident in the lowest loading concentration. However, it is interesting to note that the highest loading concentration is virtually unaffected up to 2 g/L; the effect is only evident towards the cell bottom, but in practice this region may not be observed in meniscus depletion experiments due to the inability to resolve the Rayleigh interference fringes.

The presence of material smaller than the putative monomer (for example, proteolytic fragments) leads to a characteristic upwards deflection of $\Omega(r)$ at low concentrations, most evident at the highest initial loading concentration (Fig. 1d). Once again, divergence of the three curves indicates failure of all sedimenting species

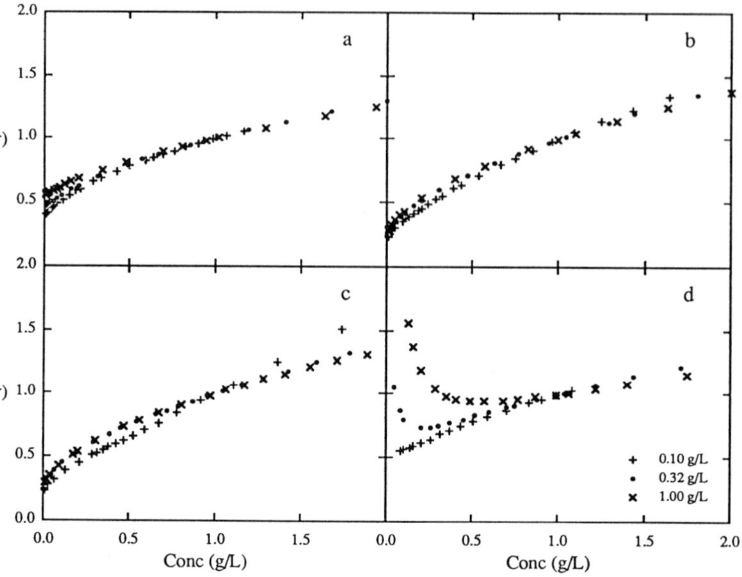

Figure 1. Simulated omega function curves for an ideal monomer-dimer reaction at sedimentation equilibrium in a meniscus-depletion experiment (M_1 = 50,000; k_2 = 10.0 L/g) in the presence of 20% contaminating material: (a) association-incompetent monomer; (b) irreversible tetramer; (c) irreversible octamer; and (d) non-participating species of M = 25,000. Simulated initial loading concentrations were (i) 0.1 g/L, (ii) 0.32 g/L, (iii) 1.0 g/L. $c(r_F)$ = 1.0 g/L.

to attain chemical equilibrium. Similar upward curvature may be seen when the value chosen for the putative monomer molecular weight is appreciably higher than the true value, or when the putative protomer of self-association can itself dissociate to smaller species; in these cases, however, where the reaction may reach equilibrium, the three curves will still overlap. Upward curvature of this nature and poor overlap were seen in a study of myelin basic protein[19,20], and in a study with human luteinizing hormone[21], and may indicate the presence of low molecular weight contaminants in these experiments.

Figs. 1(a) - 1(d) give an indication of the detectability of different types of contaminant as a function of initial loading concentration. High molecular weight contaminants are most detectable at the lowest loading concentrations; with higher loading concentrations, the effects on estimates of the equilibrium constants may be negligible. Conversely, low molecular weight contaminants are most detectable in the highest loading concentrations.

Since it is essential to be able to detect heterogeneity, such tests should also be robust, and should not be compromised by accumulation of experimental and computation errors. Experiments with lysozyme[22] showed poor overlap of $M_{w,app}$, and pressure effects were initially invoked to account for the discrepancy. However, subsequent analysis with the omega function showed excellent overlap of the $\Omega(r)$ *vs.* c(r) curves, and of the computed monomer activity distributions[23], and the conclusion was drawn that the discrepancies in the original $M_{w,app}$ data (±3.5%) arose from the numerical differentiation, rather than from heterogeneity or volume changes on association.

This phenomenon is illustrated in Fig. 2, which shows both omega and $M_{w,app}$ distributions for a simulated ideal monomer-dimer reaction. The data in Fig. 2 were simulated as c(r) *vs.* r, calculated on the basis of an ideal monomer-dimer reaction for three separate loading concentrations in the absence of contaminants. Random errors of ± 0.0027 g/L were superimposed over the concentration distribution and the perturbed data were used for the computation of the distributions of $M_{w,app}$ and $\Omega(r)$.

The failure of overlap in the $M_{w,app}$ data might lead one to reject the 'experiment' on the grounds of failure to attain chemical equilibrium, or because of the presence of contaminants. In the low concentration end of the graph, the discrepancy is as great as 25%, arising in part as a result of chance correlation of some of the errors leading to an inaccurate estimate of the meniscus concentration. This error is then propagated systematically through the rest of the data.

While similar errors are detectable in the omega function distribution computed from the same data, the errors are not magnified by numerical differentiation, and one is more prepared to accept the entire set of data as describing a system in chemical equilibrium. Estimates of the parameters obtained from direct fitting of the omega function curves (see below) are within 5% of those used in the simulation: those from $M_{w,app}$ are up to 15% in error.

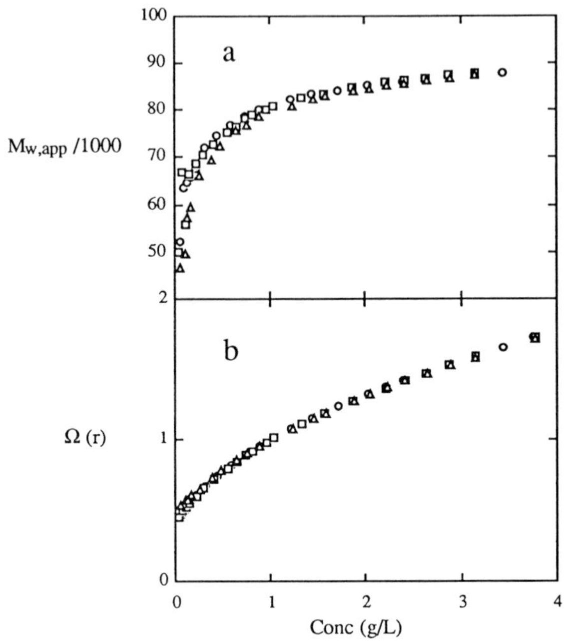

Figure 2. Simulated curves of (a) apparent weight average molecular weight (not smoothed) and (b) the omega function against concentration, based on an ideal monomer-dimer model ($k_2 = 4.0$ L/g, $M_1 = 50{,}000$). Each set of data (represented by the different symbols) was obtained from a different set of c(r) *vs.* r data, mimicking a different loading concentration of solute. Normalized random error was placed on the concentration data before determining the $M_{w,app}$ and $\Omega(r)$ values for each set. Fitting $M_{w,app}$ using nonlinear regression returned an incorrect value of $k_2 = (3.5 \pm 0.4)$ L/g; fitting to the $\Omega(r)$ data returned an acceptable value of $k_2 = (3.90 \pm 0.1)$ L/g.

With real experimental data, the failure of overlap in the $M_{w,app}$ distribution is frequently worse than is found with simulated data. Fig. 3 shows the $M_{w,app}$ and $\Omega(r)$ distributions for the self-association of human spectrin, the major protein of the red cell membrane cytoskeleton. Spectrin appears to undergo a cooperative indefinite self-association (or SEK III) reaction[24]. In the SEK III reaction[6], monomer can add to the dimer and to higher oligomers with equal free energy change, and therefore with the same equilibrium constant, K, on the molar scale. The formation of the dimer from two monomers is governed by a different equilibrium constant, K_{12}; *i.e.*:

$$\begin{array}{llll} 2A & \rightleftarrows & A_2 & K_{12} \\ A_2 + A & \rightleftarrows & A_3 & K \\ A_3 + A & \rightleftarrows & A_4 & K \\ \cdot\cdot \quad \cdot\cdot & & \cdot\cdot & \cdot\cdot \\ A_i + A & \rightleftarrows & A_{i+1} & K \end{array} \tag{6}$$

Again, the overlap in the $\Omega(r)$ data is significantly better than for the molecular weight data.

Extrapolation to Infinite Dilution

The accuracy of the analysis of self-association *via* the use of $a_1(r)$ is highly dependent on the accuracy with which the $\Omega(r)$ *vs.* c(r) curve can be extrapolated to zero concentration[25]. Incorrect extrapolation leads to a propagation of errors throughout the activity distribution. This leads to inaccurate estimates of the parameters when the activity distribution is fitted with the correct model. Furthermore, the residuals from the fit may be non-randomly distributed, leading one to reject the correct model[25].

Milthorpe *et al.*[14] demonstrated that the extrapolation of an omega function curve to infinite dilution for a nearly ideal, weakly self-associating solute is relatively unambiguous. Firstly, there is little curvature in the omega function near zero concentration. Secondly, for a small value of the reference concentration, the intercept will have a value close to 1.0. Therefore, the relative error in estimating the intercept is likely to be small and the error propagated in $a_1(r)$ will be correspondingly small. Analysis of the monomer activity curve with any particular reaction model will then give good estimates of the parameters. Several studies have shown that with relatively weak self-association, extrapolation to an intercept between 0.8 and 1.0 is relatively straightforward[16,17], although some curvature may still be apparent in some cases[18].

With stronger self-association reactions, however, the problems of extrapolation become more serious: the omega function curve becomes steeper at low concentration, the curvature in this region becomes greater, and extrapolation is more hazardous. In addition, even if a small reference concentration is chosen, the intercept

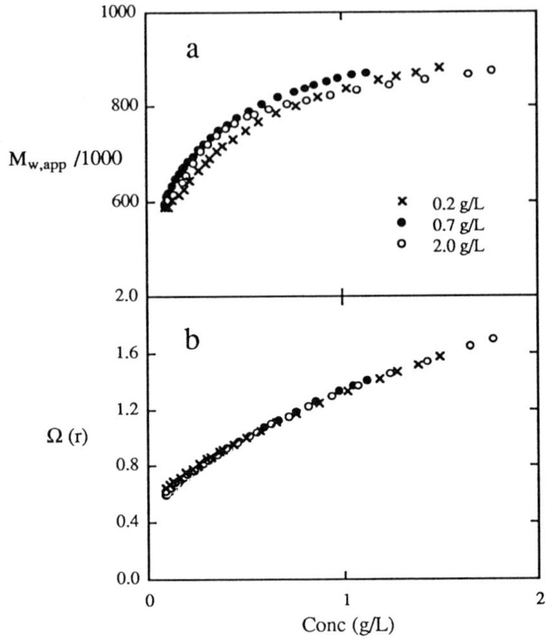

Figure 3. Experimental data for (a) smoothed apparent weight average molecular weight and (b) the omega function against concentration for spectrin centrifuged at 8,000 rev/min at pH 7.5 and 30°C. Each set of data (represented by the three different symbols) was obtained from a different loading concentration of spectrin: 0.2 g/L, 0.7 g/L and 2.0 g/L. The smoothing process appears to have superimposed additional systematic error on the $M_{w,app}$ data.

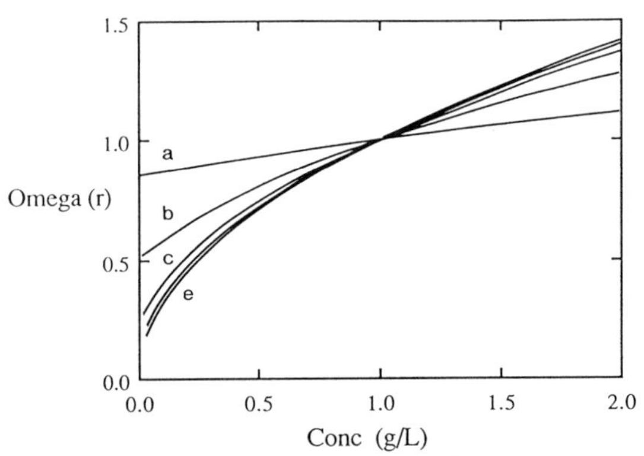

Figure 4. The effect of increasing k_2 on the omega distribution for an ideal monomer-dimer reaction. $\Omega(r)$ was simulated for an ideal monomer-dimer system with values of k_2: (a) 0.1 L/g; (b) 1.0 L/g; (c) 10 L/g; (d) 100 L/g; (e) 1000 L/g. M_1 = 50,000; $c(r_F)$ = 1.0 g/L.

value will be substantially smaller than 1.0, reflecting the low proportion of monomer in equilibrium with oligomeric species. In these cases, a small absolute error in the extrapolation will result in a large relative error in the intercept value and hence in the computed $a_1(r)$ distribution. Therefore, for strong self-associations there is a possibility of rejecting the correct self-association model because of an error made in extrapolating the omega function curve to zero concentration. The difficulty of extrapolation has been shown with simulated data[25], and has been cited as grounds for abandoning the use of the omega function in analysis of the self-association of bovine sulphatase A[26].

Turning points are possible in the omega function. On the one hand, a minimum is possible when material is present for which $M(1-\bar{v}\rho)$ is less than that used for the calculation of omega (Fig. 1d). A maximum is also possible when non-ideality is severe, as is the case in Figs. 5 and 6. Milthorpe *et al.*[14] showed that turning points were only possible when $d \ln a_1(r)/d \ln c(r) = 1.0$.

When the equilibrium constant for the first step in a series of self-association reactions is so great that no monomer is detectable in the concentration range available, then the omega function distribution will extrapolate to an intercept near zero. This has been used[27] to show that the tetramers of aldolase and glyceraldehyde-3-phosphate dehydrogenase do not dissociate appreciably in the concentration range accessible to the Rayleigh optical system.

Milthorpe *et al.*[14] suggested two ways in which errors in extrapolation might be avoided. Firstly, they suggested that an incorrect extrapolation can be detected because the monomer activity curve derived from the extrapolation will not pass through the origin. However, it has been shown that, for a self-associating system, a plot of $a_1(r)$ *vs.* c(r) derived from an incorrect extrapolation will still pass through the origin irrespective of the error in the extrapolation[25].

Secondly, Milthorpe *et al.*[14] suggested that an improved plot of the monomer activity might be achieved by averaging the values of $a_1(r)$ obtained from several curves based on different reference concentrations. However, this procedure does not necessarily lead to better estimates of $a_1(r)$. For a particular set of data, $\Omega(r)$ values at a given value of c(r) calculated for different values of $c(r_F)$ are proportional. The ratio of two such values of $\Omega(r)$ is

$$\exp[\phi_1 M_1(r_{F1}^2 - r_{F2}^2)] \frac{c(r_{F2})}{c(r_{F1})}$$

where $c(r_{F2})$ and $c(r_{F1})$ are different choices for the reference concentration. So, if a reasonable, but incorrect, extrapolation is made for one curve, it is likely that the points used to guide the extrapolation will be treated in a similar manner for a curve based on a different reference concentration. The result is a tendency to propagate a systematic error in the evaluation of the intercepts and hence in the value of $a_1(r_F)$. It would be sufficient to determine a single omega curve, using a convenient reference concentration, and to use a range of feasible extrapolations as a guide to the errors involved.

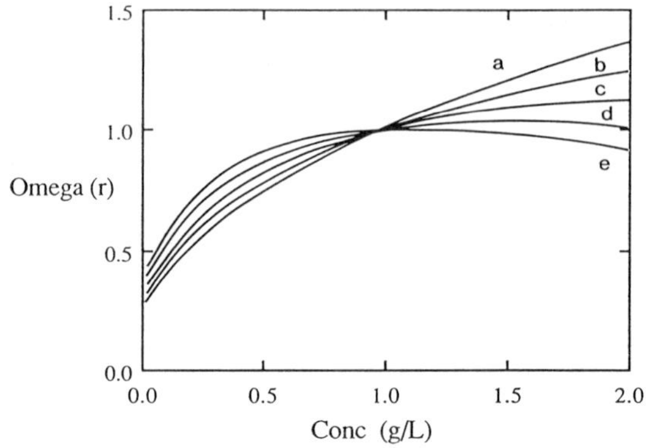

Figure 5. The effect of increasing non-ideality on the omega distribution for an ideal monomer-dimer reaction (k_2 = 10 L/g; M_1 = 50,000; $c(r_F)$ = 1.0 g/L). The values of BM_1 were: (a) 0 L/g; (b) 0.1 L/g; (c) 0.2 L/g; (d) 0.3 L/g; (e) 0.4 L/g.

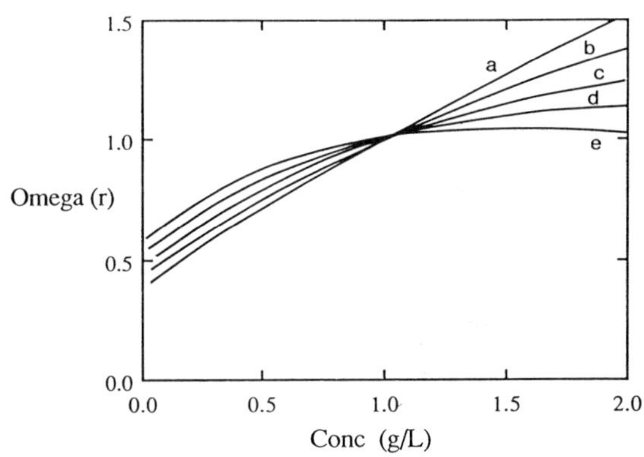

Figure 6. The effect of increasing non-ideality on the omega distribution for an ideal isodesmic reaction (k = 1.0 L/g; M_1 = 50,000; $c(r_F)$ = 1.0 g/L). The values of BM_1 were: (a) 0 L/g; (b) 0.1 L/g; (c) 0.2 L/g; (d) 0.3 L/g; (e) 0.4 L/g.

One way of overcoming the problems of extrapolation when testing simple, ideal models has been to generate the entire $[a_1(r), c(r)]$ distribution from various reasonable extrapolated intercepts, and to estimate the equilibrium constant for self-association for each distribution, *e.g.* through application of eq. (4) [28]. The most appropriate intercept, and thence the best estimate of k_{app}, is chosen as that for which the standard error in k_{app} is a minimum. It must be stressed, however, that this approximation is only valid insofar as non-ideality is negligible and the model used is appropriate. Furthermore, the distribution of $a_1(r)$ is effectively smoothed; since r can be measured effectively without error, $a_1(r)$ will be free of random errors (although it may contain systematic errors from incorrect extrapolation). The "independent" variable c(r) will contain the random errors.

Non-Ideality

The simplest approach to thermodynamic non-ideality is to ignore it. In regions of low concentration, for globular proteins near their isoelectric points, non-ideality terms may be of sufficiently low magnitude that they do not affect the concentration distribution appreciably. In this case, $a_1(r)$ may be approximated closely by $c_1(r)$, and with simple models, a closed solution for c(r) in terms of $c_1(r)$ may be obtained. The various equilibrium constants may then be estimated simply as indicated above.

However, non-ideality cannot be ignored, even at low concentrations of solute, if the solute is large, highly charged or asymmetric. In addition, high solute concentrations are often required to discriminate between possible competing models, and in higher concentration ranges non-ideality cannot be ignored, even for small, globular proteins, close to their isoelectric points.

Over a limited range of concentration, the logarithm of the activity coefficient, y_i, of each species may be approximated by a truncated power series in the total solute concentration expressed in the g/L scale[10]:

$$\ln y_i = i\, BM_1\, c(r) + \text{higher terms} \tag{7}$$

where B is termed the second virial coefficient of the solute (in units of $L.mol.g^{-2}$), taking into account non-ideal behaviour arising from excluded volume and charge repulsion. All activity coefficients are assumed to be described in terms of a single non-ideality coefficient B which is taken to be independent of concentration. In most circumstances the higher order terms are considered negligible.

The use of this "Adams-Fujita approximation" also has the effect of leaving the apparent equilibrium constants (determined from the concentrations rather than the thermodynamic activities) unaffected by non-ideality. From eq. (7), $y_n/y_1{}^n = 1$; for a monomer-n-mer reaction the thermodynamic equilibrium constant, k_n, may be expressed:

$$k_n = \frac{a_n}{a_1{}^n} = \frac{c_n}{c_1{}^n}\frac{y_n}{y_1{}^n} = \frac{c^n}{c_1{}^n} \tag{8}$$

With the use of the Adams-Fujita approximation, the value of B may become a non-specific fitting parameter, taking into account not only the thermodynamic non-ideality, but also to some extent deficiencies in the model. Returned negative values of B have no physical significance[1,4,14] but instead serve to warn the experimenter that the model being used does not take into account sufficient sources of association.

Given explicit formulation of $c_1(r)$ in terms of $c(r)$ and the relevant k values, the thermodynamic activity of the monomer may be obtained:

$$a_1(r) = c_1(r) \exp [BM_1c(r)] \tag{9}$$

For example, in the monomer-dimer case, $c_1(r)$ can be expressed as an explicit function of $c(r)$:

$$c_1(r) = \frac{-1 + \sqrt{1 + 4k_2c(r)}}{2k_2} \tag{10}$$

where $k_2 = 2K_2/M_1$ is the dimerization constant in the g/L scale. Similarly, for isodesmic self-association $c_1(r)$ can be written explicitly in terms of $c(r)$:

$$c_1(r) = \frac{1 + 2kc(r) - \sqrt{1 + 4kc(r)}}{2k^2c(r)} \tag{11}$$

where $k = K/M_1$.

Thus, from any two points $[a_1(r), c(r)]$ the parameters k (or k_2) and B may be determined by solving numerically[14]. However, to make full use of the data, k_2 and B should be estimated from a series of data point pairs, and the mean values taken. For more complex reactions, the solutions to the relevant equations are less simple, and it may not always be possible to write expressions for $c_1(r)$ explicitly in terms of $c(r)$.

3. DIRECT FITTING TO THE OMEGA FUNCTION

$\Omega(r)$, like $a_1(r)$, is dependent only on the total concentration of the associating component and can be written in terms of the self-association parameters and appropriate measures of non-ideality. Thus, in principle, full use of the data can be made by fitting reaction models to the $\Omega(r)$ distribution using non-linear regression. Direct fitting of the $\Omega(r)$ distribution also eliminates extrapolation and its associated problems.

Using the Adams-Fujita approximation (see section 2) eq. (1b) can be rewritten:

$$\Omega(r) = \frac{c(r)c_1(r_F) \exp\{BM_1[c(r_F) - c(r)]\}}{c(r_F)c_1(r)} \tag{12}$$

where B is the second virial coefficient for the self-associating solute.

Various self-association models can be used to evaluate the right-hand side of eq. (12) for comparison with the experimental $\Omega(r)$ values, provided an implicit or explicit function of $c_1(r)$ in terms of $c(r)$ and the equilibrium constant(s) is available (*e.g.*, eqs. (10) and (11)). The values of the reaction model parameters can then be determined by nonlinear regression analysis of the omega function data as a function of c(r).

Fig. 4 shows the effect of increasing k_2 on the form of the $\Omega(r)$ distribution for an ideal monomer-dimer reaction. It is apparent that for values of k_2 between 0.1 L/g and 100 L/g, the omega distribution is relatively sensitive to changes in the equilibrium constant. For higher values of k_2, the concentration of monomer becomes vanishingly small, and as a consequence, it is not possible to distinguish precisely between very large values of k_2. This effect arises primarily through the sensitivity of the optical systems of the ultracentrifuge, rather than a particular limitation of the omega analysis.

The effects of increasing non-ideality (through a single value of B) on the omega function are shown for a monomer-dimer reaction, with $k_2 = 10$ L/g, in Fig. 5, and for an isodesmic reaction, with $k = 1.0$ L/g, in Fig. 6. The effects of non-ideality are evident on either side of the reference concentration.

With only the equilibrium constants and the second virial coefficient as parameters, the fitting function is forced to pass through the reference point, since from eq. (1a), $\Omega(r) = 1$ at the reference point irrespective of the values of the parameters or the error made in measuring the reference concentration. As a result, the estimates of the parameters and the distribution of the residuals of the fit are greatly dependent on the choice of the reference concentration and the accuracy with which it can be measured. If the reference concentration is in error, incorrect parameter values are returned and the distribution of residuals is non-random, even if the correct model is fitted to the data[25].

This problem can be overcome by allowing the concentration at the reference point to be an additional parameter to be estimated in the fitting process[25]. It can be shown mathematically, and with simulated and experimental data, that the same vector of reaction parameters (the relevant equilibrium constants and non-ideality terms) and the same distribution of residuals is returned, irrespective of the error made in measuring the reference concentration, when the reference concentration is included as an additional parameter[25].

With this fitting procedure, the choice of reference point is arbitrary; any selected radial position may be chosen, and will yield the same distribution of residuals and the same vector of reaction parameters[25]. A value near 1.0 g/L is usually a convenient choice (see Section 2).

The returned value of the reference concentration parameter also gives one a check on the appropriateness of the model used for fitting. If the model is appropriate, the difference between the returned reference concentration parameter and the measured concentration at the reference point should be within the expected error of measurement. Larger discrepancies usually indicate an inappropriate model.

For more complex models, such as the non-ideal SEK III, the monomer concentration cannot be written explicitly in terms of the total concentration. In these cases, $c_1(r)$ must be solved by iterative means, such as the Newton-Raphson method[29], in order to find $c_1(r)$ for any particular value of c(r) and a given set of parameter estimates. Because complex models (such as SEK III) result in a cubic dependence of c(r) on $c_1(r)$, and other models may show higher-order dependence, there may be more than one accessible root to the solution for $c_1(r)$. The location of these roots depends on c(r) and the parameters, and sometimes the Newton-Raphson procedure may stray towards physically meaningless parameter values. In these cases a safer, though slower, procedure may be to use a bisection approach to solving for $c_1(r)$ such that reasonable limits may be put on the domain in which the root is searched[29].

In fitting to the omega distribution, it should be noted that, since the value of the reference concentration is a fitting parameter, its value may change after each cycle of iteration during non-linear regression. The value of $a_1(r_F)$ is needed to compute the fitted values of $\Omega(r)$, and so, before each cycle of iteration, a new value of $a_1(r_F)$ must be calculated from the current value of the reference concentration and the current values of the other parameters.

A potential objection to the direct fitting of the omega function is that both the dependent and independent variables are functions of c(r) and that both, thereby, are perturbed by experimental error. Such objections also apply to the fitting of $M_{w,app}$ distributions. Nevertheless, we have found fitting the omega function to be quite satisfactory, both with synthetic data, derived from simulated c(r) *vs.* r distribution in which normalized random errors had been placed on c(r), and with experimental data. When analyzed by direct fitting of the omega function, we have found consistently that the correct parameter values are returned from the synthetic data, for a variety of different models, including non-ideal monomer-dimer, non-ideal isodesmic, and non-ideal SEK III. The approximated standard errors in the parameters are usually within 10% of the values themselves (and frequently better), and are smaller than those obtained from fitting the $M_{w,app}$ distributions of the same data. Furthermore, the correct parameter values lie within the confidence limits defined by the standard errors.

Real experimental data from the self-association of spectrin return consistent values of the reaction parameters when analyzed both by fitting the omega function and by fitting the $M_{w,app}$ distribution. From a series of experiments with human spectrin over a wide range of temperature and pH, the self-association of this protein was found consistently to be described best by the SEK III model. The value of k_{12} varied over an order of magnitude with changing temperature and pH, while the value of k showed substantially less variation. The values of k_{12} estimated from fitting the SEK III model to the omega function correlated extremely well with estimates derived from fitting the same model to the $M_{w,app}$ distribution (Fig. 7). Similar good correlation was seen with the values of k from the two methods, and also with the values of B. However, the standard errors of the parameters returned from fitting the omega function were between 50% and 70% of those from fitting the $M_{w,app}$ distribution, reflecting the accumulation of uncertainty in the $M_{w,app}$ distribution.

Direct fitting to the omega function distribution, with the reference concentration as a parameter, allows the determination of the parameters of self-

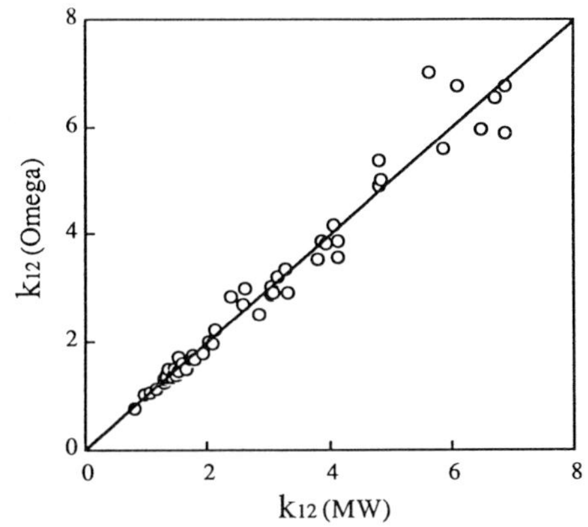

Figure 7. Correlation between values of k_{12} obtained from direct fitting of the omega function and from non-linear regression of $M_{w,app}$. Data from studies of the self-association of spectrin obtained over a range of pH from 6.1 to 8.5 and temperature from 21°C to 35°C were analysed by both methods. The slope of the fitted line is 0.995 and the correlation coefficient is 0.959.

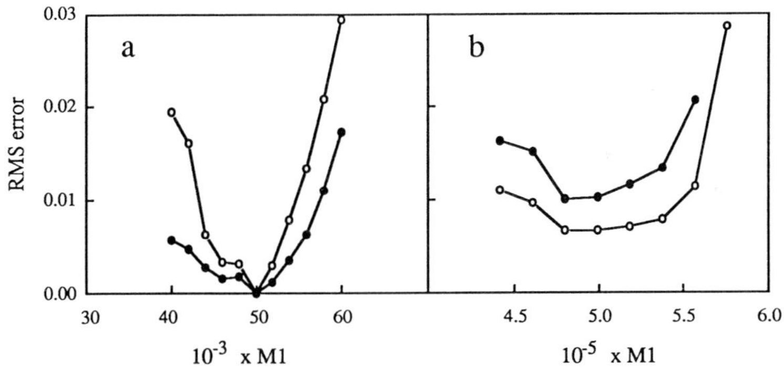

Figure 8. Estimation of M_1 from the minimum of the RMS relative deviation (solid symbols) and the RMS absolute deviation (open symbols) for: (a) simulation of an ideal monomer-dimer reaction, in the absence of errors, with k_2 = 10.0 L/g, M_1 = 50,000; and (b) an experiment with human spectrin (M_1 = 480,000) conducted at 8,000 rev/min at pH 7.5 and 30°C in 0.1 M NaCl. The absolute RMS deviation in (b) has been multiplied by 10 for clearer comparison.

association without reliance on visual extrapolation, and without the need to construct an activity distribution. In the method we have outlined above, all of the experimental data are used to make maximum use of the available information. However, there are cases where construction of the activity distribution may be desired. In such cases, the fitting process described above may be used to guide the extrapolation to infinite dilution, and thus to obtain an accurate estimate of $a_1(r_F)$. When non-ideal effects at high concentration occur that cannot be taken into account through the application of the Adams-Fujita approximation, it may still be appropriate to fit the omega distribution at low concentration (less than 1.0 g/L) to a simplified model. In this concentration range, where a limited number of species exist in detectable concentration, and where higher-order virial coefficients can be reasonably ignored, the association behaviour may be approximated by a truncated series of discrete association steps (*e.g.*, monomer-dimer-trimer) with a single value of the second virial coefficient. Once the monomer activity at the reference point is obtained, iterative procedures can be used to determine successive equilibrium constants (see Section 4, below).

The use of the omega function in determining the pattern of self-association and for estimating the parameters of self-association requires accurate knowledge of the monomer molecular weight. Simulations with ideal monomer-dimer models indicate that an underestimate of M_1 by 25% can result in an overestimate of k_2 by an order of magnitude. It is possible to estimate M_1 from the limiting slope of a plot of ln c(r) *vs.* r^2 [17]. However, with increasing strength of self-association, this approach may fail, as the concentration of monomer decreases.

We have investigated the possibility of estimating the value of M_1 from fitting the $\Omega(r)$ distributions calculated from several different initial estimates of M_1. For each of a series of values of M_1, the $\Omega(r)$ distribution is calculated from eq. (1a). Then, each distribution is fitted with the chosen model, and the root mean square (RMS) relative deviation of the fit is determined:

$$\text{RMS relative deviation} = \sqrt{\frac{\sum\left(\frac{\Omega_{obs}(r) - \Omega_{calc}(r)}{\Omega_{obs}(r)}\right)^2}{n}} \tag{13}$$

where the subscript "obs" indicates the experimental, and "calc" the fitted values of $\Omega(r)$. For a set of data for an ideal monomer-dimer reaction, in the absence of errors, a clear minimum in the RMS relative deviation was found at the true value of M_1 (Fig. 8a). It is important that the relative deviations be used, since the absolute values of $\Omega(r)$ are altered systematically by use of different values of M_1. The RMS absolute deviation is skewed, and may not even show a minimum. With real data, incorporating both non-ideal behaviour and experimental errors, however, the minimum may not always be clearly defined (Fig. 8b). This aspect is under current investigation.

4. EXPRESSION OF ACTIVITY COEFFICIENTS IN TERMS OF DETAILED SOLUTION COMPOSITION

An alternative approach to the use of the Adams-Fujita approximation is to express the activity coefficients of all species in the reaction as a function of solution composition, through the use of statistical mechanical theory[23,30]. This approach is made possible by the availability of the distribution of monomer activity, calculated from the use of the omega function.

The activity coefficient, y_i, of species i may be expressed[23]:

$$\ln y_i = \sum_j \alpha_{ij} \frac{c_j}{M_j} + \sum_{j,k} \alpha_{ijk} \frac{c_j c_k}{M_j M_k} + \ldots \tag{14}$$

where the various c/M represent molar concentrations** and each of subscripts is allowed independently to span the entire set of oligomer species.

The sets of coefficients $\{\alpha_{ij}\}$ and $\{\alpha_{ijk}\}$, corresponding to second and third virial coefficients, respectively, take into account the interactions between pairs and triplets of molecules and are calculated from excluded volumes (most simply on the basis of equivalent spheres) and from consideration of charge interactions[23].

With a suitable starting approximation (*e.g.*, $\ln y_i = \alpha_{ii}\, c(r)/M_i$), approximate values of y_i can be calculated for each oligomeric species as a function of c(r). For a suitable model, the appropriate apparent equilibrium constant(s) can be estimated from non-linear regression of c(r) on $a_1(r)$. With this first estimate of the apparent equilibrium constants, estimates of the thermodynamic activities of higher oligomers can be calculated from $a_1(r)$.

These activities, in conjunction with the initial estimates of y_i, may now be used to compute the various $c_i(r)$ in eq. (14) to provide improved estimates of y_i. This iterative procedure will converge to the best estimates of y_i appropriate to the current approximation of the equilibrium constant(s)[23]. The values of y_i are then used to refine the k values. The entire process may be repeated until best estimates of k and y_i are obtained within the required tolerance. With values available for the various k and y_i, the c(r) distribution may be back-calculated and compared with the experimental data to discriminate between competing models.

This analysis has been applied to the self-association of lysozyme[23]. On the basis of the standard deviation of the c(r) data calculated from the best estimates of the relevant parameters, the isodesmic model was the most favoured; the standard

** The molar activity coefficient has generally been written in terms of the *molal* concentrations; however, again, rigorous analysis of the appropriate definition for the chemical potential (ref. 15) have shown that the concentrations in eq. (14) should be written in the *molar* scale. This change has the effect of equating the coefficients in eq. (14) with the molar excluded volumes.

deviation of the calculated c(r), ±0.039 g/L, was within the range (±0.04 g/L) of expected experimental error.

Nevertheless, it should be pointed out that lysozyme represents a favourable case for this type of approach with an isodesmic reaction model. The equilibrium constant of 461 M^{-1} is sufficiently low, that even at 20 g/L, the decamer is the largest species contributing appreciably to the concentration distribution[23]. With higher values of the equilibrium constant, even more species would need to be included, making the analysis considerably more difficult. With simpler models, such as monomer-dimer, the number of species to be considered is limited, even at higher solute concentrations.

5. MODEL DISCRIMINATION

The most appropriate model is not usually known *a priori*. Choice of appropriate models for testing may often be guided by consideration of data available from other experimental approaches. For example, indefinite models were examined for spectrin self-association, since a series of large oligomers has been observed by means of non-denaturing gel electrophoresis.

For a given model, the most appropriate set of parameters pertaining to that model may be estimated. The most appropriate model from a set of possible models can be chosen after examination of the distribution of residuals, either visually or through the use of an objective criterion such as the runs test[31].

Further objective criteria for discriminating between alternative models can be obtained by comparing the parameter estimates derived from different sets of experimental data; consistent values of the parameter estimates from different experiments may be an indication of a reliable model. However, it should be borne in mind that because of the limitations in measuring concentrations from Rayleigh interference patterns in regions of high concentration gradients, due to inability to resolve the fringes, meniscus depletion experiments carried out with different loading concentrations will, in general, all cover a similar range of concentrations. Since the omega function (as well as $M_{w,app}$) is specified by the total concentration of self-associating solute at any point, and in the absence of contaminants the curves will all overlap, similar parameter estimates will be returned from different experiments, *even when the model chosen is incorrect.* In order to make use of consistency of parameters to discriminate between models, data from different concentration ranges need to be compared. If different parts of the concentration range require different parameter sets for their description, then fitting the combined data set at once will result in non-randomly distributed residuals.

The RMS deviation of the data from the best model would be expected to be within the experimental error expected of the data. There is little advantage in increasing the number of parameters beyond that point, since competing models may not be distinguishable from each other, and with increasing numbers of parameters, the fitting process may not converge to a unique solution.

6. HETEROGENEOUS ASSOCIATIONS

The preceding discussion has dealt almost exclusively with self-association of macromolecules; indeed, most of the application of the omega analysis to now has been devoted to homogeneous self-association. Nichol and colleagues[32], however, have applied the omega function to heterogeneous association and mixtures of non-interacting solutes. In this approach, the thermodynamic activity of the unbound component with the lowest value of $M(1-\bar{v}\rho)$ may be obtained by extrapolation, to zero total solute concentration, of the omega function distribution calculated for that component through the use of an equation analogous to eq. (1a). In heterogeneous associations, however, overlap of the omega function curves cannot be used to check for attainment of equilibrium or the absence of contaminants, since the solute is no longer a single thermodynamic component.

Relations can be formulated which permit evaluation of the equilibrium constants for systems where at least one binding site for that component exists on the second component. This approach can also be used to analyse mixtures of non-interacting components. It is interesting in this context to note that the omega function distribution for a component in the presence of another component for which $M(1-\bar{v}\rho)$ is lower, leads to turning points in plots of $\Omega(r)$ *vs.* $c(r)$, and curves that are concave upwards, such as are shown in Fig. 1d.

This type of analysis has been applied to the interaction between ovalbumin and lysozyme at low ionic strength, with the assumption that non-ideality was negligible[33]. The conclusion, that this interaction led to an array of complexes, including crosslinked complexes, was also supported by results from frontal gel chromatography experiments and from precipitation reactions.

REFERENCES

1. D.C. Teller, *Methods Enzymol.*, 1973, *27*, 346.
2. H. Kim, R.C. Deonier and J.W. Williams, *Chem. Rev.*, 1977, *77*, 659.
3. M.L. Johnson, J.J. Correia, D.A. Yphantis and H.R. Halvorson, *Biophys. J.*, 1981, *36*, 575.
4. M.S. Lewis and G.D. Knott, *Biophys. Chem.*, 1976, *5*, 171.
5. E.T. Adams, L.-H. Tang, J.L. Sarquis, G.H. Barlow and W.M. Norman, in 'Physical Aspects of Protein Interactions', N. Catsimpoolas, ed., Elsevier/North-Holland, Amsterdam, 1978, Chapter 1, p. 1.
6. L.-H. Tang, D.R. Powell, B.M. Escott and E.T. Adams, *Biophys. Chem.*, 1977, *7*, 121.
7. E.T. Adams, *Biochemistry*, 1965, *4*, 1646.
8. J.M. Beckerdite, C.C. Wan and E.T. Adams, *Biophys. Chem.*, 1980, *12*, 199.
9. K.E. van Holde, G.P. Rosetti and R.D. Dyson, *Ann. N.Y. Acad. Sci.*, 1969, *164*, 279.
10. E.T. Adams and H. Fujita, in 'Ultracentrifugal Analysis in Theory and Experiment', J.W. Wiliams, ed., Academic Press, New York, 1963, p. 119.
11. D.A. Yphantis, *Biochemistry*, 1964, *3*, 297.

12. D.E. Roark and D.A. Yphantis, *Ann. N.Y. Acad. Sci.*, 1969, *164*, 245.
13. D.C. Teller, T.A. Horbett, E.G. Richards and H.K. Schachman, *Ann. N.Y. Acad. Sci.*, 1969, *164*, 66.
14. B.K. Milthorpe, P.D. Jeffrey and L.W. Nichol, *Biophys. Chem.*, 1975, *3*, 169.
15. P.R. Wills and D.J. Winzor, Chapter 17, this volume.
16. C.M. Jackson, G.M. Brenckle, P.J. Hogg and D.J. Winzor, *J. Biol. Chem.*, 1987, *262*, 13472.
17. L.D. Ward and D.J. Winzor, *Arch. Biochem. Biophys.*, 1984, *234*, 125.
18. R. Tellam, D.J. Winzor and L.W. Nichol, *Biochem. J.*, 1978, *173*, 185.
19. R. Smith, *Biochemistry*, 1980, *19*, 1826.
20. R. Smith, *Biochemistry*, 1982, *21*, 2697.
21. A. Nureddin and P. Johnson, *Biochemistry*, 1977, *16*, 1730.
22. G. Howlett, P.D. Jeffrey and L.W. Nichol, *J. Phys. Chem.*, 1972, *76*, 777.
23. P.R. Wills, L.W. Nichol and R.J. Siezen, *Biophys. Chem.*, 1980, *11*, 71.
24. M. Morris and G.B. Ralston, *Biochemistry*, 1989, *28*, 8561.
25. M. Morris and G.B. Ralston, *Biophys. Chem.*, 1985, *23*, 49.
26. C.I. Prosser and A.B. Roy, *Biochim. Biophys. Acta*, 1980, *613*, 130.
27. C.J. Masters and D.J. Winzor, *Arch. Biochem. Biophys.*, 1981, *209*, 185.
28. K.E. Shearwin and D.J. Winzor, *Biophys. Chem.*, 1988, *31*, 287.
29. R.L. Burden, J.D. Faire and A.C. Reynolds, 'Numerical Analysis', Prindle, Weber and Schmidt, Boston, 1981, Chapter 2, p. 20.
30. P.D. Jeffrey, B.K. Milthorpe and L.W. Nichol, *Biochemistry*, 1976, *15*, 4660.
31. W.J. Dixon and F.J. Massey, 'Introduction to Statistical Analysis', McGraw-Hill, New York, second edition, 1957 p. 287.
32. L.W. Nichol, P.D. Jeffrey and B.K. Milthorpe, *Biophys. Chem.*, 1976, *4*, 259.
33. P.D. Jeffrey, L.W. Nichol and R.D. Teasdale, *Biophys. Chem.*, 1979, *10*, 379.

15

MSTAR: A FORTRAN Program for the Model Independent Molecular Weight Analysis of Macromolecules using Low Speed or High Speed Sedimentation Equilibrium

By S.E. Harding, J.C. Horton and P.J. Morgan

UNIVERSITY OF NOTTINGHAM, DEPARTMENT OF APPLIED BIOCHEMISTRY AND FOOD SCIENCE, SUTTON BONINGTON, LE12 5RD U.K.

1. INTRODUCTION

Besides conformational and homogeneity analyses, the principle use of the analytical ultracentrifuge is for the absolute (*i.e.* not requiring calibration standards) measurement of molecular weight and related parameters (such as interaction constants and thermodynamic non-ideality coefficients). There is currently a range of software that has been produced by users of the ultracentrifuge over the years for this purpose. This ranges from simple linear regression analysis of log concentration *versus* radial displacement squared plots from the sedimentation equilibrium of well-behaved single solute protein systems right through to software for the analysis of complicated self-associating and non-ideal systems. For example, the FORTRAN programs BIOSPIN and NONLIN from the laboratory of David Yphantis at Storrs, Connecticut, have been very popular for this purpose.

In our laboratory at Nottingham, we have been using various forms of a program called MSTAR for the molecular weight analysis of systems of macromolecules from sedimentation equilibrium patterns recorded using the Rayleigh interference and absorption optical systems of an analytical ultracentrifuge. It derives from an early program written by Michael Creeth - then at the University of Bristol - for a Wang desk-top calculator and later extended and adapted for mainframe FORTRAN by one of us (SEH). These versions were for the analysis of a relatively limited number of data points recorded from Rayleigh fringe profiles using manually operated microcomparators. It has now been recently adapted by us to the case of Rayleigh interference data captured semi-automatically *off-line* from the Beckman Model E *via* a laser densitometer gel scanner (see Chapter 5 of this volume) and also to the case of absorption data captured *on-line* from the Beckman Optima XL-A.

MSTAR serves two purposes:

(i) First, for a finite loading concentration, it will estimate the (apparent) *whole cell* weight average molecular weight ($M^0_{w,app}$) using a procedure involving an operational point average molecular weight, M^*: this procedure appears particularly well suited for the analysis of difficult heterogeneous systems.

(ii) Secondly, it estimates apparent *point* weight average molecular weights as a function of the square of the radial displacement (or equivalently point concentration). Providing the initial fringe or absorbance data is of sufficient quality, it will then estimate point z-average molecular weight and the Roark-Yphantis[1] M_{y2} point average (a "compound" average, free of first order effects of non-ideality).

Purpose (ii) resembles a similar objective of the Yphantis program BIOSPIN[2,3] although (i) is different. It also has a facility for evaluating meniscus concentrations, useful for the evaluation of low speed sedimentation equilibrium solute distributions recorded using the Rayleigh interference optical system. MSTAR does not assume any model (monodisperse, monomer-dimer, Gaussian distribution, *etc.*) and hence is different from programs like NONLIN[2,3] for the analysis of non-ideal self-associations and our own POLY for the analysis of non-ideal systems which are polydisperse in a discrete (as opposed to quasi-continuous) sense.

In this Chapter we will describe the two versions of MSTAR: MSTARI (for the analysis of sedimentation equilibrium solute distributions recorded using the Rayleigh interference optical system) and MSTARA (for distributions recorded using absorption optics). Although in their current form they have been written for a mainframe computer (with graphics peculiar to the IBM/3084Q at the University of Cambridge), it is expected that both forms will be available for a PC in the near future.

2. MOLECULAR WEIGHT PARAMETERS SOUGHT

Although all three principle molecular weight averages, M_n, M_w and M_z, are in theory extractable, because the bulk of our work in the past has involved the use of the Rayleigh optical system and because we have been looking at highly polydisperse systems - such as polysaccharides which are not well suited for using the meniscus depletion method - number averages have not been readily obtainable* and hence we have focussed almost exclusively on the extraction of weight average molecular weights, and - to a lesser extent - on z-average molecular weights. Although the low-speed method for many systems of macromolecules is more suitable, only the high-speed method readily yields number average molecular weight data - for the latter BIOSPIN is recommended.

---

* A method for the evaluation of $M_n(a)$, the point average molecular weight is normally required.

Point Average Molecular Weights

The simplest molecular weight parameter to come out of sedimentation equilibrium analysis, if Rayleigh or absorption optics are used, is the apparent point weight average molecular weight, $M_{w,app}(r)$ defined by[5]

$$M_{w,app}(r) = \frac{1}{k}\frac{d \ln C(r)}{d(r^2)} \tag{1a}$$

where C (r) is the concentration (g/ml) at a radial position r and k is a run constant given by

$$k = \frac{(1-\bar{v}\rho)\omega^2}{2RT} \tag{2}$$

where ω is the angular velocity of the rotor (rads/sec), R is the gas constant, T the absolute temperature, $\bar{v}$ the partial specific volume of the macromolecular solute (in ml/g) and ρ the solution density (in g/ml)†. The reason why the molecular weight in eq. (1) is an *apparent* one is that it corresponds to a finite concentration C. If absorption optics are used, then within the validity of the Lambert-Beer equation

$$A = \varepsilon_\lambda Cl \tag{3}$$

[where ε_λ is the extinction coefficient ($ml.g^{-1}.cm^{-1}$), and l the cell path length (cm)], the absorbance, A, at a given wavelength, λ is proportional to C and hence

$$M_{w,app}(r) = \frac{1}{k}\frac{d \ln A}{d(r^2)} \tag{1b}$$

If Rayleigh optics are used, the fringe displacement, J is also proportional to C:

$$J = \frac{\mathfrak{R} lC}{\lambda} \tag{4}$$

(*cf.* eqs. (2.5) and (2.6) of ref. 5) where $\mathfrak{R}$ is the specific refraction increment ($ml.g^{-1}$). Unfortunately unless the meniscus depletion method is employed, Rayleigh interference patterns yield directly fringe displacements *only relative to the meniscus*[5], and we denote these relative fringe displacements by j(r), where

$$j(r) = J(r) - J_a \tag{5}$$

J_a being the meniscus fringe number. The form of eq. (1) for Rayleigh optics is therefore[5]

$$M_{w,app}(r) = \frac{1}{k}\frac{d \ln J}{d(r^2)} \equiv \frac{1}{k}\frac{d \ln (j + J_a)}{d(r^2)} \tag{1c}$$

There are various ways of extracting J_a (or equivalently C_a): MSTARI uses a fairly simple procedure which is summarised in section 3 below.

---

† In practice, the solvent density can be used: this significantly affects only the interpretation of virial coefficient information and can be positively advantageous (see Chapter 17 this volume and ref. 4, pp. 284-285).

To obtain point z-average molecular weights, $M_{z,app}(r)$, an estimate for the meniscus concentration is *not* required, even for distributions recorded using Rayleigh optics. However a double derivative of the fringe data *is* required:

$$M_{z,app}(r) = M_{w,app}(r) + \frac{1}{k}\left\{\frac{d \ln M_{w,app}(r)}{d(r^2)}\right\} \tag{6a}$$

$$\equiv \frac{1}{k}\frac{d}{d(r^2)}\left(\ln\left\{\frac{1}{r}\frac{dC(r)}{dr}\right\}\right) \tag{6b}$$

(*cf* eqs. (2.12) and (2.42) of ref. 5) and so data of a high quality is usually needed, unless heavy data smoothing routines are employed.

It is possible also to define another derived or "compound" point average. This is the Roark-Yphantis[1] M_{y2} point average, which is free from first order effects of thermodynamic non-ideality:

$$M_{y2}(r) = \frac{M^2_{w,app}(r)}{M_{z,app}(r)} \tag{7}$$

Whole Cell Weight Average Molecular Weight

Arguably the most useful - at least the most basic - parameter to derive from sedimentation equilibrium analysis is the (apparent) average molecular weight for the distribution of solute averaged over all radial positions in the solution column of the cell, $M^0_{w,app}$. This is conventionally obtained by estimating an average slope of a plot of log C *versus* r^2, or equivalently, from determination of the concentrations of solute at the cell meniscus (r=a) and base (r=b), C_a, C_b respectively, and from knowledge of the initial loading concentration C_0:

$$M^0_{w,app} = \frac{1}{k(b^2-a^2)}\left\{\frac{C_b-C_a}{C_0}\right\} \tag{8}$$

where the concentrations could either be on a weight basis, in terms of absorbance (with the limits of the Lambert-Beer law) or in terms of absolute Rayleigh fringe displacements, J. For simple, well behaved, monodisperse, approximately ideal solutes this equation gives adequate (*i.e.* to within ±5%) estimates for $M^0_{w,app}$ (provided that the term $\bar{v}\rho$ in eq. (2) is not too close to unity). For heterogeneous solutions, particularly where there is strong curvature in the log C *versus* r^2 plots (especially if Rayleigh optics are being employed), and also if the optical records near the cell base are not clearly defined, estimates for C_b can be very difficult to obtain with any reasonable precision. We find a useful way of minimising these problems is to use an operational point average, M^*. M^* can be found from Rayleigh or absorption sedimentation equilibrium records *via* the relation

$$\frac{C(r) - C_a}{M^*(r)} = kC_a(r^2-a^2) + 2k\int_a^r r[C(r) - C_a]\, dr \tag{9}$$

Several years ago[7] Michael Creeth and one of us (SEH) explored the properties of M^* in some detail and found it to have some useful features, one of which is that at the cell base $M^* = M^0_{w,app}$: as is pointed out later in Chapter 27 of this volume, use of M^* generally provides a better way of estimating $M^0_{w,app}$ compared to the conventional use of eq. (8) (see Table 1 of Chapter 27). Further, an independent estimate of the initial cell loading concentration, C_0 is *not* required. As the name implies, it is this M^* function which forms the basis of MSTARI and MSTARA.

The M^* function provides no improvement on obtaining whole cell z-average molecular weights, $M^0_{z,app}$ so the current versions of MSTAR don't evaluate this.

Reduced Molecular Weights

Optical records from sedimentation equilibrium do not yield molecular weights directly but *reduced* (apparent) molecular weights, A_i†, defined by

$$A_i = k\, M_{app,i} \tag{10}$$

(*cf.* eqs. (1), (6) and (8)) where k is given by eq. (2), and where the subscript i can represent point number, weight, z or y2 averages, or the whole-cell weight average, or the star operational average (eq. (8)). We use the symbol A_i for reduced averages following the original convention of Rinde[8]. Another popular alternative is the use of "reduced moments", σ_i, where $\sigma_i = 2A_i$ (see *e.g.* Chapters 7 and 13 of this volume), although, since our main focus of attention has been on polydisperse macromolecules like polysaccharides and glycoconjugates, we prefer to use the Rinde notation to avoid confusion of σ with standard deviations of molecular weight distributions.

3. DETERMINATION OF MENISCII CONCENTRATIONS

For the evaluation of point weight average molecular weights and whole cell weight average molecular weights from Rayleigh interference optical records, and also for the evaluation of whole cell weight average molecular weights from absorption optical records it is necessary to estimate the solute concentration at the meniscus, C_a (or equivalently, for Rayleigh optics, J_a).

C_a from Absorption Optical Records

Fortunately if the absorption system is used this is usually trivial, since absorption is a direct record of solute concentration within the Lambert-Beer range. A simple linear extrapolation to the meniscus usually suffices: even for cases where a long extrapolation has been necessary (due to errors in cell filling), this extrapolation has given reasonable results.

---

† Note that the symbol A is used for absorbance whereas the symbol for reduced molecular weight is A_i (*i.e.* with a subscript or superscript).

C_a (or J_a) from Rayleigh Optical Records

As has been well reported in the past, this is unfortunately not as easy since the optical records are a direct record of *relative* concentrations. Creeth and Pain[5] considered a whole variety of procedures for extracting J_a. We have generally used a procedure described in ref. 7; we will give an outline of it here.

One of the forms of the fundamental equation of sedimentation equilibrium is

$$\frac{J(r)}{A_n(r)} - \frac{J_a}{A_n(a)} = 2\int_a^r rJ\, dr \qquad (11)$$

(see *e.g.* eqs. (8) and (26) of ref. 6). Creeth[9] suggested a new variable, $A^*(r)$, be identified by writing

$$\frac{j(r)}{A^*(r)} \equiv \frac{J(r)}{A_n(r)} - \frac{J_a}{A_n(a)} \qquad (12)$$

Substituting eq. (12) and eq. (5) into eq. (11) gives

$$\frac{j(r)}{A^*(r)} = J_a(r^2 - a^2) + 2\int_a^r rj\, dr \qquad (13)$$

A further re-arrangement gives

$$\frac{j(r)}{(r^2 - a^2)} = J_a A^*(r) + \frac{2A^*(r)}{(r^2 - a^2)}\int_a^r rj\, dr \qquad (14)$$

Thus a plot of $j(r)/(r^2 - a^2)$ against $I(r)/(r^2 - a^2)$, where $I(r)$ is defined by

$$I(r) \equiv \int_a^r rj\, dr \qquad (15)$$

gives an intercept at r=a of $J_aA^*(a)$ and a limiting slope at r=a of $2A^*(a)$. Hence

$$J_a = \frac{2 \times \text{intercept at r=a}}{\text{slope at r=a}} \qquad (16)$$

The integral can be evaluated using a standard numerical package (in the case of MSTARI this is the NAG (Numerical Algorithms Group, Oxford, UK) routine D01GAF as will be described below. A similar procedure has been given by Teller *et al.*[10].

For systems that are fairly ideal and monodisperse such a plot is approximately linear (Fig. 1a) and extraction of J_a is therefore easy. For other systems however it is not linear and the functional form of the integral in eq. (15) is not known (Fig. 1b). Two opposing criteria have to be taken into account: (i) because a limiting slope is required at the meniscus, *greater weight* should be paid towards those data nearer the meniscus; (ii) unfortunately, as is well known, fringe displacement data are not

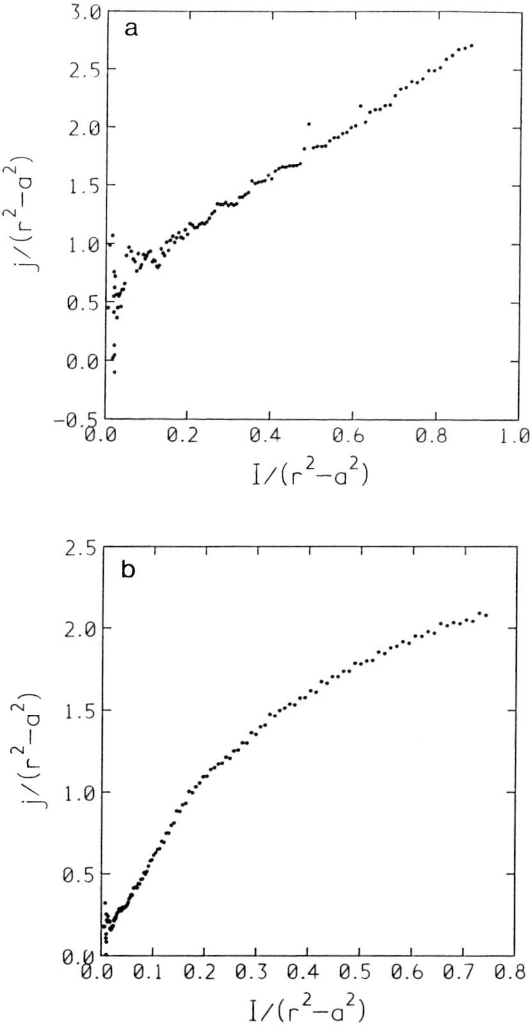

Figure 1. MSTARI $j/(r^2\text{-}a^2)$ *versus*

$$\frac{1}{(r^2\text{-}a^2)} \int_a^r rj \; dr$$

plots for sedimentation runs (recorded using the Rayleigh interference optical system on a Beckman Model E) on (a) a fairly homogeneous/ideal solution of colonic mucin "T-domains" ($J_a \sim 0.58\pm0.05$); (b) a highly non-ideal solution of xanthan ("RD") ($J_a \sim 0.01\pm0.01$ *i.e.* near meniscus depletion conditions).

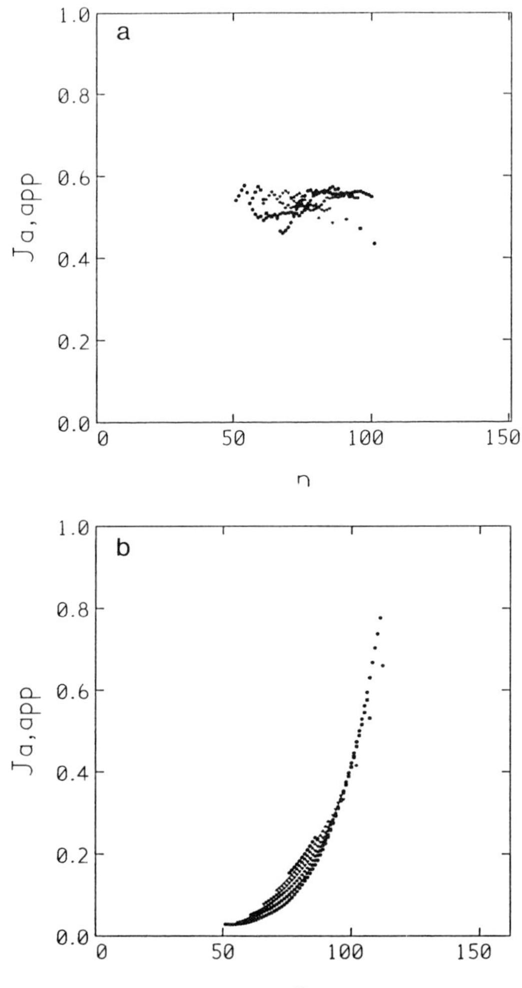

Figure 2. MSTARI plots of apparent meniscii concentrations ($J_{a,app}$) as a function of radial position number, n from sedimentation equilibrium solute distributions (in aqueous solvents) of (a) colonic mucin "T-domains" and (b) xanthan where $n = \{(r-a)/(b-a)\}\ N_{TOT}$ and N_{TOT} is the total number of experimental points (at equal intervals of radial distance, r) between a and b.

accurate near the meniscus and so *less weight* should be applied. This apparent dilemma is offset by the facility of having large amounts of fringe data from automatic (off- or on-line) data capture, and so in practice what we do is to use a sliding strip procedure which is iterated along the $j/(r^2-a^2)$ *versus* $I/(r^2-a^2)$ curve, the procedure repeated for several sliding strip lengths. These estimates for J_a ("$J_{a,app}$") can then be extrapolated to zero meniscus position to given an "ideal" J_a value (Fig. 2). The procedure usually works well even for very heterogeneous systems (Fig. 2b).

We will now consider in turn the two versions MSTARA and MSTARI of MSTAR. We will consider MSTARA first since it is easy to implement (no problems of J_a estimates).

4. MSTARA

Evaluation of the Integral, I(r)

I(r) is evaluated *via* eq. (15) by employing the NAG routine D01GAF. D01GAF is a method due to Gill and Miller[11] for integrating a function of which at least four data points are known. In the XL-A implementation of this program, the first data point is $r=r_1$ (>a). In such cases I(r) is split into two parts: from a to r_1 and from r_1 to r. The second part is evaluated with D01GAF and the first part (which, in fact, is usually quite small compared with the second part) with a simple linear extrapolation (see Appendix). The whole procedure is summarised in Fig. 3. Once I has been evaluated, M^* can then be readily calculated and an example of a plot of M^* *versus* the radial displacement squared parameter, ξ (together with the corresponding plot of ln A *versus* ξ) is given for IgM in Fig. 4 where

$$\xi = \frac{r^2 - a^2}{b^2 - a^2} \tag{17}$$

Point Weight Average Molecular Weight

We can now return to the calculation of the point (apparent) weight average molecular weight *i.e.* $M_{w,app}(r)$. Values of $M_{w,app}(r)$ are calculated by considering sliding strips of consecutive (r^2, ln J) data points and calculating the slope of the sliding strip at the middle point. (Sliding strips are always chosen so that they contain an odd number of points. This is ensured by defining their length to be 2w+1 where w is an integer. As will be seen below, this is convenient since w has a natural meaning in this method). This slope calculation is done using three NAG routines. First, E02ADF (which uses Chebyschev polynomials) is called to fit a quadratic line to the data points. This is the process referred to as FIT in Fig. 5. Next the fitted line is differentiated (in its Chebyschev representation) with E02AHF - DIFF in Fig. 5. Finally, since both the line and its differential are still in the form of sums of Chebyschev polynomials, E02AEF is called to evaluate them at the middle point of the strip. For the sake of completeness, the residuals (*i.e.* the differences between the actual data points and the fits) were also calculated (RESID in Fig. 5). Fig. 5 is a flow diagram summarising the

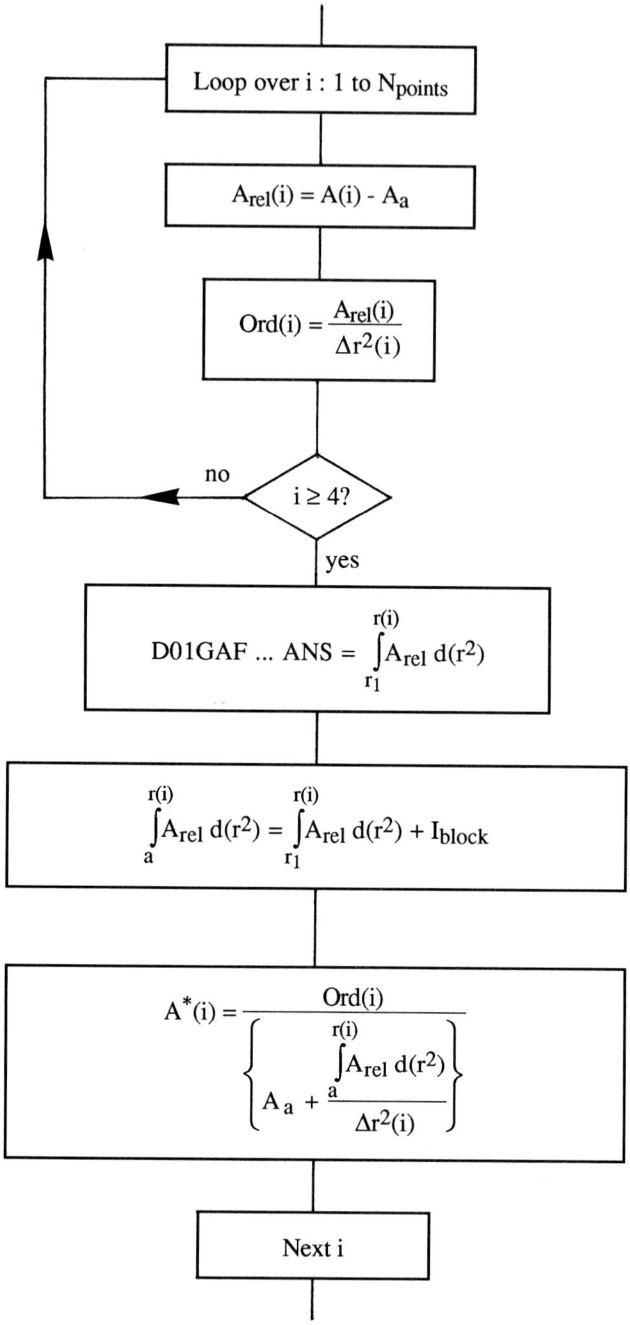

Figure 3. Calculation of A^* (and hence M^*). I_{block} is defined in the appendix.

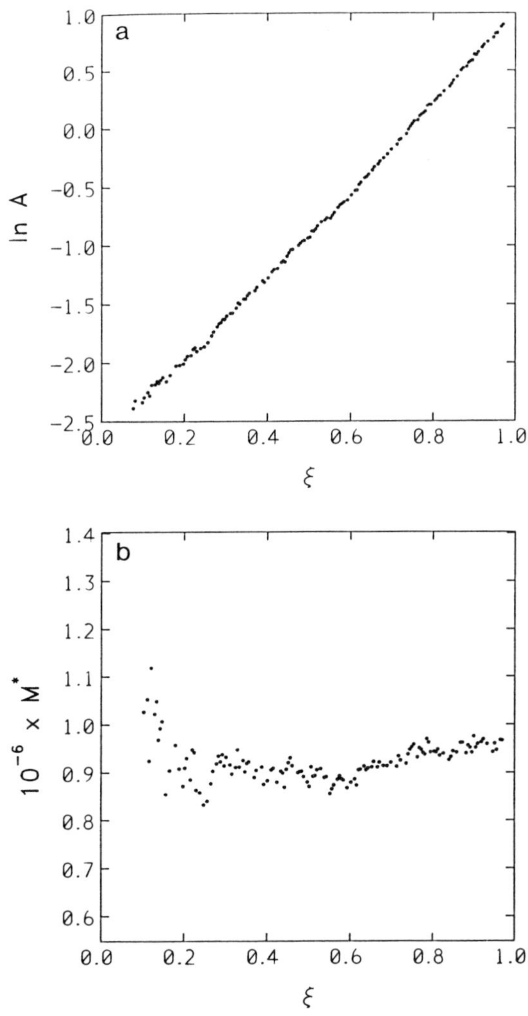

Figure 4. MSTARA (a) ln A *versus* ξ and (b) corresponding M^* *versus* ξ plots from a low speed sedimentation equilibrium experiment (recorded using the absorption optical system on the Beckman Optima XL-A) on human IgM_1 in phosphate/chloride buffer (pH=6.8, ionic strength, I=0.1) at a loading concentration of ~0.6 mg/ml. Scanning wavelength = 278 nm. Rotor speed = 5000 rev/min; temperature = 20.0°C. $M_w^0 \sim M_{w,app}^0 = M^*(\xi \rightarrow 1)$ (from plot (b)) = $(1.00 \pm 0.02) \times 10^6$ g/mol.

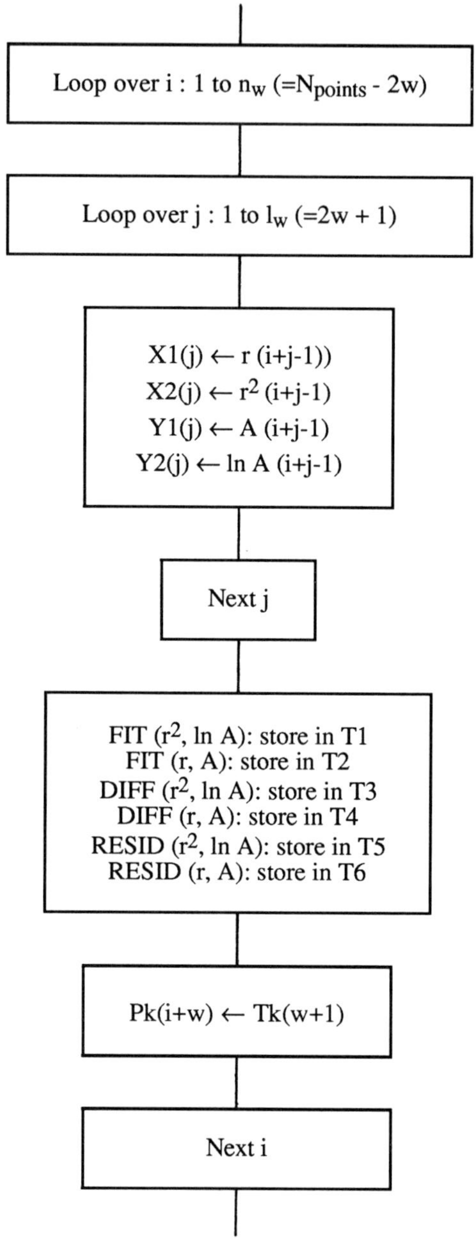

Figure 5. Sliding strip calculations. The Tk ($1 \leq k \leq 6$) are temporary matrices used inside the loop over i. At the end of the calculations (and hence just before the loop is incremented), the middle value (*i.e.* the $(w+1)^{th}$ value) from each Tk is transferred to a permanent array Pk. The FIT, DIFF and RESID operations are referred to in the text.

procedure. To reduce the diagram to as simple a form as possible, all technicalities arising solely from the NAG routines have been omitted. It is assumed that the reader is either familiar with the routines in question or can become familiar with them. By "technicalities" is meant such features as the routines working in terms of a normalised abscissa, $\bar{x}$

$$\bar{x} = \frac{2x - (x_{max} + x_{min})}{(x_{max} - x_{min})} \tag{18}$$

This sliding strip procedure is repeated for as many data points as possible. Obviously, it cannot be carried out on the points at the very beginning and very end of the data. In particular, if we use our above definition of each sliding strip consisting of 2w+1 points (*i.e.* the central point and w points on either side), the method cannot be used for the first w and the last w points. Therefore, if we start with N (r^2, ln J) points then the maximum number of (r, M_w) points we can derive is N-2w. Once evaluated the sliding strip Chebyschev polynomials can be used to generate plots of point M_w *versus* r (or equivalently ξ) or point M_w *versus* C (or absorbance). MSTARA gives both and Fig. 6 gives an example for IgM using a sliding strip length of 13 (*i.e.* w=6). As would be expected, data of this type becomes more noisy near the meniscus or, equivalently, smaller values of (C-C_a) where the concentration increments ΔC are small.

Point z-average Evaluations

The z-average (apparent) molecular weight is calculated from eq. (6b). When calculating the local slopes of ln C *vs.* r^2, opportunity is also taken of doing the same for the (r, j) data (see Fig. 3). Thus, when it comes to calculating M_z, an array of

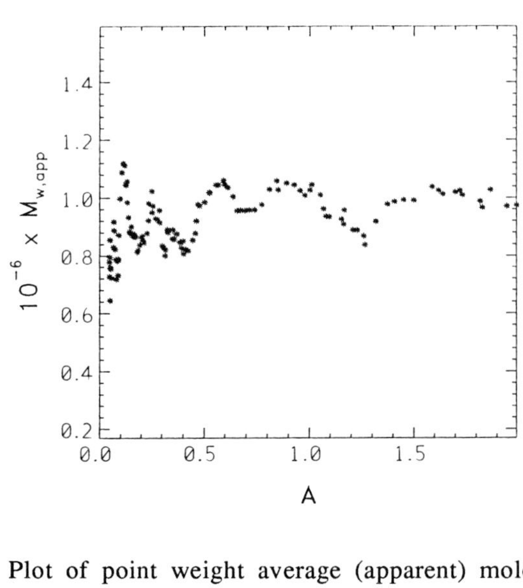

Figure 6. Plot of point weight average (apparent) molecular weight *versus* concentration (in absorbance units at 278 nm) for a low speed sedimentation equilibrium on human IgM_1 (other details as Fig. 4).

dC/dr values is already available (except that it is deficient in the first w and last w values). The second differentiation in eq. (6b) (*i.e.* with respect to r^2) is now carried out with another sliding strip procedure. For generality, the number of points used in each strip is 2z+1 where z is another integer which is, in general, not equal to w. We have been using the default value z=7.

An important feature of eq. (6b) is that it shows that the $M_z(r)$ clearly do not depend on the value of J_a. However, because of the double differentiation in eq. (6a) (or (6b)), $M_{z,app}$ is very sensitive to data error and we have found so far that absorption optics do not provide reliable enough data for $M_{z,app}$ to be evaluated unless whole curve quadratic fitting to the data is used, with the usual risks of "oversmoothing".

5. MSTARI

MSTARI is similar to MSTARA except for two features. First, the data is taken from the analytical ultracentrifuge in the form of a photographic negative showing the fringes from the Rayleigh interferometer in the optical system. After this negative has been enlarged, the data are digitised on a gel scanner controlled from a PC. As noted above, data near the meniscus are generally unavailable or, at least, very noisy. The laser densitometer ANALYSE software (Chapter 5) incorporates a procedure to extrapolate the fringe data back to the meniscus. This obviates the need for a procedure equivalent to the calculation of I_{block} (see above and the Appendix) since by the time the data is presented to MSTARI it is already complete.

The second difference is that, as pointed out above, Rayleigh interferometry only gives concentration in relative units. These "fringe" units (j) are relative to the concentration at the meniscus (J_a). Thus, before starting the calculations, it is necessary to evaluate J_a. This is done by using the method leading up to eq. (16). Despite the problem of J_a evaluation (a problem which can be avoided *if* the high speed meniscus depletion method can be applied), data captured using Rayleigh interferometry tend to be more precise than that captured using absorption optics and so the $M_{z,app}(r)$ results are usually more realistic: Fig. 7 [12] shows plots of ln J *versus* ξ, M^* *versus* ξ, M_w *versus* J and M_z *versus* J for a *relatively* "ideal" "monodisperse" system (colonic mucin T-domains); Fig. 8 [13] gives the corresponding set of plots for a very non-ideal system of xanthan.

6. MODE OF OUTPUT FROM MSTAR

The principle mode of output is graphical: For MSTARI this will be plots of $j/(r^2-a^2)$ *versus* $I/(r^2-a^2)$ and $J_{a,app}$ *versus* r in the initial running of the programme (to obtain an estimate for J_a) and plots of ln J *versus* ξ, M^* *versus* ξ, M_w *versus* J (and ξ) and M_z *versus* J (and the option of M_{y2} *versus* J). Estimates for J_a and $M^*(\xi \rightarrow 1)$ are usually best done by manual extrapolation (because of the perils of computer extrapolations of non-linear plots!). MSTARA produces plots of ln C (or A) *versus* ξ, M^* *versus* ξ, M_w *versus* ξ, M_w *versus* C (or A) and the option of M_{y2} *versus* C (or A). Both versions of MSTAR gives the option of printed output from the various calculation stages, including M^*, M_w and M_z *versus* ξ or C although with the large number of data points involved (100 - 200), this can yield an unwieldy amount of output. The routine also however prints out an estimate for C_0 (the initial loading concentration in appropriate units) based on the usual conservation of mass equation:

$$\int_a^b C(r)\,dr = \frac{C_0\,(b^2+a^2)}{2} \tag{19}$$

and the comparison of this with the "expected" C_0 provides a useful check for possible errors.

7. MSTAR AND THERMODYNAMIC NON-IDEALITY

MSTAR makes no provision for calculating virial coefficients. All molecular weights are apparent values - *i.e.* corresponding to a finite concentration, C. However, for most systems at low loading concentrations (for proteins $\lesssim$ 1.0 mg/ml; polysaccharides $\lesssim$ 0.4 mg/ml), non-ideality effects can be negligible. In cases of severe non-ideality (as manifested by strong downward curvature of M_w *versus* C, A or J plots), a crude estimate for the "ideal" molecular weight can be obtained by extrapolating M_w (or the reciprocal thereof) back to zero C, A or J. For example, in the xanthan example of Fig. 7c, an "ideal" value for $M_w(J \rightarrow 0)$ of $\approx 3 \times 10^6$ could be inferred. This procedure can lead to underestimates however especially if there is significant re-distribution of the sample in very polydisperse materials, so some caution has to be expressed. The most rigorous - albeit time consuming - procedure is to measure $M^0_{w,app}$ (from $M^*(\xi \rightarrow 1)$ - see *e.g.* ref. 14) at a series of loading concentrations C_0 and extrapolate back to zero C_0.

MSTAR *does* permit the input of a known thermodynamic (or "osmotic pressure") second virial coefficient B from which ideal M values can be calculated

$$\frac{1}{M} \approx \frac{1}{M_{app}} - 2BC \tag{20}$$

The M values in all the plots then refer to *ideal* molecular weights, not *apparent* ones.

8. MSTAR AND FLOTATION EQUILIBRIUM

For systems of macromolecules whose density is *less* than that of the solvent (*e.g.* lipoproteins in aqueous solvent or synthetic polymers in chloroform), the solute distribution at equilibrium will be opposite to that of the centrifugal field - *i.e.* flotation equilibrium. The situation for *low-speed* flotation equilibrium has been considered in ref. 15 and a version of MSTAR for this case is currently being written.

APPENDIX - CALCULATION OF 'BLOCK' FOR MSTARA

For MSTARA we need to calculate (for the evaluation of $M^*(r)$ *via* eq. (9)) the integral

$$2\int_a^{r_1} r(C(r) - C_a)\,dr \equiv \int_a^{r_1} (C(r) - C_a)\,d(r^2)$$

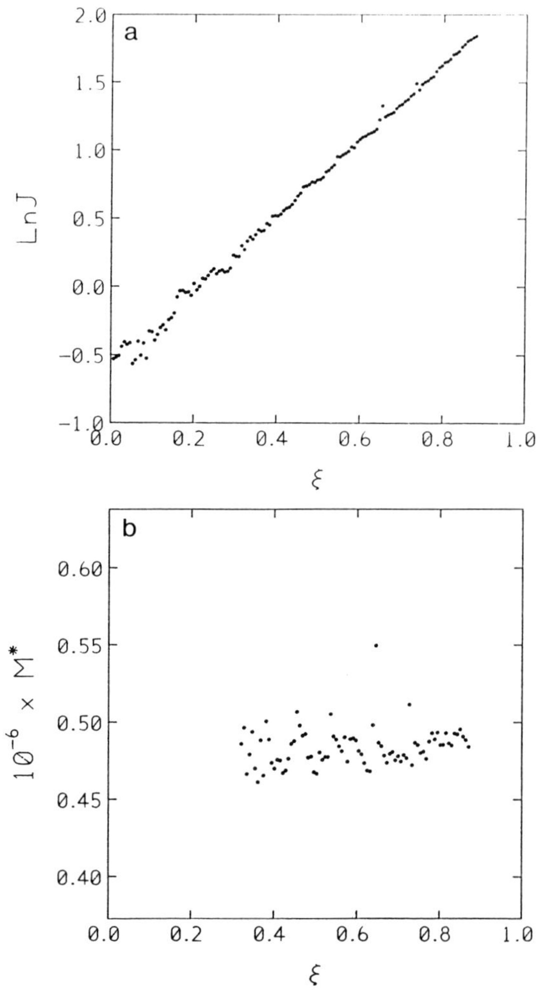

Figure 7. MSTARI analysis of solute low speed sedimentation equilibrium distribution data for colonic mucin T-domains. $M^0_{w,app}$ (from plot (b)) ~ $(0.50\pm0.02)\times10^6$ g/mol. Loading concentration ~ 1.0 mg/ml. Solvent: phosphate/chloride buffer (pH=6.8, I=0.10). Rotor speed = 5200 rev/min; temperature = 20°C. (From ref. 12).

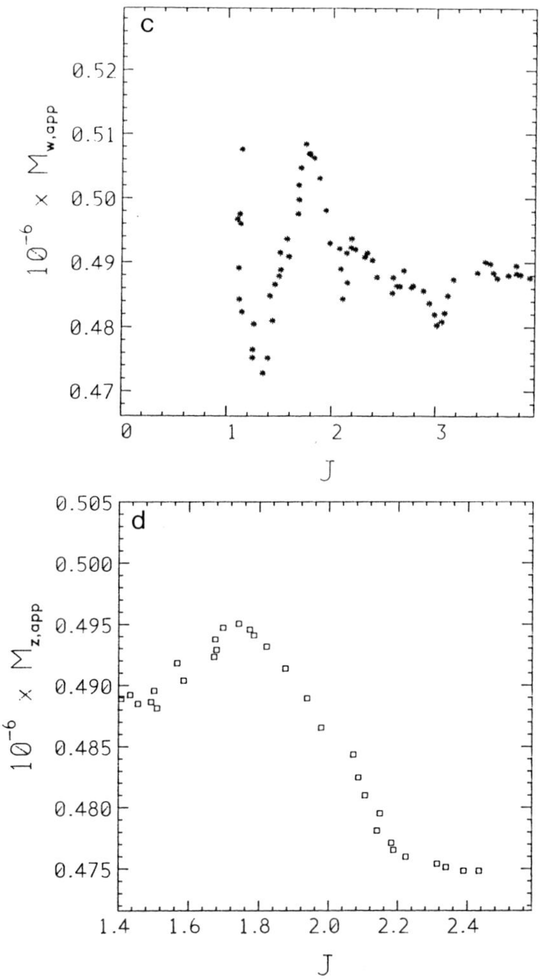

Figure 7 continued.

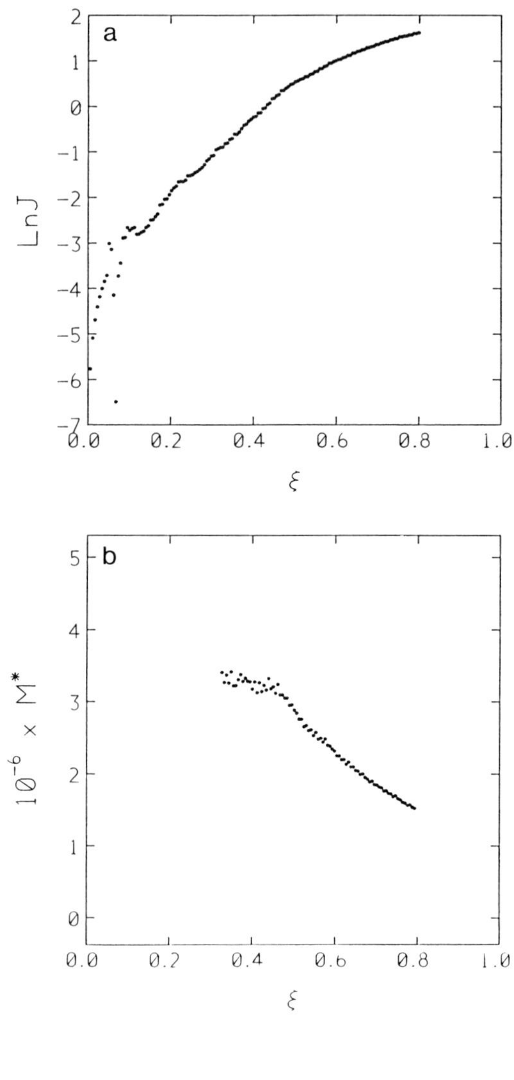

Figure 8. MSTARI analysis of solute (near meniscus depletion) sedimentation equilibrium distribution data for a highly non-ideal xanthan "RD" solution. $M^0_{w,app}$ (from plot (b)) ~ $(1.0\pm0.2)\times10^6$ g/mol. Loading concentration ~ 0.5 mg/ml. Solvent: phosphate/chloride buffer (pH=6.8, I=0.10). Rotor speed = 3000 rev/min; temperature = 20.1°C. (From ref. 13.)

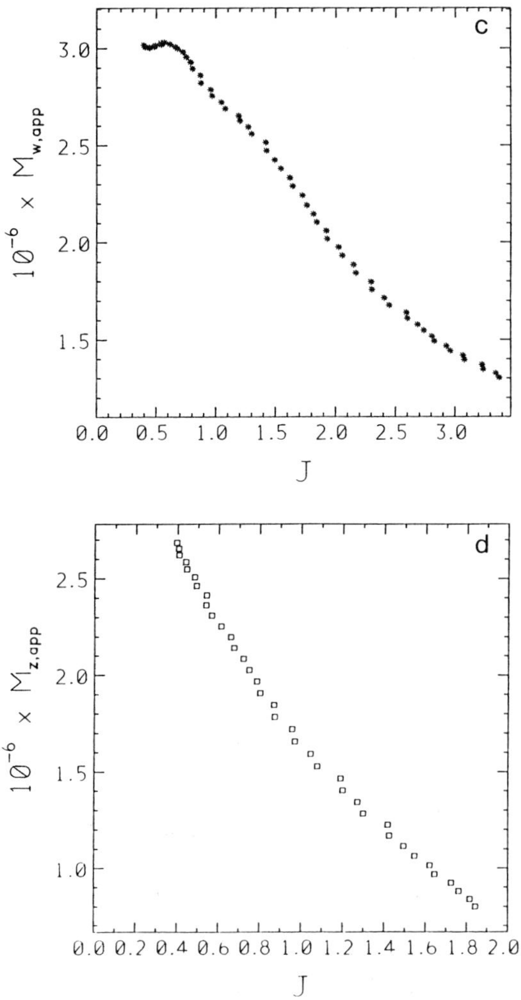

Figure 8 continued.

where C can either be in g/ml or in absorbance units.

Let $$I_{block} = \int_a^{r_1} (C(r) - C_a)\, d(r^2)$$

Assume ln C *vs.* r^2 is linear (reasonable for small r_1–a)

i.e. $\ln C(r) = X + Yr^2$

or $C(r) = \exp X \exp Yr^2$

Thus $$I_{block} = \int_a^{r_1} \{\exp X \exp Yr^2 - C_a\}\, d(r^2)$$

$$= \frac{\exp X}{Y}\{\exp Yr_1^2 - \exp Ya^2\} - C_a\{r_1^2 - a^2\}$$

Also note:

$$C_a \equiv C(r{=}a) = \exp X \exp Ya^2$$

and

$$C_b \equiv C(r{=}b) = \exp X \exp Yb^2$$

REFERENCES

1. D. Roark and D.A. Yphantis, *Ann. N.Y. Acad. Sci.*, 1969, *164*, 245.
2. M.L. Johnson, J.J. Correia, D.A. Yphantis and H.R. Halvorson, *Biophys. J.*, 1981, *36*, 575.
3. T. Laue, Chapter 7, this volume.
4. H. Fujita, 'Foundations of Ultracentrifuge Analysis', J. Wiley and Sons, New York, 1975.
5. J.M. Creeth and R.H. Pain, *Prog. Biophys. Mol. Biol.*, 1967, *17*, 217.
6. D.C. Teller, *Methods Enzymol.*, 1973, *27*, 346.
7. J.M. Creeth and S.E. Harding, *J. Biochem. Biophys. Methods*, 1982, *7*, 25.
8. H. Rinde, Ph.D. Thesis, 1928, Uppsala, Sweden.
9. J.M. Creeth, *Biochem. Soc. Trans.*, 1980, *8*, 520.
10. D.C. Teller, J.A. Horbett, E.G. Richards and H.K. Schachman, *Anal. N.Y. Acad. Sci.*, 1969, *164*, 66.
11. P.E. Gill and G.F. Miller, *Comput. J.*, 1972, *15*, 80.
12. A. Allen, F. Fogg, S.E. Harding and N. Errington, in preparation.
13. N. Errington, J. Morgan and S.E. Harding, in preparation.
14. S.E. Harding, A.J. Rowe and J.M. Creeth, *Biochem. J.*, 1983, *209*, 893.
15. S.E. Harding, P.J. Morgan and K. Petrak, *J. Phys. Chem.*, 1990, *94*, 978.

16

Determination of Molecular Weight Averages and Molecular Weight Distributions from Sedimentation Equilibrium

By M.D. Lechner

PHYSIKALISCHE CHEMIE, UNIVERSITÄT OSNABRÜCK, BARBARASTRASSE 7, D-4500 OSNABRÜCK, GERMANY

1. INTRODUCTION

The two most important advantages of the analytical ultracentrifuge (AUC) with respect to the determination of molecular weight averages and molecular weight distributions are that the AUC is an absolute method compared with relative methods *e.g.* gel permeation chromatography (GPC) and that it permits the analysis of a large range of molecular weights (from $M \approx 10^3$ to $M \approx 10^9$ g/mol) by choosing an appropriate rotor speed.

In this Chapter the main equations for the different molecular weight averages with respect to interference and schlieren optics are derived and different procedures for the determination of molecular weight distributions are presented: finally some experimental results are given.

2. BASIC EQUATIONS FOR MOLECULAR WEIGHT AVERAGES

Starting with the well known differential equation of Lamm[1]

$$\left(\frac{\partial C}{\partial t}\right)_r = \frac{1}{r}\frac{\partial}{\partial r}\left[rD\,\frac{\partial C}{\partial r} - s\omega^2 r^2 C\right] \qquad (1)$$

C = concentration of the dissolved polymer
t = time
r = distance from centre of rotation
D = translational diffusion coefficient
s = sedimentation coefficient
ω = angular velocity of the rotor

there will be no change of concentration with time at sedimentation equilibrium (*i.e.* $(\partial C/\partial t)_r = 0$): this means that one has an equilibrium between sedimentation and diffusion of all particles dissolved in the solvent. The distribution of solute at sedimentation equilibrium can be described for solutions of monodisperse polymers by the following relation[2]

$$r\,dr\,\frac{(1-\bar{v}_2\rho_1)\,\omega^2}{RT} = d(\ln C)\left\{\frac{1}{M} + 2A_2^{se}C + 3A_3^{se}C^2 + \ldots\right\} \tag{2}$$

$\bar{v}_2$ = partial specific volume of the dissolved polymer
ρ_1 = density of the solution
M = molecular weight of the dissolved polymer
A_2^{se}, A_3^{se} = virial coefficients of sedimentation equilibrium

For solutions of polydisperse polymers the following relation holds for the ith component of the polymer[2]

$$r\,dr\,\frac{(1-\bar{v}_2\rho_1)\,\omega^2}{RT} = \frac{d(\ln C_i)}{M_i} + 2\sum_{k=1}^{q} A_{2ik}\,dC_k + \ldots \tag{3}$$

i, k = number of solute component

Summation over all solute components i and division by $\Sigma C_i \equiv C(r)$ at distance r yields the weight average molecular weight at a radial displacement r

$$\frac{\Sigma C_i M_i}{\Sigma C_i} = M_{w,r,app} = \frac{1}{L_1}\frac{d(\ln C)}{d(r^2)} \tag{4}$$

$$L_1 = \frac{(1-\bar{v}_2\rho_1)\,\omega^2}{2RT}$$

$M_{w,r,app}$ includes correction terms for non-ideality. The correct $M_{w,r}$ is obtained by extrapolating the $M_{w,r,app}$ values to zero concentration. As schlieren optics measures dn/dr instead of C one has to re-arrange eq. (3) with respect to dn/dr. This gives the z-average molecular weight

$$\frac{\sum C_i M_i^2}{\Sigma C_i M_i} = M_{z,r,app} = \frac{1}{L_1}\frac{d}{dr^2}\left[\ln\left\{\frac{1}{r}\frac{dn}{dr}\right\}\right] \tag{5}$$

Integration of eqs. (4) and (5) from the meniscus to the bottom of the cell yields the molecular weight averages for the whole solution:

$$M_{w,app} = \frac{1}{L_1}\frac{(C_b - C_m)}{C_0(r_b^2 - r_m^2)} \tag{6}$$

$$M_{z,app} = \frac{1}{L_2}\frac{1}{C_b - C_m}\left[\frac{1}{r_b}\left(\frac{dC}{dr}\right)_b - \frac{1}{r_m}\left(\frac{dC}{dr}\right)_m\right] \tag{7}$$

$C_b \equiv C(r_b)$ = concentration at bottom of the cell
$C_m \equiv C(r_m)$ = concentration at meniscus of the cell

r_b = distance of bottom from centre of rotation
r_m = distance of meniscus from centre of rotation
$L_2 = 2\,L_1$

The difference in concentration between the bottom of the cell and the meniscus (C_b-C_m) can easily be determined by the number of fringes at the bottom of the cell (interference optics) or by integration (schlieren optics).

$$C_b - C_m = \Delta C(r_b) = \int_{r_m}^{r_b} dC = \int_{r_m}^{r_b} \frac{dC}{dr}\,dr \tag{8}$$

Determination of $M_{w,r,app}$ at the meniscus and the bottom of the cell according to eq. (4) allows the calculation of $M_{z,app}$

$$M_{z,app} = \frac{M_{w,b}C_b - M_{w,m}C_m}{C_b - C_m} \tag{9}$$

As has been pointed out, all M_{app} values include correction terms for non-ideality. Correct molecular weights are obtained by[2]

$$\frac{1}{M_{app}} = \frac{1}{M} + 2A_2^{se}\left\{\frac{C_b+C_m}{2}\right\} + 3A_3^{se}\left\{\frac{C_b+C_m}{2}\right\}^2 + \ldots \tag{10}$$

$$\frac{1}{M_{app}} = \frac{1}{M} + 2A_2^{se}\,C_0 + 3A_3^{se}\,C_0^2 + \ldots \tag{11}$$

The determination of number average and (z+1)-average molecular weights will in principle be possible with equations which are similar to eqs. (6) and (7). Nevertheless the applicability is very restricted due to larger experimental errors on determining intercepts, slopes and their derivatives of C and dn/dr at the bottom of the cell and at meniscus. It will be pointed out in the next Chapter that the direct or indirect transformation of the reduced concentration profile or reduced gradient concentration profile is more suitable for the determination of molecular weight averages.

3. BASIC EQUATIONS FOR MOLECULAR WEIGHT DISTRIBUTIONS

The determination of the molecular weight distribution W(M) of a polymer requires the connection of the experimental measured polydisperse reduced concentration profile $U_W(X)$ (interference optics) or gradient profile $V_W(X)$ (schlieren optics) with the molecular weight distribution function and the reduced monodisperse concentration profile U(X,M) or gradient profile V(X,M)[2,3-5]

$$U_W(X) = \left(\frac{C(X)}{C_0}\right)_{C_0\to 0} = \int_0^{\infty} W(M)\,U(X,M)\,dM \tag{12}$$

$$U(X,M) = \lambda M \frac{\exp(\lambda MX)}{\exp(\lambda M) - 1} \qquad 0<X<1 \tag{13}$$

$$V_W(X) = \left(\frac{1}{C_0}\frac{dC(X)}{dX}\right)_{C_0\to 0} = \int_0^\infty W(M)\, V(X,M)\, dM \tag{14}$$

$$V(X,M) = \lambda^2 M^2 \frac{\exp(\lambda MX)}{\exp(\lambda M) - 1} \qquad 0<X<1 \tag{15}$$

$$X = \frac{r^2 - r_m^2}{r_b^2 - r_m^2}$$

$$\lambda = \frac{(1-\bar{v}_2\rho_1)\omega^2(r_b^2 - r_m^2)}{2RT}$$

Eqs. (12) to (15) are only valid for ideal or pseudo-ideal solutions. For non-ideal solutions one has to perform a very precise extrapolation to infinite dilution or introduce correction terms. The problem of non-ideal solutions may be avoided by measuring pseudo-ideal solutions. However pseudo-ideal solutions are not available in all cases and furthermore it restricts the application.

There are two ways to get the molecular weight distribution from the experimental measured concentration profiles. The first is to perform a Laplace transformation (LT) of the Fredholm type integral in eqs. (12) and (14):

$$U_W(X) = \left(\frac{C(X)}{C_0}\right)_{C_0\to 0} = \int_0^\infty W(M)\, U(X,M)\, dM$$

$$V_z(X) = \left(\frac{1}{C_0}\frac{dC(X)}{dX}\right)_{C_0\to 0} = \int_0^\infty W(M)\, V(X,M)\, dM \tag{16}$$

$$\mathrm{LT}^{-1} \;\Big\downarrow\; \mathrm{LT}$$

$$W(M)$$

The disadvantage of this mathematically rigorous procedure is that the operators are poorly conditioned *i.e.* small changes of the experimental values exhibit large changes of the distribution function W(M) and as a result physically meaningless negative values of W(M) may occur[2-6]. Even the introduction of regularizing parameters do not avoid in principle the difficulties of the inverse transformation[6].

The other way is the direct calculation of W(M) from $U_W(X)$ and $V_W(X)$ by non-linear regression[7]. Suppose that $U_W(X)$ and $V_W(X)$ are functions of X and **k** where **k** determines the molecular weight distribution function:

$$U_W(X) = \left(\frac{C(X)}{C_0}\right)_{C_0\to 0} = \int_0^\infty W(M)\, U(X,M)\, dM$$

$$V_W(X) = \left(\frac{1}{C_0}\frac{dC(X)}{dX}\right)_{C_0\to 0} = \int_0^\infty W(M)\, V(X,M)\, dM$$

$$U_W(X) = f(X, k_1, k_2, ..., k_n)$$

$$V_W(X) = g(X, k_1, k_2, ..., k_n)$$

$$S = \sum (f(X_i, k_1, k_2, ..., k_n) - U_{w,i})^2$$

$$S = \sum (f(X_i, k_1, k_2, ... , k_n) - V_{w,i})^2$$

$k_1, k_2, ..., k_n \equiv \mathbf{k}$ = constants of a molecular weight distribution function *i.e.* $M_{n,i}$, $M_{w,i}$, $M_{z,i}$, ..., g_i; C_i, B_i, K_i, ..., g_i.

The multi-modal distribution function W(M) is connected with the mono-modal distribution function $W(M)_i$ *via* the weighting factor g_i

$$W(M) = W(M) = \sum g_i\, W(M)_i \tag{18}$$

The nonlinear regression problem is then defined by minimizing S which is the squared sum of the differences between experimental points $U_{w,i}$ and calculated points $f(X_i, \mathbf{k})$. The problem is solved in the usual way by calculating n partial derivatives of S with respect to **k** and setting them equal to zero. The constants may be $M_{n,i}$ $M_{w,i}$, $M_{z,i}$, ..., g_i or constants of a model molecular weight distribution function *e.g.* C_i, B_i, K_i, ..., g_i where the g_i are the weighting factors for the construction of a multi-modal distribution function W(M) from a couple of mono-modal distribution functions $W(M)_i$. Unlike the inverse problem, this method of calculating W(M) is well conditioned and very stable. The calculation can be performed mathematically by the multi-dimensional *Newton* method or by the multi-dimensional *Simplex* method: the Simplex method is prefered because it needs no derivatives of the integral equation. The Newton method normally converges faster but it sometimes causes problems especially if the experimental values are noisy. The calculation needs a model molecular weight distribution $W(M)_i$. The following mono-modal three-parameter distribution function $W(M)_i$ is proposed[7,8]:

$$W(M)_i = C_1\, M^K \exp(- BMC) \tag{19}$$

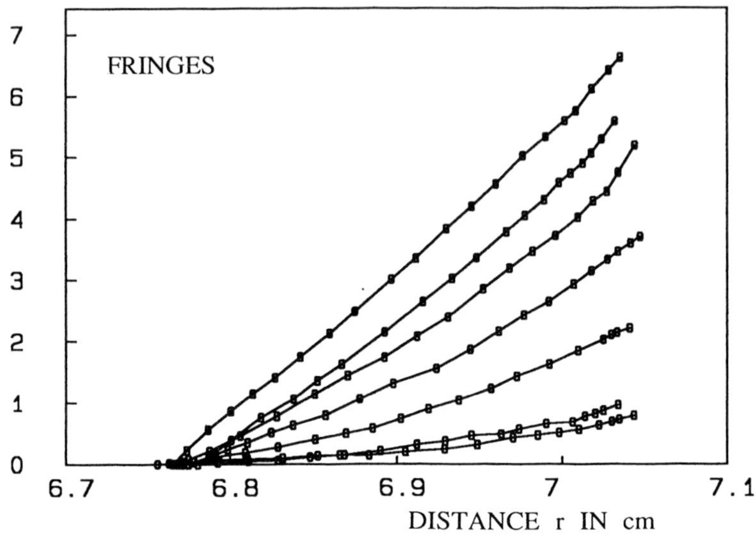

Figure 1. Fringe shift number Δj as a function of the rotor radius r for polystyrene NBS 706 in toluene at 25°C and 5,600 rev/min, interference optics.

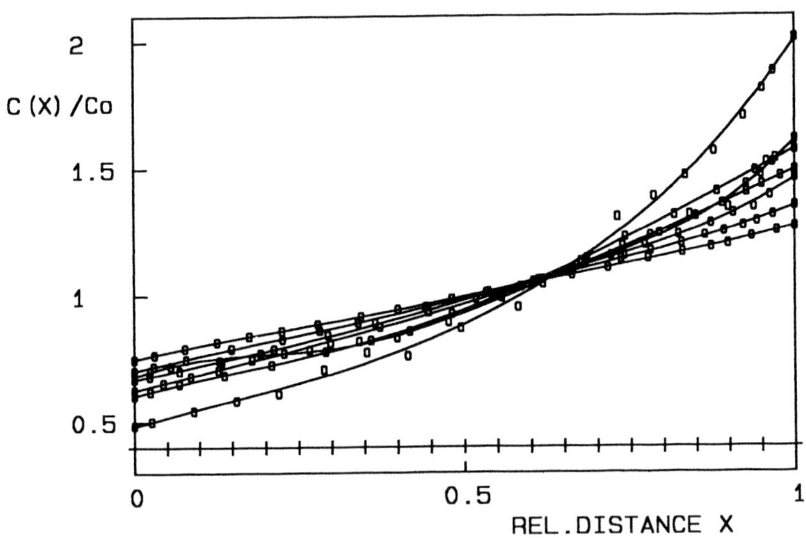

Figure 2. Reduced concentration profile $C(X)/C_0$ as a function of the relative distance X for polystyrene NBS 706 in toluene at 25°C and 5,600 rev/min.□are experimental values; continuous line is the spline interpolation curve.

$$C_1 = \frac{C\, B^{K/C + 1/C}}{\Gamma(K/C + 1/C)}$$

$$M_n = \frac{\Gamma(K/C + 1/C)}{B^{1/C}\, \Gamma(K/C)}$$

$$M_w = \frac{\Gamma(K/C + 2/C)}{B^{1/C}\, \Gamma(K/C + 1/C)}$$

$$M_z = \frac{\Gamma(K/C + 3/C)}{B^{1/C}\, \Gamma(K/C + 2/C)}$$

$$M_{z+1} = \ldots$$

..

..

..

C = 0.5 Wesslau Distribution, Log-normal Distribution;

C = 1 Schulz-Flory Distribution;

C = 2 Maxwell Distribution, Poisson Distribution.

Eq. (18) covers nearly all types of distribution functions (Poisson, Schulz-Flory, Wesslau, log-normal, *etc.*) with different skewnesses.

3. EXPERIMENTAL VERIFICATION

In the following section some experimental results are presented using the Beckman Model E analytical ultracentrifuge running with an eight-hole rotor, interference multiplexer and schlieren multiplexer with modulable laser. The evaluation of the optical records recorded on photographic film was performed with a comparator, a digitiser and a computer with 80286 processor and maths co-processor. Experimental details are given elsewhere[7,9]. (These experiments were performed whilst the author was a visitor at the Kunststofflaboratorium, BASF AG, Ludwigshafen, Germany.)

Fig. 1 demonstrates interference fringe distributions of polystyrene NBS 706 (National Bureau of Standards, Washington, D.C., U.S.A.) in toluene at 25°C and 5,600 rev/min for seven concentrations measured simultaneously in one run. From this we can obtain $M_{w,app}$, $M_{z,app}$, M_w and M_z according to eqs. (6), (7), (9), (10) and (11).

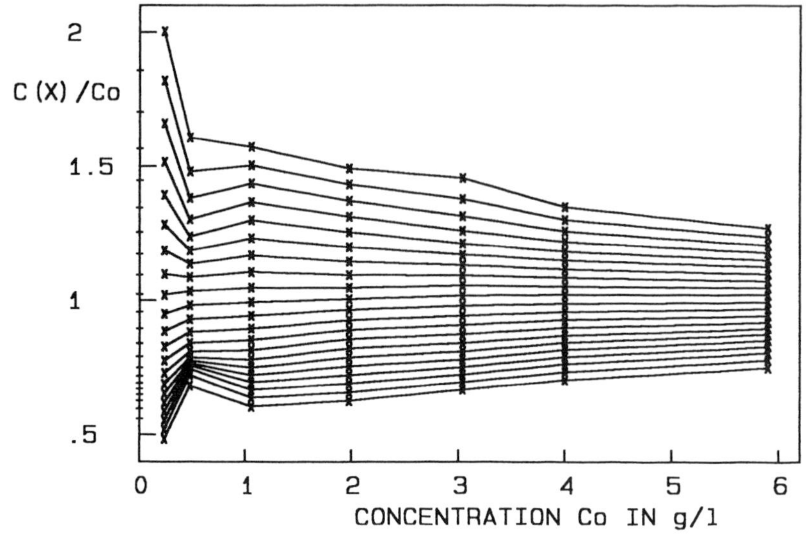

Figure 3. $C(X)/C_0$ as a function of C_0 for polystyrene NBS 706 in toluene at 25°C and 5,600 rev/min - x are interpolated values (spline interpolation); + are extrapolated values of $(C(X)/C_0)_{C_0 \to 0}$; (quadratic regression).

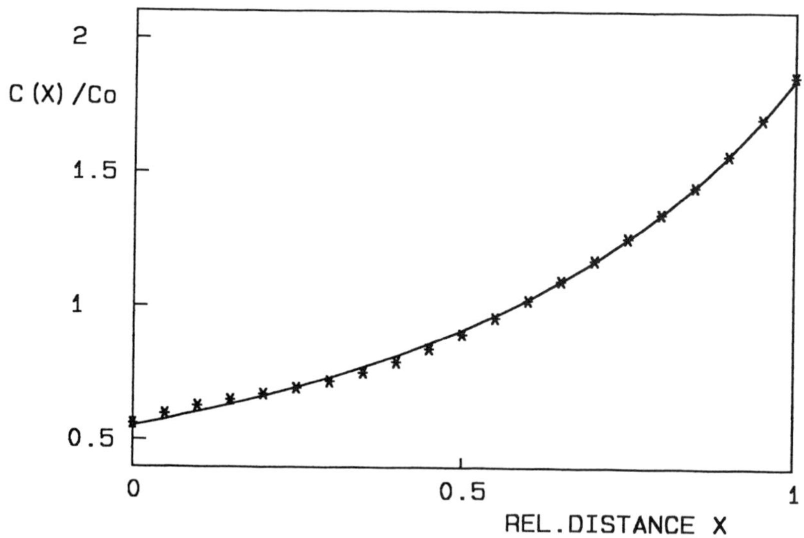

Figure 4. $(C(X)/C_0)_{C_0 \to 0}$ as a function of X for polystyrene NBS 706 in toluene at 25°C and 5,600 rev/min - * are experimental values; continuous line is calculated curve (nonlinear regression, simplex method).

Fig. 2 shows the reduced concentration profile $C(X)/C_0$ as a function of the relative distance X according to eq. (12). The measurements at seven concentrations allow a precise extrapolation to zero concentration. The interpolation to constant X values, which is necessary for extrapolation to zero concentration, may be done by different smoothing and interpolation procedures, but an approximate spline interpolation will give the best results. The extrapolation to zero concentration should be done in the case of a good solvent by quadratic regression and in the case of a poor solvent by linear regression. Fig. 3 demonstrates this extrapolation to infinite dilution. The extrapolated values $(C(X)/C_0)_{C_0 \to 0}$ are indicated as stars in Fig. 4 together with the calculated line according to the non-linear regression with the simplex method eqs. (17) and (19). Finally Fig. 5 shows the molecular weight distribution function W(M) as a function of M. The different procedures give the following results:

Polystyrene NBS 706 in toluene, 25°C, 5,600 rev.min, interference optics

(i) Extrapolated molecular weight averages and A_2 according to eqs. (6), (9), and (11):
$M_w = (273 \pm 10) \times 10^3$ g/mol
$M_z = (344 \pm 50) \times 10^3$ g/mol
$A_2^{se} = (4.3 \pm 0.3) \times 10^{-4}$ cm^3 mol g^{-2}

(ii) Calculated molecular weight distribution and averages according to eqs. (17) and (19):
$M_n = (104 \pm 10) \times 10^3$ g/mol
$M_w = (280 \pm 5) \times 10^3$ g/mol
$M_z = (460 \pm 20) \times 10^3$ g/mol
C = 0.99

Figs. 6 to 8 demonstrate the same procedure as mentioned above for the system polystyrene PCC 8b (Pressure Chemical Company, Pittsburgh, U.S.A.) in cyclohexane at 36°C and 20,000 rev/min. As cyclohexane is a poor solvent for polystyrene linear regression for extrapolation to infinite dilution is preferred.The results are:

Polystyrene PCC 8b in cyclohexane, 25°C, 20,000 rev/min, interference optics

(i) Extrapolated molecular weight averages and A_2 according to eqs. (6), (9) and (11):
$M_w = (11.7 \pm 0.5) \times 10^3$ g/mol
$M_z = (12.5 \pm 1.0) \times 10^3$ g/mol
$A_2^{se} = (1.3 \pm 0.5) \times 10^{-4}$ cm^3 mol g^{-2}

(ii) Calculated molecular weight distribution and averages according to eqs. (17) and (19):
$M_n = (9.4 \pm 0.5) \times 10^3$ g/mol
$M_w = (10.9 \pm 0.5) \times 10^3$ g/mol
$M_z = (12.3 \pm 0.5) \times 10^3$ g/mol
C = 1.02

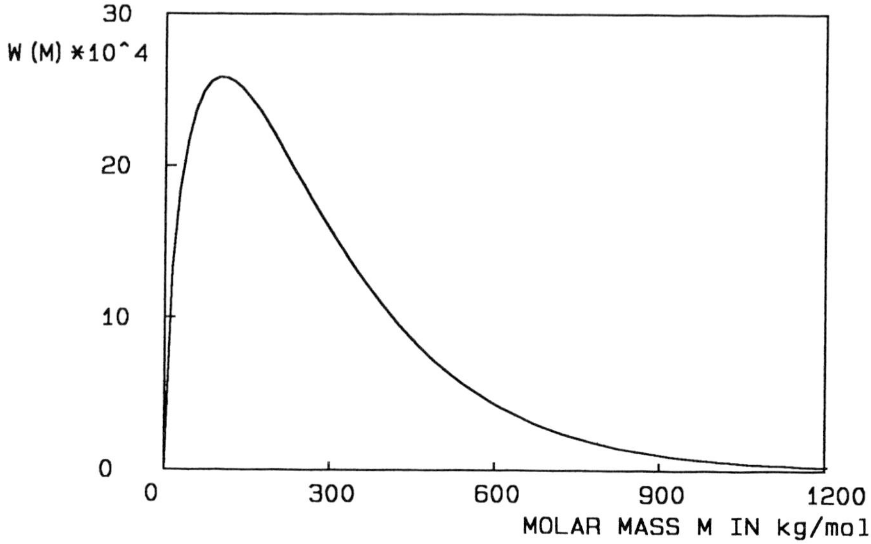

Figure 5. Model molecular weight distribution for polystyrene NBS 706 in toluene.

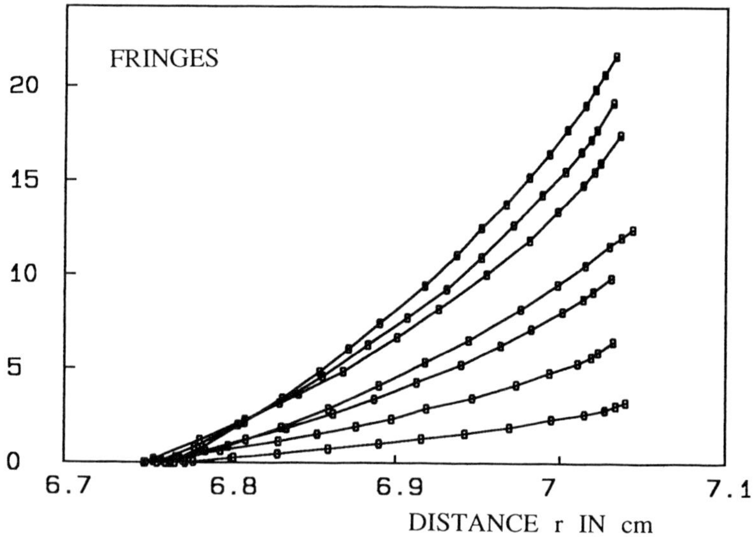

Figure 6. Fringe shift number Δj as function of the rotor radius r for polystyrene PCC 8b in cyclohexane at 36°C and 20,000 rev/min, interference optics.

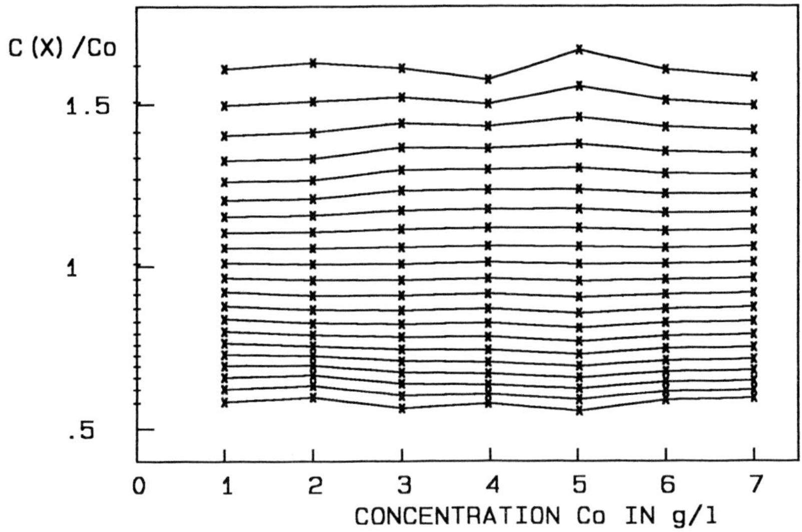

Figure 7. C(X)/C0 as a function of C0 for polystyrene PCC 8b in cyclohexane at 36°C and 20,000 rev/min - x are interpolated values (spline interpolation); + are extrapolated values of $(C(X)/C_0)_{C_0 \to 0}$ (linear regression).

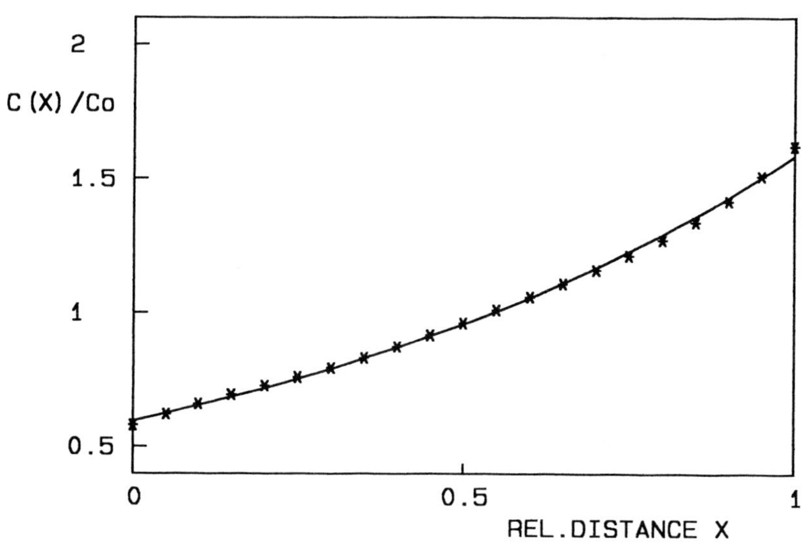

Figure 8. $(C(X)/C_0)_{C_0 \to 0}$ as a function of X for polystyrene PCC 8b in cyclohexane at 36°C and 5,600 rev/min - * are experimental values; continuous line is calculated curve (nonlinear regression, simplex method).

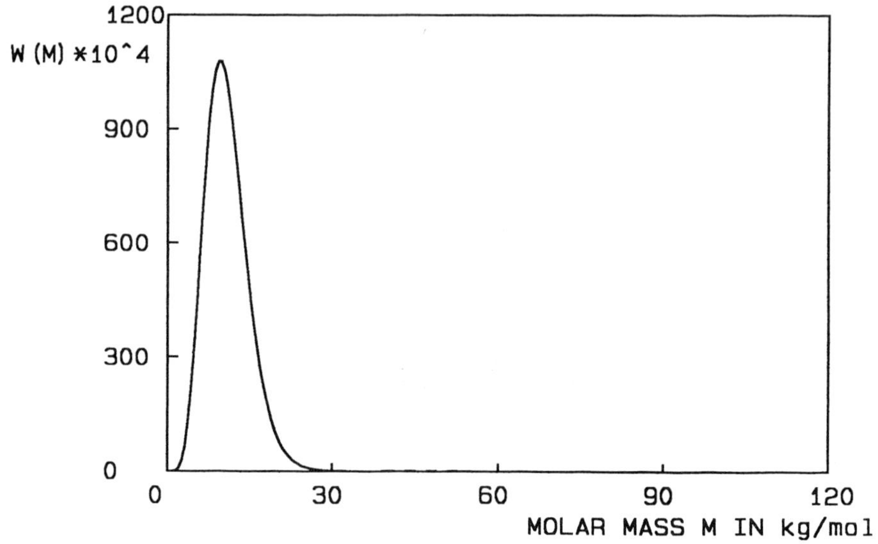

Figure 9. Model molecular weight distribution for polystyrene PCC 8b in cyclohexane.

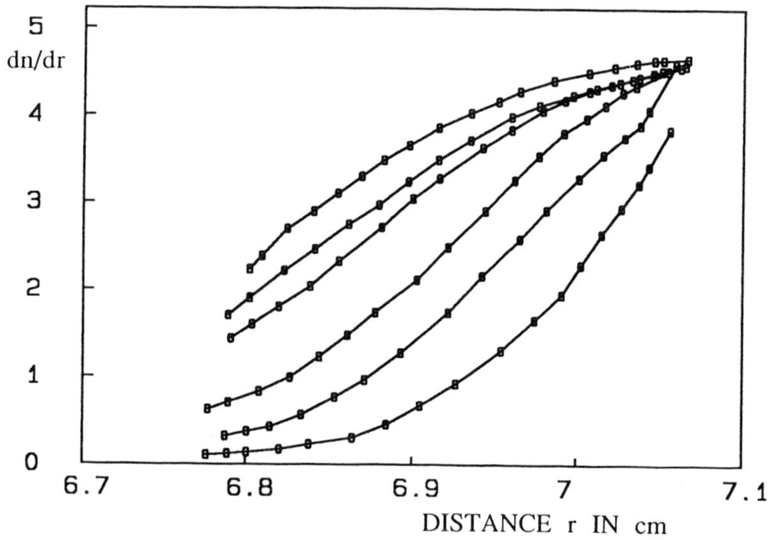

Figure 10. Refraction increment dn/dr as function of the rotor radius r for polystyrene NBS 706 in toluene at 25°C and 9,000 rev/min, schlieren optics.

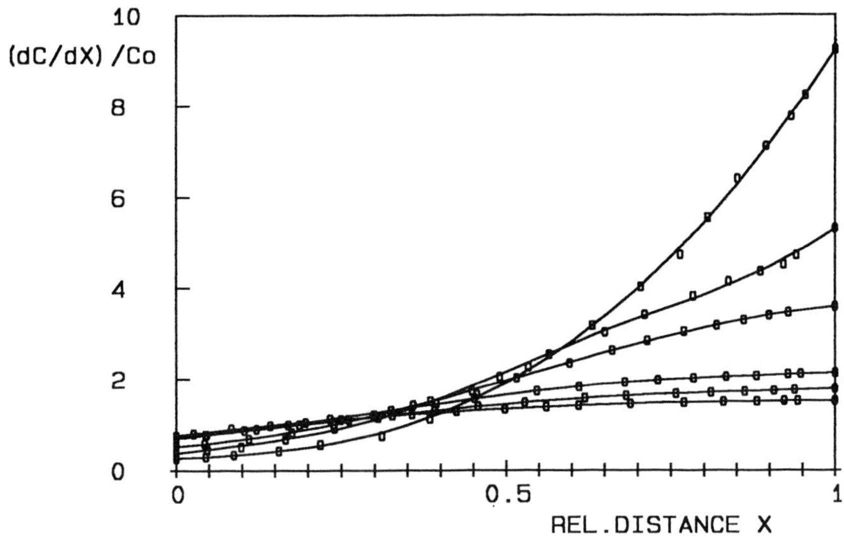

Figure 11. Reduced concentration gradient profile $(1/C_0)$ (dC(X)/dX) as a function of the relative distance X for polystyrene NBS 706 in toluene at 25°C and 9,000 rev/min - □ are experimental values; continuous line is spline interpolation curve.

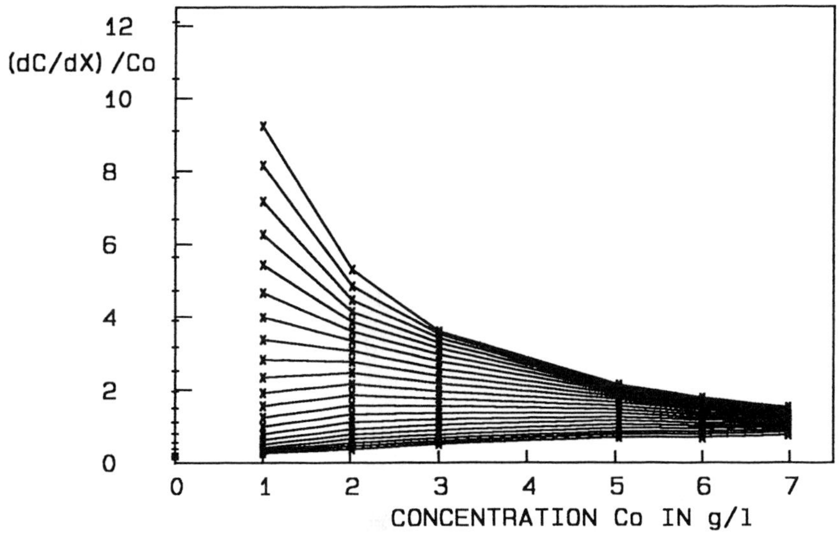

Figure 12. $(1/C_0)$ (dC(X)/dX) as a function of C_0 for polystyrene NBS 706 in toluene at 25°C and 9,000 rev/min - x are interpolated values (spline interpolation); + are extrapolated values of $(1/C_0)\ (dC(X)/dX))_{C_0 \to 0}$ (quadratic regression).

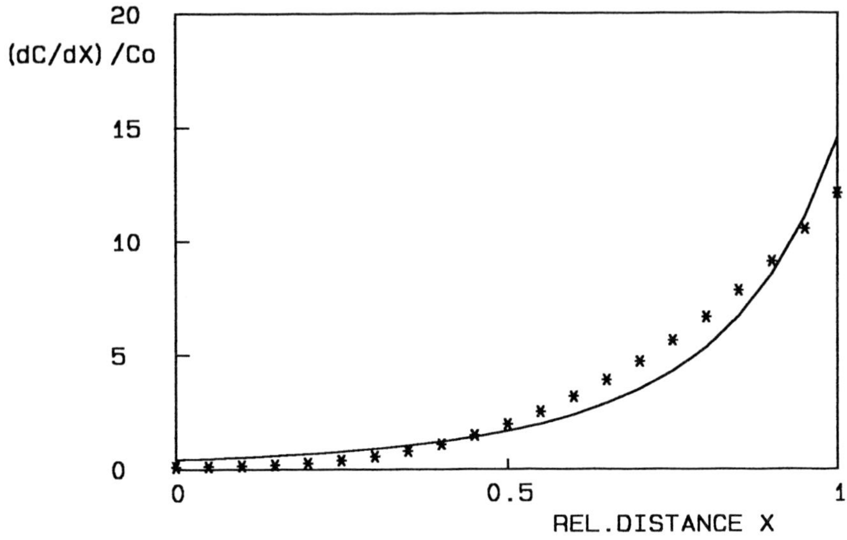

Figure 13. $(1/C_0)\ (dC(X)/dX)_{C_0 \rightarrow 0}$ as a function of X for polystyrene NBS 706 in toluene at 25°C and 9,000 rev/min - * are experimental values; continuous line is calculated curve (nonlinear regression, simplex method).

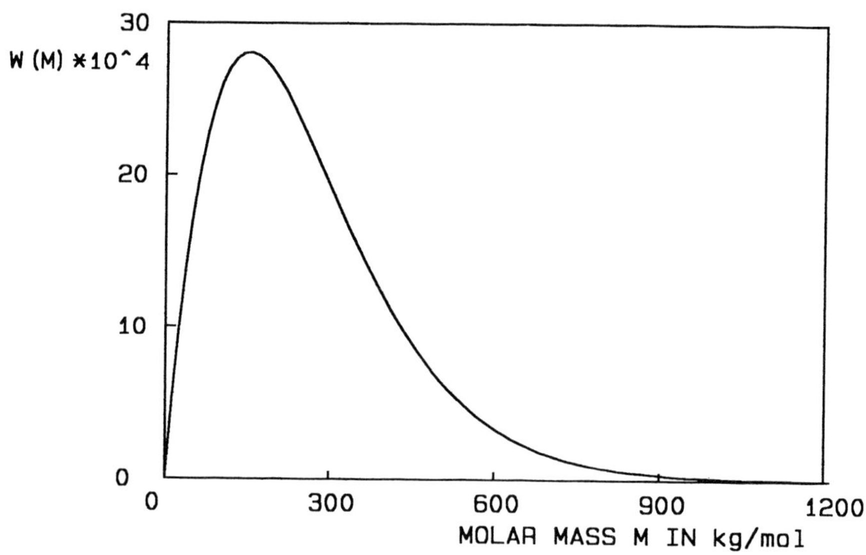

Figure 14. Model molecular weight distribution for polystyrene NBS 706 in toluene.

Figs. 9 to 13 demonstrate the procedure for determining molecular weight distributions and averages with respect to schlieren optics for the system polystyrene NBS 706 in toluene at 25°C and 9,000 rev/min. As has been well-known by AUC experimentalists, schlieren optics are less accurate than interference optics but schlieren optics give more precise M_z values compared with interference optics. Fig. 9 yields $M_{w,app}$, $M_{z,app}$, M_w and M_z according to eqs. (6), (7) and (11). The procedure for the determination of the molecular weight distribution is exactly the same as for the interference optics using eqs.(14), (15) and (17). The results are:

Polystyrene NBS 706 in toluene, 25°C, 9,000 rev/min, schlieren optics

(i) Extrapolated molecular weight averages and A_2 according to eqs. (6), (7) and (11):
$M_w = (170 \pm 50) \times 10^3$ g/mol
$M_z = (305 \pm 50) \times 10^3$ g/mol
$A_2^{se} = (7.3 \pm 2) \times 10^{-4}$ cm^3 mol g^{-2}

(ii) Calculated molecular weight distribution and averages according to eqs. (17) and (19)
$M_n = (130 \pm 20) \times 10^3$ g/mol
$M_w = (260 \pm 20) \times 10^3$ g/mol
$M_z = (365 \pm 20) \times 10^3$ g/mol
$C = 1.2$

4. CONCLUSIONS

Sedimentation equilibrium in the analytical ultracentrifuge is a powerful tool for the absolute determination of molecular weight averages and of molecular weight distributions of polymers. Particularly for complicated systems such as polyelectrolytes or polymers with microgels the AUC is a better choice than GPC or scattering methods. A thorough evaluation of data together with improved mathematical treatments yield reliable values of molecular weight averages and molecular weight distributions of polymers.

ACKNOWLEDGEMENTS

Support of this work by BASF AG, Ludwigshafen and by the Fonds der Chemischen Industrie is gratefully acknowledged. The author is indebted to the Kunststofflaboratorium of BASF AG for enabling him to stay as a guest.

REFERENCES

1. H. Fujita, 'Mathematical Theory of Sedimentation Analysis', Academic Press, New York, 1962.
2. H. Fujita, 'Foundation of Ultracentrifugal Analysis', Wiley, New York, 1975.
3. Th.G. Scholte, *J. Polym. Sci. A2*, 1968, *6*, 91; *ibid.* 111.
4. Th.G. Scholte, *Eur. Polym. J.*, 1970, *6*, 51.
5. G.S. Greschner, *Eur. Polym. J.*, 1984, *20*, 475.

6. D.R. Wiff, *J. Polym. Sci., Polym. Symp.*, 1973, *43*, 219.
7. M.D. Lechner and W. Mächtle, *Makromol. Chem.*, 1991, *192*, 1183.
8. M.D. Lechner, *Eur. Polym. J.*, 1978, *14*, 61.
9. W. Mächtle and U. Klodwig, *Makromol. Chem.*, 1979, 180, 2507.

17

Thermodynamic Non-Ideality and Sedimentation Equilibrium

By Peter R. Wills and [1]Donald J. Winzor

DEPARTMENT OF PHYSICS, UNIVERSITY OF AUCKLAND, AUCKLAND, NEW ZEALAND

[1]DEPARTMENT OF BIOCHEMISTRY, UNIVERSITY OF QUEENSLAND, BRISBANE, QUEENSLAND 4072, AUSTRALIA

1. INTRODUCTION

Current theory of sedimentation equilibrium stems from studies by Goldberg[1], who introduced the concept that the equilibrium distribution of solute should be considered solely in thermodynamic terms, rather than as a state reflecting a balance between fluxes due to sedimentation and diffusion. Early investigations[2-5] emphasized the development of procedures whereby an empirical non-ideality coefficient could be evaluated from the concentration-dependence of molecular weight for a single, non-interacting solute. Subsequently, greater advantage was taken of the fact that the sedimentation equilibrium technique had potential for accurate definition of the thermodynamic activity of solutes in incompressible solutions[6-10]. This advance then led to testing the extent to which the measured thermodynamic non-ideality might be rationalised quantitatively on the basis of excluded volume and charge-charge interactions[9,11,12]. Finally, it was realized that statistical mechanics held the key to interpretation of the experimentally derived activities and activity coefficients[13,14].

From the outset, thermodynamic activity was usually expressed in terms of molarity or the corresponding weight concentration, a scale that was pertinent to sedimentation equilibrium distributions recorded by ultracentrifuge systems. However, it now transpires that this action has led to a fundamental thermodynamic error in the derivation of the basic sedimentation equilibrium equation. The logical starting point for these considerations of thermodynamic non-ideality in sedimentation equilibrium is therefore the establishment of the correct basic expression.

2. DESCRIPTION OF THERMODYNAMIC NON-IDEALITY

Definition of Thermodynamic Activity

In general, the chemical potential, μ_A, of the solute, A, in a two-component solution is a function of temperature, T, pressure, P, and composition. Variations in μ_A due to changes in composition at fixed temperature can be conveniently expressed in terms of a thermodynamic activity if either the pressure or the chemical potential, μ_S, of the solvent, S, is held constant as the concentration of solute is varied. In an ideal solution, this variation is expressed by an equation of the form

$$\mu_A = \mu_A^0 + RT \ln c_A \quad (1)$$

where μ_A^0 is the chemical potential in a defined standard state and c_A is the solute concentration expressed on a molar, molal or mole fraction basis; or, indeed, on any linear concentration scale. Real solutions approximate to ideal behaviour in the limit of zero solute concentration, but at finite concentrations physical interactions between solute molecules give rise to deviations from eq. (1). The chemical potential of these thermodynamically non-ideal solutions is then described by an equation of the same form as eq. (1) but with the concentration replaced by a thermodynamic activity that is the product of the concentration and a dimensionless activity coefficient.

For systems in which the chemical potential of solvent is kept constant as the solute concentration is varied (osmometry and any other partition equilibrium procedure) the activity, z_A, is defined by the equation:

$$(\mu_A)_{T,\mu_s} = (\mu_A^0)_{T,\mu_s} + RT \ln z_A \quad (2)$$

and because it is a molar quantity it is usually represented as the product of the molar solute concentration C_A (moles of solute per unit volume of solution) and an activity coefficient γ_A:

$$z_A = \gamma_A C_A \quad (3)$$

The standard state chemical potential $(\mu_A^0)_{T,\mu_s}$ refers effectively to a state of the solvent, considered infinite to allow the addition of a solute molecule without change in solvent concentration. In solutions described by eqs. (2) and (3) the pressure is unconstrained: indeed, increasing the solute concentration gives rise to the osmotic pressure, Π, which is the difference in pressure between the solution and the standard state.

For systems in which the pressure is fixed (a normal laboratory constraint) the activity a_A is defined by the equation

$$(\mu_A)_{T,P} = (\mu_A^0)_{T,P} + RT \ln a_A \quad (4)$$

and it is a molal quantity, usually expressed as the product of the molal solute concentration and activity coefficient y_A:

$$a_A = y_A m_A \quad (5)$$

Once again, the standard state chemical potential, $(\mu_A^0)_{T,P}$, refers effectively to a state of the solvent, but at the same pressure as the solution. In order to give C_A and m_A the same dimensions, we choose to define a unit volume of pure solvent at temperature T and pressure P, rather than a kilogram, as the quantity of solvent defining molalities, so that $m_A = N_A/N_S V_S$ where V_S is the molar volume of pure solvent at temperature T and pressure P, and N_i represents the number of moles of substance i. (Note that the usual molal concentration, defined in terms of a kilogram of solvent, is

$1000\ m_A V_S/M_S$ where M_S is the solvent molecular weight.) When the partial molar volume of the solvent is independent of pressure and solution composition, an assumption which we shall make throughout this exposition, this change in definition does not change any thermodynamic relation involving molalities.

Virial Expansions

It is possible to express the osmotic pressure of a macromolecular solution as a virial expansion in molar concentration C_A:

$$\left(\frac{\Pi}{RT}\right)_{T,\mu_S} = C_A(1 + B_2C_A + B_3C_A^2 + \ldots) \tag{6}$$

As discussed later, the coefficients B_n have a strict statistical mechanical meaning. By a purely thermodynamic argument[13] it may be shown that the logarithm of the molar activity coefficient is described by an expansion in which the same coefficients appear, namely,

$$\ln \gamma_A = 2B_2C_A + \frac{3}{2}B_3C_A^2 + \ldots \tag{7}$$

For solutions held at fixed temperature and pressure the expressions analogous to eqs. (6) and (7) are :

$$-\frac{(\mu_S - \mu_S^0)_{T,P}}{V_S RT} = m_A(1 + C_2m_A + C_3m_A^2 + \ldots) \tag{8}$$

$$\ln y_A = 2C_2m_A + \frac{3}{2}C_3m_A^2 + \ldots \tag{9}$$

The solvent molar volume, V_S, is required in eq. (8) on account of our definition of molality. The coefficients C_n appearing in eqs. (8) and (9) cannot in general be expressed simply in terms of any statistical mechanical variables. However, when the partial molar volumes are independent of pressure and composition, the two sets of virial coefficients are related by expressions involving the partial molar volume (the product of molecular weight M_A and partial specific volume $\bar{v}_A$) of the solute. For example,

$$C_2 = B_2 - M_A\bar{v}_A \qquad B_2 = C_2 + M_A\bar{v}_A \tag{10a}$$

$$C_3 = B_3 - 2B_2M_A\bar{v}_A + M_A^2\bar{v}_A^2 \qquad B_3 = C_3 + 2C_2M_A\bar{v}_A + M_A^2\bar{v}_A^2 \tag{10b}$$

When the partial molar volumes are constants, any of the virial expansions (eqs. (6-9)) can be converted to the other concentration scale by using eq. (10) and the relationships:

$$C_A = \frac{m_A}{1 + M_A\bar{v}_A m_A} \tag{11a}$$

$$m_A = \frac{C_A}{1 - M_A\bar{v}_A C_A} \tag{11b}$$

Of significance later in this investigation is the expansion of ln z_A in terms of molal quantities, namely,

$$\ln z_A = \ln m_A + (2C_2 + M_A\bar{v}_A)m_A + \left(\frac{3}{2}C_3 + C_2 M_A\bar{v}_A\right)m_A^2 + \ldots \tag{12}$$

McMillan-Mayer Theory

By considering the interactions among clusters of n solute molecules[15], the virial coefficients B_n appearing in eq. (6) may be expressed in terms of the potential of mean force $u(r_{ij})$ between pairs of molecules (i,j) that are separated by a distance r_{ij}:

$$B_2 = -2\pi N \int_{0}^{\infty} f_{12}(r_{12})\, r_{12}^2\, dr_{12} \tag{13a}$$

$$B_3 = \frac{-8\pi^2 N^2}{3} \int_{0}^{\infty} f_{12}(r_{12}) r_{12} dr_{12} \int_{0}^{\infty} f_{23}(r_{23}) r_{23} dr_{23} \int_{|r_{12}-r_{23}|}^{r_{12}+r_{23}} f_{13}(r_{13}) r_{13} dr_{13} \tag{13b}$$

In eq. (13), N is Avogadro's number and the Mayer function f_{ij} is defined by the equation

$$f_{ij}(r_{ij}) = \exp[-u_{ij}(r_{ij})/kT] - 1 \tag{14}$$

where k is Boltzmann's constant. In the event that the solute may be modelled as an uncharged hard sphere with radius R_A, we make the substitution

$$u(r_{12}) = \begin{cases} \infty & r_{12} \leq R_A \\ 0 & r_{12} > R_A \end{cases} \tag{15}$$

and obtain the expressions

$$B_2 = \frac{16\pi}{3} N R_A^3 \tag{16a}$$

$$B_3 = \frac{160\pi^2}{9} N^2 R_A^6 \tag{16b}$$

Expressions for B_2 are available for other geometrical shapes such as ellipsoids of revolution[9,16], tri-axial ellipsoids[17], rods and cylinders[18,19].

Allowance for the effects of a symmetrically distributed net charge (valence) Z_A on spherical molecules may be accommodated by modifying eq. (15) such that

$$u(r_{12}) = \frac{(Z_A e)^2 \exp[-\kappa(r_{12}-2R_A)]}{\varepsilon r_{12}(1 + \kappa R_A)^2} \qquad r_{12} > R_A \tag{17}$$

where κ is the Debye-Hückel inverse screening length, e the electronic charge, and ε the dielectric constant. The integral in eq. (13a) may be evaluated approximately by expanding the exponential term as

$$f_{12}(r_{12}) = -\frac{u(r_{12})}{kT} + \frac{1}{2}\left(\frac{u(r_{12})}{kT}\right)^2 - \ldots \tag{18}$$

Truncation of eq. (18) after the linear term and substitution of $(8\pi Ne^2I/\varepsilon kT)^{1/2}$ for κ with the appropriate definition of molar ionic strength I then gives the following expression for B_2:

$$B_2 = \frac{16\pi}{3} N R_A^3 + \frac{Z_A^2(1 + 2\kappa R_A)}{4I(1 + \kappa R_A)^2} \tag{19}$$

Roark and Yphantis[20] have derived the corresponding expression for B_2 in the event that Donnan effects provide the dominant contribution to thermodynamic non-ideality. Although such a situation would certainly prevail for highly charged macro-ions in the absence of a supporting polyelectrolyte, the use of moderate ionic strengths in most sedimentation equilibrium studies usually assures the dominance of excluded volume effects, or charge-charge interactions which can be accounted for by using eq. (19).

Multi-Component Systems

When there is more than one macromolecular solute, the virial expansions (eqs. (6-10)) discussed above become more complicated. The fundamental relationships

$$\left(\frac{\Pi}{RT}\right)_{T,\mu_s} = \sum_i C_i + \sum_{i,j} B_{ij}C_iC_j + \sum_{i,j,k} B^{ijk}C_iC_jC_k + \ldots \tag{20}$$

and

$$-\frac{(\mu_s - \mu_s^0)_{T,P}}{V_sRT} = \sum_i m_i + \sum_{i,j} C_{ij}m_im_j + \sum_{i,j,k} C_{ijk}m_im_jm_k + \ldots \tag{21}$$

involve independent summations over all macromolecular solutes (i, j, k ... $\in$ {A, B, ...}). McMillan-Mayer theory remains valid for the evaluation of the virial coefficients B_{ij}, B_{ijk} ... appearing in the expansion for the osmotic pressure and this means that the order of the indices on the coefficients is of no significance ($B_{ij} = B_{ji}$ and so on). Conversely, by using the relationships

$$C_i = \frac{m_i}{1 + \sum_j M_j\bar{v}_jm_j} \tag{22a}$$

$$m_i = \frac{C_i}{1 - \sum_j M_j\bar{v}_j C_j} \tag{22b}$$

we find

$$C_{ij} = B_{ij} - \frac{1}{2}(M_i\bar{v}_i + M_j\bar{v}_j) \qquad B_{ij} = C_{ij} + \frac{1}{2}(M_i\bar{v}_i + M_j\bar{v}_j) \tag{23}$$

In deriving expansions for the activity coefficients it is convenient to introduce new sets of coefficients α_{ij} and β_{ij} such that

$$\ln \gamma_i = \sum_j \alpha_{ij} C_j + ... \tag{24a}$$

$$\ln y_i = \sum_j \beta_{ij} m_j + ... \tag{24b}$$

whereupon we obtain the relationships

$$\alpha_{ij} = 2B_{ij} \qquad \beta_{ij} = 2C_{ij} \tag{25}$$

For dissimilar molecules, McMillan-Mayer theory leads to the result

$$B_{ij} = \frac{16\pi}{3} N (R_i+R_j)^3 + \frac{Z_iZ_j\,[1 + \kappa(R_i+R_j)]}{4I\,(1+\kappa R_i)\,(1+\kappa R_j)} \tag{26}$$

when hard-sphere and weak electrostatic interactions between pairs of spherical molecules are taken into account. Some results are also available for hard-sphere interactions between dissimilar molecules of more complex geometry[11], but the co-volume of interaction between a globular protein and a random coil polymer molecule remains problematic[21-23].

Very Concentrated Solutions

As solutions with higher and higher concentrations are considered, the use of virial expansions (eqs. (6-9)) to account for thermodynamic non-ideality becomes increasingly difficult. In the first place, it is necessary to consider more terms in the expansion, and secondly, evaluation of statistical mechanical expressions of the higher coefficients becomes an intractable problem. One solution to this difficulty has been to model the potential of mean force between molecules in terms of an effective hard-sphere potential. This approach has been used, for example, by Minton and co-workers[12,24,25]. Its advantage is that the virial coefficients for the hard-sphere potentials are known exactly to high order, but in modelling experimental data it provides only one variable parameter - the effective radius of the sphere.

An alternative approach to the use of the virial expansion is scaled particle theory[26-28]. The advantage of scaled particle theory is that it expresses the concentration dependence of the osmotic pressure in closed form. The theory provides three variable parameters, the volume, the surface area and the mean radius of the particle (molecule), which are clearly inter-related for any specified shape. The only interactions considered in scaled particle theory are hard-sphere type excluded volume interactions, but the theory can accommodate a wide range of model shapes. With judicious choice of shape, the effects of thermodynamic non-ideality on experimental results obtained at high solute concentrations (250 g/litre) can be accurately represented[29]. Furthermore, correspondence between the predicted and experimental non-ideality is relatively insensitive to precise details of the shape assigned. Consequently, little error is likely to be introduced by assigning spherical shape to globular problems[29]. Details of the manner in which the concentration dependence of the molar activity coefficient, y_A, may be calculated on the basis of scaled particle theory have been discussed by Chatelier and Minton[30].

3. BASIC EQUATION FOR SEDIMENTATION EQUILIBRIUM

The criterion for sedimentation equilibrium in a two-component solution may be expressed as[3]

$$\frac{d\mu_A}{dr} - M_A\omega^2 r = 0 \tag{27}$$

where ω is the angular velocity and r the radial distance. At constant temperature, μ_A is a function of pressure, P, and solute molality, m_A, in which case

$$\frac{d\mu_A}{dr} = \left(\frac{\partial\mu_A}{\partial P}\right)_{T,m_A}\frac{dP}{dr} + \left(\frac{\partial\mu_A}{\partial m_A}\right)_{T,P}\frac{dm_A}{dr} \tag{28}$$

Substitution of the partial molar volume, $M_A\bar{v}_A$, for $(\partial\mu_A/\partial P)_{T,m_A}$ and $\omega^2 r\rho$ for dP/dr allows expression of eq. (27) in the form

$$M_A(1 - \bar{v}_A\rho)\omega^2 r = \left(\frac{\partial\mu_A}{\partial m_A}\right)_{T,P}\frac{dm_A}{dr} \tag{29}$$

where ρ is the solution density.

It has become generally accepted[3,4,31,32] that at this stage of the derivation it is permissible to substitute molar or weight concentration for molality in eq. (29). However this procedure is thermodynamically invalid except in the limit of infinite dilution. We therefore proceed with eq. (29) as the basic sedimentation equilibrium expression.

From eqs. (5) and (29) we obtain

$$M_A(1 - \bar{v}_A\rho)\frac{\omega^2 r}{RT} = \frac{1}{m_A}\left[1 + \left(\frac{\partial \ln y_A}{\partial m_A}\right)_{T,P}\right]\frac{dm_A}{dr} \tag{30}$$

In order to separate the variables for integration purposes it is necessary to take into account the dependence of the buoyancy factor, $(1 - \bar{v}_A\rho)$, on solute concentration. In general, $(1 - \bar{v}_A\rho)$ may depend on both composition and pressure, whereupon it cannot be split into separate factors that are independent functions of either m_A or r. However, when the partial molar volumes of the solvent and solute are independent of both composition and pressure, the solution density may be written as[4]

$$\rho = \rho_s + (1 - \bar{v}_A\rho_s) M_A C_A \tag{31}$$

where ρ_s is the solvent density. The molar concentration C_A may be converted to its molal counterpart *via* eq. (11) to give

$$M_A (1 - \bar{v}_A\rho_s)\frac{\omega^2 r}{RT} dr = \frac{1}{m_A}(1+ M_A\bar{v}_A m_A)\left[1 + m_A\left(\frac{\partial \ln y_A}{\partial m_A}\right)_{T,P}\right] dm_A \tag{32}$$

When eq. (5) is used to evaluate $(\partial \ln y_A/\partial m_A)_{T,P}$ and the right-hand side of eq. (32) is then integrated, the resulting power series is identical with that appearing in eq. (12), which is the expansion of $\ln z_A$ in terms of m_A. This means that the sedimentation equilibrium equation for incompressible solutions may now be expressed in terms of the molar activity. Accordingly we write

$$M_A (1 - \bar{v}_A\rho_s)\frac{\omega^2 r}{RT} dr = d \ln z_A = \frac{1}{C_A}\left[1 + C_A\left(\frac{\partial \ln \gamma_A}{\partial C_A}\right)_{T,\mu_s}\right] dC_A \tag{33}$$

The first point to note is that the partial differential of γ_A with respect to C_A refers to constant solvent chemical potential rather than constant pressure. Secondly, the density appearing in the buoyancy factor is the solvent density rather than the solution density at radial distance r, a point which is implicit in an early paper by Fixman[33]. Thirdly, eq. (33) provides rigorous justification for the use of sedimentation equilibrium to define thermodynamic activity in terms of molarity, or weight concentration. Furthermore, it is now seen that use of the solvent density in the buoyancy factor is not an approximation but a requirement for rigorous evaluation of thermodynamic non-ideality, or indeed molecular weight, by sedimentation equilibrium of solutions whose component partial molar volumes are effectively constant.

When there is more than one macromolecular component, the equilibrium distribution of any particular solute species may be found by considering the basic thermodynamic relation

$$M_i(1 - \bar{v}_i\rho)\omega^2 r = \left(\frac{\partial \mu_i}{\partial m_i}\right)_{T,P,\Sigma m_{j\neq i}} \frac{dm_i}{dr} + \sum_{j\neq i}\left(\frac{\partial \mu_i}{\partial m_j}\right)_{T,P,\Sigma m_{k\neq j}} \frac{dm_j}{dr} \tag{34}$$

The presence of terms involving $(\partial \mu_i/\partial m_j)_{T,P,\Sigma m_{k\neq j}}$ would make this set of equations intractable, were it not for the relationships between virial coefficients expressed in eqs. (23) and (25). By using these expressions, along with the equation for the solution density, namely

$$\rho = \rho_s + \sum_i (1 - \bar{v}_i\rho_s) M_iC_i \tag{35}$$

and working to first order in solute concentrations, it is possible to derive the simple result

$$M_i(1 - \bar{v}_i\rho_s)\frac{\omega^2 r}{RT} dr \approx d \ln z_i = d \ln C_i + \sum_j \alpha_{ij} dC_j + \ldots \tag{36}$$

in accordance with the definition of γ_i (eq. (24a)). This represents an extension of the result obtained for a single solute (eq. (33)) to the case of individual species in a multi-component solute mixture. Eq. (36) accounts for the effects of non-ideality correct to first order in solute concentrations.

4. ACTIVITY COEFFICIENT OF A SINGLE SOLUTE

In early studies of single solutes by sedimentation equilibrium the main emphasis centred on the use of the integrated form of eq. (33) to obtain the molecular weight of the solute. Accordingly, the non-ideality was incorporated into an apparent molecular weight, M_{app}, defined as

$$M_{app} = \frac{M_A}{1 + C_A\left(\frac{\partial \ln \gamma_A}{\partial C_A}\right)_{T,\mu_s}} \tag{37}$$

or, by use of eq. (7) to obtain the expanded form of the partial derivative,

$$M_{app} = \frac{M_A}{1 + 2B_2C_A + 3B_3C_A^2 + \ldots} \tag{38}$$

The dependence of M_{app} upon C_A is most readily obtained by differentiating a plot of $\ln C_A$ *versus* r^2 at any radial distance r to obtain M_{app} at the corresponding solute concentration, C_A. The molecular weight and second virial coefficient may then be determined from the ordinate intercept and limiting slope, respectively, of a plot of $1/M_{app}$ *versus* C_A. Although many non-ideality coefficients have been obtained by this type of procedure, their evaluation places extreme demands upon the accuracy with which the differentiation to obtain M_{app} can be performed. Furthermore, a much simpler and potentially more accurate procedure has been developed.

It follows from eq. (33) that $z_A(r)$, the molar activity at radial distance r, is related to $z_A(r_F)$, the corresponding activity at any reference radial distance, r_F, by the equation

$$z_A(r) = z_A(r_F) \exp\left(\frac{M_A (1-\bar{v}_A\rho_s)\, \omega^2 (r_F^2 - r^2)}{2RT}\right) \tag{39}$$

Thus, whereas the experimental record is a plot of C_A as a function of r, the thermodynamic description is in terms of $z_A(r)$. By selecting a reference point, (r_F, C_A), in any sedimentation equilibrium distribution, the function[8]

$$\Omega_A(r) = \frac{C_A(r)}{C_A(r_F)} \exp\left(\frac{M_A(1-\bar{v}_A\rho_s)\,\omega^2\,(r_F^2-r^2)}{2RT}\right) \quad (40)$$

allows representation of experimental results in a manner whose form is dictated by eq. (39). Indeed, application of L'Hopital's rule shows that the ordinate intercept of a plot of $\Omega_A(r)$ *versus* $C_A(r)$ is given by

$$\lim_{C_A(r) \to 0} \Omega_A(r) = \frac{z_A(r_F)}{C_A(r_F)} = \gamma_A(r_F) \quad (41)$$

Furthermore, this ability to define the molar activity coefficient, and hence the solute activity, at the selected reference concentration, means that it is possible to generate the entire activity coefficient distribution, since

$$\gamma_A(r) = \frac{\gamma_A(r_F)}{\Omega_A(r)} \quad (42)$$

this being a relationship obtained by combining eqs. (39) and (40).

The application of this procedure to a sedimentation equilibrium distribution for isoelectric ovalbumin is summarized in Fig. 1, where the upper panel (Fig. 1a) shows the manner in which the Ω function is used to obtain $\gamma_A(r_F)$. From Fig. 1b. the resulting concentration dependence of ln γ_A conforms with eq. (9) truncated after the linear term. A value of 500 l/mol is thus obtained for $2B_2$ [9], which is in agreement with that predicted on the basis of eq. (16a) and a radius of 2.92 mm for ovalbumin. For a charged protein, eq. (19) would be used for the macromolecular interpretation of the experimentally obtained second virial coefficient.

5. STUDIES OF SOLUTE SELF-ASSOCIATION

The self-association of a macromolecular solute may be described in terms of a series of potential chemical reactions, each of which is characterised by an equilibrium constant K_i:

monomer + monomer ⇄ dimer; $\quad K_2 = \dfrac{z_2}{z_1^2}$ (43a)

dimer + monomer ⇄ trimer; $\quad K_3 = \dfrac{z_3}{z_1 z_2}$ (43b)

and so on. We note that each equilibrium constant is a ratio of thermodynamic activities, and that the activities of all oligomers may be expressed solely in terms of the product of equilibrium constants and monomer activity raised to the appropriate power ($z_2 = K_2 z_1^2$; $z_3 = K_2 K_3 z_1^3$, *etc.*). For a self-associating solute, z_1 is the quantity obtained by the $\Omega(r)$ analysis[8] when the monomer molecular weight is substituted for M_A and the total base-molar concentration of solute, $\bar{C}$, (weight

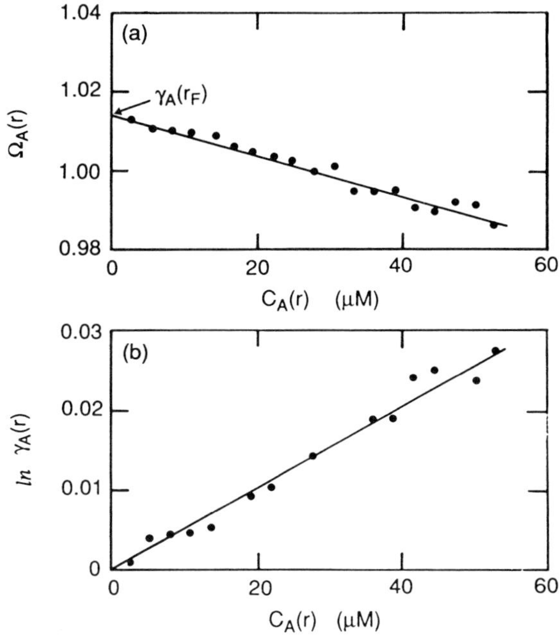

Figure 1. Use of a sedimentation equilibrium distribution to evaluate the concentration dependence of the molar activity coefficient, γ_A, for isoelectric ovalbumin.

(a) Application of the Ω-function (eq. (40)) to determine $\gamma_A(r_F)$, the activity coefficient pertaining to the selected reference concentration, $C(r_F)$.

(b) Concentration dependence of $\gamma_A(r)$, obtained by application of eq. (42) and plotted in accordance with eq. (9). [Adapted from Jeffrey *et al.*[9]]

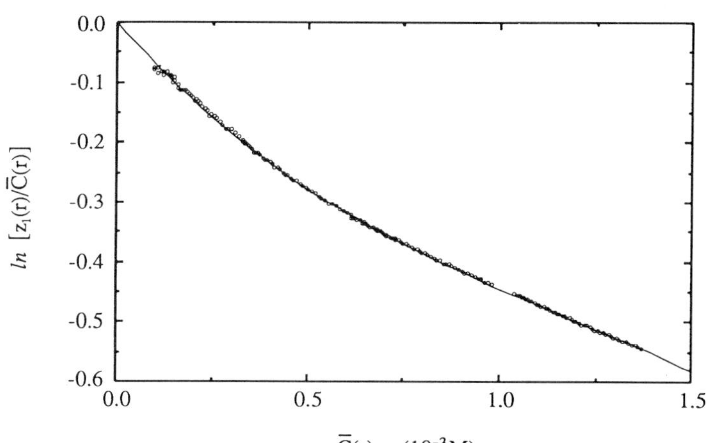

Figure 2. Analysis of the self-association of lysozyme. The solid line represents a third order polynomial which was fitted in accordance with eq. (46) to give estimates of K_2 and K_3. Data points (over 200 in all) are shown as small open circles. [Adapted from Wills *et al.*[14]]

concentration divided by monomer molecular weight) is substituted for C_A. (Note that $\bar{C} = \sum_i iC_i$ where C_i is the molar concentration of i-mer.) What is therefore required is a means of interpreting the concentration dependence of z_1, in a manner which takes into account both thermodynamic non-ideality and self-association explicitly. Several applications of this procedure have been reported[14,34], but the statistical mechanical theory derived by Hill and Chen[28] offers a more direct approach to the analysis of experimental data which expresses z_1 as a function of $\bar{C}$.

The thermodynamic description of binary solutions can be developed without any consideration of self-association phenomena, which are in these terms merely another form of non-ideality. However, when we wish to quantify the aggregate formation, the virial expansions for the osmotic pressure must be modified to include terms for the equilibrium constants that are to be evaluated. This problem has been addressed by Hill and Chen[28] who obtained the result

$$\left(\frac{\Pi}{RT}\right)_{T,\mu_s} = \bar{C} + B_2\bar{C}^2 + B_3\bar{C}^3 + \ldots \tag{44}$$

where

$$B_2 = B_{11} - K_2 \tag{45a}$$

$$B_3 = B_{111} - 4K_2(B_{11} - 2B_{12}) + K_2^2 - 2K_3K_2 \tag{45b}$$

and the coefficients B_{ij}, B_{ijk}, *etc.* are the McMillan-Mayer quantities defined in eq. (20) with numerical subscripts (1, 2, ...) used to identify aggregates (monomer, dimer, ...). Approximate procedures for evaluating these coefficients, taking into account excluded volume and charge-charge interactions between clusters comprised of up to three separate aggregates, have been discussed previously[14]. Furthermore, it has been shown that the concentration dependence of monomer activity may be expressed in the form[28]

$$\ln z_1 = \ln \bar{C} + 2B_2\bar{C} + \frac{3}{2}B_3\bar{C}^2 + \ldots \tag{46}$$

Eq. (46) has strong resemblance to a combination of eqs. (3) and (7), but in this instance B_2 and B_3 are given by eq. (45). It is noted that eqs. (44) and (46) meet the requirement that they be equivalent to those for a single solute in the event of no self-association ($K_2 = K_3 = \ldots = 0$). Eq. (44) provides the basis for polynomial curve-fitting to an experimentally derived plot of $\ln(z_1/\bar{C})$ *versus* $\bar{C}$ in order to determine the equilibrium constants, K_2, K_3, *etc.*, for any proposed pattern of self-association. Such an analysis of sedimentation equilibrium data for lysozyme is shown in Fig. 2. Fitting a cubic polynomial to the data gave values of $K_2=8.3\times10^2$ l/mol and $K_3=1.1\times10^3$ l/mol for the first two steps in the self-association of lysozyme. (The statistical uncertainty in B_4 was such that K_4 could not be accurately determined.) The advantage of this method over that used previously[14] to evaluate equilibrium constants from ($\bar{C}$, z_1) data is that it does not necessitate the iterative calculations of the concentration and activity distributions of all oligomeric species throughout the centrifuge cell.

For completeness some mention should be made of the extensive literature on the characterization of self-association equilibria from the concentration dependence of average molecular weights[5,35-40]. Most such analyses are based on an *ad hoc* definition of activity coefficients that eliminates them from the expression for the equilibrium constant ($K_2 = z_2/z_1^2$ is arbitrarily defined to be equal to C_2/C_1^2). The $\Omega(r)$ function has clearly opened the way for analyses that are free from such approximations. Alternatively, Chatelier and Minton[30,41] have recommended use of scaled particle theory to take into account the effects of thermodynamic non-ideality on the concentration dependence of apparent weight-average molecular weight. Application of their method entails the iterative evaluation of oligomeric concentrations distributions, a task that is obviated through the use of the approach we have outlined.

6. INTERACTIONS BETWEEN DISSIMILAR MOLECULES

We now wish to consider situations where non-ideality reflects, at least in part, the presence of additional solute(s). In regard to the study of such non-ideality in paucidisperse systems by sedimentation equilibrium, at least three cases are pertinent: (i) those in which the thermodynamic activity of a macromolecular solute is influenced by the presence of a small inert solute; (ii) those in which the solute of interest and the inert solute effecting thermodynamic non-ideality are both macromolecular; and (iii) those in which neither solute is inert. This section then concludes with a brief discussion of the effects of thermodynamic non-ideality in heterogeneous, inert solutions comprising a polydisperse array of chemically identical solute molecules that differ only in respect of polymer chain length. These four situations are considered in turn.

Excluded Volume Effects in the Presence of a Small Solute

Prior dialysis is used in sedimentation equilibrium studies to establish the identity of solvent chemical potentials in the solution and solvent sectors of the cell[42], this being a procedure that also ensures identity of the corresponding chemical potentials of any small dialysable solute such as sucrose or glycerol. When such a solute, B, is present at high concentration (*e.g.* 0.1 M – 1.0 M) the activity coefficient of the macromolecular component becomes dominated by the cross-interaction term ($\ln \gamma_A \approx \alpha_{AB}C_B$). If the solute, B, is sufficiently small, then its concentration, C_B, does not vary significantly across the centrifuge cell, and it may be regarded as part of the solvent. Eq. (33) or (36) may then be used to describe the sedimentation equilibrium distribution of any macromolecular solute with $\rho_0 = \rho_s + (1 - \bar{v}_B\rho_s)M_BC_B$ in place of ρ_s. The effect of the small solute on the equilibrium distribution of an inert macromolecular solute, A, may be calculated in terms of a simple correction to the buoyancy term and there is a uniform contribution, $\exp(\alpha_{AB}C_B)$, to the activity z_A of the macromolecular component at all points in the centrifuge cell.

The sedimentation equilibrium distribution of a self-associating species is greatly influenced by the presence of a small solute, because molecular crowding displaces the association equilibrium towards polymeric state[43]. This phenomenon is demonstrated in Fig. 3, where the effect of 0.15 M sucrose on the dimerization of α-

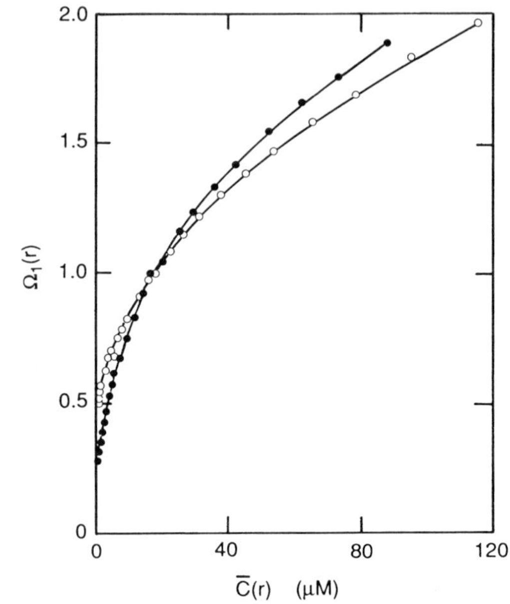

Figure 3. Quantitative evaluation of the effect of sucrose on the dimerization of α-chymotrypsin by application of the Ω-analysis to sedimentation equilibrium distributions of the enzyme in acetate-chloride buffer in the absence (open symbols) and presence (closed symbols) of 0.15 M sucrose. In each case the ordinate intercept corresponds to $C_1(r_F)/\bar{C}(r_F)$, the proportion of monomer present in the reference concentration, $\bar{C}(r_F)$, in accordance with eq. (49). [Adapted from Shearwin and Winzor[43].]

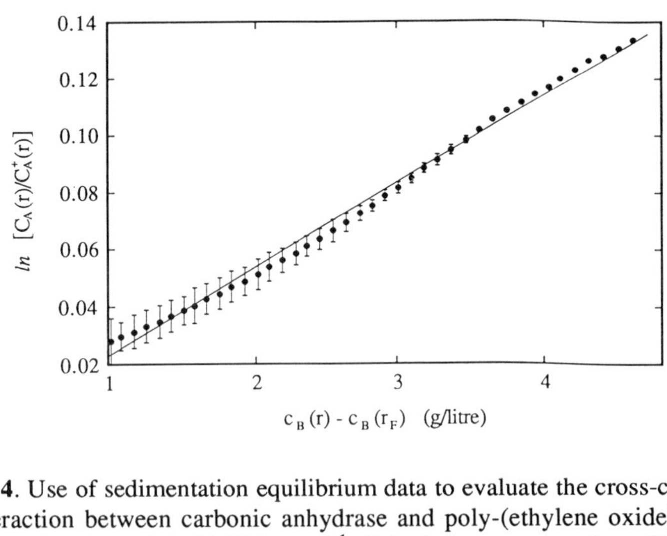

Figure 4. Use of sedimentation equilibrium data to evaluate the cross-covolume for the interaction between carbonic anhydrase and poly-(ethylene oxide) of weight-average molecular weight 21,000 g.mol^{-1}. Calculations employed eq. (52). [Adapted from Georgalis[49].]

chymotrypsin (pH=3.9, I=0.2) has been considered in terms of its consequences on the $\Omega(r)$ analysis. An interesting point arises in the application of the Ω analysis, which was used above to define the activity of monomer, $z_1(r)$, in a mixture of monomeric and polymeric species with total base molar concentration $\bar{C}(r)$. As noted in the report of the analytical procedure[8], the ordinate intercept of the $\Omega(r)$ plot is defined as

$$\lim_{\bar{C}(r)\to 0} \Omega_1(r) = \frac{\gamma_1(r_F)\, C_1(r_F)\, \bar{C}(r)}{\gamma_1(r)\, C_1(r)\, \bar{C}(r_F)} \tag{47}$$

which, in a system containing self-associating solute alone, is $z_1(r_F)/\bar{C}(r_F)$ on the grounds that $\gamma_1(r)$ and $C_1(r)/\bar{C}(r_F)$ both tend to unity as $\bar{C}(r) \to 0$. However, for a system with a high concentration, C_B, of sucrose, $C_1(r)\bar{C}(r)$ still tends to unity, whereas, in the case of a monomer-dimer system, $\gamma_1(r)$ is governed by the expression

$$\gamma_1(r) = \exp\left\{\alpha_{11}C_1(r) + \alpha_{12}C_2(r) + \alpha_{1B}C_B(r) + \ldots\right\}$$

$$\approx \exp\left\{\alpha_{1B}C_B(r)\right\} \tag{48}$$

When the solute B is sufficiently small, the monomer activity coefficient (eq. (48)) is essentially independent of radial distance, r, and eq. (47) simplifies to

$$\lim_{\bar{C}(r)\to 0} \Omega_1(r) = \frac{C_1(r_F)}{\bar{C}(r_F)} \tag{49}$$

Calculations based on eq. (33) with solute identified as sucrose showed that the concentration of sucrose varied by only 4% across the liquid column at sedimentation equilibrium[43]. The decreased ordinate intercept in Fig.3 for α-chymotrypsin in the presence of sucrose was therefore taken to reflect a decreased proportion of monomeric enzyme in the reference concentration (18.4 μM) of α-chymotrypsin. On the basis of thermodynamic ideality in the absence of sucrose a value of 5.5×10^4 l/mol is calculated for K_2, the thermodynamic dimerization constant, whereas that for the apparent dimerization constant, $K_{app} = [\bar{C}(r_F) - C_1(r_F)]/2C_1(r_F)^2$, obtained in the presence of sucrose is 23×10^4 l/mol. In terms of eq. (48) the two values are related by the expression

$$K_{app} \approx K_2 \exp[(2\alpha_{1B} - \alpha_{2B})C_B]. \tag{50}$$

Protein-Polymer Interactions

Prior to the realization that the effects of small solutes were amenable to interpretation on the statistical mechanical basis of excluded volume[44] inert macromolecular solutes such as poly(ethylene glycol) and dextran were used as space-filling solutes to incur additional thermodynamic non-ideality in protein solutions[45-47]. The use of sedimentation equilibrium to investigate protein-polymer interactions has received little attention. Here we consider an example where it is possible to make independent measurements of the sedimentation equilibrium distribution of two solutes, a protein,

A, and a polymer, B. Independent measurements of this sort are possible whenever the protein distribution in the centrifuge cell can be determined by using absorption optics and the total macromolecular distribution can be determined by using interference optics; or whenever separate analytical procedures are used to determine the concentration distributions of the individual constituents after sedimentation equilibrium experiments in a preparative ultracentrifuge[48]. In making the absorption measurements in a Model E ultracentrifuge it is desirable to choose a wavelength at which the polymer has no significant absorption and the measurements reflect only the protein distribution, $C_A(r)$. Furthermore, if the protein is relatively dilute (of the order 0.1 g/l) and the polymer is relatively concentrated (of the order 10 g/l), results obtained from interference measurements reflect the polymer distribution, $C_B(r)$, in the centrifuge cell, since any contribution from the protein is insignificant. When the partial molar volumes of all the relevant components (solvent, protein and polymer) are independent of pressure and solution composition, eq. (36) may be used to describe either distribution, $C_A(r)$ or $C_B(r)$.

Suppose two experiments are conducted, one to determine the equilibrium distribution of protein, $C_A(r)$, in the absence of polymer and the other to determine the same distribution, $C_A^\ddagger(r)$, in the presence of polymer. From eqs. (36) and (24a) it is evident that by choosing a reference radial distance r_F common to both experiments, the ratio of protein concentrations at any radial distance common to both experiments may be expressed in the form

$$\frac{C_A(r)}{C_A^\ddagger(r)} = \frac{\gamma_A^\ddagger(r)}{\gamma_A(r)} \left(\frac{C_A(r_F)\,\gamma_A(r_F)}{C_A^\ddagger(r_F)\,\gamma_A^\ddagger(r_F)} \right) \tag{51}$$

and the expression in brackets is a constant. By taking the logarithm of both sides of eq. (51) and using the approximation $\ln \gamma_A \approx \alpha_{AB} C_B$, we obtain the result

$$\ln\left(\frac{C_A(r)}{C_A^\ddagger(r)}\right) \approx \ln\left(\frac{C_A(r_F)}{C_A^\ddagger(r_F)}\right) + \alpha_{AB}[C_B^\ddagger(r) - C_B^\ddagger(r_F)] \tag{52}$$

which allows α_{AB} to be determined by plotting $\ln (C_A(r)/C_A^\ddagger(r))$ against $C_B^\ddagger(r) - C_B^\ddagger(r_F)$. The absolute concentration of the polymer need not be determined. Such an approach has been used to determine the cross-covolume between various proteins and polymers[49]. Results for carbonic anhydrase are shown in Fig. 4. The weight concentration scale (g/litre) has been used along the abscissa, meaning that the gradient, α_{AB}^*, is in fact the number-average of α_{AB} divided by the number-average molecular weight of the polymer, PEG 20000. From this graph, a value of 22±2 l/g is obtained for α_{AB}^*, consistent with the value of 22.0±0.8 l/g obtained from independent gel-chromatography experiments[49]. Interpretation of this result in terms of the dimensions of carbonic anhydrase (Stokes radius 2.4 nm) and PEG 20000 (Stokes radius 3.8 nm, radius of gyration 2.4 nm) must await the resolution of questions concerning the functional form of the mutual co-volume of exclusion between a globular protein and a random coil polymer[21-23].

The manner in which a polymer may promote the self-association of a protein has been discussed by Nichol *et al.*[50], but the application of this theory to the technique of sedimentation equilibrium has not yet been explored.

Heterogeneous Associations

It was noted above that the formation of aggregates comprised of similar species may be viewed as a manifestation of thermodynamic non-ideality. When dissimilar solute species aggregate, the analysis of sedimentation equilibrium data is greatly simplified by defining explicitly each species of aggregate present in solution. The formation of aggregates of arbitrary composition from two constituents, A and B, may be analysed by considering the reaction[51]

$$mA + nB \rightleftarrows C \qquad K = \frac{z_C}{z_A^m\, z_B^n} \tag{53}$$

When the concentrations of individual species are sufficiently low and the interactions which lead to aggregate formation are sufficiently strong, the relatively small effects of excluded volume and charge-charge interactions may be neglected. Under such circumstances, the sedimentation equilibrium distribution of each explicitly defined species may be approximated by the pseudo-ideal condition

$$C_i(r) = C_i(r_F)\exp\left(\frac{M_i(1 - \bar{v}_i\rho_s)\omega^2(r_F^2 - r^2)}{2RT}\right) \tag{54}$$

and each equilibrium constant may be approximated as a ratio of concentrations rather than activities. The interaction of ovalbumin and lysozyme has been extensively studied through approaches which make use of weight-average molecular weight distributions[52], or alternatively, a variation of the $\Omega(r)$-analysis[53,54] discussed previously. Direct non-linear fitting to an experimentally determined optical density distribution in a centrifuge cell can also be used to analyse heterogeneous associations. This method has been used to determine the thermodynamic constants for the association of plasminogen and fibrinogen[55] and for the association of ricin subunits[56]. Methods which allow for simultaneous concentration determination of dissimilar species after sedimentation equilibrium in a preparative centrifuge cell[48,57] may also be applied to the study of heterogeneous associations.

Non-Ideality in Polymer Systems

In conclusion, it is appropriate to make passing mention of the application of sedimentation equilibrium to the analysis of non-ideality in polydisperse systems. Each separate species in a polydisperse system is distributed in the centrifuge cell according to eq. (36). When the effects of non-ideality are negligible, analysis of sedimentation equilibrium data enables information to be gleaned about the polymeric molecular weight distribution[31]. When non-ideality is taken into account, it is necessary to define some "average" or "effective" virial coefficient because the individual coefficients α_{ij} for the interaction between two arbitrary polymeric species, i and j, cannot all be calculated or evaluated. Clearly the value of any "average" virial coefficient is in principle a function of radial distance, simply because the polymeric solute composition changes with radial distance. However, the only techniques so far devised[31,58] entail the evaluation of a single parameter which is comparable with similar quantities obtained by using light-scattering and other techniques.

7. SUMMARY AND CONCLUSION

The thermodynamic activity of a solute has alternative definitions depending on whether the solvent chemical potential or the pressure is held constant while the solute concentration is varied. When the solvent chemical potential is held constant, the activity, z_A, is most conveniently defined in terms of the molar concentration, C_A. On the other hand, when the pressure is held constant, the activity, a_A, is best defined in terms of the molal concentration, m_A. When the partial molar volumes of the solvent and solute are constants, independent of pressure and solution composition, simple relationships exist between the two activities, z_A and a_A, and the corresponding activity coefficients, γ_A and y_A. Virial expansions for the dependence of these quantities on solute concentration allow the relationships between activity coefficients to be expressed very simply in terms of quantities whose values can be calculated from molecular parameters.

In order to write the fundamental equation for sedimentation equilibrium in such a way that even first order deviations from ideal thermodynamic behaviour are taken into account rigorously, it is necessary to use the molal concentration scale. When the dependence of solution density on solute concentration is taken into account, the relevant variables of integration, radial distance, r, and solute concentration, m_A, are separable. Conversion to the molar concentration scale at this stage gives rise to a radical simplification. The molar activity, z_A, is found to replace the concentration variable which appears in the equation for an ideal system, but the buoyancy term, $(1 - \bar{v}_A\rho_s)$, has to be defined in terms of the density of solvent. This result applies quite generally to the description of non-ideal two-component systems for which the partial molar volumes of both solvent and solute are constants. It is also correct, at least to first order in concentration variables, for similar multi-component systems.

These theoretical results provide a context for the unified and consistent analysis of non-ideality in systems at sedimentation equilibrium. For a single solute, the virial coefficient B_2 can be determined from experimental results and related to the size, shape and charge of the molecular species, or alternatively, the effects of non-ideality can be calculated so that other effects, such as self-association, can be studied. Equilibrium constants for sequential steps in the self-association of an aggregating species can be determined without recourse to the iterative calculation of activity coefficients. A variety of interactions between dissimilar species, including interactions between proteins and small solutes or large polymer molecules, are also amenable to analysis. A small solute may be treated as a component of the solvent, but it can nevertheless have a marked effect on the self-association of a protein, just as a result of excluded volume effects alone. Larger solutes, such as polymers, also increase the activity of other macromolecular species present in solution, and in some cases sedimentation equilibrium data can be analysed to yield an estimate of the cross co-volume of interaction between a protein and polymer species. However, the analysis of sedimentation equilibrium data for systems containing a mixture of macromolecular solutes is in general a very complex undertaking.

Thermodynamic non-ideality has a marked effect on the spatial distribution of macromolecular solutes at sedimentation equilibrium. Initially, the effects of non-ideality were eliminated empirically, but it is now evident that the analysis of non-ideality can yield useful information about macromolecules and their interactions. However, from an experimenter's viewpoint, the most pleasing outcome of this consideration of thermodynamic non-ideality in *sedimentation equilibrium* is undoubtedly the demonstration that use of the solvent density in the buoyancy correction term is not the approximation that it was thought to be: for thermodynamic rigour the solvent density must be used.

REFERENCES

1. R.J. Goldberg, *J. Phys. Chem.*, 1953, *57*, 194.
2. K.E. Van Holde and R.L. Baldwin, *J. Phys. Chem.*, 1958, *62*, 734.
3. J.W. Williams, K.E. Van Holde, R.L. Baldwin and H. Fujita, *Chem. Rev.*, 1958, *58*, 715.
4. H. Fujita, 'Mathematical Theory of Sedimentation Analysis', Academic Press, New York, 1962.
5. E.T. Adams, Jr. and H. Fujita in 'Ultracentrifugal Analysis in Theory and Experiment' (J.W. Williams, ed.), Academic Press, New York, 1963, p. 119.
6. L.W. Nichol, A.G. Ogston and B.N. Preston, *Biochem. J.*, 1967, *102*, 407.
7. R.H. Haschemeyer and W.F. Bower, *Biochemistry*, 1970, *9*, 435.
8. B.K. Milthorpe, P.D. Jeffrey and L.W. Nichol, *Biophys. Chem.*, 1975, *3*, 169.
9. P.D. Jeffrey, L.W. Nichol, D.R. Turner and D.J. Winzor, *J. Phys. Chem.*, 1977, *81*, 776.
10. P.D. Jeffrey in 'Protein-Protein Interactions' (C. Frieden and L.W. Nichol, eds.), New York, 1981, p. 213.
11. A.C. Ogston and D.J. Winzor, *J. Phys. Chem.*, 1975, *79*, 2496.
12. P.D. Ross and A.P. Minton, *J. Mol. Biol.*, 1977, *112*, 437.
13. T.L. Hill, 'Thermodynamics for Chemists and Biologists', Addison-Wesley, Reading, MA, 1968.
14. P.R. Wills, L.W. Nichol and R.J. Siezen, *Biophys. Chem.*, 1980, *11*, 71.
15. W.G. McMillan, Jr. and J.E. Mayer, *J. Chem. Phys.*, 1945, *13*, 276.
16. J.O. Hirschfelder, C.F. Curtiss and R.B. Bird, 'Molecular Theory of Gases and Liquids', Wiley, New York, 1967.
17. J.M. Rallison and S.E. Harding, *J. Colloid Interface Sci.*, 1985, *103*, 284.
18. A.G. Ogston, *Trans. Faraday Soc.*, 1958, *54*, 1754.
19. A.G. Ogston, *J. Phys. Chem.*, 1970, *74*, 668.
20. D.E. Roark and D.A. Yphantis, *Biochemistry*, 1971, *10*, 3241.
21. J. Hermans, *J. Chem. Phys.*, 1982, *77*, 2193.
22. J.J. Hermans and J. Hermans, *J. Polym. Sci., Polym. Phys. Ed.*, 1984, *22*, 279.
23. K.E. Shearwin and D.J. Winzor, *Eur. J. Biochem.*, 1990, *190*, 523.
24. P.D. Ross, R.W. Briehl and A.P. Minton, *Biopolymers*, 1978, *17*, 2285.

25. A.P. Minton and H. Edelhoch, *Biopolymers*, 1982, *21*, 451.
26. H. Reiss, H.L. Frisch and J.L. Lebowitz, *J. Chem. Phys.*, 1959, *31*, 369.
27. R.M. Gibbons, *Mol. Phys.*, 1969, *17*, 81.
28. T.L. Hill and Y.D. Chen, *Biopolymers*, 1973, *12*, 1285.
29. A.P. Minton, *Mol. Cell. Biochem.*, 1983, *55*, 119.
30. R.C. Chatelier and A.P. Minton, *Biopolymers*, 1987, *26*, 507.
31. H. Fujita, 'Foundations of Ultracentrifugal Analysis', John Wiley, New York, 1975.
32. J.S. Johnson, K.A. Kraus and G. Scatchard, *J. Phys. Chem.*, 1960, *64*, 1867.
33. M. Fixman, *J. Phys. Chem.*, 1958, *62*, 374.
34. P.D. Jeffrey, B.K. Milthorpe and L.W. Nichol, *Biochim. Biophys. Acta*, 1977, *495*, 195.
35. E.T. Adams, Jr., *Biochemistry*, 1965, *4*, 1646.
36. E.T. Adams, Jr. and D.L. Filmer, *Biochemistry*, 1966, *5*, 2971.
37. E.T. Adams, Jr. and M.S. Lewis, *Biochemistry*, 1968, *7*, 1044.
38. D.C. Teller, *Biochemistry*, 1970, *9*, 4201.
39. D.A. Yphantis and D.A. Roark, *Biochemistry*, 1972, *11*, 2925.
40. G.J. Howlett , P.D. Jeffrey and L.W. Nichol, *J. Phys. Chem.*, 1972, *76*, 777.
41. R.C. Chatelier and A.P. Minton, *Biopolymers*, 1987, *26*, 1097.
42. E.F. Casassa and H. Eisenberg, *Adv. Protein Chem.*, 1964, *19*, 287.
43. K.E. Shearwin and D.J. Winzor, *Biophys. Chem.*, 1988, *31*, 287.
44. D.J. Winzor and P.R. Wills, *Biophys. Chem.*, 1986, *25*, 243.
45. E. Edmond and A.G. Ogston, *Biochem. J.*, 1968, *109*, 569.
46. T.C. Laurent, *Eur. J. Biochem.*, 1971, *21*, 498.
47. D.H. Atha and K.C. Ingham, *J. Biol. Chem.*, 1981, *256*, 12108.
48. A.K. Attri and A.P. Minton, *Anal. Biochem.*, 1987, *162*, 409.
49. Y. Georgalis, 'Physico-chemical Properties of Ribosomal Proteins from *E. coli*' [in Greek], Ph.D. Thesis, University of Athens, 1982.
50. L.W. Nichol, A.G. Ogston and P.R. Wills, *FEBS Lett.*, 1981, *126*, 18.
51. L.W. Nichol and A.G. Ogston, *J. Phys. Chem.*, 1965, *69*, 4365.
52. G.J. Howlett and L.W. Nichol, *J. Biol. Chem.*, 1973, *248*, 619.
53. L.W. Nichol, P.D. Jeffrey and B.K. Milthorpe, *Biophys. Chem.*, 1976, *4*, 259.
54. P.D. Jeffrey, L.W. Nichol and R.D. Teasdale, *Biophys. Chem.*, 1979, *10*, 379.
55. M.S. Lewis, F. Carmassi and S.I. Chung, *Biochemistry*, 1984, *23*, 3874.
56. M.S. Lewis and R.J. Youle, *J. Biol. Chem.*, 1986, *261*, 11571.
57. A. Revzin and P.H. von Hippel, *Biochemistry*, 1977, *16*, 4769.
58. P.J. Wan and E.T. Adams, in 'Fifty Years of the Ultracentrifuge', (M.S. Lewis and G.H. Weiss, eds.), North Holland, Amsterdam, 1976, 207.

Part III: Transport Methods

18

Sedimentation Coefficients of Complex Biological Particles

By José García de la Torre

DEPARTAMENTO DE QUIMICA FISICA, FACULTAD DE CIENCIAS QUIMICAS Y MATEMATICAS UNIVERSIDAD DE MURCIA, 30071 MURCIA, SPAIN

1. INTRODUCTION

Sedimentation Velocity

One of the main applications of the analytical ultracentrifuge in biochemistry and polymer science is the determination of the sedimentation coefficient of macromolecular species in solution or suspension. When the solute molecules or particles are subject to a centrifugal field, they experience three different forces. One of them is just the centrifugal force, which depends on the centrifugal field and the mass of the particle. On the other hand, there is a buoyancy force related to the volume of the macromolecule and the density of the solvent. Immediately after starting the ultracentrifugation experiment, the particles begin to move in the direction of the bottom, if the centrifugal force overcomes the buoyancy, as is usual. That motion is opposed by another force, F_f, which is simply expressed as the product of the velocity of sedimentation, v, and a friction coefficient, f, which depends on the viscosity of the solvent and is related to the size and shape of the particle:

$$F_f = f\,v \tag{1}$$

After a very short time (which is usually negligible when compared with the duration of a typical run), equilibrium between the three forces - centrifugal, buoyancy and friction - is established and the particle sediments with constant velocity. This sedimentation velocity is proportional to the square of the rotational speed, ω, and to the distance, r, between the macromolecule and the axis of rotation. The proportionality constant is the sedimentation coefficient, s, given by

$$s = \frac{v}{r\omega^2} \tag{2}$$

From the balance of the three forces, s can be expressed in terms of the properties of the solute and the solvent as

$$s = \frac{M\,(1-\bar{v}\rho)}{N_A f} \tag{3}$$

where:

M is the molecular weight of the macromolecule;
$\bar{v}$ is the partial specific volume of the macromolecule;
ρ is the density of the solution;
N_A is Avogadro's number;
f is the frictional coefficient of the macromolecule (eq. (1))

The derivation of eq. (3) and further details can be found in several monographs[1-4]. The usefulness of sedimentation velocity measurements is evident from eq. (3), which shows how the sedimentation coefficient contains information about the mass of the particle (M) as well as its size and shape (f).

Transport Coefficients

The friction and sedimentation coefficients are transport properties related to the deterministic behaviour of the macromolecule when it is moving under the action of an external force. In the absence of external agents, the macromolecular solute experiences a random or stochastic, Brownian motion, which is macroscopically detected as diffusion, and can be also monitored from the time fluctuations in the light scattered by the particle[5]. The pertinent transport property for these phenomena is the diffusion coefficient, D. The deterministic and stochastic behaviours are linked by a fundamental relationship between the coefficients, that is due to Einstein:

$$D = \frac{kT}{f} \tag{4}$$

where k is Boltzmann's constant and T the absolute temperature. From eqs. (3) and (4) we obtain an alternative expression for the sedimentation coefficient,

$$s = \frac{DM\,(1-\bar{v}\rho)}{RT} \tag{5}$$

where R is the gas constant and T is the absolute temperature. Eq. (5) provides a convenient way for the determination of the molecular weight, which can be obtained from separate measurements of s and D. Sedimentation and diffusion are concentration dependent, and therefore eqs. (4) and (5) are strictly valid for infinite dilution.

The coefficients s, f and D are mutually interconvertible using eqs. (3-5). Hereafter, only one of them (usually s or f) will be mentioned, in the understanding that the similar equations or characteristics for the two others can be trivially expressed.

2. RIGID PARTICLES

Spherical Particles

The simplest case is that of a spherical particle, for which the frictional coefficient is given in terms of the radius of the sphere, R, by Stokes law:

$$f = 6\pi\eta R \tag{6}$$

where η is the viscosity of the solvent. For a non-spherical solute, one can define a hydrodynamic radius, R_H, as the radius of a sphere having the same frictional coefficient in the same solvent

$$R_H = \frac{f}{6\pi\eta} \tag{7}$$

R_H thus gives an indication of the hydrodynamic volume of the macromolecular solute, but its interpretation requires some knowledge of its shape.

Information about macromolecular size can be obtained from the molecular weight, M. If we know the specific volume of the macromolecule, $\bar{v}$, we can estimate its volume as $V_{anh}=\bar{v}M/N_A$. This corresponds to the *anhydrous* volume of the particle, and does not include bound or associated solvent (if there is any). From this volume we can obtain a spherical radius, and then a frictional coefficient:

$$f_0 = 6\pi\eta \left\{\frac{3V_{anh}}{4\pi}\right\}^{1/3} \tag{8}$$

which is that corresponding to a sphere having the same volume as the *unhydrated* macromolecule. Both hydration and non-sphericity increase friction, so that f_0 is always a lower bound of the true friction coefficient, f. In other words, the ratio f/f_0 will be greater than unity, the deviation depending on shape, or hydration, or both.

Ellipsoids

The departure from the spherical shape can be in principle characterized by ellipsoidal models. There is a variety of biological macromolecules that exhibit fairly compact or globular shapes, and for them the ellipsoid is the proper hydrodynamic particle. The general, or tri-axial ellipsoid has three different semi-axes a, b and c, and therefore two parameters are needed to characterize its shape. When the macromolecule has the aspect of a revolution body, or simply when the scarcity of experimental information does not allow the determination of more than one shape parameter, one can analyze sedimentation coefficients in terms of an ellipsoid of revolution with semi-axes a and b=c. Thus, there is only one shape parameter, $p=a/b$ with $p>1$ for prolate ellipsoids and $p<1$ for oblate ellipsoids. An exact result for the friction coefficient for ellipsoids of revolution was given many years ago by Perrin[7]. When applied to the *frictional ratio* this gives

$$f/f_0 = \frac{p^{-1/3}\,(p^2-1)^{1/2}}{\ln\,\{p + (p^2-1)^{1/2}\}} \qquad \text{for } p>1 \tag{9a}$$

$$f/f_0 = 1 \qquad \text{for } p=1 \tag{9b}$$

$$f/f_0 = \frac{p^{-1/3}\,(1-p^2)^{1/2}}{\tan^{-1}(p^{-2}-1)^{1/2}} \qquad \text{for } p<1 \tag{9c}$$

Note that f/f_0 is just a function of the shape parameter p, and does not depend on the actual values of a and b.

For globular macromolecules, and in general when one wishes a more complete characterization of the anisometry of the particle, it is possible to employ a general, tri-axial ellipsoid, with different semi-axes $a \geq b \geq c$. The shape is then determined by two axial ratios, a/b and b/c. The tri-axial ellipsoid was not extensively used in the past because the general expression for f/f_0 contains an elliptic integral[7,8] that has to be numerically computed, and expressions for other hydrodynamic properties were not known. These difficulties have been removed recently by Harding and Rowe[8,9] who derived the viscosity of tri-axial ellipsoids and devised various numerical procedures.

In the analysis of *hydration* effects, it is accepted that the shape of the hydrated particle is the same as that of the unhydrated one, so that the shape parameter p for ellipsoids of revolution, or (a/b, b/c) for general ellipsoids, remains the same. The effect of hydration alone is to increase the frictional ratio by a factor of $(V_{hyd}/V_{anh})^{1/3}$, where the hydrated volume V_{hyd} is equal to V_{anh} plus the volume of bound water, which depends on the amount of hydration, δ, expressed as the grams of water per gram of macromolecule. Thus we have

$$f/f_0 = (1+\delta/\bar{v}\rho_1)^{1/3} \tag{10}$$

In the most general case, when both size and hydration effects are considered, the frictional ratio is the product of two factors,

$$f/f_0 = (f/f_0)_{shp}\,(f/f_0)_{hyd} \tag{11}$$

the first of which is the shape factor, given by eq. (9) and the second one is the hydration factor, given by eq. (10), and thus in general f/f_0 depends on both p or (a/b, b/c) and δ.

It is possible to obtain information about particle shape independently from hydration if one combines f (or s or D) with another transport property. For instance, the term $([\eta]M)^{1/3}$, where $[\eta]$ is the intrinsic viscosity, is influenced by hydration in the same way as f. Therefore, the combination $([\eta]M)^{1/3}/f$ is independent of the hydrodynamic volume. With some constants to make it dimensionless, this combination is the well-known Scheraga-Mandelkern function, β, which, in terms of the sedimentation coefficient is formulated as

$$\beta = \frac{N_A s[\eta]^{1/3}\eta_0}{M^{1/3}(1-\bar{v}\rho)100^{1/3}} \tag{12}$$

An important drawback of β is that it is varies very slowly with axial ratio. Thus β increases from the sphere value (p=0), 2.12×10^6, to only 2.41×10^6 for a prolate ellipsoid with p=10, and is practically equal to the sphere value for oblate ellipsoids. This insensitivity, which is used in a procedure to obtain molecular weights from measurements of sedimentation and viscosity, makes the β function, on the other hand, practically useless for the determination or macromolecular shape. Other combinations have been explored involving s and other solution properties like the molecular co-volume, the rotational relaxation time and the concentration-dependence coefficients, showing that some are quite sensitive to the shape of the particle. For tri-axial ellipsoids, two of such combinations can be simultaneously used to determine the two axial ratios.

Cylinders

A variety of biological macromolecules, ranging from short DNA fragments to filamentary viruses, exhibit an elongated shape with uniform thickness. The proper hydrodynamic model in this case is the cylindrical rod. For a rod-like particle, with length L, diameter d and aspect ratio p (≡L/d), the frictional coefficient takes the form

$$f = \frac{3\pi\eta L}{(\ln p + \gamma)} \tag{13}$$

where γ is a function of p and which has a limiting value for very long cylinders, when $p \rightarrow \infty$. Substituting eq. (13) into eq. (3), we have for the sedimentation coefficient:

$$s = \frac{M(1-\bar{v}\rho)}{3\pi\eta N_A L} \left\{ \ln \left(\frac{L}{d} \right) + \gamma \right\} \tag{14}$$

If M_L=M/L is the mass per unit length, then eq. (14) can be rewritten as

$$s = \frac{M_L(1-\bar{v}\rho)}{3\pi\eta N_A} \left\{ \ln M - \ln M_L - \ln d + \gamma \right\} \tag{15}$$

The analysis of sedimentation coefficients according to eq. (15) requires a knowledge of the γ(p) function. An early result for this function derived nearly thirty years ago by Broersma[12] was later shown to be deficient. Although a correction to that result was published later[13], the currently accepted expression for γ(p) is that proposed by Tirado and the present author[14]:

$$\gamma(p) = 0.312 + \frac{0.561}{p} + \frac{0.100}{p^2} \qquad 2<p<20 \tag{16}$$

which is actually an interpolating formula[15] valid in the range from p=2-20 approximately. It is accepted that the limiting value of γ for very long rods is[16]

$$\gamma = 0.386 \qquad \text{as } p \rightarrow \infty \tag{17}$$

When the macromolecule is elongated enough (say p>10) so that the function γ(p) can be replaced by its limiting value, then a series of data of s for samples of varying molecular weight should follow a linear trend in a semi-logarithmic plot. The slope of the straight line can yield the mass per unit length and from the slope the hydrodynamic diameter can be estimated. For shorter cylinders, the full γ(p) function can be used. An example of this type of analysis is that of the hydrodynamic behaviour of short, rod-like pieces of DNA. Fig. 1 shows experimental data along with theoretical predictions from eqs. (15) and (16) with M_L=195 daltons/Å and a rise

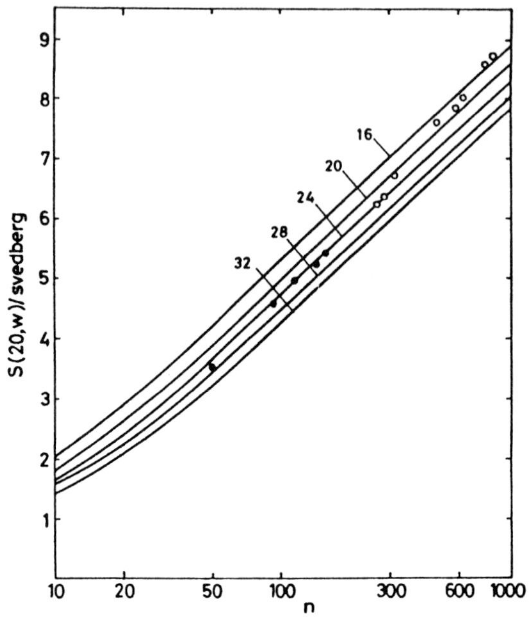

Figure 1. Sedimentation coefficients of short DNA fragments as a function of the number of base pairs, n. Filled circles are experimental data with n<200. Open circles are experimental data with n>200. The curves are calculated from the Tirado-García de la Torre theory for the indicated values of the diameter, in Å.

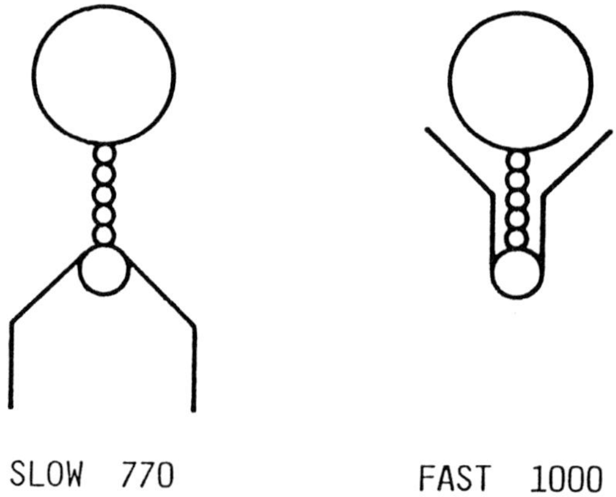

Figure 2. Bead models for the slow and fast forms of T2 bacteriophage. Each of the six tailfibres (of which only two are drawn) are modelled with 69 beads. The indicated numbers are the sedimentation coefficients.

per base pair of 3.4Å, corresponding to the B form of DNA. The data for the shortest fragments are well reproduced by the theoretical results if the hydrodynamic diameter is about 25Å. The sedimentation coefficients for the longer fragments deviate from the rod-like behaviour as a consequence of their flexibility.

From a single datum for the sedimentation coefficient of a rod-like macromolecule of known molecular weight (or known mass per unit length), one can obtain the rod length, L, from eqs. (14) or (15), if one has some estimate for the diameter, d. The choice for the latter is not critical, since d takes part in the calculations as ln d or through the weak dependence of ν on d.

The complete characterization of the cylinder dimensions (L and d) may require the knowledge of two hydrodynamic properties to be combined in a shape-dependent function. Combinations similar to those used for ellipsoids can be employed for cylinders. A particularly useful function[17], $f(\mathrm{p})$ [not to be confused with the friction coefficient], involves the translational diffusion coefficient, D (which can be obtained from sedimentation data using eq. (5)) and the rotational diffusion coefficient for end-over-end rotation, D_r, given by

$$D_r = 1/6\tau \qquad (18)$$

where τ is the longest rotational relaxation time detected in a variety of dynamic electro-optic and spectroscopic techniques[18]. This function is defined as[17]

$$f(\mathrm{p}) = \left\{\frac{9\pi\eta}{kT}\right\}^{2/3} \left\{\frac{D}{D_r^{1/3}}\right\} \qquad (19)$$

If f is experimentally determined, the shape parameter p can be obtained from[17]

$$\mathrm{p} = \exp(-6.650 + 5.489f - 0.656f^2) \qquad 2<\mathrm{p}<20 \qquad (20)$$

Knowing p, one would next obtain L and d from eq. (14).

In many cases of interest, the macromolecule is not strictly rigid and straight, but instead has some bending flexibility so that proper model for solution properties is more wormlike than rod-like. Flexibility then plays and important role in the hydrodynamic behaviour, and therefore these cases are treated along with those of typically flexible polymers in the next Chapter of this book[19].

3. BEAD MODELS

Theory

There are a number of biological macromolecules and assemblies whose shape can not be represented by simple models. A paradigmatic case is that of T-even bacteriophage, depicted in Fig. 2. These bacterial viruses have a compact head, to which a cylindrical tail is attached. At the other end of the tail, they show six long and thin fibres that can be kinked, adopting various conformations. It is evident that the hydrodynamic behaviour of these complex bioparticles cannot be represented, even as a first approach, by simple models such as ellipsoids or rods. The interpretation of the change in sedimentation coefficients of T-even bacteriophage with pH, temperature or ionic strength was indeed one of the main motivations for the development of the bead-model theory.

In 1967, Bloomfield *et al.*[22,23] proposed to model rigid macromolecular of arbitrarily complex shape using spheres (*beads*) as building elements. The spherical elements may correspond to real subunits in the assembly, although this is not necessarily the case. Actually, strings of beads had been used in polymer hydrodynamics to represent rod-like particles and flexible chains. The rationale behind bead models is that the hydrodynamics of a rigid assembly of can be described in terms of the hydrodynamic interaction concept. If the ith bead in a model of N beads is moving with velocity u_i in a solvent that would be otherwise at rest, then the force that it exerts on the solvent is given by

$$F_i = \zeta_i u_i - \sum_i T_{ij} F_j \qquad i, j = 1,2, ..., N; j \neq i \qquad (21)$$

The first term in the right hand side of eq. (21) is just the force that would result in the absence of the other spheres, given by the product of the bead velocity and the friction coefficient of the sphere. The latter is formulated as $\zeta_i = 6\pi\eta\sigma_i$, where σ_i is the radius of the sphere (as in eq. (6)). The second term is a perturbation produced by the hydrodynamic interaction between all the beads in the model, and is expressed in terms of the so-called hydrodynamic interaction tensor, T_{ij}, which depends only on the distance R_{ij} between beads.

In their early work, Bloomfield *et al.*[22] made a simplifying approximation in the treatment of eq. (21) that yields a simple formula for the frictional coefficient:

$$f = \frac{\sum_i \zeta_i}{1 + (6\pi\eta \sum_i \zeta_i)^{-1} \sum_{i,j} \zeta_i \zeta_j R_{ij}} \qquad i, j = 1, ..., n; i \neq j \qquad (22)$$

The double sum in eq. (22) requires little computational effort, and the whole equation is easily programmed. There is a slightly different version of eq. (22), due also to Bloomfield *et al.*, which is appropriate for models with beads of rather different sizes. Furthermore, another double-sum formula which includes improvements in the treatment of hydrodynamic interaction has been more recently proposed[24]. In any case, all the double-sum formulae embody approximations which may make their result somewhat erroneous. Therefore a rigorous treatment was desirable, and it was developed by Bloomfield and the present author[25-27].

If one defines B_{ij} so that B_{ii} is equal to a unit tensor, while $B_{ij}=T_{ij}$ when $i \neq j$, then eq. (21) can be rewritten as

$$\sum_i B_{ij} F_j = \zeta_i u_i \qquad (23)$$

according to which the forces are the solutions of a system of linear equations whose coefficients, the B_{ij}, depend just on the geometry of the model, *i.e.* on the size of the elements and their positions. For the particular case of the sedimentation experiment, all the bead velocities, u_i, are identical to the velocity of sedimentation.

The solution for eq. (23) is achieved by inverting the matrix of coefficients B which contains N×N blocks, the B_{ij}. If we denote the blocks of the inverse, $C=B^{-1}$, as C_{ij}, then the frictional coefficient can be calculated as

$$f = \frac{1}{3} \mathrm{Tr} \left\{ \sum_j C_{ij} \right\} \tag{24}$$

where Tr indicates the sum of the diagonal elements of a tensor. Then, from f, the sedimentation and diffusion coefficients can be calculated as usual. Further information about bead model theory, as its use for the prediction of other properties, can be found in the original references[15,25-28].

The increasing availability of computing power makes the rigorous calculation of hydrodynamic properties using the rigorous treatment feasible, so that approximate double-sum formulae such as eq. (22) are becoming obsolescent. Models with up to 100 beads can be easily calculated on modern personal computers, and the power of mainframes or vectorial computers may extend the range to the thousands. The input for the calculations is just a list of the co-ordinates and radii of the beads making up the model. A computer program that carries out the calculation of the translation (friction or sedimentation) coefficients, rotational coefficients and relaxation times, and the intrinsic viscosity has been presented in a previous review[29]. Fig. 3 shows typical output from the computer program, that corresponds in this case to a model for an immunoglobulin molecule. The output contains the input information (co-ordinates and sizes of the beads) along with the calculated properties, including translational coefficients, rotational relaxation times, intrinsic viscosity and others.

Applications

The published applications of bead model calculations of hydrodynamic properties are numerous. Usually, bead model treatments of biological structures are very specific, with details corresponding to the problem under study. So instead of a comprehensive review of applications, we just present here comments on two of them which are of particular interest.

The first one corresponds to the bacteriophage structures introduced above (see Fig. 2). It was known that, depending on pH, temperature and ionic strength, these biological entities are able to adopt two different conformations which were clearly differentiated in the ultracentrifuge. They are the *fast* form, which has the larger sedimentation coefficient and sediments faster, and the *slow* form, with the opposite characteristics. Fig. 2 shows bead models of T2 bacteriophage with their overall form based on electron micrographs. It was speculated whether the slow-fast transition was due to a change in the size of the head or a change in the disposition of the tailfibres. Bead model calculations[30] showed that the first hypothesis requires absolutely unrealistic values of the head size, and it was therefore disregarded. On the other hand, it was possible to predict quantitatively the change in sedimentation coefficients

```
SUBROUTINE TRV HAS ENTERED
VALUES OF THE CONTROL PARAMETERS
 IND   IP
   1    1

CARTESIAN COORDINATES AND RADII OF THE ELEMENTS
  1     -12.9000        0.0000        0.0000       2.0500
  2      -8.8000        0.0000        0.0000       2.0500
  3      -6.0000        0.0000        0.0000       0.7500
  4      -4.5000        0.0000        0.0000       0.7500
  5      -3.0000        0.0000        0.0000       0.7500
  6      -1.5000        0.0000        0.0000       0.7500
  7       0.0000        0.0000        0.0000       0.7500
  8       0.5000        0.0000        0.8660       0.2500
  9       0.7500        0.0000        1.2990       0.2500
 10       0.5000        0.0000       -0.8660       0.2500
 11       0.7500        0.0000       -1.2990       0.2500
 12       1.9000        0.0000       -3.2909       2.0500
 13       3.9500       -3.5507       -3.2909       2.0500
 14       1.9000        0.0000        3.2909       2.0500
 15       3.9500        3.5507        3.2909       2.0500

VOLUME=     225.62  RADIUS OF GYRATION=     7.4919

CENTER OF MASS (UNIFORM DENSITY)        -1.7162     0.0000      0.0000

CENTER OF DIFFUSION       -2.4717     0.0000      0.0000

TRANSLATIONAL DIFFUSION TENSOR
       0.189576       0.000000        0.000000
       0.000000       0.166808        0.005084
       0.000000       0.005084        0.169053

TRANSLATIONAL DIFFUSION COEFF.=    0.1751
TRANSLATIONAL FRICTION  COEFF.=    5.7095

ROTATIONAL DIFFUSION TENSOR
       0.004777       0.000000        0.000000
       0.000000       0.001757        0.000106
       0.000000       0.000106        0.001782

COUPLING TENSOR (DIFFUSION)
      -0.000425       0.000000        0.000000
       0.000000      -0.000121        0.000004
       0.000000       0.000004        0.000258

EIGENVALUES OF THE ROTATIONAL DIFFUSION TENSOR
  0.00166    0.00188    0.00478

TRACE OF ROT.DIF.TEN.=    0.00277
ANISOTROPY OF ROT.DIF.TEN.=     0.00301

RELAXATION CONSTANSTS
INDEX                    1            2            3            4            5
                     6D-2DEL      3D+3DI       3D+3D2       3D+3D3      6D+2DEL
ROTATIONAL CONST     0.01061      0.01331      0.01394      0.02265      0.02266
RELAXATION TIMES    94.27736     75.15629     71.71179     44.15849     44.13645

VISCOSITY CENTER        -2.2579     0.0000     0.0000

INTRINSIC VISCOSITY      2.577E+03

SCHERAGA-MANDELKERN PARAMETER        2.3177E+06
```

Figure 3. Model (two views) for a IgG3 antibody molecule, and output from the computer program for calculating hydrodynamic properties.

assuming a conformational change for the fibres, which would be extended in the slow form, and retracted near the head in the fast form. Considering that the fibres make a very minor fraction of the total mass of the phage particle, the change that they produce in the hydrodynamic behaviour and particularly in the sedimentation coefficients is really noteworthy.

The other application of bead modelling that is worth commenting on here is the analysis of the hydrodynamic properties of oligomeric structures made up of a few subunits of roughly spherical shapes: oligomeric proteins with well defined quaternary structure are typical examples[31]. We have calculated hydrodynamic properties of polygonal and polyhedral arrays of beads, with varying number of beads and different geometries: some results are listed in Table 1. Given the molecular weights of the oligomer, M_n and that of the monomer, M_1 (which may be accessible, for instance, from sedimentation equilibrium), then the number of subunits can be estimated as $n=M_n/M_1$: then from the ratio of sedimentation coefficients, s_n/s_1, the shape can be predicted. If sedimentation data for the monomeric subunit is unknown, one can use alternatively the f/f_0 value (from eqs. (6-8)), for which computed values are also listed in Table 1. This type of analysis for oligomeric structures can be made more comprehensive if other properties like rotational diffusion and intrinsic viscosity are included[28-29].

Table 1. Sedimentation, diffusion and frictional ratios of oligomeric structures

Structure	s_n/s_1	D_n/D_1	f_n/f_1	$(f/f_0)_n$
Monomer				
Sphere	1.0	1.0	1.0	1.0
Dimer				
Linear	1.50	0.75	1.33	1.06
Trimers				
Triangle	2.00	0.66	1.50	1.04
Linear	1.80	0.60	1.67	1.16
Tetramers				
Square	2.32	0.58	1.72	1.08
Tetrahedron	2.44	0.61	1.64	1.03
Linear	2.04	0.51	1.96	1.23
Pentamers				
Pentagon	2.60	0.52	1.92	1.12
Bipyramid	2.85	0.57	1.75	1.03
Hexamers				
Hexagon	2.82	0.47	2.13	1.17
Octahedron	3.24	0.54	1.85	1.02
Trigonal prism	3.12	0.52	1.92	1.06
Linear	2.40	0.40	2.50	1.38
Octamer				
Cube	3.68	0.46	2.17	1.09

ACKNOWLEDGMENT

The present work of the author is supported by grant PB87/0694 from the Direccion General de Investigacion Cientifica y Tecnica.

REFERENCES

1. H.K. Schachman, 'Ultracentrifugation in Biochemistry', Academic, New York, 1959.
2. K.E. Van Holde, 'Physical Biochemistry', Prentice-Hall, Englewood Clifs, 1971, Chapter 5.
3. C.R. Cantor and P.R. Schimmel, 'Biophysical Chemistry', Freeman, San Francisco, 1980, Chapter 11.
4. E.G. Richards, 'An Introduction to the Physical Properties of Large Molecules in Solution', Cambridge University Press, Cambridge, 1980, Chapter 8.
5. R. Pecora, ed., 'Dynamic Light Scattering', Plenum, New York, 1985.
6. B.J. Berne and R. Pecora, 'Dynamic Light Scattering', Wiley-Interscience, New York, 1976.
7. F. Perrin, *J. Phys. Radium*, 1936, *7*, 1.
8. S.E. Harding and A.J. Rowe, *Biopolymers*, 1983, *22*, 1813 (see also *ibid.*, 1984, *23*, 845).
9. S.E. Harding and A.J. Rowe, *Int. J. Biol. Macromol.*, 1982, *4*, 357.
10. S.E. Harding and A.J. Rowe, *Int. J. Biol. Macromol.*, 1982, *4*, 160.
11. S.E. Harding, in 'Dynamic Properties of Biomolecular Assemblies', S.E. Harding and A.J. Rowe, eds., Royal Society of Chemistry, Cambridge, 1989, p. 32.
12. S. Broersma, *J. Chem. Phys.*, 1960, *32*, 1626.
13. S. Broersma, *J. Chem. Phys.*, 1981, *74*, 6989.
14. M.M. Tirado and J. García de la Torre, *J. Chem. Phys.*, 1979, *71*, 2581.
15. J. García de la Torre and V.A. Bloomfield, *Q. Rev. Biophys.*, 1981, *14*, 81.
16. H. Yamakawa, *Macromolecules*, 1983, *16*, 1928.
17. J. García de la Torre, M.C. López Martínez and M.M. Tirado, *Biopolymers*, 1984, *23*, 611.
18. P.M. Bayley and R.E. Dale, eds., 'Spectroscopy and the Dynamics of Biological Molecular Systems', Academic, London, 1985.
19. J.J. Freire and J. García de la Torre, Chapter 19, this volume.
20. M.M. Tirado, M.C. López Martínez and J. García de la Torre, *J. Chem. Phys.*, 1984, *81*, 2047.
21. H.E. Hook, D. Beard, A.R. Taylor, D.G. Sharp and J.W. Beard, *J. Biol. Chem.*, 1946, *165*, 241, 259.
22. V.A. Bloomfield, W.O. Dalton and K.E. Van Holde, *Biopolymers*, 1967, *5*, 135.
23. V.A. Bloomfield, K.E. Van Holde and W.O. Dalton, *Biopolymers*, 1967, *5*, 149.
24. C. de Häen, R.A. Easterly and D.C. Teller, *Biopolymers*, 1983, *22*, 1133.
25. J. García de la Torre and V.A. Bloomfield, *Biopolymers*, 1977, *16*, 1747.

26. J. García de la Torre and V.A. Bloomfield, *Biopolymers*, 1977, *16*, 1779.
27. J. García de la Torre and V.A. Bloomfield, *Biopolymers*, 1978, *17*, 1605.
28. J.M. García Bernal and J. García de la Torre, *Biopolymers*, 1980, *19*, 751.
29. J. García de la Torre, in 'Dynamic Properties of Biomolecular Assemblies', S.E. Harding and A. J. Rowe, eds., Royal Society of Chemistry, Cambridge, 1989, p. 1.
30. J. García de la Torre and V.A. Bloomfield, *Biopolymers* 1977, *16*, 1779.
31. I.M. Klotz, R.R. Langerman and D.W. Darnall, *Annu. Rev. Biochem.*, 1970, *39*, 25.

19

Sedimentation Coefficients of Flexible Chain Polymers

By Juan J. Freire and [1]José García de la Torre

DEPARTAMENTO DE QUIMICA FISICA FACULTAD DE CIENCIAS QUIMICAS UNIVERSIDAD COMPLUTENSE 28040 MADRID, SPAIN

[1]DEPARTAMENTO DE QUIMICA FISICA FACULTAD DE CIENCIAS QUIMICAS Y MATEMATICAS UNIVERSIDAD DE MURCIA 30071 MURCIA, SPAIN

1. INTRODUCTION

Many polymers and biological macromolecules have a flexible molecular structure, due to the many degrees of internal rotational freedom about the bonds that constitute their backbones. Flexibility is, therefore, one of the most important characteristics which determine the properties of these molecules and, consequently, it has a crucial role in the interpretation of sedimentation coefficient measurements determined by means of the technique of sedimentation velocity in the analytical ultracentrifuge. As described in the previous Chapter of this book[1], sedimentation velocity yields the sedimentation coefficient, s, which is directly related to molecular properties such as molecular weight, M and the frictional coefficient through the relation $s = M(1-\bar{v}\rho)/N_A f$, where $(1-\bar{v}\rho)$ is the buoyancy factor and N_A is the Avogadro number. In the present Chapter we assume that the macromolecular sample is composed of chains with the same molecular weight or with narrow polydispersity so that a single value of M is used. If M is known through alternative methods (such as sedimentation equilibrium or light scattering experiments) the determination of s serves to obtain f, directly related to the mean size and shape of the chain. On the other hand, the technique allows us to obtain M, if we have an independent experimental estimation of f, usually through measurements of the diffusion coefficient in dilute solutions, D, which (as also explained in ref.1), is also inversely proportional to f through the relation $D=k_BT/f$, where k_BT is the Boltzmann factor.

The main problem for a flexible chain is that the directly related properties s, f and D reflect the behaviour of macromolecules which are continuously changing their form, as a consequence of the dynamical changes in the rotational angles which determine their instantaneous conformations. This constitutes a formidable theoretical problem, which, however, is in part alleviated because the dynamical description of long flexible chains is dominated by their large scale motions[2] so that this dynamics exhibits many universal features. Therefore, the chains admit very simple models so that the macroscopic properties only depend on a few molecular parameters.

2. MODELS

Unperturbed Chains

We can consider the polymer as composed of N segments, each one of them encompassing a certain number of real repeating units or bonds high enough to contain many rotational degrees of freedom. Then, the well known *central limit theorem* in statistics can be employed to show that the components of the vectors joining the ends of the different segments should follow a Gaussian distribution. Consequently, the chain can be envisaged as composed of N statistical segments, represented by their end-to-end vectors, of random orientations, whose lengths are Gaussianly distributed with a given root-mean-square (or statistical segment length) which we will denote as b. Of course, this Gaussian model cannot describe the local structure of the chain which depends on the bond lengths, bond angles and torsional energies of the rotational angles: more elaborate models, such as the rotational isomeric state model[3], should be employed for this purpose. In the Gaussian model, the short-range description of a particular chain is incorporated into the model only through the global parameter b. However, this model is adequate to reproduce large length-scale properties such as the frictional or sedimentation coefficients in cases where only short-range interactions need to be included. The most recent theoretical descriptions of these simple flexible chains are based on a continuous equivalent of the Gaussian model which makes use of the Wiener distribution[4]. In any case, the Gaussian model and its equivalents lead to simple analytical expressions for equilibrium properties. Thus, if one considers an appropriate characterization of the chain dimensions by means of the root-mean-square radius of gyration R_g (an average of the distance between the chain units and the centre of masses which can be directly measured from light scattering experiments at different angles), the Gaussian model leads to the following theoretical expression for long linear chains

$$R_g = \left\{\frac{N^{1/2}\ b}{6^{1/2}}\right\} \tag{1}$$

On the other hand, the transport or hydrodynamic properties are much more difficult to predict (we will give a summary of theoretical treatments in the next section). Nevertheless, we can now obtain a very crude prediction for the frictional coefficient by considering the chain as a sphere with an equivalent hydrodynamic radius R. Since, according to the Stokes law, $f=6\pi\eta R_h$, where η is the solvent viscosity, if one applies the relationship between R_g and the radius of a rigid sphere (in this case R_h)

$$R_g = (3/5)^{1/2}\ R_h \tag{2}$$

then one gets $f \sim N^{1/2}$, which is confirmed with more elaborate treatments, though this simple rigid sphere hydrodynamics is not accurate from the quantitative point of view.

Excluded Volume and Theta Conditions

The main criticism of the Gaussian model description of long flexible polymers is that it only considers short-range interactions. Real flexible chains, however, also exhibit interactions between units separated by many bonds but spatially close in a given conformation. Though these long-range interactions scarcely have an individual importance, they exert a great collective influence on the polymer size and other conformational properties. This influence is directly related to the thermodynamic solvent conditions. If the solvent is good, the polymer segments tend to disperse opposing mutual interpenetration. An adequate model for these "good" solvent conditions is to consider the chain segments as hard spheres which cannot coexist in a common space region[2,5]. Then, the model predicts an excluded volume effect which increases the chain size. A theoretical perturbation treatment[4,6] of this model shows that the appropriate parameter to describe the excluded volume is $z \sim \beta_S N^{1/2}$, where β_S represents the binary interactions between pairs of units (positive for hard spheres). An approximate dependence of β_S on temperature has the form $\beta_S \sim a_\beta + b_\beta/T$ (a_β and b_β depend on the particular polymer-solvent system) and, consequently, z takes into account simultaneously both chain length and thermodynamic conditions.

In poorer solvent systems, the segments tend to interpenetrate to expel part of the solvent: then, β_S can be small or even negative. For a given theta temperature, $T = \theta$, the binary interactions are compensated so that $\beta_S = 0$. Consequently, a polymer in the theta conditions behaves as if its binary long-range interactions were absent and can be more properly described by the Gaussian model.

The perturbation approach to the excluded volume problem leads to expansions of the type

$$\alpha^2 = \frac{R_g^2}{(R_g^2)_\theta} = 1 + cz + \ldots \tag{3}$$

where α is called the expansion coefficient, the subscript θ will be used to denote the value of a property for the same chain in the unperturbed conditions and c is a numerical constant. Though the numerical value of c and the coefficients of several other terms in the expansion are now known, the series has been shown to be divergent for moderate (realistic) values of z.

A more recent approach to the excluded volume problem is based on the application of the renormalization group theory[2,4,7]. This sophisticated treatment leads to universal scale laws, valid for long excluded volume chains (*i.e.* for $z \gg 1$). Thus, it has been shown that

$$R_g \sim N^\nu \tag{4}$$

where ν is a critical exponent whose value is $\nu = 0.588$ in three-dimensional systems. If one admits that in excluded volume chains R_g and R_h are still directly proportional, it follows that $f \sim N^\nu$.

3. LINEAR CHAINS

Theta Conditions

The hydrodynamic treatment of "equivalent hard spheres" is not able to give accurate quantitative results for linear polymers. Thus, if one defines $\rho = R_g/R_h$, eqs. (1) and (2) yield $\rho = 0.775$ for long flexible unperturbed linear chains. Although the relationships $D \sim s \sim N^{-1/2}$ have been extensively verified for different samples of long linear flexible chains in theta solvents, the most accurate experimental measurements of both R_g and R_h for practically monodisperse samples give[8] $\rho \approx 1.28$. Therefore, it seems apparent that more adequate hydrodynamic treatments should be introduced. A good estimation of ρ is important because it allows the use of the sedimentation velocity (or diffusion coefficient) measurements in determining R_g. Furthermore, this determination of R_g is very pertinent in order to understand the relationship between properties and molecular structure. Thus, according to eq. (1), $R_g/N^{1/2}$ should tend to a constant value for long chains and this constant or equivalent parameters (*e.g.* the parameter b) can be theoretically obtained through elaborate calculations which take into account the detailed local structure (for instance, using the rotational isomeric state model[3]). Then, the comparison between these theoretical calculations and results obtained from experimental properties is a useful tool to test the different assumptions introduced in the detailed models, as the description of the torsional angles and the energies of internal rotation barriers.

More complete hydrodynamic theories for flexible chain molecules in unperturbed conditions are based on dynamical equivalents of the Gaussian model where the statistical segments are also considered to be frictional beads, according to the bead model approach discussed in the previous Chapter[1]. In this case an additional parameter should be introduced in order to describe the ratio of the frictional radius of the bead to its statistical length, σ/b, which gauges the strength of the hydrodynamic interactions between beads[6]. The relevant parameter in the theory is, however, proportional to $N^{1/2}(\sigma/b)$ and, therefore, the properties turn out to be insensitive to the actual value of σ, provided that it is not very small and N is high enough (*i.e.* when we are considering long non-draining chains).

There is no rigorous dynamic theory for flexible polymers, due to the difficulties inherent in describing hydrodynamic interactions which fluctuate due to the chain flexibility. The standard existing theories make use of approximate descriptions where these interactions are pre-averaged[4,6]. In its simpler form, the pre-averaged theory gives $\rho = 1.48$. Another simple theoretical prediction is performed with approximate double-sum formulae of the type described in the preceding Chapter[1]. Thus, the popular Kirkwood double-sum formula leads to a value for ρ which does not differ much from the pre-averaged result[6]. Further progress can only be achieved through numerical (simulation) work. The pertinent simulation method consists in generating Brownian dynamics trajectories which take into account the connectivity and the frictional forces between units, together with a stochastic force[9]. These simulations are, however, very costly from the computational point of view, so that a precise asymptotic limit for very long chains is not feasible. An alternative procedure is to perform Monte Carlo calculations on equilibrium samples constituted by instantaneously rigid conformations[10]. These conformations are then treated with the techniques described in ref. 1 for rigid bodies (rigid body approximation), obtaining a lower bound estimation for ρ. This way, the value $\rho = 1.28$ is found by extrapolation for long chains. On the other hand, the pre-averaged value constitutes a rough upper bound for ρ. An improved rigid body calculation based on variational methods[11] provides $\rho = 1.38$ as the upper bound for very long chains.

Moreover, we can also consider the Scheraga-Mandelkern function, β, defined in the previous Chapter[1], which combines frictional properties, viscosity and molecular weight. This function is assumed to be weakly dependent on the molecular weight (as the ratio ρ), and also practically independent of the polymer shape. Therefore, a good estimation of β can serve to obtain the molecular weight from independent determinations of both the sedimentation coefficient (or the diffusion coefficient) and the viscosity. The most accurate experimental measurements of these properties for long monodisperse samples of flexible chains in theta conditions[8,12] yield $\rho = 2.26\times10^6$, close to the rigid sphere value ($\beta = 2.12\times10^6$). However, the simple pre-averaged theory gives a poorer value $\beta = 2.7\times10^6$. Upper and lower Monte Carlo calculations for viscosity and frictional properties lead to results for β which are finally in good accordance with the experimental data[10].

Excluded Volume

If the chain is immersed in a good solvent, the experimental determination of $R_g/N^{1/2}$ cannot be directly compared with theoretical predictions, since it turns out now to be N-dependent and, moreover, the realistic models which include short-range interactions (such as the rotational isomeric state model) do not incorporate the excluded volume effects. However, if the excluded volume is not large (*i.e.* for small values of z) the perturbation approach can be useful to avoid this problem. In this case, a frictional expansion coefficient (defined as $\alpha_f = f/f_\theta$) can be expanded as[13]

$$\alpha_f^2 = (1 + c_f z + \dots) \tag{5}$$

Since both f_θ and z are proportional to $N^{1/2}$ (or $M^{1/2}$), the representation of experimental data of $f/M^{1/2}$ against $M^{1/2}$ and further extrapolation to $M\rightarrow0$ provides an estimation of f_θ from which the equivalent unperturbed dimensions $(R_g)_\theta$ can be obtained. A similar scheme is usually employed in the treatment of intrinsic viscosity data of flexible polymers with moderate excluded volume effects (the "Stockmayer-Fixman" representation[6]) and can be confidently applied to measurements of s or D, properties which are less affected by excluded volume (the intrinsic viscosity is roughly proportional to R_g^3, and, consequently it reflects more strongly the chain expansion effects).

For higher excluded volume effects ($z \gg 1$) we know already that $f \sim M^\nu$ if ρ can be still considered as a constant. Some experimental results[14] suggest, however, a scaling law exponent smaller than ν. Adequate hydrodynamical chain models for this problem can be based on bead models used for unperturbed flexible chains, but require allowance for long-range interactions between non-neighbouring units by means of simple potentials (hard spheres or Lennard-Jones). The Monte Carlo rigid-body procedures are then able to yield upper or lower bounds for the different hydrodynamic properties[15]. A second-order renormalization group theory treatment has been also carried out[16], using the continuous equivalent of these types of models and the pre-averaged formulae. Both theoretical and simulations results show that ρ increases with the chain expansion, from its value for an unperturbed chain,

$\rho \approx 1.35$, to a higher asymptotic limit close to $\rho = 1.5$. This feature seems to be caused by a coupling between hydrodynamic interactions and excluded volume effects since the chain expansion produces an increase of its permeability. In fact, the asymptotic limit is reached very slowly for partial draining chains, *i.e.* for chains with small values of σ/b, which explains the small apparent exponents found in many real chains for the dependence of f on the molecular weight.

4. NON-LINEAR CHAINS

Branched Chains

Many polymers contain topological branching points in their chemical contour. The branched structures can adopt many different forms and, therefore, the influence of branching on the molecular properties needs to be understood in order to establish a systematic form of characterisation of the branching architecture based on experimental data and their analysis in terms of universal quantities. Thus one can consider ratios or combinations of these magnitudes, as ρ or β, which should not depend much on the chain length or short-range details, and study how they vary with the chain topology. Alternatively, it is possible to define ratios of a given property to its value for the equivalent linear structure with the same chain length and solvent conditions as

$$g = \frac{(R_g{}^b)^2}{(R_g{}^l)^2} \tag{6}$$

$$h = \frac{f^b}{f^l} \tag{7}$$

and

$$g' = \frac{[\eta]^b}{[\eta]^l} \tag{8}$$

where the superscripts b and l denote the branched and linear equivalent chains and $[\eta]$ is the intrinsic viscosity. Since the calculation of equilibrium properties such as R_g is much simpler than the prediction of hydrodynamic properties, there have been attempts to establish relationships between these ratios[6,17]. Thus, Kurata and Fukatsu[17] have made use of double-sum formulae with bead models of unperturbed chains built so that the connectivity between beads represent different types of architectures. They found that $1 < h/g^{1/2} < 1.39$. Using the standard pre-averaged theory for flexible star chains (*i.e.* for chains with a central unit and a certain number, F, of flexible arms), Zimm and Kilb[17] obtained results compatible with $g' = g^{1/2}$. Stockmayer and Fixman[17] suggested $g = h^3$. However, a systematic analysis for different star chains and comb-like structures (where different branching points are aligned along a main backbone contour) based on the pre-averaged treatment[18] points out that these relationships are not reliable in many cases.

The uniform flexible stars (*i.e.* with F arms containing equal number of units) are very interesting molecules since the coexistence of flexible arms and a more compact central core poses a challenge in the choice of models and hydrodynamic treatments while, on the other hand, the chain structure is completely defined with F and N. Moreover, adequate synthesis procedures have been designed for these types of branched polymers[19]. The experimental data for ratio g in theta conditions are consistently greater than the theoretical prediction based in unperturbed chains with the precise type of connectivity between units when F > 6. This effect is caused by the expansion of the arms due to their mutual repulsion in the region close to the core. Therefore, it seems that an adequate model for these chains in the theta state should include a long-range potential with both repulsive and attractive components. Furthermore, the theoretical values of hydrodynamic properties are shown to be more sensitive to the use of approximations than in the case of linear chains. All these problems are extensively discussed in a recent review paper[20].

Therefore, it seems that only those simulation procedures which include a proper description of long-range and hydrodynamic interactions can reproduce correctly the transport properties of star chains. In fact, recent elaborate rigid-body calculations which incorporate these characteristics have yielded results in good agreement with the experimental ratios for both the theta[21] and good solvent[15] conditions (these different conditions are represented by a Lennard-Jones potential whose parameters have been set so that $R \sim N^{1/2}$, or $R_g \sim N^{\nu}$, for sufficiently long chains). A very interesting feature shown by the different experimental and simulation ratios for different values of F is that the theoretical predictions for unperturbed chain (Gaussian) model and adequate hydrodynamic treatment are significantly different from those obtained for the theta conditions, but very close to the results for good solvents. (See a summary of all these values for h and g in Table 1).

Table 1. Theoretical predictions and experimental data for h and g'.

(a) Ratio h

Model	Hydrodynamic treatment	F = 6	F = 12	F = 18
Gaussian	Pre-averaged	0.86	0.73	0.65
Gaussian	Rigid body	0.86	0.75	0.66
LJ-θ	Rigid body	0.89	0.82	0.73
LJ-EV	Rigid body	0.89	0.78	0.66
Experimental θ		0.89	0.74-0.85	0.72-0.76
Experimental GS		0.86	0.70-0.83	0.64-0.68

(b) Ratio g'

Model	Hydrodynamic treatment	F = 6	F = 12	F = 18
Gaussian	Pre-averaged	0.69	0.51	0.42
Gaussian	Rigid body	0.58	0.38	0.24
LJ-θ	Rigid body	0.59	0.39	0.28
LJ-EV	Rigid body	0.58	0.37	0.22
Experimental θ		0.63	0.41	0.31-0.35
Experimental GS		0.58	0.35	0.22-0.26

θ: Theta conditions. GS: Good solvent conditions. LJ: model with Lennard-Jones potential. The numerical values were reported or reviewed in refs. 15, 20-21.

The ratio ρ decreases as the number of arms in the chain increases. Thus, the simulation results for a chain with F = 18 are very close to the rigid sphere limit given by eq. (2). The numerical values of ρ obtained from the simulations[15,21] are also in good agreement with experimental data. As in the case of linear chains ρ suffers a slight decrease as the chain expands when the solving conditions improve, until it reaches an asymptotic limit.

Rings

Cyclic polymers have an important role in the theoretical developments on the behaviour of flexible chains, since the absence of end effects give rise to significant simplifications in some treatments. Moreover, they can be present in many interesting processes[22]. Further recent advances of experimental techniques - particularly in sedimentation velocity analysis - makes it easier to provide an adequate characterization of these structures.

As in the case of branched chains, the dimensions and hydrodynamic properties are often described by their ratios to the equivalent linear chains. Thus, one can define

$$q_s = \frac{(R_g{}^c)^2}{(R_g{}^l)^2} \tag{9}$$

$$q_f = \frac{f^c}{f^l} \tag{10}$$

$$q_\eta = \frac{[\eta]^c}{[\eta]^l} \tag{11}$$

here the subscript c refers to the cyclic structure. The calculation of these ratios for unperturbed, or Gaussian, rings, is relatively simple when pre-averaged hydrodynamic interactions are used in the evaluation of hydrodynamic properties[6]. The results are shown in Table 2, together with a selection of experimental data for theta solvents. We also present new theoretical results obtained through rigid-body Monte Carlo simulations which avoid the pre-averaging approximation[23]. The discussion of these results is somewhat complicated by the fact that the experimental theta temperature for a given cyclic polymer-solvent system differs appreciably from the one corresponding to the linear chain, but these differences are not usually considered in the determination of the ratios. Therefore, we consider more adequate the comparison of experimental and theoretical results for combined properties of a given cyclic structure, such as those relating to the ratio ρ. The Monte Carlo results yield ρ =1.07 in good agreement with most experimental values, while the pre-averaged theory predicts ρ =1.25.

We also include in Table 2 Monte Carlo results for the ratios obtained with a model that takes into account excluded volume effects[23] and experimental data for good solvents. The agreement between the predicted and experimental values is good. The simulation results point out again the decrease of ρ with improving solvent conditions also observed for linear and star chains. The asymptotic limit for cyclic chains with excluded volume is calculated to be close to ρ =1.18.

Table 2. Ratios of cyclic to linear chain properties.

Model	Hydrodynamic treatment	q_s	q_f	q_η
Gaussian	Pre-averaged	0.50	0.85	0.66
Gaussian	Rigid body	0.50	0.86	0.55
LJ-θ	Rigid body	0.57	0.90	0.60
LJ-EV	Rigid body	0.53	0.89	0.59
Experimental θ		0.50	0.84-0.89	0.60-0.71
Experimental GS		0.52-0.55	0.83-0.90	0.56-0.79

See Table 1 for notation. The theoretical and experimental results are reported or reviewed in refs. 20 and 23.

The experimental data and rigid-body simulation values of β for non-linear chains (stars and rings) and in different solving conditions are very close to those corresponding to unperturbed linear chains. However, the pre-averaged theory yields significant differences for different types of chains, due to deficiencies in the hydrodynamic treatment.

5. SEMIFLEXIBLE POLYMERS

The chain models considered above actually correspond to a *limiting* situation, applicable to a variety of polymers, in which flexibility is manifested in a length scale of the order of a few bonds. Thus, the correlation between the orientations (or the dynamics) of two parts of the chain decreases strongly as their separation increases. The other limiting situation in that regard is that of cylindrical, rodlike particles, whose sedimentation behaviour was described in ref.1. In this case, the chain is absolutely rigid over all its contour length and, apart from the detail of the cross-sectional structure, their overall conformation can be represented by a straight cylinder.

There is a variety of polymers with an intermediate behaviour: among *synthetic polymers*, typical examples are those of poly(alkyl)isocyanates[24] and poly(γ-benzyl, L-glutamate)[25]. It is in the field of *biological macromolecules* however where semiflexible macromolecules are most frequent and relevant. In many instances, the rigidity is associated with a helical structure, as it happens, for instance, in the cases of collagen[26] and xanthan[27]. However, the archetypal example of a semiflexible biopolymer is DNA[28]. The double helical structure of DNA is essentially rigid over a contour length corresponding to a few hundred base pairs. This was indeed described in the preceding article[1] as an example of behaviour in the rodlike limit. When the DNA is longer, deviation from this behaviour takes place. DNAs much longer than that, with many thousand base pairs and molecular weights as large as 10^7 are far from the rodlike conformation, but their solution properties are still *not* described by the limit of the random coil.

The conformation and hydrodynamics of semiflexible chains is usually represented by the wormlike chain[28,29], in which the bending flexibility of the chain is represented by one parameter, the persistence length, a. Apart from its precise definition, a gives an indication of the length scale for which correlation between separate parts of the chain begin to disappear, as commented above. The persistence length gives an intrinsic or intensive measurement of flexibility; it takes a given value for a given macromolecule (say, DNA) *independent* of chain length or molecular weight. The actual conformation and behaviour of a macromolecular chain depends on the ratio of its contour length, L to the persistence length, a. If $L/a \ll 1$, the conformation is rodlike, and the hydrodynamic properties follow the equations for cylinders. If $L/a \gg 1$, the random coil limit is recovered. This is well illustrated by the chain-length dependence of the radius of gyration, given by

$$R_g^2 = \frac{La}{3}\left\{1 - \frac{3a}{L} + \frac{6a^2}{L^2} - \frac{6a^3}{L^3}(1-e^{-L/a})\right\} \tag{12}$$

When a is very small, or L is very large, so that $a/L \approx 0$, we have R_g^2 proportional to $L^{1/2}$, as in eq. (1). In the opposite case, with $a/L \gg 1$, the limit is $R_g = L/\sqrt{12}$, corresponding to a straight rod.

The hydrodynamic properties of wormlike macromolecules follow a similar trend, which varies smoothly from the results for cylinders described in the previous Chapter1 to those described in the present one for flexible chains. The sedimentation coefficient of wormlike chains was first calculated using some approximations in both conformational and dynamic aspects in a classical paper by Hearst and Stockmayer[30], and a somewhat improved treatment was later employed by Yamakawa and Fujii[31]. These theories give results that are essentially correct for wormlike chains with behaviour closer to the random coil that to the rod. Namely, the Yamakawa-Fujii treatment gives a series expansion of s in powers of $(L/2a)^{-1/2}$,

$$s = \frac{M(1-\bar{v}\rho)}{3\pi\eta N_A L}\left[1.843(L/2a)^{1/2} + A_2 + A_3(L/2a)^{-1/2} + ...\right] \tag{13}$$

where the numerical coefficients have limiting values when d « a (d is the mean diameter or cross-section of the chain): $A_2 = -\ln(d/2a)$, $A_3 = 0.1382$,... . In the original reference[31], higher terms in the series expansion are formulated as functions of d/2a. Similar results are available for the intrinsic viscosity[32]. On the other hand, when the hydrodynamic behaviour is closer to the rod limit, so that the properties show not too large a deviation with respect to the rod values, previous theories for wormlike chains are affected by the same deficiencies that provided the motivation for improvements in the theories for rigid cylinders.

Recently, García Molina *et al.*[33] have presented a study that combines the best results for rigid cylinders with a Monte Carlo simulation of the statistical aspects arising from flexibility. Indeed, the procedure employed the rigid-body treatment that we described above. The result for the sedimentation coefficient is

$$s = \frac{X_t\, M(1-\bar{v}\rho)}{6\pi\eta N_A R_g} \ln\left\{\frac{L}{d}\right\} \tag{14}$$

where R_g is given by eq. (12) and X_t is a numerical factor that depends weakly on L/a and L/d: it goes from $X_t = 0.64$ for L/a=0 and L/d=10 to $X_t = 0.45$ for L/a=10 and L/d=50. Expressions for X_t are given in ref. 33.

The recent availability of restriction fragments of DNA has made it possible to avoid problems in data analysis related to molecular weight characterization and polydispersity. A convenient set of data, for a 0.1 M salt concentration, is that reported by Kovacic and Van Holde[34], which covers the region from 50 to 1000 base pairs. Taking d = 25Å (as deduced in the rod region), García Molina *et al.* fitted those data to their theoretical results, obtaining a persistence length of 410 Å, in good agreement with other predictions at that ionic strength. The fit is shown in Fig. 1, whose most salient feature is that the deviation from the linear behaviour corresponding to the rod (number of base pairs under 200) is quite slight even for DNA as long as 1000 base pairs. This manifests the known characteristic of sedimentation coefficients as being not as sensitive to macromolecular conformation than other solution properties. On the other hand, their ease of complication-free measurement (especially compared with light scattering measurements), coupled with the many recent advances in ultracentrifugal techniques - including relevant improvements in precision - make them very promising tools for conformational studies.

ACKNOWLEDGEMENT

Present work of J.J.F. and J.G.T. is supported by Grants PB89/0093 and PB87/0694, respectively, from the Dirección General de Ciencia y Tecnología.

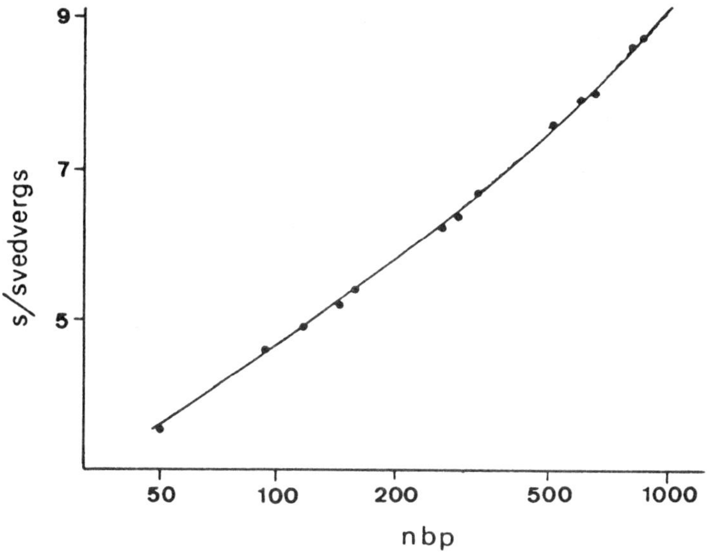

Figure 1. Sedimentation coefficients of restriction fragments of DNA as a function of number of base pairs (nbp). Closed circles are experimental data from ref. 34. The curve corresponds to the results for the wormlike chain model (ref. 33) with a persistence length of 410 Å.

REFERENCES

1. J. García de la Torre, Chapter 18 in this volume.
2. P.-G. de Gennes, 'Scaling Concepts in Polymer Physics', Cornell University, Ithaca, 1979.
3. P.J. Flory, 'Statistical Mechanics of Chain Molecules', Wiley, New York, 1969.
4. M. Doi and S.F. Edwards, 'The Theory of Polymer Dynamics', Clarendon, Oxford, 1986.
5. P.J. Flory, 'Principles of Polymer Chemistry',Cornell University, Ithaca, 1953.
6. H. Yamakawa, 'Modern Theory of Polymer Solutions', Harper and Row, New York, 1971.
7. K.F. Freed, 'Renormalization Group Theory of Macromolecules', Wiley, New York, 1987.
8. M Schmidt and W. Burchard, *Macromolecules*, 1981, *14*, 210.
9. M. Fixman, *J. Chem. Phys.*, 1983, *78*, 1594.
10. B.H. Zimm, *Macromolecules*, 1980, *13*, 592.
11. M. Fixman, *J. Chem. Phys.*, 1986, *84*, 4080.
12. Y. Miyaki, T. Einaga, H. Fujita and M. Fukuda *Macromolecules*, 1980, *16*, 588.
13. W.W. Stockmayer and A.C. Albrecht, *J. Polym. Sci.*, 1958, *32*, 215.
14. N. Nemoto, Y. Makita, Y. Tsunashima and M. Kurata, *Macromolecules*, 1984, *17*, 425.

15. A. Rey, J.J. Freire and J. García de la Torre, *Macromolecules*, 1987, *20*, 342.
16. S.Q. Wang, J.F. Douglas and K.F. Freed, *J. Chem. Phys.*, 1986, *85*, 3674.
17. P.A. Small, *Adv. Polym. Sci.*, 1975, *18*, 1 and references therein.
18. R. Prats, J. Pla and J.J. Freire, *Macromolecules*, 1983, *16*, 1701.
19. N. Hadjichristidis and J.E.L. Roovers, *J. Polym. Sci., Polym. Phys. Ed.*, 1974, *12*, 2521.
20. J.F. Douglas, J. Roovers and K.F. Freed, *Macromolecules*, 1990, *23*, 4168.
21. J.J. Freire, A. Rey and J. García de la Torre, *Macromolecules*, 1986, *19*, 457.
22. J.A. Semlyen, ed., 'Cyclic Polymers', Elsevier, London, 1986.
23. J.M. García Bernal, M.M. Tirado, J.J. Freire and J. García de la Torre, *Macromolecules*, 1990, *23*, 3357; *ibid., Macromolecules*, 1991, *24*, 593.
24. K. Kubota and B. Chu, *Macromolecules*, 1983, *16*, 105.
25. K. Okita, A. Teramoto and H. Fujita, *Biopolymers*, 1970, *9*, 717.
26. T. Saito, *Biopolymers*, 1982, *21*, 715.
27. L. Zhang, W. Liu, T. Norisuye and H. Fujita, *Biopolymers*, 1987, *26*, 233.
28. V.A. Bloomfield, D.M. Crothers and I. Tinoco, 'Physical Chemistry of Nucleic Acids', Harper and Row, New York, 1974.
29. C.R. Cantor and P.R. Schimmel, 'Biophysical Chemistry', Freeman, New York, 1979.
30. J.E. Hearst and W.H. Stockmayer, *J. Chem. Phys.*, 1962, *37*, 1425.
31. H. Yamakawa and M. Fujii, *Macromolecules*, 1973, *6*, 407.
32. H. Yamakawa and M. Fujii, *Macromolecules*, 1974, *7*, 128.
33. J.J. García Molina, M.C. Lopez Martínez and J. García de la Torre, *Biopolymers*, 1990, *29*, 883.
34. R.T. Kovacic and K.E. Van Holde, *Biochemistry*, 1977, *16*, 1490.

20

Methods for Obtaining Sedimentation Coefficient Distributions

By Walter F. Stafford, III

DEPARTMENT OF MUSCLE RESEARCH, BOSTON BIOMEDICAL RESEARCH INSTITUTE, 20 STANIFORD STREET, BOSTON, MASSACHUSETTS 02114, U.S.A.

and

DEPARTMENT OF NEUROLOGY, HARVARD MEDICAL SCHOOL, BOSTON, MASSACHUSETTS 02115, U.S.A.

1. INTRODUCTION

The determination of particle *size distributions* has been one of the goals and uses of analytical ultracentrifugation since its inception in the early 1920s by Svedberg and Nichols[1] and Svedberg and Rinde[2,3]. These early investigators first presented equations for particle diameter distributions of large spherical particles obtained from concentration profiles observed during sedimentation.

In 1942, Bridgman[4] first presented an equation for determining sedimentation coefficient distributions from concentration gradient curves. His analysis was based on the assumption that the diffusion of each component was negligible compared to sedimentation and, therefore, was restricted to the analysis of solutions of large polymers. Later, the idea of extrapolating apparent distributions to infinite time to eliminate the effects of diffusion was introduced by Baldwin and Williams[5]. They tried various methods of extrapolation to eliminate the effects of diffusion on boundary spreading[6]. Gralén and Lagermalm[7] presented a method that was used widely. Their method was extended to self-associating system by Winzor *et al.*[8] The method was later improved upon by Van Holde and Weischet[9]. Stafford[10,11] has recently devised a method for obtaining sedimentation coefficient distributions from the time derivative of the concentration profile as well as new extrapolation procedures for reducing the effects of diffusion. An extensive list of published analyses of sedimentation coefficient distributions has been given by Baldwin and Van Holde[12].

Analysis of DNA distributions has been carried out using numerically differentiated absorbance traces by Schumaker[13,14]. Particle size distributions for aqueous polymer dispersions have been studied extensively in the laboratory of Walter Mächtle[15,16,17] who has introduced a high speed eight-hole rotor combined with a multiplexer system - see W. Mächtle, Chapter 10, this volume. Analysis of lipoprotein *flotation* coefficient distributions has recently been reported by Richards *et al.*[18]

The first application of a real time system was by Runge *et al.*[19] to the association between neurofilaments and microtubules. The development of a high resolution on-line Rayleigh interferometric optical system[20] (see T.M. Laue, Chapter 6, this volume) has made it possible to acquire and analyse data rapidly and has made a significant contribution to the revitalization of the analytical ultracentrifuge. Analytical procedures that previously required a large amount of time and effort can now be carried out routinely with these automated systems. The *real time* computation of sedimentation coefficient distributions has now become possible with the Rayleigh optical system[19]. In this Chapter, we will discuss and compare several methods for obtaining apparent sedimentation distributions as well as several methods of extrapolation to deal with dispersion caused both by diffusion and by acquisition and manipulation of the data.

We define the following quantity as the differential sedimentation coefficient distribution function

$$\frac{d\{c(s)/c_0\}}{ds}\left(\frac{r}{r_m}\right)^2 \equiv g(s)_j$$

where j is "r" or "t" depending on whether we calculate it from $\partial c/\partial r$ or $\partial c/\partial t$, respectively. The function g(s) ds gives the weight fraction of material sedimenting with sedimentation coefficients between s and s+ds. The integral distribution curve is given by

$$G(s) \equiv \int_0^s g(s)\, ds$$

The concentration distribution of the sedimenting material can be viewed either as a function of radius at fixed particular instants of time or as a function of time at fixed specified values of radius. As will be shown below, apparent sedimentation coefficient distributions can be obtained from either the radial or the temporal variation of the concentration distributions. The derivations are given in the Appendix. Some advantages of using the time dependence are discussed below: the main advantage is that the precision of both the ultra-violet (UV) scanner and the Rayleigh optical system is increased by a factor of between 10 and 100. Sedimentation velocity runs can be carried out routinely with the Rayleigh optical system on solutions whose concentration is somewhat less than 0.02 mg/ml by averaging $g^*(s)$ curves. Difference patterns have been used in the past to reduce baseline contributions[19]. Analysis of the time dependence of the concentration profiles to solve the Lamm equation has been exploited by Bethune and co-workers[21, 22, 23].

An additional advantage of computing $g^*(s)$ curves, either singly or averaged, is that one can get the weight average sedimentation coefficient (assuming no component has pelleted) for the sample by simply performing the following integration:

$$\bar{s}_W = \int_0^\infty s\, g^*(s)\, ds$$

2. RELATION FOR $g^*(s)$ FROM THE TIME DERIVATIVE

The rationale for using the time derivative of the concentration profile is based on the simple idea that the raw concentration profiles, whether obtained by the UV scanner or Rayleigh optical systems, are composed of a time dependent true concentration component and a time independent baseline component plus random noise.

$$C(r,t)_{obs} = C(r,t)_{true} + C(r)_{baseline} + C(t)_{noise}$$

If the time derivative of the concentration profile is computed, the time independent part, being a function of radius only, drops out so that

$$\frac{\partial C(r,t)_{obs}}{\partial t} = \frac{\partial C(r,t)_{true}}{\partial t} + \sqrt{2}C(t)_{noise}$$

since

$$\frac{\partial C(r)_{baseline}}{\partial t} = 0$$

In this analysis we will ignore the possibility of a time dependent baseline drift. If any drift is sufficiently slow, it will not contribute significantly to the value of $\partial c/\partial t$. The source of any significant contributions should be determined and eliminated if possible.

The time derivative of the concentration profile may be estimated by two different but closely related methods depending upon the sampling frequency. For example, if the data are collected from the UV photoelectric scanner (especially if several cells are run at once), scans can be taken no more often than several minutes apart. In this case, the derivative can be estimated by simply subtracting successive scans digitised several minutes apart followed by smoothing in the radial co-ordinate. If the real-time video Rayleigh interferometer system is used, the profiles can be collected at sufficiently small time increments that the time derivative can be computed by standard numerical procedures over the time co-ordinate at each radial position.

Assuming that a sufficiently small time interval between scans is chosen so that $\Delta c(r,t)/\Delta t \approx dc(r,t)/dt$, then either this value or the value of $dc(r,t)/dt$ computed directly, can be converted to a weight fraction distribution function as shown in the appendix (eq. (A18)) to give the following relationship:

$$g^*(s)_t = \left(\frac{\partial\{c(r,t)/c_0\}}{\partial t}\right)\left(\frac{\omega^2 t^2}{\ln(r_m/r)}\right)\left(\frac{r}{r_m}\right)^2 \quad (1)$$

In cases for which diffusion can be ignored with respect to sedimentation, $g^*(s)$ is equal to the true distribution, g(s). Spreading due to diffusion goes nearly as the the square root of time whereas the spreading due to heterogeneity of sedimentation coefficient goes nearly as the first power of time. In principal, therefore, if the $g^*(s)$ patterns are extrapolated to infinite time, the spreading due to diffusion will become insignificant compared to that due to heterogeneity, and the true differential distribution function, g(s), will be obtained (see below). This relationship (eq. (1)) for $g^*(s)_t$ was first presented at the Workshop on Ultracentrifugation held at the 1989 Biophysical Society meeting in Cincinnati[10] and also at the 1990 International Biophysics Congress in Vancouver (1990)[24].

3. RELATION FOR $g^*(s)$ FROM THE RADIAL DERIVATIVE

The Bridgman equation[4] for g(s) derived from $(\partial c/\partial r)_r$ is

$$g^*(s)_r = \left(\frac{\partial c(r,t)/c_0}{\partial r}\right)\left(\omega^2 rt\right)\left(\frac{r}{r_m}\right)^2 \qquad (2)$$

A derivation is given in the Appendix (eq. (A24)). Again the asterisk is meant to indicate that in actual experimental situations, the diffusion coefficient of the sedimenting species is not identical to zero and that the distributions computed using this equation are in fact apparent distributions[5].

4. COMPUTATION OF $G^*(s)$ FROM THE CONCENTRATION PROFILE

The Gralén and Lagermalm method[7]: In this method an apparent value of s^* is computed at each level in the boundary: $s^*(c/c_p) = \ln(r^*_{(\text{at } c/c_p)}/r_m)/\omega^2 t$. If the spreading of the boundary is due only to heterogeneity the values of c *vs.* s^* will correspond to the true integral sedimentation coefficient distribution. If diffusion contributes significantly to boundary spreading, then the values of s^* will be apparent values of s. In this case, the true integral distribution, which we denote by the term G(s), is arrived at by extrapolating the values of s^* at each value of c/c_p to infinite time (*i.e.* 1/t=0). Gralén and Lagermalm originally carried out the extrapolation *versus* 1/t. Since then, this method has been improved by Van Holde and Weischet[9] who have shown that a more theoretically correct and effective procedure is to carry out the extrapolation *versus* $(1/\sqrt{t})$. The method of Gralén and Lagermalm has also been extended by Winzor *et al.*[8] and Nichol and Winzor[25] to the study of self-associating systems as a means of obtaining asymptotic boundary profiles for comparison to theoretical asymptotic patterns generated from Gilbert theory[26,27,28].

5. DATA ACQUISITION

The success of the methods described here requires that the sedimentation patterns be sampled at relatively high spatial and temporal frequency so that averaging may be carried out to reduce random noise. In this laboratory, two methods are currently employed for the acquisition of sedimentation patterns from either the UV photoelectric scanner or the Rayleigh optical system.

In the case of the scanner, the signal output from the scanner is digitised at a rate of about 30 kHz. Sets of approximately 900 sample acquisitions are initiated every $^1/_{30}$ sec, averaged and stored as a single datum point. This process produces about 450 data points from the middle of the outer reference hole to the middle of the inner reference hole (1.8 cm). These raw data are then passed through a program that finds the reference edges giving a data set with about 400 samples between reference edges. Therefore, each point is essentially the average value over each 40 μm radial interval in the cell.

In the case of the Rayleigh optical system, the Rayleigh image is refocused in both the horizontal and vertical axes onto a solid state video array (768×512 pixels) so that there are approximately 720 samples of the Rayleigh fringe pattern taken between reference edges (22 μm per sample). The video data are converted to fringe displacement at each radius by discrete, single frequency Fourier analysis.

These data are then passed to other programs that convert the data to concentration *vs.* radius. Separate programs then smooth and differentiate the data to produce plots of dc/dr *vs.* r, dc/dt *vs.* r or dc/dt *vs.* t.

The process of calculating $g^*(s)$ from either dc/dr or dc/dt requires a knowledge of the radius of the meniscus, r_m, which can be taken from the images. In addition, since the rotor does not instantaneously attain its final velocity, the analysis also requires the value of the following integral in place of the factor $\omega^2 t$ in eqs. (1) and (2):

$$\int_0^t \omega(t)^2 \, dt$$

where ω is the angular velocity of the rotor and t is the elapsed time from the beginning of acceleration. If the speed is changed or gradually increased during the run, the integral must be computed continuously and the corresponding value of the integral associated with each scan or image. If, as in the usual situation, the angular velocity is held constant after the initial acceleration, then the integral can be rearranged to compute t_{sed}, the effective time of sedimentation at the final angular velocity.

$$t_{sed} = \left(\frac{1}{\omega(a)^2}\right) \int_0^a \omega(t)^2 \, dt + (t - a) \tag{3}$$

where a is the elapsed time at the end of acceleration and the other symbols have the same meaning as in the previous equation. The value of $\omega(a)^2 t_{sed}$ is then used in the computations of $g^*(s)$.

Alternatively, if the integral during acceleration is not available, t_{sed} can be estimated by extrapolating the second moment boundary position back to the meniscus to obtain an effective time for the start of sedimentation with the assumption that the angular velocity of the rotor had been constant at its full speed value from that time onward. This procedure was advocated by Winzor *et al.*[8]

The first method is the method of choice since it allows the speed to be varied during the run and requires only one pass of the data since both r_m and $\int \omega(t)^2 dt$ are known. The second moment method is satisfactory if the angular velocity information is not available in computable form and the speed is not varied once the run is underway. Disadvantages of the second method are, firstly, that it requires two passes of the data, one to find the effective start of sedimentation at the final speed and the second to compute $g^*(s)$ and, secondly, that it requires the gradient in the plateau region to be negligible so that the second moment boundary position can be reliably calculated.

6. COMPUTATION PROCEDURES

Smoothing and differentiation of the sedimentation concentration profiles both as a function of radius and as a function of time were carried out using standard numerical techniques[29]. For the experimental cases shown here, and for most practical situations, the simple procedures described below seem to work quite satisfactorily. More sophisticated methods for smoothing have been investigated but do not seem to offer any practical advantages. Smoothing of the concentration profiles was accomplished by repeatedly passing a sliding, centred, running average spanning typically 7 to 11 points (280 to 440 µm in the cell) over the data from 5 to 10 times depending on the noise level of the data. The frequency response of the smoothing filter can be controlled easily by varying both the span and the number of passes. As expected, some resolution is sacrificed if extensive smoothing is required. Therefore, one chooses the least amount of smoothing that gives usable data.

The data were then differentiated analytically by passing a sliding 13 point centred, cubic least squares differentiating formula[29] over the data spanning about 500 µm. The derivatives were then subjected to further smoothing. The smoothed derivatives, either $\partial c/\partial r$ or $\partial c/\partial t$, were then used in the computation of $g^*(s)_r$ and $g^*(s)_t$, respectively. When the time derivative of the concentration profiles was computed by subtracting successive pairs of concentration profiles spaced at appropriate time intervals, the difference curves, $\Delta c/\Delta t$ *vs.* r, were then smoothed in the radial co-ordinate with a single pass of an 11 to 21 point centred moving average sliding strip, spanning from about 220 to 440 µm, and then used to compute $g^*(s)_t$.

This procedure has been completely automated so that one only need specify a value of t_{sed} corresponding to a given clock time, the radius at the meniscus, the speed of the run and list of the file names of each scan or frame. In a typical run, between 50 - 200 total video frames are acquired. With the current combination of hardware and software, a single video frame can be converted to concentration as a function of position in less than 6 seconds and stored on disc. The software then will process the entire set of raw sedimentation patterns from one sedimentation velocity run to produce the corresponding set of either $g^*(s)_r$ or $g^*(s)_t$ curves. A second program then, using the same list of files, collects the $g^*(s)$ files and either writes them into one large file for plotting or averages subsets of them or performs an extrapolation to infinite time, as desired. An example of the use of averaged $g^*(s)_r$ plots to analyze myosin-II filament formation using the UV photoelectric scanner has recently appeared[30].

7. COMPARISON OF $g^*(s)_r$ AND $g^*(s)_t$ USING SIMULATED DATA

Simulated sedimentation curves were computed by the finite element method of Claverie *et al.*[31] (programs kindly supplied by Dr. David Cox) for an ideal, two component non-associating system. A typical set of simulation curves is shown in Figs. 1A-1E. The case shown here is for a two species system (s_1=8 S, D_1=5 F** and s_2=10 S, D_2=5 F). Fig. 1A shows the evolution of the concentration profile with time of sedimentation. Figs. 1B and 1C show the corresponding curves of dc/dr and dc/dt *versus* r. Figs. 1D and 1E show the corresponding curves for $g^*(s)_r$ and $g^*(s)_t$. The agreement between these two methods is good. One feature of the $g^*(s)_t$ curves that needs to be addressed is the contribution from dc/dt in the plateau region in front of the boundary. As discussed in the Appendix, this contribution is small and under most circumstances can be ignored. The disadvantages associated with the contribution from the plateau region are compensated for by the large increase in signal-to-noise ratio gained by using dc/dt. Routine analyses can be carried out with the Rayleigh optical system on boundaries of less than 0.1 fringes (corresponding to less than 0.03 mg/ml in a 12 mm optical path length cell; with a He-Ne laser light source, λ=6328Å, a 1 mg/ml protein solution gives a fringe deflection of approximately 3.4 fringes). Boundaries at lower concentrations than this can be analysed if some extra care is taken and if the value of $\omega^2 s/D$ is sufficiently large.

8. DETERMINATION OF THE DIFFUSION COEFFICIENT AND MOLECULAR WEIGHT FROM $g^*(s)_{max}$

At values of $\omega^2 st/D$ sufficiently large that any component's boundary in an ideal mixture has become separated from the others, we can use the following relation to obtain the molecular weight of that component

$$g^*(s)_{max} = \left(\frac{s}{2\pi D}\right)^{1/2} (\omega^3 r_m) \left(\frac{t}{(1-e^{-2\omega^2 st})^{1/2}}\right) \qquad (4)$$

This relation was derived from the Faxén equation (see Appendix, eq. (A27)). If one plots $g^*(s)_{max}$ *vs.* $t/(1-e^{-2\omega^2 st})^{1/2}$, the slope of the plot is proportional to the square root of the ratio s/D and, therefore, also proportional to the square root of the molecular weight through the Svedberg equation, so that

$$M = \frac{2\pi RT}{(1-\bar{v}\rho)\omega^6 r_m^2} (\text{slope})^2 \qquad (5)$$

where the symbols have their usual meaning.

This relationship was tested with simulated data for the case above of s=10 S and D=5 F [M=180,000, assuming $(1-\bar{v}\rho)$ = 0.270] with r_m=6.0 cm at 60,000 rev/min. A plot of $g^*(s)_{max}$ *vs.* $t/(1-e^{-2\omega^2 st})^{1/2}$ is shown in Fig. 2. This plot gives a

** $1\ F = 10^{-7}\ cm^2\ s^{-1}$

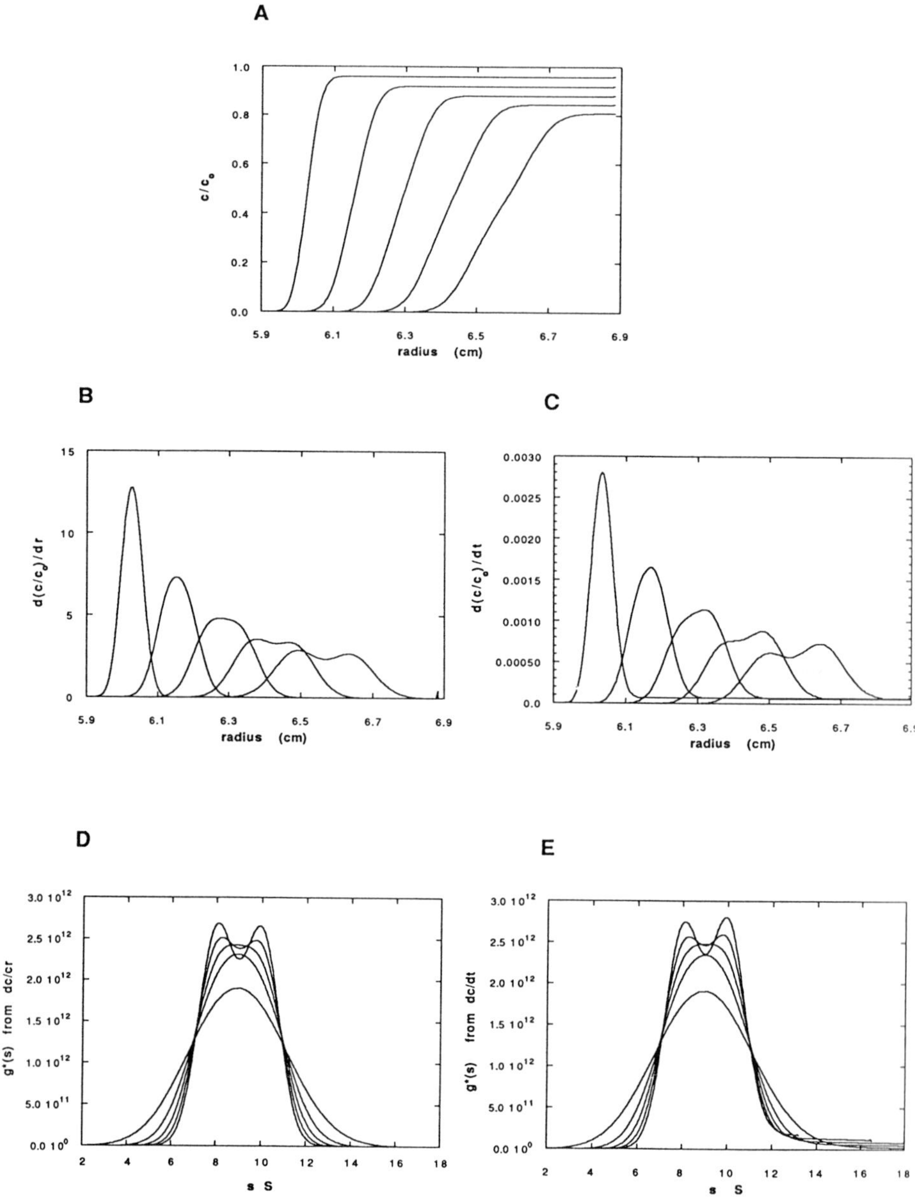

Figure 1. Simulated sedimentation curves at various times for the following system composed of 2 species: s_1=8 S, D_1=5 F and s_2=10 S, D_2=5 F. (A) c *vs.* r, (B) dc/dr *vs.* r, (C) dc/dt *vs.* r. Comparison of apparent sedimentation distribution curves: (D) $g^*(s)_r$ *vs.* s, (E) $g^*(s)_t$ *vs.* s.

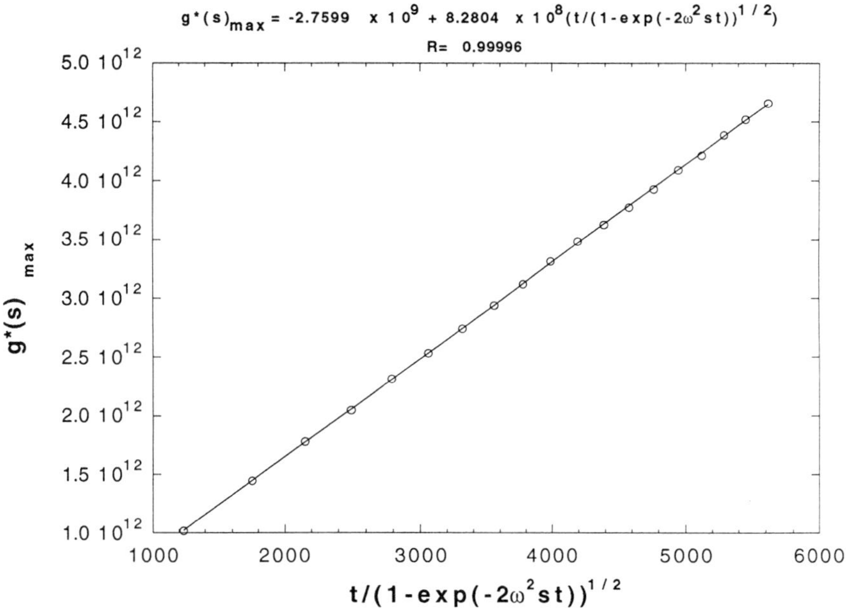

Figure 2. Plot of $g^*(s)_{max}$ *vs.* $t/(1-\exp(-2\omega^2 st))^{1/2}$. The slope is proportional to the square root of the molecular weight [see text for details (eq. (5))].

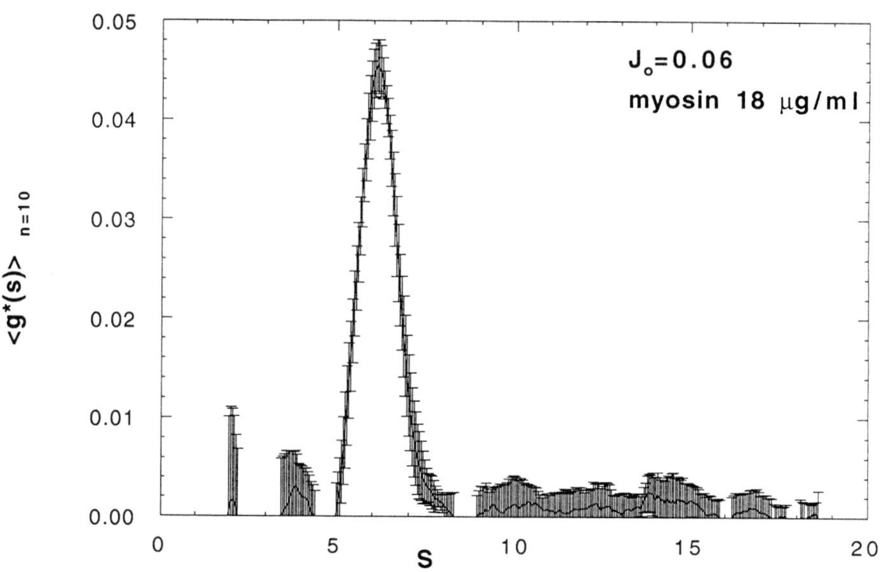

Figure 3. Plot of $<g^*(s)>$ *vs.* s (n=10) for rabbit skeletal myosin. C_0=0.06 fringes (20 μg/ml), 12 mm cell. The $g^*(s)_t$ patterns were obtained using eq. (1) by computing $\Delta c/\Delta t$ over 300 second time intervals at 52,000 rev/min ($\Delta\omega^2 st = 0.006$). Ten of these difference curves with 60 sec between each pattern were converted to $g^*(s)_t$ and averaged to give this $<g^*(s)>$ curve.

value of 8.28×10^8 sec^{-1} for the slope and a value of 176,000 for the molecular weight, agreeing with the correct molecular weight to within 2% - 3%. In practice, the estimation of M by this method is sensitive to artificial dispersion as well as to heterogeneity with both causing downward curvature in the plots leading often to 10% - 20% underestimation of the molecular weight (Stafford, unpublished observations). This result is not surprising since similar effects have been observed by Van Holde and Weischet[9] in attempts to estimate the diffusion coefficient from the time dependence of s^*.

9. AVERAGING $g^*(s)$ CURVES TO INCREASE SIGNAL-TO NOISE RATIO

If $g^*(s)_t$ and $g^*(s)_r$ curves are acquired sufficiently frequently so that several can be averaged over small time intervals, the signal-to-noise ratio of the curves can be improved considerably. In the case of $g^*(s)_t$ with the Rayleigh optical system using 30 mm cells, boundaries of less than 0.01 mg/ml can be analysed. Fig. 3 shows the results of g(s) analysis performed on a sample of myosin from rabbit skeletal muscle, at an initial concentration of ~20µg/ml (0.06 fringes in a 12 mm cell). After $\Delta c/\Delta t$ was calculated, with Δt=300 at 52000 rev/min ($\Delta\omega^2 st = 0.006$ at the peak), $g^*(s)_t$ was calculated from eq. (1). The average $<g^*(s)>$ was computed from 10 $g^*(s)_t$ curves obtained 1 min apart. The ability to compute baseline corrected $g^*(s)_t$ curves, combined with averaging has extended the precision of the Rayleigh system about two orders of magnitude.

10. REDUCING DISPERSION BY EXTRAPOLATION TO INFINITE TIME

Diffusion

The problem is to find a correct functional form for the extrapolation of $g^*(s)$ that fits the behaviour at the relatively short times ($\omega^2 st < 0.15$) before the boundary arrives at the cell bottom and also has the correct limiting behaviour at large values of $\omega^2 st$ as 1/t goes to zero. In the past, the theoretical justification for the form of the extrapolation to infinite time has been taken from an analysis of Faxén's[32] approximate solution to the Lamm equation[33] for the hydrodynamically ideal case. Faxén's equation is valid for relatively short times of sedimentation for which $\omega^2 st \ll 1$. However, since the extrapolation procedure involves the range of very large times, this form will not provide the correct limiting behaviour. According to Gosting[34], Faxén's corresponding expression for long times ($\omega^2 st > 1$) does not converge rapidly enough to be useful in the usual experiment (see Faxén's eq. (3)). It seems that there is currently no analytical solution available which can be used to obtain the correct form at very large times. Therefore, a semi-empirical approach has been taken by this writer to investigating the choice of extrapolation form by studying the behaviour of simulated sedimentation patterns. The extrapolation form will have to handle *both* diffusion *and* artificial dispersion in the $g^*(s)$ patterns.

Artificial Dispersion

Artificial dispersion of the patterns can arise from both instrumental and computational sources. For example, in the case of data obtained from the UV scanner, radial dispersion will be introduced into the data simply because a finite slit width is used at the photomultiplier. Similarly, any smoothing or differentiation procedure will introduce further radial dispersion. Since the dispersion introduced from these sources should be independent of the time of sedimentation, it will make a progressively smaller contribution to the $g^*(s)$ curves as time goes on. In principle, the relative contribution from these sources will decrease rapidly with time and should disappear when the data are extrapolated to infinite time. The effect of artificial dispersion on the extrapolation procedure is discussed below.

11. EXTRAPOLATION PROCEDURES

The requirement of the ideal extrapolation procedure is that, when given a series of simulated $g^*(s)$ profiles generated from a known distribution of g(s), it returns the original distribution. It has been difficult to find an extrapolation form that will satisfy all of our needs; however, there are some forms that provide considerable reduction in the contributions from both diffusion and artificial dispersion. The most difficult case is for mono- and pauci-disperse systems since these distribution functions are discrete and, therefore, should extrapolate to delta functions at infinite time. This is too much to ask of the analytical functions used for the extrapolation, and therefore, the best we can hope for is a compromise giving narrow peaks centred at the correct values of the sedimentation coefficient whose areas correspond to the amounts of each component. According to our requirements, in addition to artificial dispersion, the extrapolation form must accommodate the contributions from diffusion both at the peak and the off peak positions. Several extrapolation forms have been tried and will be discussed below.

The various extrapolation forms have been tested using simulated, noise-free sedimentation patterns generated by the method of Claverie *et al.*[31] to provide numerical solutions to the Lamm equation for several systems of one or more components. Both a normal length cell, with a solution column height of 1 cm, with the meniscus at 6.0 cm from the centre of rotation, and an extra-long cell with a column height of 5 cm, with the same meniscus position were simulated. The 5 cm cell was simulated to investigate what happens at larger values of $\omega^2 st$ (> 0.15) than those associated with the usual centrifuge cell for the purpose of finding a limiting form for extrapolation. The calculated $g^*(s)$ curves were interpolated onto an evenly spaced grid of $\Delta s = 0.05$ S for extrapolation at constant s.

Extrapolation of $g^*(s)$

One of the pitfalls associated with some of the forms that have been used in the past is that they can over-extrapolate the $g^*(s)$ curves so that negative values are sometimes obtained in regions where one expects the extrapolations to go to zero (Figs. 4C-E).

A

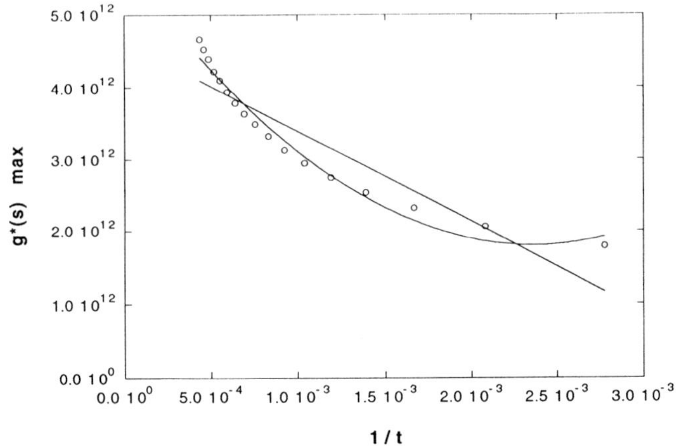

B

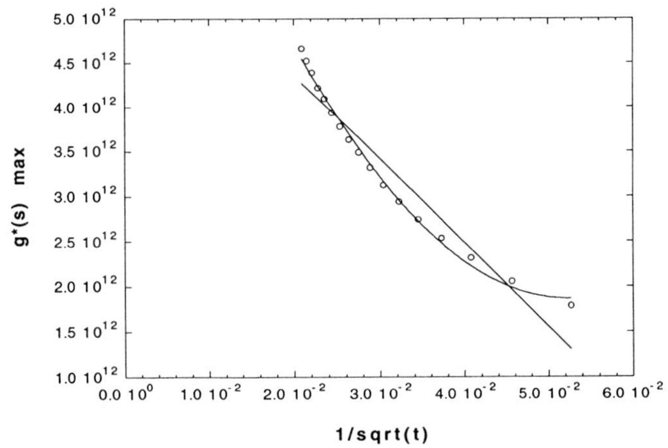

Figure 4. Plots of $g^*(s)_{max}$: (A) for the single species system shown in Fig. 6 system *vs.* 1/t, (B) *vs.* $1/\sqrt{t}$. Complete extrapolations carried out as $g^*(s)$ *vs.* (C) 1/t (D) $1/\sqrt{t}$ and (E) complete extrapolation *vs.* $1/\sqrt{t}$ for the two species system shown in Figs. 1 and 7. The arrows indicate the regions of negative g(s).

C

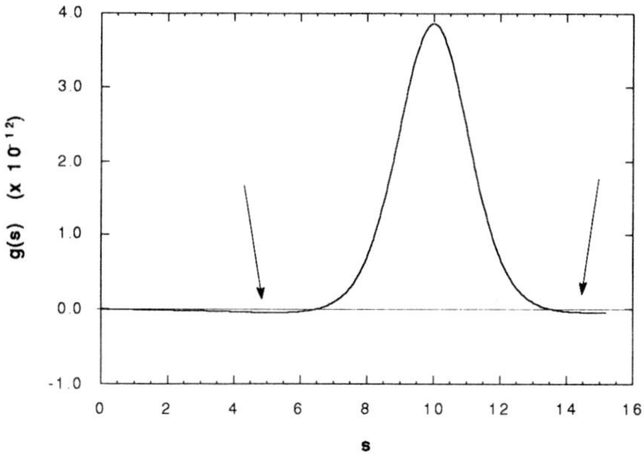
g(s) (x 10^{-12})
4.0
3.0
2.0
1.0
0.0
-1.0
0
2
4
6
8
10
12
14
16
s

D

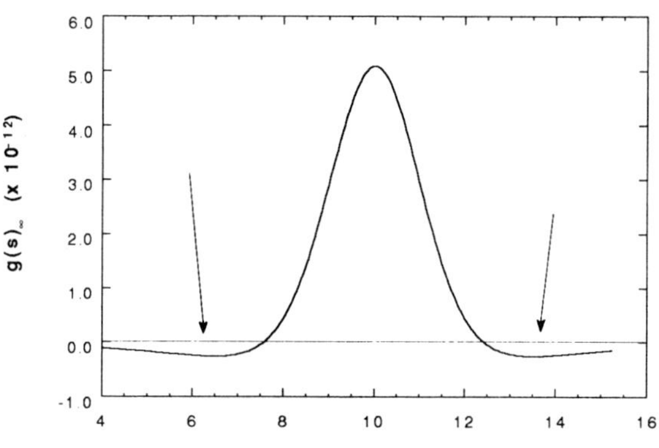
g(s)$_\infty$ (x 10^{-12})
6.0
5.0
4.0
3.0
2.0
1.0
0.0
-1.0
4
6
8
10
12
14
16
s

E

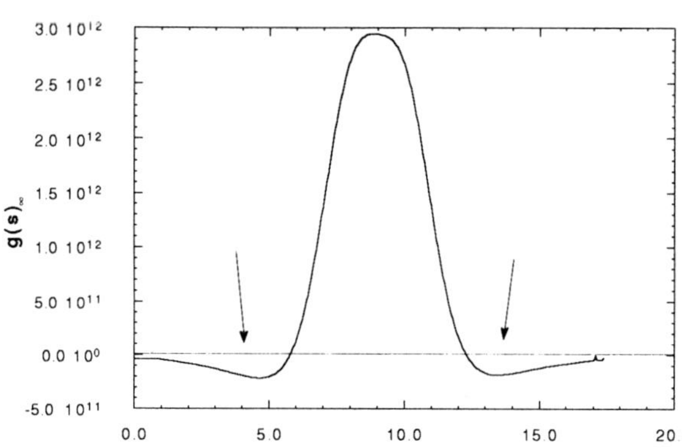
g(s)$_\infty$
3.0 10^{12}
2.5 10^{12}
2.0 10^{12}
1.5 10^{12}
1.0 10^{12}
5.0 10^{11}
0.0 10^{0}
-5.0 10^{11}
0.0
5.0
10.0
15.0
20.0
s

Specifically, procedures that employ extrapolations of $g^*(s)$, at constant s, *versus* either 1/t or $1/t^{1/2}$ have this tendency. Neither of these variables takes fully into account the time dependent curvature in the $g^*(s)$ plots even if a second order term is included (Figs. 4A and B). Therefore, several other forms (shown below) which, although not satisfactory in all respects, were chosen for the extrapolation because they offer distinct advantages over the previous forms. One of the chief advantages is an increase in resolution for pauci-disperse systems making them useful for detection of heterogeneity in spite of the lack of a firm theoretical basis for their choice.

The possibility of over-extrapolation to negative values of g(s) can occur also in the procedure of Gralén and Lagermalm[7]. However, even though it involves extrapolation of s^* and not $g^*(s)$, it apparently can over-extrapolate in another way so that negative values of dc/ds are seen in some regions of G(s) plots as shown by Winzor *et al.*[8]

Extrapolation of $1/g^*(s)$

In spite of the possibility of over-extrapolation of $g^*(s)$ to negative values, the following procedure seems to work for a single component. Eq. (4) above suggests that an extrapolation of the following form would be correct at least at the peak position:

$$\frac{1}{g^*(s)} = \frac{1}{g^*(s)_\infty} + a_1 x \tag{6}$$

where

$$x = \frac{(1-e^{-2\omega^2 st})^{1/2}}{t}$$

Indeed, for a single component, a linear extrapolation works very well (Fig. 5A) both at the peak and off peak positions. However, this form does not work very well for two closely separated components. In order to accommodate the contribution from a second component at the peak position of the first, a second term must be introduced to account for the decreasing contribution with time from the second component. This is because the derivative of $g^*(s)$ with respect to time is always positive at the peak position but is initially positive and then becomes negative at the off peak positions with increasing time. It therefore follows that an extrapolation form employing at least *two* terms must be used. A quadratic form has been tried with some success (Figs. 5B and C):

$$\frac{1}{g^*(s)} = \frac{1}{g^*(s)_\infty} + a_1 x + a_2 x^2 \tag{7}$$

However, this procedure can lead to over-extrapolation to negative values of $g^*(s)$ at the peak depending on the range of $\omega^2 st$ chosen for the extrapolation. It is also sensitive to noise.

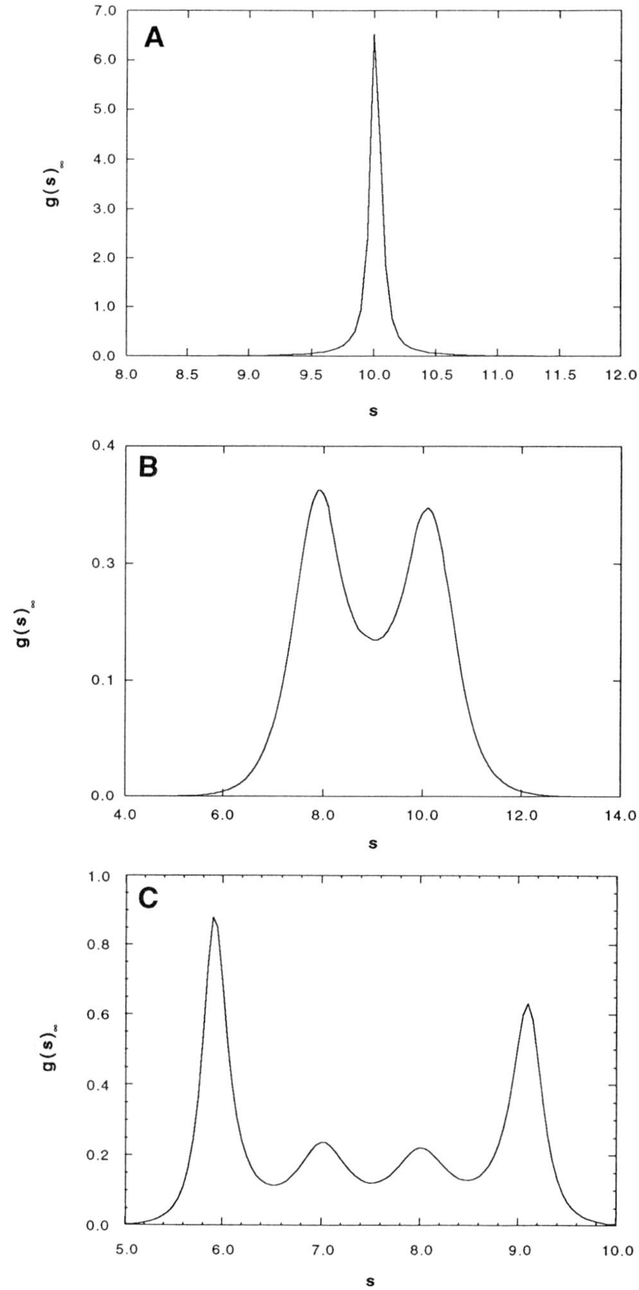

Figure 5. Extrapolation of $1/g^*(s)$ *vs.* $(1\text{-}\exp(-2\omega^2 st))^{1/2}/t$ based on eq. (7). (A) For the single component system with s=10 S, D=5 F (M=180,000, assuming $(1-\bar{v}\rho)$ = 0.270). (Same data as Fig. 6, *qv.*). (B) For the 2 component system with s_1=8 S, D_1=5 F (M_1=144,000); s_2=10 S, D_2=5 F (M_{qs}=180,000). (Same data as Fig. 8, *qv.*). (C) For the four species system shown in Fig. 9.

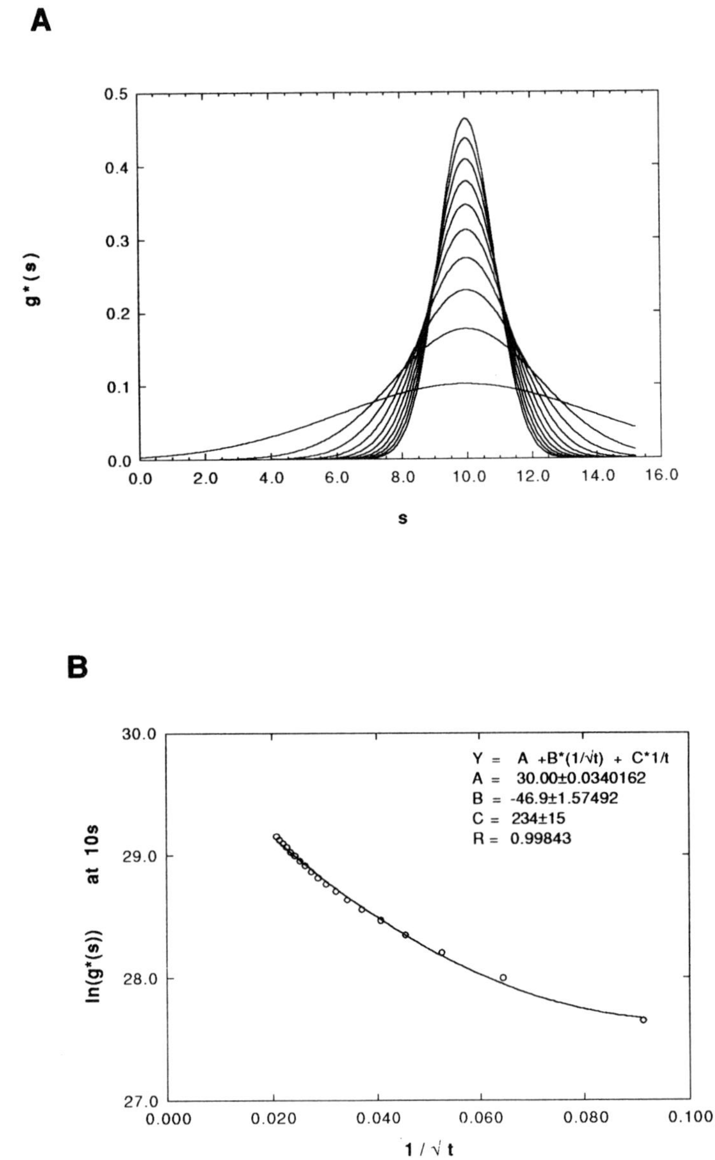

Figure 6. (A) Plots of $g^*(s)$ at various times for a single component system with s=10 S, D=5 F (M=180,000, assuming $(1-\bar{v}\rho) = 0.270$). Simulation was done on a grid of 800 points between meniscus and base with a time interval of 0.5 seconds between iterations. (B) Plot of $\ln(g^*(s)_{max})$ *vs.* $F(1/\sqrt{t},1/t)$ using eq. (8); (C) Plot of $\ln(g^*(s)_{max})$ *vs.* $F(1/\ln(t),1/\sqrt{t})$ using eq. (9); (D) Plot of $1/\ln(g^*(s)_{max})$ *vs.* $F(1/\ln(t),1/\sqrt{t})$ using eq. (10). (E) Plots showing the g(s) pattern extrapolated to infinite time using (A) eq. (8), (B) eq. (9) and (C) eq. (10).

C

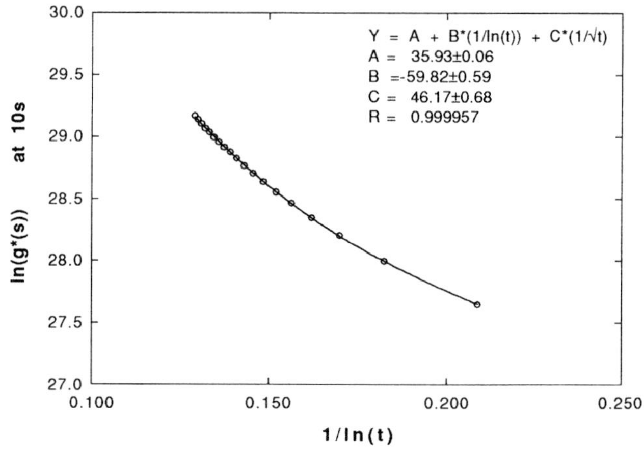

Y = A + B*(1/ln(t)) + C*(1/√t)
A = 35.93±0.06
B =-59.82±0.59
C = 46.17±0.68
R = 0.999957
ln(g*(s)) at 10s
1/ln(t)

D

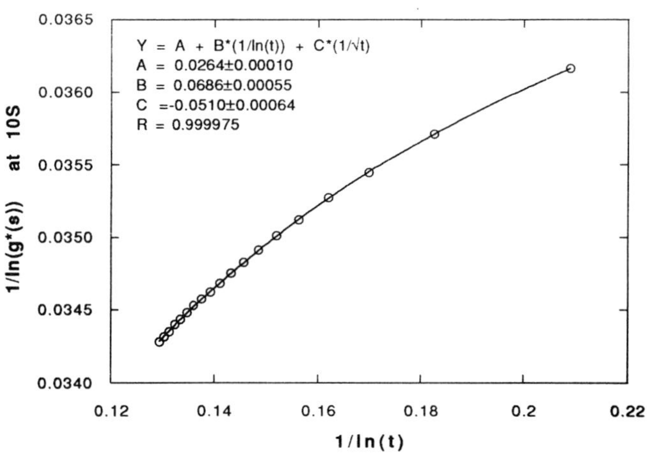

Y = A + B*(1/ln(t)) + C*(1/√t)
A = 0.0264±0.00010
B = 0.0686±0.00055
C =-0.0510±0.00064
R = 0.999975
1/ln(g*(s)) at 10S
1/ln(t)

E

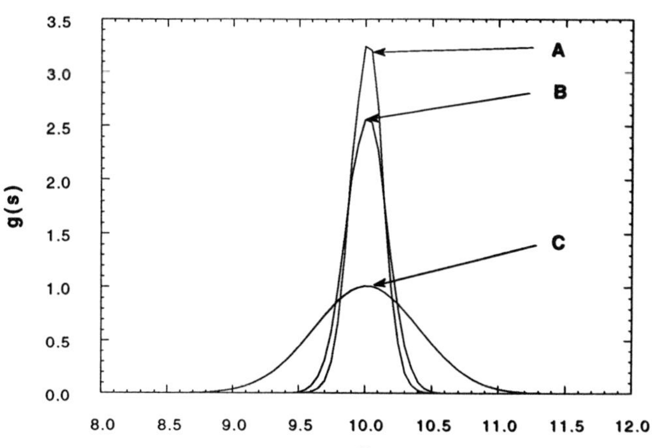

A
B
C
g(s)
s

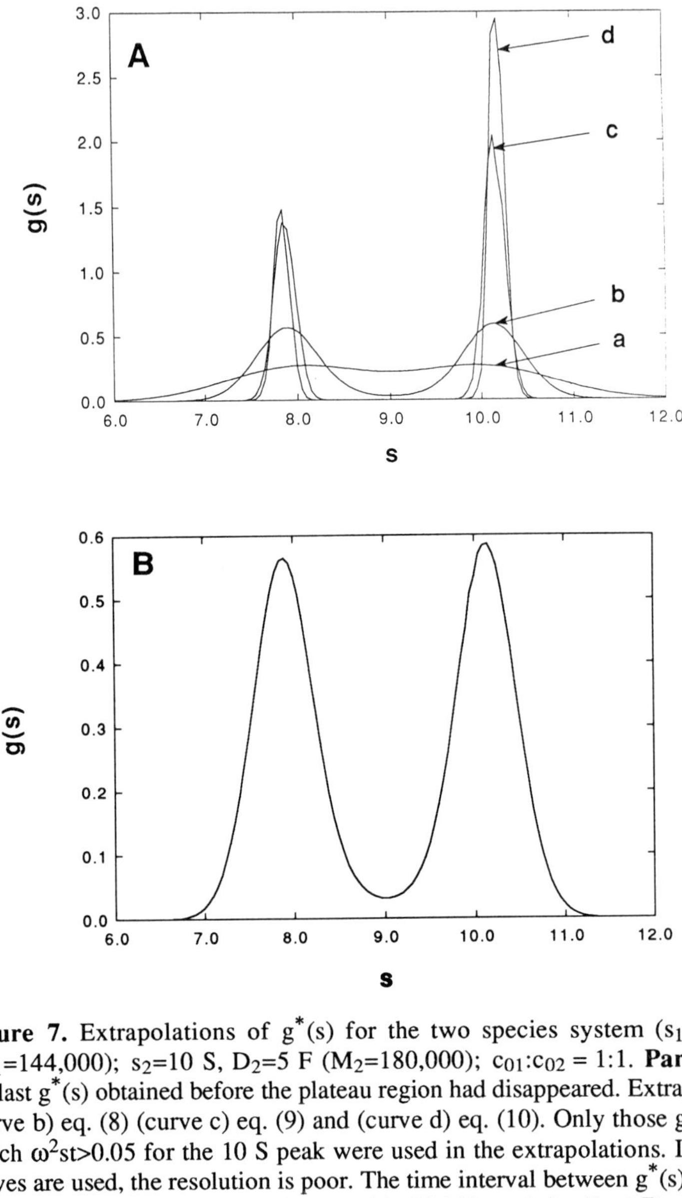

Figure 7. Extrapolations of $g^*(s)$ for the two species system (s_1=8 S, D_1=5 F (M_1=144,000); s_2=10 S, D_2=5 F (M_2=180,000); $c_{01}:c_{02}$ = 1:1. **Part A:** (curve a) the last $g^*(s)$ obtained before the plateau region had disappeared. Extrapolations using (curve b) eq. (8) (curve c) eq. (9) and (curve d) eq. (10). Only those $g^*(s)$ curves for which ω^2st>0.05 for the 10 S peak were used in the extrapolations. If all the earlier curves are used, the resolution is poor. The time interval between $g^*(s)$ curves used in the extrapolation is 120 sec. The speed is 60,000 rev/min. **Part B and C:** g(s) and G(s) curves obtained using eq. (8) in Part A showing that when the components are present in equal amount their relative amounts are preserved in the extrapolation process. **Part D and E:** g(s) and G(s) curves obtained using eq. (8) for the same system as Part A but with $c_{01}:c_{02}$ = 0.1:0.9. In this case the relative amounts of the two components is not preserved in the extrapolation process and the 8 S peak position becomes shifted to somewhat lower values. The ratio of $c_{01}:c_{02}$ becomes nearly 0.05:0.95 underestimating the smaller component by about a factor of 2. The underestimation becomes worse with eqs. (9) and (10).

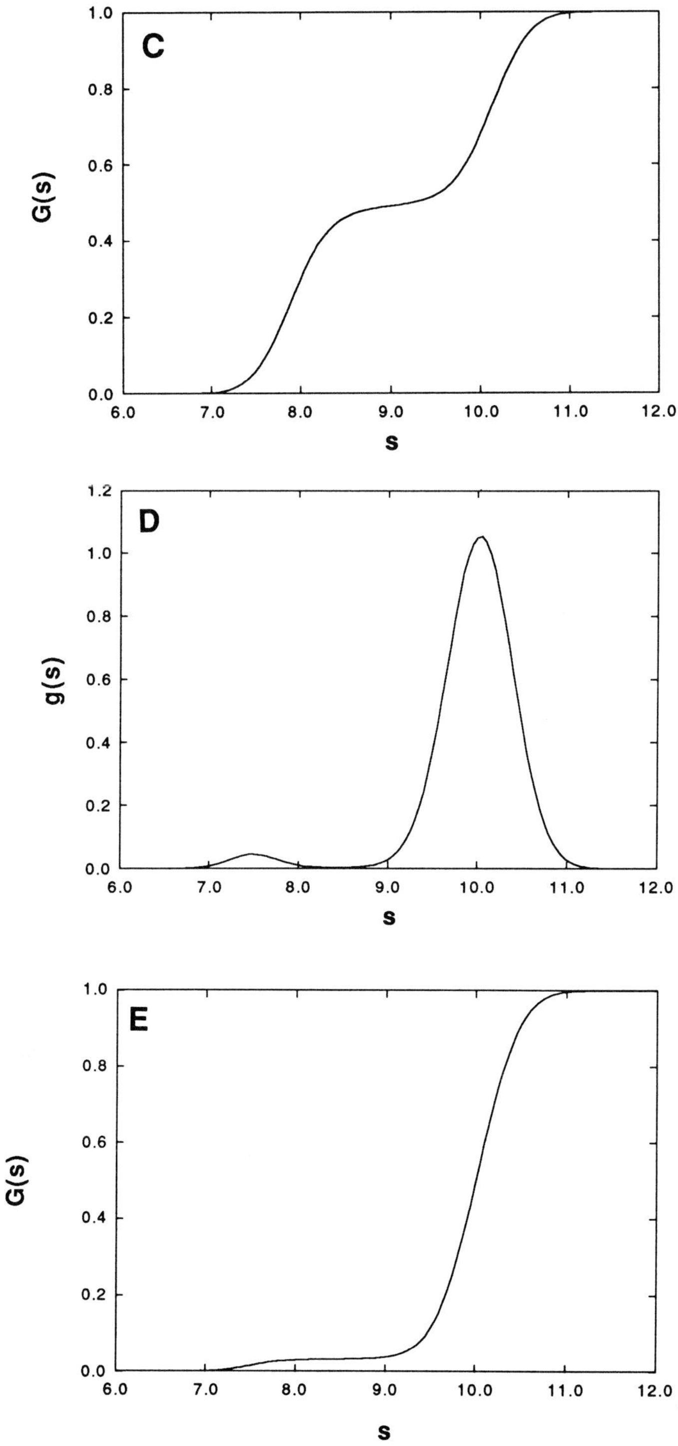
C
G(s)
s
D
g(s)
s
E
G(s)
s

Extrapolation of $\ln(g^*(s))$

Over-extrapolation to negative values of g(s) can be eliminated by carrying out the extrapolation as $\ln(g^*(s))$ *vs.* F(t) where F(t) is a function of time of some appropriate form. In addition, somewhat better resolution of pauci-disperse systems than can be obtained with eq. (7) is obtained with the forms presented next (eqs. (8-10)). Using the logarithm of $g^*(s)$ restricts the extrapolation to the positive domain of $g^*(s)$.

Several extrapolation functions of the following form were tried:

$$\ln(g^*(s)) = A + B\,\{1/\sqrt{t}\} + C\,(1/t)^x \tag{8}$$

$$\ln(g^*(s)) = A + B\,\{1/\ln(t)\} + C\,(1/t)^x \tag{9}$$

$$1/\ln(g^*(s)) = A + B\,\{1/\ln(t)\} + C\,(1/t)^x \tag{10}$$

where A, B, C are coefficients determined by least squares fitting and x, in general, can be a non-integer exponent. In these forms the coefficient, A, gives the limiting value of $g^*(s)$ as $t \rightarrow \infty$. In all the cases shown below, x was set equal to 1.0 in eq. (8) and equal to 0.5 in eqs. (9) and (10).

A Single Species with Diffusion (Monodisperse)

Fig. 6A shows simulated $g^*(s)$ curves obtained from sedimentation patterns obtained for a single species system with s=10 S and D=5 F (M=180,000) at 60,000 rev/min. These curves were generated using a grid spacing of 800 points between meniscus (6.0 cm) and base (7.0 cm) with a time difference between calculated patterns of 0.5 seconds. The correctness of the computed patterns was determined by measuring the diffusion coefficient from the patterns: clear evidence that they give the correct sedimentation coefficient can be seen directly from the $g^*(s)$ curves. The diffusion coefficient calculated from the corresponding c *vs.* r curves was 5.1 F, showing that the simulated patterns satisfactorily represent a system with the chosen parameters. The plot in Fig. 2 was obtained from this system.

The various extrapolation procedures were investigated first by looking at the behaviour of $g^*(s)_{max}$ plotted against various functions of time. Plots of $g^*(s)_{max}$ *vs.* 1/t and $g^*(s)_{max}$ *vs.* $1/\sqrt{t}$ are shown in Figs. 4A and B as mentioned above. Neither linear nor quadratic least squares fits gave a good fit to the data in the usable time range ($\omega^2 st < 0.15$). As mentioned above, another drawback of these forms is that they often result in negative values of g(s) at the off-peak positions at $t = \infty$ (Figs. 4C-E). Plots of $g^*(s)_{max}$ using the above forms are shown in Figs. 6B-D. All three of these forms give good fits to $g^*(s)_{max}$ over the observable time range.

The three forms above (eqs. (8-10)) were used to extrapolate the entire $g^*(s)$ distribution at each value of s to infinite time. The extrapolations are shown in Fig. 6E, curves a, b and c, respectively. All three forms give narrow g(s) distributions centred around the correct value of the sedimentation coefficient. However, one feature of

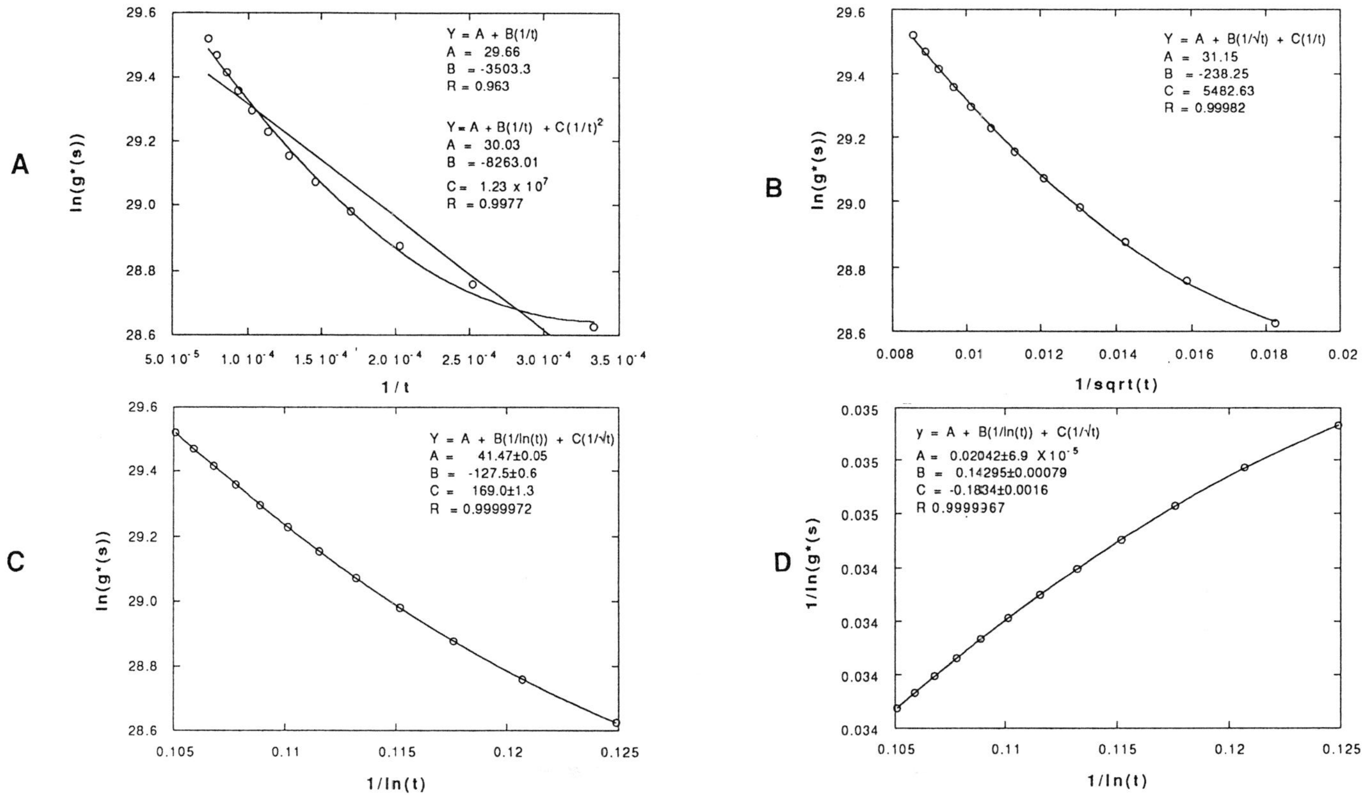

Figure 8. Plots of $\ln(g^*(s))$ (at s=10 S) *vs*. various functions of time for a two component system with s_1=8 S, D_1=5 F (M_1=144,000); s_2=10 S, D_2=5 F (M_2=180,000). *Versus* (A) 1/t (B) $1/\sqrt{t}$ (C) 1/ln(t) and (D) plot of $1/\ln(g^*(s))$ *vs*. 1/ln(t). The justification for the last form is that extrapolation to small values of $1/\ln(g^*(s))$ should be easier than to large values of $\ln(g^*(s))$.

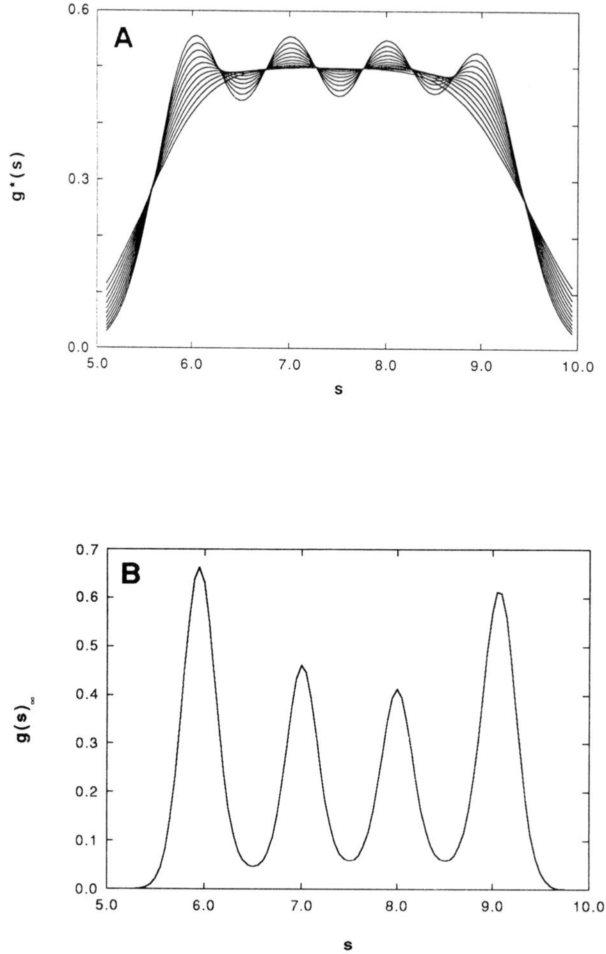

Figure 9. Plots of $g^{*}(s)$ *vs.* s for a four component system, all present in equal amounts: (A) ($s_i = 6$ S, 7 S, 8 S and 9 S and $D_i = 1.50$ F, 1.58 F, 1.63 F and 1.65 F; corresponding to M_i = 341,000, 398,000, 441,000 and 490,000, respectively, assuming $(1-\bar{v}\rho) = 0.270$). (B) g(s) and (C) G(s) extrapolations using eq. (8) and (D) g(s) and (E) G(s) extrapolations using eq. (9). The effect of neighbouring components is clearly seen in parts (D) and (E).

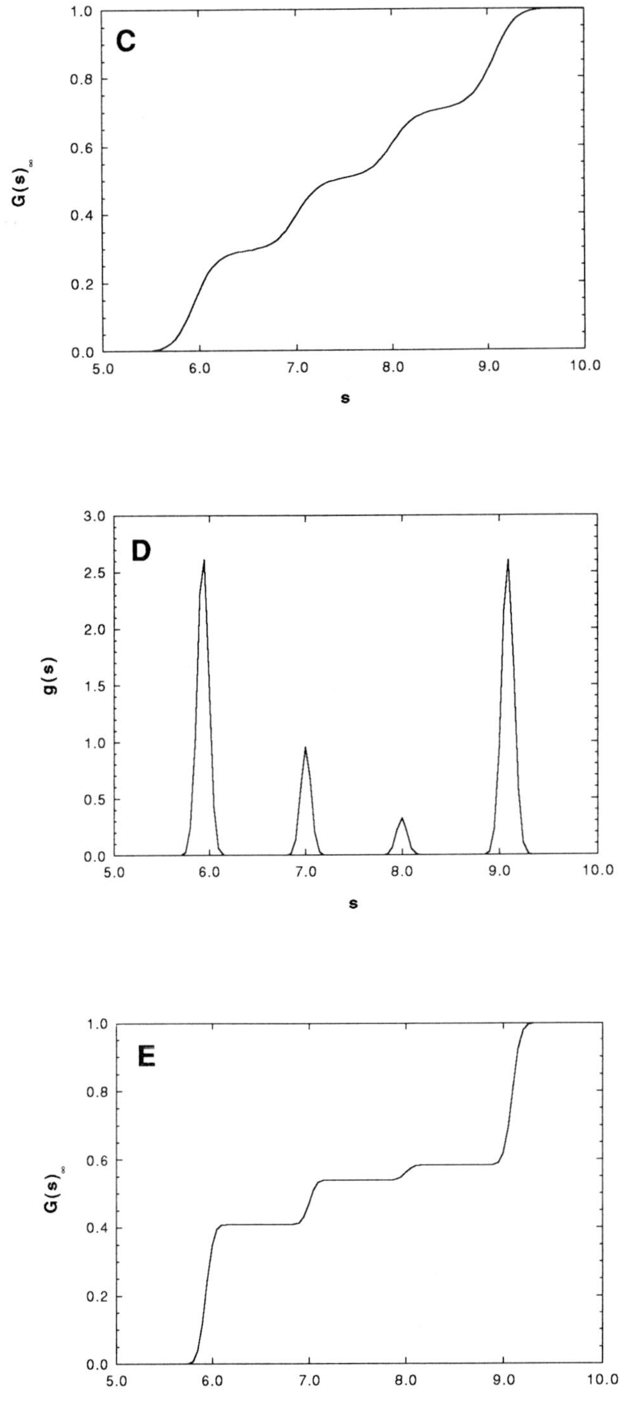
C
G(s)∞
s
D
g(s)
s
E
G(s)∞
s

these extrapolations is that they do not preserve the area of the original $g^*(s)$ curves and, therefore, must be renormalised. The renormalised curves are shown.

Two or More Species with Diffusion (Paucidisperse)

One of the requirements of an extrapolation procedure is that it at least preserve the relative amounts of each component in the original mixture. This is possible for a system composed of two components present in nearly equal amounts. However, when one of the components is present in significantly smaller amounts than the other, the component present in the lesser amount is usually underestimated. A system composed of two species (s_1=8 S, D_1=5 F, M_1=146,000 and s_2=10 S, D_2=5 F, M_2=182,000) was considered above (Fig. 1). The extrapolated g(s) curves (Fig. 7A) and corresponding G(s) curves (Fig. 7B) using several different extrapolation forms are shown for mixtures composed of different ratios of the two components (1:1 and 1:9). In the mixture of 1:9, the component present in the lesser amount was underestimated by more than a factor of two. This problem arises because as the $g^*(s)$ curves become sharper the contribution from the larger peak in the region of the maximum of the smaller peak is decreasing preventing it from growing as rapidly as it would in the absence of the larger peak. Therefore, it has extrapolated to a lower value than it otherwise would have. This problem is exacerbated in mixtures composed of more than two components. For example, in another system composed of four species (s_i = 6 S, 7 S, 8 S and 9 S and D_i = 1.50 F, 1.58 F, 1.63 F and 1.65 F; corresponding to M_i = 360,000, 399,000, 442,000 and 490,000, respectively) the middle two peaks do not grow as fast as the two outer ones even though they all attain equal height by the end of the run (Fig. 9). Therefore, the two inner components are underestimated when the curves are extrapolated. Fig. 9 shows two different extrapolations, both exhibiting this problem.

Two Species with Diffusion: the 5 cm Cell

The two species system described above was simulated for a 5 cm cell in order to investigate the behaviour of $g^*(s)$ at values of $\omega^2 st$ beyond the usual centrifuge cell. The values of $g^*(s)$ at s = 10 S were fitted to various of the extrapolation forms shown in eqs. (8-10) (Figs. 8B, C and D). Fig. 8A shows a fit to $\ln(g^*(s))$ *versus* a linear function of 1/t and a quadratic function of 1/t and shows that the functions give a poor fit. However, the ones shown in Figs. 8B, C and D give good fits indicating that these forms do represent the behaviour of $g^*(s)$ beyond the bottom of the usual 1 cm cell.

12. ARTIFICIAL DISPERSION

Single Boundary with D=0

The effect of dispersion introduced by the computational procedures was investigated by simulating a hypothetical case of sedimentation for which D=0 by generating a step function boundary whose position obeyed $r=r_m\exp(\omega^2 st)$ and whose plateau concentration obeyed $c_p=c_0\exp(-2\omega^2 st)$, where c_0 is the initial concentration (Fig.

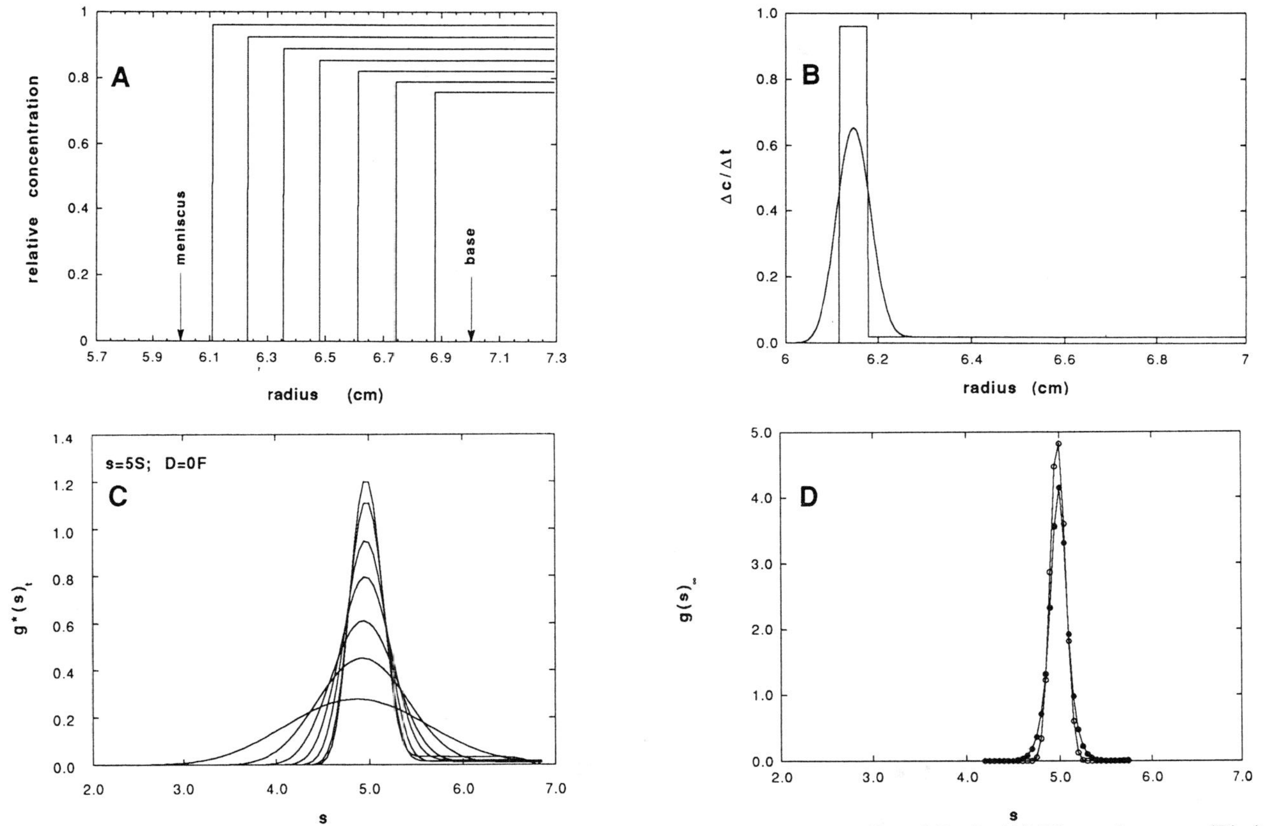

Figure 10. Effects of artificial dispersion on a single component system with s=5 S and D=0. (A) Plots of c *vs*. r. (B) A plot of $\Delta c/\Delta t$ with the smoothed curve superimposed. (C) Corresponding $g^*(s)_t$ curves and (D) extrapolated to $t=\infty$, showing that most of the effects of smoothing can be eliminated by extrapolation to infinite time. Solid circles extrapolated using eq. (8); open circles extrapolated using eq. (9).

10A). Difference patterns, $\Delta c/\Delta t$ *vs*. r, were generated by subtracting pairs of patterns (Fig. 10B). Dispersion was introduced by passing these difference curves through a typical smoothing process (nine times with an 11 point moving average spanning 360 μm). The $g^*(s)$ corresponding curves (Fig. 10C) were extrapolated to infinite time using eq. (8) (Fig. 10D). As expected, most of the artificial dispersion is eliminated in the extrapolation process.

Continuous Gaussian Distribution of s with D=0

The various extrapolation methods were tested further by applying them to a quasi-continuous Gaussian distribution of sedimentation coefficients (from 1 S to 100 S in increments of 1 S with mean of 50 S and standard deviation of 10 S). A continuous distribution was generated as a series of steps each obeying the same relations as in the previous section. The amplitude of each step was given by the Gaussian distribution function. The c *vs*. r curve generated at each time point was passed through a typical analysis procedure for scanner data as described in the legend to Fig. 11. The smoothing procedure introduces dispersion that can be nearly completely removed in the extrapolation process (Fig. 11A, curve b). The extrapolation of this distribution exemplifies the problem of asymptotic limited behaviour expected for $g^*(s)$ curves for continuous distributions. That is, as the contributions from dispersion are diminished with time, $g^*(s)$ approaches the true g(s) distribution asymptotically, implying that

$$\lim_{1/t \to 0} g^*(s) = g(s)$$

and that

$$\lim_{1/t \to 0} \left(\frac{\partial g^*(s)}{\partial (1/t)} \right) = 0$$

This tendency is clearly seen in the case of the Gaussian distribution of s for which D=0 and accounts for the observation that the extrapolation forms used tend to over-extrapolate at the maximum. If the curves are renormalised, then the values of g(s) near the tails of the distribution are underestimated. In fact, in a case of no artificial dispersion and for which the diffusion coefficient is identical to zero, $g^*(s)/g(s)$ is always equal to one and, therefore, $(\partial g^*(s)/\partial(1/t))$ is always equal to zero. The approach to an asymptote is clearly seen in Fig. 11B and C at all values of s. However, none of the forms used so far has included a term to account for this limiting behaviour.

13. NON-IDEALITY

Concentration Dependence

Hydrodynamic concentration dependence is handled in the usual way by repeating the analysis at several different concentrations and extrapolating the g(s) curves to zero concentration if necessary by either analytical[35] or graphical[36,37] means. The $\Delta c/\Delta t$ method allows the routine use of relatively low concentrations (0.02 mg/ml - 0.5

mg/ml with the Rayleigh optical system) because of its higher sensitivity; therefore, an extrapolation of $g(s)_t$ to zero concentration often is not required (Fig. 3).

Pressure Dependence

This chapter has been restricted specifically to incompressible systems (mainly, aqueous biopolymer systems); therefore, pressure dependence will not be dealt with in detail. However, there is an extensive literature describing procedures for dealing with density variation in the centrifugal field found most often in compressible organic solvents. The reader is referred to the procedure by Williams[38]. Earlier methods were presented by Eriksson[39]. Several empirical procedures have also been proposed[40,41,42,43,44].

14. CONCLUSION

The methods described here allow the rapid analysis of heterogeneous and interacting systems by sedimentation transport. Calculation of $\langle g^*(s)_t \rangle$ can now be carried out and displayed *while* the run is proceeding so that $g^*(s)$ distributions can be allowed to develop to the desired degree of resolution (within the limits allowed by diffusion and the dimensions of the cell); the run can be stopped when the necessary information has been obtained obviating the necessity of having to wait for the run to finish before the data can be analyzed. Extrapolated g(s) curves may also be computed while the run is proceeding. The use of the time derivative of the concentration profiles combined with averaging $g^*(s)$ curves has allowed an increase in the precision of the Rayleigh optical system by about two orders of magnitude.

By combining $g^*(s)$ analysis with real-time short column equilibrium methods[45], with its rapid attainment of equilibrium (*cf* Chapter 6), very rapid, precise analyses of interacting systems can now be carried out. Analytical ultracentrifugation remains the technique of choice for the detailed analysis of both interacting and heterodisperse biopolymer and synthetic polymer systems. The introduction of new instrumentation with modern computer interfacing will allow the routine investigation of macromolecular materials on a scale previously unattainable.

APPENDIX - DERIVATION OF $g^*(s)$ FROM dc/dt

First, we define the abscissa, s^*, through the following relationship for the case of a single, hydrodynamically ideal component with zero diffusion coefficient. The distinction between the sedimentation coefficient of an individual component, s, and this variable, s^*, must be kept in mind. In the case of zero diffusion coefficient, $s^* = s$, at the boundary position.

$$s^* = \frac{1}{\omega^2 t^*} \ln\left\{\frac{r^*}{r_m}\right\} \qquad \text{(A1)}$$

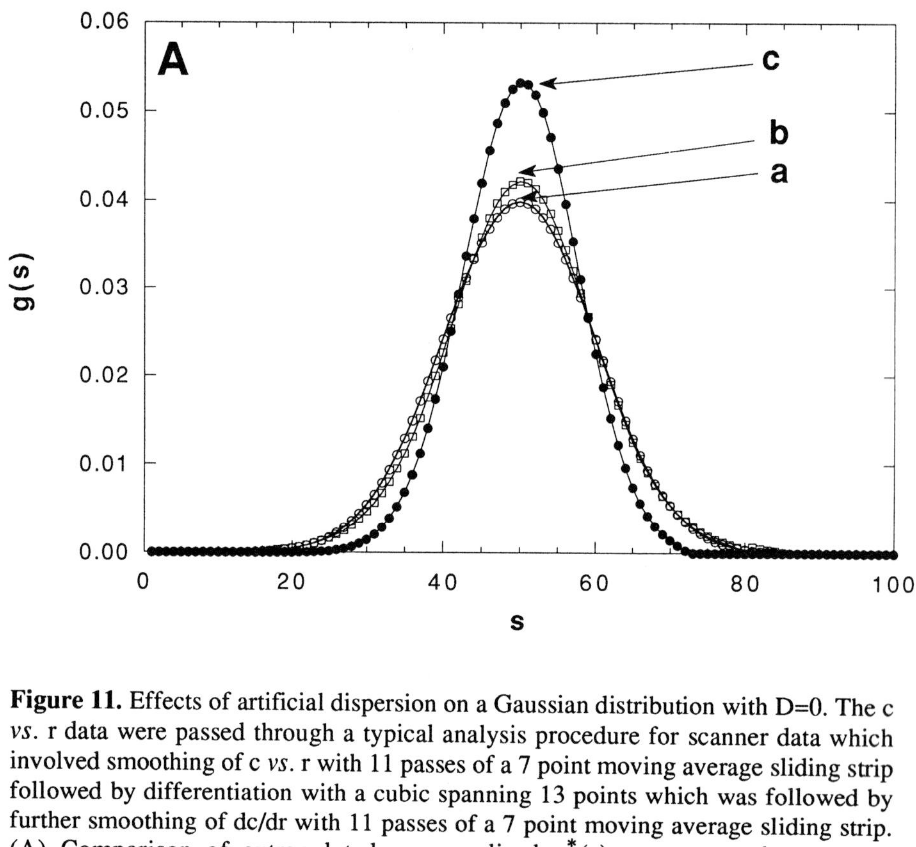

Figure 11. Effects of artificial dispersion on a Gaussian distribution with D=0. The c *vs.* r data were passed through a typical analysis procedure for scanner data which involved smoothing of c *vs.* r with 11 passes of a 7 point moving average sliding strip followed by differentiation with a cubic spanning 13 points which was followed by further smoothing of dc/dr with 11 passes of a 7 point moving average sliding strip. (A) Comparison of extrapolated, renormalised $g^*(s)_\infty$ curves to the true g(s) distribution function. (a, open circles) The true Gaussian distribution of sedimentation coefficients from 1 S to 100 S in increments of 1 S with mean 50 S and standard deviation 10 S; (b, open squares) g(s) extrapolated using eq. (7), and (c, solid circles) extrapolated using eq. (8). Extrapolations using eqs. (9) and (10) (not shown) resulted in large over-extrapolation both at the peak and at the off peak positions resulting in extrapolated curves much narrower than the true distribution. (B) and (C) Plots of $\ln(g^*(s))$ *versus* $1/\sqrt{t}$ and $1/\ln(t)$ respectively showing the limiting behaviour as $1/t \rightarrow 0$ (see text for discussion of the problem of the limiting behaviour in the limit when $D \rightarrow 0$). From (solid squares) 21 S to (open squares) 50 S with curves for 27 S, 33 S, 39 S and 45 S in between. The limiting behaviour is best seen from the slight downward curvature in the curve for 50 S as $1/t \rightarrow 0$.

B

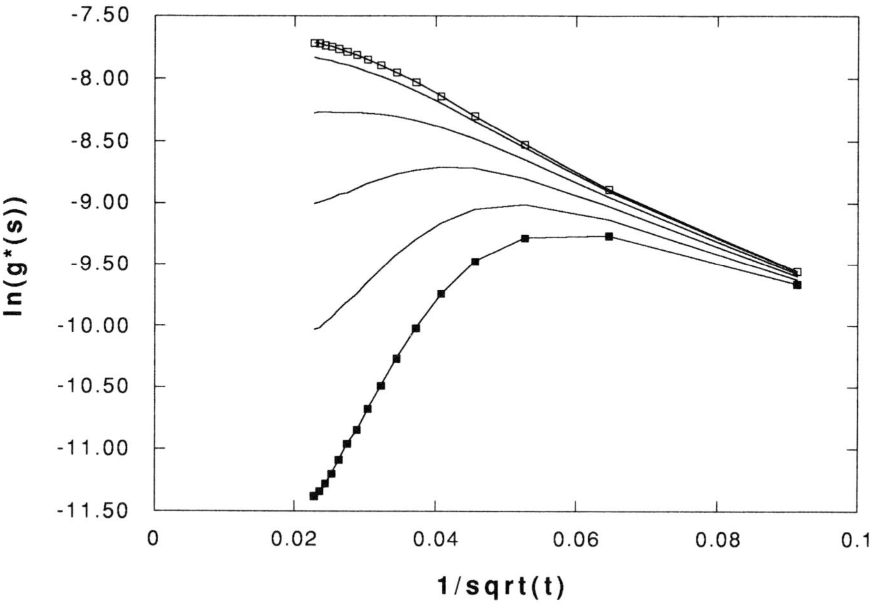

ln(g*(s))
-7.50
-8.00
-8.50
-9.00
-9.50
-10.00
-10.50
-11.00
-11.50
0
0.02
0.04
0.06
0.08
0.1
1/sqrt(t)

C

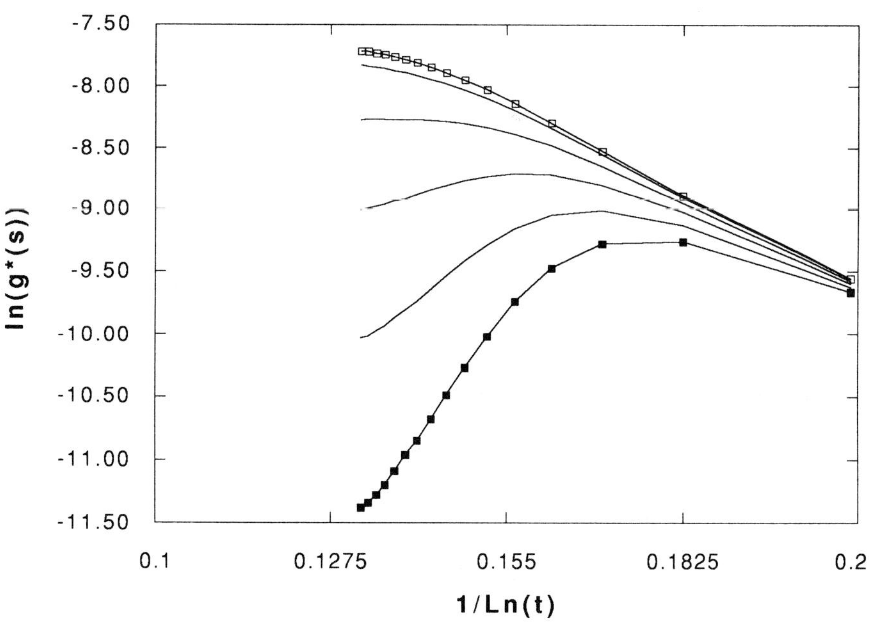

ln(g*(s))
-7.50
-8.00
-8.50
-9.00
-9.50
-10.00
-10.50
-11.00
-11.50
0.1
0.1275
0.155
0.1825
0.21
1/Ln(t)

$$F = \ln\left\{\frac{r^*}{r_m}\right\} - \omega^2 s^* t^* = 0 \tag{A2}$$

so that at any any given time, t^*, r^* corresponds to the position of the boundary or conversely, at any given radius, r^*, t^* corresponds to the time at which the boundary is located at that radius. Now, since,

$$dF = \left(\frac{\partial F}{\partial r^*}\right) dr^* + \left(\frac{\partial F}{\partial s^*}\right) ds^* + \left(\frac{\partial F}{\partial t^*}\right) dt^* \tag{A3}$$

we have

$$\frac{dr^*}{ds^*} = -\frac{\left(\frac{\partial F}{\partial s^*}\right)_{t^*}}{\left(\frac{\partial F}{\partial r^*}\right)_{t^*}}; \qquad \frac{dt^*}{ds^*} = -\frac{\left(\frac{\partial F}{\partial s^*}\right)_{r^*}}{\left(\frac{\partial F}{\partial t^*}\right)_{r^*}}$$

and

$$\left(\frac{\partial F}{\partial s^*}\right)_{t^*} = -\omega^2 t^* \qquad \text{and} \qquad \left(\frac{\partial F}{\partial r^*}\right)_{t^*} = \frac{1}{r^*};$$

so that

$$\frac{dr^*}{ds^*} = -\omega^2 r^* t^* \tag{A4}$$

Likewise,

$$\left(\frac{\partial F}{\partial s^*}\right)_{r^*} = -\omega^2 t^* \qquad \text{and} \qquad \left(\frac{\partial F}{\partial t^*}\right)_{r^*} = -\omega^2 s^*$$

so that

$$\frac{dt^*}{ds^*} = -\frac{t^*}{s^*} = \frac{\omega^2 t^2}{\ln(r_m/r)} \tag{A5}$$

These relations will be used below.

Now, assuming an infinitely long centrifuge cell, the time dependence of the concentration at an arbitrarily fixed position in the cell, for the case of D=0, for a single component is given by

$$c_r(t) = c_0 \exp(-2\omega^2 st)\, U(t^*-t) \tag{A6}$$

where $c_r(t)$ is the concentration at radial position r as a function of time; c_0 is the initial cell loading concentration; ω is the angular velocity of the rotor; s is the sedimentation coefficient; and t is time. $U(t^*-t)$ is the Heaviside unit step function defined by

$$U(t^*-t) = 1 \qquad 0 < t < t^* \tag{A7}$$

$$U(t^*\text{-}t) = 0 \qquad\qquad t^* < t$$

where t^* is the time at which the the boundary passes the point r. We can also write that the position of the boundary is given by

$$r^* = r_m \exp(\omega^2 s t^*) \tag{A8}$$

so that at t^*, $r=r^*$. Note that s is not starred in eq. (A8) since it is the actual value of s for the component under consideration.

Differentiating eq. (A6) with respect to time at constant r

$$\frac{\partial c}{\partial t} = -c_0 \exp(-2\omega^2 s t^*)\, \delta(t^*\text{-}t) - 2\omega^2 s c_0 \exp(-2\omega^2 s t)\, U(t^*\text{-}t) \tag{A9}$$

where $\delta(t^*\text{-}t)$ is the Dirac delta function defined as

$$\delta(t^*\text{-}t) = \infty; \qquad\qquad \text{for } t = t^* \tag{A10}$$

$$\delta(t^*\text{-}t) = 0; \qquad\qquad \text{for } t \neq t^*$$

The delta function also has the properties that

$$\int_{-\infty}^{+\infty} \delta(t^*\text{-}t)\, dt = 1 \tag{A11}$$

and

$$\int_{-\infty}^{+\infty} f(x)\, \delta(x\text{-}a)\, dx = f(a) \tag{A12}$$

Evidently, in the ideal case for which the delta function is an appropriate representation of the boundary, the second term of eq. (A9) is negligible with respect to the first term and we have for each component that

$$\frac{\partial c}{\partial t} = -c_0 \exp(-2\omega^2 s t^*)\, \delta(t^*\text{-}t) \tag{A13}$$

Eq. (A13) will also be correct for most practical cases for which the ratio of $s\omega^2/D$ is sufficiently large ($>80\ cm^{-2}$). The second term of eq. (A9) will also become insignificant compared to the first term if apparent distribution functions are extrapolated to infinite time to remove the effects of diffusion.

In the case of a continuous distribution of s, we arrive at the derivative of the total concentration by integrating over all values of s:

$$\frac{\partial (c/c_{0,T})}{\partial t} = -\int_0^{\infty} g(s) \exp(-2\omega^2 s t^*)\, \delta(t^*\text{-}t)\, ds \tag{A14}$$

since this applies for each species at the instant t^*, and noting that for D=0, $s=s^*$ at each boundary position, we can write also that

$$\frac{\partial(c/c_{0,T})}{\partial t} = -\int_0^\infty g(s)\exp(-2\omega^2 st^*)\,\delta(t^*-t)\,\frac{ds^*}{dt^*}\,dt \qquad \text{(A15)}$$

substituting for ds^*/dt^* from above, we have

$$\frac{\partial(c/c_{0,T})}{\partial t} = -\int_0^\infty g(s)\exp(-2\omega^2 st^*)\left(\frac{\ln(r^*/r_m)}{\omega^2 t^{*2}}\right)\delta(t^*-t)\,dt \qquad \text{(A16)}$$

Integrating and relying on eq. (A12), we have

$$\frac{\partial(c/c_{0,T})}{\partial t} = -g(s)\exp(-2\omega^2 st^*)\left(\frac{\ln(r^*/r_m)}{\omega^2 t^{*2}}\right) \qquad \text{(A17a)}$$

or

$$\frac{\partial(c/c_{0,T})}{\partial t} = -g(s)\left(\frac{r^*}{r_m}\right)^2\left(\frac{\ln(r^*/r_m)}{\omega^2 t^{*2}}\right) \qquad \text{(A17b)}$$

Rearranging and dropping the stars, but keeping them in mind, we have, finally,

$$g(s)_t = \frac{\partial(c/c_{0,T})}{\partial t}\left(\frac{\omega^2 t^2}{\ln(r_m/r)}\right)\left(\frac{r}{r_m}\right)^2 \qquad \text{(A18)}$$

Derivation of $g^*(s)$ from dc/dr

Now, for the sake of the completeness, we present a derivation of the Bridgman equation[4], using delta function notation to generate a relation between dc/dr and $g(s)_r$, following closely the derivation presented by Fujita[47] in 1975. At any given fixed instant in time, the radial distribution of concentration for a single component system, for which $D = 0$ is given by:

$$c(r) = c_0\left(\frac{r^*}{r_m}\right)^{-2} U(r-r^*) \qquad \text{(A19)}$$

where, as above, c is concentration, c_0 is the initial loading concentration, r is the radius from the centre of rotation, r_m is the radius of the meniscus, r^* is the radius of the boundary and $U(r-r^*)$ is the Heaviside unit Step function such that $U(r-r^*) = 0$ for $r < r^*$ and $U(r-r^*)=1$ for $r > r^*$. Differentiating eq. (A19) with respect to r, we have

$$\frac{\partial c}{\partial r} = c_0\left(\frac{r^*}{r_m}\right)^{-2}\delta(r-r^*) \qquad \text{(A20)}$$

$$\frac{\partial(c/c_0)}{\partial r} = \int_0^\infty g(s)\left(\frac{r^*}{r_m}\right)^{-2}\delta(r-r^*)\,ds \tag{A21}$$

$$\frac{\partial(c/c_{0,T})}{\partial r} = \int_{r_m}^{r_b} g(s)\left(\frac{r^*}{r_m}\right)^{-2}\delta(r-r^*)\,\frac{ds^*}{dr^*}\,dr \tag{A22}$$

Integrating and, again, relying eq. (A12), we have

$$\frac{\partial(c/c_{0,T})}{\partial r} = g(s)\left(\frac{r^*}{r_m}\right)^{-2}\frac{ds^*}{dr^*} \tag{A23}$$

After substituting for (ds^*/dr^*), rearranging and dropping the stars we have the Bridgman relation:

$$g(s)_r = \omega^2 rt\left(\frac{\partial(c/c_0)}{\partial r}\right)\left(\frac{r}{r_m}\right)^2 \tag{A24}$$

In the case of systems for which there is a significant contribution to boundary spreading from diffusion, we will note that g(s) distributions computed from eqs. (A18) and (A24) are apparent distributions. As has been customary in the literature, we will indicate the apparent distributions with a star (*viz* $g^*(s)_r$ or $g^*(s)_t$), not to be confused with the star introduced in eqs. (A1) and (A2).

Derivation of Relation for $g^*(s)_{max}$

For a single component, we have, from the Faxén equation (eq. (2.127) from Fujita[48]) that at the maximum gradient:

$$\left(\frac{\partial c}{\partial r}\right)_{max} = \left(\frac{c_0}{r_m}\right)\frac{\exp[-3\omega^2 st]}{\left\{\pi\,\frac{2D}{\omega^2 s r_m^2}\,(1-e^{-2\omega^2 st})\right\}^{1/2}} \tag{A25}$$

converting this to $g^*(s)_r$ using eq. (1), we have, after rearranging,

$$g^*(s)_{max} = \left(\frac{r}{r_m}\right)^3\frac{\exp[-3\omega^2 st]}{\left\{\pi\,\frac{2D}{\omega^2 s}\,(1-e^{-2\omega^2 st})\right\}^{1/2}}\,\{\omega^2 r_m t\} \tag{A26}$$

and, since $(r/r_m)^3 = \exp(+3\omega^2 st)$, we have, after simplification,

$$g^*(s)_{max} = \left(\frac{s}{2\pi D}\right)^{1/2}\{\omega^3 r_m\}\left(\frac{t}{(1-e^{-2\omega^2 st})^{1/2}}\right) \tag{A27}$$

As shown above, eq. (A27) can be used to obtain the diffusion coefficient and molecular weight under favourable circumstances.

REFERENCES

1. T. Svedberg and J.B. Nichols, *J. Am. Chem. Soc.*, 1923, *45*, 2910.
2. T. Svedberg and H. Rinde, *J. Am. Chem. Soc.*, 1923, *45*, 943.
3. T. Svedberg and H. Rinde, *J. Am. Chem. Soc.*, 1924, *46*, 2677.
4. W.B. Bridgman, *J. Am. Chem. Soc.*, 1942, *64*, 2349.
5. R.L. Baldwin and J.W. Williams, *J. Am. Chem. Soc.*, 1950, *72*, 4325.
6. J.W. Williams, R.L. Baldwin, W. Saunders and P.G. Squire, *J. Am. Chem. Soc.*, 1952, *74*, 1542.
7. N. Gralén and G. Lagermalm, *J. Phys. Chem.*, 1951, *56*, 514.
8. D.J. Winzor, R. Tellam and L.W. Nichol, *Arch. Biochem. Biophys.*, 1977, *178*, 327.
9. K.E. Van Holde and W.O. Weischet, 1978, *17*, 1387.
10. W.F. Stafford, *Biophys. J.*, 1989, *55*, special appendix to the abstracts.
11. W.F. Stafford, *J. Cell Biol.*, 1990, *109*, 1537.
12. R.L. Baldwin and K.E. Van Holde, *Fortschr. Hochpolym.-Forsch.*, 1960, *Bd. 1. S.*, 451.
13. V.N. Schumaker and H.K. Schachman, *Bioc. Biop. Acta.*, 1957, *23*, 628.
14. V.N. Schumaker, E.G. Richards and S.T. Freer, *J. Mol. Biol.*, 1965, *12*, 517.
15. W. Mächtle, *Makromol. Chem.*, 1984, *185*, 1025.
16. W. Mächtle and U. Klodwid, *Makromol. Chem.*, 1979, *180*, 2507.
17. W. Mächtle, *Die Angewandte Makromoleculare Chemie*, 1988, *162*, 35.
18. E.G. Richards, *Amer. J. Card.*, 1989, *63*, 1214.
19. M.S. Runge, T.M. Laue, D.A. Yphantis, M.R. Lifsics, A. Saito, M. Altin, K. Reinke and R.C. Williams, Jr., *Proc Natl. Acad. Sci. U.S.A.*, 1981, *78*, 1431.
20. T.M. Laue, Ph.D. Dissertation, University of Connecticut, Storrs, CT, 1981.
21. R. Cohen, J. Cluzel, H. Cohen, P. Male, M. Moignier and C. Soulie, *J. Biophys. Chem.*, 1976, *5*, 77.
22. J.L. Bethune, *Biochemistry*, 1970, *9*, 2737.
23. B.J. McNeil and J.L. Bethune, *Biochemistry*, 1973, *12*, 3254.
24. B.J. McNeil and J. Bethune, *Biochemistry*, 1973, *12*, 32.
25. W.F. Stafford, Abstracts of the International Biophysics Congress, 1990, Vancouver, 1990.
26. L.W. Nichol and D.J. Winzor, *Methods Enz.*, 1986, *130*, 6.
27. G.A. Gilbert, *Discuss. Farad. Soc.*, 1955, *20*, 68.
28. G.A. Gilbert, *Proc. Roy. Soc., Ser A*, 1959, *250*, 377.
29. G.A. Gilbert, *Proc. Roy. Soc., Ser A*, 1959, *253*, 420.
30. F.B. Hildebrand, 'Introduction to Numerical Analysis', 1956, McGraw-Hill Book Co. Inc., New York.
31. J. Sinard, W.F. Stafford and T. Pollard, *J. Cell Biol.*, 1990, *109*, 1537.
32. J.-M. Claverie, H. Dreux and R. Cohen, *Biopolymers*, 1975, *14*, 1685.
33. H. Faxén, *Arkiv Mat. Astron. Fysik.*, 1929, *21B*.
34. O. Lamm, *Arkiv Mat. Astron. Fysik.*, 1929, *21B*.
35. L.J. Gosting, *J. Am. Chem. Soc.*, 1952, *74*, 1548.
36. R.L. Baldwin, *J. Am. Chem. Soc.*, 1954, *76*, 402.
37. J.W. Williams, W.M. Saunders and J.S. Cicirelli, *J. Phys. Chem.*, 1954, *58*, 774.
38. J.W. Williams and W.M. Saunders, *J. Phys. Chem.*, 1954, *58*, 854.

39. J.W. Williams, 'Ultracentrifugation of Macromolecules, Modern Topics', 1972, Academic Press, New York.
40. A.F.V. Eriksson, *Act. Chem. Scand.*, 1956, *10*, 360.
41. T. Kotaka and N. Donkai, *J. Poly. Sci., A-2*, 1968, *6*, 1457.
42. I.H. Billick, *J. Poly. Sci.*, 1962, *62*, 167.
43. M. Wales and S.J. Rehfeld, *J. Poly. Sci.*, 1962, *62*, 179.
44. J.E. Blair and J.W. Williams, *J. Phys. Chem.*, 1964, *68*, 161.
45. M. Wales in 'Characterization of Macromolecular Structure', Natl. Acad. Sci., Washington, D.C., 1968, p. 343 ff.
46. D.A. Yphantis, *Ann. N.Y. Acad. Sci.*, 1960, *88*, 856.
47. H. Fujita, 'Foundations of Ultracentrifugal Analysis', 1975, John Wiley and Sons, New York.
48. H. Fujita, 'Mathematical Theory of Sedimentation Analysis', 1962, Academic Press, New York.

21

The Concentration Dependence of Sedimentation

By A.J. Rowe

NATIONAL CENTRE FOR MACROMOLECULAR HYDRODYNAMICS, LEICESTER LABORATORY, UNIVERSITY OF LEICESTER, LEICESTER LE1 7RH, U.K.

1. INTRODUCTION

The sedimentation coefficient (s) of a solute particle, as measured in a sector-shaped cell in a rotor, is defined as the velocity of movement (dr/dt) per unit field applied:

$$s = \frac{1}{\omega^2 r}\frac{dr}{dt} \tag{1}$$

where ω is angular velocity (rad/sec) and r is radial position. The s value so defined is usually 'normalised' *via* simple conversion formulae to standard conditions - namely water as solvent at 20.0°C.

The units of s are thus *seconds*, but as the values yielded by macromolecular solutes are very small, the general practice for many decades has been to use the 'Svedberg Unit' (S), this having the value 10^{-13} seconds. More consistent with S.I. nomenclature is the use of the picosecond (10^{-12} seconds), but this has not become general practice: presumably because 'S values' of well-known structure are well entrenched, and there is a reluctance to refer to '1.6 picosecond RNA' rather than '16 S RNA', for example.

Under the assumption that the sedimenting particles are at terminal velocity, the s value thus defined is related to the molecular weight M (g/mol) or 'relative molecular mass' (M_r), the frictional coefficient (f) and partial specific volume ($\bar{v}$) of the particle by

$$s = \frac{M_r(1 - \bar{v}\rho)}{N_A f} \tag{2}$$

where ρ is the solvent density. The impression given by eq. (2) that s is linear in M_r is misleading: since f is related to the effective hydrodynamic radius of the particle, and this itself depends on the cube root of the particle volume (and hence for *all other things being constant* to $M_r^{1/3}$), it is much more nearly true to say that s varies with $M_r^{2/3}$ for a series of particles of closely related (globular) shape, solvation and partial volume.

As only length and time measurements are involved in the determination of s, and these are inherently measurable with high precision and accuracy, s values should be both accurate and precise. This is indeed the case. For well-defined species, where a single, well-defined sedimenting boundary can be resolved, a precision of down to 0.2% and an absolute accuracy of better than 0.5% can be expected. The principal cause of uncertainty in the latter is the absolute viscosity of the solvent, which for aqueous solvents is temperature dependent to an extent that poses problems for the determination of absolute rotor temperature to an adequate accuracy.

The precision with which s values can be measured has meant that it has long been appreciated that the sedimentation rate of all solutes shows a measurable and in some cases large dependence upon solute concentration (c). From this follows the long established custom of extrapolating s values - measured of necessity at finite concentration - to 'infinite dilution' to give the extrapolated value s^0. For the purposes of extrapolation it has been assumed that either s or 1/s is linear to an adequate extent in c (eqs. (3) and (4)): *i.e.* that a limiting slope of value k_s (ml/g) *does* actually exist:

$$s = s^0 (1 - k_s c) = s^0 (1 - K_s \phi) \tag{3}$$

or

$$\frac{1}{s} = \frac{1}{s^0} (1 + k_s c) = \frac{1}{s^0} (1 + K_s \phi) \tag{4}$$

(with the volume fraction $\phi = cv_s$, where v_s is the specific volume of the 'hydrated' or 'solvated' particle and the dimensionless parameter $K_s = k_s/v_s$). We will see below that this is a formal necessity, but it has long been observed experimentally that at least for monodispersed, non-interacting solutes, the approach to the c=0 ordinate is not asymptotic. From the point of view of laboratory practice, it is accepted that especially for solutes of high particle asymmetry the reciprocal plot (eq. (4)) is linear over a wider range in c than the direct plot: but for compact, near-spherical particles the linear plot (eq. (3)) is as good and may be preferable (*vide infra*). The value for the regression coefficient k_s (or K_s) can itself be determined with moderate precision (1% - 2%) provided an adequate range in c (or ϕ) can be studied.

Values for K_s have historically been more readily measured than interpreted. There are two principle reasons why s depends upon c: hydrodynamic and charge interactions between the individual particles (two-component system) and between the particles and other solute components (more than two components). Historical explanations for the hydrodynamic interaction were fully reviewed by Schachman[1]. These were of an empirical or semi-empirical nature, and will not be further reviewed here other than by reference to the classical studies of Burgers[2,3]. Two relatively recent accounts have been published, which are reviewed below, but these offer non-identical solutions for the simplest case of non-swollen spherical particles, and the reason for the discrepancy is not clear. One account does however fit the experimental evidence very closely for a wide range of solutes. With regard to the 'charge interactions' there appears to be no published theory which describes these in even semi-rigorous fashion.

It should be noted that s values are transport coefficients determined in a system well displaced from equilibrium. There is thus no precise correlation obtained or expected with the virial coefficients which describe the concentration dependence of equilibrium distributions (Wills and Winzor, Chapter 17, this volume), although in general terms it is obvious that a large virial coefficient will be associated with a large k_s value.

2. HYDRODYNAMIC INTERACTIONS AND CONCENTRATION DEPENDENCE

If we can assume that charge effects between particles are negligible, then it follows that all effects of raising the solute concentration must be attributable to particle-particle interaction *via* the solvent medium. The question at once arises as to whether such an assumption, *i.e.* of no charge interaction, can ever be justified. Indeed the classical theory of the solubility of macromolecular solutes as accepted in colloid chemistry (DLVO theory*, see Shaw[4]) supposes that solutes exist in stable mono-dispersion in solvent as a result of a balance of forces in which charge repulsion predominates! There may however be an answer to this apparent contradiction. The ionic strength of the solvent is normally adjusted to at least 100 mM to obviate charge effects. Under these conditions the Debye-Hückel screening distance ($1/\kappa$) is about 1 nm. This is considerably smaller than the average particle-to-particle separation in dilute solutions. Even in concentrated solutions, it must be the case that many charges arising from ionised side chains will lie *within* the plane of hydrodynamic shear by reason of either (i) solvation layers or (ii) surface rugosity, in either case much reducing their contribution. The experimental evidence certainly favours this assumption, since above an ionic strength I = 100 mM there seems to be no published evidence to suggest continuing I-variance of s, as would be expected if ($1/\kappa$) continued to be important. Thus we can cautiously accept that pure hydrodynamic interaction is - at least within plausible bounds of empirical accuracy - an effect which can be conceptualised, defined and studied.

The three basic theories described below (due to Burgers, Batchelor and Rowe) all seek to evaluate the leading order term in a c- (or more properly ϕ-) expansion of s. It may be questioned whether such a term must exist. In fact, if the solute particles individually occupy finite volume and that (specific) volume is not itself concentration dependent, then return flow of displaced solvent cannot have a zero value, and hence a resultant first order term *must exist*, although its magnitude cannot be stated on this argument.

3. THE THEORY OF BURGERS

Early and important studies on this problem were published in the early 1940s by Burgers[2,3]. These followed classical hydrodynamic approaches originating from the Navier-Stokes formulations for the migration of spherical particles through a continuum. For such spherical particles, Burgers considered two effects to be

---

* DLVO theory: Derjaguin-Landau-Verwey-Overbeek theory.

operative: a 'return flow' of displaced and convected solvent, and an increased 'effective viscosity' of the solvent, both effects arising from the effects of (n-1) other particles on each individual particle. For the former, he deduced that for a solute volume fraction ϕ the velocity of movement (u_c) would be reduced by factor (1-5ϕ) relative to the 'infinite dilution' velocity, u_0 (eqs. (61a), (62a) and (64), *loc. cit.*). However, Burgers also noted that the sedimentation would be retarded by the increasing density of the solution as the solute concentration increased: an effect amounting to (1-ϕ) (eq. (15), *loc. cit.*). There is thus a reduction in the limiting coefficient from 5 to 4 *if the s values are values corrected for solution rather than solvent density*. Fujita[5] has since that time shown the correctness of this latter procedure.

Burgers then found an 'effective viscosity' term to be added to the returned solvent term, and overall gave an estimate of K_s = 6.875 for values corrected for *solvent* concentration. There seems unfortunately to have been little interest amongst either biochemists or others in confirming the correctness of this estimate for the c-regression of spherical particles.

4. THE THEORIES OF BATCHELOR AND OTHERS

Still considering spherical particles, Batchelor has over a number of years undertaken a detailed re-consideration of this problem, using more current methodology in hydrodynamics, based upon solutions to the problem of the interaction of two spheres, and assumptions of pairwise additivity. Batchelor's solution[6] published in 1972 gives K_s = 6.55 (volume/volume) to leading order in ϕ, slightly lower than the estimate of Burgers. Following Fujita[5] we should correct this for the *solvent density* effect, giving K_s = 5.55 to an adequate approximation. Among those working in the field of fluid mechanics, however, the former value is generally accepted as definitive (see for example Cichocki and Felderhof[7], Beenakker and Mazur[8] and Ladd[9]); and current work is occupied with attempts to define higher order coefficients (*e.g.* Phillips *et al.*[10]).

A problem with this solution which occurs with an analysis of the biophysical, biochemical and biopolymer literature however is that Batchelor's value of K_s = 5.55 (or the even higher earlier value of Burgers) is not supportable by empirical evidence, and indeed, appears to be in serious disagreement with it (*vide infra*) - see *e.g.* refs. 11-13.

5. THE THEORY OF ROWE

A re-examination of this problem from a totally different theoretical standpoint was carried out by Rowe[14] in 1977. The theory as published has a number of ellipticities, and a re-worded presentation is given here. A full derivation is given in Chapter 28.

The theory argues that at any solute concentration a frame of reference must be definable in which $s_c = s_0$, and that the concentration regression equation (eq. (3) - in volume units) has the form of a transformation between this novel frame of reference and the conventional cell-fixed frame. The evaluation of the coefficient k_S therefore calls only for an adequate definition of the volume fluxes associated with solute particle movement in these two frames of reference. To achieve this, it was noted that the conventionally defined frictional ratio (f/f_0) can by a simple re-arrangement of the Stokes equation be regarded as a linear phenomenological coefficient defining an effective volume flux. It can then be shown that

$$k_S = 2\bar{v}\left\{\frac{v_S}{\bar{v}} + \left[\frac{f}{f_0}\right]^3\right\} \tag{5}$$

which for the case of compact spheres yields a value $k_S = 4v_S$ or equivalently $K_S = 4$. The treatment also covers the concentration dependence of the intrinsic viscosity, and is applicable under both headings to *any* particle of any shape or solvation.

6. COMPARISON OF THE THEORIES OF BURGERS, BATCHELOR AND ROWE

It is difficult to make a comparison between the theory of Rowe and the theories based upon more classical hydrodynamics, there being no clear point of overlap. The Rowe theory in essence states that the problem is solved merely by its careful definition and by one or two particular and novel constructions: it does however make use of a construction which is equivalent to postulating that an *equivalent* solvent flux can be used to describe the total concentration dependence. This equivalent flux has the value of 3 volumes/unit particle volume for compact spheres. It may be coincidence but as noted above this is precisely the value computed by Burgers for the same property (*i.e.* 3 volumes + 1 self-volume = 4 total volumes). It has been suggested (Rowe[14]) that a reconciliation with treatments based upon classical hydrodynamics is possible if one assumes that the 'effective viscosity' effect is not an additive term, but is rather an *alternative* way of (partly) computing the total dependence. This appears to be a logical possibility, but it is hard to see how the issue can be advanced further.

7. EMPIRICAL EVIDENCE AND THE THEORIES OF THE HYDRODYNAMIC EFFECT

Since the theories of classical hydrodynamics deal only with spherical particles, we are restricted in making a direct comparison to this area. The very careful work of Cheng and Schachman[15] for polystyrene latex spheres gave a value $K_S = 4.06$ for the limiting linear coefficient, as measured in the analytical ultracentrifuge. Re-fitting the data today, one would probably not use a simple quadratic expression, but this would hardly cause change in the estimate of K_S by more than a trivial amount, and overall it is hard to see any significant sources of error in this classical study. There is in addition other and much more extensive (if slightly less precise) data available from the many studies which have been performed on 'globular' proteins[11] and viruses[12,13]. A knowledge of K_S for a solute permits one to compute the molecular mass if the solute is assumed to be spherical: and also for non-spherical solutes if the swelling is known or can be assumed (Rowe[14]). The agreement of the relative molecular mass thus computed with the known relative molecular mass is a clear test of the use of the value $K_S = 4$ for the case of spheres: and a test of the relationship (eq. (5)) in general for non-spherical particles. It was shown initially (Rowe[14]) that such an agreement was indeed found for a number of proteins, and in a more extensive study the relationship was found to be empirically valid for a wide range of macromolecular solutes (Rowe[16]). The degree of scatter found is not larger than would be expected from literature values: for 31 proteins and viruses a ratio of 1.02 (S.D. 0.06, S.E. of mean 0.01) was found for M (computed from s and k_S) to M (known value). Similarly for other solutes, such as cellulose acetate derivatives in cuprammonium sulphate and in acetone, aqueous levans and certain organic polymers in esters, the ratio was in each case within the range 0.97 to 1.05, although the standard deviations and hence standard errors of the estimate were considerably higher (up to 0.10 for the latter). Moreover, the extended theory of Rowe[16] described below is capable of describing the detailed regression over a considerable range for both spherical and non-spherical particles: clearly any basic error in the value of k_S for spheres would again be reflected here, unless one is to assume a most unlikely type of error cancellation.

In conclusion, there seems to be no real doubt that the empirical value of K_S for spheres can be given as 4.0±0.2, with a probable error lower that this conservative estimate. This value is removed by a distance totally outside any plausible experimental error from the Batchelor theoretical value of 5.55. Estimates of K_S by techniques other than analytical ultracentrifugation (*e.g.* settling under gravity) are possible, but these are of much lower precision. Even so, the range of current estimates for K_S from such approaches which extends from 5 (Kops-Werkhoven and Fijnaut[17]) down to 3 (Ham and Homsey[18]), both published values being corrected for solution density, with estimated errors of the order of ±1, probably supports the lower theoretical estimate. The fact that the theory of Rowe predicts that $K_S = 4$ does not of course prove the correctness of that theory: but there is a good degree of agreement found not only for spheres but for particles of general shape and swelling between prediction based upon this theory and experiment. In that sense the theory is successful to a level well beyond any other theory published to current date.

8. AN EXTENDED THEORY OF CONCENTRATION DEPENDENCE

An extension of the original theory of Rowe, derived and published in preliminary form (Rowe[16]), aims to cover the total range of solute concentration, up to the maximum packing fraction (ϕ_p in volume terms) obtainable. Under such conditions the system is removed from the simple limiting case, and the dependence upon concentration must now be expressed in more general form by an equation of the type

$$s = s^0 \{1 - g(c)\} \tag{7}$$

where the function g(c) is of unknown form.

The extended theory is derived on the basis of a simple concept. As the concentration of a system of sedimenting particles increases, and the inter-particle separation becomes significant in relation to the particle dimensions, so the force which must be applied *per particle* to produce a given velocity includes a component which tends to reduce that force. This effect was described by Burgers[2,3] and named the 'cloud effect'. It arises because the force required is related in conventional theory of laminar flow to the friction associated with the slippage of (notional) overlapping planes and, as particles approach, a degree of co-operativity sets in.

At first sight this merely states a familiar type of multi-body problem, insoluble in analytical terms. The extended theory circumvents this difficulty by a statistical approach, derived from the mathematics of finite probability space. Using the construction employed in the basic theory, with volume fluxes identified ultimately within the cell fixed frame of reference, a solution for g(c) of some simplicity is obtained:

$$g(c) = \frac{k_s c - \frac{c v_s^2 (2\phi_p - 1)}{\phi_p^2}}{k_s c - 2 c v_s + 1} \qquad (8)$$

This expression contains no arbitrary or empirical terms, and should apply to the hydrodynamic interaction for all solutes over the entire concentration range.

Eq. (8) is found to predict types of c-dependence which closely mirror those found in empirical practice. The regression of s upon c is linear over a modest range for compact spheres, but for asymmetrical particles a much better fit is seen with the reciprocal plot: whereas for particles of extremely high k_s (and particular if swelling is also present) the steep initial fall in s values with c leads on to an almost plateau region in which the fall becomes very much less steep (Fig. 1). We have tested the applicability of eq. (8) to a range of proteins, two examples are shown in Fig. 2.

There is again an excellent fit in each case, and the values yielded for the parameters can be checked by back-substitution into the relevant equations to yield estimates for M_r which are accurate within experimental error (Table 1).

Table 1. Estimating M_r from $s^0_{20,w}$ and k_s.

	$s^0_{20,w}$ (S)	k_s (ml/g)	v_s (ml/g)	ϕ_p	$v_s / \bar{v}$	M_r
Fibrinogen	8.54	17.7	2.14	0.40	3.01	314000
Serum albumin	4.05	7.87	1.50	0.49	2.03	65215

The value for the 'swelling' of the particle ($v_s / \bar{v}$) was also estimated independently for the same preparation of fibrinogen from the relative concentration dependence of the viscosity and the sedimentation (Rowe[14]) yielding a value of 2.57, which for this parameter is satisfactory agreement with the above tabulated value.

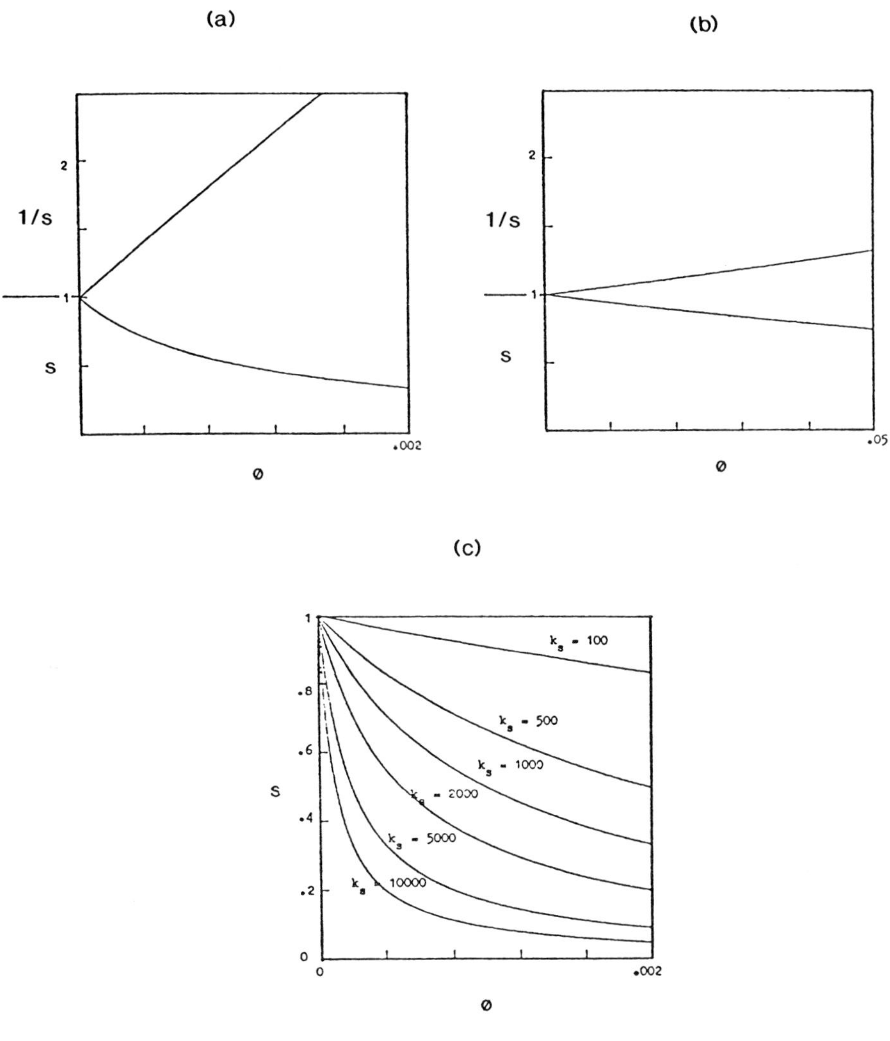

Figure 1. Curves illustrating the dependence of sedimentation on volume fraction (ϕ) as calculated from eq. (8) for different ranges of values of K_S and v_S. (a) computed for $K_S = 1000$: the reciprocal plot is linear over a much wider range than the direct plot; (b) computed for $K_S = 6$: both plots are close to linear; (c) computed for increasing values of K_S, showing the tendency to 'plateau' at higher volume fractions with higher K_S values.

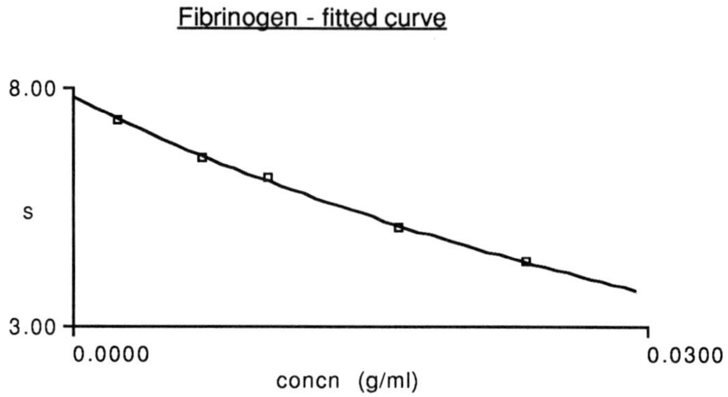

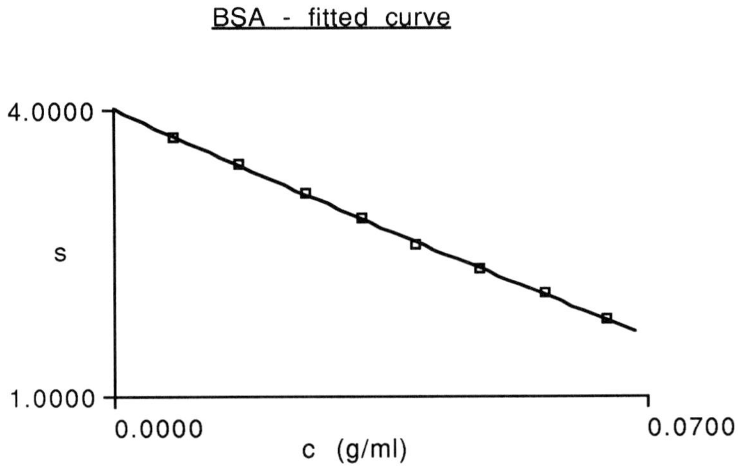

Figure 2. The measured sedimentation-concentration dependence of fibrinogen (above) and bovine serum albumin (below), with the regression lines obtained by non-linear least squares fitting of eq. (8). Note that for the case of fibrinogen the s values are *uncorrected*, and therefore differ slightly from those quoted in Table 1.

9. THE FITTING OF CONCENTRATION REGRESSION EQUATIONS

As noted earlier, over modest ranges of solute concentration simple linear regression (better usually of 1/s) in c suffices to obtain an adequate estimate for k_s or K_s (eq. (4)). It would be expected that the derivation of a general equation covering a full range in c would enable a better (or at least more valid) fit to be performed by non-linear least squares methods.

This expectation is not, in our general experience, wholly fulfilled. The reason is that *four* parameters must now be fitted: $s^0_{20,w}$, K_s, v_s and ϕ_p, as compared to *two* ($s^0_{20,w}$, K_s) in the case of eq. (4). This throws severe strain on the precision of the basic data. For spherical particles especially, detailed computer simulation shows that the stability in respect of the solution for K_s and v_s is found to be low - not surprisingly, since their contribution is very similar. The stability with respect to the maximum packing fraction ϕ_p is better, but unfortunately our knowledge with respect to values for this parameter for most macromolecular solutes is extremely limited, and hence it must be treated as a variable and not as a pre-determined constant. That being said, one would expect on general principles that values for ϕ_p would be somewhat less for proteins than for perfect spheres, because of less perfect ability to pack together, and the values estimated from eq. (8) and noted above are consistent with this.

The overall conclusion however must be that there is little or no advantage, if one is seeking to determine an experimental value for k_s, in fitting to eq. (8) rather than to eq. (4), unless one has data over an unusually large range in c.

10. WHAT PRACTICAL USE CAN BE MADE OF k_s VALUES?

Is all of the above merely abstract theory, or are there practical reasons why we should wish to determine k_s values for macromolecular solutes? There are two areas in which practical application has been found:

(i) Determination of Molecular Mass for Large, Asymmetrical Solutes

Particles with an M_r in excess of 10^6 pose certain problems for mass determination. The attainment of sedimentation equilibrium is slow, and requires very low rotor speeds. Light scattering is not simple, because of clarification problems and (where asymmetry is high) because of the breakdown of the Rayleigh-Gans approximation. Determination of the diffusion coefficient (to combine with s) *via* quasi-elastic light scattering calls in the latter case for extrapolation to zero angle, which can be tedious.

If the extrapolated s value is being measured, then k_s is determinable with no extra experimentation. For asymmetrical solutes the value of M_r determined from the conventionally derived equation :

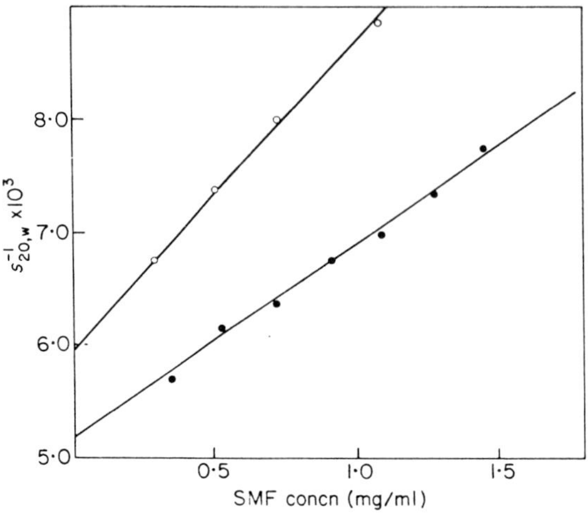

Figure 3. The sedimentation-concentration dependence of synthetic myosin filaments in the presence of 0.2 mM Mg^{2+} (open circles) and 3.0 mM Mg^{2+} (filled circles). The molecular mass values computed *via* eq. (9) were 1.78×10^6 and 1.84×10^6 respectively, showing that the very large change in k_s (from 470.5 ml/g to 334 ml/g) and in $s^0_{20,w}$ (from 169 S to 193 S) as the magnesium level is raised arises from a change in filament conformation and *not* from a change in filament mass. From Persechini and Rowe[19].

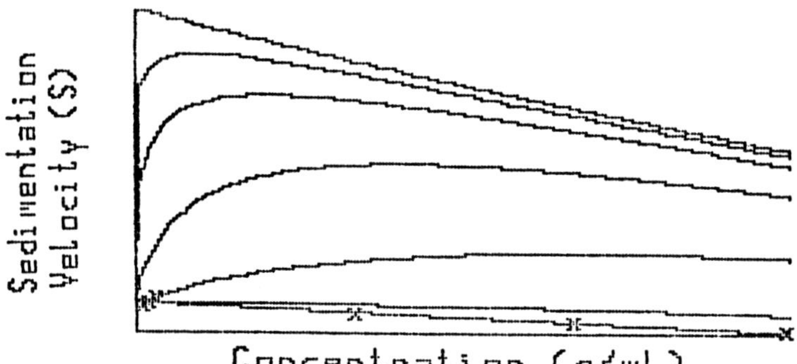

Figure 4. The computed regression curves for the sedimentation-concentration dependence of chloramphenicol acetyl transferase in the presence of 1 mM Co^{2+}, for degrees of dimerisation specified by (from lower to upper traces) K_a = 0, 10, 100, 1000, 10000, 100000 and infinity respectively. The experimental values (x) are seen to lie on the K_a = 0 line[21].

$$M_r = (6\pi\eta s)^{1.5} \left\{ \frac{3\bar{v}}{4\pi} \left[\frac{k_s}{2\bar{v}} - \frac{v_s}{\bar{v}} \right] \right\}^{0.5} \tag{9}$$

is very insensitive to the assumption of a plausible value for $v_s/\bar{v}$, since by inspection that term in the second bracketted expression is small in comparison to the first term, *i.e.* the term $k_s/2\bar{v}$. An example of the use of this approach is taken from a study by Persechini and Rowe[19] on the effect of divalent cations on the sedimentation of myosin filaments. The determination of M_r from s and k_s provided in this study a clear proof that a conformational and not a mass change was involved (Fig. 3).

(ii) Estimation of Interaction Constants for Monomer-Dimer Equilibria

The method of Gilbert and Gilbert[20], based upon a theoretical analysis which shows that a single sedimenting boundary is seen in the case of a rapidly established monomer-dimer equilibrium, allows K_a values to be estimated by iterative approaches from sedimentation velocity data over a range of solute concentrations. The approach requires a knowledge of K_s (as well as the extrapolated s value) for both monomer and dimer species. Since K_s can be computed (eq. (5)) given a knowledge of M_r and s^0, much more critical use can be made of the Gilbert approach than if K_s values must be treated as unknown parameters needing to be 'guessed'.

A simple application is shown in Fig. 4. The possible presence of dimerisation in the enzyme chloramphenicol acetyl transferase is explored on this basis, and the results show that in this case no significant dimerisation is present[21].

11. CONCLUSIONS

Although problems remain in reconciling the two theoretical approaches described above, in practice it is clear that the use of eq. (5) (and those derived from it) causes no errors significant at the level of good modern experimentation. Certain problems can be tackled on the basis of eq. (5) which would otherwise pose difficulties, and it seems likely that further theoretical and experimental work will in the near future clarify the problems noted above, and will lead on to a general methodology becoming available which will make the study of relatively concentrated solutions much more practicable than at present. This will be of some direct biological interest - cellular fluids are in general rather concentrated with respect to macromolecular solutes - as well as being of empirical importance in connection with the developing use of the modern analytical ultracentrifuge as a powerful research tool.

REFERENCES

1. H.K. Schachman, 'Ultracentrifugation in Biochemistry', Academic Press, New York and London, 1959.
2. J.M. Burgers, *Proc. Ned. Akad. Wet., Amsterdam*, 1941, *44*, 1045.

3. J.M. Burgers, *Proc. Ned. Akad. Wet., Amsterdam*, 1942, *45*, 9.
4. D.J. Shaw, 'Introduction to Colloid and Surface Chemistry', Butterworths, London, 1970.
5. H. Fujita, 'Mathematical Theory of Sedimentation Analysis', Academic Press, New York, 1962, p. 23.
6. G.K. Batchelor, *J. Fluid Mechanics*, 1972, *52*, 245.
7. B. Cichocki and B.U. Felderhof, *J. Chem. Phys.*, 1988, *89*, 1049.
8. C.W.J. Beenakker and P. Mazur, *Physica*, 1983, *120A*, 388.
9. A.J.C. Ladd, *J. Chem. Phys.*, 1990, *93*, 3484.
10. R.J. Phillips, J.F. Brady and G. Bossis, *Phys. Fluids*, 1988, *31*, 3462.
11. J.M. Creeth and C.G. Knight, *Biochim. Biophysica Acta*, 1965, *102*, 549.
12. S.E. Harding and P. Johnson, *Biochem. J.*, 1985, *231*, 543.
13. S.E. Harding and P. Johnson, *Biochem. J.*, 1985, *231*, 549.
14. A.J. Rowe, *Biopolymers*, 1977, *16*, 2595.
15. P.Y. Cheng and H.K. Schachman, *J. Polym. Sci.*, 1955, *16*, 19.
16. A.J. Rowe, Euromech 120 (Abstracts), 1979, 5.3.
17. M.M. Kops-Werkhoven and H.M. Fijnaut, *J. Chem. Phys.*, 1982, *77*, 2242.
18. J.M. Ham and G.M. Homsey, *Intl. J. Multiphase Flow*, 1988, *14*, 533.
19. A. Persechini and A.J. Rowe, *J. Mol. Biol.*, 1984, *172*, 23.
20. L.M. Gilbert and G.A. Gilbert, Methods in Enzymology (C.W. Hirs and S.N. Timasheff, eds), Academic Press, London, 1973, *27D*, 273.
21. W.V. Shaw, J. Ellis and A.J. Rowe, 1992, unpublished observations.

22

Sedimentation Coefficients of Self-Associating Species. Analysis of Monomer-Dimer-n-Mer Associations and Some Indefinite Associations

By E.T. Adams, Jr.

CHEMISTRY DEPARTMENT, TEXAS A&M UNIVERSITY, COLLEGE STATION, TEXAS 77843, U.S.A.

1. INTRODUCTION

Three types of ultracentrifuge experiments, sedimentation velocity, sedimentation equilibrium and Archibald experiments, have been used to study self-associations. Sedimentation velocity experiments (and also Archibald experiments) can be performed rapidly (1-2 hrs) compared to sedimentation equilibrium experiments (several hours to several days), and one would like to use the data from those experiments to analyze self-associations. Unfortunately, this is not an easy matter. To a first approximation sedimentation coefficients (s) depend on the molecular weight (M) *and* on the shape of the molecules. Thus, there is no one definite relationship between s and M, whereas there is a fixed relation between the molecular weight of a j-mer (M_j) and that of the monomer (M_1), since $M_j = jM_1$ (j = 2, 3, ...). Even so, there have been notable attempts to analyze self-associations from sedimentation velocity experiments using methods proposed by Gilbert and his associates[1-5]. Cox and his associates have done very elegant simulations of sedimentation velocity experiments on self-associations[6,7].

Originally the sedimentation coefficient (s) was called the sedimentation constant, but it was discovered that s varied with c when no self-association was present. Thus s could be represented by

$$s = \frac{s^0}{1 + k_s c} \tag{1}$$

or by

$$s = s^0 (1 - k_s c) \tag{1a}$$

where s^0 is the sedimentation coefficient at infinite dilution and s is the apparent sedimentation coefficient. Here k_s is the hydrodynamic concentration dependence parameter. For heterogeneous, non-associating solutes one obtains s_{wa} instead of s and $s_w{}^0$ instead of s^0 in eqs. (1) and (1a). Here $s_w{}^0$ is the weight average sedimentation coefficient and s_{wa} is its apparent value. With non-associating solutes $s_w{}^0$ is a constant whose value depends on the temperature, solute and solution conditions.

When self-associations are present, one can write[8,9]

$$s_{wa} = \frac{s_{wc}}{1 + gcs_{wc}}$$

or, (2)

$$\frac{1}{s_{wa}} = \frac{1}{s_{wc}} + gc;$$

$$s_{wa} = \frac{s_{wc}}{1 + g_*c}$$

or, (2a)

$$\frac{1}{s_{wa}} = \frac{1}{s_{wc}} + \frac{g_*c}{s_{wc}}$$

There are two models for describing s_{wa}: Model I using g is shown in eq. (2), and Model II using g_* is shown in eq. (2a). Here g or g_* is the symbol that will be used here, since k_s could be confused with equilibrium constants k_i. Either model is correct, since the concentration dependence parameter is an empirical quantity. Note that there are two factors complicating the concentration dependence of s_{wa}: the effect of the self-association is to cause s to increase as the concentration of the associating solute increases from zero, hence the symbol s_{wc} to indicate that self-association is present; the other effect due to g or g_*, which are always present and always positive, is to cause s_{wa} to decrease as c increases from zero. The result of both effects usually causes plots of s_{wa} *vs.* c to increase as c increases from zero, go through a maximum, and subsequently decrease with increasing c. Also note that

$$\lim_{c \to 0} s_{wc} = s_1$$

and that for real concentrations s_{wc} becomes[8,9]

$$s_{wc} = \frac{s_1 + k_2c_1s_2 + k_3c_1^2s_3 + \dots}{1 + k_2c_1 + k_3c_1^2 + \dots} = \frac{\Sigma c_is_i}{\Sigma c_i}$$

Here c_i is the concentration (in g/l) of self-associating species i (i = 1, 2, ...); $c_i = k_nc_1^n$, where c_1 is the monomer concentration. The s_i are the true values for the sedimentation coefficients of the various self-associating species. It is the two complications (self-association and the usual concentration dependence of s, namely, g or g_*) that both complicates the analysis and also makes it interesting and challenging. Fig. 1 shows a plot of s_{wa} *vs.* c for β-lactoglobulin A in 0.2 M glycine buffer[9] (0.1 ionic strength, pH 2.46) at 20°C; sedimentation equilibrium studies showed that β-lactoglobulin A underwent a non-ideal monomer-dimer association under these conditions[10]. The value of the dimerization constant was $k_2 = 0.350$ l/g and the non-ideal term $BM_1 = 0.0131$ l/g under these conditions.

One can combine thermodynamic experiments (sedimentation equilibrium, membrane osmometry, or light scattering) with sedimentation velocity experiments

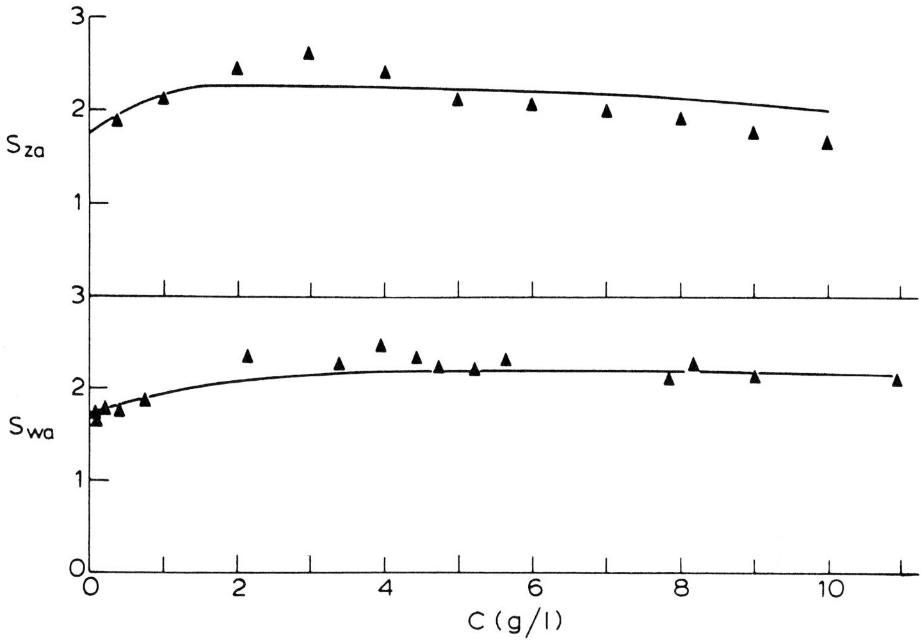

Figure 1. Plots of s_{za} *vs.* c_{av} (upper curve) and s_{wa} *vs.* c_{av} (lower curve) for β-lactoglobulin A at 20°C in 0.2 M glycine buffer, pH 2.46. Values of s_{wa} *vs.* c at lower concentrations were obtained with a photoelectric scanner at $\lambda = 280$ nm; values at higher concentrations were obtained using schlieren optics at $\lambda = 546$ nm. Values of s_{za} *vs.* c were obtained from eq. (10). The solid curve through the s_{wa} *vs.* c points was obtained from regenerated data based on model I, method III (see Table 3). These same values were used to regenerate the s_{za} *vs.* c curve using the appropriate form of eq. (10) for model I. Reproduced from ref. 9 by permission of Elsevier Science Publishers B.V. (Biomedical Division).

performed at the same temperature and under the same solution conditions (pH, ionic strength) to extract more information about the self-association than either experiment alone can provide. The Appendix shows some of the quantities available from thermodynamic experiments.

In addition to obtaining the apparent weight average sedimentation coefficient, it is possible, when self-association is present, to obtain the apparent z-average sedimentation coefficient, s_{za},[8,9,11] and also the apparent number average sedimentation coefficient, s_{Na} [8,9]. In previous communications in this area we have had limited success in developing methods to analyze self-associations that are more complicated than a monomer-n-mer association[8,9]. By combining a relation obtained from s_{Na} with a relation obtained from s_{wa}, it is possible to develop a method for obtaining s_2 and s_n (n = 3, 4, ...) the sedimentation coefficient of the dimer and the n-mer, respectively. This method can be applied to some indefinite self-associations also. In addition, it will be shown how one can obtain the usual hydrodynamic concentration dependence parameter (g or g_*) of the sedimentation coefficient.

2. EVALUATION OF s_{wa}, s_{za} AND s_{Na}

It will be assumed here that the reader is familiar with ultracentrifugation theory and procedures. Those who are unfamiliar with them should consult the books by Schachman[12], Elias[13,14], Bowen[15] or Chervenka[16] for details.

The quantity s_{wa} is obtained from the relation[12-16]

$$\frac{d(\ln r_*)}{dt} = \omega^2 s_{wa} \tag{4}$$

Here $\omega = 2\pi(\text{RPM})/60$ is the angular velocity, t is the time and $r_* = \langle r^2 \rangle^{1/2}$ is the radial position of the square root of the second moment of the boundary gradient curve. The quantity $\langle r^2 \rangle$ is defined by[17,18]

$$\langle r^2 \rangle = \frac{c_m r_m^2 + \int_{r_m}^{r_p} r^2 \left(\frac{\partial c}{\partial r}\right)_t dr}{c_p} \tag{5}$$

and the meniscus concentration, c_m, is obtained from[19]

$$c_m = c_0 - \frac{1}{r_m^2} \int_{r_m}^{r_p} r^2 \left(\frac{\partial c}{\partial r}\right)_t dr \tag{6}$$

The plateau concentration c_p is obtained from

$$c_p = c_m + \int_{r_m}^{r_p} \left(\frac{\partial c}{\partial r}\right)_t dr \tag{7}$$

In these relations r_m is the radial position of the air-solution meniscus in the ultracentrifuge cell; r_p is the radial position at which the plateau begins, the region where the concentration gradient, $(\partial c/\partial r)_t$, is zero. These relations are used with refractometric (Rayleigh/schlieren) optics. When the boundary gradient curve breaks away rapidly from the meniscus, c_m will become zero after a short time, and from that moment on c_m will drop out of eqs. (5-7). If one has a slowly sedimenting particle so that $c_m > 0$, then one must include c_m in eqs. (5-7). When absorption optics are used (assuming the Beer-Lambert law is valid), s_{wa} is obtained from[8,9]

$$\frac{d \ln c_p}{dt} = -\omega^2 s_{wa} \tag{8}$$

Experimentally values of $\ln r_*$ or $\ln c_p$ are plotted against t, and the slope of the plot gives s_{wa}. Because of the radial dilution law, this is an average value of s_{wa} at an average concentration c_{av} (not c_0). Here c_{av} is given by[8,9]

$$c_{av} = \frac{1}{2}[c_f + c_l] \tag{9}$$

where c_f is the associating solute concentration on the first photograph or scan, and c_l is the concentration of the last photograph or scan. The analysis described is only valid if a plateau region is present.

Kegeles[11] showed that one could obtain the apparent z-average sedimentation coefficient, s_{za}, from

$$s_{za} = \frac{d(cs_{wa})}{dc} \tag{10}$$

He used Model II. Weirich, Adams and Barlow[8,9] showed that one can obtain the apparent number average sedimentation coefficient from a combination of sedimentation velocity and thermodynamic studies; they obtained

$$s_{Na} = \frac{M_{nc}}{cM_1} \int_0^c s_{wa} \left(\frac{M_1}{M_{wc}}\right) dc \tag{11}$$

It should be noted that evaluation of s_{za}[8,9,11] can be done without performing thermodynamic experiments under the same solution conditions. Table 1 gives the definitions of s_{wc}, s_{Nc} and s_{zc}, as well as expressions for s_{wa}, s_{za} and s_{Na} using Models I and II.

Table 1. Definitions of s_{wc}, s_{zc}, and s_{Nc} and expressions for s_{wa}, s_{za} and s_{Na}

A. Definitions of s_{wc}, s_{zc} and s_{Nc}[a]

$$s_{wc} = \frac{\sum c_i s_i}{\sum c_i}, \ i = 1, 2, \ldots \qquad c_i = k_i c_1^{\,i}, \ i = 1, 2, \ldots; \ k_1 \equiv 1$$

$$s_{zc} = \frac{\sum i c_i s_i}{\sum i c_i} = \frac{d(c s_{wc})}{dc}$$

$$s_{Nc} = \frac{\sum n_i s_i}{\sum n_i} = \frac{\sum \left(\frac{c_i s_i}{i}\right)}{\sum \left(\frac{c_i}{i}\right)} = \left(\frac{M_{nc}}{cM_1}\right) \int_0^c s_{wc} \left(\frac{M_1}{M_{wc}}\right) dc$$

B. Expressions for s_{wa}, s_{za} and s_{Na} from Model I[b]

$$s_{wa} = \frac{s_{wc}}{1 + gcs_{wc}}$$

or

$$\frac{1}{s_{wa}} = \frac{1}{s_{wc}} + gc \qquad (4a)$$

$$s_{za} = \frac{d(cs_{wa})}{dc} = \frac{s_{zc}}{(1 + gcs_{wc})^2} = s_{zc} \left\{\frac{s_{wa}}{s_{wc}}\right\}^2$$

or

$$\frac{s_{za}}{s_{wa}^2} = \frac{s_{zc}}{s_{wc}^2} \qquad (10)$$

$$s_{Na} = \left\{\frac{M_{nc}}{cM_1}\right\} \int_0^c s_{wa} \left\{\frac{M_1}{M_{wc}}\right\} dc$$

$$= s_{Nc} - g \left\{\frac{M_{nc}}{cM_1}\right\} \int_0^c cs_{wa}s_{wc}\left\{\frac{M_1}{M_{wc}}\right\} dc \qquad (11)$$

C. Expressions for s_{wa}, s_{za} and s_{Na} from Model II[b]

$$s_{wa} = \frac{s_{wc}}{1 + g_* c}$$

or

$$\frac{1}{s_{wa}} = \frac{1}{s_{wc}} + \frac{g_*}{s_{wc}} c \tag{4b}$$

$$s_{za} = \frac{d(cs_{wa})}{dc} = s_{wa}\left[\frac{s_{zc}}{s_{wc}} - \frac{s_{wc} - s_{wa}}{s_{wc}}\right] = \frac{s_{zc} - g_* c s_{wa}}{1 + g_* c} \tag{10a}$$

$$s_{Na} = \left\{\frac{M_{nc}}{cM_1}\right\} \int_0^c s_{wa} \left\{\frac{M_1}{M_{wc}}\right\} dc$$

$$= s_{Nc} + g_* \left\{\frac{M_{nc}}{cM_1}\right\} \int_0^c c s_{wa} \left\{\frac{M_1}{M_{wc}}\right\} dc \tag{11a}$$

---

[a] These definitions are independent of the choice of models.
[b] Note that $g \neq g_*$. Model I uses g; model II uses g_*. See refs. 8 and 9 for more details.

3. MONOMER-N-MER ASSOCIATION[8,9]

This association is described by $nP_1 \rightleftarrows P_n$, n = 2, 3, ... It is also the simplest case to treat. The apparent weight average sedimentation coefficient gives the following relation with model I:

$$\frac{1}{s_{wa}} = \frac{1 + k_n c_1^{n-1}}{s_1 + k_n c_1^{n-1}} + gc \tag{12}$$

The concentration dependence parameter, g, can be eliminated using the relation:

$$\frac{1}{s_{wa}} - c\frac{d}{dc}\left(\frac{1}{s_{wa}}\right) = \frac{1}{s_{wc}} - c\frac{d}{dc}\left(\frac{1}{s_{wc}}\right) \tag{13}$$

This relation can also be shown to give

$$\frac{1}{s_{wa}} - c\frac{d}{dc}\left(\frac{1}{s_{wa}}\right) = \frac{1}{s_{wa}} + \frac{c}{s_{wa}^2}\frac{ds_{wa}}{dc} = \frac{1}{s_{wa}^2}\left\{s_{wa} + c\frac{ds_{wa}}{dc}\right\}$$

$$= \frac{s_{za}}{s_{wa}^2} = \frac{s_{zc}}{s_{wc}^2} = \frac{c^2(c_1 s_1 + nk_n c_1^n s_n)}{(c_1 + nk_n c_1^n)(c_1 s_1 + k_n c_1^n s_n)^2} \tag{14}$$

Since c_1, k_n, n and s_1 are known, there is only one unknown, s_n. This equation is quadratic in s_n, so one can readily solve it for s_n.

A second method for solving a monomer-n-mer association involves the following relation based on model I:

$$\frac{1}{s_{wa}} - \frac{2}{c}\int_0^c \frac{dc}{s_{wa}} = \frac{1}{s_{wc}} - \frac{2}{c}\int_0^c \frac{dc}{s_{wc}} \tag{15}$$

For a monomer-dimer association, this becomes

$$\frac{1}{s_{wa}} - \frac{2}{c}\int_0^c \frac{dc}{s_{wa}} = \frac{1 + k_2c_1}{s_1 + k_2c_1s_2}$$

$$-\frac{2}{c}\left\{\frac{1}{k_2s_2}\left(1 - 3\frac{s_1}{s_2} + 2\frac{s_1^2}{s_2^2}\right)\ln\left(\frac{s_1 + k_2c_1s_2}{s_1}\right) + \frac{c_1}{s_2}\left\{3 + k_2c_1 - 2\frac{s_1}{s_2}\right\}\right\} \tag{16}$$

Although this equation has only unknown s_2, it is more complicated and must be solved for by successive approximations. Nonetheless, it may be the most precise method to use, since integration is a smoothing process. Another procedure to use would be to modify an equation developed for the monomer-dimer-n-mer association, which is described subsequently. If dimer is absent $k_2=0$; if there are no species higher than dimer, then $k_n=0$ (n = 3, 4, ...) in the relation.

With model II (which uses g_*) we have for method I

$$\frac{1}{s_{wa}} = \frac{1}{s_{wc}}(1 + g_*c) \tag{17}$$

Differentiating this equation and subsequent rearrangement leads to

$$g_* = \frac{s_{wc}\left\{\frac{1}{s_{wc}^2}\frac{ds_{wc}}{dc} - \frac{1}{s_{wa}^2}\frac{ds_{wa}}{dc}\right\}}{1 - \frac{c}{s_{wc}}\frac{ds_{wc}}{dc}} \tag{18}$$

Substitution of g_* into the equation for $1/s_{wa}$ leads to

$$\frac{1}{s_{wa}} = \frac{1}{s_{wc}} + \frac{c\left\{\frac{1}{s_{wc}^2}\frac{ds_{wc}}{dc} - \frac{1}{s_{wa}^2}\frac{ds_{wa}}{dc}\right\}}{1 - \frac{c}{s_{wc}}\frac{ds_{wc}}{dc}} \tag{19}$$

Using

$$\frac{1}{s_{wc}} = \frac{1 + k_nc_1^{n-1}}{s_1 + k_nc_1^{n-1}s_n} \tag{19a}$$

and

$$\frac{d}{dc}\left(\frac{1}{s_{wc}}\right) = \frac{1}{s_{wc}^2}\frac{ds_{wc}}{dc} = \frac{(n-1)k_n c_1^{n-2}(s_1-s_n)}{(s_1+k_n c_1^{n-1}s_n)^2(1+nk_n c_1^{n-1})} \tag{19b}$$

in eq. (19) leads to one equation in one unknown, s_n; note that ds_{wa}/dc is evaluated numerically. The analysis of the monomer-dimer association of β-lactoglobulin A shown in Fig. 1 is considered in the discussion[9].

4. MONOMER-DIMER-N-MER-ASSOCIATIONS[8,9]

Here the self-association is described by $nP_1 \rightleftarrows qP_2 + rP_m$, where P represents the self-associating solute and m = 3, 4 ... One can use the monomer-dimer-trimer (1, 2, 3) association as an example. Here,

$$c = c_1 + k_2c_1^2 + k_3c_1^3 \tag{20}$$

and

$$s_{wc} = \frac{s_1 + k_2c_1s_2 + k_3c_1^2s_3}{1 + k_2c_1 + k_3c_1^2} \tag{21}$$

Although c_1 and the equilibrium constants, k_2 and k_3, are known from thermodynamic experiments, and s_1 and g can be eliminated, it would appear that one is left with complicated equations involving two unknowns s_2 and s_3 (or s_n). For example, method 1 gives

$$\frac{s_{za}}{s_{wa}^2} = \frac{s_{zc}}{s_{wc}^2} = \frac{(s_1+2k_2c_1^{\,2}+3k_2c_1^{\,2}s_3)\,(1+k_2c_1+k_3c_1^2)^2}{(1+2k_2c_1+3k_3c_1^2)\,(s_1+k_2c_1s_2+k_3c_1^2s_3)^2} \tag{22}$$

This cannot be solved directly. Equations developed using method 3, *i.e.*,

$$\frac{1}{s_{wa}} - \frac{2}{c}\int_0^c \frac{dc}{s_{wa}} = \frac{1}{s_{wc}} - \frac{2}{c}\int_0^c \frac{dc}{s_{wc}}$$

would be even more complicated, and there would be no simple way to combine methods 1 and 3 to solve for s_2 and s_3 (or s_m). If there were a known relation between s_2 and s_3, such as $s_3/s_2 = (3/2)^\alpha$, then there would be no problem.

However, there is a way to evaluate s_2 and s_3 directly! One can use the apparent weight (s_{wa}) and number (s_{Na}) average sedimentation coefficients to do this. This is how it is done (see Table 1). The number average sedimentation coefficient is defined by[8,9]

$$s_{Nc} = \frac{\Sigma M_i s_i}{\Sigma M_i} = \frac{\sum \frac{c_i s_i}{M_i}}{\sum \frac{c_i}{M_i}} = \frac{\sum \frac{c_i s_i}{i}}{\sum \frac{c_i}{i}} \tag{23}$$

since $M_i = iM_1$ (i = 1,2,...). Note that

$$\frac{cM_1}{M_{nc}} = M_1 \sum \frac{c_i}{M_i} = \sum \frac{c_i}{i} \tag{24}$$

Thus,

$$\sum \frac{c_i s_i}{i} = \left\{\frac{cM_1}{M_{nc}}\right\} s_{Nc} = c_1 s_1 + \frac{k_2 c_1^2 s_2}{2} + \frac{k_3 c_1^3 s_3}{3} + \ldots \tag{25}$$

How does one evaluate $\sum(c_i s_i/i)$? One notes that

$$\int_0^{c_1} c s_{wc} \frac{dc_1}{c_1} = \int_0^{c_1} \sum c_i s_i \frac{dc_1}{c_1} = \int_0^{c_1} (s_1 + k_2 c_1 s_2 + k_3 c_1^2 s_3 + \ldots)\, dc_1$$

$$= \sum \frac{c_i s_i}{i} = \int_0^{c} s_{wc} \frac{M_1}{M_{wc}} dc \tag{26}$$

Here

$$\frac{dc_1}{c_1} = M_1 \frac{dc}{cM_{wc}} \tag{27}$$

In the real case $(\sum c_i s_i/i)_{app}$ is measured.
For Model II,

$$c s_{wc} = c s_{wa} + g_* c^2 s_{wa} \tag{28}$$

so it follows that

$$\int_0^{c} s_{wa} \frac{M_1}{M_{wc}} dc \equiv \left(\sum \frac{c_i s_i}{i}\right)_{app}$$

$$= \left(\sum \frac{c_i s_i}{i}\right) - g_* \int_0^{c} c s_{wa} \frac{M_1}{M_{wc}} dc \tag{29}$$

Now eqs. (28) and (29) can be used as follows:

$$c s_{wa} - c_1 s_1 = k_2 c_1^2 s_2 + k_3 c_1^3 s_3 - g_* c^2 s_{wa} \tag{30}$$

$$\left(\sum \frac{c_i s_i}{i}\right)_{app} - c_1 s_1 = \frac{k_2 c_1^2 s_2}{2} + \frac{k_3 c_1^3 s_3}{3} - g_* \int_0^{c} c s_{wa} \frac{M_1}{M_{wa}} dc \tag{31}$$

Let

$$I^* = \int_0^c c s_{wa} \frac{M_1}{M_{wc}} dc \tag{31a}$$

Then the two equations above can be arranged to yield

$$\Delta \equiv I^*(cs_{wa} - c_1 s_1) - c^2 s_{wa} \left\{ \left(\sum \frac{c_i s_i}{i} \right)_{app} - c_1 s_1 \right\}$$

$$= k_2 c_1^2 s_2 \left\{ I^* - \frac{c^2 s_{wa}}{2} \right\} + k_3 c_1^3 s_3 \left\{ I^* - \frac{c^2 s_{wa}}{3} \right\} \tag{32}$$

Dividing both sides by $c_1^2[I - c^2 s_{wa}/2]$ leads to

$$\Delta^* = \frac{\Delta}{c_1^2 \left\{ I^* - \frac{c^2 s_{wa}}{2} \right\}} = k_2 s_2 + k_3 s_3 c_1 \frac{\left[I^* - \frac{c^2 s_{wa}}{3} \right]}{\left[I^* - \frac{c^2 s_{wa}}{2} \right]} \tag{33}$$

A plot of

$$\Delta^* \ vs. \ c_1 \frac{\left[I^* - \frac{c^2 s_{wa}}{3} \right]}{\left[I^* - \frac{c^2 s_{wa}}{2} \right]}$$

would have a slope = $k_3 s_3$ and an intercept = $k_2 s_2$. Since k_2, k_3, c_1, and M_1/M_{wc} are known from the thermodynamic studies, then one can readily obtain s_3 and s_2 from the slope and intercept, respectively.

One can do a similar procedure with Method I. Here

$$s_{wa} = s_{wc} - g c s_{wa} s_{wc} \tag{34}$$

Since s_{wc} is not known beforehand, one can to a first approximation replace $g c s_{wa} s_{wc}$ by $g c s_{wa}^2$; this approximation becomes better and better as $c \rightarrow 0$. The two starting equations become

$$c s_{wa} - c_1 s_1 = k_2 c_1^2 s_2 + k_3 c_1^3 s_3 - g c^2 s_{wa}^2 \tag{35}$$

$$\left(\sum \frac{c_i s_i}{i} \right)_{app} - c_1 s_1 = \frac{k_2 c_1^2 s_2}{2} + \frac{k_3 c_1^3 s_3}{3} - gI \tag{36}$$

where

$$I = \int_0^c c s_{wa}^2 \frac{M_1}{M_{wc}} dc \tag{36a}$$

Using similar procedures, one obtains

$$\delta = \left[I\ (cs_{wa} - c_1 s_1) - c^2 s_{wa}{}^2 \left\{ \left(\sum \frac{c_i s_i}{i} \right)_{app} - c_1 s_1 \right\} \right] \tag{37}$$

and

$$\delta^* = \frac{\delta}{c_1{}^2 \left\{ I - \frac{c^2 s_{wa}{}^2}{2} \right\}} = k_2 s_2 + k_3 s_3 c_1 \frac{\left[I^* - \frac{c^2 s_{wa}{}^2}{3} \right]}{\left[I^* - \frac{c^2 s_{wa}{}^2}{2} \right]} \tag{38}$$

A plot of

$$\delta^* \ vs.\ c_1 \frac{\left[I^* - \frac{c^2 s_{wa}{}^2}{3} \right]}{\left[I^* - \frac{c^2 s_{wa}{}^2}{2} \right]}$$

has an

$$\text{intercept} = k_2 s_2 \tag{38a}$$

a limiting

$$\text{slope} = k_3 s_3 \tag{38b}$$

Since k_2 and k_3 are known from the thermodynamic experiments, one readily obtains s_2 and s_3 from eqs. (38a) and (38b), respectively.

Although a monomer-dimer-trimer association has been used as an example of a monomer-dimer-n-mer association, the general equation for a monomer-dimer-n-mer equation, using Model II, becomes

$$\Delta^* = \frac{\Delta}{c_1{}^2 \left\{ I^* - \frac{c^2 s_{wa}}{2} \right\}} = k_2 s_2 + k_n s_n c_1{}^{(n-2)} \frac{\left[I^* - \frac{c^2 s_{wc}}{n} \right]}{\left[I^* - \frac{c^2 s_{wa}}{2} \right]} \tag{39}$$

Here n = 3, 4, ... With the appropriate value of n, a plot based on this equation gives

$$k_2 s_2 = \text{intercept}$$

and

$$k_n s_n = \text{slope.}$$

5. INDEFINITE SELF-ASSOCIATION[8,9]

Associations of the type

$$nP_1 \rightleftarrows qP_2 + hP_3 + \ldots \tag{40}$$

or

$$nP_1 \rightleftarrows qp_2 + mP_4 + \ldots \tag{41}$$

that appear to continue without limit are known as *indefinite* self-associations; here P_1 represents a self-associating solute. This discussion will be restricted to the first equation. There are two types we will consider: the SEK (Sequential, Equal Equilibrium Constant)[17] and the AK (Attenuated Equilibrium Constant)[18] Type I models. These are described in Table 2.

For the SEK Model Type I indefinite self-association[8]

$$s_{wc} = \frac{s_1 + 2kc_1s_2 + 3k^2c_1{}^2s_3 + \ldots}{1 + 2kc_1 + 3k^2c_1{}^2}$$

$$= (s_1 + 2kc_1s_2 + 3k^2c_1{}^2s_3 + \ldots)\,(1 - kc_1)^2 \tag{42}$$

This would appear hopeless, since there is no way to sum the series in the numerator. However, this association is also known as an open, random, linear or *isodesmic* association. With synthetic linear polymers, one notes that s and the molecular weight (M) are related by[19,20]

$$s = KM^\alpha \tag{43}$$

where K and α are constants, whose values depends on the temperature and on the solute-solvent combination. Thus, assume that one can write

$$s_i = K(iM_1)^\alpha,\ i = 1, 2, \ldots \tag{44}$$

for the indefinite self-association of the first type. With this assumption one obtains

$$s_1 = KM_1{}^\alpha \tag{44a}$$

$$s_2 = K(2M_1)^\alpha \tag{44b}$$

etc.

It follows that

$$s_2/s_1 = 2^\alpha \tag{45}$$

$$s_3/s_1 = 3^\alpha \tag{45a}$$

etc.

This leads to

$$\frac{s_{wc}}{s_1} = [1 + 2kc_1(2^\alpha) + 3k^2c_1{}^2(3^\alpha) + \ldots]\,(1 - kc_1)^2 \tag{46}$$

for the Type I SEK model[8]. If s_{wc}/s_1 were known, then there would be only one unknown on the right, namely, α. Eq. (33) for Δ^* can be modified to give

Table 2. Indefinite self-associations of the first type.

$$nP_1 \rightleftarrows qP_2 + mP_3 + \ldots$$

SEK[a] Type I Model[17]	AK[b] Type I Model[18]
$P_1 + P_{i-1} \rightleftarrows P_i,\ i = 2, 3, \ldots$	$P_1 + P_{i-1} \rightleftarrows P_i,\ i = 2, 3, \ldots$
K_m = same molar equilibrium constant for all steps	$(K_m) = \frac{K_m}{i}$ for each step
$c = c_1 + 2kc_1^2 + 3k^2c_1^3 + 4k^3c_1^3 + \ldots$	$c = c_1 + kc_1^2 + \frac{k^2}{2}c_1^3 + \frac{k^3}{2.3}c_1^4 + \ldots$
$= \frac{c_1}{(1-kc_1)^2}$ if $kc_1 < 1$	$= c_1 \exp(kc_1) \quad 0 < kc_1 < \infty$
$k = \frac{K_m}{M_1}$ = intrinsic equilibrium constant.	$k = \frac{K_m}{M_1}$ = intrinsic equilibrium constant.
$\frac{M_1}{M_{nc}} = 1-kc_1 = \sqrt{f_1}$	$\frac{M_1}{M_{nc}} = \frac{1}{kc_1}[1 - \exp(-kc_1)] = \frac{1}{x}[1 - e^{-x}]$
$\frac{M_1}{M_{wc}} = \frac{1-kc_1}{1+kc_1} = \frac{\sqrt{f_1}}{2-\sqrt{f_1}}$	$\frac{M_1}{M_{wc}} = \frac{1}{1+kc_1} = \frac{1}{1+x}$
$f_1 = \frac{c_1}{c}$ = weight fraction of monomer	$f_1 = \frac{c_1}{c}$ = weight fraction of monomer
$= (1-kc_1)^2$	$= \exp(-kc_1) = \exp(-x)$

[a]SEK = sequential, equal equilibrium constant

[b]AK = attenuated equilibrium constant

$$\frac{\Delta^*}{s_1} = 2k(2^\alpha) + 3k^2(3^\alpha)c_1 \frac{\left[I^* - \frac{c^2 s_{wa}}{3}\right]}{\left[I^* - \frac{c^2 s_{wa}}{2}\right]} + 4k^3(4^\alpha)c_1^2 \frac{\left[I^* - \frac{c^2 s_{wa}}{4}\right]}{\left[I^* - \frac{c^2 s_{wa}}{2}\right]} + \ldots \quad (47)$$

This series cannot be summed, so it would have to be truncated. As a guide to how many terms to carry before truncation, one could note how many terms are sufficient to approximate c (see Table 2) within experimental error. Suppose that six terms are sufficient for c, then it might be assumed six terms are sufficient for the truncation of Δ^*/s_1. On the other hand, note that the intercept of a plot of Δ^*/s_1 *vs.*

$$c_1 \frac{\left[I^* - \frac{c^2 s_{wa}}{3}\right]}{\left[I^* - \frac{c^2 s_{wa}}{2}\right]}$$

is $2k(2^\alpha)$ and the limiting slope is $3k^2(3^\alpha)$. Hence, one can estimate α from the intercept or the limiting slope, since k is known. With synthetic polymers in ideal, dilute solutions[20] (theta conditions) $s = KM^{1/2}$, so values of α may be $^1/_2$ or larger. Once α is known, then the value of the sedimentation coefficient of any j-mer (j = 2, 3, ...) s_j can be calculated from:

$$\frac{s_j}{s_1} = j^\alpha$$

or

$$s_j = s_1 j^\alpha \qquad j = 2, 3, \ldots \quad (48)$$

Remember that s_1 is the value of s_{wa} at c = 0.

With the AK Type I indefinite model, similar assumptions are made, such as

$$\frac{s_j}{s_1} = j^\alpha \qquad j = 2, 3, \ldots$$

The equation for Δ^*/s_1 becomes

$$\frac{\Delta^*}{s_1} = k(2^\alpha) + \frac{k^2}{2!}(3^\alpha)c_1 \frac{\left[I^* - \frac{c^2 s_{wa}}{3}\right]}{\left[I^* - \frac{c^2 s_{wa}}{2}\right]} + \ldots \quad (49)$$

A plot of Δ^*/s_1 *vs.*

$$c_1 \frac{\left[I^* - \frac{c^2 s_{wa}}{3}\right]}{\left[I^* - \frac{c^2 s_{wa}}{2}\right]}$$

has an

intercept = $k(2^\alpha)$

and a

$$\text{limiting slope} = \frac{k^2(3^\alpha)}{2!}$$

Since k is known, one can readily obtain α. Alternatively, one could truncate eq. (49) for Δ^*/s_1 after six (or more) terms and solve for α by successive approximations.

6. DISCUSSION

The analysis of sedimentation coefficients of self-associating solutes can be done successfully in many instances if thermodynamic experiments on the same solute have been run under identical solution conditions. The difficulty with self-associations is that there is usually not a definite relation between s_j (j = 2, 3, ...) and s_1, whereas with molecular weights $M_j = jM_1$ always. Another complication is the hydrodynamic concentration dependence parameter g or g_*. It is always present, and it is always positive. Although sedimentation velocity experiments on Zn^{2+}-free α-anylase at pH 7 indicate no concentration dependence of s_{wa} on c [21]; the data are rationalized by noting that the ordinary concentration dependence of the sedimentation coefficient, which would cause s to decrease with increasing c, is balanced by a self-association, which would cause s to increase with c. Indeed, sedimentation equilibrium experiments on Zn^{2+}-free α-anylase under identical solution conditions indicated that a monomer-dimer self-association was present[21].

Which is the appropriate expression:

$$s_{wa} = \frac{s_{wc}}{1 + g_* c} \tag{2}$$

or

$$s_{wa} = \frac{s_{wc}}{1 + gcs_{wc}} \tag{2a}$$

$$\frac{1}{s_{wa}} = \frac{1}{s_{wc}} + gc\ ?$$

There is no theory to state which is the correct choice. Both Model I and Model II describe the fact that s_{wa} will decrease with increasing c, if no self-association is present. When self-association is present, one can have s_{wa} being independent of c, as in the case of α-anylase, or when the self-association dominates, then the plot of s_{wa} *vs.* c will go through a maximum as c increases and then subsequently decrease with increasing c.

With monomer-n-mer associations, attempts have been made to estimate s_j (j = 2, 3, ...) by drawing a tangent to the curve of s_{wa} *vs.* c (or $1/s_{wa}$ *vs.* c) at high concentrations. With a simulated monomer-dimer model Beckerdite, Weirich *et al.* were able to show that one could obtain up to 88% of the correct answer, depending on where the tangent was drawn[9]. With some monomer-n-mer associations having n $\geq$ 3 the concentration gradient (dc/dr) curve may show two peaks, then one may be able to evaluate s_j (j = 3, 4, ...) directly. When this does not occur, then recourse must be made to an analysis like the one my associates and I have developed.

In the preceding papers[8,9], it *appeared* that the analysis of monomer-dimer-n-mer associations was quite difficult. Fortunately, this is not the case. Even though s_{wa} is the same for both models for the hydrodynamic concentration dependence parameter, the values of g and g_* are different. For example, sedimentation velocity (see Fig. 1) and sedimentation equilibrium experiments at 20°C have been performed on the monomer-dimer association of β-lactoglobulin A in 0.2 M glycine buffer (ionic strength 0.1, pH 2.46)[9]. From the sedimentation equilibrium experiments Tang and Adams[10] obtained $k_2 = 0.350$ l/g and $BM_1 = 0.0131$ l/g. Experiments at low concentrations ≤ 3 g/l were measured using a photoelectric scanner at λ = 280 nm; data at higher concentrations were obtained at λ = 546 nm using schlieren optics. The values of s_{wa} were obtained from eq. (8) with the scanner data and from eq. (4) using the schlieren data. The solid line represents a smooth curve drawn through these values of s_{wa} *vs.* c. From this smooth curve values of s_{za} were obtained using a five point method for the derivative. The values of s_{za} *vs.* c are shown in the upper curve. The values of s_1, s_2 and g or g_* are shown in Table 3 using the best fits (as judged by the variance) for both models. Both methods gave essentially the same s_2, but the values of g and g_* are quite different. The value of s_1 was obtained from

$$\lim_{c \to 0} s_{wa} = s_1$$

Table 3. Values of s_1, s_2 and g or g_* for the self-association of β-lactoglobulin A at 20°C in 0.2 M glycine buffer (ionic strength 0.1, pH 2.46)[a]

Model	Method	s_1(S)	s_2(S)	g ($l\,g^{-1}\,S^{-1}$)	g_* ($l\,g^{-1}$)	Variance
I	3	1.68	3.15±0.08	0.0068 ±0.003	-	2.38×10^{-4}
II	1	1.68	3.17±0.03	-	0.0179 ±0.008	2.57×10^{-4}

[a] Reproduced from ref. 9 by permission of Elsevier Science Publishers B.V. Only the best values are shown. The values of K_2 and BM_1 were $K_2 = 0.350$ l/g and $BM_1 = 0.0131$ l/g.

Once s_1 and s_2 are known, then they can be used in the Svedberg equation,

$$s_i = \frac{M_i\,(1-\bar{v}\rho_0)}{Nf_i} = \frac{M_iD_i(1-\bar{v}\rho_0)}{RT} \tag{50}$$

to obtain the translational diffusion coefficients D_i. Note that[24]

$$D_i = \frac{RT}{Nf_i} \tag{51}$$

Here we obtained $D_1 = (9.0 \pm 0.3)\times10^{-7}$ cm^{2}s^{-1} and $D_2 = (8.4 \pm 0.2)\times10^{-7}$ cm^{2}s^{-1}.

I have shown that one can indeed analyze a monomer-dimer-n-mer association using a combination of cs_{wa} and $\sum(c_is_i/i)_{app}$ (see eqs. (33) and (38)). This method could probably be extended to a monomer-dimer-trimer-tetramer or related associations as well. It could be applied to a monomer-dimer association or a monomer-n-mer association as well. With AK and SEK Type I indefinite self-associations, eqs (47) or (49) can be used in the analysis provided that eq. (44) applies. By using the limiting slope and the intercept one may estimate the value of α and hence s_2, s_3, *etc.*

The virtue of the procedures described here is that it allows for the unambiguous (unprejudiced) determination of s_1, s_2 and s_j ($j = 3, 4,...$). Once the sedimentation coefficients s_1, s_2, *etc.* have been determined, the Svedberg equation (see eq. (50)) can be used to evaluate the translational frictional coefficients, f_i, or the translational diffusion coefficient, D_i, from eq. (51). One can then calculate $f_i/f_{i\ min}$ and compare this ratio to that obtained from spheres, ellipsoids or other models using the Kirkwood-Riseman theory or modifications of it[22]. This has been done with haemoglobin. The observed sedimentation coefficient of the tetramer is 4.45 S, which compares very favourably with a value of (4.43 ± 0.13) S for a tetrahedral arrangement of subunits. It must be reiterated that this analysis is valid only if pressure effects are absent so that there is a plateau present in sedimentation velocity experiments; it is also assumed that any polyelectrolyte used has been dialyzed against a solution containing supporting electrolyte and buffer components. If this is done, then Eisenberg[23] has pointed out that the Svedberg equation may be valid. Once the sedimentation coefficients s_i are known, then one can calculate s_{wc} from the appropriate equation for the particular self-association and one can then evaluate g or g_* from eqs (2) or (2a), respectively[8,9].

The methods described here can be applied to other weight average properties, x_{wc}[8,9]; here x_{wc} could be a weight average partition coefficient σ_{wc} or a weight average elution volume V_{ewc}. For instance, the z-average property, x_{zc}, is given by

$$\frac{d(cx_{wc})}{dc} = x_{zc} \tag{52}$$

and the number average property, x_{Nc}, is given by

$$x_{Nc} = \left\{\frac{M_{nc}}{cM_1}\right\} \int_0^c \left\{\frac{M_1}{M_{wc}}\right\} dc$$

If there is an additional concentration dependence parameter associated with x_{wc}, then one would evaluate x_{wa}, x_{za} or x_{Na}. To do the analyses described here one must do thermodynamic experiments (sedimentation equilibrium, light scattering or membrane osmometry) under the same solution conditions (temperature, pH, ionic strength).

APPENDIX: QUANTITIES AVAILABLE FROM THERMODYNAMIC EXPERIMENTS

In order to do the analyses presented here and elsewhere[8,9], one needs to have values of c_1, k_i (i = 2, 3, ...), M_1/M_{wc} and cM_1/M_{nc}. These quantities are available from thermodynamic experiments (sedimentation, equilibrium, light-scattering, osmotic pressure) performed under the same solution conditions as the sedimentation velocity experiments[25]. If one assumes that the natural logarithm of the activity coefficient of the self-associating species obeys the relation[17,18,25,27]

$$\ln y_i = iBM_1c \qquad i = 1,2,... \tag{A1}$$

It follows that $y_n/y_1^n = 1$ and that

$$c = c_1 + k_2c_1^2 + k_3c_1^3 + ... \tag{A2}$$

The apparent weight average molecular weight, M_{wa}, is obtained as primary data from sedimentation equilibrium or light scattering experiments[25,28]. The apparent number average molecular weight, M_{na}, is obtained as primary data in osmotic pressure experiments. When eq. (A1) is obeyed and when self-association occurs, then M_{na} and M_{wa} are interrelated. Thus, from sedimentation equilibrium or light scattering experiments one obtains M_1/M_{na} from[25]

$$\frac{M_1}{M_{na}} = \frac{1}{c}\int_0^c \left\{\frac{M_1}{M_{wa}}\right\} dc = \frac{M_1}{M_{nc}} + \frac{BM_1c}{2} \tag{A3}$$

Similarly, one can obtain M_1/M_{wa} in osmotic pressure experiments from[25,26]

$$\frac{M_1}{M_{wa}} = \frac{d}{dc}\left(\frac{cM_1}{M_{na}}\right) = \frac{M_1}{M_{wc}} + BM_1c \tag{A4}$$

It is also possible to obtain the natural logarithm of the apparent weight fraction of monomer from[25]

$$\ln f_a = \int_0^c \left\{\frac{M_1}{M_{wa}} - 1\right\}\frac{dc}{c} = \ln f_1 + BM_1c \tag{A5}$$

or from[25,27]

$$\ln f_a = \int_0^c \left\{\frac{M_1}{M_{na}} - 1\right\}\frac{dc}{c} + \left\{\frac{M_1}{M_{na}} - 1\right\} \tag{A6}$$

Here $f_1 = c_1/c$ is the weight fraction of monomer. These quantities can be combined in various ways to find out what type of self-association is present, to obtain the equilibrium constant (k_i) and the non-ideal term (BM_1). Details on how this is done have been published extensively[17,18,25,27]. Some authors prefer to use nonlinear least squares for the analysis of self-associations from sedimentation equilibrium experiments[29,30]. Milthorpe, Jeffrey and Nichol[3] have shown that one can always obtain the activity of monomer, a_1, from sedimentation equilibrium experiments. It should be noted that $a_1 = y_1 c_1$, so that even though a_1 is known, neither y_1 or c_i is known. However, if eq. (A1) is obeyed, then one notes that

$$\ln\left(\frac{a_1}{c}\right) = \ln (f_1 y_1) = \ln f_1 + BM_1 c \tag{A7}$$

This is the same quantity that is obtained from eqs. (A5) or (A6).

ACKNOWLEDGMENT

The author wishes to thank Dr. D.J. Winzor of the Biochemistry Department, University of Queensland for his interest, comments and discussion about this work. This Chapter is dedicated to the memory of my teacher and friend, the late Professor J.W. Williams.

REFERENCES

1. G.A. Gilbert, *Disc. Faraday Soc.*, 1955, *20*, 68.
2. G.A. Gilbert, *Proc. Royal Soc. London, Ser. A*, 1959, *250*, 377.
3. G.A. Gilbert and R.C.Ll. Jenkins, *Proc. Royal Soc. London, Ser. A*, 1959, *253*, 420.
4. L.M. Gilbert and G.A. Gilbert, *Nature*, 1962, *194*, 1173.
5. L.W. Nichol and H.A. Mackenzie, 'Milk Proteins - Chemistry and Molecular Biology,' H.A. Mackenzie, ed., Academic Press, New York, 1970, Vol. 1, p. 280.
6. D.J. Cox, *Arch. Biochem, Biophys.*, 1971, *146*, 18.
7. D.J. Cox, *Arch. Biochem, Biophys.*, 1971, *142*, 154.
8. C.A. Weirich, E.T. Adams, Jr. and G.A. Barlow, *Biophys. Chem.*, 1973, *1*, 35.
9. J.M. Beckerdite, C.A. Weirich, E.T. Adams, Jr. and G.H. Barlow, *Biophys. Chem.*, 1983, *17*, 203.
10. L.-H. Tang and E.T. Adams, Jr., *Arch. Biochem. Biophys.*, 1973, *157*, 520.
11. G. Kegeles, *Proc. Nat. Acad. Sci. U.S.A.*, 1972, *69*, 2577.
12. H.K. Schachman, 'Ultracentrifugation in Biochemistry,' Academic Press, New York, 1959.
13. H.-G. Elias, 'Ultrazentrifugen-Methoden,' second edition, Beckman Instruments GmbH, Munich, 1961.
14. H.-G. Elias, 'Méthodes de L'Ultracentrifugation Analytique,' revised and augmented with an extra chapter by the translator, Alexandre Fritsch [French translation of ref. 13], Beckman Instruments International S.A., Geneva, 1964.
15. T.J. Bowen, 'An Introduction to Ultracentrifugation,' Wiley-Interscience, London, 1970.
16. C.H. Chervenka, 'A Manual of Methods for the Analytical Ultracentrifuge,' Spinco Division of Beckman Instruments, Palo Alto, CA, 1969.
17. L.-H. Tang, D.R. Powell, B.M. Escott and E.T. Adams, Jr., *Biophys. Chem.*, 1977, *7*, 121.
18. J.M. Beckerdite, C.C. Wan and E.T. Adams, Jr., *Biophys. Chem.* 1980, *12*,

199.
19. P.J. Wan and E.T. Adams, Jr., *Biophys. Chem.*, 1976, *5*, 207.
20. H. Fujita, 'Foundations of Ultracentrifugal Analysis,' John Wiley, New York, 1975, pp. 204-206.
21. R. Tellam, D.J. Winzor and L.W. Nichol, *Biochem. J.*, 1978, *173*, 185.
22. C.R. Cantor and P.R. Schimmel, *Biophysical Chemistry*, Freeman, San Francisco, 1980, Part II, pp. 565-570 and 610-612.
23. H. Eisenberg, *J. Chem. Phys.*, 1962, *36*, 1837.
24. T. Svedberg and K.O. Pederson, 'The Ultracentrifuge,' Clarendon Press, Oxford, 1940, p. 3, pp. 38-48.
25. E.T. Adams, Jr., L.-H. Tang, J.L. Sarquis, G.H. Barlow and W.M. Norman, 'Physical Aspects of Protein Interactions,' N. Catsimpoolas ed., Elsevier Science Publishing Co., New York, 1978, p. 1.
26. E.T. Adams, Jr., *Biochemistry*, 1965, *4*, 1655.
27. E.T. Adams, Jr., P.J. Wan and E.F. Crawford, *Methods Enzymol.*, 1978, *48*, 69.
28. C. Tanford, 'Physical Chemistry of Macromolecules,' John Wiley, New York, 1961, Chapter 5, pp. 275-316.
29. M.L. Johnson, J.J. Correia, J.J. Halvorson and D.A. Yphantis, *Biophys. J.*, 1981, *36*, 575.
30. D.A. Yphantis and T. Arakawa, *Biochemistry*, 1987, *26*, 5422.
31. B.K. Milthorpe, P.D. Jeffrey and L.W. Nichol, *Biophys. Chem.*, 1975, *3*, 169.

23

The Analytical Ultracentrifuge as a Tool for Diffusion Measurements. Cross Diffusion Effects in Ternary Polymer: Polymer: Solvent Systems

By W.D. Comper and B.N. Preston

BIOCHEMISTRY DEPARTMENT, MONASH UNIVERSITY, CLAYTON, VICTORIA, AUSTRALIA 3168

1. BACKGROUND

The importance of understanding diffusional transport of macromolecules in binary systems, as measured in the analytical ultracentrifuge, has been recognised for many years. Studies have been made particularly in terms of molecular physico-chemical characterisation in dilute solution[1] and more recently in terms of evaluating thermodynamic non-ideality and concentration dependent hydrodynamic frictional coefficients in semi-dilute solutions[2,3]. These areas have already received considerable attention in the literature and will not be reviewed here. Rather, we devote this Chapter to efforts in examining the poorly understood and less studied[4] systems of diffusion of macromolecules in ternary systems consisting of solvent: polymer (1): polymer (2) in semi-dilute solution.

The stimulus for investigating transport in these multicomponent systems comes from the need to understand transport phenomena in biological systems. Biological systems could be generally characterised as semi-dilute, multicomponent and macromolecular, such as gels and polymeric networks associated with cell membranes, cytoplasmic regions and extracellular matrices of connective tissues. The non-equilibrium status of these 'living' systems requires transport in many different contexts extending from dynamic reorganization to metabolic turnover.

At the molecular level when solutions of high molecular weight polymers exist in semi-dilute systems, it becomes increasingly evident that due to interactions between individual molecules both the static and dynamic properties of the macromolecules may be markedly altered compared to their behaviour in dilute solution. In such complex systems consideration must then be made that the diffusion of a polymer may be influenced by the presence of another polymer and in turn may influence the diffusive flow of the other polymer. These dynamic interactions, which may be quite non-specific in nature, are embodied in cross or coupled diffusion arising in ternary polymer systems; depending on the magnitude of the cross diffusion term and initial concentration profiles, the system may develop a number of distinct dynamic states.

The coupled diffusion may generate *incongruent diffusion*[5] a process which drives the flux of material against its own concentration gradient. Or, under certain conditions, the cross diffusion may drive the system into a convection regime which rapidly destroys the initial diffusion gradients. In this case the instability condition may be of two types[6]: (i) the hydrodynamic instability condition which is one where convective flows arise from gravitational driven instabilities generated by random velocity fluctuations at the initial boundary defining the concentration gradients and (ii) the hydrostatic instability condition, in which diffusive processes rapidly yield a density inversion at the boundary which is acted on by gravity to generate convective flows. The hydrostatic stability condition is a sufficient, but not a necessary condition, for a convective regime to develop, as the hydrodynamic mechanism may operate at different concentrations. It is the hydrostatic condition that has been shown, however, to operate only in ternary water: polymer: polymer systems[6].

Very few measurements have been made on the cross or coupled diffusion effects in ternary polymer systems[4]. Such direct measurements are difficult and time consuming. We take the opportunity here to present an extended analysis of work that was originally published in 1984[7] that utilised the analytical ultracentrifuge as a tool for the direct demonstration of cross diffusion effects between polymers as studied in ternary water-dextran-albumin systems. The striking schlieren patterns obtained and their semi-quantitative explanation, as described below, would suggest that the technique could be exploited to a much greater extent in quantitating multicomponent transport and cross diffusion.

2. OBSERVATIONS IN THE ANALYTICAL ULTRACENTRIFUGE

A schematic diagram of the initial distribution of dextran and albumin in the centrifuge cell is shown in Fig. 1. Albumin is initially at a uniform concentration throughout whereas there is a small concentration gradient of dextran which introduces a density gradient as well. After a period of time, the dextran diffusing under its own concentration gradient will generate a sigmoidal distribution. If cross diffusion operates, the diffusion of dextran will be accompanied by a corresponding flux of albumin which will generate a concentration inversion (this has been shown directly in the analytical ultracentrifuge using UV optics[7]) but no density inversion (large cross diffusion effects are accompanied by density inversion as described below). The variation of the concentration gradients with distance are also depicted in Fig. 1. The dextran generates a Gaussian-like profile whereas the albumin generates a split peak. A schlieren pattern could be obtained from the summation of these profiles after converting them to refractive index difference gradients. The resultant profile may also appear as a split peak as indicated.

The schlieren patterns actually obtained with the albumin-dextran system, as shown in Fig. 2, are certainly consistent with this description incorporating a significant cross diffusion effect. The photographs in Fig. 2 were obtained at approximately one hour after boundary formation. The initial dextran (molecular weight ~200,000) concentration gradient was varied left to right in the order of 5, 10

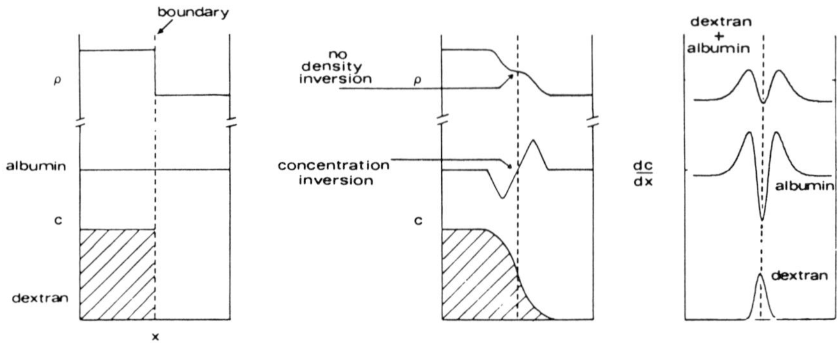

Figure 1. Schematic diagram of the putative concentration and concentration gradient profiles in a ternary system resulting from cross diffusion between albumin and dextran. ρ = macroscopic fluid density, c = concentration x = distance (reproduced from Comper *et al.*[7] with permission of publishers).

Table 1. Computed ternary diffusion coefficients in albumin (1): dextran (2): systems (from Comper *et al.*[7]).

Polymer concentration (kg m^{-3})		$10^{11} \times$ Ternary diffusion coefficients (m^2s^{-1})							
		Dextran T70[a]				Dextran T200[a]			
C_1	C_2	D_{11}	D_{12}	D_{21}	D_{22}	D_{11}	D_{12}	D_{21}	D_{22}
200.0	20.0	5.6	5.6	0.7	2.5	5.9	18.5	0.2	2.0

[a] Dextran T70 molecular weight ~70,000; dextran T200 molecular weight ~200,000.

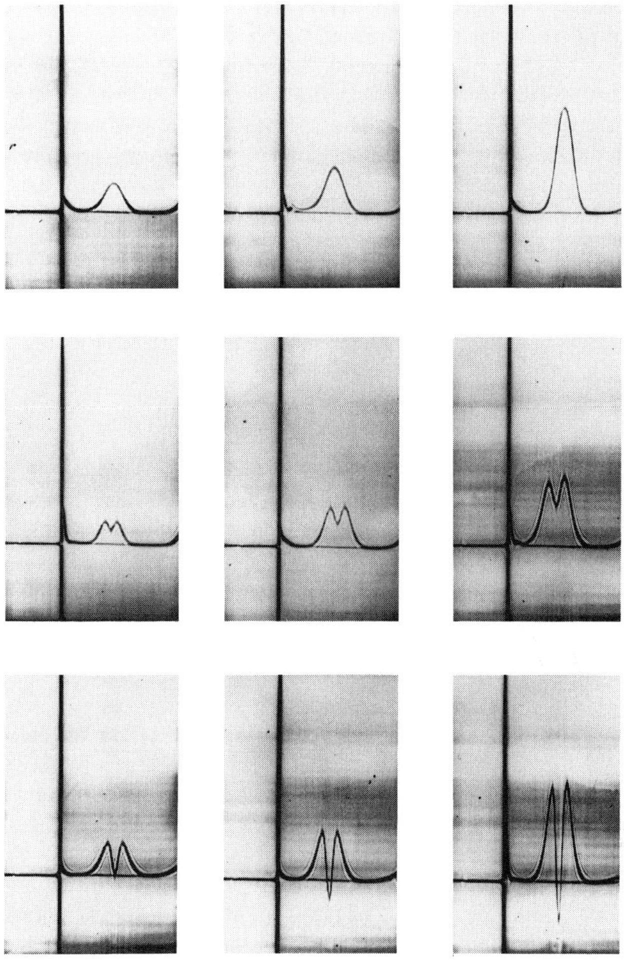

Figure 2. Schlieren patterns taken at approximately one hour after boundary formation of dextran T200 (M~2,000,000) concentration gradients in albumin solutions. The dextran concentration was varied left to right in the order of ~5, ~10 and ~20 mg g^{-1} (for exact concentrations see below) and the albumin concentration (initially uniform) was varied top to bottom in the order of ~50, ~100 and ~200 mg g^{-1}. Exact concentrations given below are in the order of left to right, top to bottom; each in the form of the initial albumin concentration of solution that was layered (right-hand-side solution in Fig. 1), albumin concentration and dextran gradient concentration, respectively in the solution furthest from the centre of the rotor (left-hand side solution in Fig. 1); all concentrations are in kg m^{-3}: 50.64/50.48, 4.99; 50.68/50.35, 10.12; 50.68/50.01, 20.29; 102.61/102.53, 5.22; 102.61/101.88, 10.48; 102.61/101.24, 20.77; 211.66/210.89, 5.21; 211.66/210.12, 10.60; 211.66/208.42, 21.79 (reproduced from Comper *et al.*[7] with permission of publishers).

and 20 mg g^{-1} and the concentration of albumin was varied from top to bottom in the order of 50, 100 and 200 mg g^{-1}. We found that as the concentration of albumin was increased the schlieren pattern transformed from a single peak at a 50 mg g^{-1} albumin concentration to a split peak at albumin concentrations of ~100 mg g^{-1} and ~200 mg g^{-1}. The nature of the schlieren profile was not significantly affected by the magnitude of the dextran concentration gradient over the 5-20 mg g^{-1} concentration range although the size of the inverted peak in the split schlieren seemed to be directly related to the magnitude of the dextran concentration gradient. We have previously shown that similar profiles were obtained for dextran with molecular weight of 70,000 generating the concentration gradient in the albumin solutions[7]. Furthermore, the refractive index difference of dextran was independent of the presence of albumin so the boundary patterns observed in Fig. 2 were not the result of anomalous variation in the refractive index difference of the polymer components in the ternary system.

3. TERNARY DIFFUSION ANALYSIS AND SIMULATED PROFILES

In an isobaric, isothermal ternary system (polymer (1) + polymer (2) + solvent (3)) the diffusional flows, J_i, of each polymer are given by the following relationships,

$$-J_1 = D_{11}\frac{\partial c_1}{\partial x} + D_{12}\frac{\partial c_2}{\partial x} \qquad (1)$$

$$-J_2 = D_{21}\frac{\partial c_1}{\partial x} + D_{22}\frac{\partial c_2}{\partial x} \qquad (2)$$

where $\partial c_i/\partial x$ is the concentration gradient of i, the D_{ii} are the main diffusion coefficients and the D_{ij} are the cross diffusion coefficients. When the polymer component 1 is initially distributed uniformly so that $\partial c_1/\partial x = 0$ in eq. (1), then its flux is determined by the concentration gradient of the other polymer (2) and the magnitude of the cross diffusion term D_{12}.

A direct quantitative analysis of the cross diffusion coefficients and the other ternary diffusion coefficients in the centrifuge has been difficult up till now due to the low sensitivity of the absorption optics in measuring albumin concentration directly and the complexity of the schlieren pattern. However, by the use of computed ternary diffusion coefficients[8] from measurements of all binary frictional and thermodynamic interactions between components we can calculate concentration and concentration gradients across the boundary. The ternary diffusion coefficients for dextran-albumin and a number of other ternary systems have been evaluated previously[7]. Some of these values have been displayed in Table 1. These analysis demonstrated that the cross diffusion coefficient D_{12} was dominated by the thermodynamic interaction (expressed as a coefficient, A^+, as obtained from osmotic pressure measurements and represents non-ideality arising from excluded volume effects) between the two polymers, rather than the frictional interaction, and could be approximated by the equation

$$D_{12} \approx \frac{RTA^+M_2C_1}{f_{13}} \qquad (3)$$

where R is the universal gas constant, T the absolute temperature, C_1 the mass concentration of 1, M_2 the molecular weight of 2 and f_{13} is the frictional coefficient of the polymer component with water. This equation bears a strong resemblance to the concentration dependent polymer mutual diffusion coefficient D_1 in a binary (polymer (1) + solvent (3)) system[2] which is defined as

$$D_1 \approx \frac{RT}{f_{13}}(1 + 2(A_2)_1 M_1 C_1) \tag{4}$$

where $(A_2)_1$ is the second virial coefficient of component 1 which, for uncharged polymers, represents the excluded volume interaction between like polymers. By analogy, therefore, the finite value of A^+ represents the fact that both dextran and albumin will mutually increase each others activity or effective concentration. Therefore, the albumin flux in Fig. 1 could alternatively be described as a result of albumin diffusing in accord with its effective concentration gradient. This will give rise to a net flux of albumin in the direction of dextran diffusion as determined by direct analysis in a diffusion cell (Table 2).

A summary of the computed ternary diffusion coefficients for the albumin - dextran system given by Comper *et al.*[7] (see also Table 1) demonstrated that the main diffusion coefficients (D_{11}, D_{22}) have very similar values to their corresponding mutual diffusion coefficient in binary solutions. The cross diffusion term D_{21} (with albumin as component 1, dextran as component 2) is insignificant whereas D_{12} is relatively large in magnitude and varies significantly with albumin concentration but not with dextran gradient concentration as embodied in eq. (3). The interaction coefficient A^+, has previously been shown to represent purely entropic or excluded-volume interaction between components; *i.e.* it does not produce any heat effects[9]. The value of this thermodynamic interaction coefficient between dextran and albumin was taken from studies on albumin interaction with cross-linked dextran and soluble chain dextrans[10,11]. It was demonstrated that this coefficient is effectively molecular weight independent (when non-ideality is expressed in a concentration scale of mass per volume). The molecular weight independence of the dextran on cross diffusion is borne by the cancellation of the molecular weight terms embodied in eq. (3) and the molar concentration scale of c_2 in eq. (1).

In Fig. 3 it is seen we can successfully generate the essential features of the schlieren pattern from the computed ternary diffusion coefficients. Superposition of the computed profile onto the schlieren pattern by matching peak heights generates a good matching over the whole profile.

4. INCONGRUENT DIFFUSION

Incongruent diffusion of a polymer may arise if the concentration gradients of polymer 1 and of polymer 2 are in opposite direction, and when

$$\left| D_{12}\frac{\partial c_2}{\partial x} \right| > \left| D_{11}\frac{\partial c_1}{\partial x} \right| \tag{5}$$

In an effort to counterbalance the flow of albumin which is coupled to dextran diffusion we have purposely constructed a system with opposing polymer

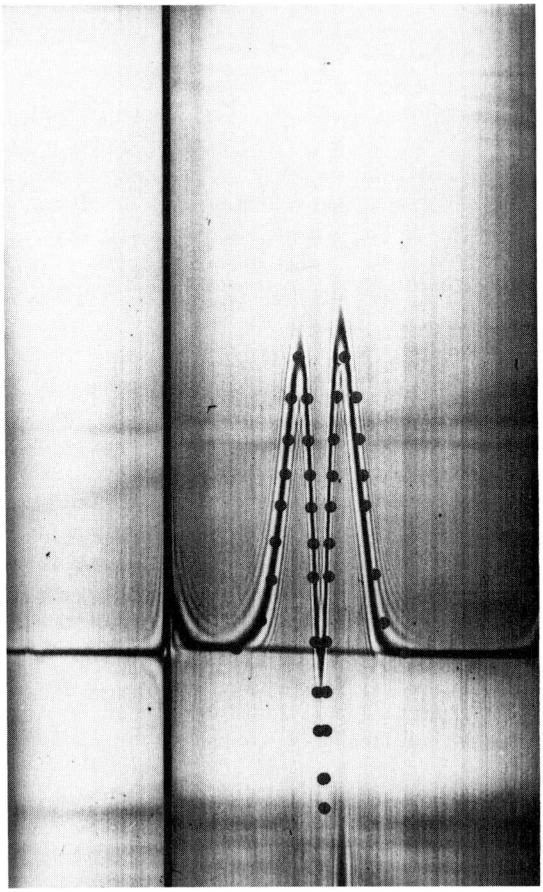

Figure 3. Schlieren profile obtained by layering an albumin solution of 209.2 kg m^{-3} onto a solution containing albumin at 210.5 and 22.5 kg m^{-3} Dextran T200. The photograph was taken approximately one hour after boundary formation. The schlieren profile has been superimposed with a computed profile (•) obtained by use of ternary diffusion coefficients in Table 1 and specific refractive increments for albumin of 0.188×10^{-3} m^3kg^{-1} and dextran 0.153×10^{-3} m^3kg^{-1}. Initial matching was made by making the left-hand peak height of the schlieren profile and the computed profile identical (reproduced from Comper *et al.*[7] with permission of publishers).

Table 2. Unidirectional diffusion coefficients[a] of $[^3H]$albumin (=Alb*) (present in trace quantites) in systems with 20 mg g^{-1} gradient of dextran (=D) with either zero or non-zero gradients of albumin (=Alb). Arrows indicate direction of diffusion for labelled albumin.

Polymer concentrations mg g^{-1} Lower solutions		Upper solutions	$10^{11}\times$Unidirectional diffusion coefficient[a] (m^2s^{-1})	$10^{11}\times$Net diffusion coefficient[b] (m^2s^{-1})
200Alb + 20D + Alb*	→	200Alb	3.27 (±0.3)	+1.79
200Alb + 20D	←	200A + Alb*	1.48 (±0.1)	
200Alb + 20D + Alb*	→	220Alb	2.09 (±0.2)	$+6.28\times10^{-2}$
200Alb + 20D	←	220A + Alb*	1.96 (±0.2)	

[a] Measurements made in an upright cell diffusion cell with shear-formed boundaries[13]. Each diffusion coefficient has been obtained from at least 16 experimental points corresponding to different times the diffusion was allowed to proceed.

[b] Defined as $(C'D' - C''D'')/\bar{C}$ where C' and C'' (with C'>C'') are the concentrations of albumin on each side of the boundary, D' and D'' are the corresponding unidirectional diffusion coefficients and $\bar{C}$ is the mean albumin concentration.

concentration gradients. Under these conditions the schlieren profile so obtained (Fig. 4) demonstrates a characteristic pattern similar to that seen in Fig. 2. The initial albumin concentration gradient would enhance the inverted middle peak and the peaks on either side indicate that significant cross diffusion was still occurring in this system. An estimate of the net flux of albumin across the boundary cannot be directly made from this schlieren profile. However, an estimate can be made through computation of the albumin concentration profile through use of the ternary diffusion coefficients given in Table 1. This analysis indicated that for a mean albumin concentration of 210 kg m^{-3} across the boundary, a net flux of albumin would occur against its concentration gradient (*i.e.* positive gradient giving a positive flux, see eq. (1)) of up to 25 kg m^{-3} with an opposing dextran gradient of 20 kg m^{-3}. For higher concentration gradients of albumin, the net flux would be in the direction of the albumin concentration gradient.

Unidirectional flux measurements of $[^3H]$albumin also demonstrate that the net flux of albumin was at the point of reversing sign near a concentration gradient of 20 mg g^{-1} albumin opposing a 20 mg g^{-1} dextran gradient as the net flux was extremely low under these conditions (Table 2).

5. CROSS DIFFUSION LEADING TO A DENSITY INVERSION

If we reverse the distributions of albumin and dextran we now have a uniform distribution of dextran with a small concentration gradient of albumin then the resulting schlieren profile seems Gaussian-like and without apparent anomaly. (These techniques have been used previously to study many similar systems in an effort to understand the restricted diffusion of one polymer in the presence of another[12].) The computer-simulated profiles (Fig. 5) do predict that a small concentration inversion of the dextran will occur and that this will affect the schlieren profile. It will result in an approximately 10% error on the apparent diffusion coefficient of albumin calculated from the spreading of the schlieren pattern. The relatively smaller magnitude of the concentration inversion of the dextran in Fig. 5 as compared to albumin studied above is, in part, due to the differences in the specific refractive increments of these polymers and the hydrodynamical frictional coefficients.

The corresponding concentration and density profiles for this dextran-albumin system from Comper *et al.*[7] are shown in Fig. 6. While a concentration inversion of the matrix is predicted, there is no accompanying density inversion. On the other hand, if we substitute the albumin concentration gradient with a poly(vinylpyrrolidone) concentration gradient, then a density inversion leading to boundary instability is predicted and with the subsequent development of a convective flow regime (see below). This is primarily due to the higher magnitude of the interaction coefficient between dextran and poly(vinylpyrrolidone) (interaction coefficient[8] on a mass per volume concentration scale *i.e.* $A^+ = 13.5\times10^{-4}$ mol g^{-2} cm^{-3}) as compared to the dextran-albumin interaction coefficient of $A^+ = 3.8\times10^{-4}$ mol g^{-2} cm^{-3}. These differences in A^+ essentially reflect the differences in the conformations or shapes of the interacting molecules with dextran[10]. Therefore, by increasing the magnitude of the excluded volume interaction coefficient

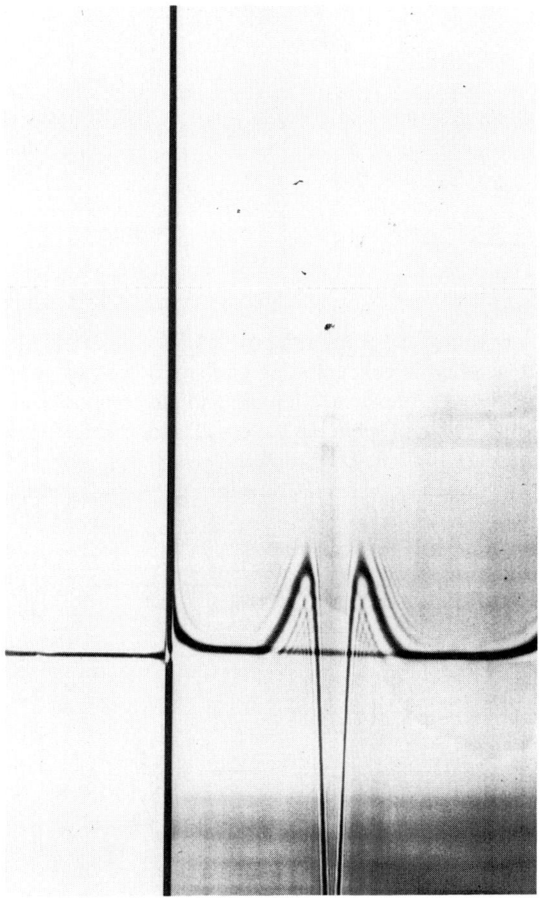

Figure 4. Schlieren pattern taken at approximately 1h of a solution of albumin at 229.7 kg m^{-3} layered onto a solution containing albumin at 203.6 kg m^{-3} and dextran T200 at 23.1 kg m^{-3}.

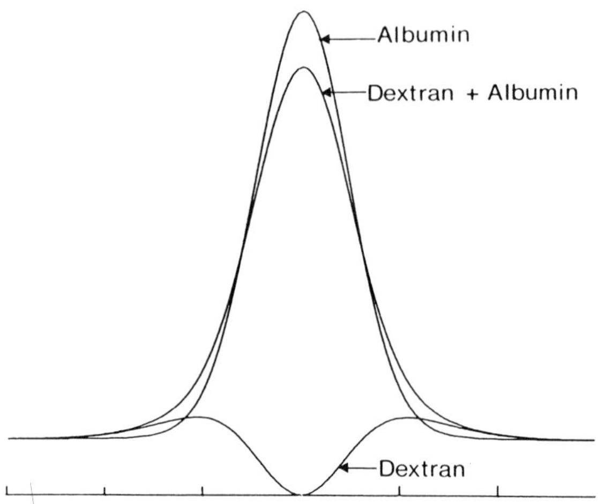

Figure 5. Computer-simulated schlieren profiles for a dextran (M~1.5×10^5) solution of 94 kg m^{-3} with an albumin concentration gradient of 5 kg m^{-3} evaluated at a time of 300 s. Profiles are given for dextran, albumin, and the composite of these two profiles corresponding to the predicted schlieren pattern. Ternary diffusion coefficients in units of 10^{-11} m^2 s^{-1} were D_{11} = 1.6, D_{12} = 0.49, D_{21} = 0.73, D_{22} = 4.8. The abscissa is in units of 2×10^{-4} m (reproduced from Comper *et al.*[7] with permission of publishers).

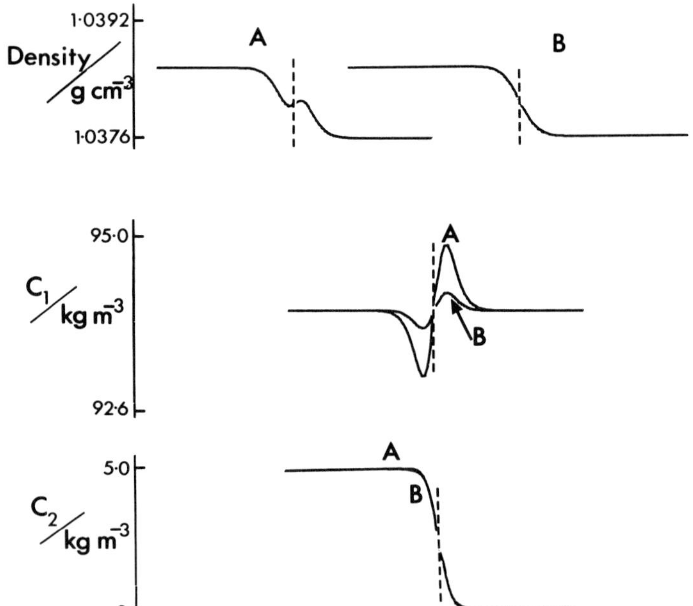

Figure 6. Computer-simulated density profiles and concentration profiles across a boundary (represented by the dashed line) of a system described in the legend of Fig. 5 (profiles designated by B) and an identical system, but with albumin gradient material substituted for a poly(vinylpyrrolidone) of identical molecular weight (designated A) (reproduced from Comper *et al.*[7] with permission of publishers).

by approximately four-fold the diffusive system converts into a convective flow regime initiated by a density instability at the boundary[6].

Convective flow regimes in poly(vinylpyrrolidone)-dextran systems described above are characterised by spectacular convective fingering at the initial boundary[14]. With time, the interdigitated, countercurrent, convective fingering flows grow in the gravitational field. Observations of these systems in the ultracentrifuge has been made by schlieren and UV absorption optics[15]. In the use of these optical systems it must be remembered that these techniques result in space average measurements in the direction parallel to the initial boundary (or at right angles to the optical axis) and thus any concentration gradients arising in this direction from the presence of convection may not be monitored (see below and Fig. 8). Such is the case in the analysis of UV absorption scans of a system with a uniform dextran distribution superimposed with a small concentration gradient of poly(vinylpyrrolidone). The time evolution of the initial absorption scans of polyvinylpyrrolidone transport monitored at 237 nm, is shown in Fig. 7. The anomalous feature of these scans is that, with time, material which absorbs at 237 nm rapidly accumulates on the left hand side of the boundary (also accounted for by depletion of absorbing material on right hand side of boundary). This material appears to be evenly distributed in this region. Absorption scans at 280 nm, which specifically detect dextran, demonstrated that no measurable gradients of this material were formed at this time. We concluded that changes in the net distribution of material as observed at 237 nm were singularly due to the net transport of poly(vinylpyrrolidone). (The rate of material transfer, as measured with the UV scan, is similar to the transport rate obtained through the specific measurement of unidirectional flux of [^{3}H] labelled poly(vinylpyrrolidone) in diffusion cells at unit gravity). For the same system an apparently normal schlieren profile is registered. This profile could be quantified, too, in terms of diffusion behaviour although the diffusion coefficients are anomalously high[15]. One further feature of this optical method with this system is that there is an apparent rapid removal of material from the concentration gradient at the boundary as evidenced by a reduction (up to 20%) in the area under the schlieren curve.

Analysis of the fingering dextran-poly(vinylpyrrolidone) in a glass cuvette by using blue dye labelled poly(vinylpyrrolidone) and colour indexing the degrees of grey as monitored through a video camera is shown in Fig. 8. The cross sectional summation of concentration gives the average concentration profile shown on the tracing which obviously has the appearance of being sigmoidal, and therefore 'diffusive', and would account for the similar 'diffusive' behaviour seen in the centrifuge.

6. CONCLUDING REMARKS

The observations reported in this Chapter highlight the importance of couple diffusion in multicomponent polymer systems not only to generate unusual diffusional fluxes against concentration gradients but also their participation in the transition of a multicomponent system from a diffusional regime to a convective flow regime. These different types of transport may arise through non-specific interactions between polymers *i.e.* excluded volume interactions (which are shape and conformation-

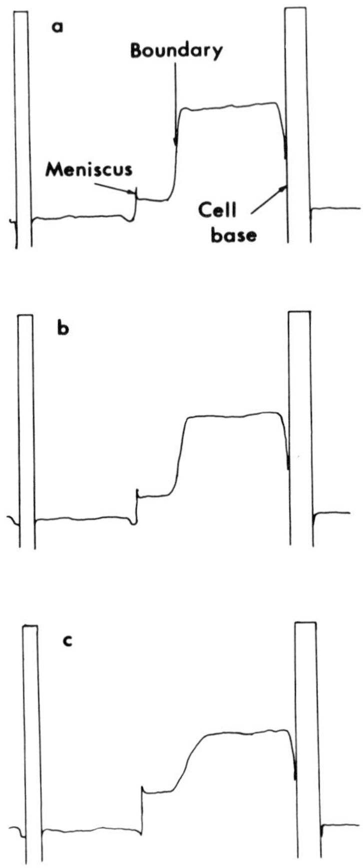

Figure 7. Time evolution of UV scans monitored at 237 nm in the analytical ultracentrifuge of a solution of dextran T10 (M~10,000) at 135.2 kg m^{-3} layered onto a solution of dextran at 135.2 kg m^{-3} and poly(vinylpyrrolidone) (M~360000) at 5.3 kg m^{-3} at times (a) 180, (b) 306, and (c) 780 s (reproduced from Preston *et al.*[15] with permission of publishers).

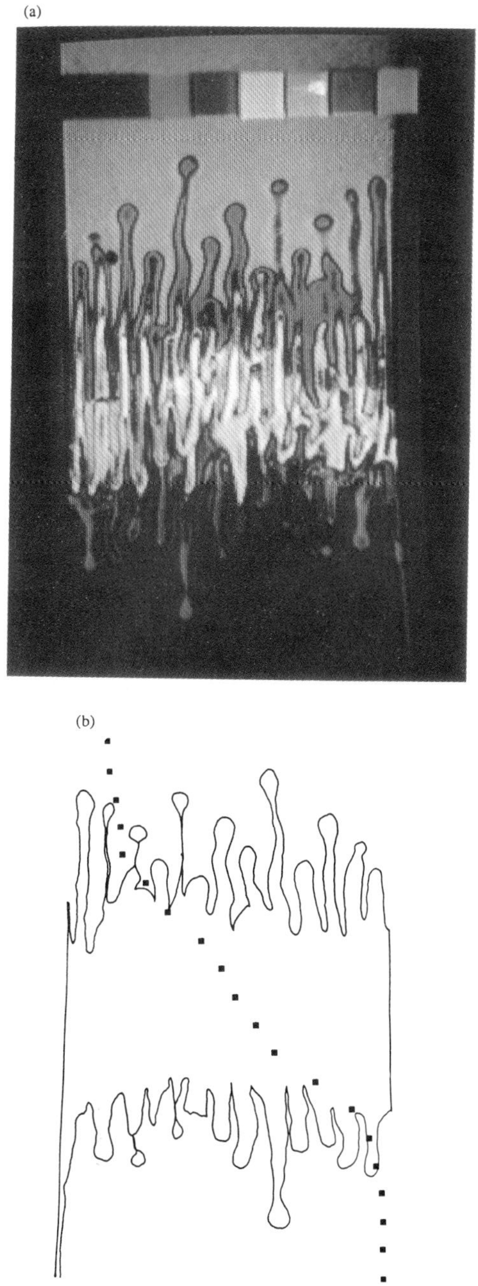

Figure 8. (a) Colour indexing of a fingering system in a thin (2 mm pathlength glass cuvette) created with solutions described in legend of Fig. 7. The poly(vinylpyrrolidone) was labelled with blue dye. Image taken at 20 min after initial boundary formation. (b) Represents a tracing of the outline of the colour indexed image in Fig. 8a and the corresponding average cross sectional concentration evaluated by summing contributions from calibrated colour indexing analysis.

dependent) and are therefore expected to occur in a wide variety of concentrated multicomponent polymer systems that are encountered in biology and technology. It is clear that the switching control in converting the diffusive regime to a convective regime in polymer systems is incorporated in the magnitude of the excluded volume interaction. The manifestation of these different regimes, in most cases, can clearly be delineated in the centrifuge. This is particularly the case in association with the striking schlieren patterns obtained in concentrated albumin solutions maintained in the diffusive regime. It would appear that this type of analysis could be exploited further in understanding both static and dynamic interactions between molecules.

ACKNOWLEDGEMENTS

This work was supported by grants from the Australian Research Council. We acknowledge the invaluable contributions by students and research assistants towards this work particularly G.J. Checkley, P.M. MacDonald and G. Wilson.

REFERENCES

1. For example C. Tanford, 'Physical Chemistry of Macromolecules', John Wiley and Sons, New York, 1961.
2. W.D. Comper, B.N. Preston and P. Daivis, *J. Phys. Chem.*, 1986, *90*, 128.
3. T.C. Laurent, L.-O. Sundelöf, K.-O. Wik and B. Wärmegård, *Eur. J. Biochem.*, 1976, *68*, 95.
4. E.L. Cussler, 'Multicomponent Diffusion', Elsevier, Amsterdam, 1976.
5. A. Katchalsky and P.F. Curran, 'Non-Equilibrium Thermodynamics in Biophysics', Harvard University Press, Cambridge, Mass., 1965.
6. W.D. Comper, R.P.W. Williams, G.J. Checkley and B.N. Preston, *J. Phys. Chem.*, 1987, *91*, 993.
7. W.D. Comper, P.M. MacDonald and B.N. Preston, *J. Phys. Chem.*, 1984, *88*, 6031.
8. W.D. Comper, G.J. Checkley and B.N. Preston, *J. Phys. Chem.*, 1984, *88*, 1068.
9. W.D. Comper and T.C. Laurent, *Biochem. J.*, 1978, *175*, 703.
10. A.G. Ogston and P. Silpananta, *Biochem. J.*, 1970, *116*, 171.
11. T.C. Laurent, *Biochem. J.*, 1964, *93*, 106.
12. B.N. Preston, T.C. Laurent and W.D. Comper, 'Molecular Biophysics of the Extracellular Matrix', S. Arnott, D.A. Rees and E.R. Morris (eds.), Humana Press, Clifton, New Jersey, 1984, Chapter 6, p. 119.
13. M.-P.I. Van Damme, W.D. Comper and B.N. Preston, *J. Chem. Soc., Faraday Trans. 1*, 1982, *78*, 3357.
14. B.N. Preston, T.C. Laurent, W.D. Comper and G.J. Checkley, *Nature*, 1980, 287, 499.
15. B.N. Preston, W.D. Comper, T.C. Laurent, G.J. Checkley and R.G. Kitchen, *J. Phys. Chem.*, 1983, *87*, 655.

Part IV: Specific Applications to Biochemical and Polymer Systems

24

Sedimentation Analysis of Seed Proteins

By V. Prakash

BIOPHYSICAL CHEMISTRY UNIT, FOOD CHEMISTRY DEPARTMENT, CENTRAL FOOD TECHNOLOGICAL RESEARCH INSTITUTE, MYSORE - 570 013, INDIA

1. INTRODUCTION

Seed proteins have in recent years attracted the attention of biochemists, plant physiologists, enzymologists, physicists, biophysical chemists and protein chemists because of their unique properties. The nature of information about these proteins which started in the early days of protein concentrates has slowly emerged to isolation of pure protein fractions, protein-protein interactions and protein-ligand interactions utilizing a number of biophysical techniques. Of particular interest in the context of the present article is the sedimentation analysis of these proteins[1-4].

Before we analyse the behaviour of these proteins in the analytical ultracentrifuge it may be in order to briefly summarize the role of these proteins in seeds. These proteins in the seeds serve as the nitrogen reserve of the seed. They supply the necessary free amino acids and amide nitrogen to the growing plant during the initial stages of germination. The location of these storage proteins is generally in distinct organelles in seed often called 'protein bodies'. This may be one of the unique nature of the seeds in that they contain substantial amounts of particle-bound proteins. It is possible that these proteins may possess some unique features which confine them to the protein bodies. Hence these seeds can be treated as a class of tissues which secrete proteins and the secretions are mostly retained within the cell[5].

Considerable amount of work has gone into understanding the protein bodies especially in the case of soybean (*Glycine max*) and groundnut or peanut (*Arachis hypogea*). It is also shown that these protein bodies contain besides storage proteins other proteins such as enzymes and lectins in much smaller amounts[6-8]. Recently the storage proteins from seeds are regarded in the same category as secretory proteins as they are shown to undergo post-translational modification[9]. Of importance to this article is the exhibition of multiplicity of quaternary structures by these seed proteins as a result of polymorphism and occurrence of several degrees of oligomerization resulting in a series of sedimentation coefficients[9-11]. The variation in the quaternary structure is affected by minor perturbations of pH, temperature, concentration of intrinsic ligands and protein concentration itself, resulting in a large amount of data on the sedimentation behaviour of these class of proteins[12]. In this article an attempt is made to consolidate the role played by individual parameters and to see whether any generality exists in such associating-dissociating protein systems in a number of seeds as a result of variation in solution conditions such as pH, temperature, ionic strength, dissociating and denaturing agents and ligands.

Number of Protein Fractions

The number of protein fractions of total proteins in many seeds are mostly recognised from the sedimentation velocity pattern. In the field of seed proteins in fact the proteins are recognised by their sedimentation coefficient such as 11 S, 7 S and 2 S proteins. The sedimentation coefficient values of the various protein fractions are summarized in Table 1. From Table 1 it can be seen that all oilseeds except mustard and rapeseed consist predominantly of four protein fractions. Based on the $s_{20,w}$ values of these protein fractions one can classify the total proteins of oilseed into four fractions. They can be designated as 2 S (low molecular weight or LMW protein), 7 S (medium molecular weight or MMW protein), 11 S (high molecular weight or HMW protein) and 15 S - 18 S (polymer-resulting from possible aggregation of 2 S, 7 S or 11 S or may be inherently present in the seed).

Table 1. Sedimentation coefficient ($s_{20,w}$) value of the various protein components from several oilseeds[12].

Oilseed	Protein fractions*			Polymer
	LMW	MMW	HMW	
Soybean	2	7	11	15
Peanut	2	7	11	18
Sesame seed	2	7	11	15
Sunflower seed	2	7	12	16
Mustard seed	2	7	12	-
Rape seed	2	7	12	-
Cotton seed	2	7	12	18
Safflower seed	2	7	12	17

* LMW, Low molecular weight; MMW, Medium molecular weight; HMW, High molecular weight.

It is important to understand that the generalization from $s_{20,w}$ values is an over simplification since the $s_{20,w}$ value of all the proteins from various oilseeds have not been determined under identical conditions of pH, solvent composition, temperature, protein concentration and other ultracentrifuge parameters. Recently in the author's laboratory an attempt was made to evaluate some of the hydrodynamic parameters (sedimentation and diffusion coefficients) of four major protein fractions namely arachin from groundnut, α-globulin from sesame seed, brassin (M) from mustard seed and helianthinin in sunflower seed in a single solvent system to see whether the hydrodynamic parameters are close enough[13]. The $s^0_{20,w}$ values indicate a close enough value for the above protein with value of (12.5 ± 0.5) S as the standard deviation for the four proteins mentioned. Similarly the $D^0_{20,w}$ values also shows an average value of $(3.9 \pm 0.6)\times10^{-7}$ cm^2 s^{-1} for these four proteins. These results indicate and confirm the earlier hypothesis that there exists a close similarity in the sedimentation and diffusion behaviour of these HMW proteins. Similar results are obtained with the LMW proteins also[14,15]. The result indicates that the LMW proteins sediment with an s value close to 2.0 S.

Fig. 1 shows a typical sedimentation velocity profile of seed protein with the various fractions designated as shown in the figure. The pattern is that of safflower seed total protein. However the percent composition of these fractions varies depending on the oil seed in question[12]. The HMW protein or 11 S protein fraction is the major protein fraction in groundnut, sesame seed, sunflower seed, poppy seed, safflower seed and linseed[12]. On the other hand, it is only present to the extent of 20% to 30% in mustard seed, rapeseed, cottonseed, soybean which has a predominance of other protein fractions such as the LMW or the MMW proteins designated as 2 S and 7 S respectively[12].

Since there is a large amount of data on these proteins and it is often referred to as '11 S' or '2 S' recently from the author's laboratory the names for the LMW and HMW proteins are proposed and accepted to a large extent in the literature. These are summarized in Table 2 along with the botanical names of the oilseed from which the names for different protein fractions are mostly derived[12,14,15].

Isolation and Characterization

Most of these seed proteins are isolated to homogeneity and characterized for their amino acid composition, biophysical and physico-chemical properties. A comprehensive review of their properties has recently appeared by Prakash and Rao[12,14]. Their behaviour in the analytical ultracentrifuge is very unique for each of these proteins which makes this technique a unique one to follow the changes in the quaternary structure of assembly-disassembly of these proteins.

The 11 S or the 2 S protein fractions from various oilseeds are isolated to homogeneity and characterized for their hydrodynamic properties especially for $s^0_{20,w}$, $D^0_{20,w}$, partial specific volume, frictional ratio, molecular weight (by approach to

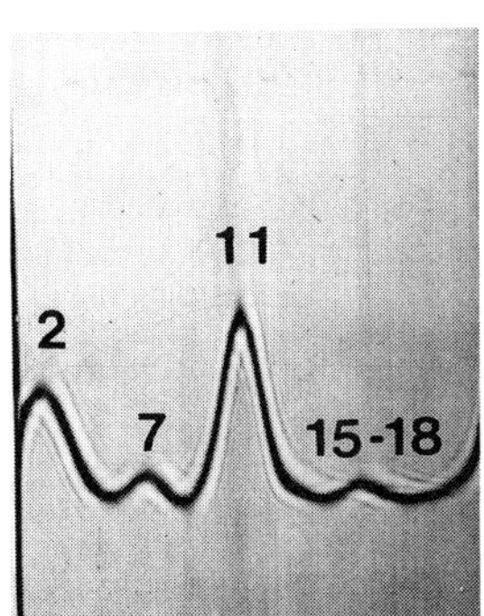

Figure 1. Typical sedimentation velocity profile of seed protein at neutral pH. The pattern is that of safflower seed total protein extracted in 0.02 M phosphate buffer pH 7.5 containing 1 M sodium chloride. The concentration of the protein is 18 mg/ml and the photograph was taken 30 min after attaining two-thirds maximum speed of 59,780 rev/min using a 12 mm, 4° single sector Kel F centrepiece at 27°C in a Beckman Model E analytical ultracentrifuge. The nominal sedimentation coefficients are indicated in the figure.

Table 2. Table of nomenclature of the high molecular weight (HMW) and low molecular weight (LMW) protein fraction from various oil seeds.

Oilseed	Botanical name	HMW protein		LMW protein	
		Existing name	Suggested name	Existing name	Suggested name
Soybean	*Glycine max*	Glycinin	Glycinin	Conglycinin	Conglycinin
Peanut	*Arachis hypogea*	Arachin	Arachin	Conarachins	Conarachins
Sesame seed	*Sesamum indicum*	α-globulin	Sesamin	α-globulin	Consesamin
Sunflower seed	*Helianthus annuus*	12 S protein	Helianthinin	Albumin fraction	Conhelianthinin
Mustard seed	*Brassica species*	12 S protein	Brassin (M)	2 S protein	Conbrassin (M)
Rape seed	*Brasicca species*	12 S protein	Brassin (R)	2 S protein	Conbrassin (R)
Cotton seed	*Gossypium species*	11 S protein	Gossypinin	Albumin	Congossypinin
Safflower seed	*Carthamus tinctorius*	Carmin	Carmin	LMW protein	Concarmin
Linseed	*Linum usitatissium*	12 S protein	Linin	LMW protein	Conlinin

sedimentation equilibrium), intrinsic viscosity measurements and sedimentation equilibrium methods of both native and denatured molecules. The data is utilised for calculation of number average, weight average, and z-average molecular weight of these proteins[14]. Table 3 summarizes such data.

In some cases such as α-globulin, carmin and brassin even the hydrated volume and hydration factors are reported[16]. In Table 3 is also summarised some of the hydrodynamic properties reported in literature for HMW proteins from several oilseeds.

2. ASSOCIATION-DISSOCIATION

The association-dissociation phenomenon of the high molecular weight proteins stems mainly from two aspects (i) their multi- subunit composition and (ii) protein-protein interaction in solution. The high molecular weight proteins from the various oilseeds have an average molecular weight of around 2.5×10^5 g/mol. They contain six acidic and six basic subunits in them stabilized by inter and intra non-covalent interactions such as hydrogen, hydrophobic and other weak interactions resulting in two 'dough nut' kind of rings sandwiched one above the other such that the acidic and basic subunits are in contact with each other to form probably a hollow cylinder[12,19]. The effect of solution parameters such as pH, ionic strength, nature of buffer ions, protein concentration, temperature, salts (cations and anions), denaturants such as urea and guanidine hydrochloride, detergents such as sodium dodecyl sulphate (SDS), cetyl trimethyl ammonium bromide (CTAB), triton and ligands such as chlorogenic acid, phytic acid, acacetin have been shown to affect the process of association-dissociation of these proteins in solutions. Some of the classical results observed in the sedimentation velocity behaviour of these proteins in presence of the above reagents are summarized below.

Effect of Ionic Strength

The effect of ionic strength on the association-dissociation phenomenon is predominantly shown by glycinin from soybean and arachin from peanut. Glycinin, is partially associated into a faster sedimenting component at 0.5 ionic strength and dissociates to monomeric 11 S at higher ionic strength[20]. At very low ionic strength of 0.001 it undergoes dissociation into its subunits. Similar observation is made with arachin of peanut which also undergoes association-dissociation between 9 S and 14 S as a function of ionic strength[21]. The nature of the association appears to be non-covalent as in most of the cases the phenomenon is reversible upon changing the experimental conditions. The equilibrium position depends on the nature of the salt, protein concentration and pH. In most of the studies of the above kind the pH has been maintained since it is well known that these proteins undergo an association-dissociation equilibrium at high or low pH values.

The changes brought about by ionic strength are found to be reversible in the case of arachin from peanut which follows the schematics[21]

$$A_n \rightleftarrows nA$$

Table 3. Hydrodynamic properties of the HMW proteins from various oil seeds[12,13,16-18].

Property	Arachin	Carmin	α-globulin	Brassin (R)	Helianthinin
Sedimentation coefficient $s^0_{20,w}$ (S)	12.8	12.7	12.8	12.7	12.8
$10^7 \times$ Diffusion coefficient $D^0_{20,w}$ ($cm^2\ s^{-1}$)	3.3	5.5	3.9	3.8	3.8
Molecular weights (g/mol):					
(a) from sedimentation and diffusion data	3.3×10^5	2.4×10^5	2.9×10^5	3.0×10^5	3.0×10^5
(b) from sedimentation and gel chromatography data	2.2×10^5	2.9×10^5	2.5×10^5	2.9×10^5	3.1×10^5
(c) sedimentation equilibrium (weight average)	-	2.6×10^5	2.7×10^5	-	-
(d) approach to sedimentation equilibrium	-	2.5×10^5	-	-	-
Frictional ratio (f/f_0)	-	1.1	1.1	1.28	1.27
Partial specific volume (ml/g)	0.716	0.735	0.725	0.725	0.714

where A_n is the 14 S component and A is the 9 S component of arachin. Johnson and Shooter[21] have also given the general equation as a function of pH and salt characterizing the monomeric and dimeric arachin molecule. Similar observations are also made with conarachin II of peanut[22]. Conarachin II exhibits association-dissociation equilibrium as monitored in the analytical ultracentrifuge clearly indicating that low ionic strength favours association of conarachin II to a 13 S component[22].

Hence, the nature of interaction of the subunits in the case of glycinin, arachin and conarachin could be predominantly hydrophobic as compared to other oilseed protein fractions. Also, the nature of inter and intra-subunit hydrophobic interaction must be different in these two classes of protein based on the sedimentation velocity profile.

Effect of pH

The role played by pH on the various properties of seed proteins as monitored in the sedimentation velocity experiments clearly indicate (i) a low extractability of the total proteins in the acidic pH region and a high-extractibility profile in the alkaline pH region; (ii) dissociation and denaturation of the homogeneous high molecular weight protein fractions initially which reaggregates at extreme acidic pH conditions and (iii) dissociation and denaturation of homogeneous high molecular protein fraction in the alkaline pH region also. These data are principally monitored in the analytical ultracentrifuge as a function of pH.

In the case of arachin from peanut several workers[21,23] have clearly shown that the protein associates in low pH and high salt concentration and dissociates in low pH and low salt concentration. The 14 S molecule in the alkaline pH progressively dissociates to the 9 S and 2 S components whereas in the acidic region, in the low salt condition, it dissociates to 8.2 S component initially and subsequently to a 2 S component . The results with arachin in the acidic pH region can be summarized as shown below[24].

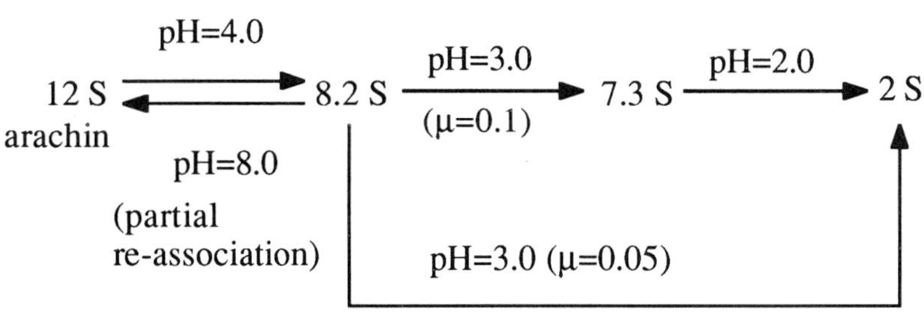

where μ = ionic strength of the solvent.

In the case of conarachin II also the 8 S molecule initially undergoes dissociation to the 2 S molecule (pH 3.0) and ultimately to a 1.8 S denatured molecule at pH 2.0[24]. The process was shown to be irreversible when dialysed back upto pH 8.0 (which is the starting pH) whereas in the case of arachin dialysing back to pH 8.0 produced a small quantity of 8.2 S component as shown earlier in the schematics.

The effect of acidic and alkaline pH on the protein α-globulin from sesame seed has been investigated by Prakash and Nandi[25]. Both acidic (pH 4.0 to 3.0) and alkali (pH 7.0 to 12.0) induced dissociation and denaturation of the protein. Schematically, it can be represented as,

$$11\,S \xrightarrow{pH \geq 3.0} 2\,S \xrightarrow{pH \leq 3.0} 2\,S + \text{"11 S"}$$

$$11\,S \xrightarrow{8.0 \leq pH \leq 10.0} 7\,S \xrightarrow{pH \geq 10.0} 2\,S$$

Where "11 S" is the reassociated molecule from 2 S protein. This has been attributed to binding of the anion which minimizes the intramolecular charge repulsion that initially was responsible for unfolding and ultimately being stabilized by hydrophobic interactions[26].

The effect of pH on the low molecular weight protein fraction, β-globulin (single polypeptide chain) from sesame seed indicates that in the acidic region the protein denatures with a reduction in the sedimentation coefficient from 2.0 S to 1.2 S [27]. Below pH 3.2 there is an increase in the sedimentation coefficient to 1.7 S possibly arising out of pH induced aggregation very similar to the mechanism explained in the case of α-globulin. In the alkaline pH there is a progressive decrease in the sedimentation coefficient from 2.0 S at pH 7.0 to 1.8 S at pH 8.2 to 1.4 S at pH 11.3, respectively. This decrease in the $s_{20,w}$ value has been explained as due to an increase in the frictional ratio of protein β-globulin due in turn to opening up of the molecule as a result of the increase in the hydroxyl ion concentration[27].

Similar results are reported for the high molecular weight protein, helianthinin from sunflower seed[28-30]. It also follows the schematics of 11 S dissociating to 2 S upto pH 2.5 below which reaggregation process produces a "12 S" component. In the alkaline pH the protein dissociates initially to a 7 S component ultimately equilibrating down to a 2 S component.

In the case of mustard seed protein brassin (M), the protein is also shown to dissociate initially which denatures later leading to refolding and reaggregation at lower pH values[31]. Schematically it can be represented as

$$12\,S \xrightarrow[H^+]{\text{upto pH 3.0}} \begin{array}{c}\text{dissociation}\\ \text{and denaturation}\end{array} \xrightarrow[H^+]{\text{below pH 3.0}} \begin{array}{c}\text{refolding and}\\ \text{reaggregation}\end{array}$$

In the alkaline region also the protein is shown to dissociate to a 2 S molecule.

The effect of acidic and alkaline pH on the protein, carmin from safflower seed has been investigated in greater detail to answer questions such as (i) is the dissociation and denaturation process simultaneous or sequential? and (ii) what is the kinetics of such a process? The results in the alkaline region of the multimetric protein, carmin as investigated in the analytical ultracentrifuge clearly demonstrates a step-wise dissociation as shown in the schematics below[32]

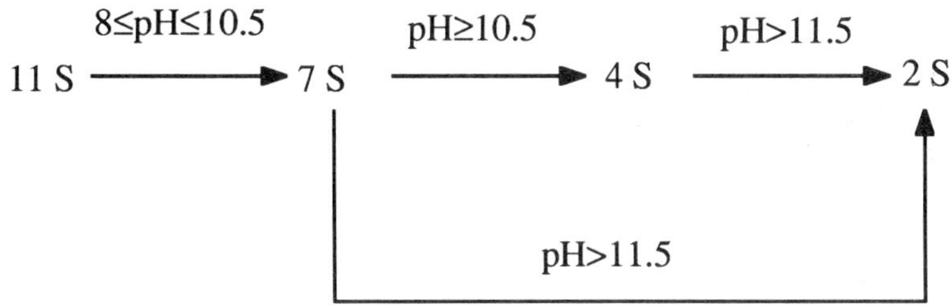

It is quite possible these reactions may not follow the same sequence, in the reverse order, when the pH is lowered. In Fig. 2 is clearly shown the effect of reversibility at different periods and conditions of experiments both with dissociated (but not denatured) carmin and denatured carmin. It is very clear from the figure that the dissociation reaction of carmin in the alkaline pH *i.e.* from 11 S to 7 S is reversible and that of dissociation of 7 S to 4 S is partially reversible. However, the denatured product, *i.e.* 2 S is not reversible to either the 4 S or 7 S component[32]. Schematically

$$11\,S \rightleftharpoons 7\,S \rightleftharpoons 4\,S \longrightarrow 2\,S$$

These results clearly indicate that the process of dissociation and denaturation is simultaneous above pH 10.0 whereas below pH 10.0 they are sequential. This probably is the first instance where it is shown with seed proteins that one can kinetically distinguish the two processes in the analytical ultracentrifuge. This is supported by stopped-flow experiments which also indicated that the dissociation step is faster (complete in about 40 msec) whereas the denaturation step appears to be at least three to four times slower than this[32].

In the acidic region carmin has been shown to undergo an initial dissociation upto pH 3.5 without any significant denaturation[33]. Below this pH the process of denaturation and dissociation is simultaneous. The protein goes through an intermediate 7 S component in the process of dissociation from 11 S to 1 S. This was trapped by judicial selection of concentration of protein and pH in the analytical ultracentrifuge. The concentration of 7 S was at no point higher than 10%. The schematics being operative in the dissociation phenomena of carmin in the acidic region can be summarised as[33]

$$11\,S \xrightarrow{\text{upto pH 4.0}} 7\,S \xrightarrow{\text{upto pH 3.0}} <1\,S \ \text{(denatured)}$$

If one carefully evaluates the sedimentation coefficient below pH 3.0 it can be seen that there is an increase in the $s_{20,w}$ value indicating reassociation of the dissociated and denatured molecule. In Fig. 3 is shown sedimentation coefficient *versus* pH for carmin in the acidic region which supports the schematics:

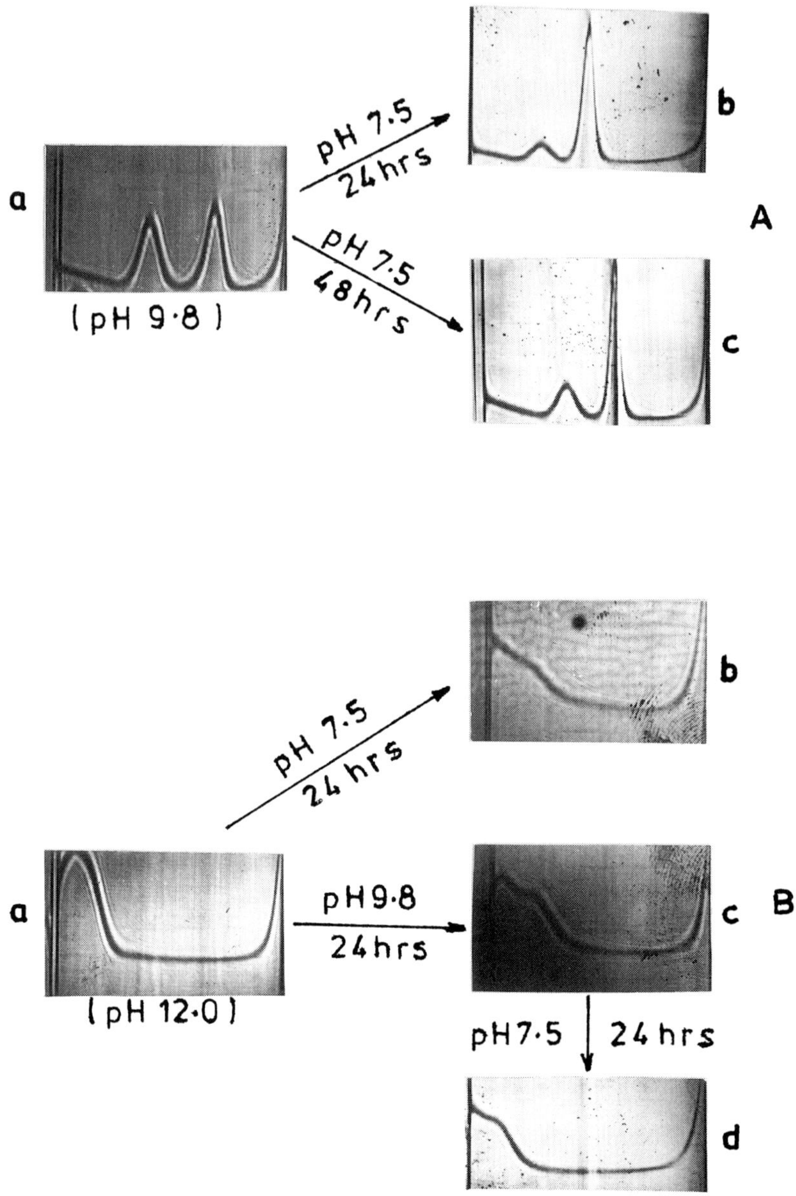

Figure 2. Sedimentation velocity pattern of carmin after decreasing the pH from either 12.0 and 9.8 to 9.8 and 7.5 respectively as indicated against each. The speed of the run was 59,780 rev/min. The pH and time of photograph are as follows. A. (a) pH 9.8, 34 min; (b) pH 9.8 to 7.5, 34 min; (c) pH 9.8 to 7.5, 40 min. B. (a) pH 12.0, 34 min; (b) pH 12.0 to 7.5, 48 min; (c) pH 12.0 to 9.8, 48 min; (d) pH 12.0 to 9.8 to 7.5, 32 min (from S. Rajendran and V. Prakash *J. Protein Chem*., 1988, *7*, 689).

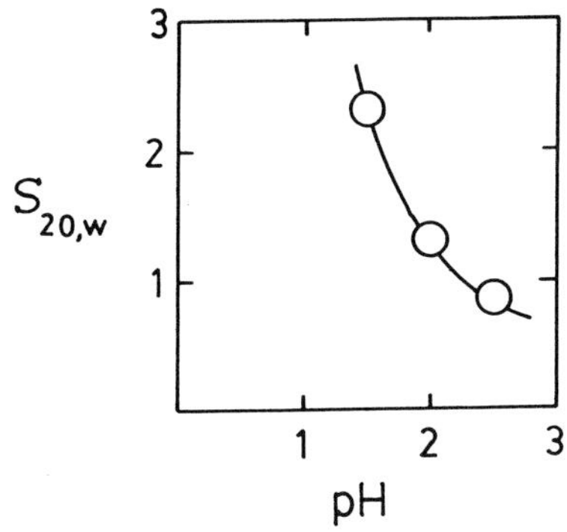

Figure 3. Variation of the sedimentation coefficient of carmin from safflower seed in the pH range of 2.5 to 1.5 (from K. Sudhindra Rao and V. Prakash, *J. Protein Chem.*, 1989, *8*, 529).

$$<1\ \mathrm{S} \xrightarrow{2.5 \leq \mathrm{pH} \leq 4.0} 2.5\ \mathrm{S}$$

(Dissociated and denatured) (Reaggregated)

The mechanism of such a reassociation process has been demonstrated as a result of reaggregation of the dissociated subunits due to anion binding and stabilized as a result of favourable cumulative hydrophobic interactions[34]. This is uniquely supported by the anomalous pK value of the carboxyl group in this region[18]. These results clearly eliminate the possibility of hydrogen bonding at this low pH.

Is such a process reversible? In Fig. 4 is shown complete schematics of the reversibility experiments that is performed in the acidic region[33]. At pH 3.5 only dissociated product was present and at pH 2.0 the denatured product was produced and both were independently reversed to pH 7.5 by (i) dialysis and (ii) readjustment immediately to pH 7.5. The result demonstrates that it is possible to generate the native carmin molecule from the dissociated molecule by the slow process of dialysis whereas the denatured protein failed to produce the original molecule upon adjusting the pH to 7.5. In fact, it produces a 9 S component instead of the 11 S - 12 S component. This has been characterized also by fast kinetic reaction analysis which clearly demonstrates an order of magnitude difference between the dissociation and denaturation process[33].

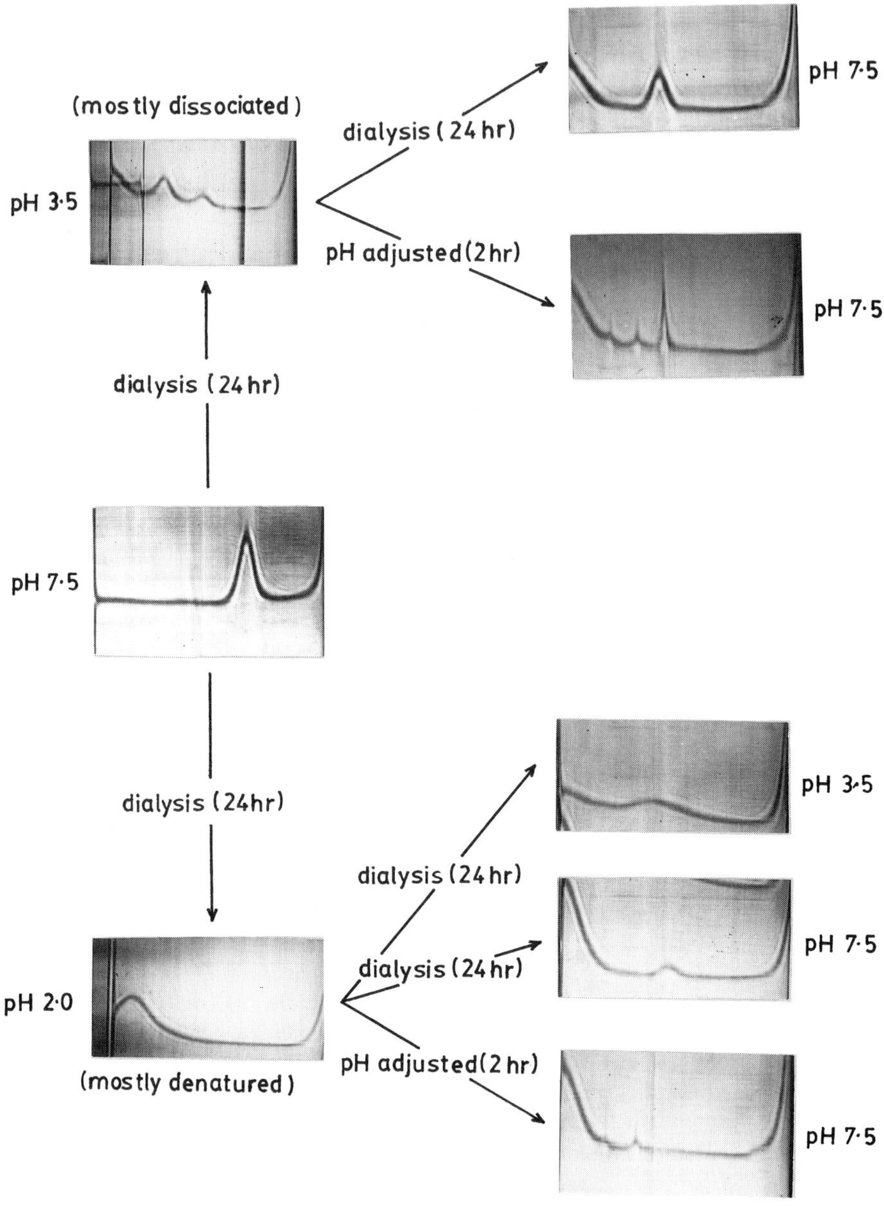

Figure 4. Sedimentation velocity pattern of carmin on reversal of pH from pH 3.5 to 7.5 (24 hr dialysis), pH 2.0 to 3.5 (24 hr dialysis), pH 2.0 to 7.5 (24 hr dialysis), pH 3.5 to 7.5 (2 hr) and pH 2.0 to 7.5 (2 hr) (from K. Sudhindra Rao and V. Prakash, *J. Protein Chem.*, 1989, *8*, 529).

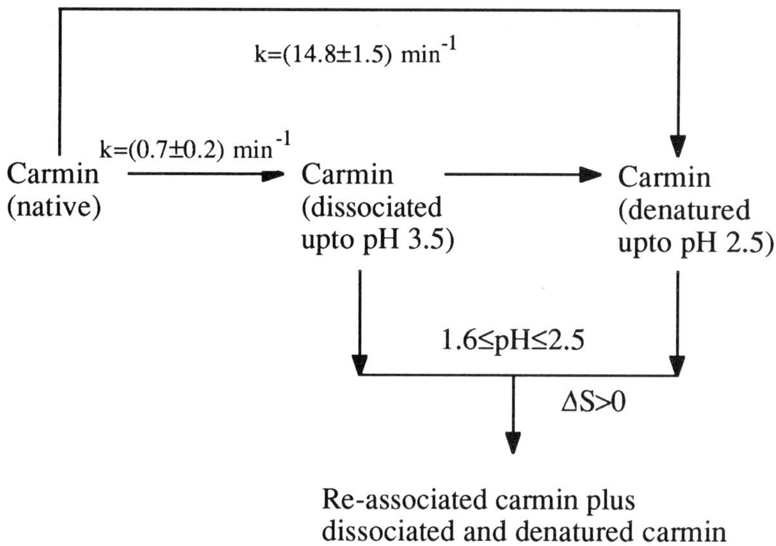

Effect of Temperature

The high molecular weight protein fractions from various oilseeds are multimeric and the subunits are mainly associated by non-covalent interactions stabilizing it to a hexamer wherein two trimers are sandwiched forming a central core[12,15,19]. The effect of temperature on α-globulin, the high molecular weight protein component of sesame seed shows that α-globulin exists as a dissociated molecule at 27°C at pH 9.0 in 0.5 M Tris-HCl buffer constituting 60% of 13 S and 40% of an 8 S component. The effect of variation of temperature on α-globulin indicates that the increase in temperature favours association and suggests hydrophobic association of the subunits[16,35]. From the ratio of concentration of 13 S to 8 S the apparent association constant K for the association process as a function of temperature is calculated[25]. From the Van't Hoff plot of such a data, a value of 13 kcals mol^{-1} for the enthalpy of the association reaction, with the ΔF for the reaction of -0.19 kcals mol^{-1} at 27°C is obtained. These results contribute a value of 42 e.u. for the entropy of the reaction at 27°C. Thus this study using the analytical ultracentrifuge indicates that the association of the subunits is predominantly governed by hydrophobic interactions quantified by thermodynamic data.

Even though a number of studies on the effect of temperature on total proteins of oilseed using various techniques are available the direct application of using the analytical ultracentrifuge using pure proteins has been very limited: possibly, more needs to be done to fill this gap.

Effect of Electrolytes

Studies on the effect of electrolytes on the depolymerization and association of multimeric proteins using the technique of analytical ultracentrifugation in the field of seed proteins, has been limited. The major work has arisen from author's laboratory in the above field with the protein α-globulin from sesame seed.

The major protein fraction α-globulin of sesame seed (*Sesamum indicum L.*) contains six acidic and six basic subunits predominantly held together by hydrophobic interactions[16]. The effect of chaotropic salts on the association-dissociation of the protein (concentrations at which there is no conformational change) has been followed using sedimentation velocity analysis[35]. The results show the following effectiveness of anions in dissociating the 13 S protein component to the 8 S component:

$$SO_4^{2-} < Cl^- < Br^- < ClO_4^- < SCN^- \leq I^- < CCl_3COO^-$$

the first two members are *association*-inducing ions. CCl_3COONa is found to be the most effective in causing *dissociation.* The effect of cations like Li^+, Na^+, K^+, Cs^+ is to induce association, the order of effectiveness being Ca^+, $K^+ \geq Na^+ > Li^+$. The low concentration of salts necessary to induce dissociation had no effect on the conformation of the protein[35]. Prakash and Nandi[35] have postulated that the association-dissociation equilibria 13S $\rightleftarrows$ 8S of α-globulin in chaotropic salt solution is in favour of dissociation due to overall favourable energetics of interaction of the chaotropic ion with the amide dipole over lesser or practically zero unfavourable/free energy of the non-polar groups of the dissociated subunits in these solutions thereby stabilizing the dissociated molecule. They also have indicated that the unfavourable positive free energy of interaction of non-polar groups in Na_2SO_4 and NaCl solutions will dominate favourable interactions of ions, which will stabilize the associated form of the protein.

Lakshmi *et al.*[36] have investigated the effect of electrolytes at high concentrations on the heat coagulating behaviour of the same protein α-globulin at pH 8.6. The salts that were used at different concentrations are Na_2SO_4, Na_2HPO_4, sodium potassium tartarate and sodium citrate. The results indicate that these anions/polyanionic electrolytes prevent the heat coagulation of α-globulin and has been explained due to dissociation of the protein (13S $\rightarrow$ 8S) reducing the overall charge of the protein molecule and binding of salts to amide groups. They also have demonstrated that binding of electrolytes to proteins would prevent the aggregation of heat denatured protein molecules due to electrostatic repulsion of the heat denatured subunits of the protein[36].

In a similar study Prakash *et al.*[37] have shown that citrate ions in the concentration range of 25 mM to 150 mM bring about dissociation of arachin the high molecular weight protein of peanut. As the concentration of citrate is increased the concentration of the 14 S arachin molecule decreases with a concomitant production of low molecular weight 2 S component, which reaches a plateau region at 150 mM concentration. This dissociation is not accompanied by denaturation as revealed by far ultraviolet circular dichroism and viscosity data[37].

The destabilizing effect of citrate ions on the compact structure of arachin is attributed to either indirect solvent effects (resulting from changes in the structure of solvent), as well as direct interaction of the counterion with the charged groups of the protein whose free energy of binding could be in the range of 1 to -5 kcals mol^{-1} [38,39].

Effects of Detergents

The role of detergents in the investigation of seed proteins has been mainly to understand the subunits using SDS-polyacrylamide gel electrophoresis[12]. Interest in the interaction of anionic, cationic and non-ionic detergents with α-globulin, the high molecular weight protein from sesame seed has been from the point of view of heat coagulation behaviour as well as conformational stability of the molecule.

The analysis of seed proteins in the presence of detergents (mainly SDS) using sedimentation velocity analysis has been limited so far to the study of glycinin from soybean, arachin from peanut, α-globulin from sesame seed, brassin from mustard and rapeseed and carmin from safflower seed. The focal point of all these studies has been to follow the dissociation and denaturation of the proteins in the presence of SDS. In some cases however, there appears to be more intermediates than in other cases, indicating the specificity of the interaction of SDS with the proteins in this class of seed proteins.

In the case of arachin (the high molecular weight protein of peanut) the protein progressively dissociates to a 7.4 S component up to a concentration of 5×10^{-3}M concentration of SDS, above which the reaction product is a 3.1 S component. Johnson and Joubert[40] observed an associating effect also with arachin: the results indicate that arachin molecule progressively dissociates in presence of SDS, equilibrating ultimately to a 3S component, the intermediate being 7.4 S component.

In the case of α-globulin from sesame seed, the presence of two more intermediates *i.e.* 2 S and 4 S has been observed by Prakash and Nandi[41] in the analytical ultracentrifuge when the SDS concentration is increased from 1×10^{-4} M to 1×10^{-2} M. The sequence of dissociation follows the order:

$$11\ S \longrightarrow 7\ S \longrightarrow 4\ S \longrightarrow 2\ S$$

The 2 S molecule is in fact denatured as shown by other biophysical techniques. The binding profile of SDS demonstrates a single class of binding site for SDS in the protein[41].

However, in the case of brassin from mustard seed and rapeseed, the protein undergoes an initial association (SDS concentration < 3.5 mM) and subsequently dissociates to 3.1 S component at higher SDS concentration[42]. In the case of glycinin from soybean the 11 S appears to dissociate directly to 2 S as the SDS concentration is increased from 0.7 mM to 3.5 mM[43]. In the case of carmin from safflower seed also one sees a direct dissociation in the SDS concentration range of 1×10^{-3} M to 1×10^{-2} M, without any change in the conformational status of the protein[44].

The above results demonstrate clearly the effectiveness of an anionic detergent like SDS at very low concentrations on the destabilization of the quaternary structure of some of the seed proteins as judged by sedimentation velocity experiments. The need for isolating the intermediate components, characterizing them and studies on reversibility is clearly indicated: for example questions arise such as does the dissociated protein molecule undergo reassociation to the native protein molecule upon complete removal of SDS? Further why do we see intermediates in one class of seed proteins whereas in other class there is a direct dissociation from polymer to monomers? Such questions need thorough investigation using several approaches of analytical ultracentrifugation techniques from a thermodynamic angle.

Effect of Urea and Guanidine Hydrochloride

The use of urea and guanidine hydrochloride for the analysis of association-dissociation results of seed proteins mainly stems from the sedimentation velocity measurements with glycinin, α-globulin, carmin, gossypin, arachin, helianthinin and brassin (M). Table 4 summaries the various sedimenting components detected in the analytical ultracentrifuge in the presence of urea, guanidine hydrochloride and SDS. Although various intermediates are produced in the course of dissociation the end product appears to be the denatured 2 S protein. This occurs either in 8 M urea, 6 M guanidine hydrochloride or at 1×10^{-2} M SDS. The dissociation and denaturation of the high molecular weight protein involve more than a single step as suggested by these intermediates. Prakash[45] has shown that in the case of protein α-globulin nearly 500 moles of guanidine hydrochloride is required to bring about dissociation and denaturation of the molecule. The dissociation and denaturation brought about by these denaturants with the various seed protein is by interacting with peptide bond as a bifunctional reagent along with interaction with aromatic residues[45]. Investigation of individual subunit interaction with the denaturant is needed to understand the overall energy of interaction and the kinetics of the dissociation of the polymer. Studies on the mode of interaction of different classes of ureas and guanidine salts are in progress in the author's laboratory in an attempt to understand the behaviour of these seed proteins and in particular their stability in various solutions. The role of analytical ultracentrifuge with a scanner can undoubtedly contribute to a significant extent in such a pursuit.

Table 4. Intermediate components ($s_{20,w}$) detected in analytical ultracentrifuge from various seed proteins in urea, guanidine hydrochloride and SDS[12].

Seed protein	Components
Glycinin	11,7,4,3,2,1
Arachin	14,12,9,7,4,3,2
α-globulin	11,120,7,4,2
Helianthinin	11,7,2
Brassin (M)	12,4,2
Gossypin	11,7,2
Carmin	11,7,4,2

Effect of Proteolysis

The effect of proteolytic enzymes such as trypsin, α-chymotrypsin and pepsin on the hydrodynamic properties of α-globulin has been investigated using the analytical ultracentrifuge as one of the probes[46]. For example, α-globulin showed definite structural changes after proteolysis as indicated by various biophysical data. The sedimentation velocity pattern of α-globulin hydrolysed by trypsin or α-chymotrypsin indicated reduction in percentage of 11 S component while the pepsin hydrolysed sample is devoid of any 11 S component indicating the absence of native protein molecule. From the analytical ultracentrifugation study it was concluded that trypsin was least effective followed by α-chymotrypsin and pepsin in hydrolysing α-globulin.

The native α-globulin at pH 7.0 has a sedimentation coefficient of (12.2±0.2) S in 0.1 M borate buffer pH 8.1. The unhydrolysed fraction of α-globulin after being hydrolysed with α-chymotrypsin has a sedimentation coefficient value of (11.7±0.2) S under similar conditions indicating the reduction in size of α-globulin after the action of the enzyme[47]. The hydrolysed peptides failed to form a boundary during sedimentation process since it consists of only low molecular weight peptides in it. The unhydrolysed fraction was characterized for its various biophysical properties and amino acid composition which shows a significant reduction in the aromatic amino acid residues in the protein[47].

These results indicate clearly the utility of the analytical ultracentrifuge for the study of partial hydrolysis of pure proteins which could be detected by a careful analysis of sedimentation data under identical conditions. With the latest methodology such studies can be taken to much more depth to evolve mechanisms and to follow precisely the hydrodynamics of proteolysis processes in proteins.

Effect of Specific Ligands

Interaction of Phytic Acid with Sesame Seed Protein Oilseeds in general contain intrinsic ligands such as phytic acid and trypsin inhibitor in soybean, phytic acid and oxalates in sesame seed, chlorogenic acid in sunflower seed and gossypol in cotton seed[12]. Sesame seed is reported to contain nearly 5% phytate in it[48]. The two major fractions of sesame seed namely, α-globulin and β-globulin interact with phytic acid. In their very recent study Rajendran and Prakash[49] have shown that β-globulin (which sediments as a 2 S protein), does not bind phytic acid significantly. On the other hand the major protein component of α-globulin has a greater degree of interaction with phytic acid[49]. The interaction has been mainly followed by analytical ultracentrifugation to characterize the aggregate which is generated by the phytic acid-α-globulin interaction[49]. Maximum interaction of phytic acid with α-globulin is shown to be at pH 2.3. At this pH α-globulin is already in the dissociated state consisting 35% of 8.3 S and 65% of 1.5 S components (Fig. 5 - inset). This is also accompanied by formation of an aggregate whose sedimentation coefficient increases with increasing concentration of phytate. In Fig. 5 is shown the percent fraction of 1.5 S component and the "aggregated" component as a function of increasing concentration of sodium phytate. The schematics being[49]:

$$8.3\ S \xrightarrow{\text{phytic acid}} 1.5\ S + \text{"aggregated" component}$$

In order to understand the process of aggregation of α-globulin in presence of different concentration of phytic acid the sedimentation coefficient is determined as a function of free phytic acid concentration. The sedimentation coefficient increases asymptotically with increase in sodium phytate concentration (Fig. 6). Schematically it can be summarized as:

$$1.5\ S \xrightarrow[\text{upto } 10^{-3}\text{M}]{\text{Phytic acid}} 15\ S \xrightarrow[\text{upto } 10^{-2}\text{M}]{\text{Phytic acid}} \text{Precipitate}$$

(Precursor of precipitate?)

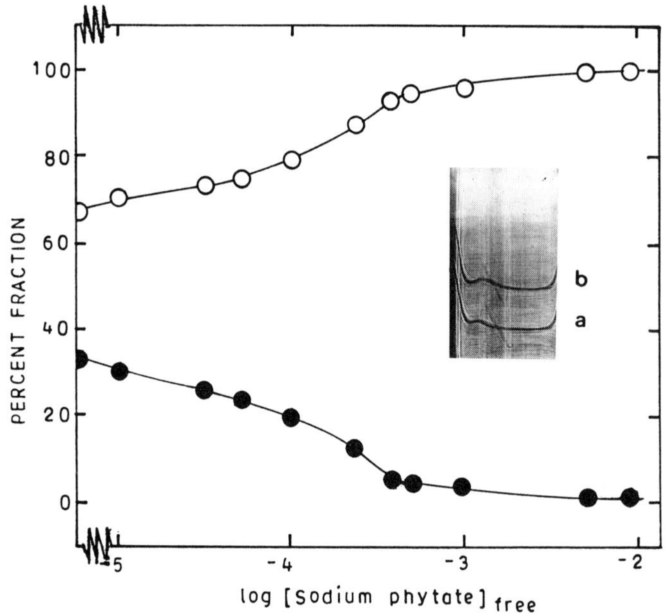

Figure 5. Percent fraction of protein components generated in presence of phytic acid from α-globulin at pH 2.3 in 0.15 M glycine-hydrochloric acid buffer as a function of increasing concentration of phytic acid (○) low molecular weight protein component and (●), the "aggregated" component. The data were calculated from sedimentation velocity patterns of α-globulin in presence of different concentrations of sodium phytate at 27°C, at a protein concentration of 15 mg/ml. Inset (a) typical sedimentation velocity pattern of α-globulin at pH 2.3, 40 min after attaining two thirds maximum speed. (b) As (a) but containing $1{\times}10^{-4}$ M sodium phytate and the pattern corresponds to 40 min after attaining two thirds maximum speed.

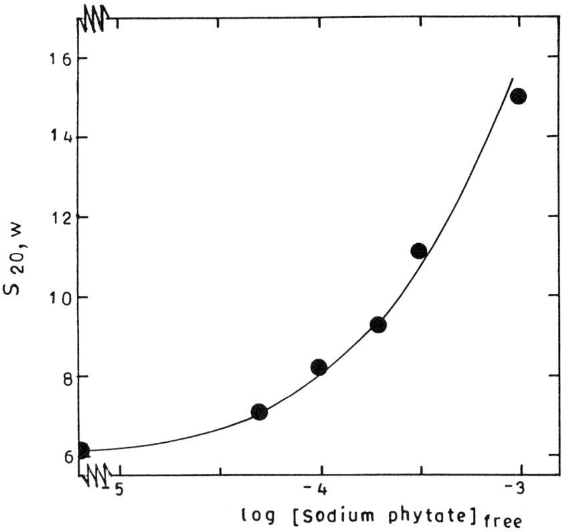

Figure 6. Effect of sodium phytate concentration on the sedimentation coefficient value of aggregated component of α-globulin in 0.15 M glycine-hydrochloric acid buffer of pH 2.3 as monitored from sedimentation velocity patterns in the analytical ultracentrifuge at 27°C. The concentration of α-globulin used was 20 mg/ml in all the experiments.

In order to characterise further whether the "aggregate" both in the absence and presence of sodium phytate are same, the effect of protein concentration on the sedimentation coefficient on (i) "aggregate" in the absence of sodium phytate and (ii) the "aggregate" in the presence of phytic acid concentration 1×10^{-4} M is determined: the results are shown in Fig. 7. The results indicate that the two "aggregates" are not the same and conforms to the standard equation of dependency of sedimentation coefficient on concentration[50]. The $s^0_{20,w}$ value for the two "aggregates" are 8.3 S and 15.3 S respectively with corresponding values for the concentration regression parameter ("g" or "k_s") of 0.09 ml/g and 0.3 ml/g respectively. The 1.5 S component did not show a significant change in the sedimentation coefficient either in the presence or absence of phytic acid as a function of protein concentration. Under these conditions the sedimentation profile of polymer of α-globulin is skewed forward as a result of the presence of soluble polymer of different sedimentation rates existing with monomer. These results clearly indicate that at pH 2.3 phytic acid interacts differently with 1.5 S component and the "aggregates" leading to polymerization.

From Fig. 8 it can be seen that the concentration of polymer increases with increasing concentration of sodium phytate and the polymer appears to move faster with increasing phytic acid concentration. What is the mechanism of such an aggregation process? Bimodality in the sedimentation velocity pattern of a homogeneous protein equilibrated with a ligand arises principally due to (i) existence of protein in more than one state of stable aggregation and (ii) rapidly re-equilibrating self-association leading to polymers greater than a dimer and (iii) rapidly re-equilibrating self association by a strong binding small ligand[51-54]. The diagnostics of

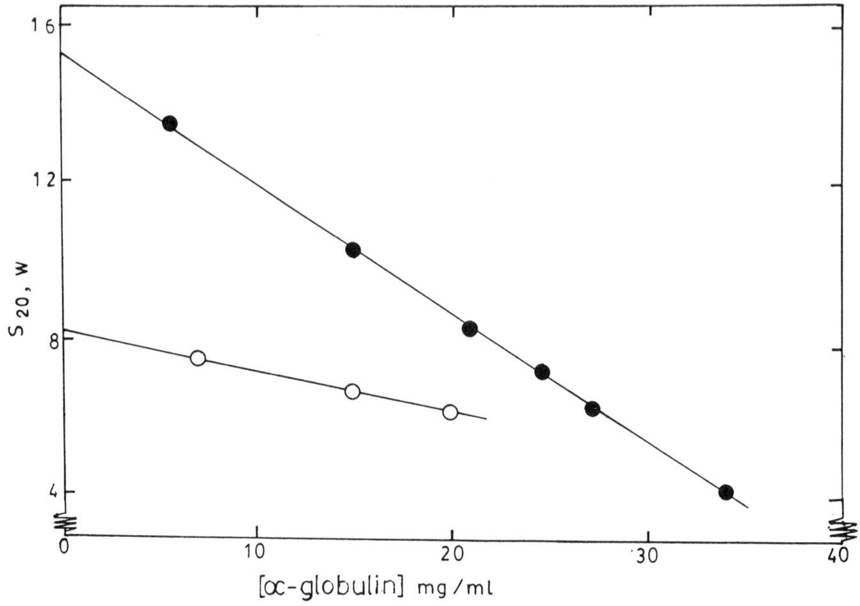

Figure 7. The effect of protein concentration on the sedimentation coefficient of the 'aggregate' (○) acid induced product of α-globulin and (●) sodium phytate induced polymer generated from α-globulin at 1×10^{-4} M sodium phytate concentration as monitored by sedimentation velocity patterns in the analytical ultracentrifuge at 27°C.

the above mechanism can be qualitatively obtained by a close observation of the schlieren pattern in the presence of ligand as described by Prakash and Timasheff[54]. Fig. 9 shows the sedimentation velocity pattern of α-globulin at 22 mg/ml protein concentration with 1×10^{-4} M sodium phytate concentration in a double sector cell. It indicates that the initial separation of the fast moving component is at 9 min (frame i) after reaching top speed. The minimum between peaks touches the baseline at 48 min after reaching top speed (frame v). The bimodality appears in the early stages of the run and subsequent descending of the minimum later in the experiment clearly demonstrates the presence of two independent components, (i) the stable 1.5 S component and (ii) the generated polymer.

The role of ligand promoted polymerization is unequivocally demonstrated by the effect of protein concentration at a constant free concentration of sodium phytate of 1×10^{-4} M. Fig. 10 shows such a plot of the percent fraction of the polymer formed as a function of α-globulin concentration. The plot indicates a *"critical concentration"* of protein required for the formation of the polymer.

The above interaction study demonstrates clearly the mechanism of precipitation of α-globulin in presence of phytic acid using the analytical ultracentrifuge as a primary tool leading to the characterization of the intermediates and to diagnose and distinguish different mechanisms choosing the right experiment above a "critical concentration" of the protein. The results are, in addition, further supported by other measurements such as binding and conformational status of the protein. It is important to note that the concentration of the ligands referred in the above study is the

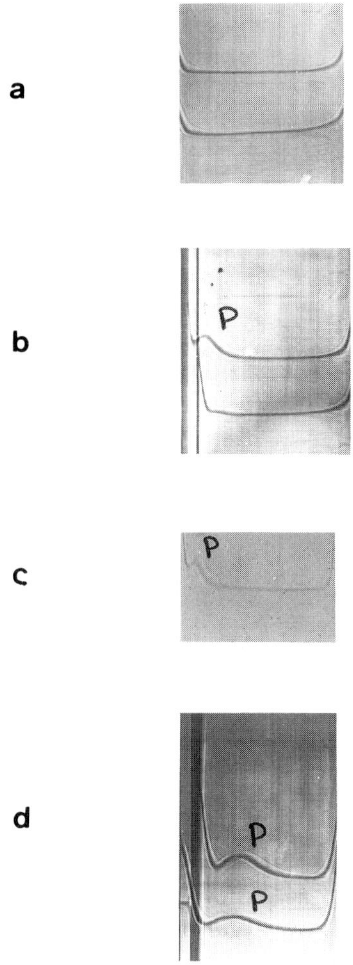

Figure 8. Effect of α-globulin concentration on the sedimentation velocity pattern of polymer formed in presence of 1×10^{-4} M sodium phytate in 0.15 M glycine-hydrochloric acid buffer of pH 2.3 at 27°C. A bar angle of 65° is used in all the photographs unless otherwise indicated. The time of photographs after attaining two-thirds maximum speed are also shown in the parentheses along with concentration of α-globulin in mg/ml. (a) 18 min, 2.7 mg/ml (upper), 1.4 mg/ml (lower); (b) 10 min, 5.4 mg/ml (upper), 3.3 mg/ml (lower); (c) 7 min, 60°, 9.7 mg/ml; (d) 40 min, 20 mg/ml (upper), 13.7 mg/ml (lower). In the case of (a) and (b) a 30 mm aluminium single sector centrepiece was used, and in the case of (c) and (d) a 12 mm Kel F single sector centrepiece was used. "P" in the photograph indicates polymer.

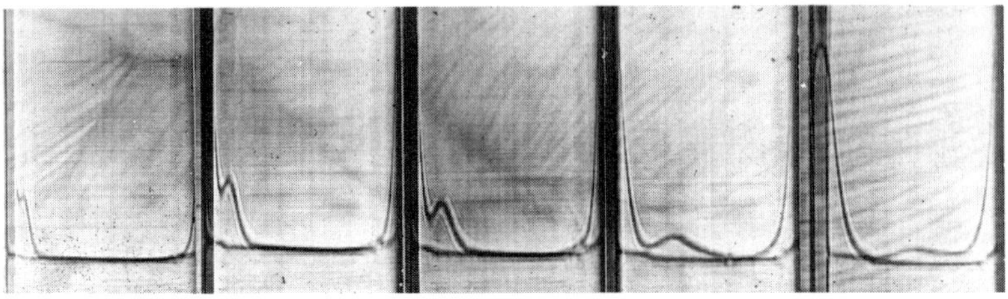

Figure 9. Effect of sedimentation time on the formation of the schlieren boundary of α-globulin in 1×10^{-4} M free concentration of sodium phytate at 27°C in 0.15 M glycine-hydrochloric acid buffer of pH 2.3. A standard 12 mm, Kel F double sector centrepiece, having protein equilibrated to 1×10^{-4} M sodium phytate in the above buffer in one sector and the buffer containing 1×10^{-4} M sodium phytate in the other sector was used. All photographs were taken at 60°C unless otherwise indicated. A protein concentration of 22 mg/ml was used. The time of photograph after attaining two-thirds maximum speed of 59,780 rev/min was (i) 9 min (70°); (ii) 13 min; (iii) 17 min; (iv) 33 min and (v) 48 min.

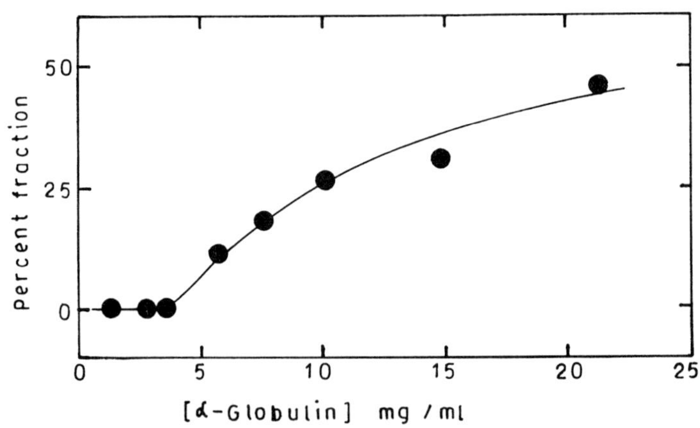

Figure 10. Percent fraction of polymer formed, equilibrated at pH 2.3 in 0.15 M glycine-hydrochloric acid buffer containing 1×10^{-4} M sodium phytate, as a function of α-globulin concentration as monitored by sedimentation velocity patterns in the analytical ultracentrifuge at 27°C.

free concentration: in other words, the macromolecule is in equilibrium with the ligand at any given macromolecule and ligand concentration and at a specified temperature.

Interaction of Other Ligands with Seed Proteins Other ligand-seed protein interactions include the interaction of allyl isothiocyanate with 12 S protein fraction from mustard; this protein has been shown not to undergo any dissociation reaction in the presence of the ligand[55]. Specific interaction of calcium ion with glycinin did not cause dissociation or association[56].

An interesting experiment of hybridization of subunits from two different oilseeds has been attempted by combining acidic subunits of sesame seed with basic subunits of soybean; this system has been characterized with a molecular weight of 265,000 and a sedimentation coefficient value of 9.6 S. On the other hand the combination of the acidic subunit of soybean and basic subunit of sesame seed results in a molecular weight of 310,000 [57].

Analytical ultracentrifugation has also been used in other areas of seed protein research such as for the investigation of the effect of wet and dry heating on the sedimentation velocity profile of total proteins of peanut protein and sunflower protein[12,58] and chemical modification such as acetylation succinylation of sunflower total protein, helianthinin and brassin[12,58,59]. The results demonstrate the dissociation of the protein under the above conditions as a result of chemical modification and heat treatments.

3. CONCLUSIONS

The role of the analytical ultracentrifuge in the study of seed proteins stems from identifying the number of protein components in the total protein and in their assignment in terms of specific sedimentation coefficient values such as 11 S and 7 S. The different protein fractions after being isolated to homogeneity are characterized in solution for their molecular weight, sedimentation and diffusion coefficient and frictional ratios using the technique. Further understanding of these homogeneous multimeric protein fractions has been facilitated by following the association-dissociation phenomena of the proteins as a function of ionic strength, pH, temperature, chaotropic salts, detergents, urea and guanidine hydrochloride and proteolysis. The interaction of specific ligands has been mainly followed using analytical ultracentrifugation by applying the principles of ligand induced self-association of protein using standard theories in the case of α-globulin-phytic acid interaction recently. The need for more detailed experiments using pure proteins in the area of seed proteins is emphasized for a deeper understanding of structure-function relationship of this unique class of proteins in the plant system by judicious application of the theory and practice of modern analytical ultracentrifugation.

REFERENCES

1. A.M. Altschul, 'Processed Plant Protein Foodstuffs', Academic Press, New York, 1978.
2. J.E. Kinsella, *CRC Crit. Rev. Food Sci. Nutr.*, 1976, *7*, 219.
3. J.W. Dieckert and M.S. Dieckert, 'Genetic Improvements of Seed Proteins', National Academy of Science, Washington D.C., 1976.
4. J.F. Thompson and H. Doll, 'Proc. Symp. Seed Protein Improvement of Cereals, Grains and Legumes', I.A.E.A., Vienna, 1979, Vol I, p. 109.
5. F.M. Ashton, *Ann. Rev. Plant Physiol.*, 1976, *27*, 95.
6. B.W. Bagley, J.H. Cherry, M.L. Rollins and A.M. Altschul, *Am. J. Bot.*, 1963, *50*, 523.
7. J.W. Dieckert, J.E. Snowden, Jr., A.T. Moore, D.C. Heinzeiman and A.M. Altschul, *J. Food Sci.*, 1962, *27*, 321.
8. C.B. Sharma and J.W. Dieckert, *Physiol. Plant*, 1975, *33*, 1.
9. J.C. Pernollet and J. Mosse, 'Seed Proteins', J. Daussant, J. Mosse, J.G. Vaughan, eds., Academic Press, London, 1983, p. 155.
10. J. Mosse, *Physiol. Veg.*, 1973, *11*, 361 - as cited in ref. 9.
11. J. Mosse and J.C. Pernollet, 'Chemistry and Biochemistry of Legumes', S.K. Arora, ed., IBH Publishing Company, Oxford, 1982, p. 111.
12. V. Prakash and M.S.N. Rao, *CRC Crit. Rev. Biochem.*, 1986, *20*, 265.
13. B. Sathyanarayana Murthy, V. Prakash and M.S. Narasinga Rao, *Ind. J. Biochem. Biophys.*, 1990, *27*, 295.
14. V. Prakash, 'Interactions in Protein Systems', K.D. Schwenke and B. Raab, eds., Akademie-Verlag, Berlin, 1989, p. 221.
15. V. Prakash, *Biochem. Rev.*, Society of Biological Chemists (India), Sarma Memorial Award Lecture, 1992.
16. V. Prakash, *J. Biosci.*, 1985, *9*, 165.
17. K.D. Schwenke, 'Interaction in Protein Systems', K.D. Schwenke and B. Raab, eds, Akademie-Verlag, Berlin, 1989, p. 215.
18. T.S. Latha and V. Prakash, *Die Nahrung*, 1986, *30*, 833.
19. V. Prakash and M.S.N. Rao, *J. Biosci.*, 1988, *13*, 171.
20. W.E.F. Naismith, *Biochim. Biophys. Acta*, 1955, *16*, 203.
21. P. Johnson and E.M. Shooter, *Biochim. Biophys. Acta*, 1950, *5*, 561.
22. J.M. Dechary, K.T. Talluto, W.J. Evans, W.B. Corney and A.M. Altschul, *Nature (London)*, 1962, *190*, 1225.
23. W.E.F. Naismith and R.K. Williams, *J. Appl. Chem.*, 1956, *6*, 421.
24. W.E.F. Naismith and K.O. Kelley, *J. Appl. Chem.*, 1958, *8*, 598.
25. V. Prakash and P.K. Nandi, *Int. J. Pept. Protein Res.*, 1977, *9*, 319.
26. V. Prakash and P.K. Nandi, Abstract in *10th International Union of Biochemistry*, Hamburg, Federal Republic of Germany, 1976.
27. S. Rajendran and V. Prakash, *J. Am. Chem. Soc.*, 1992, submitted for publication.
28. K.D. Schwenke, M. Schultz and K.J. Linow, *Die Nahrung*, 1975, *19*, 425.
29. J. Behlke and K.D. Schwenke, *Stud. Biophys.*, 1976, *597*, 55.
30. G. Sripad, Ph.D. Thesis, University of Mysore, Mysore, India, 1985.
31. N.V. Kishore Kumar Murthy and M.S.N. Rao, *Int. J. Pept. Protein Res.*, 1984, *23*, 94.

32. S. Rajendran and V. Prakash, *J. Protein Chem.*, 1988, *7*, 689.
33. K. Sudhindra Rao and V. Prakash, *J. Protein Chem.*, 1989, *8*, 529.
34. V. Prakash, Abstract in *Workshop on Structure, Interactions and Design of Proteins*, Indian Institute of Science, Bangalore, February 4-7, 1991.
35. V. Prakash and P.K. Nandi, *J. Biol Chem.*, 1977, *252*, 240.
36. T.S. Lakshmi, P.K. Nandi and V. Prakash, *Ind. J. Biochem. Biophys.*, 1985, *22*, 274.
37. V. Prakash, A.G. Appu Rao and D. Rajagopal Rao, *J. Am Oil. Chem. Soc.*, 1987, *64*, 1732.
38. I. Fridovich, *J. Biol. Chem.*, 1963, *238*, 572.
39. D. Robinson and W.P. Jencks, *J. Am. Chem. Soc.*, 1965, *87*, 2462.
40. P. Johnson and F.J. Joubert, *J. Polymer Sci.*, 1951, *605*, 1; *Chem. Abstr.*, 1954, *48*, 13365c.
41. V. Prakash and P.K. Nandi, *Int. J. Pept. Protein Res.*, 1976, *8*, 355.
42. A. Gururaja Rao and M.S.N. Rao, *J. Biosci.*, 1983, *5*, 301.
43. B.R. Sureshchandra, V. Prakash and M.S.N. Rao, *J. Biosci.*, *9*, 177.
44. K. Sudhindra Rao and V. Prakash, *Eur. J. Biochemistry*, 1992, submitted for publication.
45. V. Prakash, *J. Biosci.*, 1982, *4*, 347.
46. R. Tasneem and V. Prakash, *J. Protein Chem.*, 1989, *8*, 251.
47. R. Tasneem and V. Prakash, *J. Sci. Food Agric.*, 1992, in press.
48. A.R. Deboland, G.B. Garner and B.L. O'Dell., *J. Agric Food Chem.*, 1975, *23*, 1186.
49. S. Rajendran and V. Prakash, *Biochemistry*, submitted for publication, 1992.
50. H.K. Schachman, 'Ultracentrifugation In Biochemistry', Academic Press, New York, 1959.
51. J.R. Cann and W.B. Goad, *Arch. Biochem. Biophys.*, 1972, *153*, 603.
52. J.R. Cann and W.B. Goad, *Methods Enzymol.*, Academic Press Inc., New York, 1986, *27*, 273.
53. V. Prakash and S.N. Timasheff, *Anal. Biochem.*, 1983, *131*, 232.
54. V. Prakash and S.N. Timasheff, *Methods Enzymol.*, Academic Press Inc., New York, 1986, *130*, 3.
55. N.V. Kishore Kumar Murthy, Ph.D. Thesis, University of Mysore, Mysore, India, 1984.
56. A.G. Appu Rao and M.S.N. Rao, *J. Agric. Food Chem.*, 1975, *23*, 657.
57. K. Hasegawa, T. Tanaka and S. Tamai, *Agric. Biol. Chem.*, 1981, *45*, 809.
58. Aruna Venkatesh and V. Prakash, *J. Agric. Food Chem.*, 1992, submitted for publication.
59. K.D. Schwenke, K.J. Linow, B. Raab and J. Uhlig, *Die Nährung*, 1979, *23*, 169.

25

Sedimentation Analysis of Membrane Proteins

By Geoffrey J. Howlett

RUSSELL GRIMWADE SCHOOL OF BIOCHEMISTRY, UNIVERSITY OF MELBOURNE, PARKVILLE 3052, VICTORIA, AUSTRALIA

1. INTRODUCTION

The methods of sedimentation equilibrium and sedimentation velocity may be used to characterize the size and state of association of membrane proteins. Since most membrane proteins require detergents for solubilisation, the analytical procedures differ from those used for soluble cytoplasmic proteins. This difference arises from the need to consider the effects of bound detergent on the molecular weight and buoyant density of the solubilised protein.

Where purified membrane proteins are available, sedimentation studies may be performed using the analytical ultracentrifuge with absorption optics[1-4]. In cases where only small amounts of unpurified or semi-purified detergent-solubilised proteins are available alternate methods are required. One approach is the use of the air-driven ultracentrifuge (Airfuge) and the determination of molecular weights by sedimentation equilibrium[5-7]. The high angular velocities available using the Airfuge also permit rapid separation of membrane bound and membrane free proteins allowing rates of association and dissociation to be determined[8,9]. Density gradient sedimentation velocity measurements[10] are more widely used for the characterization of semi-purified membrane proteins. Since transport methods depend on molecular shape this approach must be combined with data on the Stokes radius to yield estimates of molecular weight[11].

Early studies on detergent-solubilised membrane proteins determined the molecular size of important functional proteins such as membrane associated enzymes[12-14], hormone receptors[15,16] and transmitter receptors[1,17]. Recent studies have analysed the subunit structure of membrane associated proteins (*e.g.* refs. 2, 18-20) and the effects of hormones and other receptor ligands on the interactions between subunits and other membrane proteins[21-23]. This Chapter reviews the procedures which have been developed for the sedimentation analysis of membrane proteins. The assumptions and limitations are indicated together with recent examples where sedimentation methods have been used to characterise the subunit structure and ligand-sensitivity of receptor interactions within the cell membrane.

2. MOLECULAR WEIGHT AND BUOYANT DENSITY ANALYSIS BY SEDIMENTATION EQUILIBRIUM

Theory

The sedimentation equilibrium behaviour of a detergent-solubilised protein depends on the molecular weight and density of both the protein and the bound detergent. One approach to the analysis of such systems is to define the sedimenting species (component 2^*) on the basis of a hypothetical dialysis experiment such that an excess or deficiency of component 3 (electrolyte, detergent or any other low molecular weight solute) on the solution side is included in the definition of component 2^* [24]. The sedimentation equilibrium distribution (concentration, c, as a function of radial distance, r) of the detergent-solubilised protein complex may then be written as :

$$\sigma_2^* = \frac{d \ln c_2^*}{dr^2} = M_2^* (1-\phi_2^* \rho) \frac{\omega^2}{2RT} \tag{1}$$

where ϕ_2^* is the apparent specific volume of component 2^*, M_2^* the molecular weight, ρ the solution density and T the absolute temperature. The use of eq. (1) implies that non-ideal effects are negligible and that the composition of component 2^* is essentially independent of the concentration of component 2^* [24].

For purified preparations of component 2 the value of ϕ_2^* may be determined experimentally from densitometric studies comparing the density of a dialysed solution with that of the dialysate. Such measurements require precise concentration measurements and relatively large amounts of protein and are not usually practical for membrane protein studies[3].

An alternate, more suitable technique for protein-detergent complexes is to separate the contributions of the protein and detergent to the sedimentation distribution and to neglect possible contributions from other low molecular weight solutes so that eq. (1) becomes[3]

$$\frac{d \ln c_2^*}{dr^2} = \frac{[M_2(1-\bar{v}_2\rho) + M_D(1-\bar{v}_D\rho)]\, \omega^2}{2RT} \tag{2}$$

where M_2 is the molecular weight and $\bar{v}_2$ the partial specific volume of the protein moiety, M_D the combined molecular weights of the bound detergent and $\bar{v}_D$ the partial specific volume of the associated detergent. For purified membrane proteins, separate detergent binding experiments may be performed to determine M_D allowing determination of M_2 from eq. (2) provided values of $\bar{v}_2$ and $\bar{v}_D$ are known or can be assumed.

In the special case where the solvent density (ρ_s) is chosen to be equal to the density of the detergent ($1/\bar{v}_D$) eq. (2) reduces to

$$\frac{d \ln c_2^*}{dr^2} = \frac{M_2(1-\bar{v}_2\rho_s)\omega^2}{2RT} \tag{3}$$

In this case M_2 may be determined directly from the sedimentation equilibrium distribution and an appropriate value for $\bar{v}_2$.

A complementary approach, first described by Edelstein and Schachman[25], involves solving for ϕ_2^* in eq. (1) by measuring σ_2^* in solutions of different densities. An assumption is that preferential interactions of component 2^* with component 1 (water) and 3 are identical in the different solvents. H_2O and D_2O have been widely used for such studies.

$$\phi_2^* = \frac{k - (\sigma_2^*)_{D_2O} / (\sigma_2^*)_{H_2O}}{\rho_{D_2O} - \rho_{H_2O} (\sigma_2^*)_{D_2O} / (\sigma_2^*)_{H_2O}} \tag{4}$$

where k is a measure of the effect of deuteration on the molecular weight and partial specific volume of component 2^*. Determination of ϕ_2^* using eq. (4) allows M_2^* to be calculated from eq. (1). Where detergent is the only significant solute associated with component 2 the amount of bound detergent can be estimated from

$$\phi_2^* = \frac{\bar{v}_2 + \delta\bar{v}_3}{1 + \delta} \tag{5}$$

where $\bar{v}_2$ and $\bar{v}_3$ are the partial specific volumes of the protein and detergent respectively and δ is the number of grams of bound detergent per gram of protein. Determination of δ permits calculation of

$$M_2 = \frac{M_2^*}{1+\delta} \tag{6}$$

where M_2 is the molecular weight of the protein alone.

Experimental

Choice of Detergent A key consideration is the ability to solubilise the protein of interest in a form which retains "native" structure. One indicator of native structure is the retention of biological activity. In a recent study the degree of solubilisation of the erythropoietin receptor varied from 0.5% to 31% for the different detergents tested while the loss of erythropoietin binding activity varied from 5% to 100% [20]. Triton X-100 was chosen for further study, giving 31% solubilisation and 83% recovery of binding activity.

Since detergent binding may vary with the concentration of free detergent another consideration is to select a detergent and centrifugal conditions to maintain a relatively constant concentration of free detergent within the centrifuge cell. The extent of sedimentation of the free detergent depends on the size and density of the detergent micelles in solution[26-28]. A further consideration concerning the density of the detergent[27] is the effect on the buoyant density of the solubilised protein. One option is to choose a detergent which differs significantly in partial specific volume compared to the protein. Determination of the density of the detergent solubilised complex then provides an indication of the proportion of bound detergent. Triton X-100, which has a partial specific volume of 0.91 ml/g [29], has been widely used for this purpose. The binding of detergents with densities equal to the protein has no effect on the density of the solubilised species. In this case the molecular weight of the complex may be determined with accuracy although further studies on detergent binding would be required to determine the molecular weight of the protein alone.

Other considerations include the availability of radioactive forms of the detergent which facilitate studies on the amount of detergent bound to the purified protein[26]. In analytical ultracentrifuge studies the use of an optically transparent detergent allows the concentration distribution of the protein to be measured independently of the bound detergent[2,4].

Measurement of Membrane Protein Concentration Both sedimentation velocity and sedimentation equilibrium experiments require determination of the concentration of the sedimenting species. For purified membrane proteins, concentrations may be determined from optical density measurements using the analytical ultracentrifuge. Studies with unpurified membrane proteins require other procedures. Measurement of enzymic activity provides a convenient marker for membrane-associated enzymes. For receptors, ligand binding may be used for detection. One method is to include radioactive ligands during centrifugation. For high ligand affinities, conditions may be chosen such that the ligand is essentially all bound and the distribution of the ligand measures the distribution of the binding protein[30]. Where dissociation is likely during the course of a sedimentation velocity experiment the ligand may be included in the solutions used to form the gradient[23]. Other possibilities include assaying the receptor concentration in aliquots collected at the conclusion of centrifugation, chemical crosslinking of the ligand to the receptor to prevent dissociation during centrifugation[18] or the use of quantitative SDS gel electrophoresis. This latter approach has been used to separately determine the sedimentation of monomers and higher polymers of membrane proteins[22].

Examples of Equilibrium Studies using the Analytical Ultracentrifuge

Sedimentation equilibrium studies using the analytical ultracentrifuge provide the most direct way of determining the molecular weights of detergent solubilised membrane proteins. The results obtained are independent of macromolecular shape and may be analysed by conventional procedures to indicate the extent of self-association or other macromolecular interaction. If the detergent absorbs light the ultracentrifuge scanner can detect both free and bound detergent allowing estimates of the amount of detergent associated with the protein[31].

The use of eqs. (1) and (2) for the analysis of sedimentation equilibrium data is demonstrated in studies of human apolipoprotein A-1 [3]. While not a membrane protein this protein binds significant amounts of detergent and may be purified in abundant quantities from serum. Sedimentation equilibrium studies of apolipoprotein A-1 in deoxycholate yielded a value of 9510 for $\sigma_2^*(2RT/\omega^2)$. Separate binding studies indicated 0.4 g of deoxycholate bound per g of protein. Insertion of this value into eq. (2) and taking the values for $\bar{v}_2$ and $\bar{v}_D$ as 0.734 ml/g and 0.778 ml/g respectively yielded a value of 26,800 for M_2, the molecular weight of the protein alone. This value agrees well with values obtained from the sequence of apolipoprotein A-1 [32]. A similar approach has been used to determine the molecular weight of Triton X-100 solubilised acetylcholine receptors[1].

More recent studies of detergent solubilised membrane proteins by sedimentation equilibrium have made use of D_2O to vary solvent density. Data obtained at different densities for purified beef heart cytochrome *c* oxidase in laurylmaltoside[4] were extrapolated to a density corresponding to that of the detergent. The molecular weight (194,000), calculated using eq. (3), gave the molecular weight of the detergent free enzyme. The value obtained supported a model composed of 12 subunits[4]. Sedimentation equilibrium studies of cytochrome oxidase were also performed by comparing the sedimentation equilibrium data obtained in solutions of H_2O and D_2O [25]. The apparent specific volume of the complex obtained using eq. (4) was used to determine the molecular weight of the detergent solubilised complex (300,000) and hence the detergent binding capacity (0.55 g/g).

Sedimentation equilibrium studies at solution densities equal to that of detergent have also been performed for chicken hepatic lectin[2] solubilised using Brij-58. These studies indicated that this lectin consists of multiple oligomers made up of aggregates of trimeric species. The state of aggregation was reduced by the addition of calcium ions or N-acetylglucosamine. These solution studies at sedimentation equilibrium agree well with results obtained from radiation inactivation[2] and illustrate the utility of this method for studies of the interactions of solubilised membrane proteins in solution.

Use of the Preparative Ultracentrifuge for Equilibrium Studies

The analytical ultracentrifuge is limited by the requirement for purified proteins. The preparative, air-driven ultracentrifuge (Chapter 4), allows the analysis of sedimentation equilibrium behaviour of proteins, without prior purification. The methods are based on the observation[33] that under conditions of density stabilization (such as a self-forming gradient of 5 mg/ml bovine serum albumin) the rotor may be stopped and the distributions formed during centrifugation sampled directly. Techniques to sequentially fractionate the tube contents provide complete sedimentation profiles[34].

The air-driven ultracentrifuge has been used to examine the sedimentation equilibrium behaviour of the human insulin receptor covalently labelled with ^{125}I-labelled insulin and solubilised by Triton X-100 [29]. Measurements in solutions of different densities, prepared using D_2O and D_2O^{18}, yielded values of 0.78 ml/g and 475,000 for $\bar{v}_2^*$ and M_2^* (eq. (1)). Taking a value of 0.71 cm^3/g for the partial specific volume of the receptor protein and 0.91 cm^3/g for the detergent, values of 0.54 g/g and 310,000 were obtained for δ and M_2 (eqs. (5) and (6)) respectively. Parallel sedimentation velocity studies indicated a single macromolecular species of sedimentation coefficient 10.5 S. Since this work, the complete sequence of the human insulin receptor precursor has been determined[35]. The precursor contains the separate α and β subunits of molecular weights 82,500 and 69,700 respectively. The sedimentation equilibrium data support a model composed of two α and two β subunits with a predicted unglycosylated molecular weight of 304,400.

The Airfuge has also been used to characterize the sedimentation equilibrium behaviour of detergent solubilised enzymes. Sedimentation equilibrium results for rat liver γ-glutamyltranspeptidase solubilised in Triton X-100 were obtained using solutions containing H_2O or D_2O as solvents[36]. For these studies small plastic caps were fitted to the Airfuge tubes to prevent the transfer of D_2O and H_2O between tubes allowing a better comparison of the effects of density on sedimentation. The linearity of the ln c *versus* r^2 plots indicated that the solubilised enzyme was homogeneous with respect to molecular weight. Values of 0.819 ml/g and 175,000 were obtained for $\bar{v}_2^*$ and M_2^* respectively. The proportion of bound detergent was 55% (using eq. (5)) and M_2 was 79,000. The value for M_2 compares with 87,000 obtained from sucrose density gradient centrifugation and 81,000 obtained from an analysis of the separate subunits by gel electrophoresis[37]. Nucleic acid sequencing of the cDNA of the human enzyme, which is homologous to the rat enzyme, yielded a molecular weight for the non-glycosylated enzyme of 58,400 [38] suggesting that the glycosylated and detergent solubilised enzyme is monomeric.

3. HYDRODYNAMIC ANALYSIS OF MEMBRANE PROTEINS BY SEDIMENTATION VELOCITY

Theory

Solubilised membrane proteins in sedimentation velocity experiments may be considered as described above for sedimentation equilibrium. Thus the sedimentation coefficient of component 2^* is given by the rate of movement (dr/dt) in a given centrifugal field divided by the field strength.

$$s_2^* = \frac{1}{\omega^2 r}\left(\frac{dr}{dt}\right) \tag{7}$$

or

$$s_2^* = \frac{1}{\omega^2}\left(\frac{d \ln r}{dt}\right) \tag{8}$$

where ω is the angular velocity in rad/sec. The value of s_2^* is related to the molecular weight, M_2^* according to the relationship[39]

$$s_2^* = M_2^* \frac{(1 - \bar{v}_2^* \rho)}{N f_2^*} \tag{9}$$

where $\bar{v}_2^*$ is the partial specific volume of the component 2^*, f_2^* the frictional coefficient, ρ the solution density and N, Avogadro's number. The frictional coefficient depends on the shape of the sedimenting complex and is related to the Stokes radius of the particle, r_2^*:

$$f_2^* = 6\pi\eta r_0 \left\{\frac{f_2^*}{f_0}\right\} \tag{10}$$

$$\frac{f_2^*}{f_0} = \frac{r_2^*}{\left\{\frac{3\bar{v}_2^* M_2^*}{4\pi N}\right\}^{1/3}} \tag{11}$$

where r_0 and f_0 are the Stokes radius and frictional coefficient of a sphere of equivalent mass and volume and f_2^*/f_0 is known as the frictional ratio[40].

From eqs. (7-11) it follows that s_2^* is a function of ρ, T, η, $\bar{v}_2^*$ and r_2^*. Comparison of sedimentation coefficients measured at different temperatures and in solvents of different density and viscosity is achieved using the relationship[39]

$$s^0_{20,w} = s\left[\frac{\eta}{\eta_{20,w}} \frac{(1-\bar{v}\rho)_{20,w}}{(1-\bar{v}\rho)}\right] \tag{12}$$

where $\eta_{20,w}$ and $(1-\bar{v}\rho)_{20,w}$ are the viscosity of water and the bouyant density term at 20°C.

Density Gradient Sedimentation Velocity of Membrane Proteins

Most recent studies on the sedimentation behaviour of solubilised membrane proteins are based on the analysis of sedimentation velocity in the preparative ultracentrifuge using pre-formed density gradients[10]. These gradients allow the application of a small zone of sample to the gradient at the beginning of the experiment and provide density stabilization for the migrating zones of macromolecular solutes during centrifugation and braking of the rotor. The method is suitable for studies using only small quantities of unpurified sample allowing a range of specific and sensitive assays to be used to locate the migration position of the protein of interest. An assumption in the use of density gradients is that effects due to preferential interactions with the solvent and density gradient material are negligible. The possible significance of this assumption in relation to membrane proteins has been discussed previously[28].

Analysis of the sedimentation behaviour of solutes in density gradients requires consideration of the variation of solution density and viscosity as a function of radial distance. Combination of eqs. (7) and (12) yields[10]

$$s^0_{20,w} = \frac{(1-\bar{v}\rho)_{20,w}}{\omega^2\, t\, \eta_{20,w}} \int_{r(0)}^{r(t)} \frac{\eta(r)\, dr}{r\,(1-\bar{v}\rho(r))} \tag{13}$$

Eq. (13) may be solved by trapezoidal integration to yield the distance migrated by macromolecules as a function of time and for various density and viscosity gradients. Such calculations[10], for linear sucrose gradients, using the rotor dimensions of the Beckman SW-39 rotor, indicate an approximately linear relationship between the distance migrated and time. This relationship and its dependence on the sedimentation coefficient and partial specific volume of the macromolecule lead to the suggestion that for macromolecules of similar partial specific volume, the sedimentation coefficient of an unknown (s_u) may be determined from

$$s_u = s_s \left\{ \frac{r_u}{r_s} \right\} \tag{14}$$

where r_u and r_s are the distances migrated by the unknown and standard respectively and s_s is the sedimentation coefficient of the standard. For detergent solubilised proteins, where the partial specific volume is unknown, the use of eq. (14) yields operationally defined values for s_u which may differ significantly from the value of the sedimentation coefficient defined in eq. (7).

Eq. (14) is based on the constant rate of migration of proteins in a density gradient[10] and therefore makes the assumption that the field strength ($r\omega^2$) is constant over the length of the tube. For the SW-39 rotor the field strength increases by approximately 60% over the length of the tube. Nevertheless, there is experimental support for a linear relationship between the distance migrated and the sedimentation coefficient of a range of standard proteins of similar partial specific volume (*e.g.* ref. 15). In addition theoretical calculations based on the geometry of the SW-40 rotor indicate an approximately linear relationship between sedimentation coefficients and distance migrated for proteins of partial specific volume 0.7 ml/g [41]. The linearity of this relationship diminishes as the value of the partial specific volume increases[41]. Clearly, as the density of the sedimenting species approaches the density of the solution the rate of sedimentation decreases toward zero. While the effects of this non-linearity are reduced by choosing standards with sedimentation characteristics similar to the unknown, more accurate determinations of the sedimentation coefficient in these cases require the use of eq. (13) [10].

The Determination of Partial Specific Volume

As indicated above the determination of sedimentation coefficients from sucrose gradient sedimentation velocity experiments requires knowledge of the partial specific volume of the sedimenting species. This parameter, for detergent-protein complexes, is a function of the amount of bound detergent. Where the amount of bound detergent is known from separate experiments, sucrose gradient sedimentation velocity experiments using D_2O and H_2O as solvents can be used to estimate the partial specific volume of the membrane protein alone and hence the sedimentation coefficient[14].

For unpurified detergent solubilised proteins it is not possible to determine the amount of detergent bound to specific proteins directly. A number of procedures for the simultaneous determination of the partial specific volume and sedimentation coefficient using density gradients prepared with H_2O and D_2O have been described. Two different approaches have emerged.

Method 1 The procedure proposed by Clarke[14] involves the use of eq. (12) written for D_2O and H_2O.

$$s_H = s_D \left[\frac{\eta_D}{\eta_H} \frac{(1-\bar{v}\rho_H)}{(1-\bar{v}\rho_D)} \right] \tag{15}$$

Where the subscripts H and D refer to sucrose gradients in H_2O and D_2O. The viscosity of the medium, η, and the solution density, ρ are defined at r_{ave} where r_{ave} is the distance from the centre of rotation to a point midway between the meniscus and the peak at radius r. The experimentally determined sedimentation coefficient is given by

$$s = \frac{1}{r_{ave}\omega^2} \frac{dr}{dt} \tag{16}$$

Determination of s_D and s_H allows the determination of $\bar{v}$ from eq. (15). The value for $s_{20,w}$ may then be calculated using eq. (12) where the densities and viscosities (ρ and η) are those at the average position of migration. In practice, the procedure proposed by Clarke involves the determination of the solution density at r_{ave} from standard tables describing the density of sucrose solutions as a function of concentration and then solving for the value of solution viscosity at r_{ave} by interpolating data obtained using a series of proteins of known sedimentation coefficient and partial specific volume.

There are a number of assumptions in the use of Method 1. The method assumes that the distance migrated in the gradient is a linear function of time. In this regard the method is similar to procedures described by others[17,42]. The assumption is supported by both theoretical and experimental studies[10,41] but is dependent on the geometry of the rotor, the characteristics of the gradient and on the partial specific volume of the sedimenting species. An analysis of the effects of non-linearity on the determination of s and $\bar{v}$ has been presented[41]. Implicit in the use of eq. (15) is that the extent of detergent binding is similar in H_2O and D_2O solutions and is independent of sucrose concentration. This assumption is supported by results which indicate little difference in the binding of detergents to a variety of membrane proteins in H_2O solutions compared to D_2O solutions[14]. A further assumption is the neglect of possible effects due to preferential hydration. Studies on the effects of concentrated sucrose on the density of proteins indicate significant preferential hydration which varies as a function of sucrose concentration[43]. The magnitude of such effects on the determination of $\bar{v}$ and s are difficult to estimate. The use of a procedure where the solution viscosity is chosen to give the correct value of s for standard proteins may minimise the effects of preferential hydration on the results obtained.

Method 2 An alternative procedure, proposed by Smigel and Fleischer[16] uses eq. (13). This equation allows values of $s_{20,w}$ to be determined from the distance migrated in a density gradient, over a particular time period, for an assigned value of the partial specific volume, $\bar{v}$. Thus a family of $s_{20,w}$ and assigned $\bar{v}$ values may be obtained from studies performed using H_2O as the solvent and a separate family of points from studies using D_2O as the solvent. The intersection of the two families of points yields the single value of $\bar{v}$ consistent with the migration of the solubilised protein in the two solvents.

The use of this procedure for standard proteins of known $s_{20,w}$ and $\bar{v}$ was initially found to give s values which were slightly in error (within 10%). The error was attributed to so-called "bang-bang" regulation of the rotor temperature leading to uncertainty in the correct value to choose for the solution viscosity. An empirical correction was therefore applied to the solution viscosity so that the standard protein running closest to the unknown assumed the correct value for s. To this extent the method is dependent on the use of standard proteins. The method assumes that the extent of detergent binding is similar in the D_2O and H_2O solvents and neglects possible effects due to preferential hydration. The procedure does not assume a linear dependence of migration on time or sedimentation coefficient and in this respect offers an advantage compared to Method 1.

Determination of Molecular Weight

The determination of M_2^* from the measurement of $(s_2^*)_{20,w}$ and $\bar{v}_2^*$ using eq. (9) requires assignment of values to f_2^*. The value of f_2^* may be determined from separate measurements of the diffusion coefficient D_0 and the relationship

$$D_0 = \frac{kT}{f_2^*} \tag{17}$$

Alternatively gel filtration experiments - using standards of known Stokes radii - may be used[11] to estimate r_2^* and eqs. (10) and (11) used to determine f_2^*. Determination of the value of M_2^*, $\bar{v}_2^*$ and f_2^* allows the frictional ratio to be determined. This value is a function of the shape or axial ratio of the molecule and the degree of hydration[40]. Determination of the axial ratio requires a value for the degree of hydration to be assumed. For these purposes the term for $\bar{v}_2^*$ in eq. (11) is expanded to include a term for hydration. Initial high values for the axial ratio of detergent solubilised proteins[17] determined from sedimentation studies have been questioned on the basis of electron microscope studies and the neglect of effects due to detergent hydration[29].

Examples of Sedimentation Velocity Studies of Membrane Proteins

Most sedimentation velocity studies of detergent solubilised membrane proteins have been performed using unpurified samples. An example using Method 1 is provided with studies of the rat insulin receptor[18]. The results indicate two species of sedimentation coefficient 10.2 S and 6.6 S. Studies in D_2O and H_2O were used to calculate the partial specific volume and molecular weight of the detergent solubilised species. The values obtained using Triton X-100 and deoxycholate as the solubilising agents were consistent with the binding of one detergent micelle per receptor species yielding molecular weights for the two forms of the receptor of (360,000 ± 65,000) and (180,000 ± 45,000). These values suggest a monomer-dimer model where the monomer is composed of an α and a β subunit. The value for the molecular weight of the larger species ($\alpha_2\beta_2$) is similar to the value reported above for the human insulin receptor from sedimentation equilibrium experiments[29].

An example using Method 2 is provided with studies on the rat liver vasopressin receptor[44]. Radiolabelled vasopressin was used to locate the receptor on sucrose gradients. The sedimentation coefficient and partial specific volume for the receptor complex solubilised by digitonin were determined to be 16.8 S and 0.736 ml/g respectively. The migration of the receptor paralleled the migration of GTPase activity suggesting that the receptor was associated with a GTP-binding protein. Other examples of the use of Method 2 include studies on the rat dihydropyridine receptor[45] and the rat lactogenic hormone receptor[46] solubilised by Triton X-100.

Sedimentation velocity studies of solubilised membrane proteins are also useful for qualitative purposes aimed at comparing the rate of movement of the protein under different conditions. Such studies can provide valuable insight into the approximate size and subunit structure of membrane proteins. For the erythrocyte glucose transport protein[19], detergent binding capacity was calculated from the lipid binding capacity of the purified protein. This allowed an estimate of the density and hence molecular weight of the solubilised protein. The results indicated a dimer-tetramer equilibrium with reductants favouring the formation of dimer. A recent study has used sucrose gradient sedimentation velocity to characterize the human transferrin receptor[22]. Recombinant DNA procedures were used to produce a functional receptor lacking the cysteine residues required for the covalent dimerization of the receptor. In the absence of added ligand (transferrin) the modified receptor sedimented, in the presence of Triton X-100, as a monomer. The addition of transferrin to the gradient lead to the formation of a dimeric form of the receptor indistinguishable in sedimentation behaviour from the naturally occurring receptor dimer. These studies indicate the utility of sedimentation analysis for the characterization of the subunit structure and interactions of membrane proteins.

4. RATES OF ASSOCIATION AND DISSOCIATION OF PROTEINS FROM CELL MEMBRANES

While many membrane proteins require detergents for solubilisation there are some which may be dissociated by simple manipulation of solution conditions. Sedimentation analysis using the preparative ultracentrifuge provides a useful tool to characterize the interactions of these proteins with the membrane. The high angular velocities available are capable of centrifuging intact cell membranes to the bottom of the centrifuge tube within minutes. The rate of dissociation and association of the cytoskeletal proteins of the erythrocyte membrane (spectrin, actin, band 2.1 and band 4.1) can be conveniently measured using the small desk-top air-driven ultracentrifuge[8,9]. These proteins are rapidly released from the membrane at low ionic strength and in the presence of ethylenediamine tetra-acetic acid (EDTA). The rate of release of spectrin/actin is decreased by adjusting the temperature from 37°C to 23°C and the ionic strength to 1.4 mM. The rate of release of spectrin/actin is even less for membranes prepared from patients with hereditary spherocytosis[9]. The air-driven ultracentrifuge has also been used to measure the rate of re-association of spectrin/actin with spectrin/actin depleted vesicles[9] and the effect of ATP and 2,3-

diphosphoglycerate on the structural stability of erythrocyte cytoskeletons prepared by detergent extraction of erythrocyte membranes[8]. In the latter case, changes in erythrocyte morphology correlated with an increased tendency for dissociation of the cytoskeletal network.

5. CONCLUDING REMARKS

The techniques of sedimentation equilibrium and sedimentation velocity provide accurate information on the size and composition of detergent solubilised membrane proteins. For unpurified membrane proteins, methods are available which allow simultaneous determination of the molecular weight and buoyant density of the solubilised protein. Buoyant density values may be used in some cases to estimate the amount of bound detergent and hence the molecular weight of the protein alone. Density gradient sedimentation velocity studies of solubilised membrane proteins can indicate ligand induced perturbations of the subunit structure of membrane proteins. Where purified membrane proteins are available the method of choice is to separately determine the extent of detergent binding and to calculate the buoyant density of the solubilised complex. Currently there are only a few studies on the sedimentation behaviour of purified membrane proteins. The development of recombinant DNA methods for the production of larger amounts of membrane proteins and the availability of a commercially available analytical ultracentrifuge with a more sensitive optical density scanner will facilitate the application of the techniques described in this chapter to the study of membrane proteins.

REFERENCES

1. S.J. Edelstein, W.B. Beyer, A.T. Eldefrawi and M.E. Eldefrawi, *J. Biol. Chem.*, 1975, *250*, 6101.
2. C.J. Steer, J.C. Osborne and E.S. Kempner, *J. Biol. Chem.*, 1990, *265*, 3744.
3. C. Tanford, Y. Nozaki, J.A. Reynolds and S. Makino, *Biochemistry*, 1974, *13*, 2369.
4. M.D. Suarez, A. Revzin, R. Narlock, E.S. Kempner, D.A. Thompson and S. Ferguson-Miller, *J. Biol. Chem.*, 1984, *259*, 13791.
5. G.J. Howlett, 'Molecular and Chemical Characterization of Membrane Receptors' (J. Craig and L. C. Harrison, eds.), Alan R. Liss, New York, 1984, Vol 3, p. 179.
6. R.J. Pollet, 'Methods in Enzymology' (C.H.W. Hirs and S.N. Timasheff, eds.), Academic Press, 1985 ,Vol. 117, p. 3.
7. G.J. Howlett, 'Methods in Enzymology' (G. Di Sabatino, ed.), Academic Press, 1987, vol 150, p. 447.
8. M.P. Scheetz and J. Casaly, *J. Biol. Chem.*, 1980, *255*, 9955.
9. J.S. Hill, W.H. Sawyer, G.J. Howlett and J.S. Wiley, *Biochem. J.*, 1982, *201*, 259.
10. R.G. Martin and B.N. Ames, *J. Biol. Chem.*, 1961, *236*, 1372.
11. L.M. Siegel and K.J. Monty, *Biochim. Biophys. Acta*, 1966, *112*, 346.
12. J.N. Umbreit and J.L. Strominger, *J. Biol. Chem.*, 1973, *248*, 6759.
13. E.J. Neer, *J. Biol. Chem.*, 1974, *249*, 6527.

14. S. Clarke, *J. Biol. Chem.*, 1975, *250*, 5459.
15. P. Cuatrecasas, *J. Biol. Chem.*, 1972, *247*, 1980.
16. M. Smigel and S. Fleischer, *J. Biol. Chem.*, 1977, *252*, 3689.
17. J.C. Meunier, R.W. Olsen and J.P. Changeux, *FEBS Letters*, 1972, *24*, 63.
18. R.A. Aiyer, *J. Biol. Chem.*, 1983, *258*, 14992.
19. D.N. Hebert and A. Carruthers, *Biochemistry*, 1991, *30*, 4654.
20. P. Mayeux, N. Casadevall, C. Lacombe, O. Muller and P. Tambourin, *Eur. J. Biochem.*, 1990, *194*, 271.
21. R. Biswas, M. Basu, A. Sen-Majumdar and M. Das, *Biochemistry*, 1985, *24*, 3795.
22. E. Alvarez, N. Girones and R.J. Davis, *EMBO J.*, 1989, *8*, 2231.
23. S. Bishayee, S. Majumdar, J. Khire and M. Das, *J. Biol. Chem.*, 1989, *264*, 11699.
24. E.F. Casassa and H. Eisenberg, 'Advances in Protein Chemistry' (C.B. Anfinson, M.L. Anson, J.T. Edsall and R.M. Richards, eds.), Academic Press, Vol 19, p. 287.
25. S.J. Edelstein and H.K. Schachman, *J. Biol. Chem.*, 1967, *242*, 306.
26. N.C. Robinson and C. Tanford, *Biochemistry*, 1975, *14*, 369.
27. J.A. Reynolds and D.R. McCaslin, 'Methods in Enzymology' (C.H.W. Hirs and S.N. Timasheff eds.), Academic Press, 1985, Vol. 117, p. 41.
28. C. Tanford and J.A. Reynolds, *Biochim. Biophys. Acta*, 1976, *457*, 133.
29. R.J. Pollet, B.A. Haase and M.L. Standaert, *J. Biol. Chem.*, 1981, *256*, 12118.
30. G.J. Howlett and G. Markov, *Arch. Biochem. Biophys.*, 1980, *202*, 507.
31. I.Z. Steinberg and H.K Schachman, *Biochemistry*, 1966, *5*, 3728.
32. N.N. Baker, A.M. Gotto and R.L. Jackson, *J. Biol Chem.*, 1975, *250*, 1015.
33. M.A. Bothwell, G.J. Howlett and H.K. Schachman, *J. Biol. Chem.*, 1978, *253*, 2073.
34. R.J. Pollet, B.A. Haase and M.L. Standeart, *J. Biol. Chem.*, 1979, *254*, 30.
35. A. Ullrich, J.R. Bell, E.Y. Chen, R. Herrera, L.M. Petruzzelli, T.J. Dull, A. Gray, L. Coussens, Y.-C. Liao, M. Tsubokawa, A. Mason, P.H. Seeburg, C. Grunfield, O.M. Rosen and J. Ramachandran, *Nature*, 1985, *313*, 756.
36. G.J. Howlett, H. Birch, P.W. Dickson and G.H. Schreiber, *Biochem. Biochem. Res. Commun.*, 1982, *105*, 895.
37. R.P. Hughey and N.P. Curthoys, *J. Biol. Chem.*, 1976, *251*, 7863.
38. D. Sakamuro, M. Yamazoe, Y. Matsuda, K. Kangawa, N. Taniguchi, H. Matzuo, H. Yoshikawa and N. Ogasawara, *Gene*, 1988, *73*, 1.
39. T. Svedberg and K.O. Pedersen, ' The Ultracentrifuge', Oxford University Press, 1940.
40. H.K. Schachman, 'Ultracentrifugation in Biochemistry', Academic Press, 1959.
41. R.D. O'Brien, C.A. Timpone and R.E. Gibson, *Anal. Biochem.*, 1978, *86*, 602.
42. J.E. Sadler, J.I. Rearick, J. Paulson and R.L. Hill, *J. Biol. Chem.*, 1979, *254*, 4434.
43. S.N. Timasheff, J.C. Lee, E.P. Pittz and N. Tweedy, *J. Colloid Interface Sci.*, 1976, *55*, 658.
44. T.J. Fitzgerald, R.J. Uhing and J.H. Exton, *J. Biol. Chem.*, 1986, *261*, 16871.

45. W.A. Horne, G.A.Weiland and R.E. Oswald, *J. Biol. Chem.*, 1986, *261*, 3588.
46. R.C. Jaffe, *Biochemistry*, 1982, *21*, 2936.

26

Disaggregation of the Membrane Protein Cytochrome P450 by Detergents

By J. Behlke*

DEPARTMENT OF HYDRODYNAMICS, INSTITUTE OF MOLECULAR BIOLOGY, BERLIN-BUCH, O-1115, GERMANY

1. INTRODUCTION

Living systems contain many membrane-integrated proteins which display their full activity. Usually these proteins possess more or less large hydrophobic sections which are occupied mostly by lipids. The separation of such membrane proteins is possible after treatment with detergents (solubilization effect). In spite of their mostly tight binding, these detergent molecules can then be removed from proteins almost completely. However, this purification procedure is associated with a '*cluster*' or '*assembly*' formation involving an average number of single protein monomers: the number of monomers in such aggregates differs depending on the quality and rigidity of the protein.

The behaviour of such 'assemblies' or aggregates as a function of different detergent treatments will be discussed by consideration of the membrane protein, cytochrome P450, which together with the NADPH-cytochrome P450 reductase and phospholipids is responsible for the catalytic activity of liver microsomes in biotransformation of organic substances such as drugs, xenobiotics or steroids[1,2]. In this ensemble the phospholipids can be replaced partially by detergents at low concentrations[3-10]. High amounts of added amphiphils cause a loss of activity, probably due to disruption of hydrophobic interactions in membrane-like structures[4,11]. Assemblies of purified proteins may dissociate in the presence of detergents, too.

The activity of reconstituted P-450 systems in the presence of detergents has been analysed in various ways. The results obtained were partially discussed with regard to dissociation including the formation of monomeric subunits as shown by sedimentation or chromatographic experiments[4-9,12]. Extending earlier studies[13] we now present results on eight different detergents concerning their ability to disaggregate P450 LM2. The aim of these experiments was to obtain information about the binding sites and efficiency of different detergents and likewise about the

---

* Present address: Max Delbrück Centre for Molecular Medicine, O-1115 Berlin-Buch, Germany.

mechanism of the interaction with proteins, especially with P450 LM2, an electrophoretically homogeneous cytochrome P450 (EC 1.14.14.1) of liver microsomes which is encoded by the rabbit P450 II B1 gene[14].

2. MATERIAL AND METHODS

Cytochrome P450 LM2 was prepared from liver microsomes of phenobarbital induced rabbits as described earlier[13]. The content of Triton N 101 from the preparation procedure and phospholipid was found to be 3 or 1.5 molecules/molecule P450, respectively[15]. Samples were dissolved in 0.1 M K-phosphate buffer, pH 7.4, containing 0.1 mM EDTA and 20% (v/v) glycerol. The concentration of P450 LM2 was determined as published by Omura and Sato[16]. According to ref. 17, from the absorption ratio (418 nm/278 nm) of 1.48 ± 0.02 a sedimentation coefficient of about 13.5 S and a molecular mass of 390,000 is to be expected for the P450 LM2 used in the experiments. Therefore the average number of P450 LM2 monomers (M=55721 [18]) in the assembly is ≈ 7.

Detergents (see Table 1) at different molar concentrations were added to the cytochrome P450 LM2 samples (final concentration usually 5 μM) 30-60 minutes before the ultracentrifuge runs.

Sedimentation velocity experiments were carried out, as previously described[13] in an analytical ultracentrifuge (Spinco Model E, Beckman Instruments) equipped with UV absorption optics, *i.e.* monochromator and a photoelectric scanner using 12 mm double sector cells and an AnG rotor. The sedimenting boundaries were recorded at 417 nm to determine only the heme-linked material with bound detergent. Free detergent molecules or micelles were not determined. Sedimentation coefficients calculated from the 50% value of the sedimenting boundary were corrected to solvent density and viscosity (20°C, water) in the standard way.

The detergent-dependent decrease of sedimentation coefficients, proceeding from the value of P450 aggregates (s_a) down to the fully dissociated or monomeric P450 molecules (s_d) is an expression of the reduction of the aggregate concentration (c_a) by increased r values (detergent/P450 ratio). It can be described by the single differential equation

$$-\frac{dc_a}{dr} = kc_a \tag{1}$$

The change in c_a given by the dimensionless value, α, can be calculated from the experimentally determined sedimentation coefficients (s_{exp}) in relation to s_a and s_d according to

$$\alpha = \frac{s_{exp} - s_d}{s_a - s_d} \tag{2}$$

Integration of eq. (1) between 0 and r and substitution leads to the following relation which reflects the detergent-dependent changes of sedimentation coefficients

$$s_{exp} = s_d + (s_a - s_d)\, e^{-kr} \tag{3}$$

Table 1. Critical micelle concentration (CMC) values, molecular masses and partial specific volumes of detergents.

Detergent	Commercial name	CMC (μM)	M	$\bar{v}$ (ml/g)
Nonylphenyl-PEG(13)ether	Emulgen 13	40	792	n.d.
Dodecyl-PEG(9-10)ether	Lubrol PX	40	582	n.d.
Cetylstearylphenyl-PEG(9-10)ether	Lubrol WX	45	1004	0.929
Nonylphenyl-PEG(9-10)ether	Triton N101	85	638	0.922
p-t-octylphenyl-PEG(9-10)ether	Triton X100	240	624	0.913-0.930
3-(3-cholamidopropyl)-dimethylammonio-1-propanesulphonate	CHAPS	4000	614	0.730
Sodium cholate	Cholate	14000	430	0.778
Octyl-β-(D)-glucopyranoside	Octylglucoside	22000	292	0.930

CMC data taken from refs. 11, 20 and 21.
Values of partial specific volumes from refs. 21 and 22.
PEG: polyoxyethylenglycol.
n.d.: not determined.

The interaction (or decay constant) k can be obtained from the relation

$$-k = \frac{\ln \alpha}{r} \tag{4}$$

where k has the dimension of the reciprocal detergent concentration. For membrane proteins consisting of n monomers we can assume also a step-by-step dissociation by analogy with the aggregation process, where different aggregates (A_n, A_{n-1}, A_{n-2}, ...) are in equilibrium at least with the monomers (A_1). Because of the globular character of P450 aggregates and dissociation products[6,12,17] the relation between monomers and different aggregates can be reasonably represented by their sedimentation coefficients using the relation

$$s_1 = \frac{s_n}{n^{2/3}} \tag{5}$$

3. RESULTS

Sedimentation coefficients of $s_{20,w} = (13.5 \pm 0.1)$ S indicate that the lipid free P450 LM2 exists in an assembled form (Fig. 1). According to ref. 17, these values correspond to molecular weights of about 390,000 meaning that, on average, P450 LM2 aggregates consist of *heptamers*. Addition of detergents leads to an average exponential decrease of sedimentation coefficients of up to about 3 S (Fig. 1).

For comparison of different titration curves with other detergents the use of semi-logarithmic plots is advantageous because (i) the parameter r is spread over several orders of magnitude and (ii) the change of sedimentation coefficients can be represented by linear sections. Without physical interpretation, the conditions in r for the decay of aggregates can be well demonstrated in this way. In Fig. 2 titration curves with eight different detergents are represented. They show large differences in their efficiency to dissociate P450 LM2. In all cases a distinct ratio r (r_{crit}) is apparently necessary to initiate the process. For the group of non-ionic detergents: Emulgen 913, Lubrol PX, Lubrol WX, Triton N101, Triton X100 low r_{crit} data of 10-20 were observed whereas for the more bulky substances CHAPS, cholate and octylglucoside high values of 300-1500 were determined. In all experiments the detergent concentrations involved in r_{crit} are lower than the critical micelle concentration (CMC) of the different substances (see also Fig. 3). That means, monomeric detergent molecules should interact at first with the P450 LM2 by penetrating the clustered P450 molecules with their hydrophobic parts, thus loosening the cluster stabilization by hydrophobic interactions. Once the r_{crit} is attained, the first monomers or some larger parts dissociate from the aggregates. This process is indicated by the subsequent decrease of the sedimentation coefficients. With the exception of Lubrol WX, all other detergents investigated yield nearly the same slope, and the lowest $s_{20,w}$ values amount to (3.6±1.0) S. The deviations can be explained by different detergent binding to the protein and different values of the partial specific volumes. As to Lubrol WX, the decrease of sedimentation coefficients stopped at $s_{20,w}$ values usually found for dimers (A_2). Presumably two monomers were bound to the long side chain (cetyl-

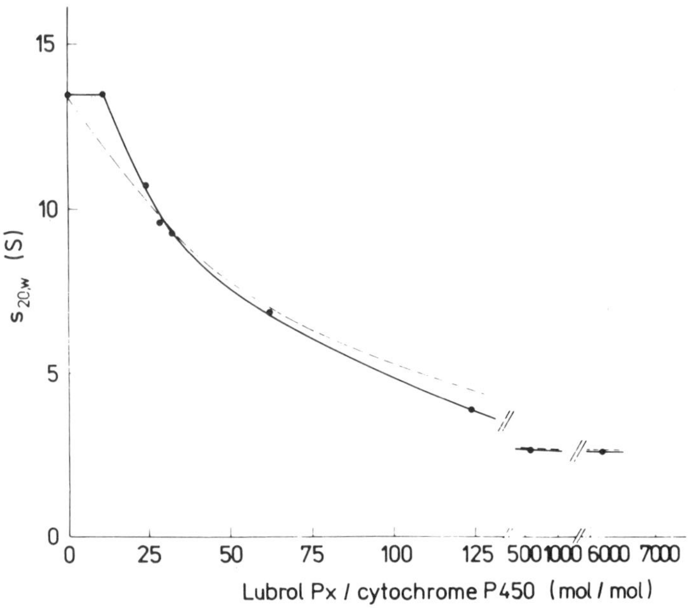

Figure 1. Sedimentation coefficients of cytochrome P450 LM2 depending on r, the molar Lubrol PX/protein ratio. Conditions: cytochrome P450: 5 μM, 0.1 M K-phosphate buffer pH 7.4, containing 0.1 mM EDTA and 20% glycerol, 20°C. Broken line: fitted curve according to eq. (3) with k = 0.0015.

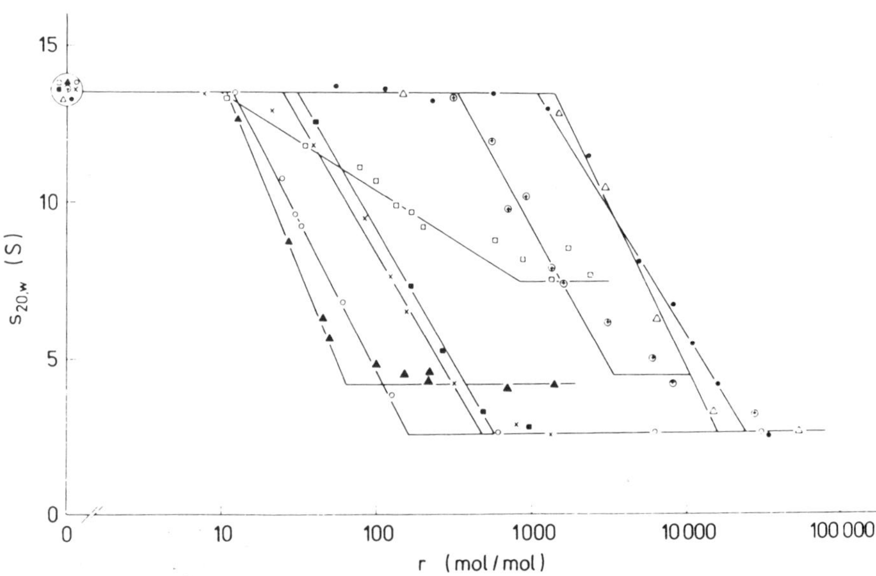

Figure 2. Semi-logarithmic representation of sedimentation coefficients of cytochrome P450 LM2 depending on the molar detergent/P450 ratio (r). Conditions as given in Fig. 1. (▲) Emulgen 913, (○) Lubrol PX, (×) Triton N101, (■) Triton X100, (□) Lubrol WX, (⊕) CHAPS, (△) octylglucoside, (●) cholate.

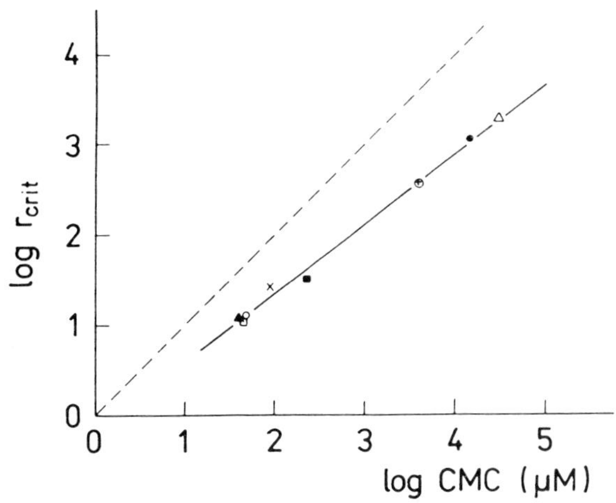

Figure 3. Influence of CMC on the r_{crit} values according to Fig. 2.

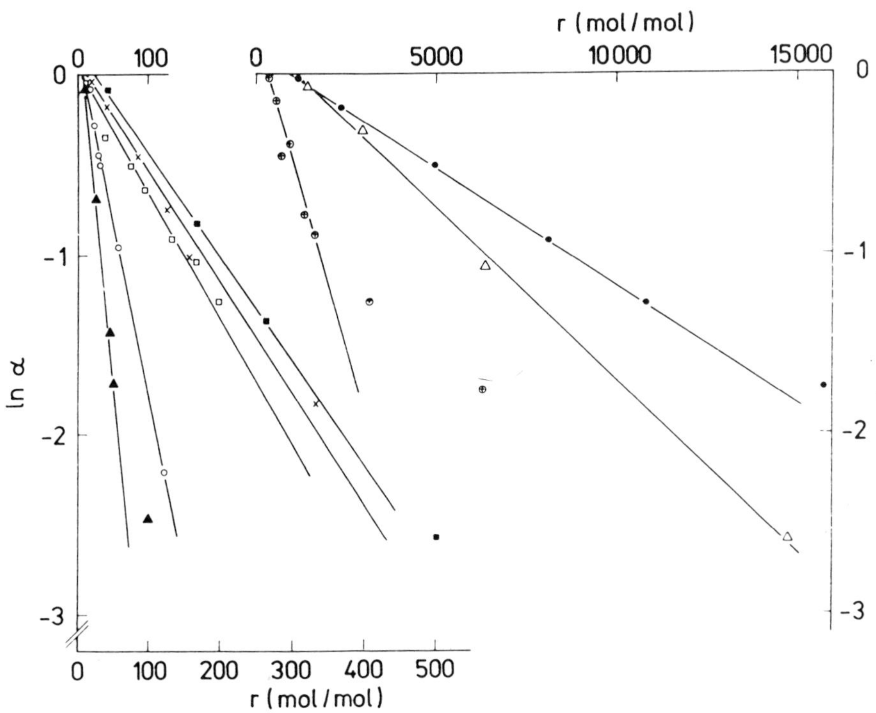

Figure 4. Plots ln α *versus* r as representation of the disappearance of P450 aggregates. Conditions as in Fig. 1.

stearylphenyl residue) of Lubrol WX, or large detergent micelles were bound to P450 resulting in $s_{20,w}$ values of about 7 S.

Generally the disaggregation process proceeds in a range of one or two orders of magnitude of r, indicating a cooperative behaviour. To measure it, the k values or decay constants have to be determined from the plots of ln α *versus* r (Fig. 4), which result in linear plots and demonstrate the disappearance of aggregates according to a first order reaction (pseudo-monomolecular reaction). Due to the fact that these linear plots do not start from the origin, the k-values vary during the titration at first stronger and, later on, less strongly. For *quantitative* consideration of the co-operativity or the increase of k values in the interaction of detergents with the protein, the amount of alteration in the sedimentation coefficients ($s_a - s_d$) was formally separated into six different steps according to the $s_{20,w}$ values calculated for the different aggregates using eq. (5). The corresponding k values describe the 50% decay of heptamers to hexamers (D_6) or hexamers to pentamers (D_5) *etc.* as can be calculated from eq. (4). As can be seen in Fig. 5a, in all experiments we can observe an increase of k for the first dissociation steps independent of the absolute data. The average change or enhancement of k during the disaggregation process amounts to about two, when taking into account the transition of D_6 to D_5 (Fig. 5b). In the further steps, the alteration in k became smaller and finally disappeared because the aggregates had apparently lost their tension (see model in Fig. 6). Further interactions of detergents with the more relaxed P450 aggregates take place with normal intensity and without limitation in space.

4. DISCUSSION

Treatment of P450 LM2 with different detergents has provided new insights into the aggregation behaviour of the membrane protein which is assembled in the purified state. The disaggregation process represented by an exponential decrease of sedimentation coefficients caused by increasing of r can be transformed into linear dependencies using a semi-logarithmic plot. This enabled us to recognize clearly the conditions for the dissociation procedure. The agreement of our results on P450 LM2 with the few literature data available on determinations of the molecular mass using the sedimentation equilibrium technique[6,7] or chromatographic procedures[9] proved the sedimentation velocity method as a suitable tool for analysing the detergent dependent disruption of membrane protein aggregation. For this process the following facts seem to be important.

(i) The efficiency of detergents in dissociating of P450 LM2 depends on their CMC, which reflects the hydrophobic free energy for the introduction of detergent molecules into the micelle. The formation of micelles, however, is an expression of the entropically unfavoured interaction of alkyl - or alkylphenyl chains with water. Therefore the transfer of monomeric detergent molecules into the assemblies is favoured because of the reduction of free energy (*i.e.* because the packing in mixed micelles [*e.g.* in comparison to protein-detergent] is energetically favoured compared to the the pure ones[19]). The above mentioned penetration process of detergent molecules into the clusters of P450 LM2 should proceed also at concentrations higher than the CMC.

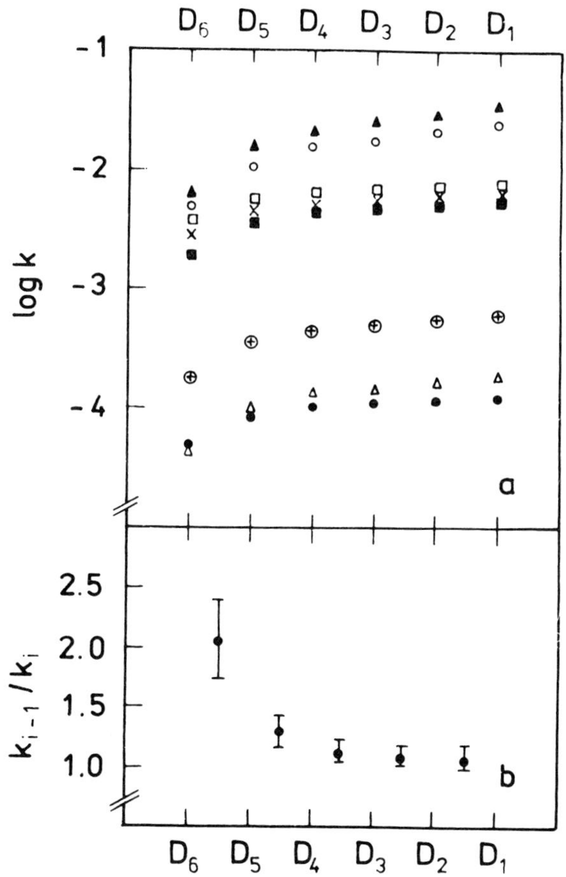

Figure 5. (a) Change of k-values during the detergent-dependent dissociation of P450. Dn represents the state where an average equivalent portion, each 50% of A_n and A_{n+1} exists. Symbols as in Fig. 2. (b) Average enhancement of k-values during the detergent-dependent dissociation of P450.

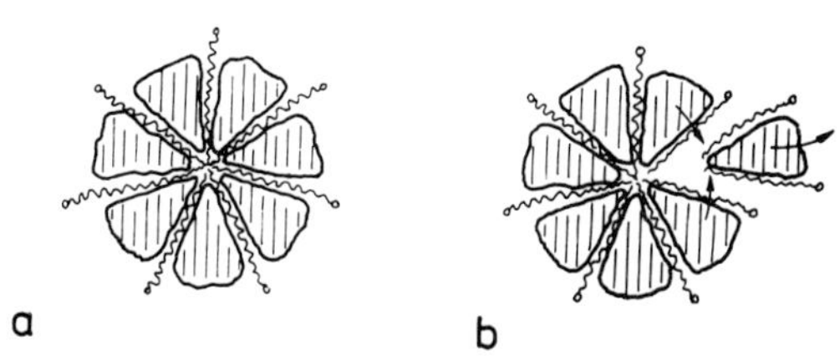

Figure 6. Model for the detergent - dependent relaxation process of P450 aggregates.

Analyzing the r_{crit}/CMC dependence (Fig. 3) from the slope smaller than unity we can conclude furthermore that detergents with high CMC have a relatively stronger ability to form mixed micelles than substances with low CMC values.

In addition, the relation between r for the beginning of disaggregation and the CMC of the detergents derived from Fig. 3 can be summarized in the following relation

$$r = k_r \, CMC^a \qquad (6)$$

This connection permits us to predict sedimentation coefficients or the degree of disaggregation by different detergents with known CMC. The data for the constants k_r and a depend on the state of aggregation as can be seen from Table 2.

Table 2. Data for the constants in eq. (6).

State of disaggregation	k_r	a
Beginning of disaggregation (r_{crit})	0.636	0.768
50% disaggregation	†2.035	†0.800
End of disaggregation	†6.400	†0.824

† Valid for all detergents without Lubrol WX.

(ii) From the data available so far it can be concluded that differences in the efficiency between non-ionic, ionic and zwitterionic detergents are caused by the hydrophobic part of the molecules, their flexibility and ability to fit the surface of the proteins rather than by the presence or absence of charges. If we compare the efficiency of detergents by the size of their hydrophobic parts, we may conclude that the 1.6 nm length of the nonylphenyl residues in Emulgen 913 or Triton N101 seems to be suitable for optimal covering of the hydrophobic areas between the monomers in aggregates, whereas the length of 2.4-2.6 nm for cetylphenyl or stearylphenyl side chains in Lubrol WX is less well suited because of the limited dimensions of such interface areas.

(iii) Although the affinity of detergents to bind the P450 LM2 aggregates differs considerably, all the substances investigated disaggregate the protein clusters of varying compositions by the same mechanism. After a lag phase in which only a defined number of detergent molecules is involved, the aggregates dissociate, on the average, step by step depending on the given detergent concentration. Obviously, in the detergent free solution the aggregates exist in a more rigid state with high tension. They are capable only of exerting weak interactions with other substances. After splitting off the first monomers from the heptamers, the conformation presumably becomes more "relaxed", and the interaction constant k increases considerably. In further dissociation steps, this process becomes insignificant, showing only weak increments of the k-values (Fig. 5). Finally the pentamers, tetramers, trimers, *etc.* possess such a relaxed structure which allows a suitable interaction with the reductase. Such particles containing partially dissociated P450 assemblies are of higher activity than those with full association[6,7]. Possibly, the connections observed between detergents and P450 LM2 are valid also for other membrane proteins.

5. SUMMARY

The membrane protein P450 in the lipid- or detergent-free state forms aggregates (containing seven monomers on the average), which are stabilized by hydrophobic forces. Increased detergent concentrations or detergent/protein ratios r provoke an interruption of the hydrophobic interface contacts and, therefore a dissociation of the P450 clusters, as can be demonstrated by the exponential decrease of sedimentation coefficients. The semi-logarithmic plot of $s_{20,w}$ values against log r permits us to estimate the detergent concentration or r values necessary for the cluster dissociation: this process proceeds by single detergent molecules. The efficiency of detergents is proportional to their CMC. Plotting the disappearance of aggregates as ln α *versus* r results in different slopes or k values as a manifestation of the strength of detergent interaction with the protein and as a reflection of a co-operative behaviour.

ACKNOWLEDGEMENT

I am grateful to Dr. G.-R. Jaenig for preparing cytochrome P450 and Prof. K. Ruckpaul for help in discussions.

REFERENCES

1. A.Y.H. Lu and M.J. Coon, *J. Biol. Chem.*, 1968,*243*, 1331.
2. W.H. Strobel, A.Y.J. Lu, J. Heidema and M.J. Coon, *J. Biol. Chem.*, 1970, *245*, 4851.
3. A.Y.H. Lu, W. Lewin and R. Kuntzman, *Biochem. Biophys. Res. Comm.*, 1974, *60*, 266.
4. D. Pfeil, G.-R. Jaenig, J. Behlke and K. Ruckpaul, *Acta Biol. med. germ.*, 1977, *36*, 1243.
5. J. Blanck, K. Rohde, J. Behlke, G.-R. Jaenig, D. Pfeil and K. Ruckpaul, *Acta biol. med. germ.*, 1979, *38*, 399.
6. W.L. Dean and R.D. Gray, *J. Biol. Chem.*, 1982, *257*, 14679.
7. S.L. Wagner, W.L. Dean and R.D. Gray, *J. Biol. Chem.*, 1984, *259*, 2390.
8. S.L. Wagner, W.L. Dean and R.D. Gray, *Biochemistry,* 1987, *26*, 2343.
9. I. Bachmanova, E.D. Skotselyas, I.P. Kanaeva, G.P. Kutznetsova, S.A. Gordeev, E.N. Kerneva, A.V. Karyakin and A.I. Archakov, *Biochem. Biophys. Res. Comm.*, 1986, *139*, 883.
10. S.L. Kaminsky, D. Dumbar, F.P. Guengerich and J.J. Lee, *Biochemistry*, 1987, *26*, 1276.
11. A. Helenius and K. Simons, *Biochim. Biophys. Acta*, 1975, *445*, 29.
12. W.L. Dean and R.D. Gray, *Biochem. Biophys. Res. Comm.*, 1982, *107*, 265.
13. J. Behlke, G.-R. Jaenig and D. Pfeil, *Acta biol. med. germ.*, 1979, *38*, 389.

14. D.W. Nebert, M. Adesnik, M.J. Coon, R.W. Estabrook, F.J. Gonzales, F.P. Guengerich, I.C. Gunsalus, E.F. Johnson, B. Kemper, W. Levin, I.R. Philips, R. Sato and M.R. Waterman, *DNA*, 1987, *6*, 1.
15. I. Wendel, J. Behlke and G.-R. Jaenig, *Biomed. Biochim. Acta*, 1983, *42*, 623.
16. T. Omura and R. Sato, *J. Biol. Chem.*, 1964, *239*, 2370.
17. I. Wendel, J. Behlke and G.-R. Jaenig, *Biomed. Biochim Acta*, 1983, *42*, 633.
18. G.E. Tarr, S.D. Black, V.S. Fujita and M.J. Coon, *Proc. Nat. Acad. U.S.A.*, 1983, *80*, 6552.
19. C. Tanford, 'The Hydrophobic Effect: Formation of Micelles and Biological Membranes', Wiley Interscience, 1980.
20. P. Mukerjee and K.J. Mysels, 'Critical Micelle Concentration in Aqueous Surfactant Systems', National Bureau of Standards NSRDS-NBS 36, Washington, 1971.
21. L.H. Hjelmeland, *Proc. Nat. Acad. Sci. U.S.A.*, 1980, *77*, 6368.
22. H. Durchschlag, in 'Thermodynamic Data for Biochemistry and Biotechnology', ed. H.J. Hinz, Springer-Verlag Berlin, 1986.

27

Sedimentation Analysis of Polysaccharides

By Stephen E. Harding

UNIVERSITY OF NOTTINGHAM, DEPARTMENT OF APPLIED BIOCHEMISTRY AND FOOD SCIENCE, SUTTON BONINGTON, LE12 5RD U.K.

1. INTRODUCTION

Polysaccharides and related glycopolymers offer a unique challenge to analytical ultracentrifugation for two clear reasons. Firstly, there is a growing demand for basic solution biophysical information on these substances: molecular weights, molecular weight distributions, conformation, conformation distributions; information which analytical ultracentrifuge can, in principle, supply. This information is important for the understanding of for example the performance of so-called "commercial" polysaccharides in the food, drug and oil industries and for understanding the behaviour of glycopolymers such as mucus glycoproteins in health and disease.

The second reason is that despite the urgent requirement for this basic information, these macromolecules - compared to so-called 'well-behaved' macromolecules like many protein systems - are very difficult to analyse, because of their high thermodynamic non-ideality, their polydispersity (not only with respect to molecular weight but also in many circumstances with respect to composition) and, in several cases, their potential to self-associate in solution. Polysaccharides place therefore a considerable strain on available ultracentrifuge methodology and in many cases satisfactory information can only be achieved by combining results from the analytical ultracentrifuge with results from other techniques, notably light scattering, gel permeation chromatography and electron microscopy.

The aim of this Chapter is therefore threefold (i) to briefly survey the contributions that analytical ultracentrifugation has made to our understanding of polysaccharide systems; (ii) to indicate those advances in methodology of particular relevance to polysaccharides; (iii) to comment on possibilities for the future in the light of recent instrumental developments. We have also included a consideration of "mucopolysaccharides" *i.e.* mucus glycoproteins, since these share many of the characteristics (*e.g.* high viscosity and gel forming properties) of unconjugated polysaccharides, and indeed have a very high carbohydrate content (~80%). Two other important classes of glycopolymer - proteoglycans and hyaluronic acid - are considered in subsequent Chapters in this volume.

2. BASIC INFORMATION SOUGHT

Molecular Weight and Heterogeneity

One of the most fundamental pieces of information describing a macromolecule is its molecular weight, M. M values for *polypeptides* can usually be evaluated without difficulty from chemical sequence information. For *polysaccharides* this is not the case, primarily due to their heterogeneity (polydispersity and self-association phenomena); an *average* molecular weight (usually the weight average, M_w) is normally specified[1]. The ratios of the z-average molecular weight, M_z to M_w or M_w to the number average, M_n are used as indices for polydispersity. M_w is usually available to better than ±10% from sedimentation equilibrium and M_n and M_z are also available but to a somewhat lower precision[2].

Molecular weight *distributions* are a more representative indicator of heterogeneity. These can be obtained in principle at least from sedimentation equilibrium directly[3]; a much easier route is to combine gel permeation chromatography with sedimentation equilibrium[4].

A good indication of *chemical heterogeneity* (*e.g.* purity) of a saccharide system can be provided by analytical isopycnic density gradient ultracentrifugation[5].

Gross Conformation

In general two levels of information are sought:

(i) Delineation between the conformations represented by the extremes of the "Haug Triangle" (Fig. 1) *viz* compact sphere, rigid rod and random coil[6,7]. In general, most polysaccharides have conformations which - in the Haug representation - appear to lie between the extremes of rigid rod and random coil.

(ii) More quantitative information. If the particle is a relatively rigid structure (as represented by side A of Fig. 1) this can be quantified in terms of either "bead"[8] or ellipsoid[9] models of the conformation. If the particle is more appropriately described by a rod, more detailed representations can be given in terms of length, diameter and mass per unit length[10]. If the particle conformation lies between rod and coil (side B), a measure of particle flexibility can be sought in terms of the ratio of the contour length to the persistence length[10,11].

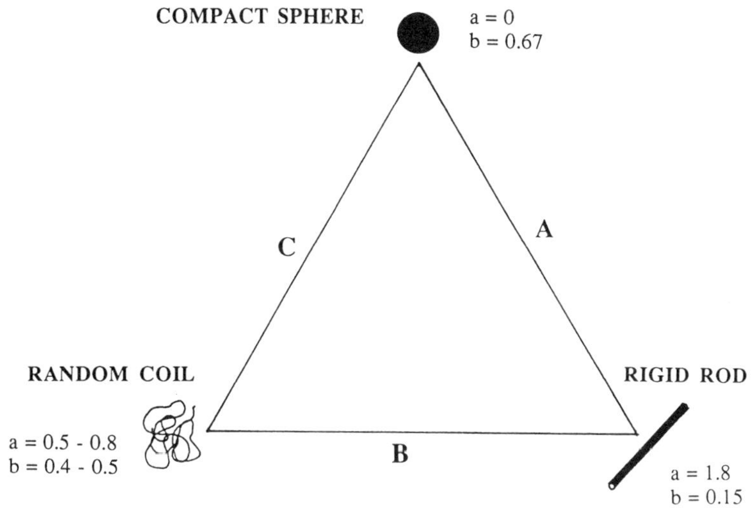

Figure 1. Haug Triangle representation of the gross conformation of macromolecules. The a and b are the exponents in the MHKS viscosity and sedimentation equations respectively (see ref. 6).

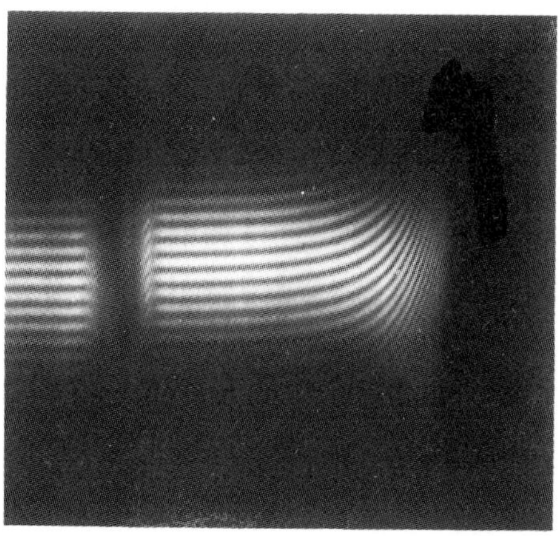

Figure 2. Rayleigh interference fringes from a low-speed sedimentation equilibrium experiment on a mucus glycoprotein or "mucopolysaccharide" from a chronic bronchitis patient ($M \sim 6.0 \times 10^6$). Rotor speed = 2233 rev/min, loading concentration $c_0 \sim 0.4$ mg/ml. Note the steep rise of the fringes at the cell base but finite slope of the fringes at the meniscus.

3. SEDIMENTATION EQUILIBRIUM

Optical System

The methodology behind recording solute distributions at sedimentation equilibrium and subsequent data capture and analysis has been dealt with earlier in this volume. Suffice to say here that most polysaccharides do not have a chromophore absorbing in the usable ultra-violet or visible spectrum and therefore to record solute distributions at equilibrium either Rayleigh or Schlieren optics have to be employed, both based on refractometric properties. Rayleigh interference is usually the optical method of choice[12] because of its greater sensitivity [concentrations as low as ~0.2 mg/ml can be used for cells of long (~30mm) optical path length] and an example of an equilibrium Rayleigh pattern is given in Fig. 2 [each solution fringe profile represents a plot of solute concentration relative to the meniscus]. For polysaccharide work the increased resolution in the Rayleigh fringes if a laser light source is used is particularly important.

Data Capture

Because of the demands on the precision and volume of data required for the analysis of polysaccharides, considerable attention has been recently afforded to the use of multiple-data-aquisition using automatic[12] or semi-automatic data capture procedures. We have been using for the last few years a semi-automatic procedure using a laser densitometer system to capture Rayleigh data off-line from photographic film[13]. Such data is then analysed *via* a frame-shift Fourier algorithm to yield the fringe concentration (relative to the meniscus) *versus* radial displacement. This data is then passed to a mainframe computer (in our case we use the IBM 3084/Q at Cambridge *via* a network link) for full molecular weight analysis.

Low Speed *versus* High Speed Sedimentation Equilibrium

An important choice has to be made prior to an ultracentrifuge run whether to attempt a meniscus depletion experiment or whether to use the more time-consuming low or intermediate speed method. The "high speed" meniscus depletion method[14] has been by far the most popular in biochemistry because it avoids the problem of evaluating the meniscus concentration or fringe number J_a. It also facilitates the evaluation of number average molecular weight data[2]. However for polysaccharides and mucus glycoproteins this method is not generally applicable because of the difficulty of obtaining proper depletion without losing optical registration of the fringes near the cell base (Fig. 2). A further difficulty in attempting meniscus-depletion conditions with polydisperse materials is that the effective thermodynamic second virial coefficient, B_{eff}, can be greatly enhanced[15]:

$$B_{eff} = B\left\{1 + \frac{\lambda^2 M_z^2}{12} + \ldots\right\} \tag{1}$$

B being the "static" virial coefficient, $\lambda = (1-\bar{v}\rho)\omega^2(b^2-a^2)/2RT$, $\bar{v}$ the partial specific volume, ρ the solution density, ω the angular velocity and a and b the radial positions at the cell meniscus and base respectively. This "speed dependence" effect can be minimised by using low speeds and short solution columns.

For polysaccharides, the low or "intermediate" speed method is therefore the more suitable: in this method the meniscus concentration remains finite and in general has to be found. A whole suite of procedures are available: Creeth and Pain[16] have given a comprehensive comparison of methods.

Whole-cell Weight Average Molecular Weights: the M* Operational Point Average

Obtaining whole-cell (apparent) average molecular weight data for polysaccharides can be particularly difficult because of problems of obtaining estimates for the solute concentration at the meniscus *and* at the cell base. It is not usually sufficient to produce simply a plot of $d\{\ln c(r)\}/dr^2$ (where c is the solute concentration at a given radial position r) and "read off a slope", as is possible with 'well-behaved', monodisperse systems of many proteins. Furthermore, application of the widely used equation (see, *e.g.* ref. 16)

$$M^0_{w,app} = \frac{1}{k}\left\{\frac{c_b - c_a}{c_0\,(b^2-a^2)}\right\} \tag{2}$$

where $k = (1-\bar{v}\rho)\omega^2/2RT$, c_0 is the initial loading concentration (g/ml) and c_a, c_b the concentrations at the cell meniscus and base respectively, or the related equation (in terms of the corresponding fringe number concentrations J_0, J_a and J_b respectively)

$$M^0_{w,app} = \frac{1}{k}\left\{\frac{J_b - J_a}{J_0\,(b^2-a^2)}\right\} \equiv \frac{1}{k}\left\{\frac{j_b}{J_0\,(b^2-a^2)}\right\} \tag{3}$$

where j_b is the fringe concentration (relative to the meniscus) at the cell base, is very difficult because of the form of the extrapolation of j(r) to the cell base; a further complication is that an accurate estimate for the initial loading concentration, J_0, is also required.

For polydisperse systems we have found an 'operational' point average molecular weight, the "star average" M^* particularly useful[17]. Although we have considered it in some detail earlier in this volume (Chapter 15), because of its particular usefulness for saccharide systems it will be constructive to briefly consider this again here but this time not in terms of reduced molecular weights. If we take the fundamental (differential) equation of sedimentation equilibrium in the form[16,2]

$$\frac{J(r)}{M_{n,app}(r)} - \frac{J_a}{M_{n,app}(a)} = 2k\int_a^r rJ\,dr \tag{4}$$

where $M_{n,app}(r)$ and $M_{n,app}(a)$ are the (apparent) number average molecular weights at a given radial position, r and the meniscus, a respectively, and then define an operational point average $M^*(r)$ by the relation

$$M^*(r) = \frac{M_{n,app}(r).M_{n,app}(a)}{[M_{n,app}(a).J(r) - M_{n,app}(r).J_a]} \cdot j(r) \tag{5}$$

eq. (4) becomes:

$$\frac{j(r)}{M^*(r)} = kJ_a(r^2\text{-}a^2) + 2k \int_a^r r\, j(r)\, dr \tag{6}$$

Eq. (6) has the limiting form

$$\lim_{r \to a} \frac{j(r)}{(r^2\text{-}a^2)} = k.M^*(a).J_a \tag{7}$$

A graph of $j(r)/(r^2\text{-}a^2)$ *vs.* $(r^2\text{-}a^2)^{-1} \times \int rj(r)dr$ therefore has a *limiting* slope of $2kM^*(a)$ and an intercept $kM^*(a).J_a$: hence J_a is determinable from the limiting slope and intercept. For ideal monodisperse systems both slope and intercept can be obtained without difficulty: for polydisperse and non-ideal systems, automatic data capture and multiple data analysis is particularly valuable for this purpose[17]. It is worth pointing out that Teller *et al.*[18] have considered other manipulations involving representations similar to eq. (6) for obtaining J_a. Other methods of obtaining J_a have been considered in detail by Creeth and Pain[16]. Once J_a has been found, M^* as a function of radial position can be defined.

A particularly useful property of the M^* function is that at the cell base (r=b),

$$M^*(b) = M^0_{w,app} \tag{8}$$

the (apparent) weight average molecular weight over the solute distribution[17]. The form of the extrapolation of M^* to the cell base is usually less severe than extrapolation of j, J (or ln J) to the cell base to obtain $M^0_{w,app}$ using eq. (3). Furthermore, an independent estimate of the initial loading concentration J_0 is *not* required (although an estimate for J_a *is* required). Table 1[17] compares the performance of the two types of extrapolation procedure (*viz via* eq. (2) or eq. (8)) for several sets of synthetic data of relevance to polysaccharides.

Table 1. Comparison of $M_w{}^0$ values obtained *via* the ln J or M^* extrapolation procedures

	I	II	III	IV	V
ln J	659000	4730000	2450000	2150000	499000
M^*	664000	4870000	2770000	2630000	500000
Theoretical	667000	4840000	3020000	3020000	500000

System I: Two component mixture, $M_2 = 3M_1$; $c_{0,2} = 0.167\ c_{0,1}$; M_1=500000;
System II: Very polydisperse (log normal, σ/log M_w=0.044; M=1-15×10^6;
System III: Isodesmic association, c=1, k=2 (M_1=500000);
System IV: As III but with fringe data with ±2 μm standard error;
System V: Single solute, M=500000, fringe data with ±2 μm standard error.

From ref. 17

Two other useful properties of M^* have been given[17]:

(i) $M^*(a) = M_{w,app}(a)$, the (apparent) point weight average at the meniscus;

(ii) $M^*(J=0) = M_{n,app}(a)$, the apparent number average at the meniscus.

Point Weight Average Molecular Weights

The differential form of the fundamental equation for sedimentation equilibrium is used[16]:

$$\frac{d \ln J(r)}{dr^2} = k M_{w,app}(r) \tag{9}$$

where $M_{w,app}(r)$ is the (apparent) point weight average molecular weight. In our hands we find the Hildebrand sliding quadratic strip formulae as considered in Teller's article[2] of particular use, although some caution has to be expressed: if the strip width is too small (particularly for short fringe increments), the data can become too noisy; conversely, if too large, important trends in the data can be missed (for example genuine maxima or minima in $M_{w,app}$ *vs*. c - or J - plots can be over-smoothed out).

Number and z-averages

Until recently, attempts to obtain accurate estimates for both point and whole-cell number and z-averages have put an almost insurmountable strain on the quality of the Rayleigh fringe data unless heavy data smoothing is employed. However, with Rayleigh fringes generated by a laser light source and with the use of either on-line[12] or off-line[13] multiple data capture procedures, reasonable estimates can be obtained without too much difficulty. To obtain both point and whole cell number average molecular weight data sets, an estimate for the number average at the meniscus, $M_{n,app}(a)$ is necessary[2,16]. As has been pointed out[14], *if* the meniscus depletion method could be used, then the $M_{n,app}(r)$ data would be much easier to obtain, but for saccharide systems this method is not always applicable as explained above. It has been suggested[17] that the identity $M^*(J=0) = M_{n,app}(a)$ may be useful in this context; and it is worth remembering that solute distributions recorded using Schlieren optics yield z-average molecular weight data directly.

Non-Ideality

For polysaccharides, non-ideality can be a particularly serious problem (Table 2[19-27]). Non-ideality affects colour the form of the dependence of the point average molecular weight on concentration or radial position; however, for many systems, measurement at a single finite, but low (0.2-0.5 mg/ml) concentration gives an estimate for the the apparent M_w^0 within a few percent of the true or "ideal" M_w^0. In extreme cases even a loading concentration as low as 0.2 mg/ml can lead to non-negligible underestimates for the true M_w^0: in Table 2, the term (1+2BMc), with B the thermodynamic or "osmotic pressure" virial coefficient, represents the factor by which the apparent molecular weight measured at a finite concentration c=0.2 mg/ml, underestimates the "true" or "ideal" molecular weight. Although in many cases the error is small, for saccharides such as alginates and hyaluronic acid, severe underestimates can result.

Table 2. Comparative non-ideality of polysaccharides.

	$10^{-6}\times M$ g/mol	$10^4\times B$ ml.mol/g^2	BM ml/g	^a1+2BMc	Ref.
Pullulan P5	0.0053	10.3	5.5	1.002	19
Pullulan P50	0.047	5.5	25.9	1.010	19
Xanthan (fraction)	0.36	2.4	86	1.035	20
β-glucan	0.17	6.1	104	1.042	21
Dextran T500	0.42	3.4	143	1.057	22
Pullulan P800	0.76	2.3	175	1.070	19
Chitosan (Protan 203)	0.44	5.1	224	1.090	23
Pullulan P1200	1.24	2.2	273	1.109	19
Mucopolysaccharide[b]	2.0	1.5	300	1.120	24
Pectin (citrus fraction)	0.045	50.0	450	1.180	25
Scleroglucan	5.7	0.50	570	1.228	26
Alginate	0.35	29.0	1015	1.406	27

[a] Based on the lowest possible loading concentration (~0.2 mg/ml in a cell with a 30 mm path length centrepiece);
[b] Bronchial mucin CFPHI from a patient with cystic fibrosis.

In these cases, measurement of apparent $M^0_{w,app}$'s over a range of concentrations is necessary followed by an extrapolation to zero concentration[19] using an equation of the form[1,7]

$$\frac{1}{M^0_{w,app}} = \frac{1}{M^0_w}\{1 + (2B_{eff}M^0_w - \bar{v})c\} \tag{10}$$

Fortunately, if solvent densities ρ_0 are used instead of solution densities ρ for evaluating k in eq. (2), the $\bar{v}$ term in eq. (10) drops out; this feature is implicit in the work of Fujita (pp. 284-285 of ref. 3) and stated explicitly by Wills and Winzor (Chapter 17 of this volume). Although this latter procedure yields an accurate $M_w{}^0$, caution has to be expressed when interpreting the second virial coefficient B from eq. (10) because of effects relating to eq. (1). In some extreme cases, third or even higher virial coefficient(s) may be necessary to adequately represent the data.

Distributions of Molecular Weight

Direct inversions of solute distribution data to obtain molecular weight distribution information are generally intractable because of complications involving thermodynamic non-ideality[3,28]. The simplest procedure[4] avoiding these complications is to use sedimentation equilibrium *in conjunction with gel permeation chromatography* (GPC). Fractions of relatively narrow (elution volume) band width are isolated from the eluate and their $M_w{}^0$ values evaluated by low speed sedimentation equilibrium in the usual way: the GPC columns can thereby be "self-calibrated" and elution volume values converted into corresponding molecular weights - a distribution can therefore be defined and in a way which avoids the problem of using inappropriate "standards" for GPC: the value of Yphantis-style multi-channel cells for this purpose[29] is clearly indicated. This procedure has been successfully applied to dextrans, alginates and pectins: for pectins excellent agreement with analogous procedures involving classical light scattering coupled to GPC have been obtained (see, *e.g.* Fig. 3 [30]).

4. SEDIMENTATION VELOCITY

Some caution has to be expressed when interpreting sedimentation coefficient data for polysaccharides because of the strong concentration dependence of the sedimentation coefficient: this feature is considered in some detail in the following chapter in this volume by Lavrenko *et al.*[31]: extrapolation to zero concentration is therefore mandatory. Once found, the sedimentation coefficient, $s^0_{20,w}$ and the concentration dependence regression coefficient, k_s, defined by the equation[32]:

$$\frac{1}{s_{20,w}} \approx \frac{1}{s^0_{20,w}} \{1 + k_s c\} \qquad (11)$$

can be used for both molecular weight and conformational analysis.

Molecular Weight Analysis

The (weight-average) sedimentation coefficient can be combined *via* the well-known Svedberg equation[33,34] with the (z-average) translational diffusion coefficient (from sedimentation analysis[35] or from dynamic light scattering) to yield the weight average molecular weight[36]. This procedure has been used to obtain molecular weights of for example galactomannans[37], β-glucans[38,39], amylopectin[40], alginate[41] and mucus glycoproteins[42]. Obtaining molecular weight *distributions* using sedimentation velocity is however not trivial. Although, after certain assumptions, estimates for distributions of sedimentation coefficients can be obtained[43,44], obtaining corresponding distributions of diffusion coefficient (and at zero concentration), and then combining this data through the Svedberg equation *with* the sedimentation coefficient distribution data is to say the very least, unwieldy.

Nevertheless attempts have been made to obtain distributions of sedimentation coefficients for polysaccharides and, after assuming a particular conformation (sphere, random coil, rigid rod) and using the appropriate coefficients in the Mark-Houwink-Kuhn-Sakurada (MHKS) sedimentation equation (see *e.g.* ref. 7 and eq. (12) below), corresponding distributions of molecular weight have been inferred. This procedure has been followed for example for pig-gastric mucus glycoprotein[45,46].

Conformational Analysis

The Wales-van Holde Ratio A useful guide to the gross conformation is the Wales-van Holde ratio[47] of the sedimentation concentration dependence regression coefficient, k_s, to the intrinsic viscosity, $[\eta]$. It is known empirically that this ratio has values of ~1.6 for random coils and compact spheres and considerably lower for more extended conformations (see Lavrenko *et al.*[31] for a discussion on this). This ratio has been used as a guide to the conformations of a whole range of saccharides and Table 3 [19,21,30,40,50-53] gives just a few representative examples.

Table 3. Values for $k_s/[\eta]$ and the MHKS b coefficient for polysaccharides.

	$k_s/[\eta]$	b	Conformation	Ref.
Dextran fractions		0.44	Random coil	50
DIT-dextrans		0.56	Semi-flexible coil	51
Pullulans	~1.4	0.45	Random coil	19
β-glucans	~0.4		Extended	21
Amylopectin[a]		1.45	Spheroidal	40
Alginates	~0.6		Extended	52
Pectins	~0.2	0.17	Rigid rod	30
Mucus glycoproteins	~1.5	~0.4	Random coil	53

All in aqueous solvents apart from "a" which is 90% DMSO, 10% H_2O.

Mark-Houwink-Kuhn-Sakurada (MHKS) Representations We have indicated above that the (sedimentation) MHKS coefficients, K" and b in the equation

$$s^0_{20,w} = K''.M^b \qquad (12)$$

can be used to obtain estimates for the molecular weight directly from sedimentation coefficients. Like $k_s/[\eta]$, the MHKS exponent b (N.B. the exponent in eq. (12) is sometimes given as 1-b - see ref. 31) is an indicator of gross conformation, having values of ~0.67, ~0.4-0.5 and ~0.15 for spheres, random-coils and rigid rods respectively (see, *e.g.*, ref. 7). MHKS b coefficients - obtained from double-log plots of $s^0_{20,w}$ *versus* M^0_w (see, below) for a range of polysaccharides are also included in in Table 3. For polysaccharides of limited flexibility (this might apply, for example to short, polyanionic species), hydrodynamic methodology now permits the modelling of macromolecular conformation in terms of bead (*i.e.* multiple-sphere) models[8,54], general triaxial ellipsoids[9] and rigid-rod models[10]. The latter has recently been used for example to estimate the mass per unit length of citrus pectins[30].

Worm-like Coil Models This type of modelling possibly provides the best representation of polysaccharide conformation in solution, and covers the range from rigid rod to random coil. Garcia Molina *et al.*[10] have recently extended the earlier theory of Yamakawa and Fujii[55,11].

5. ANALYTICAL ISOPYNIC DENSITY GRADIENT ULTRACENTRIFUGATION

W. Mächtle earlier in this volume (Chapter 10) has already stressed the power of analytical isopycnic density gradient methods for characterising the chemical heterogeneity of synthetic polymer systems. This methodology, following the classical paper of Meselson *et al.*[56] has also been of considerable value for similar studies on systems of biological macromolecules, and indeed has been widely used for assaying the purity and composition of many systems of glycoconjugates, particularly glycoproteins (see, *e.g.*, ref. 5 and 57-60). It is however fair to say that its potential has thus far largely escaped the polysaccharide community.

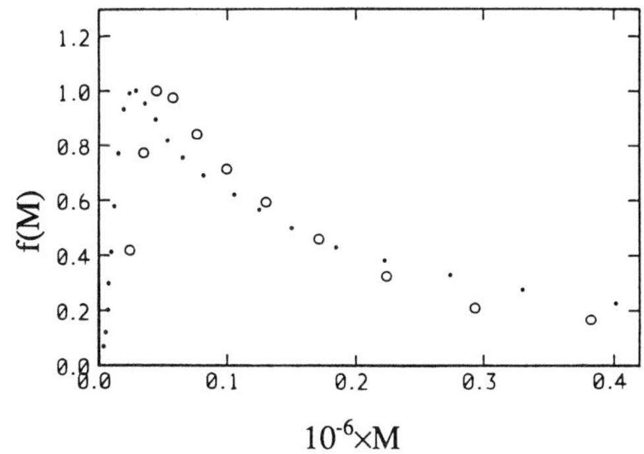

Figure 3. Distribution of molecular weight for Koch-Light citrus pectin determined by gel permeation chromatography with columns calibrated by absolute molecular weight measurements on pectin fractions by low speed sedimentation equilibrium (open circles) and total intensity or "classical" light scattering (filled circles). Reproduced, with permission, from ref. 30.

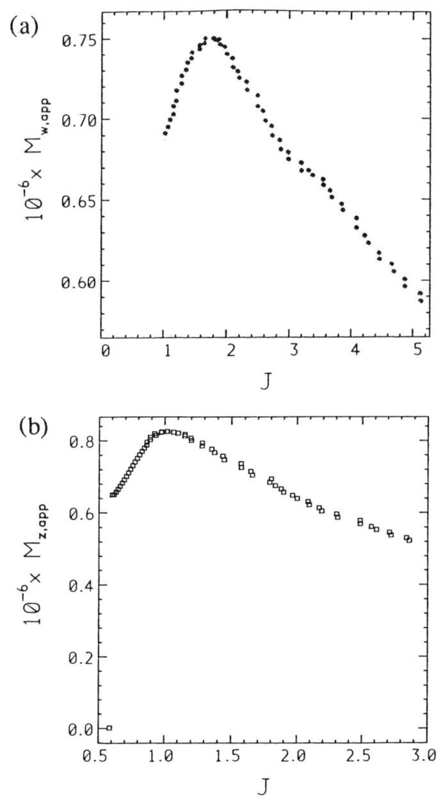

Figure 4. Plots of (apparent) point M_W, M_Z *versus* J for a purified guar preparation[60]. Rotor speed = 5200 rev/min; loading concentration, $c^0 \approx 0.7$ mg/ml. The maxima in these plots are strongly symptomatic of a polydisperse but highly non-ideal system.

Very briefly, certain salts such as CsCl, CsBr, Cs_2SO_4 and LiBr at sedimentation equilibrium distribute so that there is a significant distribution of solution density across the ultracentrifuge cell, a gradient which amongst other things will be a function of the salt, the rotor speed and the solution column length. A suspended macromolecule will, at equilibrium "band" at its isodensity point, which will either be at the meniscus, the base, or, if the density gradient has been appropriately chosen, somewhere between these two extremes. For example, the buoyant densities of proteins, polysaccharides and mucus glycoproteins are ~1.3 g/ml, ~1.6 g/ml and ~1.45 g/ml respectively.

There are three basic equations describing a density gradient (see, *e.g.* ref. 5). The first is:

$$\frac{d\rho}{dr} = \frac{\omega^2 r}{\beta} \tag{13}$$

where $d\rho/dr$ is the density gradient and β is known as the "salt parameter" (not to be confused with the Scheraga-Mandelkern shape function).

The second gives the "iso-concentration" point, r_e, where the local density equals the initial density of the salt solution, ρ_e, for sector shaped cells (14a) and for cylinders (14b):

$$r_e = \sqrt{\frac{b^2 + a^2}{2}} \tag{14a}$$

$$r_e = \sqrt{\frac{b^2 + a^2 + ab}{3}} \tag{14b}$$

The third equation gives the density at a given radial position in the cell:

$$\rho = \rho_e + E\,(r\text{-}r_e) + E\left[\frac{1}{r_e} - EF\right]\frac{(r\text{-}r_e)^2}{2} \tag{15a}$$

where both

$$E \equiv \frac{\omega^2 r_e}{\beta} \tag{15b}$$

and

$$F \equiv \left[\frac{d(\ln \beta)}{d\rho}\right]_{\rho=\rho_e} \tag{15c}$$

are independent of r. Values of β and F for the four salts commonly used in density gradient experiments are given in Table 4.

6. APPLICATIONS

We now survey some of the applications of sedimentation analysis to specific polysaccharide systems. This is by no means intended to be exhaustive, but rather to illustrate the range of polysaccharide types analysed and the diversity of molecular weights, molecular weight distributions and conformations predicted.

Dextrans

M_w^0 values for a whole series of commercial dextrans (a class of uncharged polysaccharides) have been estimated by sedimentation equilibrium and related techniques and classified in terms of "T" values (T500, T200, T50, *etc.*), the numerals representing the approximate molecular weight in kg/mol. These have subsequently been used as standard materials in, for example, gel permeation chromatography (GPC). One of the most widely studied has been T-500 dextran: good agreement has been obtained between sedimentation equilibrium and light scattering for 'whole distribution' weight average molecular weight determinations[22,7] and between GPC/sedimentation equilibrium and GPC/light scattering procedures for the determination of distributions of molecular weight[61,62].

Double logarithmic representations of sedimentation coefficient *versus* molecular weight give an MHKS b coefficient of ~0.44 suggestive of a random coil conformation[50,51].

Table 4. Isopycnic density gradient equilibrium: values of the salt parameter, β and F as a function of density for four commonly used salts.

Salt	ρ (g/ml)	$10^{-9}\times\beta$ $cm^5g^{-1}s^{-2}$	F ml/g
CsCl	1.28*	1.582	-1.61
	1.30	1.533	-1.56
	1.40	1.334	-1.19
	1.41	1.319	-1.14
	1.45†	1.265	-0.93
	1.50	1.216	-0.64
CsBr	1.30	1.023	-1.97
	1.31*	1.003	-1.95
	1.40	0.852	-1.64
	1.50†	0.744	-1.04
Cs_2SO_4	1.24*	0.890	-2.57
	1.30	0.771	-2.15
	1.33†	0.726	-1.85
	1.50	0.650	+0.69
LiBr	1.30*	8.77	+0.39
	1.40	9.31	+0.79
	1.50†	10.25	+1.11

Adapted from ref. 5 and references cited therein. Values correct at a temperature of 25.0°C

* Value regarded as characteristic of proteins

† Value regarded as characteristic of mucus glycoproteins.

The molecular weights of dextran *derivatives* have also been studied using sedimentation equilibrium, such as the cationic diethylaminoethyl (DEAE-) dextran[63] and di-iodotyrosine (DIT-) dextrans[51]. The value of ~0.56 for the MHKS b coefficient for the latter is suggestive of a somewhat less flexible configuration for DIT-dextran compared to the native species[51]. A final interesting example for dextrans: advantage has been taken of the chromophore on "blue-dextran" to evaluate both sedimentation velocity[64,65] and sedimentation equilibrium behaviour[65] using the absorption optical system on the ultracentrifuge.

Pullulans

Like dextrans, these are also now widely used as standard polysaccharides for GPC and other applications, and - like dextrans - have been classified in terms of their kg/mol molecular weights (P800, P200 *etc.*). Kawahara and co-workers[19] have determined values for M_w^0, $s^0_{20,w}$ and $k_s/[\eta]$ for a whole series of fractions. As with dextrans, both the $k_s/[\eta]$ and the MHKS b coefficient data[19] are consistent with a randomly coiled configuration in solution.

Galactomannans

This is another class of uncharged polysaccharide, but unlike dextrans and pullulans, is much more difficult to characterise because of (i) greater difficulty in solubilisation in aqueous solvents and (ii) more extreme non-ideality behaviour[66]. Fig. 4 illustrates one example of this behaviour: the maximum in a plot of point average $M_{w,app}$ (Fig. 4a) or $M_{z,app}$ (Fig. 4b) *versus* (fringe) concentration, J, even at the relatively low initial loading concentration used (~0.5 mg/ml) is strongly symptomatic of a polydisperse but very non-ideal system (similar behaviour has been observed, *e.g.* for pig gastric mucus glycoprotein)[67]. Molecular weight values have been obtained for a series of locust bean gum galactomannans using both sedimentation equilibrium[68] and sedimentation/diffusion[27] and are in agreement with each other. Earlier light scattering procedures on these substances have given much higher molecular weights, a feature however probably ascribable to problems of clarification from supramolecular aggregates[69]. These difficulties appear to have now been circumvented with the advent of on-line GPC/light scattering procedures. Satisfactory agreement has for example now been achieved between sedimentation equilibrium and results from GPC/low angle laser light scattering (GPC/LALLS)[70] and the more recent on-line GPC/multi-angle laser light scattering (GPC/MALLS) for guar[66]. Sedimentation equilibrium has been used with viscometry to monitor the thermal degradation of commercial galactomannan samples[52,71].

β-glucan

β-glucans from barley[21,38,39] and wheat[72,7] have been studied. Woodward *et al.* have obtained estimates for M_w^0 for two types of barley β-glucan ("clipper" and "commercial") from low speed sedimentation equilibrium; the values obtained (~290000 and ~160000 respectively) are comparable with values obtained earlier using sedimentation/diffusion on other preparations[28,29]. Results are also comparable for a series of wheat β-glucans also analysed by low speed sedimentation equilibrium[72,7]. Woodward *et al.*[21] also obtain relatively low values for $k_s/[\eta]$ (~0.3 for clipper and ~0.5 for commercial), which are suggestive of an extended rather than coiled or spherical conformation.

Amylopectin

Unmodified amylopectins from starch are very difficult to obtain good solution data for because of their poor solubility in aqueous solvents. Amylopectins from wheat starch have been successfully solubilised in a mixture of 90% DMSO with 10% H_2O [73] and Fronimos[40] and others have demonstrated the extreme concentration dependence of $s_{20,w}$ in this solvent. Nonetheless, reasonably consistent estimates for $s^0_{20,w}$ have been obtained: (206±13) S [40], (230±16) S [74] and (280±20) S [75].

Fronimos[40] has obtained an estimate for M_w^0 of $(85.0\pm5.0)\times10^6$ using the sedimentation/diffusion (*viz* "Svedberg equation") method. Conformationally it appears from $k_s/[\eta]$ (~1.45) to occupy a spheroidal rather than extended domain in this solvent[40].

Alginate

Alginates (from seaweed and some bacteria) are polyanionic co-polymers of mannuronic and guluronic acid residues. They are characterised by very high thermodynamic non-idealities (Table 1), and even at low loading concentrations (<0.5 mg/ml) the apparent M_w can be considerably lower than the infinite dilution value, even at relatively high ionic strength. Horton *et al.*[76] have observed that at concentrations >0.5 mg/ml, a third virial coefficient is necessary to account for the concentration behaviour. More precise estimates for the (ideal) molecular weight can be obtained by fitting molecular weights (rather than the reciprocals) directly to the concentration data (Fig. 5). Either way, results for *L. hyperborea* alginates from low speed sedimentation equilibrium are in general in good agreement with those from light scattering[77,78]. Wedlock *et al.*[27,41] have also obtained good agreement between the sedimentation/diffusion method and classical light scattering for a range of other alginates.

Conformationally, alginates appear from estimates for $k_s/[\eta]$ (<0.6) to adopt an extended rather than spheroidal conformation in solution[52].

Pectins

These are another class of polyanionic polysaccharide and a range of pectins from different sources (usually from the cell walls of fruit) have now been relatively well characterised. The application of sedimentation procedures is particularly valuable here because of problems with light scattering through the persistent contamination of pectin preparations with microgels and other supramolecular particles that even the

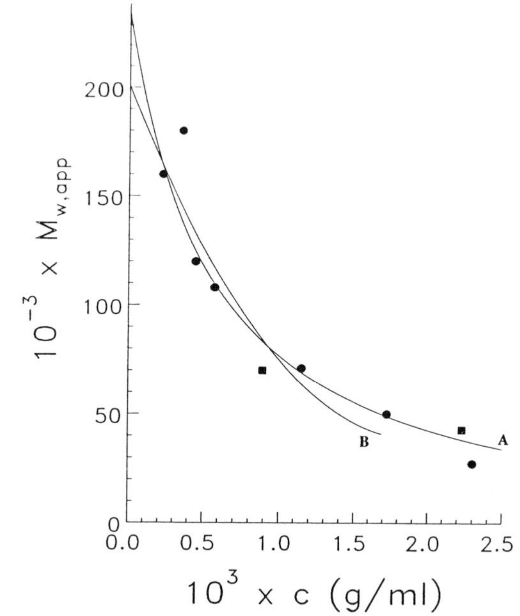

Figure 5. Plot of apparent molecular weight M_w *versus* concentration c for *L. hyperborea* alginate, I=0.3, pH=6.8. Curve A is a weighted least squares fit to $1/M_w$ *versus* c and curve B to M_w *versus* c. The different symbols (● and ■) correspond to different preparations. From ref. 76.

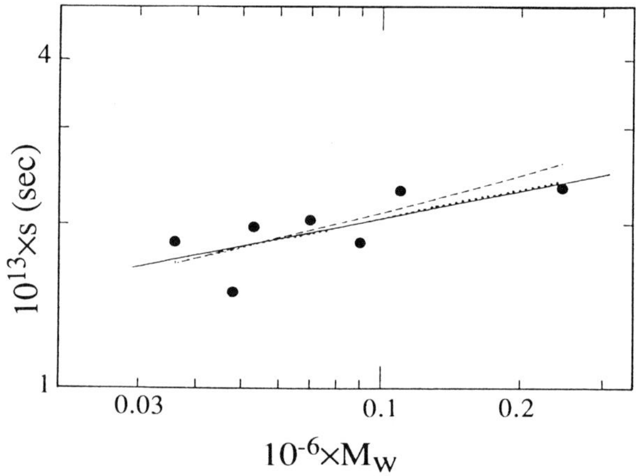

Figure 6. Double-logarithmic plot of $s^0_{20,w}$ *versus* (whole-cell) weight average molecular weight values (from low speed sedimentation equilibrium) for citrus pectin fractions. Solid line corresponds to a fit of the data to the MHKS equation (eq. (12)). Dotted line (almost coincidental with the solid line) corresponds to a rigid rod model with mass per unit length 430 $g.mol^{-1}$ nm^{-1}. Dashed line corresponds to the fit to equations representing a worm-like coil with persistence length 60 nm and mass per unit length 420 $g.mol^{-1}$ nm^{-1}. (Reproduced with permission from ref. 30.)

most rigorous clarification procedures seem difficult to remove[79]. Low speed sedimentation equilibrium has been used to monitor the changes in the molecular weight of pectins from tomato cell walls during the ripening process (a change from large, relatively non-ideal to smaller, less non-ideal but more polydisperse systems has been observed[80]), changes which can be arrested to a significant degree on transforming the genes responsible for the production of galacturonase[81]. Citrus pectins have also been studied. The molecular weight distribution from the combined GPC/sedimentation equilibrium method (see §3 above) is in excellent agreement with that obtained from GPC/light scattering, after the latter had been corrected for the presence of microgels[25]. MHKS double-log $s^0_{20,w}$ *versus* molecular weight (Fig. 6) and intrinsic viscosity *versus* molecular weight plots for a range of citrus pectin fractions are both consistent with a rod shape conformation (b=0.17±0.10) and also consistent with estimates for $k_s/[\eta]$ (~0.2). A more detailed analysis has been given in terms of both rigid rod and wormlike coil models[30]. With the latter, the ratio of the contour length, L, to the persistence length, a, is consistent only with a rod shape conformation (L~a).

"Mucopolysaccharides": Mucus Glycoproteins

These are heavily glycosylated proteins (~80% CHO), of whose physical properties are more akin to polysaccharides than proteins[76]. Mucus glycoproteins from a wide variety of sources and molecular weights (~$0.5\text{-}16\times10^6$) are thought to consist of the same hierarchical structure with basic units of M~0.5×10^6 covalently linked into a linear array to form, "subunits" of M~2.5×10^6, and in turn linked *via* disulphide bridges into further linear arrays[46]. The role of the physical properties of these substances in health and disease is undisputed[83] but so far as their characterisation is concerned they pose some additional problems compared to "pure" polysaccharides in the sense that they are polydisperse in a "discrete sense" (through variability in the numbers of "basic units") as well as a "quasi-continuous" sense (through variability in the CHO side chain lengths[46]). Furthermore, because they are highly polydisperse not only with respect to *molecular weight* but also *density* (through variable CHO/peptide content), analytical density gradient procedures have been particularly fruitful and Creeth and co-workers have pioneered much of the methodology[5,57-60]. Fig. 7 shows examples of density gradient separations/simulations recorded using either absorption optics (on an MSE Centriscan) (Figs. 7a and 7b) or using conventional Schlieren optics on a Beckman Model E (Fig. 7c).

Distributions of molecular weights for pig gastric mucus glycoprotein have been inferred from an interesting approximation involving an axiom of indistinguishability between the effects of polydispersity or isodesmic self association phenomena in *a single* sedimentation equilibrium experiment[85]. The form of these distributions are in good agreement with that expected from sedimentation velocity[45,46] - the latter assuming a random coiled configuration[46] - and electron microscopy[86]. The "discrete" polydispersity in terms of variability of basic units has been studied using a direct iterative procedure for representing non-ideal polydisperse distributions at sedimentation equilibrium and satisfactory agreement with observations from electron microscopy has been obtained[28].

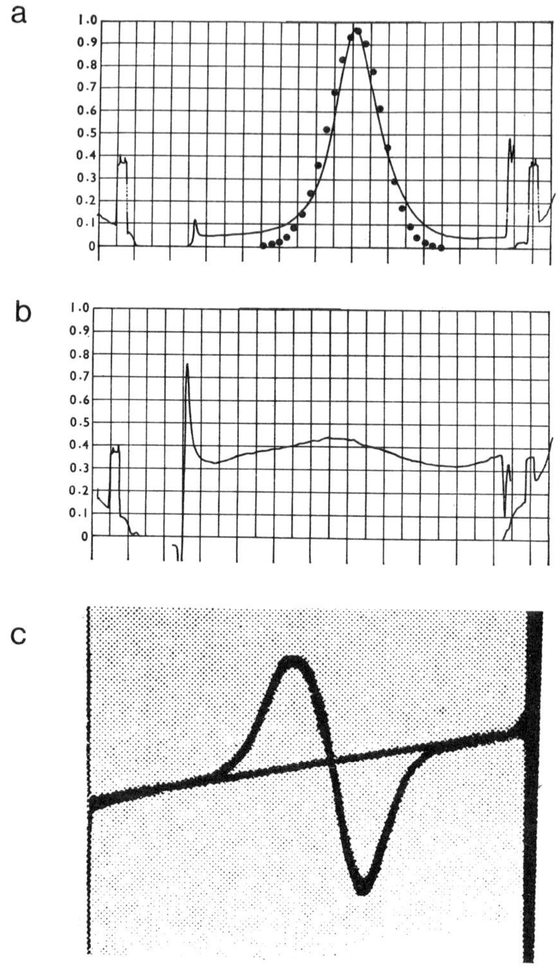

Figure 7. Isopycnic density gradient equilibrium of an ovarian cyst glycoprotein "603" (adapted from refs. 5 and 58).
(a) Fraction 50-60 (M_w=810,000 g/mol). Distribution recorded using the absorption optical system of the MSE Centriscan. Salt: Cs_2SO_4 (ρ_e=1.312 g/ml). Rotor speed=40,000 rev/min at 25.0°C. Glycoprotein c^0=0.25 (A_{280} units). The direction of the centrifugal field is from left to right. Plot is of A_{280} (absorbance at 280 nm) *versus* radial distance. Buoyant density ρ_0=1.310 g/ml. The points ● correspond to an attempted simulation of the data in terms of the same buoyant density and a solvated molecular weight $M_{s,app}$=174,000 g/mol.
(b) As (a) but in CsCl (ρ_e=1.507 g/ml), c^0=0.33.
(c) Fraction 57 ($M_w \sim 1.2 \times 10^6$ g/mol). Distribution recorded using the (phase-plate) Schlieren optical system of the Beckman Model E. Salt: Cs_2SO_4. Rotor speed=39460 rev/min at 25.0°C. Glycoprotein $c^0 \sim 1.4$ mg/ml. Buoyant density, ρ_0=1.332 g/ml. The background slope of constant gradient corresponds to the salt distribution.

Conformationally, despite their linear backbone, mucus glycoproteins from a wide variety of sources (bronchial, gastro-intestinal, ovarian, cervical) appear to adopt similar randomly coiled, spheroidal conformation in solution[42]; evidence for this has come from MHKS plots of log $s^0_{20,w}$ *versus* log M and also double log MHKS plots of the translational coefficient and intrinsic viscosity *versus* molecular weight, supported by values for $k_s/[\eta]$ of ~1.6 [53,86,87].

ACKNOWLEDGEMENT

The initial help and encouragement of my mentor Dr. J.M. Creeth, who showed me ways of avoiding the minefield associated with these substances, is gratefully appreciated.

REFERENCES

1. C. Tanford, 'Physical Chemistry of Macromolecules', Chapter 3, Wiley and Sons, New York, 1961.
2. D.C. Teller, *Methods Enzymol.*, 1973, *27*, 346.
3. H. Fujita, 'Foundations of Ultracentrifuge Analysis', Chapter 5, Wiley and Sons, New York, 1975.
4. A. Ball, S.E. Harding and J.R. Mitchell, *Int. J. Biol. Macromol.*, 1988, *10*, 259.
5. J.M. Creeth and J.R. Horton, *Biochem. J.*, 1977, *161*, 449.
6. O. Smidsrød and I.L. Andresen, 'Biopolymerkjemi', 1979, Tapir, Trondheim.
7. S.E. Harding, K.M. Vårum, B.T. Stokke and O. Smidsrød, *Advances in Carbohydrate Analysis* (C. White ed.), 1991, *1*, 63.
8. J. García de la Torre, in 'Dynamic Properties of Biomolecular Assemblies' (S.E. Harding and A.J. Rowe eds.), Chapter 1, Royal Society of Chemistry, Cambridge, 1989.
9. S.E. Harding, in 'Dynamic Properties of Biomolecular Assemblies' (S.E. Harding and A.J. Rowe eds.), Chapter 2, Royal Society of Chemistry, Cambridge, 1989.
10. M.C. Garcia Molina, M.C. Lopez Martinez and J. García de la Torre, *Biopolymers*, 1990, *29*, 883.
11. H. Yamakawa and M. Fujii, *Macromolecules*, 1974, *7*, 128.
12. T. Laue, Chapter 6, this volume.
13. A.J. Rowe, S. Wynne-Jones, D.G. Thomas and S.E. Harding, Chapter 5, this volume.
14. D.A. Yphantis, *Biochemistry*, 1964, *3*, 297.
15. See *e.g.* H. Fujita, 'Foundations of Ultracentrifuge Analysis', pp. 290-293, Wiley and Sons, New York, 1975, for a discussion on this method.
16. J.M. Creeth and R.H. Pain, *Prog. Biophys. Mol. Biol.*, 1967, *17*, 217.
17. J.M. Creeth and S.E. Harding, *J. Biochem. Biophys. Meth.*, 1982, *7*, 25.
18. D.C. Teller, J.A. Horbett, E.G. Richards and H.K. Schachman, *Ann. N.Y. Acad. Sci.*, 1969, *164*, 66.
19. K. Kawahara, K. Ohta, H. Miyamoto and S. Nakamura, *Carbohyd. Polym.*, 1984, *4*, 335.

20. T. Sato, T. Norisuye and H. Fujita, *Macromolecules*, 1984, *17*, 2696.
21. J.R. Woodward, D.R. Phillips and G.B. Fincher, *Carbohyd. Polym.*, 1983, *3*, 143.
22. E. Edmond, S. Farquhar, J.R. Dunstone and A.G. Ogston, *Biochem. J.*, 1968, *108*, 755.
23. R.A.A. Muzzarelli, C. Lough and M. Emanuelli, *Carbohyd. Res.*, 1987, *164*, 433.
24. S.E. Harding, A.J. Rowe and J.M. Creeth, *Biochem. J.*, 1983, *209*, 893.
25. G. Berth, H. Dautzenberg, D. Lexow and G. Rother, *Carbohyd. Polym.*, 1990, *12*, 39.
26. D. Lecacheux, Y. Mustiere, R. Panaras and G. Brigand, *Carbohyd. Polym.*, 1986, *6*, 477. See also T. Yanaki, T. Kojima and T. Norisuye, *Polym. J.*, 1981, *13*, 1135.
27. D.J. Wedlock, B.A. Baruddin and G.O. Phillips, *Int. J. Biol. Macromol.*, 1986, *8*, 57.
28. S.E. Harding, *Biophys. J.*, 1985, *47*, 247.
29. D.A. Yphantis, *Ann. N.Y. Acad. Sci.*, 1960, *88*, 586.
30. S.E. Harding, G. Berth, A. Ball, J.R. Mitchell and J. García de la Torre, *Carbohyd. Polym.*, 1991, *15*, 1.
31. P.N. Lavrenko, K.J. Linow and E. Görnitz, Chapter 28, this volume.
32. H.K. Schachman, 'Ultracentrifugation in Biochemistry', Academic Press, N.Y., 1959.
33. T. Svedberg and K.O. Pedersen, 'The Ultracentrifuge', Oxford University Press, 1940.
34. C. Tanford, 'Physical Chemistry of Macromolecules', p. 380, Wiley and Sons, New York, 1961.
35. W.D. Comper and B.N. Preston, Chapter 23, this volume.
36. P.N. Pusey, in 'Photon Correlation and Light Beating Spectroscopy' (H.Z. Cummings and E.R. Pike, eds.), p. 387, Plenum Press, New York.
37. W.R. Sharman, E.L. Richards and G.N. Malcolm, *Biopolymers*, 1978, *17*, 2817.
38. R. Djurtoft and K.L. Rasmussen, *Eur. Brew. Conv. Congr.*, p. 17, 1955.
39. O. Igarishi and Y. Sakurai, *Agr. Biol. Chem.*, 1965, *29*, 678.
40. P. Fronimos, M.Phil. Thesis, University of Leicester, 1991.
41. D.J. Wedlock, B.A. Fasihuddin and G.O. Phillips, *Food Hydrocolloids*, 1987, *1*, 207.
42. J.K. Sheehan and I. Carlstedt, in 'Dynamic Properties of Biomolecular Assemblies' (S.E. Harding and A.J. Rowe eds.), Chapter 17, Royal Society of Chemistry, Cambridge, 1989.
43. H. Fujita, 'Foundations of Ultracentrifuge Analysis', Chapter 3, Wiley and Sons, New York, 1975.
44. W.F. Stafford, Chapter 20, this volume.
45. R.H. Pain, *Symp. Soc. Exp. Biol.*, 1980, *34*, 359.
46. S.E. Harding, *Adv. Carbohyd. Chem. and Biochem.*, 1989, *47*, 345.
47. M. Wales and K.E. van Holde, *J. Polym. Sci.*, 1954, *14*, 81.
48. J.M. Creeth and C.G. Knight, *Biochim. Biophys. Acta*, 1965, *102*, 549.
49. P.Y. Cheng and H.K. Schachman, *J. Polym. Sci.*, 1955, *16*, 19.

50. F.R. Senti, N.N. Hellman, N.H. Ludwig, G.E. Babcock, R. Tobin, C.A. Glass and B.L. Lamberts, *J. Polym. Sci.*, 1955, *17*, 527.
51. N. Errington, S.E. Harding, E. Schacht and L. Illum, *Carbohyd. Polym.*, 1992, *17*, in press.
52. A. Ball, Ph.D. Thesis, University of Nottingham.
53. J.K. Sheehan and I. Carlstedt, *Biochem. J.*, 1984, *217*, 93.
54. J. García de la Torre, Chapter 18, this volume.
55. H. Yamakawa and M. Fujii, *Macromolecules*, 1973, *6*, 407.
56. M. Meselson, F.W. Stahl and J. Vinograd, *Proc. Nat. Acad. Sci.*, 1957, *43*, 581.
57. J.M. Creeth, K.R. Bhaskar, A.S.R. Donald and W.T.J. Morgan, *Biochem. J.*, 1974, *143*, 159.
58. K.R. Bhaskar and J.M. Creeth, *Biochem. J.*, 1974, *143*, 669.
59. J.M. Creeth, K.R. Bhaskar, J.R. Horton, I.R. Das, M.-T. Lopez-Vidriero and L. Reid, *Biochem. J.*, 1977, *167*, 557.
60. J.M. Creeth, *Modern Problems in Paediatrics*, 1977, *19*, 34.
61. A. Ball, S.E. Harding and N.J. Simpkin, in 'Gums and Stabilisers for the Food Industry' (G.O. Phillips, D.J. Wedlock and P.A. Williams, eds.), Vol. 5, pp. 447-450, IRL Press, Oxford, 1990.
62. Dawn F Application Note 7/8/88 No.3, Wyatt Technology Corp. (Santa Barbara USA), 1988.
63. M.T. Anderson, S.E. Harding and S.S. Davis, *Biochem. Soc. Trans.*, 1989, *17*, 1101.
64. N. Errington, S.E. Harding and A.J. Rowe, *Carbohyd. Polym.*, 1992, *17*, 151.
65. R. Acosta, B. Nyström and L.O. Sundelof, *Chemica Scripta*, 1981, *18*, 7.
66. K. Jumel, J.R. Mitchell and S.E. Harding, in preparation.
67. J.M. Creeth and B. Cooper, *Biochem. Soc. Trans.*, 1984, *12*, 618.
68. S.E. Gaisford, S.E. Harding, J.R. Mitchell and T.D. Bradley, *Carbohyd. Polym.*, 1986, *6*, 423.
69. G. Robinson, S.B. Ross-Murphy and E.R. Morris, *Carbohyd. Res.*, 1982, 107, 17.
70. B.R. Vijayendran and T. Bone, *Carbohyd. Polym.*, 1984, *4*, 299.
71. T.D. Bradley, A. Ball, S.E. Harding and J.R. Mitchell, *Carbohyd. Polym.*, 1989, *10*, 205.
72. K.M. Vårum, S.E. Harding and O. Smidsrød, unpublished observations.
73. E. Dickinson, J. Lelievre, G. Stainsby and S. Wright, in 'Gums and Stabilisers for the Food Industry' (G.O. Phillips, D.J. Wedlock and P.A. Williams eds.), Vol. 2, pp. 447-450, Pergamon Press, Oxford, 1984.
74. J. Lelievre, J.A. Lewis and K. Marsden, *Carbohyd. Res.*, 1986, *153*, 195.
75. W. Banks and C.T. Greenwood, 'Starch and its Components', Edinburgh University Press, 1975.
76. J.C. Horton, S.E. Harding, J.R. Mitchell and D.F. Morton-Holmes, *Food Hydrocolloids*, 1991, *5*, 125.
77. A. Martinsen, G. Skjåk-Bræk, O. Smidsrød, F. Zanetti and S. Paoletti, *Carbohyd. Polym.*, 1991, *15*, 171.
78. J.C. Horton, S.E. Harding and J.R. Mitchell, *Biochem. Soc. Trans.*, 1991, *19*, 510.
79. G. Berth, *Carbohyd. Polym.*, 1992, in press.

80. G.B. Seymour and S.E. Harding, *Biochem. J.*, 1987, *245*, 463.
81. C.J. Smith, C.F. Watson, P.C. Morris, C.R. Bird, G.B. Seymour, J.E. Gray, C. Arnold, G.A. Tucker, W. Schuch, S.E. Harding and D. Grierson, *Plant. Mol. Biol.*, 1990, *14*, 369.
82. J.M. Creeth, *Br. Med. Bull.*, 1978, *34*, 17.
83. F. Avery-Jones, *Br. Med. Bull.*, 1978, *34*, 1.
84. See, for example, J.M. Creeth, K.R. Bhaskar, J.R. Horton, I. Das, M.T. Lopez-Vidriero and L. Reid, *Biochem. J.*, 1977, *167*, 557.
85. D. Roark and D.A. Yphantis, *Ann. N.Y. Acad. Sci.*, 1969, *164*, 245.
86. J.K. Sheehan, K. Oates and I. Carlstedt, *Biochem. J.*, 1986, *239*, 147.
87. J.M. Creeth and C.G. Knight, *Biochem. J.*, 1967, *105*, 1135.

28

The Concentration Dependence of the Sedimentation Coefficient of some Polysaccharides in Very Dilute Solution

By P.N. Lavrenko, K.J. Linow* and E. Görnitz*

INSTITUTE OF MACROMOLECULAR COMPOUNDS OF THE ACADEMY OF SCIENCES, BOLSHOI PR. 31, ST. PETERSBURG, RUSSIA

*INSTITUTE OF POLYMER CHEMISTRY, KANTSTR. 55, TELTOW-SEEHOF, GERMANY O-1530

1. INTRODUCTION

The experimental values of sedimentation coefficients s_c of polymers at finite concentrations c are usually extrapolated to the conditions of infinite dilution (c=0) to exclude concentration effects. A detailed experimental investigation of these concentration effects is of considerable importance for at least two reasons. First, if the form of the dependence of s_c on c is known, it is possible to carry out the extrapolation to c=0 and to calculate the "infinite dilution" sedimentation coefficient $s= \lim_{(c\to 0)} s_c$ more reliably. Secondly, it is possible *in principle* to obtain supplementary information about the properties of macromolecules and their interactions in solution analogous to information provided by viscometric data.

2. THEORY

With increasing polymer concentration in solution, the value of s_c usually *decreases*[1]. Since the sedimentation coefficient is related to the translational frictional coefficient f_c of the macromolecules in solution by the equation

$$s_c = \frac{M\,(1-\bar{v}\rho_0)}{N_A f_c} \qquad (1)$$

it follows that the dependence of s_c on c reflects the concentration dependence of the *frictional coefficient* and hence we can write

$$f_c \approx f\,(1 + k_s c) \qquad (2)$$

correct to first order in concentration. Here $f= \lim_{(c\to 0)} f_c$, M is the molecular weight of the polymer, $\bar{v}$ is its partial specific volume, ρ_0 is the solvent density, N_A is Avogadro's number and k_s is the concentration dependence parameter.

According to Burgers[2,3], the decrease in the sedimentation velocity of the molecules in solution with increasing concentration may be interpreted by an increase in the density and effective viscosity of the solution and the existence of the counter flow of the solvent. If we denote by ϕ the volume fraction of the polymer in solution, then the effect of the two last factors is described by the equation

$$\frac{s}{s_c} = 1 + k_\phi \phi \tag{3}$$

where k_ϕ is the numerical coefficient which for rigid spherical particles is equal to $k_\phi = {}^{55}/_{8}$.

For solvated particles of an arbitrary shape, ϕ can be calculated from the equations

$$\phi = \frac{cV}{m} = c\,\frac{N_A}{M}\,\frac{4\pi a^3}{3} \tag{4}$$

$$f = 6\pi\eta_0 a \tag{5}$$

where m and V are the mass and the volume of the particle, a is the radius of the hydrodynamically equivalent sphere and η_0 is the solvent viscosity. Eqs (4) and (5) give

$$\phi = c\,\frac{N_A}{M}\,\frac{4\pi}{3}\left\{\frac{f}{6\pi\eta_0}\right\}^3 \tag{6}$$

Wales and Van Holde[4] have proposed to express f through the independently measured intrinsic viscosity $[\eta]$. They have used for this purpose the Flory equations

$$f = P\,\eta_0 \langle h^2\rangle^{1/2} \tag{7}$$

$$[\eta] = \frac{\Phi \langle h^2\rangle^{3/2}}{M} \tag{8}$$

where $\langle h^2\rangle^{1/2}$ is the mean square end-to-end distance of the dissolved macromolecules, $[\eta]$ is the intrinsic viscosity and P and Φ are constants taken to be equal to $P = 5.2$ and $\Phi = 2.1\times10^{23}$ mol^{-1}. It follows from eqs. (7) and (8) that

$$f = P\,\eta_0\left(\frac{M[\eta]}{\Phi}\right)^{1/3} \tag{9}$$

Substituting f into eq. (6) and taking $k_\phi = {}^{55}/_{8}$ these authors instead of eq. (2) have obtained the equation

$$\frac{1}{s_c} = \frac{1}{s}\{1 + 1.66\,[\eta]\,c\} \tag{10}$$

In spite of the rough approximation of these assumptions, this result has surprisingly coincided with a large number of experimental data obtained not only for globular particles[5] but also for vinyl polymers in thermodynamically good solvents[4] (when the particles shape is far from spherical). This coincidence is still more surprising if we take into account that when this approach has been used with the application of the values of P=5.11 and $\Phi = 2.87\times10^{23}$ mol^{-1} [6,7] or P = 5.99 and $\Phi = 2.51\times10^{23}$ mol^{-1} [8] which are not less reliable, it yields quite different values of $k_s/[\eta]$ (= 1.24 or 2.2). It shows that the result of eq. (10) is *fortuitous*. However, this approach to $s_c(c)$ has been developed further[9-11].

In the application to *polysaccharides*, an important requirement is the evaluation of the draining effects of macromolecules, the property characteristic of most polymers of this class[12]. It has been established experimentally that the parameter $k_s/[\eta]$ decreases with decreasing degree of coiling of the macromolecule and with increasing ratio of the length of the Kuhn segment A to the chain diameter d [13,14].

The theoretical evaluation of draining carried out by Freed[15] gives the following expression relating the ratio $k_s/[\eta]$ to the parameter of hydrodynamic interaction in the Gaussian chain X:

$$\frac{k_s}{[\eta]} = \frac{8X}{3+8X} \qquad (11)$$

Eq. (11) shows that $k_s/[\eta]$ may take the values from 0 (in the case of the absence of hydrodynamic interaction, $X\rightarrow0$) to 1 (non-draining macromolecules, $X\rightarrow\infty$). So eq. (11) allows us to calculate the value of X from experimental value of $k_s/[\eta]$.

The conclusion that $k_s/[\eta] < 1$ also follows from the theory[16] proceeding from the concept that the sedimentation coefficient is proportional to the permeability of the moving particles to solvent molecules.

Rowe's theory[17] of the dependence of s_c on c is based on other concepts. Considering this theory to be grounded on the most strict physical fundamentals we repeat its main positions in more detail.

The Stokes Law for the motion of spheres may be written in the form applicable to the particles of any shape. A particle moving at a velocity U in a medium with the viscosity η experiences the resistance

$$F = 6\pi\eta a\left[\frac{f}{f_0}\right]U \qquad (12)$$

where $a(f/f_0)$ is the radius of an equivalent sphere, f/f_0 is the frictional ratio (a dimensionless ratio characterizing the shape and/or solvation) and a is the radius of an equivalent compact spherical particle given by

$$a = \left\{\frac{3M}{4\pi N_A}\bar{v}\right\}^{1/3} \qquad (13)$$

The elimination of a from eq. (12) gives

$$F = 6\pi\eta\left\{\frac{3M}{4\pi N_A}\right\}^{1/3}\left\{\bar{v}\left(\frac{f}{f_0}\right)^3\right\}^{1/3}U \qquad (14)$$

where $\bar{v}(f/f_0)^3$ is the effective partial specific volume. The apparent volume flow of the moving component is obtained by multiplying the velocity by volume concentration

$$J_f' = c\bar{v}\left(\frac{f}{f_0}\right)^3 U \tag{15}$$

A part of the solvation solvent entrained by the macromolecules migrates with them so that the total flow moving at a velocity U with respect to the cell walls is given by

$$J_f = J_f' + cv_sU = c\bar{v}\left\{\frac{v_s}{\bar{v}} + \left(\frac{f}{f_0}\right)^3\right\} U \tag{16}$$

where v_s is the specific volume (per mass unit of the polymer) occupied by the moving component (including accompanying solvent *and* entrained solvent).

Since the cell is closed, the sedimenting flow induces the counterflow of the solvent equal to it in volume. Hence, in the co-ordinate system related to the counterflow, the migrating flow is twice larger, *i.e.*, the following equation is valid

$$J_f = 2c\bar{v}\left\{\frac{v_s}{\bar{v}} + \left(\frac{f}{f_0}\right)^3\right\} U \tag{17}$$

On the other hand, the centrifugal field $\omega^2 r$, where ω is the angular velocity and r is the radial displacement, ensures the sedimentation velocity of the macromolecules U_0 with respect to the solvent moving in the opposite direction. In an immobile co-ordinate system, this velocity is lower

$$U = U_0(1\text{-}k_sc) = \frac{U_0}{1+k_sc}$$

Here k_s is the effective specific volume of the sedimenting component. In a *mobile* co-ordinate system, the effective centrifugal field is lower than $\omega^2 r$ by a factor of U/U_0. Hence, the sedimentation flow is given by

$$J_s = \text{volume} \times \text{velocity} = k_sc\ s\ \omega^2 r\ (1\text{-}k_s\ c) \tag{18}$$

Using the definition $s_c = s(1\text{-}k_sc)$, the expression $s_c = U/\omega^2 r$ and equating J_f and J_s, we obtain the main equation of Rowe's theory

$$\frac{k_s}{\bar{v}} = 2\left\{\frac{v_s}{\bar{v}} + \left(\frac{f}{f_0}\right)^3\right\} \tag{19}$$

This equation relates k_s, the "entrainment" $v_s/\bar{v}$ and the frictional ratio f/f_0. For non-hydrated ($v_s/\bar{v} = 1$) spheres ($f/f_0 = 1$) which are not accompanied by the solvent we have $k_s/\bar{v} = 4$. This value is in good agreement with the experimental value of 4.06 obtained for a latex of polystyrene spheres in water[18] and with Burgers' result (*without* the assumption that the effective viscosity of the solvent changes).

Let us enumerate the concepts assumed for the derivation of eq. (19):

(i) The migration velocity of particles with respect to the solvent is independent of c at infinitely low c.

(ii) Interactions and interactions with additional components are neglected.

(iii) The value of $\bar{v}$ may be taken to be equal to the inverse density of the polymer.
(iv) The system is incompressible.

The value of f_0 in eq. (19) may be replaced by M and v_s

$$f_0 = 6\pi\eta \left\{\frac{3Mv_s}{4\pi N_A}\right\}^{1/3} \tag{20}$$

and $v_s/\bar{v}$ may be expressed through k_s and the Huggins constant k' [17]

$$\frac{v_s}{\bar{v}} = \frac{k'[\eta]}{k_s} \tag{21}$$

The substitution of f from eq. (20) into eq. (19) gives

$$M = N_A \left\{\frac{6\pi\eta s}{1-\bar{v}\rho_0}\right\}^{3/2} \left\{\frac{3k'[\eta]\bar{v}}{4\pi k_s} \left(\frac{k_s}{2\bar{v}} - \frac{k'[\eta]}{k_s}\right)\right\}^{1/2} \tag{22}$$

Eq. (22) makes it possible to calculate the molecular weight of the polymer (which is close to M_w in its type of averaging) by using the values of k_s, s, $[\eta]$ and $\bar{v}$.

3. EXPERIMENTAL DATA AND DISCUSSION

At low concentrations ($c < 1/[\eta]$), the dependence of s_c on concentration is usually reasonably approximated by the linear dependence of $1/s_c$ on c [19-21] characterized by the parameter k_s

$$\frac{1}{s_c} \approx \frac{1}{s}(1+k_s c) \tag{23}$$

The parameter k_s is usually calculated from the slope of the dependence of $1/s_c$ on c obtained from the data of several sedimentation experiments carried out at different but relatively low concentrations. These experiments can be facilitated by using double-sector cells and multi-cell rotors[22] or by using Vinograd's method[23] which facilitates the measurement of s and k_s from a single experiment.

The s and k_s values can also be estimated in an experiment with a differential sedimentation boundary when a solution at a concentration c_1 is layered on a solution at a different concentration c_2. The velocity of motion of the boundary between these solutions (called the differential boundary) is characterized by the sedimentation coefficient s_d [24]

$$s_d = \frac{s_1 c_1 - s_2 c_2}{c_1 - c_2} \tag{24}$$

where s_1 and s_2 are the sedimentation coefficients of the molecules in the concentration ranges c_1 and c_2. For solutions with the concentration dependence of the type (23), the following equation is obeyed[25]:

$$s_d = \frac{s_1 s_2}{s} \tag{25}$$

where the coefficient s_d corresponds to the usual sedimentation of macromolecules at a concentration $c_d = c_1 + c_2 + k_s c_1 c_2$.

Many papers (also discussed in a number of reviews) have been published dealing with the investigation of the concentration dependence of the sedimentation of *flexible-chain* macromolecules in solution[4,10,26-28]. The experimental data on the dependence of s_c on c for such rigid-chain polymers as cellulose derivatives and other polysaccharides are, however, much more scarce. In Fig. 1, the sedimentation characteristics for cellulose and some cellulose derivatives in different solvents are shown in the form of the dependence of k_s on the *characteristic sedimentation coefficient* $[s] \equiv s\eta_0/(1-\bar{v}\rho_0)$.

It follows from the experimental data that for each specific polymer the value of k_s increases monotonically with s. It is convenient to relate this increase in k_s in a polymer homologous series to s by a power function

$$k_s = B(s\times10^{13})^{æ} \tag{26}$$

The values of the coefficients B and æ are listed in Table 1.

If the character of these dependencies is known, this facilitates the problem of the determination of s from several values of s_c obtained at finite concentrations and sometimes makes it possible to exclude the effect of concentration on s_c according to the data obtained at one solution concentration. In the latter case the method of *successive approximations*[73] is used. To a zeroeth approximation, the resulting coefficient s_c is taken to be equal to s. For s the value of k_s is found from the dependence $k_s(s)$ with the aid of which the value of s to a first approximation is calculated according to eq. (23). (Two or three approximations are usually sufficient.)

The same problem may be solved with the aid of the equation

$$\frac{1}{s_c} = \frac{1}{s}(1 + \text{Const}\,[\eta]c), \qquad \text{Const} \equiv \frac{k_s}{[\eta]} \tag{27}$$

if the ratio $k_s/[\eta]$ is invariant to changes in M. An important question follows: may we consider this justified with respect to cellulose derivatives? Searching for an answer, let us examine Table 1. In most cases (within a considerable degree of random error) the value of $k_s/[\eta]$ does *not* change with M. In ten cases an increase in $k_s/[\eta]$ with increasing M is observed, whereas in seven cases $k_s/[\eta]$ decreases.

The reasons for this behaviour of $k_s/[\eta]$ are because of the different ways the draining of the macromolecule changes with increasing M in each specific case. However, the experimental conclusion that the parameter $k_s/[\eta]$ is not constant cautions us against using eq. (27) with a universal value of $k_s/[\eta]$ for calculating s of a cellulose derivative from the coefficient s_c determined at just one concentration. This caution is even more justified because the change in $k_s/[\eta]$ with increasing M has not yet been satisfactorily explained theoretically. We require also a critical attitude and care in the application of methods for the calculation of molecular weight and for the establishment of the conformational characteristics for cellulose derivatives molecules

Table 1. Values of sedimentation parameters in equations $k_s = B(s\times10^{13})^{æ}$ and $s = KM^{1-b}$ for a selection of polysaccharides.

Polymer	DS[a]	Solvent[b]	T (°C)	$10^{-3}\times M$	$10^2\times\frac{\eta_0}{1-\bar{v}\rho_0}$	B	æ	$\frac{k_s}{[\eta]}$	æ(1-b)[c]	Ref.
Cellulose		Cadoxen+H_2O (1:1)	20	7.5-380	2.88	46	2.0	1.3±0.7	0.74	29
Cellulose		Cadoxen+H_2O (1:1)	20	500-680	-	-	-	1.66	-	45
Cellulose		Cadoxen+H_2O (1:1)	25	43-745	2.88	46	2.0	0.96	0.80	46
Cellulose		Cadoxen+H_2O (1:1)	25	10.1-33.7	2.797	73	1.6	1.0±0.1	0.71	47
Cellulose		FeTNa	21	6-635	-	-	-	0.9	-	48
Cellulose		FeTNa	20	111-539	-	162	1.46	0.5±0.2	0.63	82
Cellulose		Cuoxame	20	27-8200	-	9.86	1.86	0.7±0.3	0.71	19
Cellulose acetate	2.92	DMAA	25	36.7-500	3.1526	14	1.94	1.02-0.47	0.47	49
Cellulose acetate	2.46	Acetone	25	24.2-249	0.689	0.69	2.25	0.76±0.08	0.88	49
Cellulose acetate	2.3	Acetone	27	10-194	0.65	0.21	2.8	0.26-0.63	0.95	41
Cellulose acetate-hydroxypropyl		DMAA	26	56-202	4.10	10	2.9	0.56±0.08	1.28	30
Cellulose acetobenzoate	2.2[d]; 0.55[e]	1,4-dioxane	21	7.5-1500	6.19	-	-	0.87	-	50
Cellulose benzoate	2.3	1,4-dioxane	26.6	194-660	3.51	2.0	2.0	2.7±0.1	0.67	31
Cellulose carboxymethyl	0.9	Cadoxen+H_2O (1:1)	26	28-380	-	6.0	2.7	0.90±0.08	0.94	32
Cellulose carboxymethyl	0.9	Cadoxen+H_2O (1:10)	26	28-380	1.96	13	2.2	0.84±0.13	0.88	51
Cellulose cyanoethyl	2.6	Acetone	26	54-317	0.738	0.36	2.7	0.53±0.04	0.89	34
Cellulose dimethylphosphonocarbamate	2.0	0.2 M NaCl	22	200-2400	2.93	1.6	2.53	0.4-0.9	0.89	43

Cellulose diphenylphosphonocarbamate	2.3	1,4-dioxane	26.6	450-1700	3.56	1.03	2.6	0.8±0.5	1.2	31
Cellulose hydroxyethyl		Water	25	150	-	-	-	1.0	-	52
Cellulose hydroxyethyl	0.88	Water	25	125-545	2.97	46	2.04	1.0-1.5	0.94	53
Cellulose hydroxyethyl		Water			-	-	2	-	-	61
Cellulose hydroxypropyl		DMAA	26	36-415	4.48	24	3.6	0.27-0.77	1.66	54
Cellulose hydroxypropyl		Water	25	73	-	-	-	0.45	-	52
Cellulose monophenylacetate	2.6	Benzene	26	65-5410	1.83	6.6	1.6	1.25±0.27	0.64	31
Cellulose nitrate	3	Ethyl acetate	20	93-1500	0.92	0.43	3.1	0.82-0.21	0.93	55
Cellulose nitrate	2.95	Acetone	20	112-2000	0.566	0.43	2.38	0.65-0.37	0.88	81
Cellulose nitrate	2.84	Ethyl acetate	30	41-573	0.78	0.9	2.7	0.81	0.78	56
Cellulose nitrate	2.7	Ethyl acetate	21	56-983	0.894	0.88	2.8	0.56±0.06	0.95	57
Cellulose nitrate	2.6	Ethyl acetate	25	24.8-1726	0.933	1.1	2.5	0.40	0.92	58
Cellulose nitrate	2.4	Acetone	20	27.8-495	0.568	1.2	2.0	1.1±0.4	0.74	59
Cellulose nitrate	2.4	Ethyl acetate	25	48-169	0.826	0.53	2.8	0.6	0.87	60
Cellulose nitrate	2.3	Ethyl acetate	26	22-813	0.98	0.11	2.55	0.23-0.85	1.02	35
Cellulose nitrate	2.15	Ethyl acetate	26	6.4-40.5	0.98	0.11	2.55	1.09±0.18	0.92	36
Cellulose nitrate	2.0	Ethyl acetate	26	5.4-51	0.98	0.11	2.55	0.51-1.28	0.92	36
Cellulose nitrate	1.1	DMAA+6% LiCl	25	131-566	8.077	2.8	1.8	1.3	0.70	60
Cellulose triacetatemethylol		DMFA	25	18-1100	2.61	7.4	2.44	0.6	0.84	62
Cellulose tributyrate	2.94	Methyl ethyl ketone	20	74-1700	1.27	1.86	2.2	0.8-1.1	0.88	42
Cellulose phenylcarbamate	3	1,4-dioxane	20	80.8-2580	5.16	30	1.95	1.8-0.36	0.72	63
Cellulose phenylcarbamate	3	Cyclohexanone	25	430-2170	4.0	59	1.63	1.4±0.1	-	64
Cellulose phenylcarbamate	2.2, 2.6	Ethyl acetate	25	280-2500	1.13	0.66	2.3	1.16±0.47	0.74	37,38
Cellulose carbanilate	2.9	Acetone	20	320-7530	0.851	0.99	1.68	0.6±0.2	0.66	81
Cellulose xanthate		0.2 N NaOH + 1 M NaCl	20	35-126	-	5.07	2.78	-	1.31	19
Amylose carbanilate	2.9	1,4-dioxane		83.5-2730	-	7.92	2.06	0.37-0.25	0.64	63
Ethylcellulose	2.75	Acetone	20	11.45-78.8	1.08	0.68	2.9	0.56±0.11	0.90	44

Ethylcellulose	2.27	Ethyl acetate	24	19-29.5	1.70	6.2	3	0.97±0.24	1.1	39
Co-polymers of cellulose and polyacrylamide		Cadoxen+H_2O (1:1)	20	730-1910	-	8.4	2.4	2.0	1.2	45
Chitine nitrate	1.75	DMFA	21	11-153	2.12	6.1	2.4	0.69±0.26	0.68	65
Chitosan		Acetate buffer	21	14-172	2.76	52	2.5	1.3-0.6	0.88	77
Dextran		Water	20	17.7-975	2.618	8.1	1.20	1.02-2.65	0.6	66-68
Dextran		Water	25	64.2	2.49	-	-	1.86	-	69
Potato amylose		0.33 M KCl	22.5	160-2290	-	1.18	0.95	0.10±0.01	0.48	70
Polysaccharide synthetic *Ficoll*		Water	26	15-1300	2.6	4.6	0.87	2.1-3.2	0.49	40
Polysaccharide native Schizophyllum		Water	25	96-5700	0.894	0.022	4.3	0.20-0.85	0.99	71
Polysaccharide xanthan		0.1 M NaCl	25	7.4-740	2.494	0.047	4.2	1.77-0.33	1.13	72

[a] DS: degree of substitution;
[b] DMAA: N,N-dimethylacetamide; DMFA: dimethylformamide; FeTNa: iron sodium tartrate;
[c] The term (1-b) is equivalent to the term "b" used elsewhere in this volume (Chapter 27);
[d] DS_{benz};
[e] DS_{ac}.

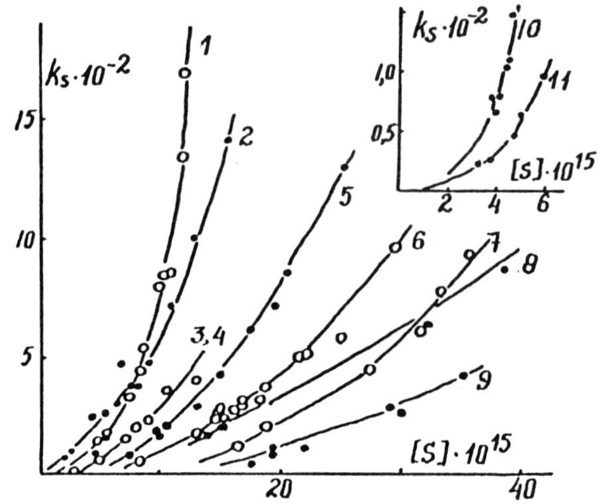

Figure 1. Dependence of k_s on [s] for cellulose and cellulose derivatives

1	cellulose nitrate in ethyl acetate[35]
2	cellulose in Cadoxen:water = 1:1 [29]
3	cellulose acetate in acetone[41]
4	cellulose benzoate in dioxane[31]
5	cellulose tributyrate in methyl ethyl ketone[42]
6	cellulose phenylcarbamate in ethyl acetate[37,38]
7	cellulose dimethylphosphonocarbamate in 0.2 M NaCl[43]
8	cellulose monophenyl acetate in benzene[31]
9	cellulose diphenylphosphonocarbamate in dioxane[31]
10,11	ethylcellulose in ethyl acetate and acetone[39,44]

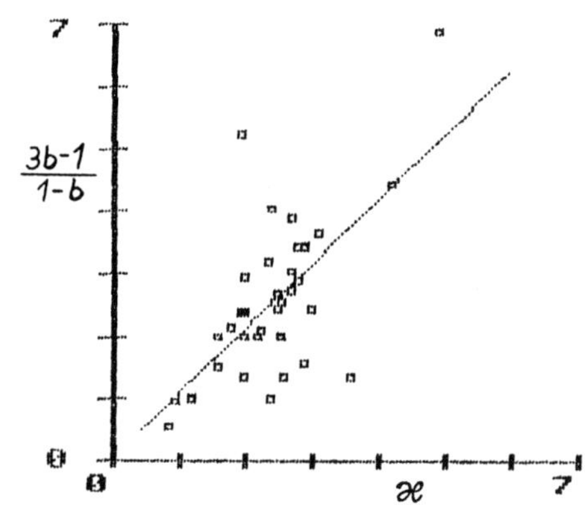

Figure 2. Comparison of the parameter æ from eq. (26) with the ratio (3b-1)/(1-b) for the values from Table 1.

based on the invariance of $k_s/[\eta]$ (insensitivity to M changes) or the assumptions equivalent to it.

Particularly from sedimentation-viscometry data the value of M can be calculated as

$$M = \left\{\frac{R}{A_0}\right\}^{3/2} \left\{\frac{[\eta]}{100}\right\}^{1/2} [s]^{3/2} \qquad (28)$$

Here R is the gas constant and A_0 is the "hydrodynamic constant" (see ref. 74). We can re-write equation (28) in the form

$$M = \left\{\frac{R}{A_0}\right\}^{3/2} \left\{\frac{[\eta]}{100k_s}\right\}^{1/2} k_s^{1/2} [s]^{3/2}$$

If A_0 and $k_s/[\eta]$ values (or the ratio $A_0^3[\eta]/k_s$) are invariant to changes in M we have the following expression applicable to the determination of M

$$M = \text{Const} \times k_s^{1/2}[s]^{3/2} \qquad (29)$$

According to Wales-Van Holde[4] the constant is

$$\text{Const} = \pi N_A \left\{\frac{16.2}{k_\phi}\right\}^{1/2} \qquad (30)$$

with k_ϕ as a numerical coefficient equal to $^{55}/_8$ (see above discussion and ref. 3). According to Pavlov[75]

$$\text{Const} = \left(\frac{N_A}{\beta_s}\right)^{3/2} \qquad (31)$$

with β_s as the sedimentation parameter introduced in refs. 76 and 77.

$$\beta_s = N_A\, k_s^{1/3}\, [s]\, M^{-2/3} \qquad (32)$$

We have to stress that the assumptions made above to obtain eq. (29) lead to the conclusion that the parameter æ in eq. (26) has to coincide with the value of ratio (3b-1)/(1-b). One should observe, however, the experimental data for polysaccharides collected in Table 1 shows that such a coincidence does not always occur. The check made in Fig. 2 gives the slope of the regression line to be (1.04±0.51) with a great degree of scatter of the points.

The choice of a specific value of $k_s/[\eta]$ is also difficult because of a considerable experimental error in its determination (Table 1). For polysaccharides this error is larger than that for other polymers. Hence, the application of the relationships of $k_s(s)$ established above to the determination of sedimentation coefficients should be considered to be more reliable than calculation with the aid of $k_s/[\eta]$ and $[\eta]$.

Finally, the molecular weight dependence of k_s may be related to the thermodynamic "strength" of the solvent characterized by the thermodynamic or "osmotic pressure" second virial coefficient A_2. The values of A_2, k_s and the parameter k_D in the concentration dependence of the diffusion coefficient $D_c = D(1+k_Dc)$ are known to be related to each other by the equation[78]

$$k_D + k_s = 2A_2M - \bar{v} \approx 2A_2M \tag{33}$$

Applying the equations $s = K M^{1-b}$ and $k_s = B s^{æ}$ we obtain

$$k_D + B K^{æ} M^{æ(1-b)} \approx 2A_2 M \tag{34}$$

It follows from Table 1 that for many cellulose derivatives the value of æ(1-b) is close to unity. (The exact statistical calculation from all values of Table 1 gives æ(1-b) = 0.86±0.23.) For the same systems it has been established experimentally[79] that at infinite dilutions the diffusion coefficient D is virtually independent of concentration, *i.e.* $k_D = 0$. For systems satisfying these two conditions, it is possible to write instead of eq. (34) the equation

$$A_2 \approx {}^1/_2\, B K^{æ} \tag{35}$$

The conclusion that A_2 is a constant drawn from eq. (35) is confirmed by experimental observations showing that the dependence of the value of A_2 on M for cellulose derivatives in thermodynamically strong solvents is very weak (or absent)[80]. Consequently, for the systems discussed here eq. (35) makes it possible to estimate the values of A_2 from the concentration parameters B and æ.

REFERENCES

1. H.K. Schachman, 'Ultracentrifugation in Biochemistry', Academic Press, Inc., N.Y., 1959, p. 272.
2. J.M. Burgers, *Proc. Ned. Akad. Wet., Amsterdam*, 1941, *44*, 1045.
3. J.M. Burgers, *Proc. Ned. Akad. Wet., Amsterdam*, 1942, *45*, 9.
4. M. Wales and K.E. Holde, *J. Polym. Sci.*, 1954, *14*, 81.
5. J.M. Creeth and C.G. Knight, *Biochim. Biophys. Acta*, 1965, *102*, 549.
6. H. Yamakawa and M. Fujii, *Macromolecules*, 1973, *6*, 407.
7. H. Yamakawa and M. Fujii, *Macromolecules*, 1974, *7*, 128.
8. B.H. Zimm, *Macromolecules*, 1980, *13*, 592.
9. K. Kakide and T. Kawai, *Chem. High Polym.*, 1962, *19*, 441.
10. H. Yamakawa, *J. Chem. Phys.*, 1962, *36*, 2995.
11. S. Imai, *J. Chem. Phys.*, 1969, *50*, 2116.
12. V.N. Tsvetkov, 'Rigid-chain polymers', Plenum, New York and London, 1989, p. 490.
13. W.S. Skazka and W.M. Yamtschikov, *Plaste u. Kautschuk*, 1973, *1*, 24.
14. P.N. Lavrenko, E.U. Urinov and A.A. Gorbunov, *Vysokomol. Soedin.*, 1977 Ser. A, *19*, 859.
15. K.F. Freed, *J. Chem. Phys.*, 1976, *65*, 4103.
16. H.J. Elmgren, *J. Polym. Sci.: Polym. Lett. Ed.*, 1981, *19*, 567.
17. A.J. Rowe, *Biopolymers*, 1977, *16*, 2595.
18. P.Y. Cheng and H.K. Schachman, *J. Polym. Sci.*, 1955, *16*, 19.
19. N. Gralen, 'Sedimentation and diffusion measurements on cellulose and cellulose derivatives', Inaugural. Diss., Almqvist a. Wiksells, Uppsala, 1944.

20. L. Mandelkern, W.R. Krigbaum, H.A. Scheraga and P.J. Flory, *J. Chem. Phys.*, 1952, *20*, 1392.
21. R.J. Goldberg, *J. Phys. Chem.*, 1953, *57*, 194.
22. T.J. Bowen, 'An introduction to ultracentrifugation', Wiley, London and New York, 1971.
23. J. Vinograd, R. Bruner, R. Kent and J. Weigle, *Proc. Natl. Acad. Sci. (U.S.A.)*, 1963, *49*, 902.
24. R. Hersh and H.K. Schachman, *J. Am. Chem. Soc.*, 1955, *77*, 5228.
25. P.N. Lavrenko and A.A. Gorbunov, *Vysokomol. Soedin.*, Ser. A, 1973, *15*, 2310.
26. R.L. Baldwin and K.E. van Holde, *Fortschr. Hochpolym.-Forsch.*, 1960, *1*, 451.
27. S. Imai, *J. Chem. Phys.*, 1970, *52*, 4212.
28. P.P. Nefedov and P.N. Lavrenko, 'Transport methods in the analytical chemistry of the polymers', Khimija, Leningrad, 1979, p. 232.
29. S.Ya. Lubina, S.I. Klenin, I.A. Strelina, A.V. Troitskaya, A.K. Khripunov and E.U. Urinov, *Vysokomol. Soedin.*, Ser. A, 1977, *19*, 244.
30. E.V. Korneeva, I.N. Shtennikova, V.P. Shibaev, I.A. Strelina, G.F. Kolbina, I.V.Ekaeva, S.A. Didenko and P.N. Lavrenko, *Vysokomol. Soedin.*, Ser. A 1991, *33* (in preparation).
31. E.V. Korneeva, P.N. Lavrenko, E.U. Urinov, A.K. Khripunov, L.I. Kutsenko and V.N. Tsvetkov, *Vysokomol. Soedin.*, Ser. A 1979, *21*, 1547.
32. P.N. Lavrenko, O.V. Okatova, T.V. Filippova, I.N. Shtennikova, V.N. Tsvetkov, H. Dautzenberg and B. Philipp, *Acta Polymerica*, 1986, *37*, 663.
33. P.N. Lavrenko, O.V. Okatova, V.N. Tsvetkov, H. Dautzenberg and B. Philipp, *Polymer*, 1990, *31*, 348.
34. V.N. Tsvetkov, P.N. Lavrenko, L.N. Andreeva, A.I. Mashoshin, O.V. Okatova, O.I. Mikriukova and L.I. Kutsenko, *Eur. Polym. J.*, 1984, *20*, 823.
35. N.V. Pogodina, P.N. Lavrenko, K.S. Pozivilko, A.B. Melnikov, T.A. Kolobova, G.N. Marchenko and V.N. Tsvetkov, *Vysokomol. Soedin.*, Ser. A, 1982, *24*, 332.
36. N.V. Pogodina, A.B. Melnikov, O.I. Mikriukova, S.A. Didenko and G.N. Marchenko, *Vysokomol. Soedin.*, Ser. A 1984, *26*, 2515.
37. L.N. Andreeva, E.U. Urinov, P.N. Lavrenko, K.-J. Linow, H. Dautzenberg and B. Philipp, *Faserforsch. u. Textiltechnik*, 1977, *28*, 117.
38. L.N. Andreeva, P.N. Lavrenko, E.U. Urinov, L.I. Kutsenko and V.N. Tsvetkov, *Vysokomol. Soedin.*, Ser. B, 1975, *17*, 326.
39. E.I. Ryumtsev, F.M. Aliev, M.G. Vitovskaya, E.U. Urinov and V.N. Tsvetkov, *Vysokomol. Soedin.*, Ser. A, 1975, *17*, 2676.
40. P.N. Lavrenko, O.I. Mikriukova and O.V. Okatova, *Anal. Biochem.*, 1987, *166*, 287.
41. S.J. Singer, *J. Chem. Phys.*, 1947, *15*, 341.
42. S.Ya. Lubina, S.I. Klenin, I.A. Strelina, A.V. Troitskaya, V.I. Kurliankina and V.N. Tsvetkov, *Vysokomol. Soedin.*, Ser. A, 1973, *15*, 691.
43. E.N. Zakharova, L.I. Kutsenko, V.N. Tsvetkov, V.S. Skazka, G.V. Tarasova and V.M. Jamtschikov, *Vestnik Leningrad Univ.*, 1970, *16*, 55.
44. G. Meyerhoff and N. Sütterlin, *Makromol. Chem.*, 1965, *87*, 258.

45. S.I. Klenin, S.Ya. Lubina, A.V. Troitskaya, I.A. Strelina, V.I. Kurliankina and V.A. Molotkov, *Vysokomol. Soedin.*, Ser. A, 1975, *17*, 197.
46. D. Henley, *Arkiv Kemi*, 1962, *18*, 327.
47. W. Brown and R. Wikström, *Eur. Polym. J.*, 1965, *1*, 1.
48. G.M. Pavlov and N.A. Shil'dyaeva, *Khimiya Drevesini* (in Russian), 1988, No.4, 10.
49. S.-I. Ishida, H. Komatsu, H. Katch, M. Saito, Y. Miyazaki and K. Kamide, *Makromol. Chem.*, 1982, *183*, 3075.
50. V.N. Tsvetkov, S.V. Bushin, E.P. Astapenko, M.A. Bezrukova, S.A. Didenko and A.K. Khripunov, *Vysokomol. Soedin.*, Ser. A, 1990, *32* (in press).
51. P.N. Lavrenko, O.V. Okatova, H. Dautzenberg and B. Philipp, *Vysokomol. Soedin.*, Ser. A, 1991, *33* (in preparation).
52. L.-O. Sundelöf and B. Nyström, *J. Polym. Sci.: Polym. Lett. Ed.*, 1977, *15*, 377.
53. W. Brown, *Arkiv Kemi*, 1961, *18*, 227.
54. E.V. Korneeva, I.N. Shtennikova, V.P. Shibaev, S.I. Klenin, G.F. Kolbina, I.V. Ekaeva and S.A. Didenko, *Vysokomol. Soedin.*, Ser. A, 1990, *32*, 43.
55. S. Newman, L. Loeb and C. Conrad, *J. Polym. Sci.*, 1953, *10*, 463.
56. M.L. Hunt, S. Newman, H.A. Scheraga and P.J. Flory, *J. Phys. Chem.*, 1956, *60*, 1278.
57. G.M. Pavlov, A.N. Kozlov, G.N. Marchenko and V.N. Tsvetkov, *Vysokomol. Soedin.*, Ser. B, 1982, *24*, 284.
58. N.V. Pogodina, S.V. Bushin, A.B. Melnikov, E.B. Lysenko, T.A. Kolobova, G.N. Marchenko, O.T. Zipina and V.N. Tsvetkov, *Vysokomol. Soedin.*, Ser. A, 1987, *29*, 299.
59. I. Jullander, *Arkiv Kemi, Min. Geol.*, 1945, *21* (3-4), 1.
60. S.V. Bushin, E.B. Lysenko, V.A. Cherkasov, S.A. Didenko, G.N. Marchenko and V.N. Tsvetkov, *Vysokomol. Soedin.*, Ser. A, 1983, *25*, 1899.
61. Polymer Handbook, ed. by J. Brandrup and E.H. Immergut, Wiley Intersci., N.Y., 1975, second edition, part IV.
62. N.V. Pogodina, A.B. Melnikov, I.N. Bogatova, T.A. Stepchenkova, V.M. Denisov, I.L.Kozlov, B.I. Zakharov and V.N. Tsvetkov, *Vysokomol. Soedin.*, Ser. A, 1989, *31*, 107.
63. W. Sutter and W. Burchard, *Makromol. Chem.*, 1978, *179*, 1961.
64. V.P. Shanbhag, *Arkiv Kemi*, 1968, *29*, 1.
65. G.M. Pavlov, N.V. Pogodina, S.V. Bushin, A.B. Melnikov, E.B. Lysenko, V.N. Marsheva, G.N. Marchenko and V.N.Tsvetkov, *Vysokomol. Soedin.*, Ser. A, 1986, *28*, 688.
66. F.R. Senti, N.N. Hellman, N.H. Ludwig, G.E. Badcock, R. Tobin, C.A. Glass and B.L. Lamberts, *J. Polym. Sci.*, 1955, *17*, 527.
67. K.A. Granath, *J. Colloid Sci.*, 1958, *13*, 308.
68. T.C. Laurent and K.A. Granath, *Biochim. Biophys. Acta*, 1967, *136*, 191.
69. W. Brown, P. Stilbs and R.M. Johnsen, *J. Polym. Sci.: Polym. Phys. Ed.*, 1982, *20*, 1771.
70. W. Banks and C.T. Greenwood, *Makromol. Chem.*, 1963, *67*, 49.
71. T. Yanaki, T. Norisuye and H. Fujita, *Macromolecules*, 1980, *13*, 1462.
72. T. Sato, T. Norisuye and H. Fujita, *Macromolecules*, 1984, *17*, 2696.

73. V.M. Jamtschikov and V.S. Skazka, *Vestnik of Leningrad Univ.*, 1973, *10*, 144.
74. V.N. Tsvetkov, P.N. Lavrenko and S.V. Bushin, *J. Polym. Sci.: Polym. Chem. Ed.*, 1984, *22*, 3447.
75. G.M. Pavlov and S.Ya. Frenkel, *Acta Polymerica*, 1988, *39*, 107.
76. G.M. Pavlov and S.Ya. Frenkel, *Vysokomol. Soedin.*, Ser. B, 1986, *28*, 353.
77. G.M. Pavlov and S.G. Selunin, *Vysokomol. Soedin.*, Ser. A, 1986, *28*, 1727.
78. H. Yamakava, 'Modern Theory of Polymer Solutions', Harper and Row, N.Y., 1971, p. 419.
79. P.N. Lavrenko, 'Hydrodynamic properties and conformations of rigid-chain polymer molecules', Doct. B. Thesis, Leningrad, 1987.
80. G. Meyerhoff and N. Sütterlin, *J. Polym. Sci., Part C*, 1972, *42*, 943.
81. K.-J. Linow, Dissertation, TH Magdeburg, 1970.
82. S. Claesson, W. Bergmann and G. Jayme, *Svensk Papperstidn.*, 1959, *62*, 141.

29

Sedimentation Analysis of Proteoglycans

By W.D. Comper and O. Zamparo

BIOCHEMISTRY DEPARTMENT, MONASH UNIVERSITY, CLAYTON, VICTORIA, AUSTRALIA 3168

1. OVERVIEW

Proteoglycans commonly exist in the extracellular regions of connective tissues. They are multifunctional molecules whose properties can be classified on the basis of their concentration in the tissue. For proteoglycans in dilute concentration their function is mainly associated with enthalpic, close proximity binding of cationic molecules or macromolecules with cationic domains; hydrophobic and hydrogen-bonding mediated interactions may also occur. In concentrated solutions (>5 mg/ml) proteoglycans are, because of their glycosaminoglycan content, highly non-ideal, and thereby influence the water distribution in tissues.

Contemporary interest in the sedimentation analysis of proteoglycans therefore stems from the ease with which this technique directly provides parameters describing the dynamic motion of water with respect to the proteoglycan in concentrated solutions, *i.e.* the specific hydraulic conductivity. This is in contrast to the type of information that can be provided from filtration studies performed with a porous membrane and a hydrostatic pressure gradient; these types of studies suffer from experimental artifacts, not the least of which is the concentration polarisation at the membrane surface, and do not provide absolute quantative parameters for water flow through the solution (see below). This chapter will, therefore, focus on sedimentation analysis of *concentrated* solutions of proteoglycan[1].

The analogy between sedimentation and filtration of solvent through a porous plug of polymer material had first been recognised by Signer and Egli[2] and by Fessler and Ogston[3] and developed theoretically by Mijnlieff and Jaspers[4]. The "porous plug" concept is based on the idea that in concentrated solutions, transient molecular networks are formed through frequent contacts between individual molecules. Gradients in molecular concentration and osmotic pressure that exist in dilute solution then disappear at high concentrations so that the polymer can essentially be regarded as being represented by its constituent segments. These segments - which have yet to be quantified in many cases - then determine the nature of water flow through the polymer solution.

In this chapter we review the sedimentation analysis, expressed here in terms of specific hydraulic conductivity, of proteoglycans, or model proteoglycans, directly involved in governing water distribution in tissues such as those from cartilage, glomerular basement membrane and synovial fluid. We will also include an extensive comparative study in order to delineate the primary structure features of these molecules that act as the major determinants for the viscous dissipation of water.

2. PROTEOGLYCAN STRUCTURE

Proteoglycans are conjugated proteins that contain one or more covalently bound polyanionic glycosaminoglycan (GAG) chains[5]. Classification of these molecules has proved difficult. Generally their description has been based on the type of GAG chain present although, more recently, efforts are being made to classify according to their protein structure[6].

Most often the GAG chain dominates the proteoglycan structure by mass and endows the proteoglycan with much of its 'segmental' properties in concentrated solutions. There are five major types of GAG chains whose physical and structural properties are shown in Table 1. Within each family of GAG chains, there exists a considerable degree of structural heterogeneity, particularly with respect to molecular weight (due to metabolic activity), the ratio of glucuronate to iduronate (for dermatan sulphate and the heparin-like polysaccharides) and the degree and position of sulphation. The invariant characteristics of each GAG-type are that they are linear polysaccharides with repeat disaccharide sugar units with characteristic glycosidic linkages. Other minor sugar components found in GAG chains are generally those associated with the specific linkage region that covalently links the GAG chain to the protein. Major subgroups exist for chondroitin sulphate with its two isomers of 4-sulphate or 6-sulphated hexosamine residue. For the heparin-like polysaccharides, the two major types of GAG included in this group will be heparin and heparin sulphate; the difference between these polysaccharides is that heparin is more highly sulphated and has a higher content of iduronate as compared to heparin sulphate although there is a range of molecules in between these two extremes. The polyanionic nature of the GAG chains comes from the fully ionised sulphate groups ($pK_a \leq 2$) and carboxyl groups of iduronate and glucuronate (pK_as in the range of 3.0-4.5)[7].

All the GAGs exist in their native form as chains covalently bound through a specific oligosaccharide binding region to a protein core, with the exception of hyaluronate which exists in the linear chain form only. Proteoglycans, too, exhibit a wide range of heterogeneity. Apart from heterogeneity of the GAG chains, proteoglycan heterogeneity is further exemplified in the population of chains within the proteoglycan, the different types of GAG chain within the proteoglycan, other carbohydrate attachments, heterogeneity in the size of the protein core (due to processing while residing in the tissue), and they may aggregate through self association or through interaction with other extracellular macromolecules. In many cases, the exact details of these structures are still being determined.

Proteoglycans that are involved in influencing water distribution in tissues must necessarily be present in concentrated form: such is the case for hyaluronate whose concentrations may range from 3.5-7.5 mg/ml for tissues including umbilical cord, rooster comb and synovial fluid[8]. The large aggregating chondroitin sulphate proteoglycan is a major contributor to the GAG content of cartilage which may

Table 1. Structural and physical properties of glycosaminoglycan chains.

Type of GAG	Major disaccharide repeat[a]	Sulphate position on sugar residue	$10^{-3} \times$ Molecular weight range
Hyaluronate	→4)-β-D-GlcA-(1-	0	200-8000
	→3)-β-D-GlcNAc-(1-	0	
Chondroitin	→4)-β-D-GlcA-(1-	0	5-50
sulphate	→3)-β-D-GalNAc-(1-	4 or 6 or both	
Keratan	→3)-β-D-Gal-(1-	0 or 6	4-20
sulphate	→4)-β-D-GalNAc-(1-	0 or 6	
Dermatan	→4)-α-L-IdoA-(1- or	0 or 2	15-40
sulphate	→4)-β-D-GlcA-(1-	0	
	→3)-β-D-GalNAc-(1-	4 or 4+6	
Heparin-like	→4)-β-D-GlcA-(1- or	0 or 2	4-50
polysaccharides[b]	→4)-α-L-IdoA-(1-,	0 or 2	
	→4)-α-D-GlcNAc-(1- or	0 or 6	
	→4)-α-D-GlcNSO$_3^-$-(1-	0 or 6 or 6+3	

[a] The polysaccharide-protein linkage regions have been omitted. Although these structures represent most of the disaccharide units found in vertebrate glycosaminoglycans the list is by no means complete because the detailed structures of the heparin-like polysaccharides and keratan sulphate are still incomplete. NSO_3, N-sulphamino; Gal, galactose; GlcA, glucuronic acid; GlcNAc, N-acetyl-glucosamine; GalNAc, N-acetyl-galactosamine; IdoA, iduronic acid;

[b] Includes heparin and heparan sulphate.

approach 100 mg/ml [9]. Similar molecules are also known to exist in aorta, tendon and intervertebral disc. The proteoglycan monomer (molecular weight (M) ~$1\text{-}4\times10^6$) consists of a large core protein (M in range of 200,000-340,000). On average there are 100 pendant chondroitin sulphate chains (M ~15,000-20,000), 50 keratan sulphate chains (M ~5,000-8,000) and a number of low molecular weight O-linked and N-linked oligosaccharide chains (for reviews see refs. 10 and 11). The various GAG chains and other carbohydrate units have a particular distribution along the protein core. These proteoglycan monomers can interact noncovalently but specifically with hyaluronate, *via* a globular region at one end of its protein core, to form an enormous aggregate (M ~200×10^6). The binding between hyaluronate and the proteoglycan protein core is stabilised by glycoproteins called link proteins. The overall carbohydrate content of the proteoglycan may amount to ~90% of the whole structure.

A quite different structure is encountered for the heparin sulphate proteoglycan isolated from glomerular basement membrane (GBM). The proteoglycan isolated from rat GBM (M ~130,000) has been shown to have a protein core of molecular weight of 26,000 and only four heparin sulphate chains (M ~26000)[12]. The structure may vary from species to species. In any case, the carbohydrate content of the proteoglycan is again dominant (up to 90%) and clearly contributes to the anionic charge properties of the membrane. The concentration of this proteoglycan in the GBM, however, has been subject to widely varying estimates. There has been the suggestion that GBM charge concentration is 160 mequiv l^{-1} (ref. 13) which would translate to a heparin sulphate concentration of 40-45 mg/ml.

3. OPERATIONAL PARAMETERS

Irreversible thermodynamics provides an expression for solute flux (J_1) in a two-component system associated with sedimentation velocity in the ultracentrifuge with respect to a volume (assumed to be the same as the cell)-fixed frame of reference[14]

$$J_1 = (S_1)_v C_1 \omega^2 r - (D_1)_v \frac{\partial C_1}{\partial r} \tag{1}$$

where ω is the angular speed of the rotor (radians/s), r is the distance from the centre of the rotor, $(S_1)_v$ is the sedimentation coefficient and $(D_1)_v$ is the mutual diffusion coefficient of solute (defined as component 1) in a volume-fixed frame of reference and C_1 is the concentration of component 1 in mass/volume units. Expressions for $(S_1)_v$ and $(D_1)_v$ have been derived previously, with due caution to reference frame and thermodynamic states[15], such that:

$$(S_1)_v = (1-\rho\bar{v}_1)(1-\phi_1)\frac{M_1}{f_{12}} \tag{2}$$

$$(D_1)_v = (1-\phi_1)^2 \frac{M_1}{f_{12}} \left.\frac{\partial \Pi^*}{\partial C_1}\right|_{T,\mu_2} \tag{3}$$

Substituting eqs. (2) and (3) into (1) gives:

$$J_1 = (1-\phi_1)\frac{M_1}{f_{12}}\left\{C_1\omega^2(1-\rho_1)\frac{\partial(r^2/2)}{\partial r} - (1-\phi_1)\left.\frac{\partial\Pi^*}{\partial C_1}\right|_{T,\mu_2}\right\} \tag{4}$$

where ρ is the solution density, $\bar{v}_1$ is the partial specific volume of component 1, ϕ_1 is the volume fraction of component 1, M_1 is the molecular mass of component 1, f_{12} is the frame-independent hydrodynamic frictional interaction between 1 mole of solute and solvent (component 2)[16], Π^* is the osmotic pressure at constant temperature and chemical potential of solvent, T is the temperature and μ_2 is the chemical potential of solvent. This equation then represents a balance of a mechanical force $[C_1\omega^2(1-\rho\bar{v}_1)\partial r^2/2]$ and an osmotic force $[(1-\phi_1)\partial\Pi^*]$ on the solute. At high speeds in the ultracentrifuge, the mechanical force will far outweigh the osmotic force for concentrations up to 80 mg/ml proteoglycan.

At high rotor speeds the flux J_1 is determined primarily by the sedimentation coefficient. A direct relationship between specific hydraulic conductivity, k, and the sedimentation coefficient has been derived by Mijnlieff and Jaspers[4]

$$k = \frac{\eta_2(S_1)_v}{C_1\left(1 - \left[\frac{\bar{v}_1}{\bar{v}_2}\right]\right)} \tag{5}$$

where η_2 is the viscosity of the solvent. We express our sedimentation data in this chapter in terms of k as it is the directly important physiological parameter describing permeation/filtration. For example, the volume flow (J_v) of fluids through porous beds is described by D'Arcy's law (1856)

$$J_v = \frac{k\Delta P}{l\eta_2} \tag{6}$$

where A is the area, l the length and ΔP the pressure gradient. This type of equation is used in describing permeation of water through polymer plugs or solutions. The value of sedimentation analysis is that it gives an unambiguous estimate of k whereas in permeation or filtration studies more often the distance parameter l is not known; this distance is where relative movement of solute to solvent occurs. The equations described above have been derived for two component systems whereas with proteoglycans we are dealing with polyelectrolytes in salt solutions (normally physiological saline buffered with phosphate buffer pH 7.4). The complication is not great as we have shown (see below) that electrolyte-dissipation is low for proteoglycans in concentrated solutions which validates the use of eqs. (4) and (5) where the solvent can be regarded as combination of water and NaCl.

It is necessary to measure the partial specific volume of the sedimenting material in order to evaluate k. These values shown in Table 2, appear generally lower than values often quoted in the literature. Yet, a closer examination of the latter reveals a degree of incompleteness or just a paucity of investigation. Mathews[19] reported a $\bar{v}_1$ for sodium chondroitin sulphate in 0.15 M phosphate buffer, pH 7.0 of 0.5 ml/g (a value close to that given in Table 2) which was based on the measured dry weight of the material. Eyring and Yang[20] suggested a value of 0.59 ml/g for chondroitin sulphate on the basis of unpublished work but this was not confirmed in later publications. Further, they used the $\bar{v}$ for protein of 0.70 ml/g to give an estimated $\bar{v}$ for chondroitin sulphate proteoglycan of 0.61 ml/g. Hascall and Sajdera[21] estimated a $\bar{v}$ for chondroitin sulphate of 0.54 ml/g on the basis of the values given in the paper by Eyring and Yang[20]. This, together with a $\bar{v}$ for the core protein of bovine nasal cartilage proteoglycan, assumed to be 0.72 ml/g, gave an estimated $\bar{v}$ for the proteoglycan of ~0.55 ml/g as compared to the value of 0.49 ml/g given in Table 2. The $\bar{v}$ for hyaluronic acid at physiological salt concentration was found to be in the range of 0.54 to 0.56 ml/g [22] and in agreement with values obtained for chemically similar chondroitin as shown as Table 2: this value for hyaluronate was also lower than previously reported values[23] which is presumably due to a better purified product.

Table 2. Partial specific volumes of polysaccharide preparations (from Comper *et al.*[17,18])[a]

Material	Partial specific volume (ml/g)
Chondroitin	0.54
Chondroitin sulphate	0.47
Chondroitin sulphate in 1 mol dm^{-3} NaCl	0.47
Swarm rat chondrosarcoma proteoglycan subunit	0.49
Heparin	0.46
Desulphated heparin	0.51
Keratan sulphate	0.49

[a] The solvent was phosphate buffered saline* unless otherwise stated.

4. SPECIFIC HYDRAULIC CONDUCTIVITY

Much of the stimulus for analysing the sedimentation of proteoglycans arose with the finding that chondroitin sulphate had a considerably lower specific hydraulic conductivity than other polysaccharides such as dextran and dextran sulphate, and other polymers such as albumin, poly(vinyl alcohol) and poly(ethyleneglycol) (Fig. 1). The relatively low conductivity or high resistance to flow per unit volume (or mass) for chondroitin sulphate seems to identify a specific structure-biomechanical functional relationship as its conductivity would generally account for the low conductivity found in cartilage[25]. In this tissue, low conductivity is required for resistance to compression.

* Phosphate buffered saline consists of 0.14 mol dm^{-3} NaCl, 2.68×10^{-3} mol dm^{-3} KCl, 1.5×10^{-3} mol dm^{-3} KH_2PO_4, 8.1×10^{-3} mol dm^{-3} Na_2HPO_4, pH 7.5.

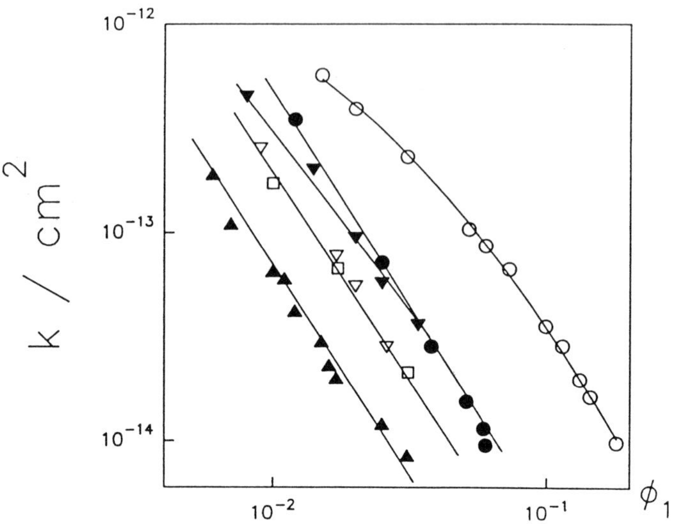

Figure 1. The specific hydraulic conductivity, k, as a function of volume fraction ϕ_1, for chondroitin sulphate (▲), dextran sulphate (□), poly(ethyleneglycol) (∇), dextran (●), poly(vinyl alcohol) (▼) and bovine serum albumin (o) in phosphate buffered saline. Lines drawn are those of best fit. Modified from Comper and Zamparo *et al.*[17,24]. Reproduced with permission of publishers.

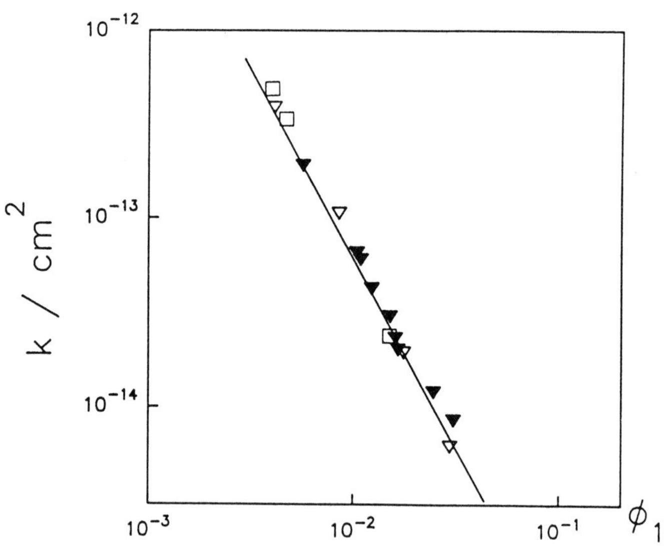

Figure 2. The specific hydraulic conductivity, k, as a function of volume fraction ϕ_1, for chondroitin sulphate (▼), Swarm rat chondrosarcoma proteoglycan subunit (monomer) (∇) and aggregate (□) in phosphate buffered saline. Lines drawn are those of best fit.

Concentration Dependence

The concentration dependence of the specific hydraulic conductivity of the proteoglycan form of chondroitin sulphate as a function of volume fraction is shown in Fig. 2. This proteoglycan, isolated from Swarm rat chondrosarcoma, is known to have similar structural and physical properties to the aggregating proteoglycans isolated from cartilage except that it does not contain keratan sulphate. There are a number of immediate features of the data in Fig. 2. First, there seems to be little difference in the concentration dependence of proteoglycan as compared to its constituent chondroitin sulphate chains. The contribution of protein to the partial specific volume and volume fraction parameters essentially normalizes out the k-values when the data are expressed in terms of volume fraction. The protein core, the branched structure of the proteoglycan, the vast differences in the macroscopic viscosity and molecular weight of the different proteoglycan forms (Table 3) then appear to have little influence on flow resistance. Further we have previously established that there was little difference in the k-values for the chondroitin sulphate isomers namely chondroitin 4-sulphate and chondroitin 6-sulphate[25].

Table 3. Molecular weight and intrinsic viscosity of swarm rat chondrosarcoma proteoglycan preparations measured in phosphate-buffered saline[24]

Material	Intrinsic viscosity (ml/g)	$10^{-6}\times M$ (g/mol)
Proteoglycan aggregate	950	211
Proteoglycan subunit	265	2.6
Chondroitin sulphate chain	40[a]	0.02

[a] From A.Wasteson, *Biochem. J.*, 1971, *122*, 477 for chondroitin sulphate of M ~20,000 g/mol.

The molecular weight independence of k demonstrates that it is the interaction of water with segments of the polysaccharide chain that determines the magnitude of k. This is not surprising in view of the earlier data on the molecular weight independence of dextran sedimentation in semi-dilute solutions[15,26] and theoretical treatments involving scaling laws[27]. The latter was not pursued in our analysis of proteoglycan sedimentation but rather we have investigated the salient primary structural features of the GAG chains that influence sedimentation.

Comparative Studies

Variation in the GAG content of the proteoglycans would appear to make little difference in the magnitude of k as data in Fig. 3 would suggest that chondroitin sulphate, hyaluronate, dermatan sulphate and keratan sulphate have similar specific hydraulic conductivities. This data clearly indicates the insensitivity of k to sulphation (as hyaluronate is not sulphated) and to the linear charge density of the GAG chain. Further evidence of the lack of influence of sulphation on the GAG chain is seen for k-

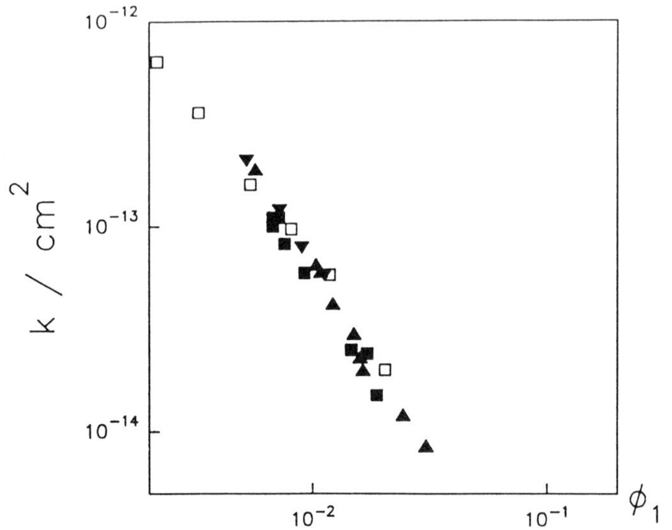

Figure 3. The specific hydraulic conductivity, k, as a function of volume fraction ϕ_1, for chondroitin sulphate (▲), hyaluronate (□), dermatan sulphate (▼) and keratan sulphate (■) in phosphate buffered saline. Data from Comper and Zamparo[18].

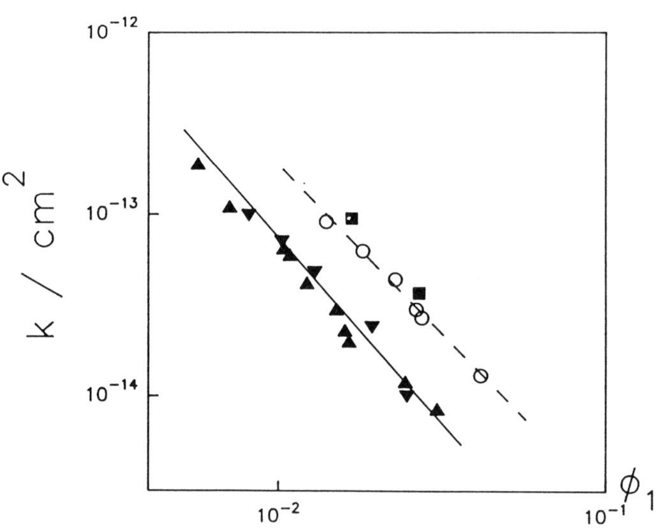

Figure 4. The specific hydraulic conductivity, k, as a function of volume fraction ϕ_1, for chondroitin sulphate (▲), chondroitin (desulphated chondroitin sulphate) (▼) , heparin (o) and desulphated heparin (■) in phosphate buffered saline. Data from Comper and Zamparo[18,25]. Lines drawn are those of best fit.

data in Fig. 4 for the comparison of desulphated chondroitin sulphate (chondroitin) to chondroitin sulphate and desulphated heparin to heparin.

The lack of influence of sulphation on the heparin chain would suggest that the range of heparin-like polysaccharides found in nature most likely have similar conductivities. The other major variable in their structures is the iduronate content although preliminary studies would suggest that it is not a major influence on k (*vide infra*). It is also clear from data in Fig. 4 that the heparin-like polysaccharides have a considerably higher specific conductivity than the chondroitin sulphate/chondroitin polysaccharides. This could be viewed as consistent with biomechanical function as regions where the heparin-like polysaccharides are retained (such as the glomerular basement membrane) require high flows of water as distinct from flow resistance in cartilage which contains chondroitin sulphate proteoglycans. Recent studies by us have demonstrated that the glomerular basement membrane heparin-like polysaccharides, given that they were controlling water flow, would not be at sufficient concentration to exert a charge selective effect seen in kidney ultrafiltration[28]. Rather, it turns out that the glomerular heparin-like polysaccharides exist at a low concentration (<1 mg/ml)[29], probably do not control water flow in this system and do not exert a direct charge selective effect on kinetic charged probes as originally proposed by Deen *et al.*[13].

Polysaccharide Flexibility

The difference seen for the polysaccharides represented in Fig. 4 add to the discrimination we originally observed with chondroitin sulphate and dextran (Fig. 1). These relative differences between various polysaccharide groups occur in spite of variations introduced through chemical modification such as the degree of sulphation. The overriding factor associated with discriminating between these polysaccharides appears to be associated with the nature of their major glycosidic linkage (a number of the polysaccharides studied have minor structural heterogeneity with respect to the type of glycosidic linkage but this issue is not raised here). This is demonstrated in Fig. 5 where there appears to be three groups of data. The group showing the relatively lowest k values correspond to polysaccharides with alternating β-1,3 and β-1,4 linkages including dermatan sulphate, chondroitin sulphate, chondroitin and also from Fig. 3 keratan sulphate and hyaluronate. The next group, with relatively higher values of k, correspond to polysaccharides with β-1,4 linkages or alternating α-1,4 and β-1,4 (or α-L-1,4) linkages as for the heparin-like polysaccharides (see also Fig. 4), carboxymethyl cellulose and hydroxyethyl cellulose[30]. The third group of data, generally exhibiting the highest value of k for any volume concentration, corresponds to polysaccharides with either α-1,6 linkages, like dextran, and α-1,4 and α-1,6 linkages for dextrin. The differentiation of these three major groups, based on the type of the glycosidic linkage, appears to be irrespective of quite major differences in the nature of the chemical groups substituted onto the native polysaccharide chain. For example, a comparison of the celluloses reveals that there is little difference between hydroxyethyl cellulose and carboxymethyl cellulose. Even more striking, in this context, is the similarity of the cellulose derivatives to the heparin-like polysaccharides, indicating the lack of influence of the iduronate group.

Details of the chain conformation of these polysaccharides, particularly with

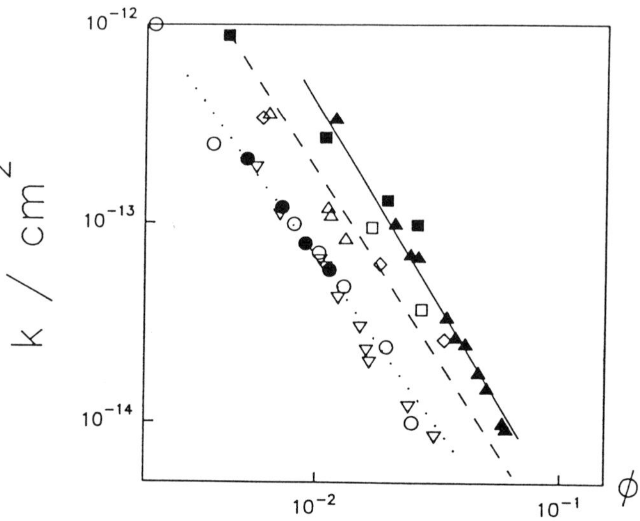

Figure 5. The specific hydraulic conductivity, k, as a function of volume fraction ϕ_1, for groups of polysaccharides defined by their major glycosidic linkage type. For the group with alternating β1,3 and β1,4 linkages we have dermatan sulphate (●), chondroitin sulphate (∇) and chondroitin (o). For the group with β1,4 linkages or alternating α1,4 and β1,4 (or α-L-1,4) linkages we have included desulphated heparin (□), carboxymethyl cellulose (Δ) and hydroxyethyl cellulose (◊) (from ref. 30). For the group with α1,4 linkages and/or α1,6 linkages we have included dextran (▲) and dextrin (■). Data compiled from Comper and Zamparo[18,31] unless otherwise stated. Lines drawn are those of best fit.

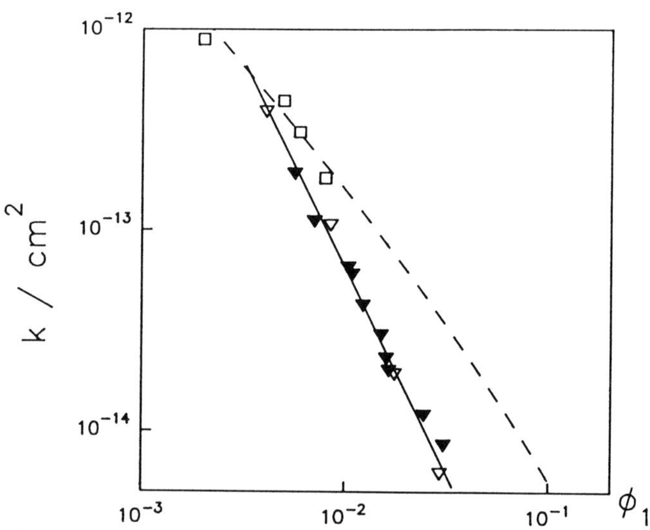

Figure 6. The specific conductivity, k, as a function of volume fraction ϕ_1 for proteoglycan subunit (∇), chondroitin sulphate (▼) in phosphate buffered saline and hyaluronate (□) in thermodynamic equilibrium with buffered 0.2 M NaCl (from Preston *et al.*[23]). The dashed line represents the predicted value of k from eq. (7) for a cylindrical radius $r_1 = 0.55$ nm.

respect to conformational flexibility, unperturbed distances and water interaction are incomplete. However, some interesting data is being generated with regard to inter-residue H-bonding especially for the mammalian GAGs. For β-1,3 and β-1,4 alternate linked GAGs hydrogen-bonding groups, as well as the glycosidic oxygens, are equatorial. This allows formation of a co-operative H-bonding pattern between residues. For hyaluronate, four different hydrogen bonds within a repeat trisaccharide unit, have been identified;namely glucosamine $C_{(4)}OH$ to the glucuronic ring oxygen atom, from the acetamido C=O to HO-2 of glucuronate, from the HO-3 of glucuronate to the hexosamine ring oxygen atom and from the amino NH to $O=C-O^-$ carboxyl of glucuronate[32-35]. For chondroitin sulphates only three hydrogen bonds can form due to the fact that the bond from the axial galactosamine HO-4 to the glucuronate ring oxygen cannot form[34]. In any case, these inter-residue H-bondings would clearly be expected to endow the critical hydrodynamic segments with decreased flexibility and create a stiffer chain segment. This may be a major discriminating factor in determining the correlation found between the type of polysaccharide glycosidic linkage and the magnitude of specific hydraulic conductivity. A possible exception to this interpretation is keratan sulphate as no inter-residue H-bonds have yet been identified for this GAG. We also know that the polysaccharides that have the relatively highest value of specific hydraulic conductivity, like α-1,6 dextrans, also have an extremely flexible type structure. For β-1,4 linked polysaccharides it has been suggested that for heparin sulphate that two possible inter-residue H-bonds may form[33] yet the structure may have increased conformational flexibility through the presence of the iduronate residue[36]. Cellulose, too, may have inter-residue H-bonds at the OH-3...O-5 and OH-2...OH-6 position[37].

Steric Hindrance

It is apparent, then, that linear polysaccharides, including mammalian GAGs exhibit specific primary structural influences on the magnitude of their specific hydraulic conductivity. This makes it difficult to predict the magnitude of k through simple steric hindrance modelling. Such efforts have previously been made through the use of the theoretical analysis of the movement of water through a matrix of neutral fibres of radius r_1 such that[38]

$$k = \frac{r_1^2(\phi_1)^3}{4G(1-\phi_1)^2} \tag{7}$$

where G is a dimensionless term called the Kozeny factor. The theoretical dependence of G on ϕ_1 was analysed by Happel and Brenner[39] for beds of uniform cylindrical fibres. Levick[40] has claimed good agreement of the Happel-Brenner (H-B) model, as embodied in eq. (7), with the early data on the sedimentation of hyaluronate[23] using the radius of hyaluronate of 0.55 nm (Fig. 6). This agreement seems coincidental as at higher concentrations the H-B theory offers a poor prediction of not only of hyaluronate (see Fig. 3) but also chondroitin sulphate. The redeeming feature of the model is that it does predict the general behaviour of the variation of k with ϕ_1. (The hyaluronate data of Preston *et. al.*[23] appear to yield relatively higher values of k which is probably a result of the use of their higher $\bar{v}$ of 0.65 ml/g).

Electrolyte Dissipation

The influence of high NaCl concentrations on hydraulic conductivity of the proteoglycan is small, with an increase in k of 1.2-1.6 fold (Fig. 7), and was smaller for chondroitin sulphate (similar results have also been obtained for heparin[18]). Analysis of k for chondroitin sulphate at 36 mg/ml reveals also that there are only small changes in k over the NaCl concentration range of 0.035 to 1.0 mol dm^{-3}. These results are consistent with the lack of influence of the linear charge density, and particularly the degree of sulphation of the GAG chain on k as described previously. The low degree of electrolyte dissipation seen with GAG chains clearly demonstrates the major influence that chain conformation and flexibility has in determining the generally low values of k obtained for these materials. We have noted however, for reasons that are unclear at present, that the more flexible dextran sulphate does exhibit significant electrolyte dissipation in semi-dilute solutions[17].

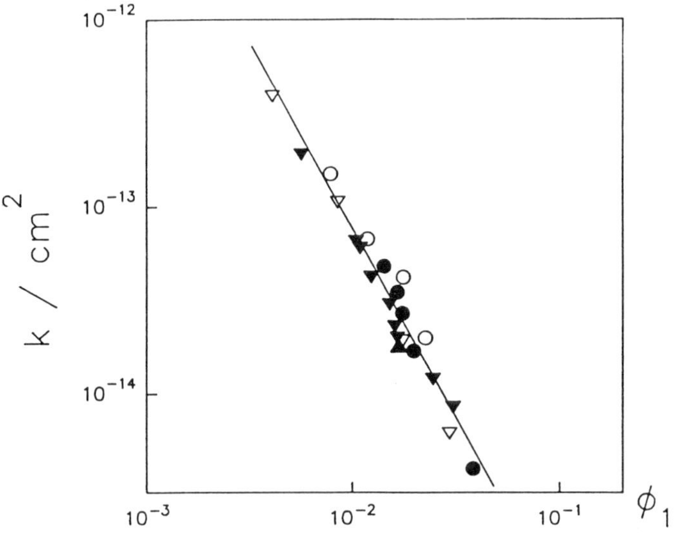

Figure 7. The specific hydraulic conductivity, k, as a function of volume fraction ϕ_1, for proteoglycan subunit in phosphate buffered saline (∇) and 1 mol dm^{-3} NaCl (o), and for chondroitin sulphate in phosphate buffered saline (▼), 1 mol dm^{-3} NaCl (•), 0.07 mol dm^{-3} NaCl (▲), and 0.035 mol dm^{-3} NaCl (Δ) (almost superimposed by the ▲ symbol). Data from Comper and Zamparo[18,25].

5. WATER TRANSPORT

The viscous dissipation of water over the polymer chain is governed by the hydrodynamical frictional coefficient as embodied in eq. (4). The molecular weight independence of k is then due to the molecular weight independence of the M_1/f_{12} term and suggests that dynamic water interaction with mass units of the polymer chain occurs over a finite segment length. This molecular weight independence is consistent with studies of water diffusion, as studied through the use of tritrated water (HTO), in polymer solutions which also demonstrated that the polymer concentration dependent HTO diffusion coefficient is molecular weight independent[41]. This has been demonstrated through studies of HTO diffusion in dextran solutions. In fact eq. (4) implies that it is the same hydrodynamical frictional coefficient that governs sedimentation that also governs the diffusive motion of the polymer relative to solvent or solvent relative to polymer. Therefore, an estimate of f_{12} may be made through the measurement of the diffusion coefficient of HTO designated D_{2*} (where 2* represents labelled water, HTO) such that[42]

$$D_{2*} = \frac{RT}{(f_{2*1} + f_{2*2})} \tag{8}$$

Using the Onsager reciprocal relationship we can rearrange eq. (8) to give[41]

$$D_{2*} = RT/(f_{2*2} + (c_1 f_{12}/c_2)) \tag{9}$$

where c_2 is the molar concentration of 2. Therefore, measurement of D_{2*} may give an estimate of f_{12} as embodied in eq. (9); such measurements have been made in dextran solutions[41]. It turns out that if f_{12} is evaluated from polymer mutual diffusion studies then estimates of D_{2*} may yield the effective molecular weight of the segment; the values obtained for dextran (M ~150,000) over a concentration of 0-200 kg m^3 were in the range of 800-10000 whereas for dextran (M ~10000) over the same concentration range segment molecular weight values of 100-1000 were obtained.

This type of extensive analysis has not been extended to the proteoglycan as yet although limited studies have suggested an unusual effect[17]. We have made comparative studies on the HTO diffusion in various polymer systems. The nature of the concentration dependent reduction in D_{2*} depends on the polymer studied[17]. Although we have found no major differences in D_{2*} for polymers at $\phi_1<0.05$ the expected differences for $\phi_1>0.05$ in the value of D_{2*} for polymers is probably best represented in terms of the corresponding ϕ_1 values for a given k value (which should also give an identical D_{2*}). Table 4 demonstrates that a k value of 1.0×10^{-14} cm^2 corresponds to ϕ_1 for PGS, poly(ethyleneglycol), dextran and albumin of 0.024, 0.046, 0.062 and 0.17 respectively. Assuming that the ϕ_1 value for D_{2*} for albumin corresponds to the k value then the corresponding ϕ_1 values obtained from HTO diffusion are shown in Table 4.

The fact that the ϕ_1 values agree for dextran and albumin for both HTO diffusion and sedimentation would appear to demonstrate that there is reasonable agreement (and internal consistency) with the obstruction effects of water with these two forms of measurement. An anomalous situation occurs, however, with poly(ethyleneglycol) and chondroitin sulphate proteoglycan whose corresponding interpolated ϕ_1 values from HTO diffusion lie well below that obtained from hydraulic flow. This demonstrates that other factors may be contributing to the enhanced resistance to hydraulic flow exhibited by these polymers. A possible contributing factor that is currently being investigated is the osmotic activity of these materials *i.e.* the osmotic activity of poly(ethyleneglycol) and proteoglycan (even in high salt) is high as compared to the other polymers, such as albumin and dextran. These results may indicate a situation that, while the overall macroscopic osmotic pressures are small in relation to the centripetal forces, the osmotic environment at microscopic dimensions around the polymer chain may be influential in determining k. This osmotic environment would not be detected by HTO exchange as the exchange process does not recognise chemical potential gradients of water. This, too, may imply that osmotic gradients may extend from the polymer chain which will be determined, in part, by the intrinsic segmental mobility and flexibility of the molecule.

Table 4. Comparison of ϕ_1 values for various polymers obtained (by interpolation) from $k = 1.0\times10^{-14}$ cm^2 and $D_{2*}/(D_{2*})^{0\ a} = 0.78$. Data from Comper and Zamparo[17].

	ϕ_1 (from sedimentation)	ϕ_1 (from HTO diffusion)
Albumin	0.17	0.17
Dextran	0.062	0.06
Poly(ethyleneglycol)	0.046	>0.09
Proteoglycan	0.024	>0.045

[a] $(D_{2*})^0$ represents the intra-diffusion coefficient in water alone.

6. CONCLUDING REMARKS

Sedimentation velocity analysis provides an unambiguous way of evaluating the flow of water over polymer chains in semi-dilute solutions. This is especially important in studies of proteoglycans as their biological function is often associated with controlling water permeability and filtration. Sedimentation analysis obviates the need to know the distance function in filtration or permeability over which relative displacement of water to solute takes place. Our comparative studies of various GAG types and other polymers have demonstrated that the material with the lowest conductivity per unit volume or mass is chondroitin sulphate proteoglycan; this identifies a structure-function relationship for this material as it is required to resist water flow in cartilagenous tissues that resist compressive load. The major structural feature of the GAG chains that determines hydraulic conductivity is the polysaccharide chain segment and its intrinsic flexibility. The size of the segments and the nature of interchain and intrachain interactions governing chain flexibility have yet to be fully evaluated. However, a strong correlation is made with the nature of the glycosidic linkage and the presence of intrachain inter-residue H-bonding in governing chain flexibility and therefore the magnitude of specific hydraulic conductivity.

ACKNOWLEDGEMENTS

This work was supported by grants from the Australian Research Council and the National Health and Medical Research Council of Australia.

REFERENCES

1. For studies in dilute solution see ref. 7 and J.C. Pita, F.J. Muller, T. Oegema and V.C. Hascall, *Arch. Biochem. Biophys.*, 1978, *186*, 66 and references therein.
2. R. Signer and H. Egli, *Rec. trav. chim.*, 1950, *69*, 45.
3. J.H. Fessler and A.G. Ogston, *Trans. Faraday Soc.*, 1951, *47*, 667.
4. P.F. Mijnlieff and W.J.M. Jaspers, *Trans. Faraday Soc.*, 1971, *67*, 1837.
5. For review of the structures see, for example, A.R. Poole, *Biochem. J.*, 1986, *236*, 1.
6. E. Ruoslahti, *J. Biol. Chem.*, 1989, *264*, 13369.
7. For review W.D. Comper, 'Heparin (and Related Polysaccharides)', Gordon and Breach, New York, 1981, pp 120-121.
8. T.C. Laurent and J.R.E. Fraser, 'Functions of Proteoglycans', Wiley, Chichester, 1986 (Ciba Foundation Symposium 124), p. 9.
9. A. Maroudas, 'The Joints and Synovial Fluid', L. Sokoloff (ed.), Academic Press, London, 1980, Vol II, p. 239.
10. S.L. Carney and H. Muir, *Physiol. Rev.*, 1988, *68*, 858.
11. V.C. Hascall, *ISI Atlas of Sci. Biochemistry*, 1988, 189.
12. Y.S. Kanwar, A. Veis, J.H. Kimura and M.L. Jakubowski, *Proc. Natl. Acad. Sci.*, 1984, *81*, 762.
13. W.M. Deen, B. Satvat and J.M. Jamieson, *Am. J. Physiol.*, 1980, *238*, F126.
14. H. Fujita, 'Mathematical Theory of Sedimentation', Academic Press, New York, 1962.
15. W.D. Comper, B.N. Preston and P. Daivis, *J. Phys. Chem.*, 1986, *90*, 128.
16. K.S. Speigler, *Trans. Faraday Soc.*, 1958, *54*, 1409.
17. W.D. Comper and O. Zamparo, *Biophys. Chem.*, 1989, *34*, 127.
18. W.D. Comper and O. Zamparo, *Biochem. J.*, 1990, *269*, 561.
19. M.B. Mathews, *Arch. Biochem. Biophys.*, 1956, *61*, 367.
20. E.J. Eyring and J.T. Yang, *J. Biol. Chem.*, 1968, *243*, 1306.
21. V.C. Hascall and S.W. Sajdera, *J. Biol. Chem.*, 1970, *245*, 4920.
22. K.-O.Wik, 'Physicochemical Studies on Hyaluronate', Acta Universitatis Upsaliensis, Uppsala, 1979.
23. See, for example, B.N. Preston, M. Davies and A.G. Ogston, *Biochem. J.*, 1965, *96*, 449.
24. W.D. Comper and R.P.W. Williams, *J. Biol. Chem.*, 1987, *262*, 13464 and unpublished data.
25. O. Zamparo and W.D. Comper, *Arch. Biochem. Biophys.*, 1989, *274*, 259.
26. A.G. Ogston and E.F. Woods, *Trans. Faraday Soc.*, 1954, *50*, 635.
27. P.G. de Gennes, 'Scaling Concepts in Polymer Physics', Cornell University Press, Ithaca, 1979.
28. O. Zamparo and W.D. Comper, *Biophys. Chem.*, 1990, *38*, 167.
29. A.S.N. Lee, M. Tay and W.D. Comper (submitted for publication).
30. L.-O. Sundelöf and B. Nyström, *J. Polym. Sci., Polym. Lett. Ed.*, 1977, *15*, 377.
31. O. Zamparo and W.D. Comper, *Carbohydr. Res.*, 1991, *212*, 193.
32. E.D.T. Atkins, D. Meader and J.E. Scott, *Int. J. Biol. Macromol.*, 1980, *2*, 318.
33. J.E. Scott and F. Heatley, *Biochem. J.*, 1982, *207*, 139.

34. J.E. Scott, F. Heatley and W.E. Hull, *Biochem. J.*, 1984, *220*, 197.
35. F. Heatley and J.E. Scott, *Biochem. J.*, 1988, *254*, 489.
36. For review, see B. Casu, M. Petitou, M. Provasoli and P. Sinaÿ, *Tr. Biochem. Sci.*, 1988, *13*, 221.
37. For review, see R.H. Marchessault and P.R. Sundararajan, 'The Polysaccharides', G.O. Aspinall (ed.), Academic Press, New York, 1983, Vol 2, Chap. 2, p. 11.
38. F.E. Curry, 'Handbook of Physiology: The Cardiovascular System, Vol IV, Microcirculation', E.M. Renkin and C.C. Michel, eds., Amer. Physiol. Soc., Bethesda, 1984, Part 1, p. 309.
39. J. Happel and J. Brenner, 'Low Reynolds Number Hydrodynamics', Prentice-Hall, Englewood Cliffs, NJ, 1965.
40. J.R. Levick, *Q. J. Exp. Physiol.*, 1987, *72*, 409.
41. W.D. Comper, M.-P.I. Van Damme and B.N. Preston, *J. Chem. Soc. Faraday Trans. I*, 1982, *78*, 3369.
42. R.J. Bearman, *J. Phys. Chem.*, 1961, *65*, 1961.

30

Non-Ideality of Hyaluronan as Determined by Sedimentation Equilibrium Studies

By B.N. Preston and [1]K.-O. Wik

DEPARTMENT OF BIOCHEMISTRY, MONASH UNIVERSITY, CLAYTON, VICTORIA 3168, AUSTRALIA

[1]PHARMACIA BIOMATERIAL, S-75-182, UPPSALA, SWEDEN

1. INTRODUCTION

The non-ideal behaviour of hyaluronan (hyaluronic acid) is usually described quantitatively in terms of the virial coefficients A_2, A_3 *etc.* as expressed in an equation relating osmotic pressure to concentration. Owing to the considerable technical difficulties in making direct osmotic pressure measurements at high concentrations of hyaluronan, Nichol *et al.*[1] suggested in 1967 that the non-ideality terms could be determined by sedimentation equilibrium. This method has been extended in this Chapter to include a polynomial regression analysis of the entire concentration distribution in the ultracentrifuge cell in high speed-high concentration experiments. This technique allows an estimate not only of the second and third virial coefficients but also, at low ionic strengths, the fourth and fifth.

The physico-chemical characterization of hyaluronan in solution at concentrations comparable to those existing in micro-environments of tissues is important for understanding the functional role of this material. There is considerable support for the concept that the osmotic contribution of hyaluronan and other connective tissue polysaccharides is an important factor in homeostasis and in controlling the movement of water in tissues. It has been suggested[2] that owing to the very strong concentration dependence of osmotic pressure exhibited by these polyions that they can act as osmotic buffers. Furthermore, solutions of hyaluronan have been shown to exhibit marked steric exclusion phenomena[3,4,5] and effect transport of material through such solutions[6,7,8].

It is therefore of considerable interest to describe quantitatively the behaviour of hyaluronan at high concentrations. This polyion displays typical polyelectrolyte behaviour in that it exhibits strong dependence of its non ideal behaviour on ionic strength of the solvent[9].

Several years ago, we were attracted to the use of sedimentation equilibrium of hyaluronan as a technique for estimating the thermodynamic properties of the polymer[1]. Our reasons for using this approach were related to the technical difficulties of making direct osmotic pressure measurements on this polymer. In that study, the limitations of using simplified equations to describe the sedimentation equilibrium of polydisperse material were clearly stated. However, the analysis appeared to give satisfactory results. The method has later been used on synthetic polymers by Chu and Munk[10,11].

We have characterised a degraded sample of hyaluronan with a molecular weight less than 10^5; the virial coefficients have been estimated at various ionic strengths using a modification of the high speed sedimentation equilibrium technique described earlier[1]. The data from the entire schlieren curve were used in a curve fitting procedure to evaluate terms up to and including the fifth virial coefficient. Experiments were performed at varying speed and concentration and the validity of certain aspects of the approach taken was checked by use of narrow molecular weight fractions of dextran. The second virial coefficient of hyaluronan was also evaluated by low speed sedimentation equilibrium[12,13] and by light scattering.

2. MATERIALS AND METHODS

Polymer Samples

Rooster comb hyaluronan was obtained from Pharmacia AB (Uppsala, Sweden) as 1% solution in 0.145 M NaCl and 0.002 M Na-phosphate, pH 7.3. The protein content was about 0.1% of the organic material.

An acid hydrolysed sample for sedimentation equilibrium experiments was prepared by addition of 13.2 ml 6 M HCl to 66 ml of a hyaluronan stock solution (10 mg/ml) and the reaction allowed to proceed for 80 min at 50°C. The degradation was terminated by removal of the sample from the water bath and adjustment of pH to 6.0 with 5 M NaOH. The sample was dialysed exhaustively against 0.15 M NaCl and stored at -20°C at a hyaluronan concentration of 6 mg/ml.

Determination of the refractive index increment and the partial specific volume was performed on a sample from the same batch of hyaluronan as above. Degradation to approximately the same molecular weight was accomplished by autoclaving the hyaluronan stock solution (10 mg/ml) in 10 ml glass vials at 120°C for 150 min.

Fractionated samples of dextran characterised by gel chromatography were kindly supplied by Dr. K. Granath.

Determination of Concentration

The concentration of hyaluronan used in sedimentation equilibrium experiments was determined by uronic acid analysis[14]. The automated carbazole method[15] was employed for all other purposes. In the conversion of the standard (D-glucuronolactone) to hyaluronan the equivalent weight of the repeating disaccharide of the polymer was taken as 401.3. Dilutions before the analyses were performed by weighing.

The concentration of solutions used for determination of the partial specific volume and the refractive index increment was also obtained as dry weight. Weighed solutions (not less than 4 g) were dried in an oven at 110°C to 120°C for 2 × 24 h. Weight to volume conversion was made using the solution densities determined in a Precision Density Meter DMA 02C (Anton Parr, Graz, Austria). Within experimental error there was no difference in dry weight between solution and dialysate. The uronic acid analysis and dry weight determinations were in excellent agreement. A p-value (two-sided) less than 0.05 was obtained when comparing the determinations by one-sample t-test applied to the differences between the values obtained from the two methods.

Refractive Index Increment

A series of six to eight solutions of hyaluronan with concentrations between 1 and 10 mg/ml was dialysed for more than 12 days against 0.2 M, 0.01 M and 0.001 M NaCl, respectively. The concentration of hyaluronan was determined on each sample as described above. The refractive index increment (dn/dc) was measured at 436 nm and 546 nm in a Rayleigh Interference Refractometer (Hilger and Watts Ltd., London, U.K.) at 20.0±0.1°C. A cell with 1 cm light path was used and dialysate placed in the solvent compartment. The dn/dc at 633 nm was extrapolated from values at 436 nm and 546 nm using the relation for simple Cauchy dispersion (dn/dc $\propto$ $1/\lambda^2$) suggested by Huglin[16]. The value of dn/dc for dextran at 546 nm, 0.150, was used[17].

Partial Specific Volume

The solutions and dialysates used for studies of the refractive index increment, were analysed for density in an Anton Paar (Graz, Austria) Precision Density Meter at 20.0±0.01°C. The values were then used to calculate the partial specific volume ($\bar{v}$) in 0.2 M, 0.01 M and 0.001 M NaCl. The value for dextran was taken as 0.60 [18].

Osmometry

Osmotic pressure was measured in an electronic high-speed osmometer (Hewlett Packard model 501, F&M Scientific Corp., Avondale, PA, USA). Three solutions with hyaluronan concentrations <10 mg/ml were analysed in 0.2 M NaCl after exhaustive dialysis and the number-average molecular weight calculated by extrapolation of the reduced osmotic pressure to zero concentration.

Light Scattering

Light scattering was carried out in a low angle light scattering photometer (KMX-6, Chromatix, Mountain View, CA, USA) at 633 nm.

Sedimentation Equilibrium Experiments

Theoretical Considerations There are many theoretical and technical problems which deter the application of the sedimentation equilibrium technique to investigate polydisperse concentration dependent systems[19,20]. For example, at any fixed point in the cell the distribution of molecular weights will differ from that in the original solution and thermodynamic non-ideality will be dependent both on the total concentration and the particular distribution of species at any fixed point. In a previous study of the sedimentation equilibrium of hyaluronan[1] the relevance of these parameters was discussed. Nevertheless the authors were successful in determining virial coefficients which agreed with osmotic measurements.

The sedimentation equilibrium of a non-ideal homogeneous solute is given by

$$\frac{M\,(1-\bar{v}\rho_0)\,\omega^2 rc}{RT} = \left\{1 + c\,\frac{d\ln\gamma}{dc}\right\}\frac{dc}{dr} \qquad (1)$$

where c is the concentration in g/ml, r is the radius of rotation, ω is the angular velocity, M the molecular weight, $\bar{v}$ the partial specific volume, γ the activity coefficient of the solute and ρ_0 the density of the solvent.

The activity coefficient can be expressed as a virial expansion in the form

$$\ln\gamma = M\left\{2A_2c + (3/2)\,A_3c^2 + (4/3)\,A_4c^3 + \dots\right\} \qquad (2)$$

where A_2, A_3, A_4 *etc.* are the conventional virial coefficients in the osmotic pressure equation.

On differentiation of eq. (2) and putting into eq. (1) we obtain

$$\frac{\omega^2 c(1-\bar{v}\rho_0)}{2RT\,(dc/dr^2)} = \frac{1}{M} + 2A_2c + 3A_3c^2 + 4A_4c^3 + \dots \qquad (3)$$

The equation is written in this form since in these studies our objective is not to determine the molecular weight or the molecular weight distribution of the sample, but to evaluate the virial coefficients describing the non-ideality of the hyaluronan.

Eq. (3) shows the effects of A_2, A_3, ... on the equilibrium gradient (schlieren) pattern whose ordinate measures dc/dr. As has been pointed out earlier[1], if $A_2 = A_3 = A_4 = \dots = 0$ then dc/dr^2 and dc/dr increases indefinitely with c (and thus r^2), the curvature of dc/dr^2 (and thus c) against r^2 is everywhere positive (Fig. 1, Curve I). If $A_2 > 0$ and $A_3 = A_4 = \dots = 0$ then dc/dr^2 approaches a limiting value as c (and r) $\rightarrow \infty$ (Fig. 1, Curve 2). If all $A_2, A_3, A_4, \dots > 0$, dc/dr^2 passes through a maximum (Fig. 1, Curve 3) and the plot of c against r^2 has an inflection beyond which the curvature is negative. However, as we shall report later, the plot of dc/dr^2 observed with hyaluronan displayed more complex behaviour in that the plot exhibited a second inflection (Fig 1, Curve 4) at increasing concentration beyond which the curvature is once again positive. Such behaviour can be explained if we allow the value of A_4 or A_5 or higher terms to be < 0.

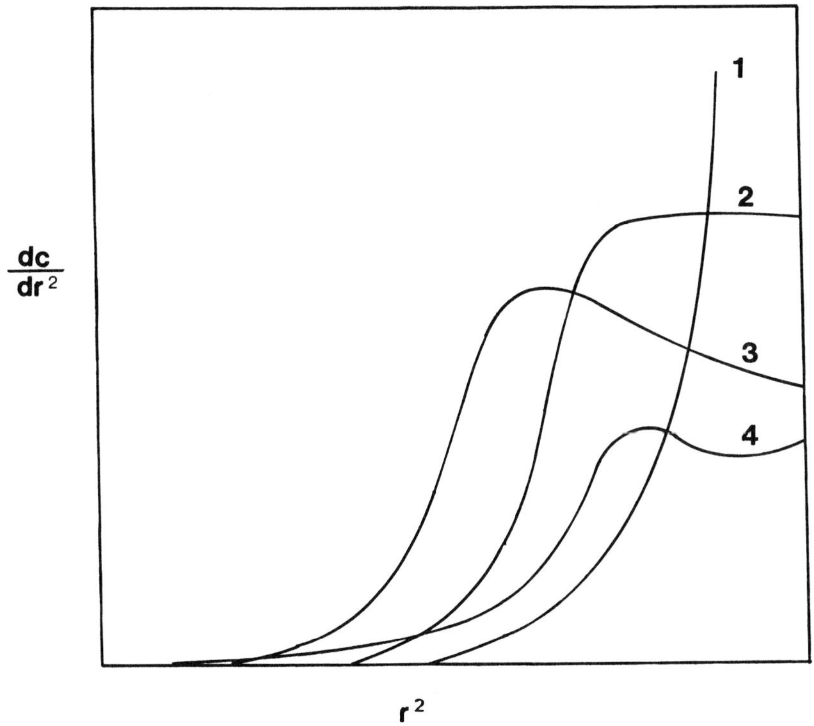

Figure 1. Diagram to show the effects of non-ideality coefficients on the variation of dc/dr^2 with r^2 (and hence with c) in sedimentation equilibrium. Curve 1: $A_2 = A_3 = A_4 = ... = 0$; curve 2: $A_2 > 0$, $A_3 = A_4 = ... = 0$; curve 3: $A_2 > 0$, A_3 *etc.* > 0: curve 4: $A_2 > 0$ $A_3 > 0$ $A_4 < 0$.

The application of eq. (3) to a study of polydisperse solutes requires extensive approximations. However, as has been pointed out in detail[1], the approximations in context with the present investigation are not too serious. Eq. (3) shows that at higher concentrations, provided A_2 is high in relation to 1/M, the right-hand side of the equation is dominated by the non-ideality terms. However, in the present studies, we had to adopt a compromise course because of technical difficulties associated with the experimental technique. The use of high molecular weight hyaluronan, where 1/M is much less than A_2, gives solutions of very high viscosity and low diffusion coefficient. This results in manipulative problems and in very long periods of time needed to attain equilibrium. To overcome this we have used an acid degraded sample of lower molecular weight. Such an action, however, places a restriction on use of eq. (3) to evaluate A_2 since the value of 1/M for the polydisperse sample is no longer small compared to A_2.

In order to evaluate to what extent the polydispersity of our sample affected the value of A_2 determined by high speed - high concentration sedimentation equilibrium we have also determined this parameter by two independent techniques, *i.e.* light scattering and low speed - low concentration sedimentation equilibrium[12,13]. At low speeds there is little fractionation with regard to molecular weight in the cell. It will be shown below that the three techniques essentially give the same values of A_2. Knowing this, we are able to use the high speed - high concentration sedimentation equilibrium technique to evaluate the higher order virial coefficients in eq. (3) with certain confidence.

Experimental Considerations All solutions studied by sedimentation equilibrium had been exhaustively dialysed against 0.2 M, 0.05 M, 0.01 M, 0.005 M or 0.001 M NaCl. The dialysate was used as solvent in the ultracentrifuge cells. The experiments were performed at 20°C in a Spinco Model E ultracentrifuge equipped with both interference and schlieren optics.

The technique employed for the high speed - high concentration runs was essentially that described by Nichol *et al.*[1] where in most cases the solute concentration at the meniscus was very close to zero. The An-D rotor and schlieren optics were used in conjunction with various cells of path lengths between 1.5 mm and 30 mm. The choice of cell was dependent mainly upon the initial concentration of the solute as described earlier[1]. Where single sector cells were used the base line was determined in separate experiments with solvent alone. In all cases a thin layer of an inert fluorocarbon oil (FC 43, Beckman Instruments, Spinco Division, Palo Alto, CA, USA) was included to define the base of the cell. The runs were continued until equilibrium had been reached as judged by successive photographs of the equilibrium pattern some six hours apart. In many runs the equilibrium pattern was measured at two or more rotor speeds. In order to reduce the time taken to attain equilibrium, higher speeds were used during the initial parts of the run. In the case of 0.2 M NaCl as solvent, it was necessary to cover a wide range in concentration of hyaluronan in order to estimate the higher order virial terms. This required sedimentation conditions where the concentration at the meniscus was not zero.

For the low speed - low concentration runs Rayleigh interference optics were used in conjunction with a 30 mm double sector cell to facilitate the study at low concentrations. The An-J rotor was used at speeds less than 16,000 rev/min. As suggested[13] the column height of the solutions was adjusted to about 3 mm. FC 43 oil was used to give a sharp definition of the base.

Measurements and Calculations The plates were read on a microcomparator (Nikon 6TC2). In the case of the low speed - low concentrations runs, in which the fringe displacement was small, considerable care was taken in the alignment of the plates in the radial directions.

For the high speed - high concentration runs, the differences of the ordinate readings of the schlieren patterns of the experimental and base lines were measured at a series of points between the position at which the trace and base line coincided (or in some cases the meniscus) and the bottom of the column. With the use of the appropriate cell and optical constants and the refractive index increment, values of dc/dr at each point could be calculated.

In earlier experiments[1] a trapezoidal integration of dc/dr *versus* r was used to give a value of c representing the concentration at each point in excess of that at the meniscus. However, in the present study, it was our intention to fit the dc/dr data of the entire schlieren trace into eq. (3) by polynomial regression. The value of the left-hand side of eq. (3) is very sensitive in the initial range to errors in c. It was found that in this region a simple trapezoidal integration to estimate c was unsatisfactory due to the low precision in the readings of the first points on the schlieren curve. A better estimate of the concentration at the first point was obtained as follows. The readings from the first point up to the inflection point on the schlieren curve (a typical schlieren curve is demonstrated in Fig. 2) were converted to $\log_{10}(dc/dr)$ and plotted *versus* r. The points were correlated by linear regression and the resulting function was assumed to describe the variation in the concentration gradient from the meniscus up to the inflection point. The exponential function of dc/dr was integrated from the meniscus to give the value of c at the point where the readings could begin. The rest of the curve was evaluated by trapezoidal integration of the schlieren pattern.

However, the improvement obtained with this procedure was not altogether satisfactory. Schlieren curves from all experiments except those performed on hyaluronan in 0.001 M NaCl increased slowly with increasing radius. Therefore not only the first but also the next points suffered from low precision in reading the ordinate, which resulted in scatter in the value of the left-hand side of eq. (3) at low concentration.

Moreover, as pointed out by Nichol *et al.*[1], polydispersity effects will be of greater importance in the region of low concentration. Therefore, in order to get a uniform evaluation, the polynomial regression analysis was performed from the inflection point onwards. The schlieren curves at 0.001 M NaCl were much steeper (very high A_2 values) and the whole curve could therefore be utilized in the regression analysis. The molecular weight was regarded as a variable in the regression analysis since the value obtained with this technique would be dependent upon the speed of

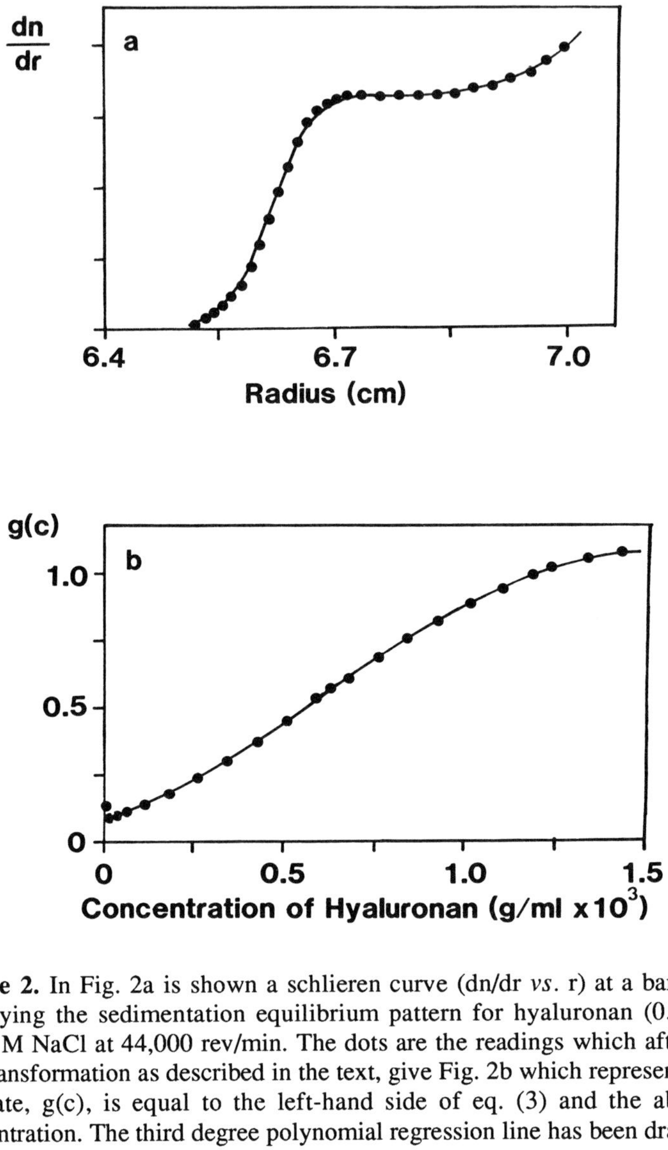

Figure 2. In Fig. 2a is shown a schlieren curve (dn/dr *vs*. r) at a bar angle of 50° displaying the sedimentation equilibrium pattern for hyaluronan (0.49 mg/ml) in 0.001 M NaCl at 44,000 rev/min. The dots are the readings which after integration and transformation as described in the text, give Fig. 2b which represents eq. (3); the ordinate, g(c), is equal to the left-hand side of eq. (3) and the abscissa is the concentration. The third degree polynomial regression line has been drawn in Fig. 2a to demonstrate how well the regression analysis fit to the actual readings.

rotation and the polydispersity. An example of a schlieren curve and the function representing eq. (3) together with polynomial regression curves are shown in Fig. 2.

In the original method suggested by Nichol *et al.*[1] only one point on the schlieren curve was utilized, *i.e.* the point where dc/dr^2 reaches its maximum value. Eqs. (6a) and (7a) in the paper of Nichol *et al.*[1] are in our notations:

$$A_2 = \frac{\omega^2 (1-\bar{v}\rho_0)}{4RT\,(dc/dr^2)_{max}} - \frac{1}{Mc_{max}} \tag{4}$$

$$A_3 = \frac{1}{3Mc_{max}^2} \tag{5}$$

where c_{max} and $(dc/dr^2)_{max}$ refer to the maximum in the (dc/dr^2) function. When employing these equations M was taken to be equal to M_w~70,000. The use of eqs. (4) and (5) in the calculations will for convenience be denoted the 'single point method'.

The procedure to calculate A_2 from the low speed - low concentration experiments was that described by Williams and co-workers[12,13] (see also Wan and Adams[21]). Three concentrations of hyaluronan (0.5 mg/ml - 1.7 mg/ml) were studied at each ionic strength. Equilibrium was attained for each concentration at three rotor speeds. The lowest speed was chosen to get a fringe displacement between meniscus and bottom of ~100 μm. The concentration difference between the meniscus and the bottom of the cell (Δc) was measured as fringe displacement and the original concentration (c_0) similarly determined by fringe displacement by a separate run in a 12 mm synthetic boundary cell[21]. For each concentration an apparent molecular weight, M_{app}, was calculated from c/c_0 as described[13]. Extrapolation of $1/M_{app}$ to zero concentration gives the weight-average molecular weight and the concentration dependence gives A_2. To obtain absolute values for c_0 for the final extrapolation the fringe displacement was transformed to concentration with the aid of the measured dn/dc at 546 nm.

3. RESULTS

Refractive Index Increment (dn/dc) and Partial Specific Volume ($\bar{v}$)

To evaluate our ultracentrifugation and light scattering data we must know dn/dc and $\bar{v}$ of hyaluronan at the different solvent conditions used as was carried out by Cleland[9]. $\bar{v}$ is a parameter in eqs. (3) and (4) and dn/dc is used in the conversion of schlieren and interference patterns to concentration gradients, as well as in the light scattering work. The results of our analyses are shown in Table 1.

Characterization of the Hyaluronan Preparation

The acid degraded hyaluronan sample was characterized by light scattering, osmometry, and low speed - low concentration sedimentation equilibrium. The results are given in Table 2. The weight-average molecular weight of the sample is in the order of 70,000 and the number-average molecular weight by osmometry ~50,000 indicating a moderate polydispersity. Table 2 also contains molecular data on the three dextran fractions.

Table 1. Partial specific volume and refractive index increment of hyaluronan, M~10^5 at 20°C determined at different salt concentrations.

Solvent NaCl (mol l^{-1})	Partial specific volume (ml/g)	Refractive index increment (ml/g) 436 nm	546 nm	633 nm
0.25	0.57	0.167	0.165	0.164
0.20	0.57	0.167	0.165	0.164
0.05	0.56	0.168	0.166	0.165
0.01	0.53	0.171	0.168	0.166
0.005	0.51	0.172	0.169	0.167
0.001	0.47	0.178	0.175	0.173

Values at 0.005 M, 0.05 M and 0.25 M NaCl were obtained by graphical extrapolation and interpolation. The simple Cauchy dispersion[16] was used to calculate the refractive index increment at 633 nm.

Table 2. Weight-average molecular weight (M_w) and number-average molecular weight (M_n) for the polymer samples used.

Polymer	$10^{-3}\times M_w$	$10^{-3}\times M_n$	M_w/M_n
Hyaluronan	73[a]	53[b]	1.4
	74[c]		
	82[d]		
Dextrans[e]:			
FDR 7314.3207/22	41.2	38.0	1.08
T500/3207/9	316	257	1.23
T500/3207/8	373	303	1.23

[a] Low angle light scattering;
[b] Osmometry;
[c] Low speed - low concentration sedimentation equilibrium;
[d] Molecular weight calculated from $[\eta]$ = 250 ml/g in 0.25 M NaCl using relationship $[\eta] = 0.12\ M^{0.86}$ derived by Wik (unpublished observations);
[e] Data from gel chromatography supplied by Dr. K. Granath.

Table 3. Virial coefficients of dextran in water determined by different methods[a].

M_w	2nd virial coefficient, $10^4 \times A_2$ (mol ml g^{-2}) Sedimentation equilibrium		2nd virial coefficient, Osmometry[b]	3rd virial coefficient, $10^3 \times A_3$ (mol ml^2 g^{-3}) Sedimentation equilibrium		3rd virial coefficient, Osmometry[b]
	(c)	(d)		(c)	(d)	
41200	4.1	6.1	5.5	3.5	1.1	2.8
316000	5.2	4.8	4.5	3.8	2.0	3.1
373000	4.2	4.0	4.4	3.0	1.2	3.1

[a] Experiments were performed at rotor speeds between 22000 and 34000 rev/min. Initial concentration varied between 22 and 56 mg/ml and concentration at bottom of the cell varied between 120 and 200 mg/ml. Equilibrium was attained after 96 h;
[b] Calculated values from virial coefficients determined by osmometry by interpolation applying a log-log relationship between number-average molecular weight and virial coefficients[41];
[c] Calculated by polynomial regression (see text);
[d] Calculated by single point method (see text).

Determination of Virial Coefficients of Dextran

The technique to determine virial coefficients by high speed - high concentration sedimentation equilibrium and polynomial regression was first tested on three well-characterized dextran fractions. The results are presented in Table 3 and compared with data obtained by the single point method and by osmometry. The very good agreement between the results from different techniques provides confidence in the present method. It is notable that polynomial regression gives much better agreement with osmometry data for A_3 than the single point method.

Determination of A_2 of Hyaluronan by Different Techniques

As discussed above errors in determination of the non-ideality term by high speed - high concentration sedimentation equilibrium may be introduced by polydispersity. It was therefore of prime interest to determine to what extent the polydispersity of the present hyaluronan sample would influence the results. A_2 was therefore determined by light scattering and low speed - low concentration sedimentation equilibrium and the values compared with the high speed - high concentration data (Table 4). The values are in general agreement with each other, with the exception that the light scattering data seem to deviate from the expected pattern at the lowest ionic strengths. The two techniques to calculate A_2 from high speed - high concentration experiments do also give different results in 0.001 M NaCl although the order of magnitude is the same.

The data in Table 4 suggest that the polydispersity effects cannot have any large influence on the determination of the non-ideality term in the high speed - high concentration experiments. It is therefore reasonable to use the technique to calculate also A_3 and higher order virial coefficients.

Table 4. Second virial coefficient, A_2, of hyaluronan determined by different methods. (For experimental conditions see Table 5.)

Solvent NaCl (mol l^{-1})	No. of expts[a]	2nd virial coefficient, $10^4 \times A_2$ (mol ml g^{-2})			
		Sedimentation equilibrium			Light scattering
		High speed[b]		Low speed[c]	
		(d)	(e)		
0.2	10	41±6	47±5	50	44
0.05	2	86±27	110±8	-	110
0.01	7	370±150	390±69	530	400
0.005	6	430±110	690±91	-	390
0.001	9	1700±1000	3500±470	3200	78

[a] Refer to high speed sedimentation equilibrium;
[b] Values are given with standard deviation;
[c] The concentration of hyaluronan and rotor speed were varied as follows: in 0.2 M NaCl: 0.4 - 1.5 mg/ml and 3200 - 7200 rev/min; in 0.01 M NaCl: 0.8 - 1.8 mg/ml and 5600 - 11000 rev/min; in 0.001 M NaCl: 0.8 - 1.7 mg/ml and 7200 - 14000 rev/min;
[d] Calculated by polynomial regression (see text);
[e] Calculated by single point method (see text).

Determination of the Higher Virial Coefficients by High Speed - High Concentration Sedimentation Equilibrium

A summary of the experimental data is given in Table 5. A large number of experiments were performed at different ionic strengths. The rotor speed and the concentration varied as described in the table, and the virial coefficients are given as mean values ±standard deviation. It was found that at lower ionic strengths it was necessary to include a fourth and fifth virial term at higher concentrations to describe the sedimentation equilibrium by polynomial regression. The table also gives the values of A_2 and A_3 calculated by the single point method. The latter calculations give a lower value of A_3 than the polynomial regression but this is readily explained in view of the fact that negligible contribution of higher order terms was assumed in the derivation of eqs. (4) and (5). A striking feature in Table 5 is the negative value of A_4. Possible interpretations will be discussed below.

Table 5. Virial coefficients of hyaluronan determined by high speed sedimentation equilibrium. Values are calculated by polynomial regression and are given as mean ± standard deviation. Values in italics were determined by the single point method.

Solvent NaCl (mol l^{-1})	No. of expts	$10^3 \times A_2$ (mol ml g^{-2})	$10^2 \times A_3$ (mol ml^{2}g^{-3})	$10^{-3} \times A_4$ (mol ml^3 g^{-4})	$10^{-6} \times A_5$ (mol ml^4 g^{-5})
0.2	10	4.1 ± 0.61	6.3 ± 3.0	-	-
		4.7 ± 0.48	*4.8 ± 2.8*		
0.05	2	8.6 ± 2.7	42 ± 30	-	-
		11.0 ± 0.8	*32 ± 20*		
0.01	7	37 ± 15	$(2.5 \pm 1.5) \times 10^2$	-4.7×10^{-2}	-
		39 ± 6.9	*$(6.8 \pm 3.0) \times 10^2$*		
0.005	6	43 ± 11	$(2.4 \pm 1.1) \times 10^3$	-3.3 ± 1.2	-
		69 ± 9.1	*$(6.8 \pm 3.0) \times 10^2$*		
0.001	9	170 ± 100	$(2.0 \pm 1.3) \times 10^4$	-81 ± 62	3.0 ± 2.3
		350 ± 47	*$(1.4 \pm 1.0) \times 10^3$*		

Experiments in 0.2 M NaCl were performed at rotor speeds between 15000 and 52000 rev/min; the initial concentration varied between 1 and 19 mg/ml and concentration at bottom of the cell from 9 to 54 mg/ml. At lower ionic strengths rotor speed varied from 30000 to 60000 rev/min and initial concentration from 0.4 to 4.5 mg/ml The concentration at bottom of the cell was 1.2 - 4.8 mg/ml, 2.3 - 5.9 mg/ml, 3.1 - 18 mg/ml and 5.9 - 17 mg/ml for experiments in 0.001 M, 0.005 M, 0.01 M and 0.05 M NaCl, respectively. For the initial and highest rotor speed equilibrium was attained after 72 - 144 h. Equilibrium times for successively lower speeds was 36-96 h.

There is relatively larger standard deviation in the values at low ionic strength and for the higher order virial coefficients. This is not surprising since the errors are magnified as more terms are needed in the curve fitting. The data is, however, consistent and is believed to give at least a semi-quantitative description of the non-ideality of hyaluronan at different ionic strengths. The value of the total non-ideality term of hyaluronan at different ionic strengths is given in Fig. 3.

In order to check that the evaluation of the virial coefficients accounted for all the material in the sample we have regularly calculated the total area under the schlieren curves and converted it to amount of hyaluronan with the aid of dn/dc. Generally, the calculated amount differed from the amount contained in the cell, as determined by uronic acid analyses, by less than 10%-20%. There was a tendency for an increasing difference with increasing rotor speed (less material in the cell) indicating that some loss of material was due to difficulties in observing material in the bottom region.

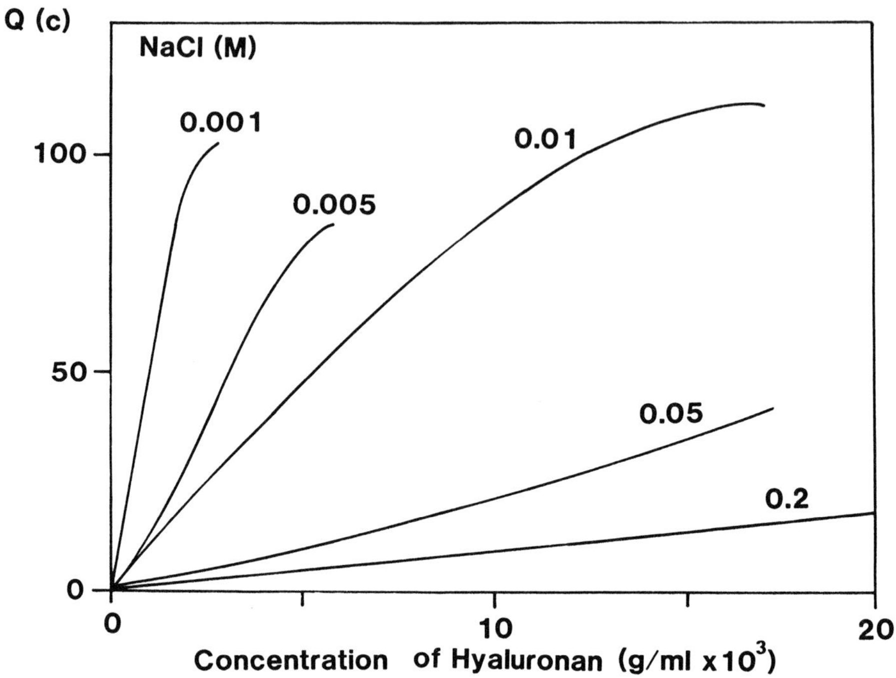

Figure 3. The non ideality term, Q(c), of hyaluronan at different ionic strengths. Q(c) is equal to $1 + 2A_2Mc + 3A_3Mc^2 + 4A_4Mc^3 + 5A_5Mc^4 + \ldots$. Q(c) has been calculated using virial coefficients from single experiments at the highest polymer concentration at the bottom of the cell at each ionic strength.

The Effect of Rotor Speed on the Calculated Virial Coefficients

Many runs were performed at different rotor speeds. It was found that within the errors of the method the apparent molecular weight slightly decreased while the value of A_2 slightly increased with increasing speed. This is an effect expected due to polydispersity.

4. DISCUSSION

Reliability of the Technique

Although the application of the present method to dextran and hyaluronan seems quite successful, two inherent short-comings need comments. First, the readings of low values of dc/dr suffer from low accuracy and therefore it is hard to obtain correct values of the expression of the left-hand side of eq. (3) at low concentrations. This problem was at least partly overcome by using an extrapolation procedure before calculating the concentration by integration and by only using points from the inflection point onwards for the regression analysis. At the inflection point (see Fig. 1a) the ordinate is quite large (about 50% of the maximum value) and can be read with high accuracy. Using this procedure almost all scatter of experimental points was avoided. However, the disregard of data at low concentrations makes impossible the determination of 1/M in eq. (3) with any kind of accuracy and we have regarded M as a variable in the regression analysis.

Secondly, even after having deleted the points at low concentrations, there is still the problem of polydispersity. The observation that A_2 was slightly dependent on rotor speed indicates that polydispersity does influence the results.

A comparison between the single point method and the polynomial regression shows (Table 5) that in the determination of A_2 there is a larger standard deviation in the latter type of calculation. This is an expression for the fact that there is a larger scatter of experimental points at lower concentration. The single point method relies on the values of $(dc/dr^2)_{max}$ and these values occur at quite high concentrations. The value of A_3 calculated by the two methods differs considerably (Tables 3 and 5), especially for hyaluronan at low ionic strength. From experiments on dextran it is apparent that the polynomial regression gives better agreement with values determined by osmometry. This is also readily explained by the approximations included in the derivation of eq. (5) used to determine A_3 by the single point method; higher terms in the virial expansion were assumed to be negligible[1]. The polynomial regression should therefore be superior to the single point method in determining A_3.

Comparison with our Results with Previously Published Data

The Second and Third Virial Coefficients The measured value of A_2 in 0.2 M NaCl is about twice as high as that generally recorded[1,23,24,25]. Most studies have, however, been performed on high molecular weight hyaluronan. Previously published values of A_2 at lower ionic strengths are generally surprisingly low, with the exception of that obtained by Cleland[23] who found A_2 to be 7.7×10^{-3} in 0.01 M NaCl by light scattering. This value is still only 20% of our value. It is notable that many other determinations of virial coefficients of hyaluronan at low ionic strength have been performed by light scattering[9,24,26,27]. Also our light scattering value (Table 4) at the lowest ionic strength is markedly low. There may be an inherent difficulty in studying polyelectrolytes at low ionic strength by light scattering. As pointed out by earlier workers[28,29,30] intermolecular interference in the scattering appears in polyelectrolyte solutions when electrostatic repulsions between the molecules lead to a non-random distribution; this may interfere in the normal extrapolation procedures to obtain molecular parameters. However, recently Reed and co-workers[27] obtained repeatable and reasonable values for the molecular parameters of hyaluronan in water by scattering technique.

The value of A_3 varies considerably in those few instances when previous investigators have tried to estimate its value. In most cases this must be ascribed to inadequate methodology. The most reliable data[9] are, however, in good agreement with ours. The estimates of A_3 in 0.5 M, 0.1 M and 0.01 M NaCl are 7×10^{-3}, (20-30)$\times10^{-3}$ and 250×10^{-3} mol ml^2 g^{-3}, respectively (compare with Table 5).

The Partial Specific Volume ($\bar{v}$) The only values of $\bar{v}$ for hyaluronan which to our knowledge have been published previously are 0.66 ml/g, recorded by pycnometry[31], and 0.653 ml/g, estimated from density gradient centrifugations[25]. Both values are considerably higher than those in Table 1. The value from, pycnometry does, however, not state the solvent used and essentially pycnometry to determine the density is also a less sensitive technique than that used by us. The value from density gradient centrifugations is a calculated value assuming a certain caesium ion binding to the polysaccharide.

The Refractive Index Increment (dn/dc) Several authors have published values of the refractive index increment of hyaluronan[9,24,26,27,32,33,34] and values between 0.14 ml/g and 0.18 ml/g have been reported. The large variation is probably due to differences in the methods used to determine the concentration of hyaluronan to varying purity of the samples. The high reproducibility in the analysis and the good agreement between two independent methods used for the concentration determinations make us believe that the present values for our sample cannot be much in error.

Interpretation of Data

The virial coefficients describe the interaction between the solute molecules. A_2 has in general terms been ascribed to the excluded volume effect and has been interpreted in terms of an effective size of each molecule. According to Flory[35] A_2 is related to the effective diameter, d, by

$$A_2 = \frac{Nu}{2M^2} = \frac{2\pi N d^3}{3M^2} \qquad (6)$$

where u is the volume excluded around each individual molecule, M is the molecular weight and N is Avogadro's number. The values of A_2 in Table 5 have been used to calculate d and the data are given in Table 6. The effective diameter increases, as expected, with decreasing ionic strength. This is due to the expansion of the coil at low salt concentrations and to increasing repulsions between individual molecules. The value at 0.2 M NaCl can be compared with the size of the hyaluronan molecule from light scattering. Wik (unpublished data) has derived a relationship between molecular weight and mean square radius of gyration, R_g. According to this relation, hyaluronate with molecular weight 730,000 should have a mean square radius of gyration of 29 nm. The radius of the hydrodynamic unit of the molecule is 0.67 R_g [36] and the diameter of the hydrodynamic unit thus 39 nm. Similarly, one can calculate a diameter of the hydrodynamic unit from the limiting viscosity number (250 ml/g) [36] and get a value of 29 nm. Our A_2 value therefore agrees with what is expected from the mutual volume exclusion by large solvated coil molecules.

Table 6. Effective diameters of the hyaluronan molecule, calculated from the second virial coefficient (A_2), at different NaCl concentrations.

Solvent NaCl (mol l^{-1})	$10^3 \times A_2$ (mol ml g^{-2})	Effective diameter (nm)
0.2	4.1	26
0.05	8.6	33
0.01	37	54
0.005	43	57
0.001	170	90

The observation that A_4 is negative at lower ionic strength deserves a comment. It is usually assumed that the virial coefficients are independent of concentration. The evaluation of the coefficients by polynomial regression in this paper is based on this assumption. The assumption that the virial coefficients are concentration independent is valid as long as they are interpreted as *thermodynamic* parameters without being *physically* interpreted. If, however, A_2 is related to the effective volume of the molecule as discussed above (eq. (6)), then it would not be surprising if A_2 were concentration dependent. The negative A_4 observed may actually be a compensation for a decrease in A_2 with increasing concentration. There are several experimental results which indicate that the effective volume of the hyaluronan coil decreases with increasing concentration. Ogston and Preston[37] have shown that dextran molecules are osmotically compressed at high concentrations. Harper and Preston[38] have performed viscosity and gel chromatographic experiments on proteoglycans which confirm this concept. Ogston[39] has provided a theoretical framework for this effect in discussions on effect of concentration on the configuration of chain polymers.

The fact that the negative A_4 is most pronounced at the lowest ionic strengths can also be the result of an increase in solution ionic strength at high hyaluronan concentrations due to the contribution of the polymer and its counterions. As is apparent from Table 5, an increase in ionic strength will dramatically decrease the value of A_2. It may be appropriate to compare the present data with the polyelectrolyte pattern of the reduced viscosity of hyaluronan[39]. The reduced viscosity at low salt concentration decreases dramatically with increasing concentration of hyaluronan which can be interpreted as a decrease in the hydrodynamic volume of the molecule with increasing concentration.

The negative A_4 can also be interpreted as an aggregation of the hyaluronan molecules at high concentration. Such behaviour may be considered in view of the recent suggestions of Scott[40] on the structure of hyaluronan.

Concluding Remarks

The present work suggests that the non-ideality in concentrated solutions of hyaluronate and other polymers can be determined by sedimentation equilibrium. There is at present no other technique which can be easily utilised in the concentration regime studied here.

ACKNOWLEDGEMENTS

We are indebted to Prof. T.C. Laurent and Prof. L.-O. Sundelof for their advice and interest in this project.

REFERENCES

1. L.W. Nichol, A.G. Ogston and B.N. Preston, *Biochem. J.*, 1967, *102*, 407.
2. T.C. Laurent, *Acta Otolaryngol (Stockh.)*, 1987, *Suppl. 442*, 7.
3. T.C. Laurent, *Biochem. J.*, 1964, *93*, 106.
4. T.C. Laurent, B.N. Preston and B. Carlsson, *Eur. J. Biochem.*, 1974, *43*, 231.
5. T.C. Laurent, *Eur. J. Biochem.*, 1971, *21*, 298.
6. T.C. Laurent, I. Bjork, A. Pietruszkiewicz and H. Persson, *Biochem. Biophys. Acta*, 1963, *78*. 351.
7. A.G. Ogston, B.N. Preston and J.D. Wells, *Proc. Roy. Soc. Lond. Ser. A*, 1973, *333*, 297.
8. B.N. Preston, B. Obrink and T.C. Laurent, *Eur. J. Biochem.*, 1973, *33*, 401.
9. R.L. Cleland, *Biopolymers*, 1968, *6*, 1519.
10. S.G. Chu and P. Monk, *J. Polym. Sci., Polym. Phys. Ed.*, 1977, *15*, 1163.
11. S.G. Chu and P. Munk, *Macromolecules*, 1978, *11*, 101.
12. H.W. Osterhoudt and J.W. Williams, *J. Phys. Chem.*, 1965, *69*, 1050.
13. D.A. Albright and J.W. Williams, *J. Phys. Chem.*, 1967, *71*, 2780.
14. T. Bitter and H.M. Muir, *Anal. Biochem.*, 1962, *4*, 330.
15. E.A. Balazs, K.O. Berntsen, J. Karossa and D.A. Swann, *Anal. Biochem.*, 1965, *12*, 547.
16. M.B. Huglin, 'Light Scattering in Polymer Solutions', Academic Press, New York, 1972.
17. P.-H. Norberg and L.-O. Sundelof, *Makromol. Chem.*, 1964, *77*, 77.

18. K.A. Granath, *J. Colloid Sci.*, 1958, *13*, 308.
19. H. Fujita, 'Mathematical Theory of Sedimentation Analysis', Academic Press, New York, 1960.
20. S.E. Harding, K.M. Vårum, B.T. Stokke and O. Smidsrød in 'Advances in Carbohydrate Analysis', C.A. White ed., 1991, pp 63-144, JAI Press, London.
21. P.J. Wan and E.T. Adams, *Biophys. Chem.*, 1976, *5*, 207.
22. C.H. Chervenka, 'A Manual of Methods', 1969, Spinco Division of Beckman Instruments, Palo Alto, U.S.A.
23. R.L. Cleland and J.L. Wang, *Biopolymers*, 1970, *9*, 799.
24. B.N. Preston, M. Davies and A.G. Ogston, *Biochem. J.*, 1965, *96*, 449.
25. P. Silpananta, J.R. Dunstone and A.G. Ogston, *Biochem.J.*, 1968, *109*, 43.
26. J. Schutz, H. Hemmetsberger, F. Sasshofer, M. Tomiska and H. Tretthart, *Hoppe-Seyler's Z. Physiol. Chem.*, 1967, *348*, 711.
27. C.E. Reed, X. Li and W.F. Reed, *Biopolymers*, 1989, *28*, 1981.
28. P. Doty and R.F. Steiner, *J. Chem. Phys.*, 1949, *17*, 743.
29. P. Doty and R.F. Steiner, *J. Chem. Phys.*, 1952, *20*, 85.
30. A. Takahashi, T. Kato and M. Nagasawa, *J. Phys. Chem.*, 1967, *71*, 2001.
31. L. Vara, *J. Biol. Chem.*, 1955, *217*, 651.
32. B.S. Blumberg and G. Oster, *Science*, 1954, *120*, 432.
33. T.C. Laurent and J. Gergely, *J. Biol. Chem.*, 1955, *212*, 325.
34. T.C. Laurent, M. Ryan and A. Pietruszkiewicz, *Biochem. Biophys. Acta*, 1960, *42*, 476.
35. P.J. Flory 'Principles of Polymer Chemistry', Cornell University Press, Ithaca, New York, 1953.
36. C. Tanford, 'Physical Chemistry of Macromolecules', John Wiley and Sons, New York, 1961.
37. A.G. Ogston and B.N. Preston, *Biochem. J.*, 1979, *183*, 1.
38. G.S. Harper and B.N. Preston, *J. Biol. Chem.*, 1987, *262*, 8088.
39. A.G. Ogston in 'Chemistry and Technology of Water Soluble Polymers', C.A. Finch, ed., Plenum Publishers Corp., New York, 1981, p. 203.
40. J.E. Scott in 'The Biology of Hyaluronan', D. Evered and J. Whelan eds., Ciba Foundation Symposium No. 124, John Wiley and Sons, London 1989, p. 6.
41. H. Vink, *Eur. Polym. J.*, 1971, *7*, 1411.

31

Sedimentation Analysis of Synthetic Polymers

By Hidematsu Suzuki

DEPARTMENT OF BIOENGINEERING, NAGAOKA UNIVERSITY OF TECHNOLOGY, NAGAOKA, NIIGATA-KEN 940-21, JAPAN

1. INTRODUCTION

This chapter might be expected - as its title might imply - to deal specifically with applications of ultracentrifugation to polymer science. However, it considers in detail some fundamental aspects of data analysis which are common to a wide range of macromolecular systems. Indeed, for the last three decades, the analytical ultracentrifuge has been indispensable in some fields and useful in many others of polymer science. In particular, the following characteristics of polymers (flexible or wormlike) have been determined using the technique:

(a) the sedimentation coefficient and its distribution;
(b) the average diffusion coefficient;
(c) the sedimentation characteristics of a polymer in a mixed medium;
(d) average molecular weights and solute-solute interactions;
(e) the molecular weight distribution;
(f) the chemical potential;
(g) the distribution of chemical composition of copolymers.

If I can deal with each of these briefly in turn. The first three of these can be studied by sedimentation velocity or synthetic boundary experiments. The *distribution of sedimentation coefficient* of a sample of flexible polymer has been determined in the past mainly for estimating its *molecular weight* distribution[1]. This conversion to a molecular weight distribution is most conveniently performed for the sedimentation coefficient, s, measured under "theta" conditions and for a randomly coiled polymer where s is approximately proportional to the square root of the molecular weight. However, in practice, the procedure is much more involved and tedious; the sedimentation coefficient obtained at finite concentration must be corrected for the pressure effect and extrapolated to zero concentration. In addition, this method is based on the assumption that each species sediments independently of the others: this latter assumption is known to be in variance with experimental facts[2]. As a consequence, applications for this particular purpose have not been made (with a few exceptions) since the advent of the alternative - and easier to use - procedure of size-exclusion chromatographly.

From ordinary moving-boundary sedimentation runs, the weight-average *translational diffusion coefficient* may be determined[3] by means of Schumaker's method[4]. However, synthetic boundary runs are more direct and better suited for this purpose (the reader is referred to Chervenka (i)* for a good general reference for this.) An ultracentrifuge is a convenient apparatus but no longer the method of choice: it is difficult to establish exact experimental conditions on this apparatus, and its applicability is, in practice, limited to a certain range of diffusion coefficient. Sedimentation of one polymer in a *mixed medium* of another polymer and a common solvent was investigated by Ogston *et al.*[5] and lately by Nemoto *et al.*[6] The latter group used as a medium an isorefractive and isopycnic pair of polymer and solvent.

Phenomena (d) to (f) are best studied by sedimentation equilibrium, a powerful absolute method for determining average molecular weights and characterising solute-solute interactions. In practice, however, a great deal of care must be taken in both experiment and data analysis particularly with regard to non-ideality and polydispersity. Since these are major and serious problems intrinsic to non-ideal polydisperse systems, words and figures are definitely required in order to understand underlying equations for data analysis.

Molecular weight distributions are routinely measured by size-exclusion chromatography these days, and researchers hardly use an ultracentrifuge for this purpose. Determination of the *chemical potential* of a given solution was a classical consideration in the Svedberg era and this method has been applied successfully to non-ideal polydisperse systems by, for example, Scholte[7] and others[8,9] (these studies will not be covered here, and readers can refer to their original papers). Finally the *distribution of chemical composition* of a given copolymer sample can be investigated best by density gradient sedimentation equilibrium (this topic has been treated by Mächtle earlier in this book - Chapter 10 - and again will not be considered here.)

What I propose to focus on, for the rest of the Chapter, is the subject of how to determine the various types of average molecular weights and solute-solute interaction parameters of a given polydisperse polymer sample. In this respect, this Chapter is intended to supplement earlier works by Williams (Chapter 1 of ref. (ii)), and by Fujita (Chapter 5, Section IV of ref. (iii)) to both of which the reader is referred.

2. CONVENTIONAL ANALYSES REVISITED

$1/M_w^{app}$ *vs.* $\bar{c}$ Plots

There are two conventional methods used for non-ideal polydisperse systems. One of them is to plot the reciprocal of the apparent weight-average molecular weight, M_w^{app}, against the average concentration, $\bar{c}$ [10]. This concentration is defined as the arithmetical mean of the equilibrium concentrations at both ends of the solution column.

This method of analysis is known to be exact for monodisperse solutes and to

* General references are indicated by Roman numerals (i), (ii) and (iii).

yield a linear plot unless the non-ideality is considerable. In addition, as a single rotor-speed run on solutions of different concentrations is sufficient, the method has frequently been adopted to the analysis of synthetic polymers having a narrow molecular weight distribution. An example is shown in Fig. 1: here, the sample examined by Sakato and Kurata[11] is a nearly monodisperse poly(α-methylstyrene) dissolved in two solvents (one "good" and the other "poor"). Although the data in the good solvent, benzene, deviate from linearity at higher concentration, the two extrapolations to $\bar{c} = 0$ of the $1/M_w^{app}$ data obtained in the different solvents are seen to yield a common intercept of the ordinate. This implies that for solutes like this, the present method *does* work irrespective of the quality of the solvent used; the $\bar{c}$ can be considered as a concentration variable sufficient to take rotor speed effects into account. Accordingly, the initial slope of each plot may be approximated to the light-scattering second virial coefficient, B_2, a measure of the strength of solute-solute interaction in a given solvent.

Applications of this method for molecular weight analysis to polydisperse solutes[12] have been supported theoretically (although caution has to be expressed in dealing with estimates for B_2) by Fujita[13]: he claims the method works provided that the solute is not too polydisperse nor its solution too non-ideal. In actuality, it is rather difficult to know prior to experiments whether a given solution meets these conditions or not. If the solute were very polydisperse, then this method might yield wrong results or would not work at all, even though the deviation from non-ideality might be very little: such an example is given in Fig. 2. Here, solutions of a polycondensate sample were examined at different, low rotor speeds[14]. Since its weight-average molecular weight is naturally unique, concentration extrapolations must yield a common intercept of the ordinate. Obviously, it is very difficult to make such extrapolations, however. It can be recognized, therefore, that the $\bar{c}$ method has a limited applicability, although this polycondensate sample is *exceptionally* polydisperse.

$1/M_w^{app}$ *vs.* λ^2 Plots

The other method is to plot $1/M_w^{app}$ against the square of the generalized speed parameter, λ, as defined by eq. (7a) below. Requirements may be satisfied by examining solutions of different concentrations at several different rotor speeds. Extrapolations to $\lambda^2 = 0$ allow us to determine $1/M_w{}^{app}$ values free from rotor speed effects, followed by a usual plot of those *versus* the initial concentration.

Although these double extrapolations are demanding and time consuming, this is the standard way of data analysis, proposed by Fujita[15], for non-ideal polydisperse solutes. An example is shown in Fig. 3. Here, the sample examined by Utiyama *et al.*[16] is a binary mixture of standard polystyrene samples different in molecular weight, both having narrow molecular weight distribution and known molecular weights. From this figure, it can be seen that the dependence of $1/M_w{}^{app}$ on λ^2 is not necessarily linear especially for solutions of higher concentrations.

This nonlinear tendency will be strongly marked for very polydisperse solutes, as can be seen in Fig. 4. Here the same data as those in Fig. 3 are replotted against λ^2.

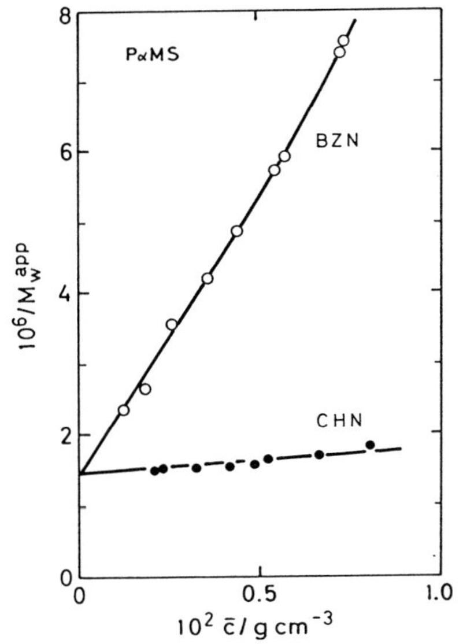

Figure 1. Plots of $1/M_w^{app}$ against $\bar{c}$ for monodisperse poly(α-methylstyrene) in benzene (○) at 30°C and in cyclohexane (●) at 39°C. Reprinted with modification and permission from Sakato and Kurata, *Polym. J.*, 1970, *1*, 260. Copyright (1970) Society of Polymer Science, Japan.

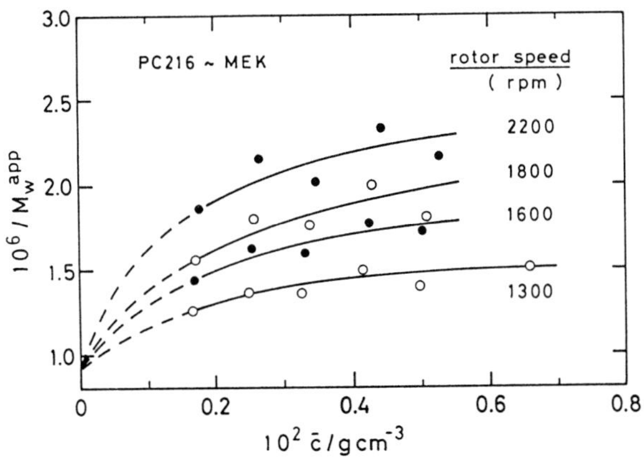

Figure 2. Plots of $1/M_w^{app}$ against $\bar{c}$ for extremely polydisperse polycondensate PC216 in 2-butanone at 25°C and four different rotor speeds. Reprinted with modification and permission from Gordon, Leonis and Suzuki, *Proc. Roy. Soc., Lond.*, 1975, A *345*, 207. Copyright (1975) Royal Society, London.

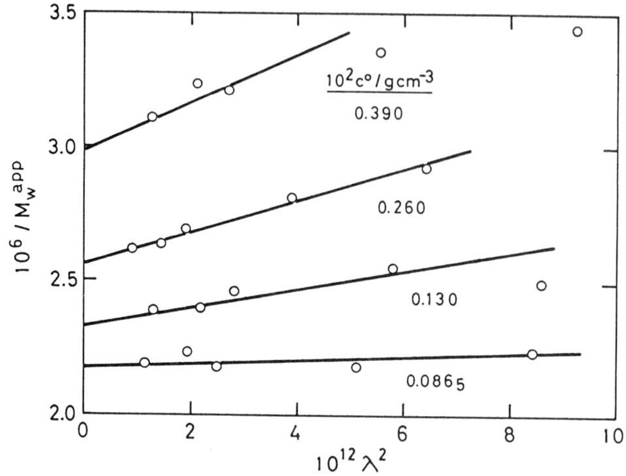

Figure 3. Plots of $1/M_w^{app}$ against λ^2 for a binary mixture of polystyrene in 2-butanone at 25°C. Reprinted with permission from Utiyama, Tagata and Kurata, *J. Phys. Chem.*, 1969, *73*, 1448. Copyright (1969) American Chemical Society.

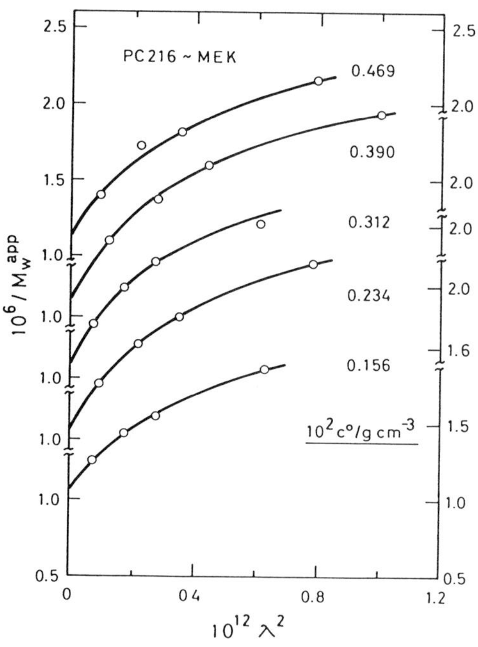

Figure 4. Plots of $1/M_w^{app}$ against λ^2 for polycondensate PC216 in 2-butanone at 25°C. Reprinted with modification and permission from Gordon, Leonis and Suzuki, *Proc. Roy. Soc., Lond.*, 1975, A *345*, 207. Copyright (1975) Royal Society, London.

Again, it is still difficult to make such non-linear extrapolations. Although new ways are definitely needed for successful analysis of situations like this, it cannot be forgotten that the present method is the standard way for determining both M_w and B_2: the problem in this particular case is lack of data at lower values of λ.

From a brief discussion given above, it may be recognized that, in ultracentrifugation, data treatment for non-ideal polydisperse systems is not as straightforward as in light-scattering photometry or in osmometry. This unfavourable feature of sedimentation equilibrium stems from the non-uniform spatial distribution of solute molecules. Since the degree of the non-uniform distribution decreases at low values of λ and the non-ideality factor is linked to λ, too, as will be seen below, it is advisable to carry out measurements at as *low* a λ value as possible: this means that the rotor speed be lower and the solution column height be shorter. As an example, a photographic record of such experiments is given in Fig. 5[14]. If, for some practical reasons, this advice cannot be followed, a great deal of difficulty will be met, as has been seen in Figs. 2 and 4. The theoretical background behind such unfavourable features and how to overcome them will be described in the following section.

So far, we have "revisited" the conventional methods for determining the weight-average molecular weight, but not for other averages like the number- and the z-average molecular weights: it goes without saying that these molecular weights are also of considerable importance in molecular characterisation. Moreover, it is *analytical ultracentrifugation* that allows us to determine these average molecular weights without reference materials and without limits inherent to the methods. Newer procedures for determining these average molecular weights will be explained with equations and illustrations after reviewing the theory for the conventional methods.

3. SEDIMENTATION EQUILIBRIUM METHODS FOR CHARACTERISING NON-IDEAL POLYDISPERSE SYSTEMS

Principles

Consider an incompressible solution containing a single solvent (component 0) and polymeric solutes of q species (components 1, 2, ..., q) differing in molecular weight, M, but with negligible difference in partial specific volume, $\bar{v}$. For equilibrium in such a system exposed to a centrifugal field at constant temperature, T, it is necessary that the total potential of a component is uniform throughout the system. That is

$$\mu_i + M_i\phi = \text{const} \tag{1}$$

Here, μ_i is the chemical potential per mol at a given radial position, r (cm), and ϕ the centrifugal field potential, $\phi = -(1/2)\omega^2 r$ (ω: the angular velocity of rotation). On differentiating with respect to r, we get

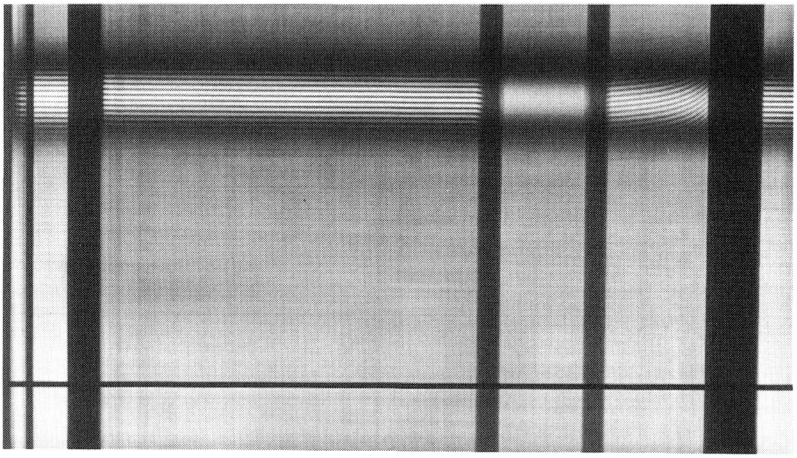

Figure 5. Typical "combined optics" pattern for sedimentation equilibrium experiments. Schlieren phase angle 90°; 900 rev/min (polycondensate PC214, c^0 = 0.606 g dm^{-3}, 25°C). Reprinted with permission from Gordon, Leonis and Suzuki, *Proc. Roy. Soc., Lond.*, 1975, A *345*, 207. Copyright (1975) Royal Society, London.

$$\frac{d\mu_i}{dr} = \bar{v} M_i \frac{dP}{dr} + \sum_{k=1}^{q} \left(\frac{d\mu_i}{dc_k}\right)_{T,P,c_j} \frac{dc_k}{dr} \tag{2}$$

where P is the pressure and c the equilibrium concentration (in g/cm^3) at the position, r. On the other hand, from the Gibbs-Duhem equation and eq. (1), we obtain

$$\frac{dP}{dr} = \rho\omega^2 r$$

where ρ is the solution density. Substitution of this equation into eq. (2) gives us

$$M_i (1-\bar{v}\rho)\, \omega^2 r = \sum_{k=1}^{q} \left(\frac{d\mu_i}{dc_k}\right) \frac{dc_k}{dr} \tag{3}$$

This is the basic equation for sedimentation equilibrium.

Working Equations

When the solution non-ideality is sufficiently small the activity coefficient, y, on the c scale may be expressed as

$$\ln y_i = M_i \sum_{k=1}^{q} B_{ik} c_k + O(c_k c_m) \tag{4}$$

where the B_{ik} are the interaction parameters. For a solution adequately dilute, its density may be expressed as a linear equation of concentration:

$$\rho = \rho_0 + (1-\bar{v}\rho_0) \sum_{k=1}^{q} c_k + O(c^2) \tag{5}$$

where ρ_0 is the solvent density. Substitution of these equations into eq. (3) yields, after re-arrangement of terms

$$\lambda M_i c_i = \frac{dc_{qs}}{d\xi} + M_i c_i \sum_{k=1}^{q} B_{ik} \frac{dc_k}{d\xi} + \bar{v} \sum_{k=1}^{q} c_k \frac{dc_i}{d\xi} \tag{6}$$

with

$$\lambda = \frac{(1-\bar{v}\rho_0)\omega^2(b^2-a^2)}{2RT} \tag{7a}$$

and

$$\xi = \frac{r^2-a^2}{b^2-a^2} \tag{7b}$$

where λ is a generalized speed parameter, ξ a reduced radial distance, and a and b the actual radial distances, respectively, to the meniscus and to the bottom of the solution column. For later convenience, two parameters are introduced here: the weight fraction, $g_i = c_i^0/c^0$ and the relative equilibrium concentration, $\theta_i = c_i/c_i^0$, where $c^0 = \Sigma c_i^0$ is called the initial concentration. With these parameters, eq. (6) may be expressed as

$$\lambda M_i\theta_i = \frac{d\theta_i}{d\xi} + c^0\left\{M_i\theta_i\sum_{k=1}^{q} B_{ik}\frac{d\theta_k}{d\xi}g_k + \bar{v}\sum_{k=1}^{q}\theta_k\frac{d\theta_i}{d\xi}g_k\right\} \qquad (i = 1, 2, ..., q) \tag{8}$$

The solution of the above differential equation θ_i may be expressed as a series function in c_0 (see ref. iii):

$$\theta_i = \theta_{i0} + c^0\theta_{i1} + ... \tag{9}$$

under the conservation of mass conditions as expressed by

$$\int_0^1 \theta_{i0}\, d\xi = 1; \qquad \int_0^1 \theta_{ij}\, d\xi = 0 \qquad (i = 1, 2, ..., q;\ j \geq 1) \tag{10}$$

Here, the θ_{ij} are unknown functions of ξ to be determined. Substitution of eq. (9) into eq. (8), followed by comparison of terms of equal powers in c_0, yields a set of equations for θ_{ij}. The first two equations and their solutions are as follows[12]:

$$\lambda M_i\theta_{i0} = \frac{d\theta_{i0}}{d\xi} \tag{11a}$$

$$\theta_{i0} = A_i \exp(\lambda M_i\xi) \tag{11b}$$

with

$$A_i = \frac{\lambda M_i}{\exp(\lambda M_i) - 1}$$

$$\lambda M_i\theta_{il} = \frac{d\theta_{il}}{d\xi} + \sum_{k=1}^{q}\left\{M_iB_{ik}\theta_{i0}\frac{d\theta_{k0}}{d\xi} + \bar{v}\theta_{k0}\frac{d\theta_{i0}}{d\xi}\right\}g_k \tag{11c}$$

$$\theta_{il} = \theta_{i0}\sum_{k=1}^{q} M_i\bar{B}_{ik}\left\{\frac{A_iA_j}{A_{ij}} - \theta_{k0}\right\}g_k \tag{11d}$$

$$\bar{B}_{ik} = B_{ik} + \frac{\bar{v}}{M_k} \tag{12}$$

with

$$A_{ik} = \frac{\lambda(M_i + M_k)}{\exp \lambda(M_i + M_k) + 1}$$

The θ_{il} will be required later on. Here θ_{i0} and $d\theta_{i0}/d\xi$ replace the θ_i and $d\theta_i/d\xi$, respectively, which appear in the right-hand-side of eq. (8). Under this pseudo-ideal solution approximation, eq. (8) may be expressed as

$$\lambda M_i \theta_i = \frac{d\theta_i}{d\xi} + c^0 \lambda \sum_{k=1}^{q} M_i M_k \bar{B}_{ik}(\theta_{i0}\theta_{k0}) g_i g_k \tag{13}$$

The integration indicated and the summation over all species, after multiplying g_i, yield the following result:

$$M_w^{app} = M_w - c^0 \sum_{i=1,k=1}^{q,q} M_i M_k \bar{B}_{ik} \left\{ \int_0^1 \theta_{i0}\theta_{k0}\, d\xi \right\} g_i g_k \tag{14}$$

$$M_w^{app} = \frac{\Delta c}{\lambda c^0} \tag{15}$$

with

$$\int_0^1 \theta_{i0}\theta_{k0}\, d\xi = \frac{A_i A_k}{A_{ik}}$$

$$= \frac{\lambda M_i M_k \exp[\lambda(M_i + M_k) - 1]}{(M_i + M_k) \{\exp(\lambda M_i) - 1\} \{\exp(\lambda M_k) - 1\}} \tag{16}$$

Here, M_w is the weight-average molecular weight and Δ in ΔX is the operator to take the difference in quantity X between the positions a and b. To extrapolate M_w^{app} one often uses its reciprocal:

$$\frac{1}{M_w^{app}} = \frac{1}{M_w} + c^0 \frac{1}{M_w^2} \sum_{i=1,k=1}^{q,q} M_i M_k \bar{B}_{ik} \left\{ \int_0^1 \theta_{i0}\theta_{k0}\, d\xi \right\} g_i g_k + \ldots \tag{17}$$

The initial slope on a plot of $1/M_w^{app}$ against c^0 depends on ξ through the factor

$$\int_0^1 \theta_{i0}\theta_{k0}\, d\xi$$

which may be approximated as

$$\int_0^1 \theta_{i0}\theta_{k0}\, d\xi \approx 1 + \frac{\lambda^2}{12} M_i M_k + \frac{\lambda^4}{720} M_i^2 M_k^2 (M_i M_k - M_i^2 - M_k^2) + \ldots$$

$$(\lambda M_i \ll 1;\ i=1, 2, \ldots, q) \qquad (18)$$

It is seen from eq. (18) that at the limit of $\lambda \rightarrow 0$, the initial slope approaches to the light-scattering virial coefficient, B_2, as defined by

$$B_2 = \frac{1}{M_w^2} \sum_{i,k} M_i M_k \bar{B}_{ik}\, g_i g_k \qquad (19)$$

The standard method as shown in Figs. 3 and 4 is based on eq. (17) together with the approximation, eq. (18). This approximation is quite safe only for the cases that solutes having the narrow molecular weight distribution are examined under the experimental condition of $\lambda M_w \ll 1$. Otherwise, the expansion of exponentials in eq. (16) cannot be allowed, and in those extreme cases where higher molecular weight species are contaminated and/or λ values are large enough, eq. (16) will reduce to an asymptotic form[14]:

$$\int_0^1 \theta_{i0}\theta_{k0}\, d\xi \approx \frac{\lambda M_i M_k}{M_i + M_k} \qquad (\lambda M_i \gg 1;\ i = 1, 2, \ldots, q) \qquad (20)$$

This is the detail behind Figs. 3 and 4.

The data shown in Fig. 4, when replotted against λ, yielded reasonable straight lines, allowing one to estimate the M_w value[14]. This type of plot is deduced as stated above and supported by numerical results on a hypothetical mixture[14]. Here, results of more direct experimental investigation will be presented in the next two figures.

The sample used is a binary mixture of monodisperse polystyrenes, whose molecular characteristics determined by light scattering are summarized in Table 1. These data together with the B_{12} value estimated are enough to calculate the λ-dependence of $1/M_w^{app}$, which is represented, for each concentration, in dot-dash lines in Figs. 6 and 7.

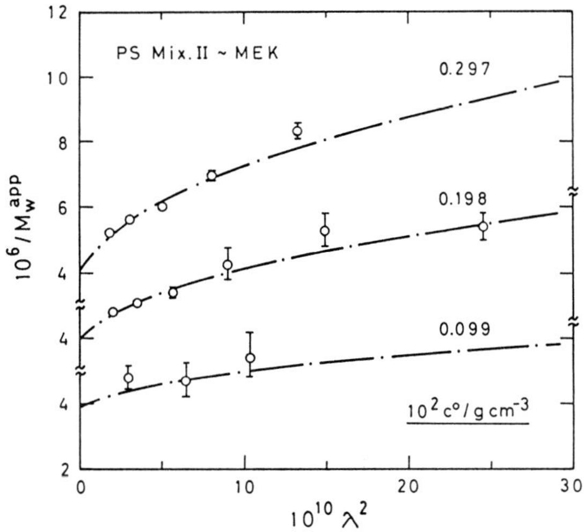

Figure 6. Plots of $1/M_w^{app}$ against λ^2 for comparison of theoretical predictions (dot-dash lines) with experimental results on polystyrene mixture II (see Table 1) in 2-butanone at 30°C.

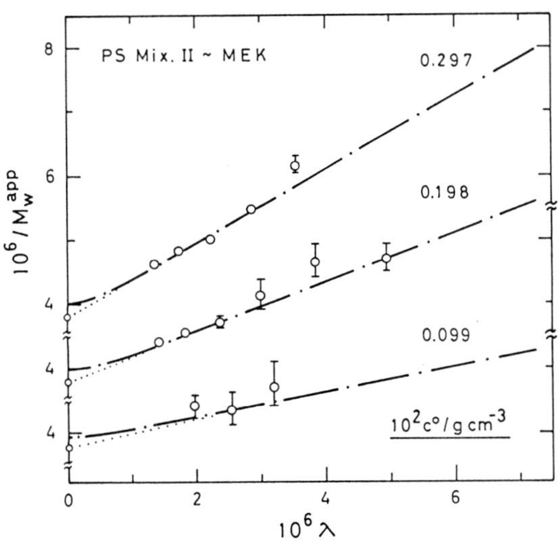

Figure 7. Plots of $1/M_w^{app}$ against λ for the comparison stated in the legend to Fig. 6. Open circles on the ordinate denote the reciprocal of the true weight-average molecular weight of polystyrene mixture II.

Table 1. Molecular characteristics of monodisperse polystyrenes and their binary mixture II in 2-butanone.

Species	M_w (g/mol)	B_2 (ml.mol.g^{-2})	g_i for Mix. II
1	1.1×10^4	8.0×10^{-4}	0.951
2	5.2×10^6	$1.5_3\times10^{-4}$	0.049
Mix. II	$2.6_5\times10^5$	$0.7_2\times10^{-4}$	...

($M_n : M_w : M_z = 0.04 : 1.00 : 18.8$)

The data obtained in three different concentrations are seen to fit each line, implying that the theoretical predictions of the λ-dependence of M_w^{app} are confirmed by experiments. From Fig. 6, plots are seen to be nonlinear for polydisperse solutes like this and the ordinate intercepts are the $1/M_w^{app}$ values at the given initial concentrations. From Fig. 7, on the other hand, plots are seen to have practically linear portions experimentally accessible and the intercepts as obtained by linear extrapolations are seen to give directly the reciprocal of the true weight-average molecular weight, which is marked with open circles on the ordinate.

These theoretical predictions of the λ-dependence of $1/M_w^{app}$ may also be presented in another plot of $\Delta c/c^0$ against λ, as has been used by Albright and Williams[17]. The M_w^{app} value may be determined from the initial slope of the plot. Applying their method to the data of polystyrene mixture II again, we get the results as shown in Fig. 8. The predicted lines are convected upward and the data are located in the parts far deviated from the initial slope of each line. Accordingly, there is no advantage in using this type of plot for determining the M_w^{app} value for polydisperse solutes like these.

Next we consider the *average* concentration $\bar{c}$ as a concentration variable. It is known that c^0 may be correlated with $\bar{c}$ by the following identity[12]:

$$c^0 = \frac{c(a)+c(b)}{2}\,\frac{\Delta c}{\lambda c^0}\,\frac{2\lambda (c^0)^2}{\Delta c^2}$$

$$= \frac{\bar{c}\,M_w^{app}}{\sum\limits_{i,k} M_k \left\{ \int_0^1 J_{ik}\,d\xi \right\} g_i g_k} \tag{21}$$

where

$$J_{ik} = \frac{1}{\lambda M_k}\,\theta_i\,\frac{d\theta_k}{d\xi} \tag{22}$$

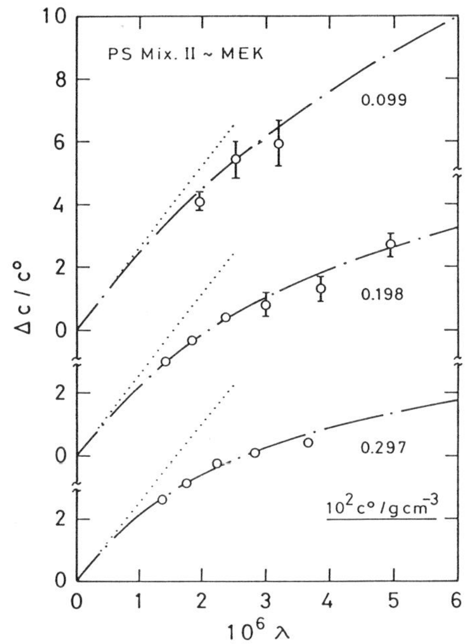

Figure 8. Plots of $\Delta c/c^0$ against λ for the comparison stated in the legend to Fig. 6. Dotted lines denote the initial slope of each theoretical curve.

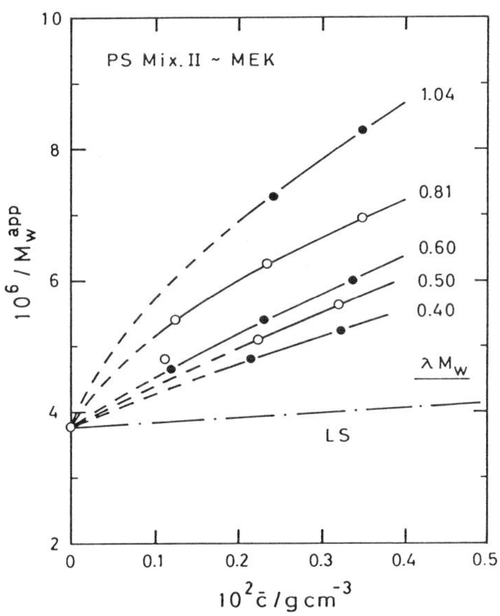

Figure 9. Plots of $1/M_w^{app}$ against $\bar{c}$ for polystyrene mixture II in 2-butanone at 30°C. Dot-dash line denotes the light-scattering result (plotted against c^0).

Substituting eq. (21) into eq. (16), one finds

$$M_w^{app} = M_w - \bar{c}M_w^{app} \times \left\{ \frac{\sum_{i,k} M_i M_k \bar{B}_{ik} \left\{ \int_0^1 \theta_{i0}\theta_{k0}\, d\xi \right\} g_i g_k}{\sum_{i,k} M_k \left\{ \int_0^1 J_{ik}\, d\xi \right\} g_i g_k} \right\} \tag{23}$$

or in reciprocal form

$$\frac{1}{M_w^{app}} = \frac{1}{M_w} + \bar{c}\, B^{app} \tag{24}$$

$$B^{app} = \frac{\sum_{i,k} M_i M_k \bar{B}_{ik} \left\{ \int_0^1 \theta_{i0}\theta_{k0}\, d\xi \right\} g_i g_k}{M_w \sum_{i,k} M_k \left\{ \int_0^1 J_{ik}\, d\xi \right\} g_i g_k} \tag{25}$$

Here, B^{app} is seen to depend on λ through the factors including the integration. However, these factors are actually equal to each other: without the pseudo-ideal solution approximation in deriving eq. (13), the factor in the numerator of eq. (23) can be replaced by

$$\int_0^1 J_{ik}\, d\xi$$

but $\bar{B}_{ik}$ should read $B_{ik} + \bar{v}/M_i$. This means that the λ-dependence of B^{app} is reduced to some extent, the degree of which depends first on the polydispersity of a given solute and secondly on the applied conditions through the quantity

$$\int_0^1 J_{ik}\, d\xi$$

It is clear that for monodisperse solutes, these λ-dependent factors cancel out completely.

Fig. 9 shows plots of $1/M_w^{app}$ against $\bar{c}$ for polystyrene mixture II. It is seen that B^{app} is always larger than B_2, the light-scattering second virial coefficient. The higher the λM_w values, the steeper the slope. So M_w determinations are more difficult. Up to the λM_w value of 0.6, however, these plots are seen to work even for such polydisperse solutes that the ratio of M_w to M_z (the z-average molecular weight) is 1.0 : 18.8. It may be concluded that apart from forgoing a rigorous determination of B_2, this type of plot does work even for considerably polydisperse solutes, as long as experiments are carried out at lower rotor speeds and with shorter solution columns. In equilibrium experiments, the λ value must be set as *low* as possible, as has been stressed elsewhere[18] and first noticed by Pedersen in the early days of ultracentrifugation[19].

In order to calculate the average concentration, one has to determine the equilibrium concentration at the meniscus, c(a), in terms of the relation:

$$c(a) = c^0 - \frac{b^2 \Delta c}{b^2 - a^2} + \int_a^b \frac{r^2\, dc}{b^2 - a^2} \tag{26}$$

It is noted that the second and the third terms on the right-hand side of eq. (26) are much larger in an absolute sense than the first one. The ratio of the first to the second term is as large as 18:1, when c^0 and Δc are comparable and a 2 mm solution column is used (b is fixed to *ca.* 7.1 cm). Accordingly, the integration of

$$\int_a^b r^2\, dc$$

must be carried out with such an accuracy that Δc can be determined experimentally.

Another concentration variable is c^{0*} first introduced by Van Holde and Williams[29] as early as in 1953 and later used by Scholte[21]; the initial concentration is modified by such factors as can be seen in the equation

$$\frac{1}{M_w^{app}} = \frac{1}{M_w} + c^{0*}\left\{B + \frac{\bar{v}}{M_w}\right\} \tag{27}$$

$$c^{0*} = c^0 \frac{\int_0^1 \left(\frac{dc}{d\xi}\right)^2 d\xi}{\left\{\int_0^1 \frac{dc}{d\xi}\, d\xi\right\}^2} \tag{28}$$

This working equation has been derived under the approximation that B_{ik} is species independent, taking a constant value of B. The M_w^{app} is the same as defined by eq. (15) and is given alternatively by

$$\frac{1}{\lambda c^0} \int_0^1 \frac{dc}{d\xi}\, d\xi$$

The modification factors may be calculated from the observable dc/dr in the Schlieren optics. An example of this type of plot is given in Fig. 10. Scholte has extended this theory to the M_z determination[21]. However, as his equation is lengthy and complicated, other types of equations for higher average molecular weights will be described below.

The z- and the (z+1)-average molecular weights, M_z and M_{z+1}, and their apparent quantities obtainable from observables can be defined as

$$M_w M_z = \sum_i M_i^2 g_i \tag{29a}$$

$$(M_w M_z)^{app} = \frac{\Delta \dfrac{dc}{d\xi}}{\lambda^2 c^0} \tag{29b}$$

$$M_w M_z M_{z+1} = \sum_i M_i^3 g_i \tag{30a}$$

$$(M_w M_z M_{z+1})^{app} = \frac{\Delta \dfrac{d^2c}{d\xi^2}}{\lambda^3 c^0} \tag{30b}$$

As a preparation for calculating these quantities, we consider the $\Delta\theta_i$, like θ_i itself, as a series function of c^0

$$\Delta\theta_i = \Delta\theta_{i0} + c^0\Delta\theta_{i1} + \dots \tag{31}$$

With the results given by eqs. (11b) and (11d), we obtain

$$\Delta\theta_i = \lambda M_i - c^0\lambda \sum_k M_i \{(M_i+M_k) - M_i\}\bar{B}_{ik} \left\{\int_0^1 \theta_{i0}\theta_{k0}\, d\xi\right\} g_k \tag{32}$$

$$\Delta(\theta_{i0}\theta_{k0}) = \lambda(M_i + M_k) \int_0^1 \theta_{i0}\theta_{k0}\, d\xi \tag{33}$$

Now we return to eq. (13). The summation over i and the operation of Δ to the equation yields, with the help of eqs. (32) and (33), the equation for M_z:

$$(M_w M_z)^{app} = M_w M_z - c^0 \sum_{i,k} M_i [(M_i+M_k)^2 - M_i^2]\, \bar{B}_{ik} \left\{\int_0^1 \theta_{i0}\theta_{k0}\, d\xi\right\} g_i g_k \tag{34}$$

or in reciprocal form

$$\frac{1}{(M_wM_z)^{app}} = \frac{1}{M_wM_z}$$

$$+ \frac{c^0}{(M_wM_z)^2} \sum_{i,k} M_i\{(M_i+M_k)^2-M_i^2\} \bar{B}_{ik} \left\{ \int_0^1 \theta_{i0}\theta_{k0}\, d\xi \right\} g_i g_k + \ldots \quad (35)$$

For the (z+1)-average molecular weight, first differentiate eq. (13) with respect to ξ, and then carry out the analogous calculations. The result obtained is as follows:

$$(M_wM_zM_{z+1})^{app} = M_wM_zM_{z+1}$$

$$+ c^0 \sum_{i,k} M_i[(M_i+M_k)^3-M_i^3] \bar{B}_{ik} \left\{ \int_0^1 \theta_{i0}\theta_{k0}\, d\xi \right\} g_i g_k \quad (36)$$

or

$$\frac{1}{(M_wM_zM_{z+1})^{app}} = \frac{1}{M_wM_zM_{z+1}} + \frac{c^0}{(M_wM_zM_{z+1})^2}$$

$$\times \sum_{i,k} [(M_i + M_k)^3-M_i^3] \bar{B}_{ik} \left\{ \int_0^1 \theta_{i0}\theta_{k0}\, d\xi \right\} g_i g_k + \ldots \quad (37)$$

This way of deriving the working equations for higher average molecular weights is straightforward[22]. In practical applications, it may be better to take, for an example, $M_w/(M_wM_z)^{app}$ instead of $1/(M_wM_z)^{app}$. This is because the initial slope of the plot can be reduced and the M_w value is usually known prior to the data analysis for M_z. The above are extensions of eq. (17) for non-ideal polydisperse solutes, requiring the double extrapolations with respect to λ^2 and to c^0 [23].

An extension of eq. (24) may also be made as follows[24]:

$$\frac{(\Delta c)^2}{c^0 \Delta r^2 \Delta\left(\frac{dc}{dr^2}\right)} = \frac{M_w}{M_z} (1 + BM_w\bar{c} + \ldots) \quad (38)$$

The derivative, dc/dr^2, at both ends of the solution column may be obtained by first making a plot of ln c(r) against r^2 and then taking the slope at the desired positions, followed by the multiplication of c(a) or c(b).

Another average molecular weight we have not mentioned above is the number-average molecular weight, M_n. The working equation for M_n may be derived from eq. (13). After multiplication by $d\xi/M_i$, followed by the summation and the integration indicated, the equation below is obtained:

$$\frac{1}{M_n^{app}} = \frac{1}{M_n} - \sum_i \frac{\theta_i(0) g_i}{M_i}$$

$$+ c^0 \sum_{i,k} \bar{B}_{ik} \left\{ \int_0^1 \theta_{i0}\theta_{k0} \, d\xi - \theta_{k0} \right\} g_i g_k \tag{39}$$

where

$$\frac{1}{M_n^{app}} = \lambda \int_0^1 \left\{ \int_0^{\xi} \theta \, d\xi \right\} d\xi = \frac{\lambda}{c^0} \int_0^1 \left\{ \int_0^{\xi} c \, d\xi \right\} d\xi \tag{40}$$

and

$$\frac{1}{M_n} = \sum_i \frac{g_i}{M_i} \tag{41}$$

The third term on the right-hand-side of eq. (39) was estimated, as before, under the pseudo-ideal solution approximation. In order to perform the integration in eq. (40), the concentration at $\xi = 0$ (the meniscus), must be effectively zero. Accordingly, this type of sedimentation equilibrium situation is usually called *meniscus depletion.* Since higher acceleration must be applied for this purpose, eq. (39) may take a simple form such as

$$\frac{1}{M_n^{app}} \approx \frac{1}{M_n} - \sum_i \frac{\theta_i(0) g_i}{M_i} + c^0 \lambda \sum_{i,k} \bar{B}_{ik} \frac{M_i M_k}{(M_i + M_k)} g_i g_k \tag{42}$$

The above equation shows that since the rotor-speed dependent factor is simplified to be solely represented by λ, a single extrapolation with respect to c^0 is enough to obtain the $1/M_n$ value.

This meniscus depletion method has two principal shortcomings. The one is unresolvable fringes near the cell bottom. This is often encountered in polydisperse solutes and reflects a high concentration gradient yielded due to high acceleration. The other is optically *undetectable* concentration at the meniscus. This concentration is surely low but might lead to an appreciable error in the final results, as will be discussed below.

In order to avoid the latter shortcoming, such a modification has been proposed that a relatively large volume of pure solvent is layered onto a small volume of sample solution in a capillary synthetic boundary cell[25]. This results in a long solution column (*ca.* 5 mm), but with equilibrium attained in the shorter time corresponding to a short column height.

The meniscus depletion method thus modified has been applied to NBS polystyrene sample 706 in toluene with the values of M_n, M_w and M_z all being determined[26]. As for these determinations, it should be noted here that the slopes of the conventional reciprocal plots against c^0 do depend on the experimental conditions, as has been seen above, but cannot be expected to have different signs for different plots. According to eq. (42), the slope of a plot of $1/M_n^{app}$ against c^0 *must* be positive (as is the case with the other averages) *unless* $\overline{B}_{lk}$ is negative.

Evidently, the potential applicabilities of meniscus depletion may be revealed for high molecular weight solutes, for no other experimental means enables us to absolutely determine M_n values exceeding a million. We will now consider this situation in more detail. For theoretical simplicity and experimental convenience, the consideration below is confined however to the case of *pseudo-ideal* solutions, where the third term of the right-hand side of eq. (39) vanishes.

In view of the problem of unresolvable fringes near the cell bottom, we define the observable number-average molecular weight, $M_n(\xi)$, at a given position, ξ, as

$$\frac{1}{M_n(\xi)} = \frac{\lambda \int_0^{\xi} \left\{ \int_0^{\xi} c \; d\xi \right\} d\xi}{\int_0^{\xi} c \; d\xi} \tag{43a}$$

$$= \lambda \left\{ \xi - \frac{\int_0^{\xi} \xi c \; d\xi}{\int_0^{\xi} c \; d\xi} \right\} \tag{43b}$$

Differentiation of eq. (43) with respect to ξ gives us

$$\frac{d}{d\xi}\left[\frac{1}{M_n(\xi)}\right] = \lambda \left\{ 1 - \frac{M_w(\xi)}{M_n(\xi)} \right\} \tag{44}$$

where

$$M_w(\xi) = \frac{c}{\lambda} \int_0^{\xi} c \; d\xi \tag{45}$$

Eq. (45) is the definition of $M_w(\xi)$ at a given position, ξ, which when integrated up to $\xi = 1$, becomes the true weight-average molecular weight of the whole sample: now we are ready to construct a new plot. First, compute the $1/M_n(\xi)$ in the resolvable fringe region and plot them against ξ. Then the extrapolation to $\xi = 1.0$ yields the true value for $1/M_n$ from the intercept and the slope of the plot reflects, as eq. (44) suggests, the polydispersity in the solute at a given position, ξ.

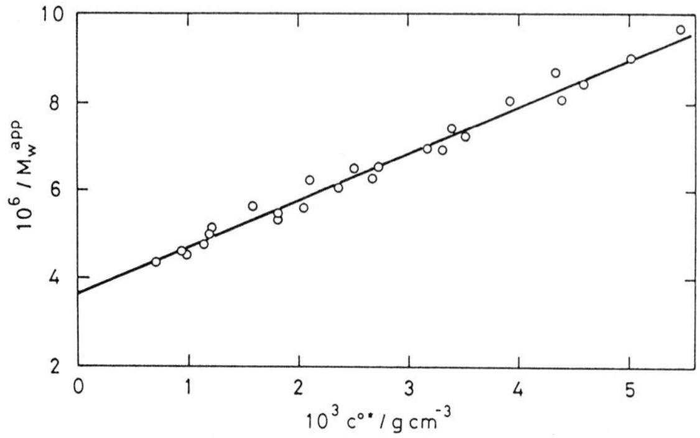

Figure 10. Plot of $1/M_w^{app}$ against c^{0*} for NBS polystyrene 706 in toluene at 25°C. Redrawn with modification from Scholte[21] with his permission.

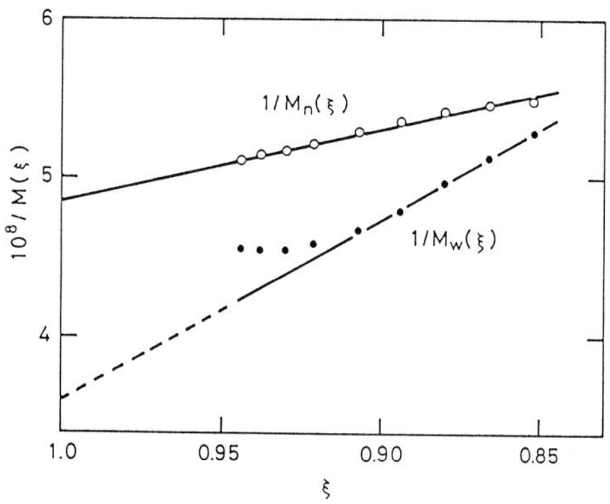

Figure 11. Plots of $1/M_n(\xi)$ and $1/M_w(\xi)$ against ξ for high molecular weight polystyrene in *trans*-decalin at 20.4°C. Reprinted with permission from Suzuki, *Brit. Polym. J.*, 1979, *11*, 91. Copyright (1979) Society of Chemical Industry.

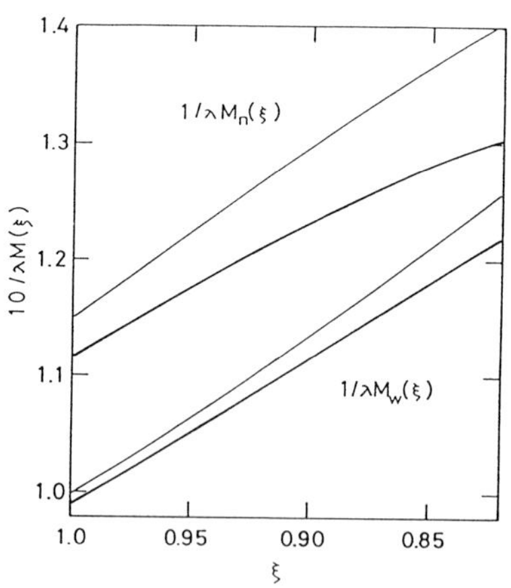

Figure 12. Theoretical predictions for $1/M(\xi)$ for a hypothetical mixture. For details, see the text. Reprinted with permission from Suzuki, *Brit. Polym. J.*, 1979, *11*, 91. Copyright (1979) Society of Chemical Industry.

Fig. 11 shows plots of $1/M_n(\xi)$ and $1/M_w(\xi)$ against ξ near unity using the results obtained on a *trans*-decalin solution of a linear polystyrene sample having the M_w value of 2.7×10^7 [27]. The column height, (b – a), was 0.167 cm and the actual rotor speed was found to be 1,952 rev/min. The sample fraction, on which each computation of the molecular weights was based were 17.2% at $\xi = 0.852$ and 38.2% at $\xi = 0.946$. The latter reduced radial position was the upper limit of the resolvable region of fringes. A linear extrapolation of the $1/M_n(\xi)$ data to $\xi = 1.0$ yields the value $M_n = 2.0_6\times10^7$.

The main purpose of Fig. 11 is to estimate the M_n value of the given sample which cannot be measured with an ordinary membrane osmometer. Additionally, the figure indicates the possibility of determining its M_w value as well from the same equilibrium distribution of the sample. Since the fringes are faint and eventually disappear near the cell bottom, it is subject to a large error in determining the fringe displacements at higher ξ. As can be seen from Fig. 11, systematic deviation of the data, $1/M_w(\xi)$, near $\xi = 1.0$ reflects such a situation. Using low weights in these data, a plot of $1/M_w(\xi)$ against ξ is linearly extrapolated to $\xi = 1.0$, yielding the value, $2.6_0\times10^7$ for the M_w of the whole sample. This agrees fairly well with the light-scattering result of 2.7×10^7 (ref. 28).

As eq. (11b) states, the relative concentration at the meniscus, $\theta(0)$, decreases exponentially with increase in the generalized speed parameter. Thus, the prerequisite of zero concentration cannot be brought about exactly. An application of higher acceleration or use of a longer solution column is expected to be favourable for this prerequisite, whereas the upper limit of the sample fraction experimentally analysable, decreases rapidly. Since the decrease in the sample fraction is a vital defect, we never completely remove the effects of non-zero concentration at the meniscus. We have to assess, therefore, the error in determining the average molecular weights due to incomplete fulfillment of the prerequisite.

Using calculation results on a hypothetical sample, we examine the further insights of the plots. The hypothetical sample used is a mixture of three monodisperse macromolecular solutes, which may be characterised as $g_1 = g_3 = 0.25$ and $g_2 = 0.50$; $\lambda M_1 = 5$, $\lambda M_2 = 10$ and $\lambda M_3 = 15$. The actual order of their molecular weights are not important, just their product with λ being essential. Under the pseudo-ideal condition, the equilibrium distribution of the hypothetical mixture may be given as

$$\sum_{i=1}^{3} \theta_{i0} g_i = \sum_{i=1}^{3} A_{i0} g_i \exp(\lambda M_i \xi) \tag{46}$$

With this equation, the $M_n(\xi)$ and $M_w(\xi)$ were calculated as a function of ξ. The results obtained are shown in Table 2 and in Fig. 12.

Table 2. Dependence of $M_n(\xi)$ and of the sample fraction on ξ for the hypothetical mixture.

ξ	$\int_0^{\xi} \{\theta_0(\xi) - \theta_0(0)\}\, d\xi$	$\dfrac{M_n(\xi)}{M_n(\xi=1)}$
1.000	0.991	1.000
0.940	0.553	0.938
0.920	0.459	0.921
0.900	0.383	0.904
0.880	0.321	0.886
0.860	0.270	0.876
0.840	0.228	0.865
0.820	0.193	0.855

The middle column of Table 2 represents the sample fraction which is counted from the meniscus up to a given position, ξ. At $\xi = 0.90$, the fraction counted is as low as 38.3%, but the number average molecular weight calculated at this position deviates from the $M_n(\xi=1.0)$ value by only 9.6%. This simply reflects the fact that in the calculation of M_n, low molecular weight components in a given sample have weights much higher than high molecular weight components have. In meniscus depletion, the counting is performed from the low molecular weight tail, and hence it can be said that the extrapolation of the $1/M_n(\xi)$ to $\xi = 1.0$ is a correct procedure.

In Fig. 12, the thin lines represent the results directly derived from eq. (46), while the thick lines represent the results obtained in such a way that the contribution of a non-zero value for $\theta(0)$ was subtracted in order to make a fair comparison with experiment. These lines are almost linear with ξ in the region shown in this figure, which is favourable for extrapolations. However, the differences between the two lines are seen to be appreciable especially for $1/M_n(\xi)$. In the case of the hypothetical mixture which is subjected to acceleration such that $\lambda M_w = 10$, the value for $\theta_0(0)$ amounts to 0.871×10^{-2}. Such a small error of 0.871% looks trivial at first sight. This view is valid only for the calculation of the total concentration, *i.e.*, for the single integral of

$$\int_0^{\xi} \theta_{i0}\, d\xi$$

For the double integral which appears as the definition of the observable number average molecular weight, it is not the case.

For the results of thick lines, the integrand, θ, in eq. (40) has been replaced by $\{\theta - \theta(0)\}$. Hence, eq. (39) with neglect of the non-ideal term can be modified as:

$$\lim_{\xi\to 1}\frac{1}{M_n(\xi)} \equiv \frac{1}{M_n(1)} = \frac{1}{M_n} - \sum\frac{\theta_{i0}(0)g_i}{M_i} - \frac{\lambda}{2}\theta_0(0) \tag{47}$$

under the pseudo-ideal condition. The comparison of the present equation with eq. (39) shows that the true correction term necessary for the limit is not $M_n\Sigma\theta_{i0}(0)g_i/M_i$ but the value of $(\lambda M_n/2)\theta_0(0)$. It is understood, therefore, that meniscus depletion *always overestimates the M_n* value by $(\lambda M_n/2)\theta_0(0)$, which helps to explain the discrepancy between the thick and thin lines in Fig. 12: this discrepancy amounts to 4.35% for this particular hypothetical mixture. The values for λM_w are usually ~10 for the present type of experiment, and the error inherent in this method, $(\lambda M_n/2)\theta_0(0)$, becomes of the order of $5\theta(0)$, when both M_w and M_n values are of the same order.

Unfortunately, the actual value for $\theta_0(0)$ is not ascertainable unless the fringes are resolvable over the entire solution column. What *is* experimentally accessible is only the state where a fringe is horizontal near the meniscus within experimental error. Even for an extremely dilute solution whose initial concentration is only two in fringe units, with the resolution of about one fiftieth of a fringe, it is assured that the $\theta_0(0)$ value is no worse than 1×10^{-2}. For a monodisperse solute, this corresponds to the acceleration such that the value for λM is as large as 6.47. In the case shown in Fig. 12, the λM_w value equals 11.0: so the $M_n(1)$ value obtained is corrected by an error of 5%, yielding a final result of $1.9_6\times10^7$.

The effect of non-zero concentration at the meniscus with regards to M_w is not as significant as in the case of M_n. The observable weight-average molecular weight is written as

$$\lim_{\xi\to 1}\frac{1}{M_w(\xi)} \equiv \frac{1}{M_w(1)} = \lambda\frac{1-\theta_0(0)}{\theta_0(1)-\theta_0(0)} \tag{48}$$

On the other hand, the true weight-average molecular weight is given by $[\theta_0(1) - \theta_0(0)]/\lambda$. It is concluded, therefore, that the experimentally determinable $M_w(1)$ value differs from the true M_w value only by $\theta_0(0)$, which is surely a minor deviation.

In the above, the working equations for various types of average molecular weights were derived and explained with illustrations. Some emphasis was placed upon the description of how to determine the number average molecular weight, since, to our knowledge, no other books describe the method in detail. It goes without saying that all applications must be accurate, otherwise unambiguous results are by no means obtained for non-ideal polydisperse systems.

ACKNOWLEDGEMENTS

We are indebted to the copyright owners, cited in the legend to each figure, who granted permission for us to reproduce them.

GENERAL REFERENCES

(i) C.H. Chervenka, 'A Manual of Methods for the Analytical Ultracentrifuge', Beckman, Palo Alto, 1963/1973.

(ii) J.W. Williams, 'Ultracentrifugation of Macromolecules, modern topics', Academic Press, New York, 1972.

(iii) H. Fujita, 'Foundations of Ultracentrifugal Analysis', Wiley-Interscience, New York, 1975.

REFERENCES

1. See, for an example, T. Kotaka and N. Donkai, *J. Polym. Sci.*, 1968, *A-2*, 6, 1457.
2. J.P. Johnston and A.G. Ogston, *Trans. Faraday Soc.*, 1946, *42*, 789.
3. H. Suzuki, C.G. Leonis and M. Gordon, *Makromol. Chem.*, 1973, *172*, 227.
4. V.N. Schumaker, *Arch. Biophys.*, 1963, *103*, 139.
5. A.G. Ogston, B.N. Preston and J.D. Wells, *Proc. Roy. Soc. Lond. Ser. A*, 1973, *333*, 297.
6. N. Nemoto, S. Okada, T. Inoue, T. Tsunashima and M. Kurata, *Macromolecules*, 1986, *19*, 2305; *ibid.*, 1988, *21*, 1502.
7. Th.G. Scholte, *J. Polym. Sci.*, 1970, *A-2*, 8, 841.
8. H. Murakami, T. Norisuye and H. Fujita, *Polym. J.*, 1975, *7*, 248.
9. P. Munk, *Macromolecules*, 1978, *11*, 101.
10. J.W. Williams, K.E. Van Holde, R.L. Baldwin and H. Fujita, *Chem. Rev.*, 1958, *58*, 715.
11. K. Sakato and M. Kurata, *Polym. J.*, 1970, *1*, 260.
12. See, for example, T. Kotaka, N. Donkai and H. lnagaki, *J. Polym. Sci.*, 1971, *A-2*, 9, 1379.
13. H. Fujita, *J. Phys. Chem.*, 1969, *73*, 1759.
14. M. Gordon, C.G. Leonis and H. Suzuki, *Proc. Roy. Soc. Lond. Ser. A*, 1975, *345*, 207.
15. H. Fujita, *J. Phys. Chem.*, 1959, *63*, 1326.
16. H. Utiyama, N. Tagata and M. Kurata, *J. Phys. Chem.*, 1969, *73*, 1448.
17. D.A. Albright and J.W. Williams, *J. Phys. Chem.*, 1967, *71*, 2780.
18. H. Suzuki, in 'Physical Chemistry of Colloids and Macromolecules', B. Rånby, ed., Blackwell Sci. Pub., Oxford, 1987. p.101.
19. K.O. Pedersen, in 'The Ultracentrifuge', by T. Svedberg and K.O. Pedersen, Oxford University Press, Oxford, 1940, p.304.
20. K.E. Van Holde and J.W. Williams, *J. Polym. Sci.*, 1953, *11*, 243.
21. Th.G. Scholte, Paper presented before the IUPAC Symposium, Helsinki, 1972.
22. H. Suzuki, *Bull. Inst. Chem. Res.*, Kyoto Univ., 1978, *56*, 89.
23. C.G. Leonis, H. Suzuki and M. Gordon, *Makromol. Chem.*, 1977, *178*, 2867.
24. T. Norisuye, T. Yanaki and H. Fujita, *J. Polym. Sci.: Polym. Phys. Ed.*, 1980, *18*, 547.
25. C.H. Chervenka, *Anal. Biochem.*, 1970, *34*, 24.
26. P.M. Budd, *J. Polym. Sci.: B. Polym. Phys.*, 1988, *26*, 1143.
27. H. Suzuki, *Brit. Polym. J.*, 1979, *11*, 91.
28. H. Suzuki, *ibid.*, 1979, *11*, 41.

32

Sedimentation Analysis of Synthetic Polyelectrolytes

By Peter M. Budd

DEPARTMENT OF CHEMISTRY, UNIVERSITY OF MANCHESTER, OXFORD ROAD, MANCHESTER. M13 9PL U.K.

1. INTRODUCTION

Polymers which possess many ionisable groups (polyelectrolytes) behave in aqueous solution in ways which are markedly different both from non-ionic polymers and from simple electrolytes[1]. The unique properties of polyelectrolytes lead to their use in a variety of industrial applications: as viscosity modifiers, flocculants, dispersants, water softeners, *etc*. Sedimentation methods[2] are of value for characterizing these materials and investigating their solution behaviour.

This chapter provides an outline of polyelectrolyte properties and discusses the sedimentation velocity and sedimentation equilibrium behaviour of flexible, linear polyelectrolytes, with particular reference to sodium poly(styrene sulphonate) (NaPSS):

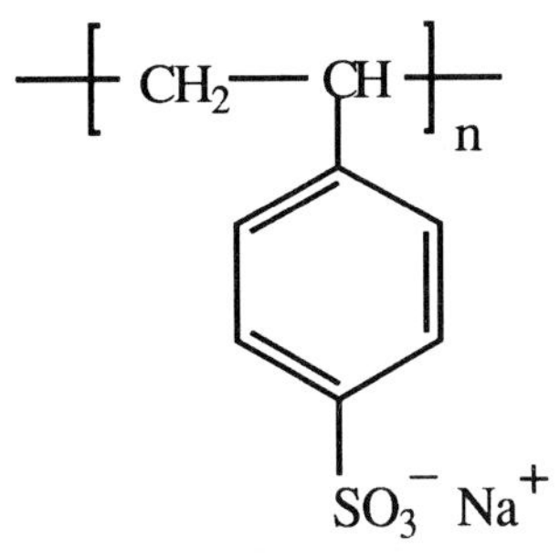

2. POLYELECTROLYTES IN SOLUTION

Polyions and Counterions

Ionization in solution gives rise to a multiply charged polyion, together with many small counterions. The electrostatic field around a polyion leads to strong interactions between the polyion and its counterions. Counterions tend to be clustered around the polyion, forming an 'ionic atmosphere', and counterion activity coefficients tend to be rather low. Addition of a simple salt to a polyelectrolyte solution screens the polyion charges, so that the thickness of the ionic atmosphere decreases, and counterion activity coefficients increase. Experiments with polyelectrolytes generally require a simple salt to be present in order to minimise charge effects.

Chain Expansion

A non-ionic, flexible polymer in solution typically adopts a coiled conformation, the precise form of which changes with time. A flexible polyion is generally more expanded than an equivalent non-ionic polymer because of repulsion between charged groups along the chain. At low concentrations in pure water a polyelectrolyte may be almost rod-like. Addition of a simple salt shields the charges and the degree of expansion of the polyion decreases. This affects the frictional properties of the polyelectrolyte chain and thereby its sedimentation behaviour.

Charge Effects in Sedimentation and Diffusion

In addition to the influences of chain expansion and the ionic atmosphere on frictional properties, charge effects arise in sedimentation and diffusion because of the different mobilities of a polyion and its counterions. For a salt-free polyelectrolyte solution, the sedimentation coefficient, s, is much lower, and the diffusion coefficient, D, is much higher than for an equivalent non-ionic polymer. This has been called the 'primary charge effect' and has been attributed to the electric field which would be set up if a rapidly sedimenting (or slowly diffusing) polyion were separated from slowly sedimenting (or rapidly diffusing) counterions[3]. Polyion and counterions are forced to move together.

Addition of a simple salt reduces the primary charge effect, but a secondary charge effect arises if the constituent ions of the added salt differ in their mobilities. The secondary charge effect is small for salts like NaCl for which cation and anion have similar frictional coefficients.

The Donnan Effect

The Donnan effect occurs when a polyelectrolyte solution is separated from a solution of a simple salt by a semi-permeable membrane. The polyion cannot pass through the membrane but the counterions and the constituent ions of the simple salt can. At equilibrium, there is a lower concentration of salt in the presence of polyion than in its absence, in other words salt is excluded from the polyelectrolyte solution to some extent[4]. This may be described in terms of a Donnan membrane distribution parameter, Γ, which is defined at infinite dilution of the polyelectrolyte to avoid the complication of interactions between polyions:

$$\Gamma = \lim_{C_u \to 0} \left(\frac{C_S' - C_S}{C_u} \right) \tag{1}$$

where C_S' and C_S are the concentrations (in mol dm^{-3}) of salt on the polyion-free and polyion-containing sides of the membrane respectively, and C_u is the concentration of polyelectrolyte per charged unit.

Given that electroneutrality must be preserved in solution and that, at equilibrium, the activities of the salt must be equal on the two sides of the membrane, it can be shown that[5]

$$\Gamma = 0.5 - \lim_{C_u \to 0} \left[\frac{C_S'}{C_u} \left(\frac{\gamma_\pm'}{\gamma_\pm} - 1 \right) \right] \tag{2}$$

where $\gamma_\pm'$ and $\gamma_\pm$ are mean activity coefficients for the salt ions in the absence and presence of polyelectrolyte respectively. In the 'ideal' case, where $\gamma_\pm' = \gamma_\pm$, the value of Γ is 0.5. In practice, $\gamma_\pm' > \gamma_\pm$ at low salt concentrations and Γ is rather less than the ideal value but increases with increasing salt concentration.

The situation of membrane equilibria may seem far removed from sedimentation analysis, but in fact the Donnan effect arises wherever there is a concentration gradient of polyelectrolyte, as in an ultracentrifuge cell. Furthermore, Donnan equilibrium represents a situation of constant chemical potential of diffusible components, which, as will be discussed, defines an appropriate density increment for use in the sedimentation analysis of polyelectrolytes.

3. SEDIMENTATION VELOCITY

The considerable influence of salt concentration on the sedimentation velocity of a polyelectrolyte is illustrated in Figs. 1 and 2 for NaPSS of molecular weight 1.0×10^5 and 7.8×10^5 respectively. It can be seen that the reciprocal sedimentation coefficient, 1/s, is linearly dependent on polymer concentration and that the slope of the plot decreases as the salt concentration increases. Fig. 3 shows that the limiting, infinite dilution, sedimentation coefficient, s_0, increases with increasing salt concentration up to 0.5 mol dm^{-3}, then appears to decrease.

Theory

There have been a number of theoretical treatments of sedimentation in polyelectrolyte solutions. The most rigorous approach involves the application of irreversible thermodynamics[6,7]. The classical approach considers the actual velocity of a polyion as having two components; there is a contribution from the centrifugal field and a contribution from the internally created electric field which would arise if the polyion were separated from its counterions[3,8,9].

For a polyelectrolyte PX_Z in a salt-free solution, s_0 is given by[10]

$$s_0 = \frac{M(1-\bar{v}\rho)}{f + iZf_X} \tag{3}$$

where M is the molecular weight and $\bar{v}$ is the partial specific volume of the polyelectrolyte, ρ is the density of the solution, f and f_X are the molar translational frictional coefficients of the polyion P and counterion X respectively and i is an effective charge parameter which takes account of interactions between the polyion and its counterions. It is assumed that, on average, (1-i)Z counterions interact in some way with the polyion so that it carries an effective charge of iZ. The magnitude of i will depend both on the number of any specifically bound counterions and on the extent of atmospheric screening.

Eq. (3) is to be compared with the corresponding expression for a non-ionic polymer:

$$s_0 = \frac{M(1-\bar{v}\rho)}{f} \tag{4}$$

In the absence of salt, s_0 for a polyelectrolyte is predicted to be smaller than for an equivalent non-ionic polymer, as is observed experimentally.

For a polyelectrolyte PX_Z in the presence of a simple salt XY, s_0 is given by[10]

$$s_0 = \frac{M(1-\bar{v}\rho) - 0.5iZM_s(1-\bar{v}_s\rho)(1+f_Y f_X^{-1})^{-1}}{f} \tag{5}$$

where M_s and $\bar{v}_s$ are the molar mass and partial specific volume respectively of the salt and f_Y is the frictional coefficient of the co-ion Y. In the absence of secondary charge effects, when $f_X = f_Y$, this reduces to

$$s_0 = \frac{M(1-\bar{v}\rho) - 0.5iZM_s(1-\bar{v}_s\rho)}{f} \tag{6}$$

When dealing with multicomponent systems such as a polyelectrolyte with added salt, it is often advantageous to utilize a density increment, $(d\rho/dc)$, rather than a buoyancy factor, $(1-\bar{v}\rho)$. However, it should be remembered that density increments may be determined under a variety of conditions. The density increment for the condition of constant salt concentration $(d\rho/dc)_{C_s}$ is directly equivalent to $(1-\bar{v}\rho)$, but a more useful increment for polyelectrolyte studies is that at constant chemical potential of diffusible components, $(d\rho/dc)_\mu$. Strictly speaking, the correct increment to use is that at constant chemical potential of the salt[10,11], but $(d\rho/dc)_\mu$ is virtually the same and is conveniently determined by density measurements on solutions which have been brought into dialysis (Donnan membrane) equilibrium. The density increments differ because of the Donnan effect. If the salt has an ion in common with the polyelectrolyte, they are approximately related by

$$\left(\frac{d\rho}{dc}\right)_\mu \approx \left(\frac{d\rho}{dc}\right)_{C_s} - (1-\bar{v}_s\rho)\left\{\frac{M_s}{M_u}\Gamma\right\} \tag{7}$$

where $M_u = M/Z$ is the molecular weight per charged unit of the polyelectrolyte.

The Donnan parameter Γ may, in equilibrium situations, be related to an effective charge parameter i by[10]

$$\Gamma = 0.5\,i \tag{8}$$

If the same identification is made in the case of sedimentation velocity, eq. (6) becomes

$$s_0 = \frac{M(d\rho/dc)_\mu}{f} \tag{9}$$

which shows that in some respects one can treat a polyelectrolyte with added salt similarly to a non-ionic polymer, provided that a density increment for the condition of dialysis equilibrium is used. Eq. (9) may be combined with the corresponding equation for the limiting diffusion coefficient and a molecular weight evaluated. For a more detailed analysis of sedimentation behaviour, however, one should bear in mind that both $(d\rho/dc)_\mu$ and f vary with salt concentration, reflecting variations in Γ, in the degree of expansion of the chain and in the thickness of the ionic atmosphere.

Concentration Dependence of s

It may be seen from Figs. 1 and 2 that 1/s is linearly dependent on the polymer concentration, c (in g cm^{-3}). According to theory[10], charge effects give rise to a slope, k, in these plots given by

$$k \approx \frac{10^3 i^2 f_Y}{M_u C_s\{M_u(1-\bar{v}\rho)(1+f_Y f_X^{-1}) - iM_s(1-\bar{v}_s\rho)\}} \times \left\{1 - \frac{M_s(1-\bar{v}_s\rho)}{f_X(1+f_Y f_X^{-1})s_0}\right\} \tag{10}$$

Rearrangement of eq. (10) enables i to be calculated, as[12,13]

$$\alpha i^2 + \beta i + \gamma = 0 \tag{11}$$

where

$$\alpha = \frac{10^3}{kM_uC_s}\left\{f_Y - \frac{M_s(1-\bar{v}_s\rho)f_Y}{f_X(1+f_Y f_X^{-1})s_0}\right\}$$

$$\beta = M_s(1-\bar{v}_s\rho)$$

$$\gamma = -\,M_u(1-\bar{v}\rho)\,(1+f_Y f_X^{-1})$$

The problem remains, however, of isolating the concentration dependence due to charge effects from the normal concentration dependence exhibited by any polymer. To a first approximation we may assume that the experimental concentration dependence is

$$k_e = k_{ni} + k \tag{12}$$

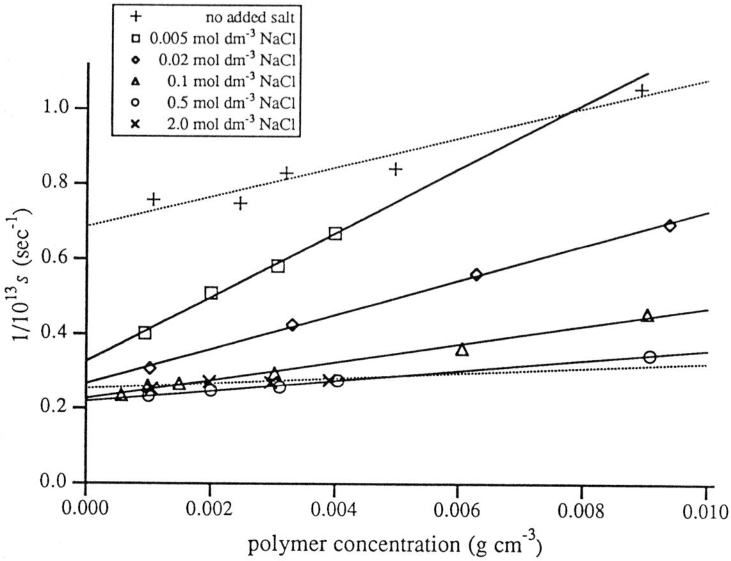

Figure 1. The dependence of the reciprocal sedimentation coefficient on polymer concentration for NaPSS of molecular weight 1.0×10^5 at various concentrations of added NaCl.

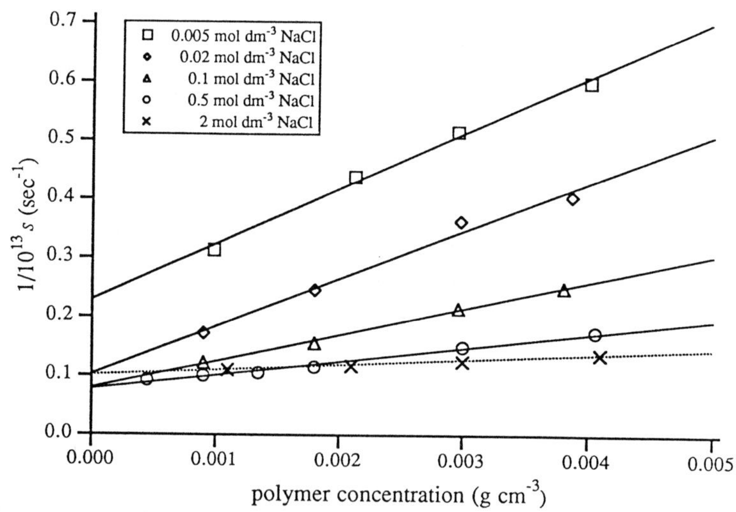

Figure 2. The dependence of the reciprocal sedimentation coefficient on polymer concentration for NaPSS of molecular weight 7.8×10^5 at various concentrations of added NaCl.

where k_{ni} is the slope expected for an equivalent non-ionic polymer. For many flexible, non-ionic polymers in good solvents it has been found that[14]

$$k_{ni} = \frac{1.6\,[\eta]}{s_0} \tag{13}$$

where $[\eta]$ is the intrinsic viscosity (limiting viscosity number).

Values for the effective charge parameter, i, calculated from the data of Figs. 1 and 2 using eqs. (11) to (13), are shown in Fig. 4. Also included in Fig. 4 are values of i calculated from the data of Nagasawa and Eguchi[15] for NaPSS of molecular weight 3.5×10^5. The value of i increases with increasing salt concentration, as expected from its relationship to the Donnan parameter, Γ.

The Frictional Coefficient

Fig. 5 illustrates the dependence on salt concentration of the frictional coefficient, corrected to the viscosity of water to eliminate the effect of the increasing viscosity of the medium, for two samples investigated by the present author (NaPSS 1.0×10^5 and 7.8×10^5) and for two samples of Nagasawa and Eguchi[15] (NaPSS 3.5×10^5 and 1.0×10^6).

The corrected frictional coefficient decreases with increasing salt concentration. It is the dominant effect of the decrease in f which leads to the increase in s_0 with increasing salt concentration up to about 0.5 mol dm^{-3} (Fig. 3). At higher salt concentrations, s_0 decreases as the increase in solvent viscosity and the increase in i (Fig. 4) become more significant.

The decrease in f with increasing salt concentration may at least largely be attributed to the decreasing degree of expansion expected for a polyelectrolyte. According to simple theory[16,17] f should be proportional to an expansion coefficient α, and hence to $[\eta]^{1/3}$. Experimentally, $[\eta]$ for a polyelectrolyte is linearly dependent on $C_s^{-1/2}$ [18] so it might be expected that a plot of f^3 against $C_s^{-1/2}$ would be linear. For molecular weights up to 3.5×10^5 g/mol such a plot does seem to be linear, as shown in Fig. 6, but at higher molecular weights the plot is curved, as illustrated in Fig. 7.

Two factors in addition to chain expansion may affect the dependence of f on salt concentration. First there is the influence of the ionic atmosphere, which will increase f and will be particularly marked at low salt concentrations. Secondly, there is the possibility of drainage. Whereas typical non-ionic polymers are non-draining, for a polyelectrolyte at high degrees of expansion at least partial drainage may occur, which would lead to a decrease in f. This situation is most likely to arise at low molecular weights.

Tentatively, one may suggest that the curvature at high molecular weight (Fig. 7) reflects the influence of the ionic atmosphere, whereas the apparently linear

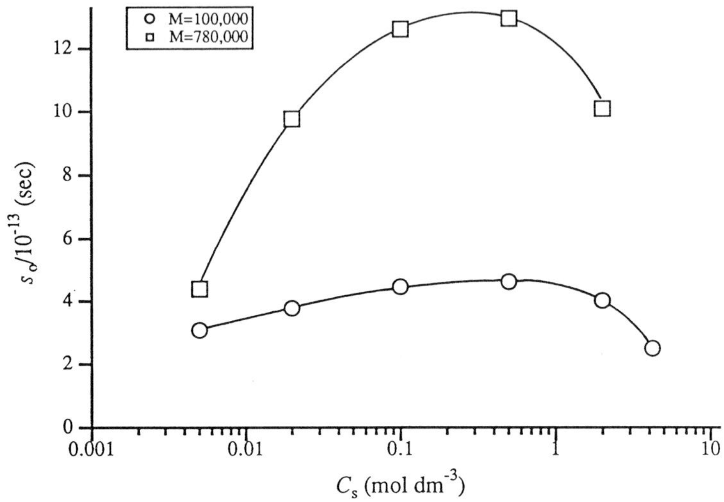

Figure 3. The dependence of the limiting sedimentation coefficient on concentration of added NaCl for two samples of NaPSS.

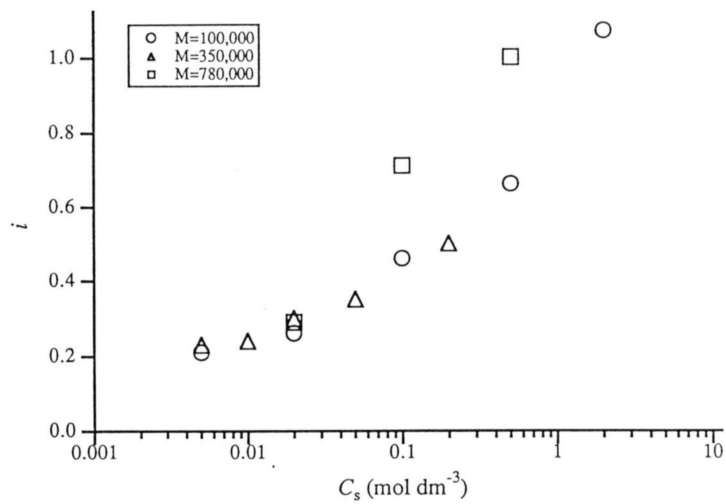

Figure 4. The dependence of the effective charge parameter on concentration of added NaCl, calculated from sedimentation velocity data for three samples of NaPSS.

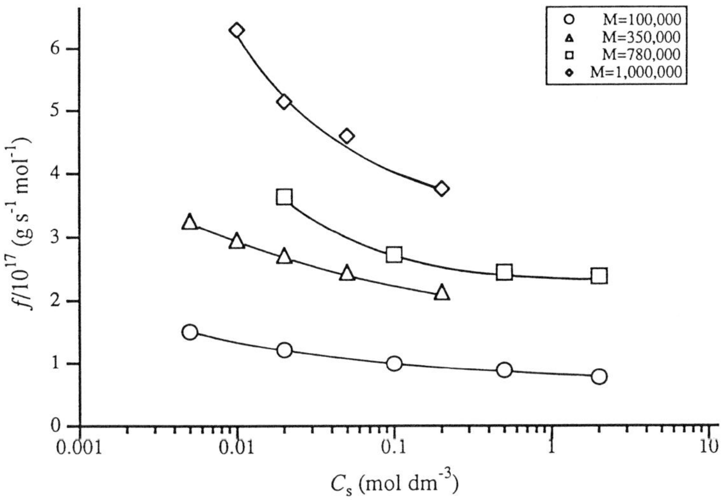

Figure 5. The dependence of the frictional coefficient, corrected to the viscosity of water, on concentration of added NaCl for four samples of NaPSS.

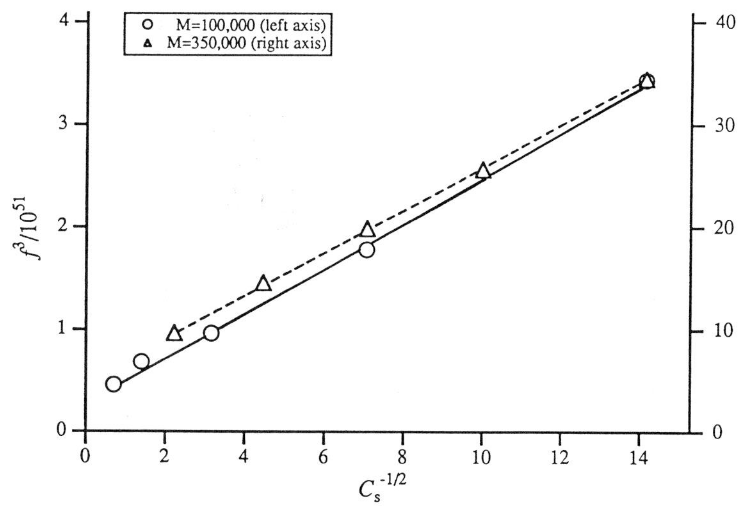

Figure 6. The dependence of the cube of the corrected frictional coefficient on $C_s^{-1/2}$ for NaPSS of molecular weight 1.0×10^5 and 3.5×10^5.

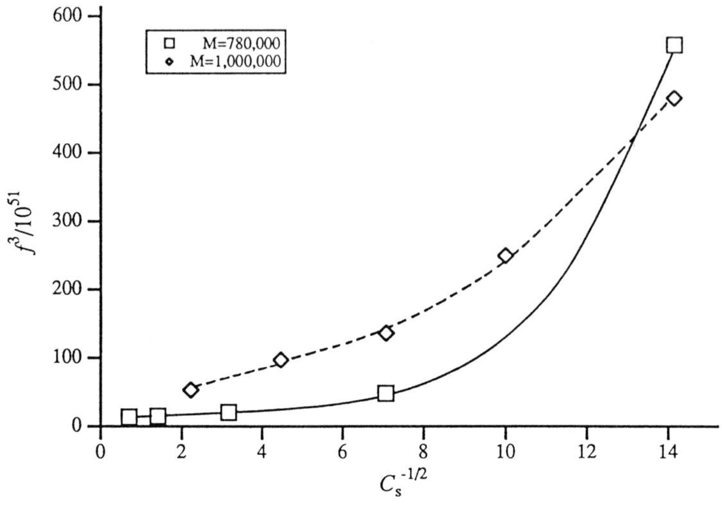

Figure 7. The dependence of the cube of the corrected frictional coefficient on $C_s^{-1/2}$ for NaPSS of molecular weight 7.8×10^5 and 1.0×10^6.

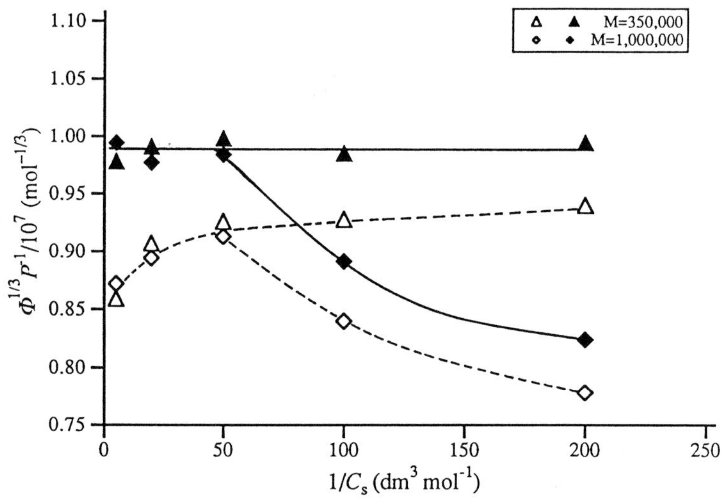

Figure 8. The dependence of the Mandelkern-Flory parameter, calculated according to eq. (14) (open symbols) and eq. (15) (filled symbols), on the reciprocal salt concentration for two samples of NaPSS.

behaviour at low molecular weight results from the fortuitous cancellation of the effects of the ionic atmosphere and of partial drainage.

Mandelkern-Flory Parameter

According to Mandelkern and Flory[16], all polymers should have much the same value of a parameter $\Phi^{1/3}P^{-1}$, given by

$$\Phi^{1/3}P^{-1} = \frac{s_0 N_A \eta_0 [\eta]^{1/3}}{M^{2/3}(1-\bar{v}\rho)} \tag{14}$$

where N_A is the Avogadro number and η_0 is the solvent viscosity. This expression may be used to determine molecular weights by a combination of sedimentation velocity and viscosity data. For non-ionic polymers the value of $\Phi^{1/3}P^{-1}$ is generally closc to 1.1×10^7 mol$^{-1/3}$ (2.5×10^6 if $[\eta]$ is expressed in dl g^{-1})[19]. For polyelectrolytes, lower values have been found[15,20,21].

Nagasawa and Eguchi[15] used eq. (14) to evaluate $\Phi^{1/3}P^{-1}$ in a study of NaPSS. Their results for two molecular weights are illustrated by the open symbols in Fig. 8, in which $\Phi^{1/3}P^{-1}$ is plotted against the reciprocal of salt concentration. It can be seen that $\Phi^{1/3}P^{-1}$ appears to decrease at high salt concentrations. This is due to the increase in i with increasing salt concentration. For polyelectrolytes the appropriate expression for $\Phi^{1/3}P^{-1}$ is

$$\Phi^{1/3}P^{-1} = \frac{s_0 N_A \eta_0 [\eta]^{1/3}}{M^{2/3}\left\{(1-\bar{v}\rho) - \dfrac{0.5iM_s(1-\bar{v}_s\rho)}{M_u}\right\}} \equiv \frac{s_0 N_A \eta_0 [\eta]^{1/3}}{M^{2/3}\left(\dfrac{d\rho}{dc}\right)_\mu} \tag{15}$$

The filled symbols in Fig. 8 show the values of $\Phi^{1/3}P^{-1}$ calculated using eq. (15), with i evaluated by means of eqs. (11–13). The decrease at high salt concentrations is eliminated. The limiting value, however, is still lower than for a typical non-ionic polymer, and at high molecular weight there is a decrease in $\Phi^{1/3}P^{-1}$ at low salt concentrations.

Referring to the discussion of the frictional coefficient above, the low value of $\Phi^{1/3}P^{-1}$ for polyelectrolytes may be attributed to the influence of the ionic atmosphere. The ionic atmosphere will have little effect on the rotational frictional coefficient operative in viscosity but will affect the translational frictional coefficient significant in sedimentation. The increasing thickness of the ionic atmosphere with decreasing salt concentration would lead to the observed decrease in $\Phi^{1/3}P^{-1}$ for the higher molecular weight sample in Fig. 8. At the lower molecular weight, this effect is likely to be counterbalanced by increased drainage.

4. SEDIMENTATION EQUILIBRIUM

Sedimentation equilibrium provides a theoretically well-grounded, absolute method for the determination of average molecular weights of polymers. For a polyelectrolyte in a salt-free solution, the attainment of equilibrium is extremely slow and the concentration dependence very large. In practice, polyelectrolytes must be studied in the presence of a simple salt.

At sedimentation equilibrium there is a concentration gradient of polyelectrolyte and therefore, because of the Donnan effect, an opposing concentration gradient of salt. A refractometric optical system will observe a composite gradient curve. In most cases this does not matter, as errors tend to cancel out in the sedimentation equilibrium equations. However, if a separate synthetic boundary experiment is performed for the determination of initial concentration, it is desirable to use solutions which have been dialysed to equilibrium, with the dialysate as reference.

As mentioned earlier, for a multicomponent system such as a polyelectrolyte with added salt it is advantageous to utilise a density increment rather than a buoyancy factor. An appropriate density increment is that for the condition of constant chemical potential of diffusible components, $(d\rho/dc)_\mu$[10,11]. The increment $(d\rho/dc)_\mu$ may be determined by density measurements on solutions which have been dialysed to equilibrium. However, it is not always feasible to perform equilibrium dialysis, so it is appropriate to ask what error is incurred if the increment $(d\rho/dc)_{C_s}$ is used instead of $(d\rho/dc)_\mu$, or whether there are any alternative experimental procedures.

Inspection of eq. (7) reveals that the difference between $(d\rho/dc)_\mu$ and $(d\rho/dc)_{C_s}$ should disappear for the specific case of a salt for which $(1-\bar{v}_s\rho)$ is zero. Alternatively, if Γ is known from some other technique, $(d\rho/dc)_\mu$ may be calculated from $(d\rho/dc)_{C_s}$.

Salt with Zero Buoyancy Factor

A salt for which the buoyancy factor is close to zero in water at room temperature is tetramethylammonium chloride[22,23]. Table 1 gives values of $(d\rho/dc)_{C_s}$ and $(d\rho/dc)_\mu$ measured for a sample of NaPSS of nominal molecular weight 1.2×10^6 in aqueous NaCl (0.5 mol dm^{-3}) and for a sample of $(CH_3)_4NPSS$, which was prepared from the NaPSS by ion exchange, in aqueous $(CH_3)_4NCl$ (0.5 mol dm^{-3})[13]. For NaPSS with NaCl there is a significant difference between the two increments; a 20% error in molecular weight would be incurred if $(d\rho/dc)_{C_s}$ were to be used rather than $(d\rho/dc)_\mu$. For $(CH_3)_4NPSS$ with $(CH_3)_4NCl$, on the other hand, the two increments are the same within experimental error.

Table 1. Density increments.

Polymer	Solvent	$\left(\frac{d\rho}{dc}\right)_{C_s}$	$\left(\frac{d\rho}{dc}\right)_{\mu}$
NaPSS	aqueous NaCl (0.5 mol dm^{-3})	0.464 ± 0.01	0.376 ± 0.01
$(CH_3)_4NPSS$	aqueous $(CH_3)_4NCl$ (0.5 mol dm^{-3})	0.210 ± 0.01	0.203 ± 0.01

Average molecular weights and degrees of polymerisation determined for these samples by a meniscus-depletion sedimentation equilibrium method are given in Table 2[13].

Table 2. Average molecular weights (M) and degrees of polymerisation (DP).

Average	NaPSS		$(CH_3)_4NPSS$	
	10^{-6}×M	DP	10^{-6}×M	DP
Number-	0.787 ± 0.08	3 900	1.026 ± 0.09	4 000
Weight-	1.014 ± 0.08	5 000	1.253 ± 0.09	5 000
z-	1.449 ± 0.18	7 100	1.637 ± 0.20	6 500

Use of Γ

A number of experimental procedures, in addition to membrane distribution studies[24-27], have been employed to evaluate Γ, including isopiestic distillation[28,29], gel permeation chromatography[30,31] and electromotive force measurements using ion-selective electrodes[32]. Sedimentation velocity may also be able to provide a value of Γ, utilising eqs. (8) and (11–13). Data obtained for NaPSS of nominal molecular weight 1.2×10^6 in aqueous NaCl (0.5 mol dm^{-3}) are summarised in Table 3 [13]. A value for Γ of 0.41 is obtained, which when put into eq. (7) gives $(d\rho/dc)_\mu = 0.385$, very close to the measured value (Table 1).

It should be noted that as i, and hence Γ, increases with increasing salt concentration (Fig. 4), the error incurred if the incorrect increment is used is greater at higher salt concentrations. It is often stated that the Donnan effect is swamped by the addition of salt, which is true in that the amount of salt excluded becomes small as a proportion of the total salt concentration. However, in the present situation the important factor is not the proportion but the actual amount of salt excluded, which increases with increasing salt concentration.

Table 3. Results for NaPSS ($M \approx 1.2\times10^6$) in aqueous NaCl (0.5 mol dm^{-3})

Parameter	Value
s_0	11.2×10^{-13} sec
k_e	2.457×10^{14}
$[\eta]^*$	129.8 ml g^{-1}
f_X^{**}	1.864×10^{15}
f_Y^{**}	1.189×10^{15}
i	0.82

* estimated from $[\eta] = 1.86\times10^{-2}\,\bar{M}_w^{0.64}$ (ref. 33)
** calculated from data of Pederson (ref. 3)

Molecular Weight

A question which may be asked concerns the calculated molecular weight; whether it is that for the polyion alone or whether the counterions are included. The answer is that it depends on how the concentration is measured when the density increment is determined. Molecular *weights* are ultimately obtained because *weight* concentrations are used. If the concentration measurement includes the counterions, for example if dry NaPSS is weighed out and dissolved, then the molecular weight includes the counterions.

In this context it is worth remembering that polyelectrolytes are generally hygroscopic, for example NaPSS powder typically contains about 10% by weight of moisture. The moisture content must be allowed for when determining the density increment; any error in the concentration is reflected in the molecular weight ultimately obtained. Commercial polymers are also frequently impure; a supposedly 'standard' sample of NaPSS received in our laboratory was found to contain 40% inorganic sulphate.

5. CONCLUSIONS

Sedimentation methods provide an invaluable insight into the molecular characteristics and properties of synthetic polyelectrolytes.

In pure water, polyelectrolytes undergo very slow sedimentation, a consequence of the primary charge effect and of the high degree of expansion of a polyion under these conditions. Addition of a simple salt leads to behaviour akin to that of a non-ionic polymer. Molecular weights may be determined from limiting sedimentation and diffusion coefficients, or by sedimentation equilibrium, provided that an appropriate density increment is used. A detailed study, however, reveals a number of factors which underlie the observed sedimentation behaviour.

An effective charge parameter may be derived from sedimentation velocity data. This parameter, which may be equated to 2Γ, increases with increasing salt concentration. The frictional coefficient, on the other hand, decreases with increasing salt concentration, reflecting a decreasing degree of expansion of the polyion, modified by other factors.

A density increment for the condition of dialysis equilibrium, $(d\rho/dc)_\mu$, should be employed in the sedimentation analysis of polyelectrolytes in the presence of a simple salt. Indiscriminate use of $(d\rho/dc)_{C_s}$ instead of $(d\rho/dc)_\mu$ may give rise to significant errors, but the two increments are identical for the special case of a salt for which the buoyancy factor is zero. In other cases it is feasible to calculate $(d\rho/dc)_\mu$ from $(d\rho/dc)_{C_s}$, using values of Γ derived from sedimentation velocity or other data.

REFERENCES

1. P.M. Budd, 'Comprehensive Polymer Science', eds. G. Allen, J.C. Bevington, C. Booth and C. Price, Pergamon, Oxford, 1989, Vol. 1, Chapter 11, p. 215.
2. P.M. Budd, 'Comprehensive Polymer Science', eds. G. Allen, J.C. Bevington, C. Booth and C. Price, Pergamon, Oxford, 1989, Vol. 1, Chapter 10, p. 199.
3. K.O. Pedersen, *J. Phys. Chem.*, 1958, *62*, 1282.
4. F.G. Donnan, *Z. Elektrochem.*, 1911, *17*, 572.
5. R.W. Armstrong and U.P. Strauss, 'Encyclopedia of Polymer Science and Technology', eds. H.F. Mark *et al.*, Wiley-Interscience, New York, 1964, Vol. 10, p. 781.
6. P.F. Mijnlieff, 'Ultracentrifugal Analysis', ed. J.W. Williams, Academic, New York, 1963, p. 81.
7. R. Varoqui and A. Schmitt, *Biopolymers*, 1972, *11*, 1119.
8. Z. Alexandrowicz and E. Daniel, *Biopolymers*, 1963, *1*, 447.
9. H. Eisenberg, *Biophys. Chem.*, 1976, *5*, 243.
10. H. Eisenberg, 'Biological Macromolecules and Polyelectrolytes in Solution', Clarendon, Oxford, 1976.
11. E.F. Casassa and H. Eisenberg, *Adv. Protein Chem.*, 1964, *19*, 287.
12. P.M. Budd, *Polymer*, 1985, *26*, 1519.
13. P.M. Budd, *Brit. Polym. J.*, 1988, *20*, 33.
14. M. Wales and K.E. Van Holde, *J. Polym. Sci.*, 1954, *14*, 81.
15. M. Nagasawa and Y. Eguchi, *J. Phys. Chem.*, 1967, *71*, 880.
16. L. Mandelkern and P.J. Flory, *J. Chem. Phys.*, 1952, *20*, 212.
17. L. Mandelkern, W.R. Krigbaum, H.A. Scheraga and P.J. Flory, *J. Chem. Phys.*, 1952, *20*, 1392.
18. I. Noda, T. Tsuge and M. Nagasawa, *J. Phys. Chem.*, 1970, *74*, 710.
19. V. Petrus, B. Porsch, B. Nyström and L.-O. Sundelöf, *Makromol. Chem.*, 1983, *184*, 295.
20. M. Stickler, *Angew. Makromol. Chem.*, 1984, *123/124*, 85.
21. H. Eisenberg and D. Woodside, *J. Chem. Phys.*, 1962, *36*, 1844.
22. R. Ziccardi and V. Schumaker, *Biopolymers*, 1971, *10*, 1701.
23. J.P. Kratohvil, *J. Coll. Interf. Sci.*, 1980, *75*, 271.
24. U.P. Strauss and P. Ander, *J. Amer. Chem. Soc.*, 1958, *80*, 6494.
25. H. Eisenberg and E.F. Casassa, *J. Polym. Sci.*, 1960, *47*, 29.

26. Z. Alexandrowicz and A. Katchalsky, *J. Polym. Sci.*, 1963, *A1*, 3231.
27. U.P. Strauss, C. Helfgott and H. Pink, *J. Phys. Chem.*, 1967, *71*, 2550.
28. G. Cohen and H. Eisenberg, *Biopolymers*, 1968, *6*, 1077.
29. D.A. Doughty, *J. Phys. Chem.*, 1979, *83*, 2621.
30. S. Kadokura, T. Miyamoto and H. Inagaki, *Makromol. Chem.*, 1983, *184*, 2593.
31. T. Lindström, A. DeRuvo and C. Söremark, *J. Polym. Sci., Polym. Chem. Edn.*, 1977, *15*, 2029.
32. N. Imai and H. Eisenberg, *J. Chem. Phys.*, 1966, *44*, 130.
33. A. Takahashi, I. Kato and M. Nagasawa, *J. Phys. Chem.*, 1967, *71*, 2001.

33

Sedimentation Analysis of Micelle-Forming Systems

By Robert W. Roxby

DEPARTMENT OF BIOCHEMISTRY, MICROBIOLOGY AND MOLECULAR BIOLOGY, UNIVERSITY OF MAINE, ORONO, MAINE 04469 , U.S.A.

1. INTRODUCTION

Micelle formation, a property characteristic of but not entirely limited to single chain amphiphiles, is ordinarily regarded as a reversible, self-association reaction and, as such, is subject to the general principles of the analysis of such systems by sedimentation methods that are described elsewhere in this volume. As for true macromolecular systems, these measurements lead to the determination of association constants and physical properties of the aggregates. This Chapter will be limited primarily to aspects of these analyses that are unique to micelle formation, which ordinarily involves the reversible, high-order association of molecules of low molecular weight, and to the analysis of products of micelle-mediated solubilization of small molecules, excluding studies of detergent solubilized proteins of biological membranes, which is taken up elsewhere.

2. GENERAL ASPECTS

There are some unique properties of micelle forming systems which influence the techniques used and the interpretation of the results, that it is well to recognize beforehand. One complication of the analysis of micelles, not encountered with true macromolecules, is the ambiguity in correcting for preferential interactions with solvent components. Using the Scatchard notation in which the subscript 1 refers to water, subscript 2 refers to the macromolecular component of interest and subscripts 3 and higher refer to other solvent components which may interact with the macromolecular species, the distribution of macromolecular component in a centrifugal field depends on the effective buoyant mass, B_2, defined by[1]

$$B_2 = M_2\left[(1 - \bar{v}_2\rho_0) + \sum_{i>2} (1 - \bar{v}_i\rho_0)\left(\frac{\partial g_i}{\partial g_2}\right)_{\mu_i}\right] \tag{1}$$

Consideration will be limited here to systems in which there is at most one additional solvent component and the summation therefore will contain only one term. It may be useful to point out, however, that this equation illustrates the reason that studies of micelle forming systems in density gradients are particularly intractable and prone to misinterpretation, since multiple terms in the summation can not be neglected. The quantity $(\partial g_i/\partial g_2)_{\mu_i}$ is the binding of component 3 to component 2 as it would be operationally measured in an equilibrium dialysis experiment. Fortunately, it need not be known explicitly, since[1]

$$\left(\frac{\partial \rho}{\partial C_2}\right)_{\mu_3} = (1-\bar{v}_2\rho_0) + (1-\bar{v}_3\rho_0)\left(\frac{\partial g_3}{\partial g_2}\right)_{\mu_3} \quad (2)$$

Casassa and Eisenberg defined an effective specific volume, ϕ', such that

$$\left(\frac{\partial \rho}{\partial C_2}\right)_{\mu_3} = 1 - \phi'\rho_0$$

For true macromolecular solutes, sedimentation studies with an auxiliary measurement of ϕ', both made against a reference solvent consisting of equilibrium dialysate, permit an unambiguous interpretation of the buoyant mass, B_2, in terms of the molecular weight of the unsolvated component 2. This approach cannot be applied with confidence to micelle forming systems, however, because the monomer is usually small enough to pass though the membrane, preventing the attainment of dialysis equilibrium. For amphiphiles with low CMCs, the slow equilibration of amphiphile relative to other solvent components might be exploited to produce solutions approximating dialysis equilibrium, but some means of independently measuring amphiphile concentration in the solutions would then be required. Alternatively, the more detailed knowledge often available concerning the chemical nature of the amphiphile solute, compared to that ordinarily available for a true macromolecule, permits justifiable assumptions in some cases about preferential interactions. This is illustrated in the examples discussed below.

For non-ionic amphiphiles in dilute salt solutions, preferential interactions are likely to be negligible. This has repeatedly been demonstrated to be so for true macromolecules near the isoelectric point and has been shown true for the detergent octyl glucoside, for which molecular weights obtained in water and in undialyzed solutions of 0.05M NaCl[2] were found to be the same within experimental error.

Ionic amphiphiles present a more difficult situation, for which preferential interactions can be estimated only to a first approximation. Since the exact charge of the amphiphile is usually known exactly, a precise calculation could be made on the assumption that the ionic amphiphile is fully dissociated, both as monomer and in the micelle, and that preferential interactions arise only from the Donnan effect. An amphiphile of charge Z dissolved as the salt AX_Z in water at infinite dilution and without supporting electrolyte will have an effective mass, M_{eff}[3]

$$M_{eff} = \frac{M_2}{Z + 1} \quad (4)$$

Non-ideality arising from electrostatic interactions is likely to be severe in such studies[39], which would make it difficult to apply this equation at finite concentrations. If AX_Z were dissolved at a concentration C_2 in a solvent containing a 1:1 electrolyte, BX, having the common anion, X, at an initial concentration $C_3{}^0$, the combined requirements of electrical neutrality and equal chemical potential of bound and free BX combine to produce negative binding (preferential hydration) of the salt to the amphiphile. If such a solution could be dialyzed against the solvent, the concentration of BX in the solvent would increase by - and the concentration of BX in the amphiphile solution would decrease by - the amount

$$\Delta C_3 = \frac{ZC_3^0C_2M_3}{M_2(4C_3^0 + ZC_2)} \tag{5}$$

This equation can be used to calculate the composition of a reference solvent to be used against the amphiphile solution. The difference between the salt concentration in the solvent and the solution at the hypothetical dialysis equilibrium is twice the amount transferred and represents the amount bound to the amphiphile. Thus,

$$\left(\frac{\partial g_3}{\partial g_2}\right)_{\mu_3} = -\frac{2ZC_3^0M_3}{M_2(4C_3^0 + ZC_2)} \tag{6}$$

Extrapolating to zero concentration of C_2 (C_2 will usually be negligible in comparison with C_3) and substituting into eq. (1) gives[3]

$$M_{eff} = M_2 - \frac{ZM_3(1-\bar{v}_3\rho_0)}{2(1-\bar{v}_2\rho_0)} \tag{7}$$

A potential flaw in the above treatment is the assumption that the amphiphile is fully ionized in the micelle. In fact some condensation of counterion at the micelle surface is likely to occur, decreasing Z and bringing M_{eff} for amphiphile in micelles closer to M_2. From isopiestic distillation measurements, Doughty[5] estimated values of Z for sodium dodecyl sulphate micelles varying from 0.35 to 0.42 as the NaCl concentration was increased from 0.1 M to 0.3 M.

While the effective mass of the particles is a consideration in any sedimentation experiment, there are a number of additional significant factors applying to transport experiments on micelle forming systems. In the general situation of a reversible equilibrium between sedimenting species, the Gilbert[6,7] theory predicts a complex profile in which the sedimentation rates of concentration gradient maxima are not constant over time and do not, in general, correspond to the sedimentation coefficients of species participating in the equilibrium. In a simple micelle forming system in which monomer is in equilibrium with micelles of a single size, such maxima provide useful approximations to the micelle and monomer sedimentation coefficients, but other modes of association can give similar profiles[8] and the assumption of the simplest model may not always be justified[2]. There may be operational difficulties as well. Because of their low densities, micelles may not sediment sufficiently rapidly to form a boundary separate from the meniscus, which will influence the measurement of the sedimentation coefficient. Micelle formation will, in general, involve significant volume changes, since the process involves transfer of nonpolar groups from water to a nonpolar environment, which means that the equilibrium will be pressure dependent, leading to further distortion of concentration profiles as a function of position in the cell. The shorter solution columns and lower rotor speeds used in equilibrium studies minimize pressure dependence problems. Pressure effects may be particularly severe when micelles have lower densities than solvent, in which case boundaries form at the bottom of the centrifuge cell, where pressures can easily exceed 100 atm [37]. Finally, the sedimentation coefficients of asymmetric particles exhibit strong dependencies upon concentration.

3. APPLICATIONS

All of the broad array of techniques routinely applied to the study of true macromolecules have been applied over a long history of measurements of micelle-forming systems reported in the literature. These include studies of micelle forming systems that are homogeneous, or effectively homogeneous[9-21], of micelle formation by inhomogeneous amphiphiles[22,23] and of complexes of micelles with solubilized products[27-31,38,45]. Theoretical papers have dealt with the mathematical treatment of multiple equilibria applicable to micelle forming systems[8,32,33] and with the prediction of structural properties of micelles from thermodynamic considerations and the geometric properties of the monomers[34,35,44]. Partial molar volumes have been reported in many of these studies and an extensive compilation of literature data has been given by Steele *et al.*[36].

Sedimentation Equilibrium Studies[2,4,5,9,13,17,18,21]

The most frequently applied method of analysis of micellar equilibria has been through the use of sedimentation equilibrium measurements using the refractometric optical systems of the Beckman/Spinco Model E analytical ultracentrifuge. One feature distinguishing micellar equilibria from those of true macromolecules is that the monomers are small and there is a large difference in molecular weight between the monomer and the smallest micelle. This has the result that, at any rotor speed, monomer will be present at all points in the cell, potentially complicating the analysis of the micellar aggregates by making a disproportionate contribution to molecular weight averages in which all species are included. The molecular weight averages as measured directly, therefore, do not even approximately correspond to the weight of any solution component. This problem has been dealt with in various ways. In some cases monomer concentration and buoyant mass may be sufficiently low as not to contribute significantly to concentration gradients at sedimentation equilibrium and thus may be neglected in the analysis of the micellar fraction. Even if monomer concentration is high enough to be detectable in the experiment, the low buoyant mass and consequently minimal redistribution may make it a satisfactory assumption to treat monomer concentration as a baseline in the sedimentation experiments[9,18]. However, it should be recognized that the "critical micelle concentration" (CMC), while a useful concept, is only an approximate one, particularly at total concentrations near that value[42]. Therefore, taking the CMC as a subtractive baseline may introduce significant error. A more exact and rigorous method is available for determining monomer concentration at any total concentration in an equilibrating system[2]. Knowledge of monomer concentration allows the contribution of monomer to be subtracted from the weight averages, allowing the contribution of the micelles to be determined. If the concentration of monomer - or, for that matter, of any component - is known at any total concentration in a sedimentation equilibrium experiment, it can be determined at any other from the relationship[40,41]

$$\ln\left[\frac{C_{mon,c}}{C_{mon,r}}\right] = \int_{C_r}^{C} \frac{B_{mon}}{B_{w,c}}\, d\ln C \tag{8}$$

where

$$B_{mon} = [M(1-\bar{v}\rho_0)]_{mon} \tag{9}$$

$$B_{w,c} = [M(1-\bar{v}\rho_0)]_w \tag{10}$$

$C_{mon,r}$ is the concentration of monomer at the reference point, $C_{mon,c}$ is the concentration of monomer at any other total concentration, B represents the measured buoyant mass and the subscript w indcates the weight-average. Eqs. (9) and (10) represent the buoyant molecular weights of monomer and micelle, the latter being the weight average at total concentration C, which are the quantities obtained most directly and without assumptions from a sedimentation equilibrium measurement. Eq. (8) may be appreciated intuitively by recognizing (i) that each sedimenting species must obey the basic equation for sedimentation equilibrium whether or not it is in equilibrium with other species and (ii) that in an equilibrating system, the monomer concentration will be a unique function of the total concentration. Eq. (8) can be applied by numerical integration of data collected from a series of experiments and plotted as B_w *vs*. C, as long as the data from different experiments overlap, which demonstrates that the system is an equilibrating one. Treatment of the data first in terms of the buoyant masses, separation of the contributions of monomer and micelles, and then insertion of the appropriate values of $\bar{v}$ allows an accurate treatment when partial molar volumes of monomer and micelle are different, as will usually be the case.

Other approaches to separating the contributions of monomer and micelle have been taken. Nichol *et al.*[4] estimated micelle molecular weight of chlorpromazine hydrochloride in an experiment carried out at high concentration and analyzed to give the z-average molecular weight, which was assumed to approximate that of a homogeneous micelle species. A similar approach was used by Ikeda and Kakuichi[43], who assumed polyoxyethylene octylphenyl ether micelles to be homogeneous and estimated monomer concentration from the baseline needed to give linear plots of ln C *vs*. r^2.

The power of the sedimentation equilibrium technique is illustrated by two recent studies of the nonionic detergent octyl-β-D-glucopyranoside. Kameyama and Takagi[9] used the techniques of sedimentation equilibrium and light scattering to investigate micelle shape and size as a function of temperature in a concentration range close to the CMC. Roxby and Mills[2] measured micelle size and free monomer concentration over a very wide concentration range using sedimentation equilibrium alone. The conclusions of these studies are entirely consistent. Octyl glucoside micelles become detectable in the sedimentation equilibrium experiments between 6 and 7 mg/ml (21-24 mM), initially with an aggregation number of 70. Although essentially homogeneous in size at the CMC, micelles increase markedly in size with concentration above the CMC. The minor quantitative differences in the two studies can be attributed to different temperatures and methods of data analysis. Kameyama and Takagi in effect averaged micelle molecular weights over a narrow concentration range, but one in which some large micelles were present, whereas Roxby and Mills extrapolated micelle molecular weights to the point of vanishing micelle concentration.

The approach to dealing with the more complicated case of ionic micelles is illustrated by the study of Nichol *et al.*[4] in which eq. (6) was used to predict the effective mass of an ionic species, chlorpromazine hydrochloride, which has a charge Z=1, in sedimentation equilibrium experiments. Chlorpromazine hydrochloride has a formula molecular weight of 355.4. At a concentration of 0.0038 M (C_2 in eq. (6)) in 0.154 M NaCl (C_3^0) the predicted M_{eff} for the monomer is 326, in excellent agreement with the measured value below the CMC of 319. The effects of decreased effective charge upon chlorpromazine micelles (if of the magnitude estimated by Doughty[5] for sodium dodecyl sulphate as described above) would be quite small in this system. The M_{eff} of the micellar amphiphile would be about 5% higher and the estimated aggregation numbers lower by the same factor, than those obtained assuming no condensation of charge, a discrepancy that would be close to the experimental error.

Transport Studies[10-15,17,19,20,22-25]

Transport studies are required in order to obtain information about micelle shape and size and have often been used as well to measure molecular weights, although they have the disadvantage of not readily providing information about size heterogeneity. An analysis of micelle size and shape from combined sedimentation equilibrium and velocity measurements has been made for n-alkyl polyoxyethylene ethers, $C_{12}E_8$ and Lubrol WX [17]. Both of these amphiphiles form micelles of sufficiently restricted size distribution as to appear homogeneous in sedimentation equilibrium experiments, notwithstanding the chemical heterogeneity of the Lubrol. Therefore the hydrodynamic analysis (by sedimentation velocity and viscosity) was straightforward and indicated that micelles are not spherical. The data are developed further on the basis of a theory of micelle shape developed by Tanford and others[17,44] which is based on the assumption that the micelle core behaves like a droplet of liquid hydrocarbon with a volume equal or approximately equal to the sum of the hydrocarbon chain volumes that it contains, excluding solvent from all but the periphery. Under this assumption the micelle core must have at least one dimension no larger than twice the statistical average length of the hydrocarbon plus the hydrophilic moieties of the amphiphile, quantities that can be reasonably estimated. Within that constraint, dimensions for prolate and oblate ellipsoidal micelles consistent with the known volumes can be calculated and compared with the hydrodynamic data. Better agreement was found with the oblate model for micelle shape than with the prolate one. Since the micelle shape predictions depend only on the geometric properties of the monomers and the aggregation numbers, with no assumption being made about hydration, it is possible then to estimate the hydration consistent with a model as the difference between the total micelle volume and the volume of the amphiphile molecules alone, as estimated from $\bar{v}$. Estimated hydration, as would be expected, increases with increasing polyoxyethylene chain length, being in the range of 2-4 molecules of water per oxyethylene unit.

In contrast to $C_{12}E_8$, n-alkyl polyoxyethylene ethers with shorter hydrophilic chains represent a more complex system in which, at higher temperatures, micelles grow in size with increasing concentration[17,44]. $C_{12}E_6$ association has been investigated by sedimentation velocity combined with diffusion coefficients measured by quasielastic light scattering[11,12]. Although the monomer concentration is low enough that it could be neglected in these experiments, the density of the micelle is close to that of water, requiring the use of D_2O solutions for the measurements, in which the micelles float. At 25°C and above, sedimentation rates first increase and then decrease with concentration, with the effect becoming more pronounced with increasing temperature. At higher concentrations, unexplained rapidly moving boundaries appear. Thus, it is difficult to make an exact analysis of the micellization equilibria of this amphiphile by transport methods alone.

A somewhat extreme example of the complexity that may be encountered is seen in a recent analysis of dodecyldimethylammonium chloride (DDAC) micelle shape by sedimentation velocity methods[10]. The interpretation was assisted by previously obtained light-scattering data, which indicated that the spherical micelles initially formed appear to undergo a secondary association to form rodlike higher aggregates at high amphiphile and NaCl concentrations. The light scattering studies also provided an independent estimate of micelle molecular weights. Sedimentation coefficients are negative in this system and the shapes of the upwardly migrating boundaries are consistent with the predictions of the Gilbert theory for a rapid, monomer:micelle:n-mer equilibrium, as had been indicated by light scattering. Estimates of weight fractions and hydrodynamic properties of spherical and rod-like micelles were obtained, but depended upon a number of extrapolations and assumptions which, the authors acknowledge, are uncertain.

Mixed Systems[27-31,38,45]

One of the most important and useful properties of amphiphiles is their ability to disperse insoluble components in solution. The principle is the same whether the amphiphile mediates the dispersal of hydrophobic material in water or hydrophilic material in a nonpolar solvent and ultracentrifugation studies that have been made of both kinds of systems show some interesting and incompletely explained parallels.

Mixed micelles of sphingomyelin and Triton-X-100[38] do not form with a sphingomyelin content lower than a mole fraction of ~0.21. If the overall composition of the solution is below this mole fraction of the lipid, two types of micelles coexist, the mixed micelle of minimum sphingomyelin content and micelles containing only Triton-X-100. As the mole fraction of sphingomyelin is increased above 0.21, the mixed micelles become larger primarily by the addition of sphingomyelin, the amphiphile aggregation number remaining approximately constant. The amphiphile aggregation number in the mixed micelles is higher than that of the pure Triton micelle. The hydrodynamic properties of the mixed micelles indicate either increased assymetry or increased hydration with increasing size and sphingomyelin content. Solubilization of sphingomyelin with Triton-X-305 [45], which has a hydrophilic polyoxyethylene chain of thirty units (compared to 10 for Triton-X-100), produces mixed micelles of two size classes at all lipid:detergent ratios, with no pure detergent micelles present at *any* solution composition. These mixed micelles again contain roughly similar numbers of Triton molecules and increasing numbers of lipid molecules as lipid is added. These results are interpreted in terms of a model in which the mixed micelles are disk shaped, with sphingomyelin molecules occupying regions of low curvature and the Triton occupying regions of high curvature near the edge. The data for the two systems suggest that there are distinct permissible states of mixed micelles, perhaps dictated by geometric considerations.

Similar effects are seen in the solubilization of hydrophilic materials, proteins[46,47] and basic oligopeptides[48], in organic solvents by the amphiphile bis(2-ethylhexyl)sodium sulfosuccinate (referred to as aerosol OT or AOT). This amphiphile can emulsify large quantities of water, which may contain solute, in reversed micelles in the organic phase, forming optically clear solutions. Small amounts of water soluble dye were included in the reversed micelles, which were demonstrated not to affect micellar properties and allowed the sedimentation experiments to be monitored by absorption optics[47]. One sedimenting boundary of reversed micelles is seen in the absence of solute, the sedimentation coefficient of which increases with the molar ratio of water to amphiphile. When hydrophilic solute is added to the system, a second, more rapidly sedimenting boundary appears, which is attributed to reversed micelles containing solute. There is a corresponding decrease in the proportion of material in, but no change in the sedimentation velocity of, the more slowly moving boundary. The incremental effect of solubilized protein on the sedimentation rate of the reversed micelles is greater for small micelles than large ones, which is not surprising since the solute represents a smaller proportion of the mass of the larger micelles. There is also a proportion of micelles which *cannot* be induced to take up solute at any water mole fraction. This may be simply a kinetic limitation in the process of preparing the solutions which do not subsequently equilibrate or it may again indicate structural constraints on the mixed micelles. It is not clear from these studies whether the presence of solute has an influence on micelle size or whether amphiphile and micelles are in equilibrium. Regardless, it is an interesting and potentially useful system that deserves further investigation.

ACKNOWLEDGEMENT

This is publication number 1618 of the Maine Agricultural Experiment Station.

REFERENCES

1. E. Casassa and H. Eisenberg, *Adv. Protein Chem.*, 1964, *19*, 287.
2. R. Roxby and B. Mills, *J. Phys. Chem.*, 1990, *94*, 456.
3. C. Tanford, 'Physical Chemistry of Macromolecules', Wiley, New York, 1961, Chapter 4, pp. 265-269.
4. L. Nichol, E. Owen and D. Winzor, *J. Phys. Chem.*, 1982, *86*, 5015.
5. D. Doughty, *J. Phys. Chem.*, 1979, *83*, 2621.
6. G. Gilbert, *Disc. Faraday Soc.*, 1955, *20*, 68.
7. G. Gilbert, *Proc. Royal Soc. Ser. A*, 1959, *250*, 377.
8. M. Tai and G. Kegeles, *Biophys. Chem.*, 1984, *20*, 81.

9. K. Kameyama and T. Takagi, *J. Coll. Interface Sci.*, 1990, *137*, 1.
10. S. Ikeda, S. Ozeki and K. Kakiuchi, *J. Phys. Chem.*, 1988, *92*, 3465.
11. W. Brown and R. Rymdén, *J. Phys. Chem.*, 1987, *91*, 3965.
12. W. Brown, R. Johnson and P. Stilbs, *J. Phys. Chem.*, 1983, *87*, 4548.
13. H.-G. Elias, D. Chung, N. Donkai, G. Hellman, K. Solc and K. Nagai, *Makromol. Chem.*, 1987, *188*, 537.
14. J. Kratohvil, T. Aminabhavi, W. Hsu, S. Fujima, A. Patkowski, F. Chen and B. Chu, *Croat. Chem. Acta*, 1983, *56*, 789.
15. J. Kratohvil and T. Aminabhavl, *J. Phys. Chem.*, 1982, *86*, 1254.
16. L. Nichol, E. Owen and D. Winzor, *J. Phys. Chem.*, 1982, *86*, 5015.
17. C. Tanford, Y. Nozaki and M.F. Rohde, *J. Phys. Chem.*, 1977, *81*, 1555.
18. C. Biaselle and D. Millar, *Biophys. Chem.*, 1975, *3*, 355.
19. J. Kratohvil, *Colloid Polym. Sci.*, 1975, *253*, 251.
20. G. Blauer and B. Zvilichovsky, *Arch. Biochem. Biophys.*, 1968, *127*, 749.
21. S. Ikeda and K. Kakiuchi, *J. Coll. Interface Sci.*, 1967, *23*, 134.
22. J. Hwang, H. Ji, S. Chen, *J. Chin. Chem. Soc. (Taipei)*, 1990, *37*, 157.
23. Z. Tuzar, V. Petrus and P. Kratochvil, *Makromol. Chem*, 1974, *175*, 3181.
24. G. Ebert, M. Plachky, M. Seno and S. Shoji, *Prog. in Colloid and Polym. Sci.*, 1988, *76*, 113.
25. Y. Sakaiguchi and H. Hirata, *Bull. Chem. Soc. Japan*, 1988, *61*, 2743.
26. H. Schott and A. Royce, *J. Pharm. Sci.*, 1985, *74*, 957.
27. S. Yedgar and G. Cooper, *Arch. Biochem. Biophys.*, 1985, *240*, 191.
28. A. Levishov, Y. Khmelnitsky, N. Klyachko, V. Chernyak and K. Martinek, *J. Coll. Interface Sci.*, 1982, *88*, 444.
29. A. Levishov, Y. Khmelnitsky, N. Klyachko, V. Chernyak and K. Martinek, *Anal. Biochem.*, 1981, *118*, 42.
30. N. Incardona, K. Miles and B. Baker, *Nature New Biol.*, 1971, *229*, 250.
31. A. Fryar and S. Kaufman, *J. Coll. Interface Sci.*, 1969, *29*, 444.
32. P. Mukerjee, in 'Micellization, solubilization and microemulsions', Vol. 1, K. Mittal, Ed., Plenum, 1977, 171.
33. P. Mukerjee, *ibid.*, p. 171.
34. P. Mukerjee *J. Phys. Chem.*, 1972, *76*, 565.
35. C. Tanford, In 'Micellization, solubilization and microemulsions', Vol. 1, K. Mittal, Ed., Plenum, 1977, 119.
35. C. Tanford, *J. Phys. Chem.*, 1974, *76*, 3020.
36. J. Steele, Jr., C. Tanford and J. Reynolds, *Meth. Enzymol.*, 1978, *48*, 11.
37. G. Howlett, P. Jeffrey and L. Nichol, *J. Phys. Chem.*, 1970, *74*, 3607.
38. S. Yedgar, Y. Barenholz and V. Cooper, *Biochim. Biophys. Acta*, 1974, *363*, 98.
39. D. Roark and D. Yphantis, *Biochemistry*, 1971, *10*, 3241.
40. R. Steiner, *Arch. Biochem. Biphys.*, 1953, *49*, 400.
41. E. Adams and J. Williams, *J. Am. Chem. Soc.*, 1964, *164*, 3454.
42. L. Nichol and A. Ogston, *J. Phys. Chem.*, 1981, *85*, 1173.
43. S. Ikeda and K. Kakiuchi, *J. Coll. Interface Sci.*, 1967, *23*, 134.
44. C. Tanford, 'The Hydrophobic Effect: Formation of Micelles and Biological Membranes', second edition, Wiley, New York, 1980.
45. S. Yedgar and G. Cooper, *Arch. Biochem. Biophys.*, 1985, *240*, 191.

46. A. Levishov, Y. Khmelnitsky, N. Klyachko, V. Chernyak and K. Martinek, *Anal. Biochem.*, 1981, *118*, 42.
47. A. Levishov, Y. Khmelnitsky, N. Klyachko, V. Chernyak and K. Martinek, *J. Coll. Interface Sci.*, 1982, *88*, 444.
48. G. Ebert, M. Plachky, M. Seno and S. Shoji, *Prog. in Colloid and Polym. Sci.*, 1988, *76*, 113.

Subject Index

W

Index of Solutes

This index lists different solutes referred to in the text. Solutes are divided into the following classes.

- Antibodies
- Bioparticulates
- Micellar Systems
- Nucleic Acids
- Polymers
- Polysaccharides and Mucins
- Proteoglycans
- Proteins
- Viruses and Viriods

An asterisk (*) accompanying a page number indicates that the reference takes the form of a table or of a diagram.

ANTIBODIES

BIOPARTICULATES

MICELLAR SYSTEMS

Mixtures

NUCLEIC ACIDS

POLYMERS

POLYSACCHARIDES AND MUCINS

PROTEINS

PROTEOGLYCANS

VIRUSES AND VIRIODS